D0908204

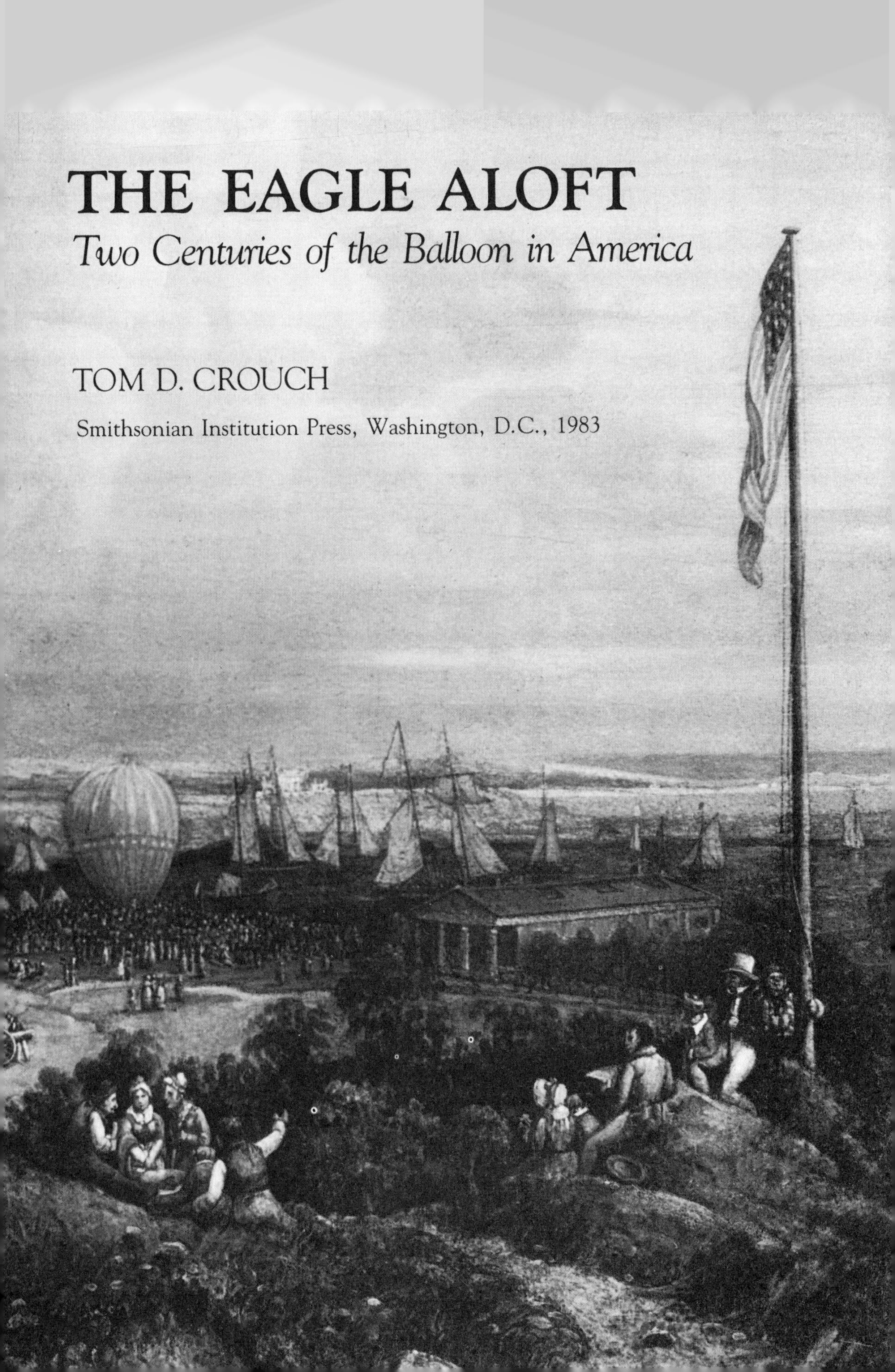

THE EAGLE ALOFT

Two Centuries of the Balloon in America

TOM D. CROUCH

Smithsonian Institution Press, Washington, D.C., 1983

Copyright © Smithsonian Institution, 1983.
All rights reserved.

Library of Congress Cataloging in Publication Data

Crouch, Tom D.
The eagle aloft.

Bibliography: p.
Includes index.
Supt. of Docs. no.: SI 1.2:Ea3
1. Balloons—United States—History. I. Title.
TL618.C76 1983 629.133'22 83-17079
ISBN 0-87474-346-X

Printed in the United States of America
Designed by Carol Hare Beehler

The paper in this book meets the guidelines for permanence and durability of the Committee on Production Guidelines for Book Longevity of the Council on Library Resources.

All photographs are from the National Air and Space Museum Library, unless otherwise noted.

The United States Organizing Committee of the Air and Space Bicentennial has selected *The Eagle Aloft* as an official book commemorating the 200th anniversary of human flight.

To the Memory of Charles Harvard Gibbs-Smith (1909–1981)

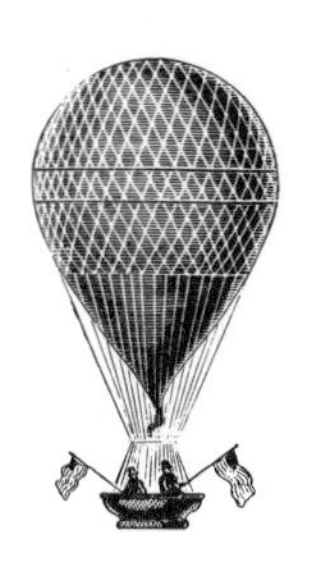

Contents

Acknowledgments

THE AUTHOR OF ANY BOOK AS LONG AS THIS ONE ACCUMULATES SUCH an incredible number of debts that any attempt to list them seems a bit ridiculous. Still, some of these obligations are so overwhelming that they cannot be overlooked. The first of these are owed to previous writers in the field. Chief among these are the late Charles Dollfus and my late friend and mentor, Charles Harvard Gibbs-Smith; Jeremiah Milbank; Joseph Jenkins Cornish III; F. S. Haydon; Roger Pineau; J. Gordon Vaeth; Kurt R. Stehling; L. T. C. Rolt; Dr. Dale Topping; and Dr. Russell J. Parkinson.

During the course of research for this book, the author drew material from a great many libraries and archives. My special thanks go to the staff of the following institutions. Don Barrett and the staff of the USAF Academy Library, Colorado Springs, Colorado, were gracious hosts and knowledgeable guides through the treasures of the Col. Richard C. Gimball Collection. Phyllis Kihn of the Connecticut Historical Society generously made her marvelous collection of New England newspaper articles on ballooning available to me. Staff members of other institutions answered questions, steered me to published sources, and photocopied reams of priceless material from their own collections to save me travel expense. These included: the Adams Family Papers, Massachusetts Historical Society; the Jay Papers, Columbia University; the Houghton Library, Harvard University; the Sterling Library, Yale University; the Library of the American Philosophical Society; the Museum of the City of New York; the Lancaster County [Pa.] Historical Society; the New York State Historical Society; the New Jersey Historical Society; the New York Public Library; the Virginia State Library; the Research Division, Colonial Williamsburg; Archives, College of William and Mary; the Maryland Room, McKeldon Library, University of Maryland; the Montana Historical Society; the Denver Public Library.

My most spectacular debt is to Fred Coker and the staff of the Manuscript Division, Library of Congress. In a very real sense, this volume is the result of their dedication to serving the scholarly researcher. The Newspaper, Rare Book, Prints and Photographs, Stack and Reader Services, Circulation, and Science and Technology Divisions of the Library of Congress provided consistent exemplary service as well.

The staff of the Old Navy and Army Division of the National Archives offered advice and counsel above and beyond the call of duty.

Katherine Scott, Frank Pietropaoli, Mimi Scharf, Pete Suthard, Phil Edwards, Lee Jones, Robert Dreeson, Greg Bryant, Liz Hand, and Mary Pavlovich, all of the National Air and Space Museum Library, offered their usual high quality of support.

My colleagues at the NASM have also contributed to the task of finishing and improving this volume. Mel Zisfein, Walt Boyne, and Noel Hinners, each of whom served as either director or assistant director of the museum during my very slow progress through this manuscript, offered all help and assistance. My special thanks go to Donald Lopez, chairman of the Department of Aeronautics of the NASM, who maintains the sort of work environment that encourages the pursuit of significant, long-range research projects.

Claudia Oakes, Dom Pisano, Jay Spencer, Karl Schneide, Kathleen Brooks-Pazmany, Glen Sweeting, Tim Wooldridge, Robert Mikesh, Dorothy Cochrane, Natalie Rowan, and Susan Owen all deserve my thanks for having carried a greater share of the office work load than would have been the case had this book not been in progress. Robert van der Linden deserves all thanks and praise for the many hours he has spent in proofreading and correcting the text and calling the author's attention to manifold inconsistencies and inaccuracies in the manuscript. As always, my thanks to Dr. Howard Wolko, who seems constantly ready to listen and comment.

Mark Hodak, Mischa Prothik, and William Maclean researched, respectively, Chapters 1 and 2, Chapter 19, and Chapters 17 and 18. The author, who is unaccustomed to working with research assistants, owes an enormous debt to these three men who eased his path by reading their way through everything from eighteenth-century newspapers to twentieth-century scientific reports.

Typing the manuscript was a long and thankless chore. Brett Fletcher, Valerie Banks, and Gladys Waters did the job cheerfully, accurately, and most professionally.

Harvey Lippincott, George Fuller, Russell Parkinson, Claudia Oakes, and Adrianne Noe each read and commented on portions of the manuscript. Russell Parkinson's advice was especially helpful. Harvey Lippincott introduced me to a number of important contacts, including Rev. Nathaniel Lauriat and Phyllis Kihn. Robert Rechs, balloonist and balloon historian extraordinaire, opened his files on American aeronauts and his records of the National Balloon Races. Rev. Nathaniel Lauriat and Mr. William Foulke deserve special thanks for opening priceless family archives to the author.

All quotations from the Adams family correspondence are with the kind permission of the Adams Family Papers project, Massachusetts Historical Society. Portions of this book originally appeared as a series of articles in *Aviation Quarterly*, *Ohio History*, and the *Bulletin of the Cincinnati Historical Society*.

The author, of course, accepts all responsibility for errors in the text. I can only hope that the final product justifies the amount of time, trouble, and energy so many individuals have contributed to it.

Particular thanks go to Felix Lowe, director of the Smithsonian Institution Press, who saw the value of the project at the outset; Frances Kianka, who edited and indexed the book; and Carol Hare Beehler, the Smithsonian Institution Press designer. My thanks also to Nancy and Michael DeCarlo for their support and encouragement. Finally, my gratitude to Nancy, Bruce, Abigail, and Nathan Crouch, who put up with it all.

Tom D. Crouch
Smithsonian Institution
December 13, 1982

Chapter 1 Flight and the Founding Fathers

SIXTEEN-YEAR-OLD JOHN QUINCY ADAMS SPENT THE MORNING OF AUGUST 27, 1783, touring the Louvre. "There are," he commented, "some good paintings there amongst a great number of indifferent ones."[1] His afternoon was to prove far more exciting. Shortly after lunch he joined the enormous throng that jammed the streets leading toward the Champ de Mars. An English observer later recalled that "there were such numbers of carriages along the Boulevards that they could not pass."[2]

The venerable duc de Richelieu, ninety-year-old senior maréchal de France, was forced to abandon his coach and walk, preceded by a guard to clear the road. The duc de Chartres made his way through the crowd on horseback, dressed, as were his servants, in the latest English fashion. The comte de Provence, popularly known simply as Monsieur, represented the royal family in this glittering assemblage. The king's younger brother, the comte would one day rule as Louis XVIII.

John Quincy Adams was not the only American in the crowd. Benjamin Franklin, the seventy-seven-year-old leader of the American diplomatic community in Paris, was also on the scene. "It is supposed that not less than 50,000 people were assembled," Franklin remarked, "the Champ de Mars being surrounded by Multitudes, and vast numbers on the opposite side of the river."[3]

Now the site of the Eiffel Tower, the Champ de Mars was a military parade ground in the eighteenth century. A large, open field bordered by neat rows of trees, it extended from the Ecole Militaire to the left bank of the Seine. The dome of the Hôtel Royal des Invalides was visible to the right as one faced the river, while Franklin's home, the Hôtel de Valentinois, could be seen across the Seine on the heights of suburban Passy.

The attention of the crowd was focused on a delicate taffeta sphere bobbing up and down within a fenced enclosure constructed near the center of the Champ de Mars. It was the *Globe*, the world's first gas balloon. Only 12 feet in diameter, it was filled with 943 cubic feet of a gas that John Quincy Adams identified as "inflammable air." Benjamin Franklin's friend Antoine Laurent Lavoisier had recently given the gas a new name, hydrogen.

Four thousand ticket holders who had paid to enter the enclosure watched as a small party of men worked to complete the process of inflating the balloon. Jacques Alexandre César Charles, a well-known natural philosopher who had conceived and designed the *Globe*, was supervising the operation. He was assisted by Jean and Noel Robert, who had constructed the balloon to Charles's specifications.

The Robert brothers, proprietors of a small workshop on the Place des Victoires, had developed a process by which fabric could be coated with a natural rubber gum suspended in linseed oil. Thus sealed, Charles's *Globe* was able to retain the hydrogen with which it was filled.

Charles had not chosen the most pleasant of days for his first flight. Gusty winds sent clouds scudding across an overcast sky. Intermittent rain had fallen earlier that afternoon, making, as one observer commented, "fine work for the milliners and mercers" who would be called upon to replace the drenched finery of the spectators.[4] Far from dampening Franklin's enthusiasm, the downpour seems to have enhanced his appreciation of the scene. The rain, he remarked, had wet the balloon, "so that it shone, and made an Agreeable appearance."[5]

Two cannons were fired promptly at 5:00 P.M. to signal the beginning of the experiment. John Quincy Adams watched as the balloon "rose at once, for some time perpendicular, and then slanted. The weather was unluckily very cloudy, so that in less than two minutes it was out of sight; it went up very regularly and with great swiftness."[6] Benjamin Franklin noted that "it diminished in Apparent Magnitude as it rose, till it enter'd the Clouds, when it seem'd to me scarce bigger than an Orange."[7]

Completely sealed to prevent the escape of any gas, the little balloon burst after forty-five minutes in the air. It fell near a group of peasants from the village of Gonesse, some 15 miles from the take-off point. Franklin described the fate of the first hydrogen aerostat to Sir Joseph Banks, president of the Royal Society: "It is said that the country people who saw it fall were frightened, [and] conceived from its bounding a little, when it touched the Ground, that there was some living Animal in it, and attack'd it with Stones and Knives, so it was much mangled."[8]

At the conclusion of the demonstration, Franklin joined a group of spectators who were discussing the various uses to which the new invention might be put. One scoffer remarked that the balloon was nothing more than a useless toy. Franklin countered with the query, "What is the use of a new born babe?"[9]

An artist's conception of the inflation of the first hydrogen balloon

This most famous of his many epigrams was immediately taken up by admiring Parisians. Baron Friedrich Melchior Grimm, a Paris philosopher and wit, lost little time in passing the remark on to his many European correspondents. By October Franklin was receiving letters congratulating him on having compared the balloon to "an infant just coming to birth."[10]

The subject of new births was much on the minds of the Americans living in Paris during the summer and fall of 1783. John and Sarah Livingston Jay, their one-year-old daughter Maria, and their sixteen-year-old nephew Peter Jay Munro were sharing quarters with Benjamin Franklin and his grandsons, William Temple Franklin and Benjamin Bache. The Jays' second daughter, Ann, was born there on August 13, only two weeks before the flight of the *Globe*.

But there was a much more significant new beginning to be celebrated. The war that had begun with a single shot fired eight long years before in the tiny village of Lexington, Massachusetts, was over at last. On the morning of September 3, 1783, Franklin, Jay, and John Adams met British peace commissioner David Hartley at the Hôtel d'York in the Quartier Latin to sign the definitive Treaty of Paris.

The treaty was nothing short of a diplomatic triumph. It not only confirmed the absolute independence of the United States but also guaranteed room for national growth. Under the terms of the agreement, the new nation took possession of the vast forested wilderness between the Appalachian Mountains and the Mississippi River. John Adams voiced the justifiable pride felt by all the American peace commissioners: "Undisciplined marines as we were, we were better tacticians than was imagined."[11]

The American community regarded the invention of the balloon as an event perfectly calculated to celebrate their hard-won independence. The colorful globes that rose, one after another, above the Paris skyline that autumn symbolized the coming of a new age. It seemed only fitting that a new nation that promised unparalleled freedom and equality of opportunity should be born at the very moment when human beings took their first faltering steps toward achieving the freedom of the skies.

The excitement had begun in June 1783 when Parisians first learned that Joseph Michel and Jacques Etienne Montgolfier had sent a strange contrivance into the sky. The flight was said to have taken place on June 4 in Annonay, a provincial market and industrial town in the Vivarais, an administrative area roughly corresponding to the modern department of the Ardèche. The experiment had apparently been witnessed by a large number of people, including the members of a local legislative assembly.

The first reports were intriguing, but contained few technical details. An

The frightened citizens of Gonesse attack and destroy the Globe.

anonymous letter from Annonay that appeared in a Paris newspaper on July 10 was typical:

> *The* Montgolfiers *have just performed a really curious spectacle here, that of a machine made of cloth and covered with paper which had the shape of a house, being 36 feet long and 16 feet wide, and about as high. They made it ascend by means of a fire to a prodigious height, until it seemed no bigger than a drum. It could be seen three leagues from the city. My peasants who saw it, frightened at first, believed it was the moon falling from the sky; they regarded this terrible phenomenon as the forerunner of Judgement Day. I watched the machine rise; it had difficulty getting started, but when once filled with smoke it left as quickly as a rocket. It stayed in the air about 7 to 8 minutes, and then fell down a quarter of a league from the place where it had started, on top of a small raised wall from which it was impossible to pull it down and save it from the fire. Therefore it was completely burned. That was perhaps a good thing, for one of the* Montgolfiers *talked seriously of ascending in it. Only one thing stopped him, which was the difficulty of making it descend where he wished; but he had no doubts he could go very far, by means of fire, without any danger. It is said that this accident did not discourage the gentlemen and that they are busy making another machine. However, the first one cost them more than 900 pounds. According to their*

reputed character, they won't give up their enterprise until one of them has broken his neck.[12]

Benjamin Franklin was apparently the first American to take note of these reports. As early as July 27, 1783, he had begun to inform his network of scientific correspondents in England and on the continent of "a vast Globe sent up into the Air, much talked of here, and which, if prosecuted, may furnish [a] means of [obtaining] new knowledge."[13] Then, like the rest of Paris, Franklin was forced to wait for the Montgolfier brothers to unveil their invention in the capital, so that he could witness the miracle of flight for himself.

In fact, the hot air balloon was not so much a miracle as it at first appeared. It was so simple a device that it is difficult to speak of it as having been "invented" in the popular sense of that term. The Montgolfiers had not been required to make any great intellectual leaps or to uncover any new bits of scientific or technical information in order to construct their balloon. They had simply taken advantage of principles and materials that had been readily available for centuries. A balloon capable of carrying human beings into the air could have been constructed by any of the ancient peoples that produced relatively lightweight, tightly woven fabrics. Proof of this fact came in 1973 when a group of balloonists built and flew a simple hot air balloon using only the materials and techniques available to the Incas. While the experiment did not prove that the Incas had flown, it demonstrated that they could have done so.[14]

Archimedes laid the theoretical foundation for the balloon when he stated the law of buoyancy in the third century B.C., yet the first hint of interest in lighter-than-air flight was entirely unrelated to theory. A second-century B.C. Chinese text describes a magic trick in which an empty egg is made to fly by burning a bit of mugwort stuffed through a hole in the shell.[15] Whether by independent invention or by diffusion, a variant of the flying egg appeared in Europe as early as 1579 when the English writer Thomas Lupton suggested that an eggshell filled with a few drops of dew would rise into the air when placed in the morning sun. Gaspar Schott, Lauretus Laurus, and other seventeenth-century authors also gave instructions for performing the trick. The fanciful tale in which Cyrano de Bergerac is carried aloft when bottles of dew tied to his belt are drawn up toward the sun was clearly inspired by the flying eggs. It is interesting to note that the Montgolfiers' father, who had taken little prior interest in the balloon experiments, described the flying egg trick to his sons in the early summer of 1783.[16]

The flying egg was an indication of interest in buoyant flight, but it is possible that small balloons had been flown in the Orient long before the trick appeared in Europe. Eighteenth- and nineteenth-century travelers' accounts of China and Southeast Asia describe hot air balloons constructed of rough paper stretched over bamboo frames. Joseph Needham, an authority on ancient Chinese technology, has suggested that these simple aerostats may have been in use at a much earlier date.[17]

The first European proposal for a genuine balloon appeared as early as the fifteenth century. Giovanni da Fontana, a well-known Italian commentator on technology, referred to an unnamed inventor who had "conceived of making, from thick cloth and rings of wood, a pyramid of very great size whose point would be uppermost, and of firmly tying across . . . at the base a board on which a man might sit or ride, holding in his hands burning brands made of pitch and tallow or other material producing intense fire which is long-lasting and creates a great deal of thick smoke." He suggested that the fire would heat the air enclosed in the pyramid making it lighter and rarer. Consequently, "the pyramid and the man sitting in it would be raised."[18] Fontana regarded the author of this scheme as an inventor who "reasoned badly," but, with benefit of hindsight, we can applaud the unknown dreamer for having provided the earliest description of a hot air balloon.

Given this lengthy interest in the possibility of lighter-than-air flight, and at least one perfectly straightforward description of a workable balloon, one is left with the question why so many centuries were to pass before the ancient dream of flight was realized. The answer is to be found in the fact that the balloon did not develop from this prior tradition but from a scientific revolution that began in the seventeenth century. It was one of those rare times in history when a new vision of the universe truly recast the shape of the future. A new alliance forged between science and technology lay at the core of the revolution. The attempt to understand the structure and operation of the universe was no longer to be regarded as a philosophical exercise devoted solely to the glorification of God. Rather, science was recognized as an essentially practical enterprise directly linked to the development of new machines and techniques.

The new relationship between science and technology was clearly evident in the atmospheric studies of seventeenth- and eighteenth-century natural philosophers. Galileo's early discussion of the specific gravity of air in his *Discourses upon Two New Sciences* had been, in part, sparked by contemporary efforts to develop a suction pump to remove water from deep mines. Galileo's studies, in turn, led Torricelli to the development of the barometer, an instrument that Marin Mersenne, Blaise Pascal, Adrien Auzout, Jean Pecquet, and others used to conduct experiments relating atmospheric pressure to altitude.

The air pump, developed by Otto von Guericke in the 1650s, was a second major technological product of early scientific studies of the atmosphere. The great English scientist Robert Boyle had read of the invention of the air pump and of von Guericke's famous experiment in which two evacuated copper hemispheres could not be pulled apart by horses. Boyle and his laboratory assistant Robert Hooke developed their own improved air pump with which they conducted experiments that led to an explanation of the operation of the barometer as well as the relationship between volume and pressure that was to become known as Boyle's Law.

Now it was the turn of the pioneering analytical chemists. Having arrived at some understanding of the physical properties of air, Robert Boyle pointed to the importance of identifying the constituent elements of this and other compounds. Joseph Black, a Scottish chemist, was the first to identify one of these separate gases, carbon dioxide, publishing his results in 1756. Henry Cavendish announced the discovery of hydrogen in 1766. Joseph Priestley followed with oxygen in 1775. Antoine Laurent Lavoisier completed the cycle of atmospheric studies begun a century before by demonstrating the role of these gases in burning and respiration.

The aeronautical implications of this scientific research had long been apparent. Francesco de Lana's well-known scheme of 1670, which called for the use of four evacuated copper spheres to carry a flying machine aloft, was based on the work of von Guericke, Schott, Boyle, Hooke, and others. The idea was intriguing but obviously impractical. Two other possibilities offered much more hope for success.

First, Cavendish had announced that his "inflammable air" (hydrogen) was seven times lighter than atmospheric air. Within a year or two of his announcement, it had occurred to Joseph Black that a lightweight, airtight membrane filled with the gas might rise in the air in accordance with Archimedes' law. Black went so far as to obtain a calf's intestine from a local butcher but did not actually conduct the experiment. By 1782 Tiberius Cavallo, an Italian experimenter living in England, had tested the notion, only to discover that the bladder he employed was too heavy. He did produce soap bubbles filled with hydrogen, however, and offered an account of these experiments to the Royal Society on June 20, 1782.

The other possibility was derived from the fact that Boyle's studies of the effect of changes in temperature and pressure on air had focused scientific attention on the age-old observation that hot air rises. Had Fontana's unknown engineer been correct after all? Would hot air enclosed in a bag provide sufficient lift to carry a man aloft?

There is much evidence to suggest that the Montgolfiers were not the first men inspired to attempt the experiment. An eighteenth-century issue of the newspaper *La Gaceta de Mexico* noted that, a century earlier, in 1667, a citizen of Las Mendarios del Perro, Veracruz, broke his leg in a fall following an ascent in "a strange device with fire."[19]

There is far more evidence to support the case of Bartolomeu Laurenco de Gusmao. Born in the Sao Paulo province of Brazil in 1685, Bartolomeu immigrated to Portugal in 1705. He became intrigued by aeronautics after reading of the work of Boyle and other contemporary natural philosophers, and in 1709 he petitioned King John V of Portugal for permission to build "an instrument to move in the air." True to his name, John the Magnanimous agreed to fund the project.

Gusmao's first attempt came on August 3, 1709, when a small paper balloon bearing a fire box ("an earthen bowl encrusted in a waxed wooden tray") accidentally caught fire just prior to launch. On August 8 Gusmao flew a similar balloon to a height of twelve feet in the ambassador's drawing room of the royal palace in Lisbon. The demonstration, which took place in the presence of many courtiers and diplomats, reportedly came to an abrupt end when the tiny craft was snatched from the air by two servants who feared that it would set fire to the draperies. During the course of a later test flight in the courtyard of the Casa de India, a similar small balloon reached a more impressive altitude. A fourth flight may also have been made with a much larger balloon.[20]

There is no evidence to suggest that Gusmao himself made a flight, but there can be no doubt that he did fly small balloons in the presence of many reputable witnesses. The Portuguese regarded the flights as nothing more than stunts, and only garbled reports of the trials reached European scientific journals. The entire episode was quickly forgotten, with the exception of a famous print of a bird-shaped flying machine, presumably intended to serve as the car of Gusmao's balloon, which circulated widely in western Europe after 1709. The drawing, identified as the *Passarola*, did not include the balloon, and thus was regarded as nothing more than a fanciful proposal for a heavier-than-air machine.

Yet another report of a pre-Montgolfier balloon comes from Russia in 1731. A young officer named Kria Kutnoi, attached to the staff of the military government stationed in the town of Ryazan, some 120 miles south of Moscow, is reputed to have flown a primitive balloon over a grove of birch trees straight into the tower of a neighboring church. The flight is said to have been made with a balloon constructed of hides, which was filled with "evil smelling smoke." According to local tradition, Kria Kutnoi was threatened with the possibility of being burned or buried alive by villagers who regarded his attempt to fly as a sacrilege, but escaped with the lighter penalty of excommunication.[21] It is clear, therefore, that while the Montgolfiers have been awarded credit for the invention of the balloon, others had responded to the same scientific stimulus during the previous century.

Joseph Michel and Jacques Etienne were, respectively, the twelfth and fifteenth of sixteen children born to Pierre Montgolfier, a prosperous paper manufacturer of Annonay. Beyond their joint heredity and a shared interest in science and technology, the brothers had little in common. Forty-three years old in 1783, Joseph was an absentminded, largely self-educated dreamer. Unable to adjust to the routine of formal education, he had run away in his teens, working his way across the south of France as an agricultural laborer before returning for other, scarcely more successful, tries at school. He had failed repeatedly in business as well. Deeply in debt, he seemed well on his way to becoming the black sheep of the family.

Etienne was quite a different fellow. Born in 1745, he was trained as an architect, a profession that provided him with the formal background in science and technology that his elder brother lacked. Steady and dependable, Etienne assumed leadership of the family enterprise following the death of an older brother and his father's retirement in 1772. He proved to be an innovative and imaginative manager. Always willing to experiment with new techniques and processes, Etienne brought a new prosperity to the Montgolfier mills during a decade that saw Joseph struggling simply to survive.[22]

And yet it was Joseph, the rootless failure, who would immortalize the family name. The origins of Joseph's interest in the balloon are lost in a welter of legends dealing with soap bubbles, rising smoke, and billowing undergarments drying before a fire. In truth, it is far less important to pinpoint Joseph's moment of inspiration than to realize that the real roots of his balloon experiments are to be found in his reading of contemporary science. Joseph had begun calculating the lift to be derived from both hot air and the newly discovered hydrogen while living in Avignon in the late 1770s.

He flew his first small balloons soon thereafter. Returning to Annonay in the fall of 1782, he communicated his enthusiasm to Etienne. Together they flew a series of aerostats constructed of paper reinforced with taffeta. During the early spring of 1783, they flew still larger balloons for distances of up to a half mile or more. Their first public flight followed in Annonay on June 4.

Etienne, the well-educated brother with friends in Paris, was chosen to carry the invention to the capital. He set out from Annonay on July 12, 1783, with two cases of equipment to be used in the construction of a large balloon.

The Montgolfiers were no longer the only inventors who hoped to fly a balloon in Paris. A rival aeronautical enterprise was already under way. Barthé-

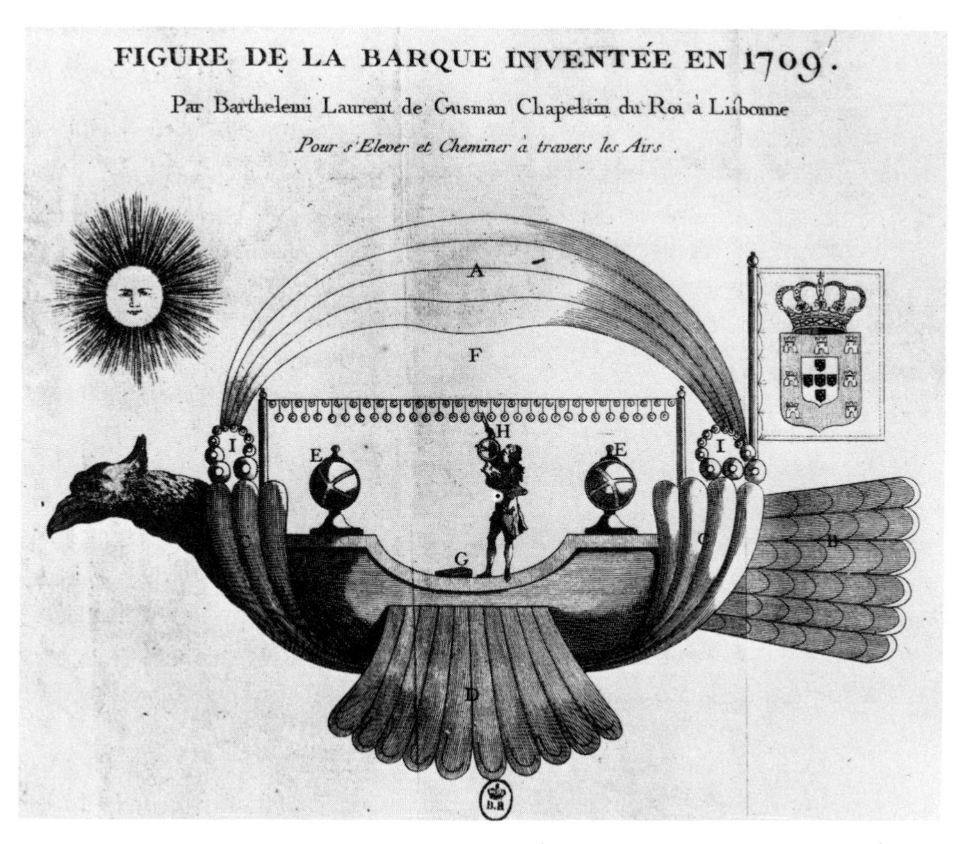

FIGURE DE LA BARQUE INVENTÉE EN 1709.
Par Barthelemi Laurent de Gusman Chapelain du Roi à Lisbonne
Pour s'Elever et Cheminer à travers les Airs .
A
F
H
I
E
E
I
G
B
D

B. Laurenco de Gusmao's famous Passarola *of 1709, for many years thought to have been a fanciful design for a heavier-than-air flying machine, may actually have been a proposed balloon car.*

lemy Faujas de Saint-Fond, a friend of Franklin's and a member of the Academy who had published a study of ancient volcanism in the Vivarais, had apparently witnessed the Montgolfier flight of June 4. Immediately after his return to Paris, Saint-Fond began collecting money to fund the construction of a new balloon. That August, John Quincy Adams noted that "a Subscription was opened some time ago and filled at once for making a globe."[23]

J. A. C. Charles took charge of the enterprise. Unaware of the technical details of the Montgolfier balloons, Charles did not even consider the possibility of a hot air aerostat, planning instead to enclose a bubble of hydrogen in a ball of lightweight, impermeable fabric.

As one of the best-known natural philosophers in Paris, Benjamin Franklin was close to the center of both the Charles and Montgolfier balloon enterprises from the outset. Franklin, the leader of the "undisciplined marines" who had negotiated their nation's independence, was one of the most honored figures in Paris. As Thomas Jefferson noted when he arrived in the city in 1784, "more respect and veneration [are] attached to the character of Dr. Franklin in France than to that of any other person, foreign or native."[24]

Franklin, who had first visited Paris as a colonial agent in 1767 and 1769, took up permanent residence as chief American diplomatic representative in the spring of 1777. Seventy-one years of age, he found himself an instant celebrity. Nobles, diplomats, philosophers, and socialites showered him with accolades. John Adams recalled that "Franklin's reputation was more universal than that of Leibnitz, Newton, Frederick or Voltaire; and his character was more beloved and esteemed than any of them."[25]

His witticisms made the rounds of society. His friendships with the ladies delighted all of Paris. On those relatively rare occasions when he traveled into Paris from his home at the Hôtel de Valentinois in Passy, he went wigless, wore his famous fur hat and spectacles, and invariably drew a crowd of admirers.

Fashionable houses were equipped with Franklin lightning rods and stoves. His picture appeared on everything from snuffboxes to teacups and fans. Franklin described the situation to his daughter Sarah Franklin Bache with typical self-mockery:

> *The clay medallion of me you say you gave to Mr. Francis Hopkinson was the first of its kind made in France. A variety of others have been made since of different sizes; some to be set in the lids of snuffboxes, and some so small as to be worn in rings; and the numbers sold are incredible. These, with the pictures, busts and prints (of which copies upon copies are sold everywhere) have made your father's face as well known as that of the moon, so that he durst not do anything that would oblige him to run away, as his phiz would discover him wherever he should venture to show it.*[26]

One Parisian commentator, Félix Nogaret, observed that Franklin's image appeared over so many French mantles that he seemed "revered . . . no less than a Penate God."[27]

Franklin appeared as a virtual god to the young scientific enthusiasts of Paris as well. Several years before, J. A. C. Charles had abandoned a promising career as a government administrator after reading Franklin's account of his early experiments with electricity and magnetism. Establishing himself as a popular lecturer on scientific topics, Charles had prospered. Franklin attended a number of the young man's public demonstrations and had been most favorably impressed.

"Nature," he remarked, "cannot say no to him."[28]

Franklin must have been particularly pleased to learn that the young man whose scientific career had been inspired by his own efforts had taken command of the Paris balloon project. Certainly he remained fully informed on the progress being made in the Robert workshop.

The fabrication of the *Globe* was a fairly simple task. Generating the unprecedented quantity of hydrogen required to inflate the aerostat proved far more difficult. Initially the envelope was filled with hydrogen produced by adding sulphuric acid to iron scraps contained in an apparatus resembling a chest of drawers. When this device proved so leaky as to be worthless, Charles and the Roberts began using a series of barrels in which the gas was generated, then conveyed to the balloon through leather pipes.

Having developed a means of producing sufficient gas, the experimenters had to deal with the heat resulting from the chemical reactions. The inflated balloon grew so hot that Charles feared the possibility of a catastrophic explosion. When he cooled the envelope with water, water vapor condensed inside the bag. This highly acidic vapor not only forced hydrogen out of the balloon but attacked the delicate rubberized coating on the fabric, causing leaks.

While the Robert brothers and Charles struggled to overcome these difficulties, the crowds in the Place des Victoires grew ever larger. By August 20, when the first attempt was made to inflate the *Globe*, the curiosity seekers milling around the statue of Louis XIV, which dominated the area in front of Maison Robert et Robert, grew so boisterous that archers were posted to preserve order.

The partially inflated *Globe* had been tied to a cart and moved to the Champ de Mars during the predawn hours of August 27. Preceded by torchbearers and surrounded by city guardsmen mounted and on foot, the nocturnal parade was an extraordinary sight. Coachmen encountered on the way to the launch site were so awed by the spectacle that they pulled up their vehicles and knelt on the pavement, hat in hand, while the cortege passed.

Etienne Montgolfier was in the crowd to view J. A. C. Charles's triumph on August 27. As Benjamin Franklin was aware, Montgolfier was building a 38-foot balloon "at the Expence of the Academy." It was, Franklin had heard, "constructed of Linen and Paper, and is to be filled with a different Air, not yet made Public, but cheaper than that produc'd by the Oil of Vitriol [sulphuric acid], of which 200 Paris pints were consum'd in filling the other."[29]

Montgolfier had made his headquarters at the Reveillon paper works in the Faubourg Saint-Germain. On September 12, 1783, Etienne inflated his large hot air balloon in the presence of the members of the Academy. This balloon, which was intended for use in a demonstration before the king and queen at

Versailles only a week later, was destroyed in a sudden downpour.

Over the next few days Montgolfier and the Reveillon workmen struggled to complete work on a brand-new balloon measuring 41 feet in diameter and 57 feet tall. Tested on Thursday, September 18, 1783, it floated "in a most majestic fashion for five or six minutes."[30]

The flight before the court at Versailles on September 19 was an unqualified success. When released, the balloon rose to an altitude of 1,450 feet, carrying with it the first three aerial travelers, a sheep, a duck, and a cock. The report of the Academy of Sciences noted: "Never before has an experiment been carried out with more brilliance and splendour, never before have there been such eminent spectators, nor in such great numbers for a like event."[31]

Franklin, who had not been able to attend the Versailles flight, was nevertheless able to keep Sir Joseph Banks well informed. "So vast a Bulk, when it first began to rise so Majestically in the Air struck the spectators with surprise and Admiration." "The basket," he continued, "contained a sheep, a duck, and a Cock, who, except the Cock, received no hurt by the fall."[32]

The balloon excitement was now building toward a peak in the city. Hair and clothing styles, jewelry, snuffboxes, wallpaper, chandeliers, bird cages, fans, clocks, chairs, armoires, hats, and other items were designed with balloon motifs. Party guests sipped *Crème Aérostatique* liqueur and danced the *Contradanse de Gonesse* in honor of Charles's *Globe*.[33]

"All the conversation here at present turns upon the Balloons . . . and the means of managing them so as to give Men the Advantage of Flying," Franklin informed Richard Price, an English correspondent.[34] His friend Baron Grimm concurred. "Among all our circle of friends," he wrote, "at all our meals, in the antechambers of our lovely women, as in the academic schools, all one hears is talk of experiments, atmospheric air, inflammable gas, flying cars, journeys in the sky."[35]

Even the taciturn John Adams had caught a touch of the balloon fever. Adams, who was not normally interested in scientific questions, noted balloon flights in his correspondence and speculated on the principles of aerostation. Unaware of Tiberius Cavallo's experiments, he suggested blowing hydrogen through a soap solution to produce "balloons of soap."[36]

The small balloons which, Franklin noted, were now being flown "everyday in every quarter" were yet another indication of the extent to which "balloonamania" permeated Paris.[37] His invitations now included the request that he visit a friend's home for "tea and balloons." Franklin attended one fête at which the duc de Chartres distributed "little phaloid balloonlets" to the delighted guests. At another memorable entertainment staged by the duc de Crillon in

The flight of the sheep, duck, and rooster from Versailles, September 19, 1783

the Bois de Boulogne, Franklin witnessed the launch of a hydrogen balloon some 5 feet in diameter that kept a lantern aloft for over eleven hours.[38]

John Quincy Adams also took note of the small balloons offered for sale by street vendors. "The flying Globes are still much in Vogue," he wrote on September 22. "They have advertised a small one of eight inches in diameter, at 6 livres a piece without air [hydrogen] and 8 livres with it, . . . several accidents have happened to persons who have attempted to make inflammable air, which is a dangerous operation, so that the government have prohibited them."[39]

Benjamin Franklin had purchased one of these small balloons as a present for his grandson and secretary, William Temple Franklin, on September 15. Released in a bed chamber that evening, "it went up to the ceiling and remained rolling around there for some time."[40] Franklin emptied the membrane of hydrogen and forwarded it to Richard Price so that he and Sir Joseph Banks might repeat the experiment. The delightful little toy was thus not only the first balloon to be owned by an American but also the first to reach England. Both Benjamin and William Temple Franklin were soon receiving requests for these little balloons from friends on the Continent.[41]

A growing rivalry between the Montgolfier and Charles camps added to the excitement. Both groups courted Franklin's favor. Etienne Montgolfier had invited the aging statesman and a group of his friends to visit the Reveillon workshop in September and October. Charles, of course, considered himself Franklin's disciple. For his part, Franklin attempted to straddle the fence, suggesting that if the balloon was a "new born babe," then Montgolfier was the father and Charles the wet nurse.[42]

The next step in this informal race was obvious. The time had come for human beings to venture aloft. John Quincy Adams noted the growing talk of the possibility of human flight in the pages of his diary: "The enthusiasm of the People of Paris for the Flying Globes is very great; several Propositions have been made from Persons, who to enjoy the honour of being the first Travelers through the air, are willing to go up in them and run ten risques to one of breaking their necks."[43]

Rumors to the effect that secret manned flights had already been made were circulating as early as October 1783. One such story appearing in a London newspaper suggested that Franklin's assistant, one Joseph Fathom, had died during just such a secret balloon ascent. According to this account, Fathom "often suffered hairbreadth escapes from the dangers into which he was betrayed by his eagerness in the imitation of his master's trials." Fathom, it seems, had constructed "an air balloon of immense size," complete with a valve "by means of which he could introduce such a portion of atmospheric air as to moderate the

"La mode au ballon"

velocity with which the balloon ascended or bring it gradually to ground again." During a secret test flight, the would-be aeronaut lost his hold on the valve line and was carried into the upper atmosphere where "air utterly unfit for respiration . . . put an immediate end to his existence." The story, of course, was a complete hoax, but it did indicate the extent to which Franklin's name was already linked to the balloon.[44]

Pilâtre de Rozier, not Joseph Fathom, was the first man to fly. A thirty-year-old scientific lecturer and keeper of a natural history cabinet and laboratory for the comte de Provence, de Rozier had become a balloon enthusiast and was, in fact, the first man to reach the Montgolfier balloon when it came to earth following the animal flight of September 19. He had begun to petition the Academy for permission to fly as early as August 30, but had been refused until the safe return of the sheep, duck, and cock demonstrated that life could survive above the surface of the earth.

De Rozier was to venture aloft in a very large balloon that Etienne Montgolfier was then constructing in the Reveillon shops. Measuring 46 feet in diameter and standing 70 feet tall, the new craft was gaily decorated with signs of the zodiac, royal suns, and the king's initials. The balloon featured a low, circular passenger gallery at its base. A brazier suspended inside the mouth could be fed with straw from the gallery.

The first tethered flight with de Rozier aboard was made from the Reveillon garden on October 15. A second tethered ascent, this time to an altitude of 200 feet, followed on October 17. It was now clear that a second man would be required aboard the craft, both to balance it during the flight and to assist in stoking the fire.

The marquis d'Arlandes was the second volunteer. The son of a minor noble family, he had grown up near Annonay and had known Joseph Montgolfier for a number of years. On October 19 d'Arlandes and de Rozier made a final tethered ascent together. All seemed to be in readiness for the great moment when two human beings would break their final ties to the earth and soar freely into space.

Peter Jay Munro was an enthusiastic observer at the tethered Montgolfier ascents in the Faubourg Saint-Germain. The Jays had followed the progress of the balloons with particular interest. John Jay had sent word of the Annonay flight to friends in America as early as July 1783.[45] He probably attended the *Globe* flight, and lost little time in informing William Livingston of the animal flight of September 19. "A Monsieur Montgolfier has invented Globes which he fills with inflammable air so much lighter than common air as that they rise above the Clouds with great Velocity and are capable of carrying up with them a very considerable weight," he wrote.[46]

Now, with John Jay temporarily detailed to London, his nephew Peter dispatched a series of letters keeping him up to date on the latest aeronautical developments: "Mr. Montgolfier has sent up a Globe tied to a Rope, and two men with it, they tried to fill it with Smoke while in the air, and succeeded. Then these new Navigators instead of <u>spare Masts</u>, <u>spare Sails</u>, <u>Cordage</u>, etc. etc. will take with them a supply of <u>Straw</u>, <u>Old Shoes</u>, <u>Rabbits Skins</u> and <u>other animal substances</u>."[47]

As Peter Munro observed, the Montgolfiers did burn small quantities of such assorted "animal substances" in the belief that they produced a lighter "gas" than the smoke obtained from straw or wool alone. There were, however, disadvantages to be faced in burning rabbit skins and old shoes. On the occasion of the ascent from Versailles, the stench of the fire was reported to have forced Marie Antoinette to cut short her tour of the furnace area.

John Jay complimented his nephew on his "just" observations and inquired as to whether or not "the men who ascended made any Experiments on the State of the air, etc.?"[48] But Peter had much more important news to report on the morning of November 21, 1783. The great moment when human beings would cut their tether and make a free flight through the air had arrived at last.

When Benjamin Franklin arrived at the Chateau de la Muette that morning, a "vast concourse of Gentry" had already gathered in the garden. The distinguished American had ample opportunity to inspect the great balloon. It was, he noted in yet another letter to Sir Joseph Banks, "larger than that which went up from Versailles and carried the Sheep, & c. Its bottom was open, and in the middle of the Opening was fixed a kind of Basket Grate in which Faggots and Sheaves of Straw were burnt. The Air rarified in passing thro' this Flame rose in the Balloon, swell'd out its sides and filled it."[49] Franklin took particular note of the crew accommodations: "The Persons . . . were plac'd in the Gallery made of Wicker, and attached to the Outside near the Bottom, had each of them a Port thro' which they could pass Sheaves of Straw to keep up the Flame, & thereby keep the Balloon full."[50]

Pilâtre de Rozier and the marquis d'Arlandes made their first attempt to ascend at about 1:50 P.M. Franklin watched in horror as the giant balloon was blown into the trees that lined the garden walks: "The Gallery hitched among the top Boughs . . . while the Body of the Balloon lean'd beyond and seemed likely to overset. I was then in Great Pain for the Men, thinking them in danger of being thrown out, or burnt for I expected that the Balloon being not upright the Flame would have laid hold of the inside that leaned over it. But by means of some cords that were still attach'd to it, it was soon brought upright again, and made to descend, & carried back to its place. It was, however, much damaged."[51]

After some hurried repairs, de Rozier and d'Arlandes climbed back into the gallery and were off once again, this time with complete success. "When it went over our heads," Franklin informed Banks, "we could see the Fire which was very considerable."[52] Further: "When they were as high as they chose to be, they made less Flame and suffered the Machine to drive Horizontally with the Wind, of which however they felt very little, as they went with it, and as fast.[53]

For twenty-five minutes the aeronauts moved sedately across the suburban rooftops, finally settling to a soft landing between two mills some 5 miles from their take-off point. The Montgolfiers had placed second in the race to fly the first balloon in Paris, but they had succeeded in sending the first human beings into the sky.

Franklin was honored by a visit from Etienne Montgolfier and the marquis d'Arlandes on the evening of November 21. He gladly added his signature to the official report on the flight that was to be submitted to the Academy of Sciences. In exchange, the marquis provided Franklin with a detailed account of mankind's first aerial voyage: "They say they had a charming view of Paris & its Environs, the Course of the River, & c. but that they were once lost, not knowing what Part they were over, till they saw the Dome of the Invalids, which rectified their ideas. . . . He informed me that they lit gently without the least shock, and the Balloon was very little damaged.[54]

In spite of his special opportunities to obtain a knowledge of the Montgolfier technology, Franklin was forced to admit to Banks that "there is a little mystery made, concerning the kind of air with which the Balloon is filled. I conceive it to be nothing more than hot smoke or common air rarify'd, tho' in this I may be mistaken."[55]

While Franklin made every effort not to choose sides in the friendly rivalry that developed between the Montgolfier and Charles camps, it was apparent that he believed the gas balloon to be far more promising. Charles and the Roberts were, he remarked, "men of science and mechanic Dexterity."[56] He added his own name to the list of subscribers underwriting the construction of the large hydrogen balloon that would carry Charles and the elder Robert aloft.

In his letter to Banks describing the de Rozier and d'Arlandes flight, Franklin had uncharacteristically failed to provide such basic details as the dimensions of the balloon, the distance traveled, or the length of time it remained in the air, yet in this same report he offered a detailed description of the Charles balloon then under construction. It was, he noted, 26 feet in diameter and constructed of alternate red and white silk gores, "so that it makes a very beautiful appearance."[57]

The first to fly, November 21, 1783

Charles had furnished the "very handsome triumphal Car" in which the aeronauts would ride with a small table and a variety of scientific instruments. The expense of filling the balloon with hydrogen would, he estimated, exceed 10,000 livres.

The first flight of a manned gas balloon illustrated both the advantages and disadvantages of this alternative to the Montgolfier hot air technique. Even a large Montgolfier could be inflated in only a few minutes. The Charles project continued to be plagued with problems in the generation and transfer of the hydrogen.

The difficulties of controlling the altitude of a hydrogen balloon also required serious thought. In the case of the Montgolfier balloon, the aeronaut had only to add fuel to the fire or allow it to decrease in order to rise or fall. The hydrogen balloon was quite another matter. Franklin had wondered aloud to Richard Price if it might not be possible to "create a mechanical Contrivance to compress the Globe at pleasure" to make it descend and allow it to expand to gain altitude.[58]

Charles developed a more practical solution, the valve and ballast system. When the balloon was inflated, ballast was loaded aboard until the craft was too heavy to ascend. Weight would then be dropped until the balloon began to rise. Once in the air, the pilot would simply drop ballast to rise and valve gas to descend.

The Charles balloon was an extraordinary device. Only the second large gas balloon to be constructed, it epitomized the finished technology. Nothing more than minor refinements would be made to the gas balloon over the next two hundred years.

The Charles flight had originally been scheduled for November 29, but inflation problems forced a postponement until December 1. Franklin, who was not feeling well on that cool, damp morning, chose not to go inside the enclosure in the Tuileries Gardens from which the ascent would be made. Instead, he stayed in his carriage near the large equestrian statue in the Place Louis XV.

From this vantage point, between one and two o'clock that afternoon, he saw Charles and the elder Robert "rise majestically from among the trees and ascend gradually above the buildings." It was, he noted, "a most beautiful spectacle!"[59]

Other Americans in the crowd at the Tuileries that day were just as impressed. Daniel Lathrop Coit found himself at a loss for words in attempting to describe this "prodigy not before heard of in the world" to his family in Connecticut.[60] A native of New London, born on September 20, 1754, Coit was apprenticed at an early age to an uncle involved in the import trade. He departed on his first European business trip aboard the brig *Iris* on June 7, 1783. After meeting

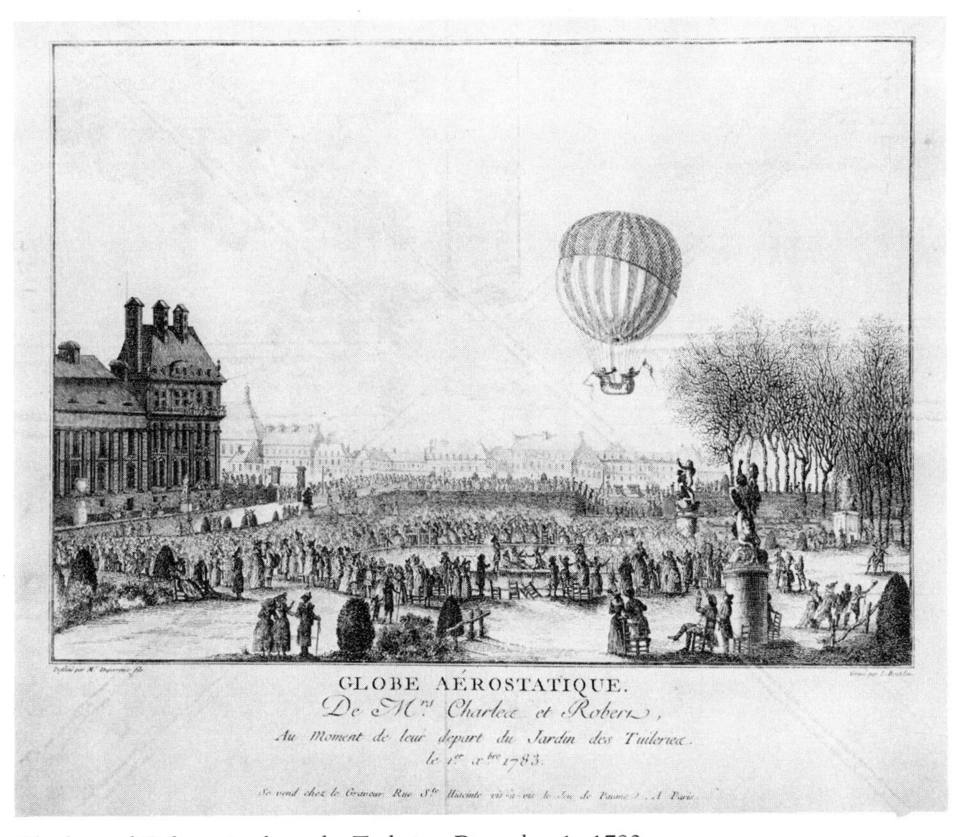

GLOBE AÉROSTATIQUE.
De M.rs Charles et Robert,
Au Moment de leur depart du Jardin des Tuileries.
le 1er xbre 1783.

Charles and Robert rise from the Tuileries, December 1, 1783.

with suppliers in Portsmouth, Birmingham, London, and Holland, he left for Paris via Dover and Calais on October 25. Like most Americans visiting the French capital, his first stop was at the Hôtel de Valentinois where he met both Franklin and the Marquis de Lafayette.

Coit arrived in Paris at the height of the balloon fever. "The novelty of the thing is so great that it engrosses half the talk and attention of the city," he remarked in a letter to his father. Although Coit attended the Charles and Robert flight, he had difficulty accepting the evidence of his senses and was careful to insure that his family and friends would not regard his account as a joke. "Had it occurred in antiquity," he remarked to his father, "we could not have believed but with pretty good evidence. . . . I presume you would even call my veracity into question, tho' I am not used to telling stories that cannot easily be believed."

In spite of his enthusiastic response to the flight, Coit was quick to recognize the major problem of the free balloon. "The basic difficulty they labor under is to steer themselves, which they are not yet able to do and I apprehend will find it a difficult matter ever to attain too, as the fluid in which they move always goes as fast as they do themselves."[61]

Peter Jay Munro and his aunt Sally Jay had been just as impressed by the sight of the sparkling red-and-white balloon rising into the air. Sally informed her absent husband, John, that the illustration on the ticket of admission which she had sent to him, "correctly represented" the scene. The aeronauts, she continued, were housed in a car "which was supported by Cords that fasten'd it to a net that was thrown over the upper part of the Globe."[62] Sally Jay had obtained a thorough knowledge of gas balloon technology, which she hastened to share with her husband:

> *Thro' the Globe pass'd a tube that descended into the Carr, which by means of a string serv'd to let out the gas from the top [of the balloon] at the discretion of the gentlemen. The Balloon as before was made of elastic gum—Nothing could have better answered their expectations, for they found themselves sufficiently masters of it, to make it rise or descend at pleasure, by means of the ballast they provided, for they took up so many sacks of sand as the Globe could raise which made their ascent very gradual, [and] when they chose to mount still higher they threw out some of the sand. It was 40 minutes before two when they left the Garden and as the wind was very gentle it [the balloon] remain'd a long time in sight.*[63]

Like Sally Jay, Benjamin Franklin felt some apprehension for the safety of the aeronauts as he watched the balloon shrink to the size of a walnut through his pocket telescope. Mrs. Jay was considerably relieved the next day to learn that the airmen had landed safely in an open field 27 miles from Paris.

Anxious to obtain additional details of the flight, she dispatched a servant to fetch a copy of the *Journal de Paris* from Benjamin Franklin, in whose home the Jays had lived until recently. William Temple Franklin appeared on her doorstep before the servant could return. Unwilling to walk the remaining distance home in order to learn the outcome of the flight, William had called to borrow Mrs. Jay's newspaper. Both of them were able to satisfy their curiosity when the servant returned from Dr. Franklin's home. What must have been a very dog-eared copy of the *Journal de Paris* was then dispatched to Sir Joseph Banks in England.[64]

Sally Jay quickly passed the good news on to John, who was still absent in London: "The Dukes of Chartres and Fitz-James, I am told, by taking horses followed its course, & arriv'd before it at a Village 9 leagues from Paris at the inst. the Balloon descended—then Mr. Robert alighted & Mr. Charles remaining in the Carr . . . after taking in some earth to replace the weight it had lost by the descent of Mr. Robert, tho' not an equal proportion, Mr. Charles remounted to make some observations with the instruments he had provided for that purpose and went a league and a half further, but with much greater velocity."[65]

Peter Jay Munro provided his uncle with additional information, including a comparison of the altitudes attained and the distances traveled by the two rival teams of aeronauts. Having described the latest French aeronautical developments, Peter asked if "the English [are] making any experiments of that kind?"[66]

Benjamin Franklin was also puzzled as to why his English colleagues had yet to take to the sky. "I am sorry this experiment is so neglected in England where mechanic Genius is so strong," he chided Banks. "Your philosophy seems too bashful. In this country we are not so much afraid of being laughed at."[67] Writing to the Dutch experimenter Jan Ingenhousz, Franklin voiced his fear that "national jealousy" had "prevented the English from prosecuting the Experiment." It was a pity, he added, for "they are such ingenious mechanicians that in their hands it might have made a more rapid progress toward perfection, and all the utility it is capable of affording."[68]

For himself, Franklin was more than pleased with the progress that had been made in France. Now that human beings had flown farther than the distance from Calais to Dover, he felt that his initial enthusiasm had been confirmed. "We think of nothing here but of flying," he informed Henry Laurens. "A few months since," he remarked to Sir Joseph Banks, "the idea of Witches riding thro' the air upon a broomstick, and that of Philosophers upon a bag of smoke would have appeared equally impossible and ridiculous."[69]

Franklin had given the old question of utility a great deal of thought since the *Globe* flight. While the balloon would "hardly become a common carriage" in his lifetime, he could foresee important military applications: "Convincing sovereigns of the Folly of wars may perhaps be one effect of it, since it will be impractical for the most potent of them to guard his dominions. Where was the Prince who could afford to field an army sufficiently large to defend his nation against a balloon borne invasion force of 10,000 men who might land anywhere?"[70] Science would also benefit from the improved understanding of the upper atmosphere that would come with extended balloon voyages.

In fact, the question of potential utility was of little importance to Franklin. The invention would be worth supporting even if there were no immediate practical benefits to be enjoyed. "It does not seem to me a good reason to decline prosecuting a new Experiment which apparently increases the Power of Man over Matter, till we can see to what use that power may be applied. When we have learnt to manage it, we may hope some time or other to find uses for it, as men have done for magnetism and electricity, of which the first experiments were mere Matters of Amusement."[71]

But Franklin was not blind to the dangers of ballooning. When a friend informed him of his intention to attempt a Channel flight, Franklin urged him not to risk his life: "It is said by some . . . who have had experience, that as yet they have not found means to keep up a balloon more than two hours; for that, by now and then discharging ballast as to avoid descending too low, these means of regulation are exhausted. Besides this, all circumstances of danger . . . in the operation of the soupapes [valves] &c. &c., seem not to be yet well known, and therefore not easily provided against."[72]

Jan Ingenhousz also expressed a desire to build and fly a balloon. Once again Franklin urged caution: "If you undertake to make one, I think it extremely proper that you send an ingenious man here . . . otherwise, for want of attention to some particular circumstance, . . . the Experiment may miscarry . . . [and] draw upon you a good deal of censure, and affect your reputation."[73] But Franklin's enthusiasm overcame his caution. He had no doubt that "important consequences that no one can foresee" would flow from this "glorious object," the balloon.[74]

John Quincy Adams shared Franklin's confidence in the future of the balloon. "This discovery is a very important one," he noted in his diary, adding that "it may become very useful to mankind."[75] Sally Jay agreed. She noted that the Charles and Robert flight "seems to have confirm'd the general idea of great utility resulting from the discovery."[76]

Even taciturn John Adams could see some romantic possibilities in the invention of the balloon. Writing to his wife, Abigail, from whom he had long been separated, he remarked that "if the balloon should be carried to such perfection . . . as to give Mankind the safe navigation of the air, I will fly [to you] in one of them at the Rate of 30 Knots an hour."[77]

John Quincy Adams shared a similar lighthearted hope for the future of the balloon with his fellow enthusiast Peter Jay Munro. "I heartily wish they would bring the balloons to such perfection, as that I might go to New York, Philadelphia, or Boston in five days time."[78] In a slightly more practical vein, young Adams confessed his desire to take a *tour aérostatique* aboard a tethered balloon the Robert brothers promised to establish in Paris. For a small price, Adams noted, "any curious person may mount as high as he pleases, and so look down upon the pendant world."[79]

The American residents of Paris continued their infatuation with the balloon during the next year. New faces were now on the scene. Dr. John Foulke, a Philadelphia physician studying in Europe, gathered material for the lectures and demonstrations of balloon flight he planned to offer once he returned home. M. G. Fontaine, who was later to become a prominent New York auctioneer, was present in Lyons on January 19, 1784, when Joseph Montgolfier unveiled his giant balloon *La Flesselles*. Standing 126 feet tall and measuring 100 feet in diameter, the monster ripped and burned at take-off.[80]

For the Americans, the pioneer period in the history of aeronautics can be said to have ended with an ascension by the Robert brothers and their cousin Colin Hullin on September 19, 1784. The trio began their journey from the Tuileries at 11:59 A.M. They landed safely at the village of Beauvray, near Béthune, at 6:40 that evening, having established a new distance record.

John Quincy Adams escorted his mother, Abigail, and his sister Abigail (Nabby) to the ascension. Nabby Adams was impressed by the fact that eight to ten thousand persons had gathered to witness the ascent. "These people are more attentive to their amusements than anything else," she noted in her diary, adding, however, that "as we were upon the same errand, it is unjust to reflect upon others whose curiosity was undoubtedly as well founded."[81] Mrs. Piozzi, an English visitor, also made note of the enormous crowd gathered in the Tuileries that day, remarking that she had seen "ten times the bustle and ten times the difficulty at a crowded playhouse in London, than the Parisians made when all the city was gathered together. Nobody was hurt, nobody was frightened—nobody was even incommoded."[82]

Nabby noted that the new Robert balloon was not a sphere but an oval built "in the form of an egg, if both ends were large." She watched as the balloon was walked into position in an open area of the garden by "some of the greatest men in the kingdom," and thrilled to the sight of the three intrepid aeronauts climbing toward the overhanging clouds a few minutes later.[83]

The man who was to replace Benjamin Franklin as chief American diplomat and balloon enthusiast in Paris was also a part of the enormous crowd that had gathered for the Robert-Hullin flight that day. Thomas Jefferson had arrived in Paris in August 1784 with his daughter Martha (Patsy) and James Hemings, a black servant.

Throughout his stay in France, Jefferson's letters home were filled with the latest bits of aeronautical information as well as advice to friends in America who were constructing small balloons. It was through Jefferson that Americans would first learn of the death of Pilâtre de Rozier, the world's first airman, and

The Robert brothers and Colin Hullin fly from the Tuileries, September 19, 1784.

his companion Pierre Romain in the crash of a mixed hot air and hydrogen balloon on June 15, 1785. Like Franklin before him, he passed on the latest prints and pamphlets on balloons to correspondents in Europe and America.

Unlike his predecessors, however, Jefferson had been introduced to the balloon long before he arrived in France. While still in the United States, he had been able to remain current on the latest balloon news and had shared his expertise with scientifically inclined correspondents. The Robert-Hullin flight of September 19, 1784, was Jefferson's first opportunity to see human beings in the air, but he had seen the large paper balloons flown by Dr. John Foulke in Philadelphia that May. Moreover, while Jefferson had not witnessed the event, Edward Warren, a thirteen-year-old Baltimore lad, had made the first tethered flight in the United States on June 24, 1784, eleven days before Jefferson sailed from Boston aboard the brig *Ceres*. The balloon had arrived in America before Jefferson arrived in France.

Chapter 2 The "New Born Babe" Comes to America

GOUVERNEUR MORRIS, THE THIRTY-ONE-YEAR-OLD ASSISTANT SUPERintendent of finance for the fledgling Continental government, was one of the first residents of the United States to learn of the invention of the balloon. Morris and his longtime friend John Jay had worked together to frame the New York state constitution and had shared in the unsuccessful struggle to abolish slavery in the state. The two men had maintained their close contact during Jay's absence in France, so the receipt of a letter from Paris dated July 20, 1783, can hardly have come as a surprise to Morris. The contents, which included mention of the Montgolfier flight in Annonay, were quite another matter however.[1]

Throughout the summer and fall, Jay continued to keep his American friends informed of the latest aeronautical developments. "All the people are running after air globes," he remarked to New York congressman Robert Livingston. "The invention of them may have many consequences, and who knows but travelers may hereafter literally pass from country to country on the wings of the wind."[2]

Jay's letters and the prints and pamphlets he forwarded to the United States aroused much curiosity. Robert Livingston informed Jay that he was considering becoming an "aerial architect" himself, now that these "first castles in the air . . . promise to have some stable use." In a more serious vein, he requested "a fuller account of their composition and the uses proposed to be made of them."[3] Robert Morris, another of Jay's correspondents, inquired if the French balloonists planned "to send passengers with a man to steer the course, so as to make them the means of conveyance for dispatches from one country to another or must they only be sent for intelligence to the moon and clouds?"[4]

Over the next few years letters passing from France to America would continue to fuel enthusiasm for the balloon in the United States. Some correspondents, Franklin and Jefferson, for example, concentrated on providing up-to-the-minute accounts of the latest aeronautical triumphs. Others transmitted a sense of the excitement created by the balloon. William Vans Murray described the impact of the balloon on London society in a letter to a Baltimore friend, Henry Maynaidier, in February 1789: "What strange relation everything stands in to the eye of caprice! The hats now worn by many of fashion are the Air Balloon hats—and the curiosity that engages the philosophy of the day is the Air Balloon or Aerostatic Globe."[5]

Secondhand accounts of balloon activity from Europeans whose friends or relatives had witnessed flights were also important in building American enthusiasm for aeronautics. John Coakley Lettsom, a West Indian-born Quaker physician living in London, was more than happy to pass reports from his contacts in Paris on to Dr. Benjamin Rush and his scientific colleagues in Philadelphia. "The French seem to have acquired a new spirit of . . . discovery in almost every department," Lettsom informed Rush in February 1784, "but it is in the aerostatic machines that they have exceeded all nations." He described the Montgolfier and Charles flights and discussed both the "great secrecy" that continued to shroud the hot air technology and the "jealousy" that had grown up between the two camps. Lettsom was also careful to note the plans the Montgolfiers and J. A. C. Charles had developed for future flights. The Montgolfiers were constructing a large balloon capable of carrying twelve people "on a voyage of 3 months without lighting for provisions." Charles, Lettsom had been informed, was hard at work on a new gas balloon that "is to astonish the world." Lettsom's Paris correspondent had himself "discovered the art of guiding these balloons and proposes to visit some neighboring kingdoms as soon as his [balloon] is perfected."[6]

Thomas Jefferson was introduced to the balloon through secondhand accounts offered by the chevalier de la Luzerne, the French minister to the United States, whose brother had witnessed a flight in Paris. Jefferson, in turn, spread the word within his broad circle of acquaintances. "What do you think of the balloons?" he queried the Philadelphia poet, musician, and patriot Francis Hopkinson: "They are really beginning to assume a serious face. . . . This discovery seems to threaten the prostration of fortified works . . . and the destruction of fleets and what not. The French may now run over their laces, wines & c. to England duty free. The whole system of British statutes made on the supposition of goods being brought into some port must be revised."[7]

Jefferson's remarks on the balloon were much appreciated by his scientific friends. James McClurg, a theologian and scientific lecturer at Jefferson's alma mater, the College of William and Mary in Williamsburg, was particularly grateful. Before hearing from Jefferson, McClurg had rejected scattered rumors of flights in France as "monstrous exaggerations."[8]

After reading an imported copy of Barthélemy Faujas de Saint-Fond's encyclopedic treatment of early ballooning in the spring of 1784, Jefferson was in a better position to share a detailed knowledge of aerostatic technology with philosophically inclined correspondents in isolated areas. In a letter to his cousin Dr. Philip Turpin, written in April 1784, Jefferson provided a virtual abstract of the history of the balloon to date, complete with comparative tables and dimensioned drawings. In addition, he added his own thoughts on the uses to which the balloons would be put:

> *The uses of this discovery are suggested to be 1. transportation of commodities under some circumstances. 2. traversing deserts, countries possessed by an enemy, or ravaged by infectious disorders, pathless & inaccessible mountains. 3. conveying intelligence into a besieged place, or perhaps enterprising upon it, reconnoitering an army & c. 4. throwing new light on the thermometer, barometer, rain, snow, hail, wind, & other phenomena of which the atmosphere is the theatre. 5. the discovery of the pole, which is but one days journey in a balloon, from where the ice has hitherto stopped adventurers. 6. raising weights; lightening ships. . . . 7. house-breaking, smuggling & c.*[9]

While Jefferson regarded some of these objectives as "ludicrous," others, he believed, were "serious, important, and probable."

Important as personal correspondence was in spreading word of the advent of the balloon, most Americans first learned of the discovery in the pages of a newspaper. George Washington was typical of thousands of his fellow citizens who were fascinated by the first published reports of aeronautical activity. "I have only news paper accts. of Air Balloons, to which I do not know what credence to give," he remarked to his friend Louis Le Bèque du Portail in April 1784. "The tales related of them are marvelous, and lead us to expect that our friends at Paris, in a little time, will come flying thro' the air, instead of ploughing the ocean to get to America."[10]

American newspaper coverage of the balloon story had begun as early as November 5, 1783, when the *Salem Gazette,* a Massachusetts journal, had carried a confused account that mixed details of the Charles *Globe* and the Montgolfier animal flight. But if the editor of the *Gazette* was a bit fuzzy in his presentation of balloon technology, he was extraordinarily successful in portraying the reaction of the French peasants to the landing of the first Montgolfier balloon launched from Versailles: "Upon the discovery of so uncommon an inhabitant of the air making its solemn descent, the country people stood aghast; nor did their panic soon subside even after it had reached the ground, its fiery particles sputtering an unknown language, & still keeping them at an awful distance."[11]

By late November, the quality of the reporting had improved and the *Gazette* published a very detailed account of the Charles *Globe* flight, including an accurate treatment of the work of Black, Cavendish, and others leading up to the invention of the balloon. The editor of the *Gazette* also included his own thoughts on the utility of the balloon, which were very much in accord with Franklin's point of view:

> *Every fact established . . . may in some distant place and time, and in unexpected hands, produce another fact; so that it should be a datum, that all philosophical facts well established, are of use. The present machine may teach us many new circumstances relative to the upper regions of the air, as to their density, temperature, electricity, moisture, currents, . . . and thence discover to us other truths or detect unsuspected errors, that may have given us much trouble. It may be applied to purposes of innocent luxury, and even to astronomical and nautical, as well as terrestrial uses: It may be applied by navigators and travelers to reconnoitre traits of country, and by others to reconnoitre the position of fleets and armies, and prevent surprises from them: it may be used as a landmark, or a signal, and it may serve in another way as a signal to animate ambition in science, and spread a desire of knowledge: In short, nobody at first sight should condemn any facts as being barren, and still less those who pursue them, as frivolous persons.*[12]

At any rate, the editor concluded, experimentation with balloons was considerably "more innocent than cutting each others throats for absurd trifles by sea and land, and . . . quite as profitable as . . . a common game of whist."[13]

Projecting new uses for the balloon became a favorite pastime for American newspaper editors. The *Essex Journal and Massachusetts and New Hampshire General Advertiser*, for example, speculated that aerostats could be employed to lift heavy weights over mountains, raise signal lights at sea and over land, and carry electroscopes and meteorological instruments into the upper atmosphere.[14]

Many of the news stories appearing in American journals were based on accounts from European newspapers and magazines. Letters from Americans living abroad were another major source of balloon news. A number of these correspondents shared Daniel Lathrop Coit's fear that his account of the Charles and Robert flight would not be believed. "Be as satirical as you please," noted an anonymous correspondent whose letter was published in Boston's *Independent Ledger*, "certain it is that all good Parisians are petrified with wonder."[15]

There were a few scoffers in the American press however. Papers such as the *Independent Ledger* and the *Massachusetts Spy or Worcester Gazette* referred to the Montgolfiers as geniuses "from whom we may soon expect to make journeys to the moon, or [to] cross over . . . from Dover to Calais and back, with as little trouble as to go on dry land."[16]

Throughout 1784 and 1785 reasonably accurate reports of European flights continued to appear in American newspapers. Readers were able to follow the rapid progress of the balloon in France, including Jean Baptiste Marie Meusnier's plan for a dirigible balloon that would move "with great velocity."[17] Reprints of the latest aeronautical pamphlets, such as Jean Pierre Blanchard's *Description of a Machine Proper to be Navigated through the Air*, which was carried in the *Massachusetts Centinel* along with the first illustration of a balloon to appear in a U.S. newspaper, helped keep Americans up to date on the latest aeronautical thinking.[18]

Balloon activity in other European nations was chronicled as well. Paolo Andreani's first flight in Italy on February 25, 1784, was widely reported in the American press, as were the pioneer efforts in the Low Countries and the German states.[19]

Naturally, American readers were particularly interested in the status of aeronautical research in England. The editor of the *Connecticut Courant* seemed to take a special delight in chiding British scientists for their apparent lack of interest in the field. "The English pride has been too apt to affect a contempt for the productions of their neighbors, particularly in their favorite science of philosophy," he noted. "Their literati have lately poured their ridicule very liberally on the French, for the attempts of several gentlemen of that nation to invent a mode of aerial navigation."[20] A writer for a Massachusetts newspaper had more faith in the English: "Though the French claim the honour of the new invention of air-globes, there is little doubt that the English will be the men to make a rational use of them. Dr. Priestley is said already to have made some collateral discoveries on the subject, which will excite greater astonishment than the original invention itself."[21]

For reasons known only to themselves, American editors seemed ever ready to open their columns to the many amateur poets anxious to celebrate the birth of the age of flight with an outpouring of "humorous" verse. The *Gazette of the State of Georgia* published what was, unfortunately, a typical example of the genre:

How add this whim, to mount
on air stuf't pillows,

T'will ruin all our coachmen
and postillions.[22]

A poet writing in the *Independent*, a Philadelphia journal, hoped to make an interplanetary tour aboard a balloon:

To see all my friends
in the stars,
Take breakfast with Mercury,
Jupiter, Saturn and Mars,
And should I fatigued or
wearisome prove,
While from planet to planet
I'm dodging
With Venus I'm welcome
to tarry all night
Where on earth can you
find such a lodging?[23]

The *Massachusetts Centinel* published yet another bit of prophetic doggerel, "The Progress of Balloons," by Philip Freneau. Freneau's satyric verses offered a lighthearted vision of the skies of the future, filled with balloons. Transportation, interstate commerce, and national defense would, he predicted, be revolutionized.

The poet's comments on aerial agriculture are typical:

To market the farmers
shall shortly repair
With their hogs and potatoes,
wholesale thro' the air,
Skim over the water as light
as a feather,
Themselves and their turkies
conversing together.[24]

Not all of the balloon humor appearing in American newspapers was poetic. One-liners such as, "John, fill the large balloon, my lady and I want to take an airing" abounded.[25]

Balloon hoaxes were also to be found. The *Providence Gazette* reprinted the story of Joseph Fathom, Franklin's mythical amanuensis. Another article in the *Massachusetts Spy* told of an ill-fated French experimenter who floated into oblivion without the aid of a balloon after inadvertently swallowing a dose of hydrogen.[26] The editor of the *Connecticut Courant* attempted to dupe his readers with a tongue-in-cheek report that J. A. C. Charles and Robert had literally been carried away by their enthusiasm. During a recent ascent the intrepid aeronauts had flown so high that they were unable to return to earth and had resorted to dropping weighted messages describing their harrowing adventures "in equilibrio at the edge of the terrestrial atmosphere."[27]

The *Gazette of the State of Georgia* afforded a home-grown tall tale in its issue of September 30, 1784. A few days earlier, it seems, "a beautiful Air Balloon" had been launched in Savannah. Bound for Jamaica, the marvelous craft had carried a crew of six, six hundred bushels of corn, and "necessaries sufficient for the trip."[28] If true, the story would certainly have pleased an anonymous New Jersey gentleman who, according to the *Massachusetts Centinel*, had offered to wager that goods would be imported by balloon within two years at an insurance rate of over five percent.[29]

"An ingenious gentleman" perpetrated another sort of hoax on the citizens of Kingston, Jamaica, that was widely reported in American newspapers. A large crowd that had gathered to witness a balloon ascent announced by this fellow were treated to nothing more than the appearance of a large black cloud of smoke emitted by a greased cannon.[30]

American magazine publishers were also quick to treat aeronautical subjects. The *Boston Magazine* began a discussion of the balloon in February 1784. The article, entitled "Explanation of the Air Balloon," was, at best, highly imaginative. The author reported, for example, that a pistol had to be discharged inside a small wooden keg attached to the bag, "by which means the Balloon is supplied with a sufficient quantity of vapour from burnt powder, to produce the necessary rarefaction, and sustain it in EQUILIBRIO."[31] During the next six months the magazine featured lengthy excerpts from European balloon accounts as well as a letter from an anonymous American correspondent describing the Charles and Robert launch. The magazine also commissioned John Norman, an associate of Paul Revere and one of the leading engravers of the Revolutionary era, to produce a fanciful illustration of the landing of the sheep, duck, and cock following the first Montgolfier flight from Versailles.

The extensive coverage of balloon activity in American newspapers and magazines helped to create a modest wave of "balloonamania" in the United States to match that which was currently sweeping across Europe. A Boston reporter noted in July 1784 that "the taste for Air Balloon matters has grown to such an extravagant pitch that nothing can pretend to have any intrinsic value in it, unless it has this name as an appendage."[32]

New Yorkers were also caught up in the balloon craze. "Air-Balloon dress is so much the fashion in this city and so generally favored," remarked one wag, "that some ingenious seamstresses have it in contemplation to establish a balloon petticoat, so constructed that every person may go up in it with safety."[33]

As a Boston journal noted, "The ladies and gentlemen of the Bon Ton are not the only objects that can boast of this aerial bombastick insignia." In American cities, as in Europe, household items and personal accessories, such as fans and jewelry, were also decorated with balloon motifs.[34]

The *Freeman's Journal*, a Philadelphia newspaper, reported that balloon enthusiasm had spread from the cities into the countryside. A farmer who came to town to sell his produce in a street stall was heard to cry, "Fine balloon string beans, fine balloon string beans!"[35]

American thespians were quick to take advantage of the growing balloon fever. The *Maryland Gazette* carried a very favorable review of a local offering entitled *A Mogul Tale*. The play was a comedy dealing with the adventures of a physician who hires a poor cobbler and his wife to accompany him on a balloon voyage that ends in the garden of the Seraglio of the Great Mogul. *A Mogul Tale*, and similar offerings, would do much to enliven the limited American theatrical circuit during the next decade.[36]

Clearly, the observation of a Providence, Rhode Island, editor that "the Balloon rage increases very rapidly in this town" was true of a great many other American cities as well.[37] The citizens of the new republic obviously shared the hope expressed by a writer in the *Freeman's Journal* that "this country which has hitherto discovered so much genius will dare to meet European philosophers above the clouds and there chart new modes of happiness for their fellow mortals below."[38]

If an impartial observer had been asked to select the American city in which the balloon would be most likely to appear, Philadelphia would have been the obvious choice. Far from being a colonial backwater, Philadelphia was the third largest city in the British Empire in 1776 (after London and Edinburgh) and the largest in the new United States. An estimated 35,000 people lived and worked in a single square mile of the city bounded by Vine, Lombard, and Seventh streets.

In fact, an anonymous hoaxer writing in a French newspaper had already set his fictitious account of the first American balloon ascent in that city. The article, which appeared in the *Journal de Paris* on May 13, 1784, is worth quoting in full:

> *Philadelphia, Dec. 29, 1783–*
>
> *No sooner was the extraordinary discovery of M. Montgolfier known here, about a month ago, than a similar experiment was attempted; not, indeed, on so large a scale, for want of means; and this circumstance has led us to perhaps the happiest application of this phenomena. A man raised himself to a height of ninety-seven English feet, and came down again, but with too much ease.*
>
> *Messrs. Ritnose and Opquisne began their experiments with bladders; and then with somewhat larger machines; they joined several of them together, and fastened them around a cage into which they put animals. The whole ascended, and was drawn down again by a rope. The next day, which was yesterday, a man offered to get into the cage, provided the rope was not let go. He rose about 15 feet and would not suffer them to let him go higher. Gimes Ouilcoxe, a carpenter, engaged to go in it for a little money. He rose 20 feet or more before he signaled to be lowered. After being instructed by Messrs. Ritnose and Opquisne, and some practice on the ground, he agreed to have the rope cut for fifty dollars. Dr. Jarvise, the principal physician in the city, attended in case of accident.*
>
> *The crowd was excited and shouted as they saw Ouilcoxe in the cage surrounded by forty-seven balloons fastened to it, with astonishing coolness, nodding his head to express satisfaction and composure. He was unable to raise higher than 97 feet, according to measures taken by two gentlemen of the Philosophical Academy. He was at least five minutes in the air, when, noticing that the wind was blowing him toward the Schuylkill River, he became frightened and, as instructed, used a knife to puncture three of the balloons. This was not sufficient, though we saw him descend a little. He pierced three more, and seeing that the machine was still not descending, his fear increased. He quickly slashed five more, unfortunately all on the same side. He then seemed to swing around and fell on a fence at the edge of a ditch. Dr. Jarvise ran up and found that the poor man had sprained his wrist, but was otherwise uninjured. He was taken care of, the machine repaired, and it is hoped it will be more successful.*[39]

Unlike the other balloon hoaxes that appeared in European and American newspapers, this story reads like a straight news account. The broad, tongue-in-cheek humor is absent, replaced by exactly the sort of detail one would expect in an eyewitness account. The author provides a graphic description of the aerial vehicle as well as a precise record of flight times and the altitudes achieved. Gimes Ouilcoxe, the hero of the piece, is a far cry from that two-dimensional caricature, Joseph Fathom. Rather, he is a believable human being—greedy, frightened, and prone to error.

The story was especially convincing to readers familiar with Philadelphia. The names Ritnose and Opquisne could easily be read as French phonetic spellings of Rittenhouse and Hopkinson. David Rittenhouse and Francis Hopkinson were leading members of the Philadelphia-based American Philosophical Society. Rittenhouse's reputation as a scientific experimenter and technician was second only to that of his friend Benjamin Franklin, while Hopkinson, a poet, musician, and scientific dilettante, was one of the best-known figures in the city.

Gimes Ouilcoxe probably represents a similar phonetic spelling of James Wilcox. Three men of this name are listed in the Philadelphia directory for 1785. As no occupations are given in the directory, it is not clear whether any of these men were carpenters. Nor does the directory list a Dr. Jarvise.

The story of the Ouilcoxe "flight" has such charm and appeal that one wishes it were true. Just imagine, an untethered flight in the United States accomplished only a few months after the first de Rozier and d'Arlandes ascent! Unfortunately, the story is not true. A careful survey of Philadelphia records as well as the papers and published biographies of Rittenhouse, Hopkinson, and others fails to produce any evidence in support of such a flight. The final proof that the story was a hoax comes from Hopkinson himself. Replying to Thomas Jefferson's inquiry as to the state of aeronautical studies in Philadelphia, Hopkinson remarked, on March 12, 1784, that "we have not yet taken the affairs of the Ballons in hand."[40]

It is impossible to discover the identity of the author of the Philadelphia hoax, but it is interesting to speculate that he may have been a member of the American community in Paris. The editor of the *Journal de Paris* would have had little reason to doubt the word of an American who claimed to have received the "news" in a letter from home. This would explain the appearance of the story in a Paris newspaper rather than in an English-language journal in the United States or Great Britain. It would also account for the use of well-known Philadelphia names and locations and phonetic spelling of the names involved. This could easily have occurred as a French writer took notes on a conversation with an American informant.

In fact, a number of witty and high-spirited Philadelphians had attached themselves to Benjamin and William Temple Franklin's household at the Hôtel de Valentinois. While this circle of Philadelphians in Passy may have perpetrated the hoax, one of them certainly was responsible for flying the first balloon in the United States.

Dr. John Foulke was a native Philadelphian, born in 1757. His father, Judah Foulke, was the head of one of the city's great Quaker families and had filled a variety of civic posts ranging from collector of excise to sheriff of Philadelphia County and keeper of the standards of brass weights and measures.

Young John received his early education at Robert Proud's school, then entered the College of Philadelphia (now the University of Pennsylvania) to study medicine. Foulke, a Quaker, did not serve in the Continental Army, but he did sympathize with the cause, and while still an undergraduate volunteered his services as a physician to the beleaguered troops encamped at Valley Forge. On one occasion when American officials asked Foulke to assist in treating enemy wounded, the young man astounded visiting British surgeons by performing in only twenty minutes amputations that normally took twice as long.[41]

During this period a fellow surgeon characterized Foulke as a "pedantic young Quaker," but this scarcely seems fair. He was outgoing and athletic, one of the finest ice skaters in Philadelphia. Nor were his activities in Europe those of a stodgy pedant.

Foulke sailed for France with his friend and classmate George Fox aboard the brig *Duke of Leicester* on May 4, 1780. Like so many young Philadelphia physicians, Foulke planned to continue his medical studies at one of the great European universities. Joseph Wharton, a leading Philadelphia citizen, and Dr. Thomas Bond, a founder of the Pennsylvania Hospital, the medical department of the College of Philadelphia, and the American Philosophical Society, provided the two young men with glowing letters of recommendation to Benjamin Franklin. As citizens of Pennsylvania, natives of Philadelphia, graduates of the University of Pennsylvania, and devotees of natural philosophy, the two young men were, as Bond noted, to be numbered among Franklin's "American Children."[42]

Upon arrival in Paris, Foulke and Fox were immediately accepted into Franklin's circle. Foulke's attitude toward the senior American diplomat can only be regarded as worshipful. After Foulke left Paris for study in Holland, his letters to Franklin opened with the salutation "Reverend Father." On one occasion Foulke remarked that he owed Franklin "more than that duty and veneration which would be due to a natural father."[43] He was especially pleased to report how well known Franklin was on the Continent: "There is a school in Leipsig . . . where I was shown by a friend—when a young pupil of fifteen inquired what countryman I was. . . . The Master told him I was from North America and I asked him if he knew what country that was. The pupil answered, yes, it was Doctor Franklin's country and that it lay there, pointing to the West."[44]

For Foulke, Franklin would always remain "to his country, a firm and affectionate Father, an honorable Patron to Science, an ornament and useful Citizen to the Universe, and a Parent to the children of America."[45]

Although Foulke, Fox, and other American scholars scattered to universities across Europe, they remained in touch through correspondence with their friend William Temple Franklin in Paris. Their letters were filled with lighthearted discourses on topics ranging from the latest in men's fashion ("You will therefore be so obliging as to enlarge, in your next, upon the subject of dress, not forgetting the lining and the buttons")[46] to Foulke's extraordinary success with the ladies of France, Holland, Germany, and England. Even William Temple Franklin, the illegitimate son of an illegitimate son who would soon present Franklin with his first illegitimate great-grandchild, seemed impressed by Foulke's romantic prowess.

Dr. Foulke's European sojourn was coming to an end in the fall of 1783. He had returned to Paris and taken rooms on the Rue du Maille when news of the balloon reached the city. Unfortunately, Foulke left no letters from this period that indicate his interest in aerostation. In view of his close association with the Franklins, however, and the fact that Foulke was still in Paris as late as November 7, it is difficult to believe that he was not present for at least one of the launches preceding the first manned flight on November 21. Certainly his lectures on the subject and the small balloons he flew in the United States indicate that he had made a careful study of balloon technology during his final months in France.

When Foulke returned to his native Philadelphia that winter he found a city full of knowledgeable balloon enthusiasts who were anxious to learn more about the subject. Much of the early correspondence on balloons from Franklin, Jay, Lettsom, and others in Europe had been directed to Philadelphians. As early as December 1783 Benjamin Franklin had sent a first edition of B. Faujas de Saint-Fond's encyclopedic *Description des Expériences de la Machine Aérostatique* to the library of the American Philosophical Society, remarking that it was the best available "account of one of the most extraordinary discoveries this age has produced, by which men are enabled to rise in the air and travel with the wind."[47] Franklin, Lafayette, and others were careful to keep the Philadelphia savants supplied with the latest and most authoritative French books and pamphlets on ballooning as soon as they appeared in the bookstalls.[48]

Like other Americans, Philadelphians often expressed their interest in balloons through humor. Francis Hopkinson, for example, had compared the balloon to a "high flying politician [who] is full of inflammability, he is driven along by every current of wind, and those who suffer themselves to be carried up by them run a great risk that the Bubble may burst and let them fall down

Dr. John Foulke. Photo courtesy of the Mutter Museum, Philadelphia, Pennsylvania.

from the Height to which the principle of Levity has raised them."[49]

Franklin would certainly have approved of his friend and fellow townsman's jest. While still living in Paris, he had offered a similar comment, remarking that great quantities of an element ten times lighter than inflammable air could be found "ready made in the promises of lovers and of courtiers and in the sighs of our widowers; in the good resolutions taken during a storm at sea, or on land, during an illness; and especially in the praise to be found in letters of recommendation."[50]

The Montgolfiers were held up as the preeminent philosophers of the age at University of Pennsylvania graduation ceremonies on June 2, 1784.[51] Talk of balloons could be heard everywhere in Philadelphia in the spring of that year. Francis Hopkinson described the situation to Jefferson, being careful to include yet another jab at politicians: "Congress imagined that when they removed to Annapolis . . . we should all be in deep distress . . . but the event is far otherwise. The Name of Congress is almost forgotten, and for one Person that will mention that respectable Body, a hundred will talk of an Air Balloon."[52]

It was a situation made to order for John Foulke, who was not only eager to establish his reputation as an amateur natural philosopher but as something of an authority on the "Air Balloon." Soon after his return to Philadelphia, Foulke obtained permission from university officials to offer a lecture on pneumatics in the "Old Hall" of the college on Fourth Street between Arch and High. In spite of all evidence to the contrary, it must be admitted that the title, "Pneumatics," the study of the mechanical properties of gases, is a bit pedantic for a talk on balloons.[53]

In addition to selling tickets for his lecture, which was scheduled for May 17, 1784, Foulke distributed a number of passes to leading citizens, including George Washington. Washington was unable to attend as a result of a previous commitment to preside over a meeting of a new veterans' organization, the Society of the Cincinnati, but his ticket and his note of regret to Foulke are now prized possessions of the Smithsonian Institution's National Air and Space Museum in Washington, D.C.[54]

Washington's inability to attend Foulke's talk did not mean that he was uninterested in the subject. In a letter to Sir Edward Newenham he had remarked on the need to explore the "upper regions." "The observations there made may serve to ascertain the utility of the first discovery and how far it may be applied to valuable purposes. To such alone I think these voyages ought at present to be consigned."[55] Moreover, Washington believed that "young men of science and spirit" should be offered "handsome public encouragements . . . for the risks they run in ascertaining its [the balloon's] usefulness, or the inutility of the pursuit."[56]

Even before he gave his lecture, Foulke offered Washington and other residents of Philadelphia an opportunity to see a balloon in flight for the first time. On March 31, 1784, Francis Hopkinson reported to Jefferson that "a Gentleman in town is making an Air Balloon of 6 feet in Diameter; it is now almost completed—what the success will be time must show."[57]

Hopkinson provided more information on the balloon, which was now definitely identified as Foulke's, in a letter to Jefferson on May 12. It was a hot air aerostat made entirely of paper and featuring a large opening at the base. This would suggest that the device may have been shaped more like a Japanese temple bell than a classic spherical balloon.

Foulke sent this first American balloon aloft from the garden of Peter John Van Berckel, Dutch minister to the United States, on May 10. The following morning he flew the same balloon from "Mr. [Gouverneur?] Morris's garden."[58] That evening he launched a second balloon from the home of the chevalier de la Luzerne, "to the great amusement of the spectators." Hopkinson reported that each of the balloons had reached an altitude of "perhaps three times the height of the houses, and then gently descended without Damage."[59]

Hopkinson was particularly pleased to be able to inform Franklin of the fact that balloons, albeit very small ones, had been flown in Philadelphia. "We have been diverting ourselves with raising paper balloons by means of burnt straw to the great astonishment of the populace," he noted.[60] Hopkinson admitted that, like Franklin's early electrical experiments, the balloon seemed "only amusing now," but he was confident that "its uses and applications will hereafter unfold themselves."[61] He suggested, for example, that a navigable balloon, an oblong gasbag powered by paddle wheels, might revolutionize transportation. For all his jokes, Hopkinson obviously saw the possibility of a bright future for the balloon.[62]

Flight had arrived in Philadelphia at a most propitious moment. At noon on Monday, May 10, the sheriff, "accompanied by the proper officers," had officially announced that the Treaty of Paris had gone into effect. Flags were flown all over the city, and a gigantic arch, measuring 50½ feet wide by 35½ feet high, "capped with a ballustrade and suitably decorated," was constructed at the Market Street wharf to serve as a focal point for the festivities that followed. Charles Willson Peale exhibited his back-lit "transparent paintings" to thousands of spectators as part of the celebration.[63]

Foulke's balloons added an additional note of gaiety to the events of the week. Thomas Jefferson, arriving in Philadelphia too late to witness the first three flights of May 10-11, nevertheless saw three additional "balloons" launched before May 21. The largest of these, he reported to James Monroe, was eight feet in diameter and rose to an altitude of 300 feet.[64]

Philadelphians were enthralled by the sight of these paper "Air Globes" rising out of the back lots and gardens of the city. Some citizens were content to remain armchair aeronauts, offering comment and advice in the columns of local newspapers. One such correspondent recommended "to the young gentlemen now making experiments with small balloons, to try the smoke of hot lime slaked."[65] There were others, however, including some of the leading citizens of the city, who were anxious to realize the Philadelphia editor's dream of sending some enterprising American aloft "to meet European philosophers above the clouds."[66]

Plans for the construction of the first manned balloon in the United States had been set in motion in late May. Little time was lost in convincing the most influential men in the city to accept subscriptions from local residents who wished to support the venture. This list of eighty-five sponsors includes some very impressive names, including those of Matthew Clarkson, who would serve as Philadelphia's mayor during a dreadful series of yellow-fever epidemics; John Morgan, Benjamin Rush, and other leading physicians; Haym Solomon, who had played such an important role in helping to finance the revolution; Joshua Humphries, the naval architect who designed both the *Constellation* and the *Constitution*; Peter Le Maigre, a leader of the city's large French and West Indian community; and Charles Willson Peale, Philadelphia's leading resident artist. The name of at least one signer of the Declaration of Independence was included, as were those of every professor on the faculty of the University of Pennsylvania and members of such old Philadelphia merchant families as the Whartons, Benezets, Harrisons, and Irwins. Strangely, John Foulke's name is not listed, nor is there any indication that he might have been involved in the scheme.[67]

Dr. John Morgan, the controversial physician, was a major figure in the enterprise however. On June 11, 1784, Morgan requested that the members of the American Philosophical Society endorse the plan for the construction of "an *Air Balloon* . . . of such magnitude as to be capable of raising great weights, men and other animals into the region of the atmosphere and returning with safety to the earth."[68] At the next meeting, on June 19, with Foulke, Fox, Hopkinson, and the Marquis de Lafayette present, John Morgan read a paper "on the construction of air balloons and the several uses to which they may be applied."[69] At the same time, Morgan withdrew his motion for a resolution supporting the balloon, having discovered that the rules of the Society precluded the group from offering "their opinions as a body upon any subject of a nature that may come before them."[70]

Failure to obtain the support of the Philosophical Society did little to dis-

The Versailles animal flight of September 19, 1783, as visualized by Connecticut engraver John Norman. This illustration, one of the first representations of a balloon published in the United States, appeared in the Boston Magazine *for August 1784.*

courage Morgan and his associates. By June 26 advertisements announcing the opening of a subscription to underwrite the construction of "a large and elegant Air Balloon" were appearing in Philadelphia newspapers. The proposal called for a silk balloon, "properly lined, covered, varnished, and painted by the best artists." The balloon was to be a hot air model standing sixty feet tall "and of proportionate diameter; to be strengthened with net-work; to have a car or boat appended to it."

In justifying their project, the authors of the scheme drew on the now standard list of uses to which a balloon could be put. In addition, they suggested that their aerostats might even serve the needs of religion by offering "new proofs of the sublime workmanship of the great Architect of the Universe, which will have a direct tendency to excite suitable ideas of the government of the world by the all-wise and omnipotent Creator of the Universe [and] to call forth our grateful admiration of his beneficence to mankind."[71]

In fact, the Philadelphia savants were about to be upstaged. Just as their fund-raising effort was getting underway, word was received that Peter Carnes, a tavern owner and lawyer from Bladensburg, Maryland, would fly his "American Aerostatic Balloon" in Philadelphia on July 4, 1784.

Peter Carnes had little in common with urbane, wealthy, well-established, and well-educated city dwellers like Foulke and Morgan, save his interest in ballooning. He represented a very different side of American life.

Born, probably, in Middlesex County, New Jersey, in 1749, Carnes was rootless and ambitious. He spent his early years moving from place to place and from job to job in a restless search for economic security and social status. Many years later, while working as an attorney in South Carolina, he informed companions that he had been variously employed as a house carpenter, Methodist exhorter, planter, barkeep, and balloon builder prior to turning to the law. Carnes's early peregrinations are difficult to trace, but we do know that he was operating a tobacco plantation in Charles County, Maryland, shortly before the outbreak of the American Revolution. At this point, he began building business connections in Bladensburg, a tobacco shipping port on the Anacostia River in neighboring Prince George's County.

Carnes became a particular friend of Jacob Wirt, the owner of the Indian Queen Tavern and an associated business complex that included a large, two-story brick house, a billiard hall, counting house, blacksmith shop, and several outbuildings in Bladensburg. Carnes rented the complex from the heirs following Wirt's death in October 1774.

Vague family references indicate that Carnes had been married to one Mary Eden Briscoe, the daughter of a Charles County planter. We can assume that she was dead by 1776, however, for in that year he married Jacob Wirt's widow, Henrietta. Over the next several years Bladensburg's new innkeeper began to make a name for himself in the community. He served as a sergeant in a county militia unit in 1777 and was apparently also reading law during this period, for he was licensed to practice before Prince George's County courts in 1780.[72]

No portraits of Carnes have survived, but we do know that he was a large man, weighing 234 pounds in the summer of 1784. Judging from accounts of his subsequent legal career in South Carolina and Georgia, he must have been an extraordinarily ebullient and outgoing fellow. One historian of the back-country southern courts put the matter as gently as possible when he remarked that "this gentleman is better known for his humor than anything else."[73]

In truth, Carnes was a practical joker of legendary proportions. He was once seated in court next to a frequent legal opponent. As his rival stood addressing the jury, Carnes quietly reached over and buttoned their jackets together. He then stood up and walked from the courtroom, dragging the sputtering and incoherent lawyer behind him.[74] On another occasion, after being chastised by a judge for appearing in court in his shirt-sleeves, Carnes strode angrily out of the courthouse, returning a few minutes later wearing a tattered blanket draped across his shoulders in lieu of legal robes.[75]

Clearly this sort of rough-and-ready humor has its limits, and it is apparent that Carnes occasionally pushed rival lawyers beyond the limits of endurance. During the course of a trial in which the legal definition of the terms assault and battery was at issue, Carnes faced an opponent who had been the repeated butt of his jokes. When asked to explain his understanding of the terms, the rival lawyer shook his fist in Carnes's face to illustrate assault, then punched him in the stomach. Carnes's comment that he "did not think that the fellow had so much sense" drew many a chuckle for years thereafter when South Carolina lawyers met to swap tales.[76]

In essence, Carnes seems to have been just the man to operate a riverfront inn or to entertain the spectators in a packed courtroom, but he can hardly have been regarded by his friends and neighbors as a very likely candidate for honors in science or technology. Nevertheless, he was to produce the first full-scale, man-carrying balloon in America within a year of the first public flight of a small aerostat in Annonay and a scant seven months after human beings had ventured aloft for the first time.

Carnes later told reporters that "the first Hint he had of the Balloon was by hearing of Monsieur Montgolfier's experiments."[77] Unfortunately, he did not tell them how he had learned of the Montgolfiers, but it seems likely that Carnes, like George Washington and so many other Americans, first heard of the balloon through a local newspaper. The *Maryland Journal and Baltimore General Advertiser*, one of the newspapers widely read in the Bladensburg area, began its coverage of the balloon story early in 1784. By March of that year the newspaper had carried a number of long articles concentrating on the Montgolfier flights, complete with reasonably detailed descriptions of these first aerial craft.[78]

The early chronology of Carnes's work is also a bit uncertain. If Carnes began planning his balloon on the basis of the news accounts, which peaked in March, he must have worked very quickly indeed, for his balloon was first exhibited in Bladensburg only three months later.

On June 15, 1784, Carnes published an advertisement in the same *Maryland Journal* that had taken such an interest in European ballooning earlier that spring. Prefaced with the usual bit of doggerel ("On vent'rous Wing in Quest of Praise I go, and leave the gazing multitude below"), Carnes announced that he would begin exhibiting an "American Aerostatic Balloon" in a field near Baltimore on Thursday, June 24.[79]

The balloon was 35 feet in diameter, stood 30 feet tall, and was constructed of "silk of various colors." Further, "a splendid chariot . . . fitted for the reception of two persons" was suspended beneath the gasbag. A cylindrical iron stove was positioned within reach of the passengers, for "Mr. Carnes prefers the common rarefied air [hot air] to the inflammable air [hydrogen] now so much in use in Europe."

Carnes promised not only to "ascend above the clouds" but to offer a short lecture "on the great uses to which this important discovery may be applied, for the convenience and delight of human life." The presentations would be made each day at 8:00 A.M. and 5:00 P.M. prior to the first flight on June 24.

The announcement also provides some definite clues about Carnes's motives for undertaking the experiment. This was clearly a hardnosed business venture. A tall fence had been constructed at the launch site. Tickets of admission to the lectures and the final ascent were $2.00 "for the first place and for the second place ten shillings each." Armed guards stationed around the fence "will be justifiable in taking the life of any person who attempts to force his way into the field." Carnes promised that anyone eluding the guards and sneaking into the enclosure without paying would be met with a lawsuit.[80]

In order to allay any suspicion on the part of potential ticket buyers that the advertisement might be a hoax, Carnes called attention to the fact that this would not be his first flight. Several leading "Gentlemen of Bladensburg" had witnessed his preliminary experiments and stood ready to testify to his ability to perform as advertised.

In spite of the fact that no accounts of the early tests referred to in Carnes's announcement have survived, we can imagine what an impression they must have made on the citizens of Bladensburg. Carnes had undoubtedly begun by launching small paper or silk balloons from one of the lots surrounding the Indian Queen Tavern on Water Street. By the time he had worked his way up to unmanned, tethered test flights of the large balloon, the whole town must have been buzzing with excitement.

Carnes's wife and his two step-children, Elizabeth and William Wirt, were probably present at the trials, beaming with pride. Carnes, who was considerably younger than his wife, would marry Elizabeth several years after Henrietta's death. William Wirt, who remained very close to his step-father until Carnes's death in 1794, became attorney general of the United States under Pres. John Quincy Adams and made an unsuccessful bid for the presidency as an anti-Masonic candidate in 1831.

Mr. Sydebotham, one of the town's busiest and most successful merchants, headed Carnes's list of those who had attended his early trials. He was, William Wirt recalled, a man "rosy from good living, who, in the fashion of Maryland, had his bowl of toddy every day—a thorough John Bull, proud, rough, absolute, and kind."

Robert Dick, another successful merchant who maintained one of the finest homes in Bladensburg on a ridge overlooking Carnes's tavern, was also present, as was Dr. Ross, the town physician. Carnes also mentioned the name of two other witnesses, Messrs. Tolforth and Miller.

But these were only the leading citizens of Bladensburg. Large crowds of seamen, workers from the tobacco warehouse, and ship outfitters from the dock area on lower Water Street, as well as women and children from all over town, must have gathered to observe Carnes's progress. Blacks as well as whites would have been present, for Prince George's County had a large black population, and the town of Bladensburg was still the site of an occasional slave auction.[81]

We do know that crowds of "several hundred" spectators were gathering to watch trial ascents of the large balloon by mid-June. On June 19 the craft became the object of a cross-country chase when it broke loose from a ground tether at an altitude of several hundred feet. Slightly damaged in landing, it was repaired and ready for flight again the following day.[82]

Some time before, Carnes had announced that he would make a public flight on June 19. There was a suggestion that he might take this opportunity to attempt the first manned ascent with his balloon. The real purpose of this final Bladensburg flight, however, was to generate additional publicity for the upcoming Baltimore ascent, which was now only five days away.

Fortunately for both Carnes and historians, the flight did draw attention to the project. A long letter from an anonymous correspondent describing the scene in Bladensburg on June 19 appeared in the *Maryland Journal* three days later. As the first eyewitness account of the flight of a large balloon in the United States, it merits quotation in full:

> *Yesterday I made one of a Party to Bladensburg, to attend the Exhibition of Mr. Carnes' Aerostatic Balloon. When we arrived, we learnt that he was then filling it in a retired Part of a Wood, about Three Miles distant, as much to avoid the Intrusion of the Ignorant, as to guard against those who would satisfy their curiosity without making an adequate Compensation. As he had informed the Public, by Advertisement, of his Intention of ascending in his Aerial Chariot from a Field near the Town, a general Murmur of Discontent seemed to pervade all Ranks, especially those who were destitute of Horses. Your humble Servant, and some other of his Friends, requested him to move it to Town, which he did, through the Air, about Three o'Clock in the Afternoon. Being thus obliged to deviate from his Morning's Plan, there was some Difficulty in getting the Balloon through the Woods, especially where they were so thick as to prevent the Persons who held the pendant Ropes from guiding it, while it was raised above the Trees. When they crossed the Fields, it was brought within about 5 or 6 Feet of the Ground, being thus more manageable than when higher up. Passing through the Road, a sudden Blast of Wind drove it against a Fence, by which Means it was much damaged, the Hoop which surrounded it, and to which it was attached by Pieces of Linen, to prevent its collapsing, was unfortunately broken; yet, notwithstanding this Accident, Mr. Carnes brought it, filled, to a Field near the Town, where he raised it about 70 Yards still keeping hold of the Cords to prevent its further Flight. While it remained thus suspended, several Gentlemen expressed a Desire that it might be brought down and a Preparation made for their ascending with it; but the Sides being collapsed by the Breaking of the Hoop, the Air could not be rarefied, without Danger of setting Fire to the Machine. So novel an appearance had an odd Effect on the "gazing Multitude"—Silent Admiration—the Stare of Surprize—the Grin of Ignorance—and a wild tumultuous Joy—sufficiently indicated that all were pleasingly disappointed.*
>
> *I felt happy that the vent'rous Philosopher had so far succeeded in his attempt. I am pleased in reflecting how much our countrymen have done to improve the various Branches of Science, and doubt not our being as much distinguished for Works of Genius in Time of Peace, as our Patriot Army have been for their Success and Sufferings during the War. For my own Part, I should not be surprized should this Discovery be very soon applied to navigate the Productions of the Earth through the Air, having seen Mr. Carnes' Balloon raise a very considerable Weight*

in private Experiments. . . . Mr. Carnes not ascending in it agreeable to Advertisement gave umbrage to some, who would not admit the Accident as an Apology, but expected the sublime Pleasure of seeing him tumble from his Aerial Charriot like another Phaeton. He, however, gratified the Philosopher, and all who were capable of judging by small Experiments, what Time and Labour may effect. He purposes being in Baltimore on Thursday, where I hope, for the Honour of American Science, he will meet with so liberal a Subscription, as may Encourage him in his bold Pursuit of Fame, not doubting, in that Case, but he will soon aloof, as he has promised, and leave the gazing Multitude below.[83]

The *Virginia Journal*, an Alexandria newspaper, also carried a lengthy first-person account of the "Gentleman of Maryland" and his "Aerial ellipses." The Virginia editor heaped praise on Carnes. The flight, he believed, had provided "the philosophic mind . . . as rich a repast as the intellectual faculties are capable of receiving." The editor was full of advice as well. Carnes was "cautioned relative to his fire" and advised to launch early in the cool of the morning, for "every European experiment of any consequence, has been made in the cool seasons."

Carnes was also urged to proceed slowly with his plans to make a flight himself. He should, noted the editor, "let no temptation, no love of fame transport him to such enthusiastic ideas as to venture himself in the Chariot of the Sun, lest the fate of the ambitious Phaeton should be his portion—the rivers Po and Potomack in this Case might be Synonymous." After all, the would-be aeronaut was reminded, "the ingenious philosophic Mind can be as well ascertained of the Powers of his Balloon, by a simple experiment, as if it carried up a Dozen Men, Women and Children."

Carnes must have been especially pleased by the editor's advice that his readers attend the Baltimore ascent and support "the Rich Genius who, regardless of Labour and Expence has prepared so grand a Spectacle to feast the Eye of the Scholar and Philosopher."[84]

As advertised, Carnes and his balloon were in Baltimore on June 24. Howard Park, a wooded area immediately to the north of the city, had been chosen as the launch site. The publicity had attracted "a numerous and respectable Congress of People, whom the Fame of his superb Balloon had drawn together from East, West, North and South."[85] Baltimore had gone "Balloon Mad," remarked a disgruntled sales clerk who was forced to remain at his post while the rest of the city flocked to Howard Park. "Every store but our own and a few others were shut," he noted to a friend.[86]

All of the flights made early on June 24 were unmanned, tethered ascents. It is quite possible that Carnes had already attempted a secret ascent himself, only to discover that the balloon would not support his own considerable bulk. Nevertheless, the crowd "generally appeared highly delighted with the awful Grandeur of so novel a Scene, as a large Globe making repeated Voyages into the Airy Regions, which Mr. Carnes's Machine actually performed, in a manner that reflected Honour on his Character as a Man of Genius, and could not fail to inspire solemn and exalted Ideas in every reflecting Mind."[87]

As Carnes was preparing the balloon for the final ascent of the afternoon, Edward Warren, a thirteen-year-old Baltimore lad, made his way through the crowd and volunteered to be sent aloft on a tethered ascent. Carnes must have known that his balloon, with a lift of only 469 pounds, would scarcely leave the ground with his own 234-pound bulk on board. The sudden appearance of a lightweight, young volunteer provided Carnes with the perfect way to cap his exhibition.

Warren "behaved with the steady fortitude of an old voyager." He "soared aloof" to the loud applause and cheers of the crowd, all of which "he politely acknowledged by a significant wave of his hat." When Warren returned "to our terrene element," a few minutes later, a collection was taken up among the spectators so that he might have a reward with "a *solid* instead of an *airy* foundation and of a species which is ever acceptable to the residents of this *lower world*."[88]

This was a significant moment in American history. For the first time a citizen of the United States had left the ground in a flying machine. We ought to know more about him. What impelled him to step out of the crowd that day, and where did he go when his short flight was over? But we do not know these things. The afternoon of June 24, 1784, was Edward Warren's single moment on the stage of history.[89]

With the successful Baltimore exhibition behind him, Carnes was anxious to repeat his performance before a larger crowd of paying spectators in a major city. As early as June 17, a full week before the Baltimore exhibition, he had placed advertisements for a Philadelphia ascent in the *Pennsylvania Packet*. The advertisement, which set July 4 as the date for the Philadelphia flight, was almost a duplicate of the earlier Baltimore ads. Once again armed "CENTINELS" with a financial interest in the venture would patrol the fence with instructions to shoot to kill.[90]

By June 21 Carnes's plans for the Philadelphia ascent were made final. Benjamin S. Coxe had arranged for Carnes to stage his flight in the "New Work-House Yard," at the corner of Sixth and Plum streets. By launching the balloon from within the walls of the city jail, Carnes saved the expense of constructing the wooden fence that both protected his craft from ground winds and enabled him to mask his preparations from non-ticket holders.[91]

Carnes was determined to upgrade his aging balloon so that he could make a flight himself. The once-splendid "American Aerostatick Balloon" was now disfigured by rips and burn holes. These were patched as carefully as possible, although a number of holes remained at the time of the Philadelphia exhibition.

Carnes also modified the "brazier" that kept the balloon aloft. A "tube" now extended from the 150-pound iron furnace to the appendix of the balloon. A triangular frame hung below the furnace on chains. In view of the fact that this frame was described as a "scaffold," it seems probable that it was a simple platform with no sides. The furnace and "scaffold" for the pilot were suspended directly from a circular hoop at the equator of the balloon. Since a later Philadelphia account mentions that the balloon had no "network," we can assume that the hoop was held in place by a number of lines passing over the top of the envelope.[92]

Carnes hoped that his modified balloon was now capable of lifting as much as 600 pounds. Considering the remaining holes and the additional weight, however, this scarcely seems possible. Even Carnes must have realized that the veteran balloon was rapidly approaching the end of its career.

When preparations for the ascent were not complete by July 4, the flight was rescheduled for July 19. By five that afternoon, "thousands of admiring spectators" had gathered in the workhouse yard and in nearby Potter's Field, now Washington Square, to witness the miracle of man in flight.

Precisely at 6:00 P.M. Carnes climbed aboard his "scaffold" and loaded more sticks of hickory into the brazier. The balloon had scarcely begun its rise out of the courtyard amid the cheers and applause of the crowd when, at an altitude of 10 to 20 feet a gust of wind blew it against the wall. The edge of the platform on which Carnes was seated caught beneath a projecting eve, snapping the suspension chain and dropping Carnes ignominiously, but safely, back to earth. Freed of its burden, the balloon then "rushed into the air with astonishing velocity." It moved to the south, rising to an altitude of several hundred feet, so that it appeared to be "the size of a barrel." It seemed to hang suspended in the air, then, suddenly, it was engulfed in flames. Apparently the accident had brought the braziers into contact with the fabric.[93]

The spectators outside the jail wall, who had not seen Carnes knocked from his perch, watched in horror as the remains of the furnace and platform, which they feared was Carnes's body, fell onto the roof of the New Playhouse.

The rumors that swept through the city to the effect that Carnes had died in the ascent were impossible to dispel. For many years thereafter, Americans who had read the initial garbled reports of the flight were certain that Peter

Carnes had been the first American martyr to the cause of flight.

Congressman John Page, for example, fixed the image of Carnes's death in a flaming balloon in his poem "The Balloon," published in New York about 1790. It is interesting to note that Page also gave credence to the Joseph Fathom hoax:

The Balloon is an old device,
And us'd long since, who now denies?
For Phoebus Phaeton's kind fire,
Lent him a Balloon which took fire.
So Carnes's balloon in a blaze,
Dropt him whilst thousands at it gaze.
Fathom from Phoebus took the thought,
And yok'd his horse, 'twas hint dear bought,
His horse poor fellow was so little,
And tackle made of ware to brittle,
His traces broke away he flew,
And went where no one ever knew.[94]

Peter Carnes abandoned the balloon business following the abortive Philadelphia ascent. He returned to Bladensburg only to discover that he was being sued by Jaspar Wirt, Jacob Wirt's brother and executor, for nonpayment of rent. Carnes had apparently believed that he was not obliged to pay rent on the Indian Queen Tavern complex as a result of his marriage to Jacob's widow. Jaspar Wirt, pointing out that the money was due him, not Henrietta, pressed his suit, and Carnes fled south. As noted, he enjoyed a successful legal career in the area of Newberry, South Carolina, before settling near the city of Augusta, in rural Richmond County, Georgia.

Carnes prospered in Georgia. Some notion of the esteem in which he was held can be seen in the fact that he was selected as one of five leading citizens to greet Pres. George Washington when he visited Augusta in 1791.[95]

At the time of Carnes's death at age forty-five in August 1794, his young widow Elizabeth and his two infant sons, Patrick and John Peter Carnes, inherited an estate that included large tracts of land in Richmond County and 2,300 acres in neighboring Franklin County. It is appropriate that Carnes's obituary in the *Augusta Chronicle* included a list of poetry quite as bad as that which had been employed to advertise his balloon ascents a decade before:

Speak not his name; but that here the great Carnes lies,

He needs no epitaph, who never dies.
Cease envy to disturb an honest name,
And dare not trample on the Dead.[96]

Peter Carnes's contribution to American aerostation was considerable. He worked in isolation, far from the great urban centers that offered support to other balloon pioneers, and had only limited access to available technical information. Carnes overcame all these obstacles, built the first large, serviceable balloon in the United States, and staged exhibitions that thrilled and delighted thousands of citizens of the young republic.

Carnes's real achievement, however, is to be found in the sense of pride and self-confidence his work had given to all Americans. His flights were reported and applauded in newspapers from Maine to Georgia. In a very real sense, many seemed to feel that the success of Carnes's balloon symbolized the potential of the new nation. As the editor of the *Massachusetts Centinel* noted: "The opinion that philosophy will never spread her rays in the new world, can certainly have no foundations in truth." Peter Carnes had flown.[97]

John Morgan and the members of the Philadelphia balloon syndicate had suspended their fund-raising efforts during the Carnes episode. They were aware that his had been a private venture and, in the spirit of goodwill, they withdrew so that the "Gentleman from Maryland" could realize the maximum profit.

Within a day or two of Carnes's Philadelphia ascent, the subscription books for the syndicate "Air Balloon" were open once again. In announcing the resumption of their effort, the Philadelphians were careful to praise Carnes as a man "whose talents for philosophy are considerable." They took equal pains to explain that the balloon they planned would differ significantly from that which had been destroyed by fire over the Philadelphia rooftops. Carnes's balloon had been too flimsy, too small, and too dangerous. Their balloon, on the other hand, would be built "on a scale befitting the metropolis of one of the first states of the Union."[98]

The Philadelphia balloon, 60 feet in diameter, would be able to lift an estimated 3,373 pounds. The "flimsy" silk used in the Carnes envelop would be replaced by fine linen, with a complete network to support "a fine car."[99] But all the explanations, calls to patriotism, and lists of useful purposes to which the magnificent aerostat might be put were to no avail. Carnes had apparently satisfied the city's curiosity, and the new project collapsed for lack of financial support.

And so Philadelphians returned to flying the small balloons introduced by John Foulke. Dr. John Morgan was one of those who refused to allow his disappointment at the failure of the large balloon project to daunt his enthusiasm for flight. Morgan's continued interest in aerostation was so well known that in the fall of 1786 Mrs. Charles Thomson, a Philadelphia matron, would comment that a friend's trip from New York had been made with such speed that "he must have used Dr. Morgan's balloon."[100]

Peter Carnes's Philadelphia ascent had occurred precisely fifty-one weeks to the day after John Jay had sent the short note to Gouverneur Morris announcing the invention of the balloon. It had been an extraordinary year which had seen Franklin's "new born babe" journey across the Atlantic to establish itself on American soil. The fact remained, however, that no American had yet ventured to make an untethered aerial voyage. It was a situation that would soon be remedied.

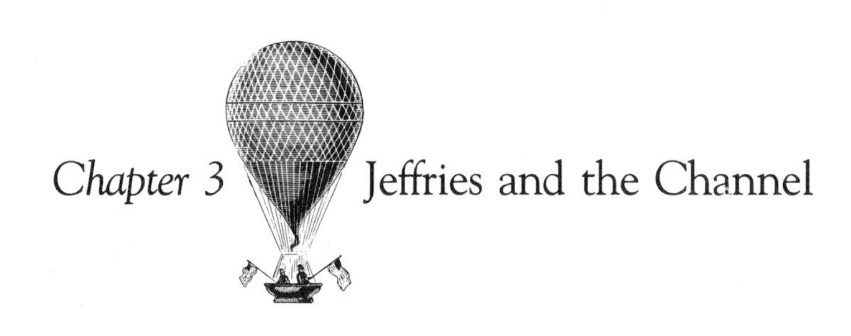

Chapter 3 Jeffries and the Channel

JOHN JEFFRIES SAW HIS FIRST BALLOON AT THE HONOURABLE ARTILLERY Company's training ground in the London suburb of Moorfields on the morning of November 25, 1783. It was a gorgeous thing, a gas balloon some ten feet in diameter, constructed of oiled silk and covered with gilt.

The balloon was the work of Count Francesco Zambeccari, an Italian seaman who had served in the Spanish navy, fought the Turks, and spent three years as a slave in Constantinople before arriving in England. Zambeccari and Michael Biaggini, a maker of artificial flowers, had startled the entire city a few weeks earlier when they launched the first small balloon in England from Biaggini's home in Cheapside on November 4.

Now Jeffries and the other "elegant ladies and gentlemen" in the crowd watched in fascination as the new balloon rose into a cloudless blue sky precisely at noon. "In about twenty minutes," Jeffries noted in his diary, "it had ascended so high, so far, as to be scarcely visible, appearing like a black speck, and almost as high as the eye could discern." It was a sight John Jeffries would not soon forget.[1]

The third son of David and Sarah Jaffrey Jeffries, he had been born in Boston thirty-nine years before, on February 5, 1744. Extraordinarily well-educated, he was a graduate of both Harvard University (B.A., 1763; M.A., 1766) and the Marshall College of the University of Aberdeen (Doctor of Physic, 1768). In addition, he had studied medicine under the tutelage of one of Boston's finest physicians, Dr. James Lloyd, served as a dresser in surgery at London's Guys Hospital, and attended lectures by leading English surgeons and physicians.

Within a few months after returning to Boston in 1769, he had established a thriving private practice and was chosen by Massachusetts officials to care for "extraordinarily difficult cases" in remote parts of the colony. In May 1774 he accepted an appointment as physician and surgeon to the Boston Almshouse, a local charity hospital. Jeffries seemed to be the most fortunate of young men, standing on the threshold of a brilliant career.

All this began to change on March 5, 1775. Jeffries was visiting his father that evening when a distraught neighbor informed the men that an angry mob was attacking the British troops stationed at nearby Murray's barracks. Jeffries rushed to the scene, arriving just in time to see the barracks gates barred behind the retreating soldiers. An hour or so before, he learned, a guard at the Custom House on King Street had been pelted with rocks and snowballs. The situation had degenerated until Capt. Thomas Preston had been forced to march to the relief of the guards. Surrounded and threatened, the troops had opened fire, killing four townsmen and wounding several others.

At eleven o'clock that night, Jeffries received word that his services were requested by Patrick Carr, an Irish immigrant who had been wounded in the fighting on King Street. Carr lingered under the doctor's care for ten days, during which he repeatedly gave Jeffries an account of the "massacre" that cleared Preston and his men of blame. Jeffries's testimony and his recital of Carr's deathbed statement were key factors in Captain Preston's subsequent acquittal.

As a result of the incident, John Jeffries was forced to reassess his political stance. Having seen an American mob in action, he was not at all certain he could continue to support the patriot cause. His changing attitude created much tension in his family. His father, the treasurer of Boston, was an ardent patriot and a close friend of radical leaders like Sam Adams, who had provided a glowing letter of recommendation when John Jeffries went abroad in 1766. David Jeffries had been a founding member of the Sons of Liberty and was a central figure in the committee of correspondence formed to solicit food and supplies from sympathizers in other colonies when the port of Boston was closed following the Tea Party in December 1773.

As events moved toward a crisis in the early summer of 1775, John Jeffries moved further from his father's position and closer to the loyalist colonial government. The final break from family and friends came on June 17, 1775, when the first major battle of the Revolution was fought on the slopes of Breed's Hill.

Following the battle, Jeffries assisted British military surgeons in dealing with their own casualties as well as the American wounded who had been left on the field. He also had the unhappy task of identifying the body of his friend Dr. Joseph Warren, the fallen patriot hero of the battle, for Gen. Sir William Howe.

When the British finally abandoned Boston in March 1776, Jeffries, his wife, and his two small children were evacuated to Halifax, Nova Scotia, with other Massachusetts loyalists. They remained in Canada for three years. Jeffries treated the ills of his fellow loyalists, British soldiers, and American prisoners of war but was unable to obtain what he regarded as a suitable commission in the army or navy.

Arriving in England, Jeffries became embroiled in the political machinations that had become the lifeblood of the expatriate American community. Benjamin Thompson, an eccentric genius from Massachusetts who had become a favorite of the secretary for colonial affairs, Lord George Germain, took a particular interest in the doctor's case. A hearing was arranged before a panel of distinguished physicians, as a result of which Jeffries was able to return to America in September 1779 with dual commissions as apothecary to the forces in Nova Scotia and surgeon-major to the general military hospitals in Canada and New York.

Over the next eight months, Jeffries had an opportunity to observe the American war at firsthand. He assisted in the capture of two American privateers, served at the sieges of Charleston and Savannah, and met the British officers, from Sir Henry Clinton and Lord Cornwallis to Maj. John André, who were directing the war effort.

The adventure came to an end for Jeffries on September 26, 1780, when he received word from Benjamin Thompson that his wife had died suddenly in London. Anxious to return to England to arrange for the care of his children, Jeffries now discovered that it was as difficult to resign his post as it had been to obtain it. He was finally ordered to army headquarters in New York, where he was able to sell his commissions, an action that would eventually lead British military authorities to exclude medical posts from the brisk trade in military commissions.[2]

Disgusted and discouraged, Jeffries returned to England, placed his children in boarding schools, and attempted to rebuild his life among the exiled loyalists of London. In spite of a number of attractive offers, one of which would have led to his appointment as head of a medical staff in India, he chose to remain in London, specializing in gynecology and obstetrics while continuing to serve as physician in residence to his expatriate countrymen.

After five years of hardship and work, Jeffries was ready to relax and savor the delights of London. An inveterate theatergoer, he haunted the Drury Lane, the Haymarket, and the Strand, enjoying performances by great actresses like Mrs. Sarah Siddons as well as his favorite soubrettes. There was Miss Wheeler, "a graceful woman who sings very well," and Miss Stuart, "a gentle, good, handsome girl," whom he saw in a performance of *The Positive Man* at the Drury Lane in September 1784.[3]

John Singleton Copley was always willing to tour the city's art galleries with his friend Jeffries.[4] The doctor attended balls and spent hours at the zoo and in the various pleasure spas scattered around the city, where one could walk through exotic gardens, purchase refreshments, meet friends, and enjoy the evening firework exhibitions. He took his children to Astley's Amphitheater to see "midgets tumbling and riding and dancing upon horseback."[5]

Jeffries saw another side of London life as well. He attended a number of public executions. On one such occasion, a convicted French spy was hung, beheaded, drawn, and quartered. "An awesome and tremendous sight," Jeffries noted.[6]

The young widower took a particular interest in the "charming, alluring Women of the Town." Fanny Hill and Moll Flanders come to life in the pages of his diaries. Jeffries knew the most fashionable madams in the city, including Miss Johnson, who kept "an elegant house in Jermyn St., St. James," where he attended rousing amateur theatricals staged by the veiled and scantily clad residents.[7] His liaisons ranged from single evenings spent in the company of a prostitute whom he had met at the theater to longer associations with young women such as Maria Callen, who "put herself under my sole protection." He later abandoned Maria after being informed that she was using the lodging he provided for her on Moors Place to entertain other gentleman callers, but not before offering her "his future services as a gentleman."[8]

Jeffries's taste in women was eclectic. He enjoyed the company of American ladies, including Miss Elizabeth Bacon and Mrs. Joshua Loring, the wife of a leading loyalist who had caught the eye of General Howe during the British occupation of Boston. But he seemed to prefer English women. Miss Fanny Vassal, Miss Stewart, and Miss Betsy Spencer were particular favorites.

Jeffries was also quick to comment on the notable beauties of the day. He met Mrs. Mary Robinson, "late the favorite Demoiselle of the Prince of Wales, now, I believe, under the Wing of Lt. Col. [Banastrae] Tarleton, the well known partisan in the cavalry on American service."[9] "Perdita" Robinson, one of Gainsborough's favorite subjects, was, Jeffries remarked, "cute," with a "lovely bosom."[10]

A religious man who was much more accustomed to being viewed as a pillar of the community than as a rake, Jeffries felt occasional pangs of conscience over his excursions into the London demimonde. For the most part, however, he was able to rationalize his behavior as "only a compliance to the laws of nature."[11]

The coming of the balloon brought a new sense of direction and purpose for John Jeffries. Fascinated by the sight of Zambeccari's golden aerostat, he took every opportunity to learn more about the subject. And there would soon be opportunities aplenty.

As Franklin had noted, the balloon was slow in coming to England. When

Dr. John Jeffries

his letters describing the invention first reached Sir Joseph Banks, most knowledgeable Englishmen agreed with Horace Walpole's view that ballooning "was as childish as the kite-flying of school boys."[12] It proved difficult to maintain this point of view in the face of flights as long as the Charles and Robert ascent of December 1783.

Certainly the performance of the first large gas balloon forced Sir Joseph Banks to alter his opinion. "I laughed when balloons, of scarce more importance than soap bubbles, occupied the attention of France," he noted, "but when men can with safety pass . . . more than five miles in the first experiment. . . . "[13] By June 1785 even Horace Walpole had some belated second thoughts: "How posterity will laugh at us one way or the other! If half-a-dozen break their necks, and balloonism is exploded, we shall be called fools for having imagined it could be brought to use: if it should be turned to account, we shall be ridiculed for having doubted."[14]

In the wake of Zambeccari's experiments, the number of small balloons being flown in England grew rapidly. Aimé Argand, a Swiss scientist, flew a small gas balloon for George III and his family at Windsor Palace on November 26, 1783. The following year saw small hot air and gas balloons launched from London, Aberdeen, Cork, and other cities in Great Britain.[15]

James Sadler launched one of the largest of these, a 30-foot aerostat, from Oxford on February 9, 1784. A 25-foot specimen complete with a "triumphal car" was sent up from Astley's Amphitheater in London on March 12.

Small, home-built balloons were so common in London that spring and summer that the editor of the *London Morning Herald* could argue that the "balloon madness" had gone past the point of "fashionable folly": "When these aerostatic machines are let off in every street in town, and at *night* too, it is high time to look to . . . safety . . . for should these globes alight upon hay ricks, or corn stalks, or thatched out houses, no man can foresee the extent of the mischief that might ensue: houses burnt down, and families reduced to beggary might be the melancholy consequences."[16]

But there was no stemming the enthusiastic tide. As early as February 1784, London papers were reporting the craze for balloon fashions: "Balloon hats now adorn the heads of such of the *parading impures* as can afford them. Whilst the more inferior tribes have invented a hat which is, not improperly, called the *bastard balloon*!—being a humble imitation and destitute of feathers."[17] Advertisements announced that "AIR BALLOON HATS, either trimmed or plain, or the wares to make them, may be had, in the greatest variety of colours, at Hartshan and Dydes on Wigmire St."[18]

The editor who proclaimed in February 1784 that "the balloon fashions are at their zenith and must soon be forgotten" was sadly mistaken.[19] Two years later the passion for balloon adornments had yet to reach a peak. In July 1786 the *London Morning Post* could still poke fun at the impact of the balloon on ladies dress. "To be balloonified is all the *ton*, especially among the *belles* of twenty, whose enormous protuberances present to the mind an idea of extreme rarification."[20]

But for all this excitement, the British were slow to attempt a manned ascent. Reports that a Jewish barber had reached an altitude of 116 yards at Oxford in February 1784 were generally discounted, and the efforts of James Tytler provided more mirth than edification. "Balloon" Tytler, as he was dubbed in the press, had flown small balloons in Edinburgh in March 1784. On August 7, over a month after young Edward Warren had made his tethered ascent in Baltimore, Tytler attempted to fly an ungainly, barrel-shaped Montgolfier from Comely Gardens in Edinburgh. The balloon was partially destroyed by fire during the inflation, but the intrepid aeronaut, goaded by a disappointed crowd that was rapidly transforming itself into an unruly mob, did make a short hop to an altitude of less than 500 feet seated on a makeshift crossbar. Tytler subsequently made other short "leaps," as he characterized them, but his "Edinburgh Fire Balloon" must obviously have been much inferior to Carnes's grand "American Aerostat." Tytler emigrated to Salem, Massachusetts, in 1795, where he died nine years later, the victim of a fall into a salt pit while wandering about in a drunken stupor.[21]

It remained for Vincenzo Lunardi, secretary to the Neapolitan ambassador in London, to make the first unequivocal manned flight in Great Britain on September 15, 1784. Lunardi created enormous excitement. Two unsuccessful attempts had been made to ascend in London during the month prior to his flight. As a result, many potential ticket buyers were convinced that Lunardi was yet another charlatan.

Nevertheless, on the morning of September 15, 150,000 Londoners streamed back to the artillery ground at Moorfields to watch Lunardi ascend, or fail to ascend, in his lovely red-and-white balloon. As the British historian L. T. C. Rolt has noted, the scene can only be described as Hogarthian: "Certainly the situation hovered on the brink of chaos all morning. First the rumour that a mad bullock was loose caused a panic stampede. Next a wooden scaffold crowded with spectators collapsed with a rending crash. Then a furious affray broke out as people began pelting the occupants of cabs which had parked in front of them and blocked their view. There was more scuffling, shouting, and excitement when a pick-pocket was seized and ducked in a nearby pond."[22]

At two o'clock that afternoon, Lunardi stepped into his car, grasped the twin silken oars that were to be used in maneuvering the aerial craft, and ordered his release. John Jeffries was present in the crowd to record the scene: "At first ascending, it appeared rather too heavy and disposed to come down, but on casting out part of the sand gravel, which he had for ballast it ascended most elegantly and with great beauty and magnificence. In about five minutes he waved his flag as a signal that he was well, and soon after that I saw his oars or wings in motion, and in a few minutes after the flag fell down and almost immediately after that . . . one oar fell.[23] Jeffries observed Lunardi's balloon for an hour and a quarter, by which time "it had ascended so far as to appear of the size of the crown of a hat and was with difficulty visible."[24]

The excitement that followed the flight was almost as impressive as the near riot that preceded it. The broken oar that dropped from the balloon fell on Ray Street, where it was retrieved by Mrs. Godfrey, a servant at the Baptist's Head in St. Johns Lane. A crowd immediately grabbed the fragile paddle from her hands and ripped it into small pieces which they divided among themselves. The *Gazette of the State of Georgia* reported that "the poor woman, with wringing hands and streaming eyes, declared that the loss of her husband or one of her children would scarcely have given her more affliction than she felt at being so cruelly disposed of the signal of good luck that the flying conjurer had thrown in her lap."[25] Mrs. Saunders of nearby Good Street was even less fortunate. Mistaking the falling oar for Lunardi's body, she collapsed and died.

John Jeffries was finding it difficult to contain his enthusiasm for aeronautics. A month after Lunardi's ascent, on October 16, 1784, he was present at Lochees Military Academy to witness a flight by a visiting French aeronaut, Jean Pierre Blanchard, and his English sponsor, John Sheldon. Once again, an enormous crowd, which Jeffries estimated at 250,000 spectators and two thousand carriages, was present for the event. At 2:05 that afternoon Blanchard's "grand air balloon, with its elegant flying boat hanging to it," ascended, then came back to earth in the enclosure after only a minute and a half in the air. The two men cast off a second time, only to descend once more outside the enclosure. On the third try, having removed some items of equipment from the basket, they remained in the air, moving southwest toward Finchley Common. It was, Jeffries noted, "a most noble appearance."[26]

Jeffries was particularly impressed by the fact that John Sheldon, an anatomist with no prior aeronautical experience, had been able to make a flight simply by agreeing to bear the costs of the venture. The notion that he might make an ascent himself was born as he watched Blanchard and Sheldon disappear in the distance, "waving their flags and . . . wing-like apparatus affixed to the boat." "I resolved," he later recalled, "to gratify this, which had finally become my ruling passion."[27]

Jeffries reasoned that the pioneer balloonists, who had "opened and pointed out the new road," had done well simply to ascend and descend with safety. Now it was time for natural philosophers to take advantage of this "new and immense field." Jeffries viewed himself, an adventurous amateur with scientific training, as the perfect man to begin the serious study of the atmosphere.[28]

The day after the Blanchard and Sheldon ascent, Jeffries began to pressure two friends who were acquainted with Blanchard, Mme Bernard and Mme De Can, for an introduction to the French aeronaut. The two men met at Mme De Can's home on November 6 and struck an immediate bargain. Jeffries was to pay Blanchard 100 guineas, in return for which he would be allowed to accompany the Frenchman on his next flight. In addition, Jeffries was to receive a dozen free tickets of admission for his friends.

It is unfortunate that Jeffries provided only a barebones account of this first meeting with Blanchard. In view of the violent enmity that grew between the two men during the next decade, it would be interesting to know a bit more about their initial reaction to one another.

Certainly they stood poles apart in almost every conceivable way. Jeffries was, as one observer noted, "a straight-built figure" standing some 5 feet 9 inches tall and weighing roughly 165 pounds. He was handsome and impeccably dressed, the very picture of an English gentleman. In politics he was a convinced monarchist with distinctly elitist social views.[29]

Blanchard, on the other hand, was aptly described by a contemporary as an "unpleasant creature—a petulant little fellow not many inches over five feet, and physically suited for vaporish regions."[30] He was, as Rolt has remarked, "a ruthless egotist, a mean-spirited and jealous man, a prima donna of the air who begrudged others the smallest share of his limelight."[31] But the feisty, jockey-sized Frenchman who dabbled in radical politics had two great virtues, his enormous courage and his extraordinary mechanical skill.

Blanchard was born in the Norman village of Petit Andelys on July 4, 1750. He inherited his mechanical skills from a father who worked as a carpenter, gunsmith, and machinist. Blanchard produced his first invention, a rat trap, at age twelve. Four years later he constructed a primitive velocipede, which he rode from Andelys to Rouen. By age twenty-nine he was working as a professional engineer, having designed the hydraulic system employed to raise water 133 meters from the Seine to the Chateau Gaillard.

In 1781 Blanchard unveiled his *Vaisseau Volant*, a heavier-than-air flying machine based on his velocipede. With the invention of the balloon, he transferred the flapping wings of the *Vaisseau Volant* to an aerostat in the hope that he could produce a navigable airship. His first ascent from the Champ de Mars on March 2, 1784, was followed by two more from Rouen on May 23 and July 18.[32] While the flight with John Sheldon from London on October 16 was only his fourth ascent, Blanchard was already one of the most experienced and skilled aeronauts in Europe and well on his way to becoming the first man to pursue ballooning as a profession.

Jeffries and Blanchard hoped to make their joint ascent within a week of their agreement of November 6. Their first problem was to locate a suitable launch site. Armed with letters of introduction from Lord North, the two men, occasionally accompanied by Mme De Can, visited the Charter House School, the Christ Hospital arena, the College Square at Westminster School, the yard at Westminster Abbey, the King's Muse Courtyard, the Foundling's Hospital, Busby Park, Canenbury House, the Artillery Ground, the Lyceum, Ashton Lever's yard, Tattersall's, the French Ambassador's Residence, Coleman's Theater, and the duchess of Devonshire's home before deciding to exhibit the balloon at and make the ascent from Mackenzie's Rhedarium.

Jeffries agreed to pay 20 guineas a day for the use of the exhibition rooms. In addition, he was charged £10 for each day the Rhedarium Yard was scheduled for an ascent. He made a £50 cash deposit, part of which was to be returned if the weather permitted an ascent on the first or second day scheduled.

The venture was already proving expensive for Jeffries in other ways as well. By the end of November he had loaned Blanchard £105 sterling above and beyond the actual expenses of the flight.

Clearly, Jeffries saw his role as being at least as important as that of Blanchard. He was the intrepid philosopher, risking danger to investigate the unknown. As a result he spent a great deal of time conferring with Dr. Joseph Blagdon, secretary of the Royal Society, and Henry Cavendish, one of England's most distinguished chemists. Accompanied by his friend William Franklin (son of Benjamin, father of William Temple), a loyalist and the former royal governor of New Jersey, he sought advice on the kind of scientific experiments that could most profitably be conducted aboard a balloon.

When Jeffries made his first ascent on November 29, 1784, he went armed with a thermometer, barometer, Nairn and Blount pocket electrometer, hydrometer, precision timepiece, compass, and small telescope. In addition, Cavendish provided him with seven small sealed vials that were to be opened at various altitudes to obtain air samples.

Preparations for the ascent began in earnest at the Rhedarium on November 28. Until the introduction of illuminating gas to ballooning in the mid-nineteenth century, the process of inflating a hydrogen aerostat was costly and dangerous. Blanchard's balloon measured some 26 feet in diameter and had a capacity of 9,026 cubic feet. The task of generating so much hydrogen required an industrial process that was very difficult to manage.

Blanchard hired Aimé Argand, the famous Swiss chemist, to supervise the inflation. Argand had not only flown small balloons of his own for the king and queen of England, but had assisted the Montgolfiers and J. A. C. Charles and had been in charge of the inflation for the Blanchard-Sheldon flight.

Argand began preparations for the Blanchard-Jeffries ascent on November 28, the day before the scheduled launch. The first step was to erect two large, upright poles to support the balloon during inflation. Twenty-eight 108-gallon barrels were required for the inflation. Two of these remained whole, while the remaining twenty-six were sawed in half and arranged in two groups around each of the large barrels. An area roughly 100 feet in diameter was required for the completed gas generator.

To begin generating hydrogen, each of the half-barrel tubs was filled with water and a smaller tub, with three legs on its lip, was inverted in it. Each upended tub had a 6-inch tube extending out of the botton through which hydrogen would be piped to the large barrels for purification and cooling before transfer to the balloon. A fifty-pound bed "of the parings of iron plates" was laid in the bottom of each of the water-filled tubs. One hundred pounds of cast iron trimmings were then added.[33]

At ten o'clock on the morning of the launch, each of the twenty-six small generators grouped around the large barrel receptacles was charged with 100 pounds of oil of vitriol (sulphuric acid), poured through a second 6-inch opening in the bottom of the upended tub, which was then stopped with clay. Occasionally during the day, Argand would order the clay plugs knocked out and the filings stirred with iron rods to expose fresh surface area to the action of the acid. Experience had taught Blanchard and Argand that some 3,000 pounds of iron and an equivalent weight of acid would be required to produce 9,000 cubic feet of hydrogen. Most of the 100 guineas supplied by Jeffries had been spent on the most expensive item, the acid.

High winds forced a postponement on November 29. Activity began early on November 30, in spite of less than ideal weather conditions. Jeffries arrived at the Rhedarium dressed in a warm flannel shirt, trousers, and a coat. In addition to his instruments, he carried a silver pen ("I did not like to trust to a common pen or pencils, they being liable to accidents"), a blank book of quarto sheets for recording his instrument readings, and a map of England cut and pasted on cardboard sheets for easy reference.[34]

After calibrating his instruments at ground level, he busied himself mounting his equipment in the "triumphal car." There were other problems to be solved before the launch. Because of the questionable weather, gas generation had not begun until after 11:00 A.M. Although Blanchard did his best to speed things along, the inflation process proceeded slowly. Jeffries blamed the situation on the confined work area at the Rhedarium and the clumsiness of the workmen, although the possibility of sabotage had occurred to him as well.

Jeffries and Blanchard took advantage of the extra time to launch a small balloon to check wind direction and to greet the distinguished guests who were present to observe the ascent. The Prince of Wales and the duchess of Devonshire were on the scene and, in fact, asked Jeffries to give up his place in the car to the duchess. His response was a firm no, softened by the comment that "there is nothing else I would not lay at her Royal Highness' feet, but this I could not give up."[35] In fact, Jeffries had already rejected substantial sums of money from John Sheldon and two others who had made similar requests.

The car was finally attached to the inflated balloon at 2:00 P.M. As Jeffries adjusted his instruments and stowed the light provisions that would be carried on the flight, Blanchard lingered with the prince, the duchess, and the French ambassador. When he finally clambered into the car and ordered its release at 2:34, he was clutching a large banner emblazoned with the arms of the duchess of Devonshire.

The two men rose gently into the air as ground handlers moved the balloon beyond the pole fence enclosure. Back on the ground in a large open area outside the fence, they attached the wings and moulinet, a silk air screw with which Blanchard hoped to maneuver the balloon in the air. Friends and well-wishers pushed forward as Jeffries's young daughter Anna was brought to the basket in tears to bid her father farewell.

The balloon, now very much overweight, could only be coaxed into the air by immediately dropping all ballast. As they rose to the level of the rooftops, the balloon was blown against a building, sending a chimney pot crashing to the street and ripping a hole in the basket. They passed up and over Grosvenor Square and flew just to the left of St. Paul's. "The whole city," Jeffries noted, "had a most lovely and beautiful appearance." Ships on the Thames appeared "like Canoes on a smoky, foggy creek." Moving away from the city, Jeffries was startled to find that "the land looks exactly like a beautiful coloured map—not the least appearance of Hills and Building—all flat."[36]

While Blanchard busied himself with the management of the balloon, Jeffries filled his vials with air and wrote notes to be dropped from the balloon for delivery to friends. One of these notes, addressed to fellow loyalist Arodi Thayer, has been preserved in the collection of Amherst College: "From the balloon above the clouds, let this afford one proof, my dear Mrs. Thayer that no separation shall make me unmindful of you—have confidence—happily I hope that happier days await me—pray you, my dear Mr. T. I salute. . . ."[37]

Jeffries addressed other notes to Margaret Vassall, Mrs. Joshua Loring, and a Mr. Geyer. Miss Vassall took violent exception to her note, the contents of which have not survived. It is not clear whether she objected to being singly honored by Jeffries or whether she disliked the idea of sharing the occasion with the well-known, and infamous, Mrs. Betsy Loring.

Blanchard and Jeffries remained in the air for two hours, traveling the whole of London, Westminster, and Greenwich, then passing over Shooter's Hill and between Dartford and Guyford before coming to earth on the banks of the Thames in the Stone Marshes of Kent.

While Jeffries removed his flannel shirt after landing, Blanchard was "as busy as a little B. can be in forcing out the inflammable air" from the envelope.[38] The two men then loaded their equipment on a post chaise and repaired to Mr. Wharton's Bull Inn in Dartford. By 2:00 P.M. the following day they were back at the Rhedarium exhibition rooms where they unloaded the basket, net, and envelope as well as Jeffries's scientific apparatus. Then they were off to Devonshire House for a party celebrating their exploits.

Jeffries was quick to apologize to the duchess once more for having refused her request to fly in his place. "Her grace spoke and looked [with] so much condescension that I am sure she forgave me." The Prince of Wales was also present, but Jeffries's attention was fully absorbed by the duchess and another lady "who but from her loveliness I think must be allied to her Grace."[39]

Jeffries spent the next several days accepting the congratulations of friends, particularly Sir Joseph Banks, who "complimented me so much and did me the honours to say I had done very well and done more in my attention, etc., to observations in my aerial voyage than had been done by anyone before."[40]

The fact that Jeffries was now the first native American to have made a free flight in a balloon was a matter of little importance to him. He was, in fact, particularly upset when the *London Morning Herald* reported that he had waved a flag "emblazoned with the Stars and Stripes as a symbol of the American standard during the flight." Jeffries penned angry letters to the press, insisting that the story was an "invidious misrepresentation." He accused the editors of "having injured my character in their annexing a political character which did not belong to me—for I had been and was a loyal American—I had done myself the honours to take up to me on our Aerial Voyage the British flag only, which had enjoyed the honour of waving over the whole City of London and more villages."[41]

Jean Pierre Blanchard had announced his intention of flying the English Channel prior to his ascent with Jeffries. During the months since the invention of the balloon, the Channel had been seen as the first obvious natural barrier to be overcome by air. Blanchard had informed Jeffries on November 17 that their joint flight would have to be postponed until after the Channel crossing. The Frenchman relented a few days later when it became apparent that he could not finance the Channel flight. It seems probable that Blanchard saw in Jeffries the perfect sponsor for the great flight over the Channel.

Jeffries, in spite of much contemporary opinion to the contrary, was not an extraordinarily wealthy man. For the past several months he had been petitioning the government for a stipend to offset his property losses in Massachusetts and New Hampshire as a result of his loyalty to the Crown. Moreover, much of his practice in London was among the impoverished American loyalists. Still, he was far from destitute and could look forward to the prospect of a considerable inheritance from a wealthy uncle, a loyalist sympathizer who had remained in America. Blanchard was offering him an opportunity to achieve immortality. Jeffries could not resist, but he was resolved to keep the expenses of the venture to a minimum.

Jeffries and Blanchard signed their final agreement on December 14, 1784. Jeffries agreed to pay all expenses associated with the Channel flight in exchange for the privilege of making the voyage. Both men agreed to post a £500 bond to insure that each would keep the agreement.

Jeffries soon realized that his "honourable little Captain" had no intention of keeping to a tight budget. Blanchard's personal debt to his American backer was also rising at an alarming rate. On December 16 Jeffries finally agreed to consolidate the Frenchman's debt to him, and to a number of others, in a single £100 note that would be a bone of contention for the next decade.[42]

Jeffries was also concerned about the size of the party he would have to support before the ascent at Dover. There was, of course, no question about the need for technical personnel to supervise the inflation. James Deeker, who had previously made ascents, had been chosen to head this crew. Jeffries was much less certain, however, about the role Blanchard's friend Mrs. John Sheldon would play at Dover. The bill for Mrs. Sheldon's room and board prior to the ascent would loom large in the later arguments between the two aeronauts.

Blanchard, for his part, was upset by what he regarded as the doctor's parsimony and was rapidly beginning to question the motives of a partner who insisted on referring to him as "my amiable little Captain."[43] By the time the unhappy group departed London for Dover on December 17, Blanchard was already afraid that Jeffries viewed himself as the leader of the group.

The trip to Dover was slow, with stops at Dartford to dine with Mr. Wharton of the Bull Inn and at Rose Inn, Sittingbourne, to retrieve a small balloon launched by the duchess of Devonshire before the flight from the Rhedarium. Arriving in Dover on December 19, Blanchard and Jeffries paid their respects to Ralph Lane, the lieutenant governor of Dover Castle, and immediately began preparations for the flight. Following one of the disagreements that were becoming ever more commonplace, the two men chose a launch site west of the castle, near an enormous gun known as Queen Anne's Pocket Pistol. They hired a local man, Mr. Kisby, to cart their equipment and prepare the site.

The three weeks from December 19 to January 7 were filled with discouragement. The workmen were slow and the weather dismal and unpromising. Jeffries struggled to meet the "great and unexpected" expenses. After paying Mr. Marie's bill for board and lodging on December 31, he was moved to pray that "kind providence in which I trust [will] favor me with a prosperous wind and weather, or, what I much prefer, bless me with dutiful, soulful resignation."[44]

There were diversions. They launched small Montgolfier balloons to check the winds over the Channel and greeted distinguished visitors like Pilâtre de Rozier, "a genteel, thoughtful, temperate man," who was returning to France for his own cross-Channel attempt.

Blanchard and Jeffries seem to have spent most of their time in constant bickering however. By January 5 the situation had become so difficult that Ralph Lane was forced to call at Marie's Hotel to arbitrate the disagreements. Blanchard, it seems, was convinced that Jeffries planned to force him out of the basket at the last minute so that he would have the honor of making the flight by himself. Jeffries informed Lane that he had no such intention and complained of "various artifices" by which Blanchard had "clandestinely attempted to deceive, deter, and prevent me from the enterprise, and to prejudice the minds of some of the principal Gentlemen of the County Kent, and of the City of Dover, insinuating, that from the incapacity of the Balloon it was madness to attempt the experiment with two persons, unless the Balloon could carry one hundred pounds weight of ballast. The pretended friends of M. Blanchard, his Countrymen, publicly circulated such reports of my having declined the enterprise, as occasioned by my being insulted while preparing for our experiment."[45]

Jeffries told Lane that Blanchard had gone so far as to obtain a lead belt, which he planned to wear beneath his clothing on the day of the flight. This would make the balloon so overweight that Jeffries would have to leave the car. The American "very reluctantly" outlined the extent of Blanchard's financial distress and informed Lane that he was so eager to make this flight that he would promise to leap from the balloon if it became necessary "for his [Blanchard's] preservation."[46]

As a result of Lane's intervention, the two aeronauts were temporarily reconciled, but strong winds, rain, and cold continued to prevent a launch. Finally, at six o'clock on the morning of January 7, Blanchard entered Jeffries's room to inform him that "the wind and weather were fair, and would do for our intended voyage from the cliff below the royal castle of Dover, for the continent of France."[47] The two men at once called on Mr. Hugget at the local pilot's lookout for his opinion on whether the winds would carry them across the Channel. The pilot was uncertain, but observing the clouds being blown across, and after testing the winds with a kite and a balloon, Jeffries and Blanchard decided to proceed.

Three cannons were fired at 9:00 A.M. and flags were hoisted from the castle to inform the people of Dover and the outlying communities that a launch would be attempted. While the car and the net were being transported to the site and the balloon was being hoisted on the poles, the two aeronauts breakfasted with Ralph Lane. Jeffries then retreated to Capt. James Campbell's quarters in the castle to dress for his great adventure.

He had given a great deal of thought to the problem of appropriate clothing for a balloon ascent since his flight from the Rhedarium. The cold had been a major problem on that occasion. This time he would wear layers of clothing, including a fur cap, two waistcoats, a coat, an undervest with detachable sleeves, stockings, trousers, and a kerchief. Several of these items, notably the undervest (undershirt), were specially designed by Jeffries to meet the particular need of the aeronaut for warmth and mobility. An oil silk greatcoat, a silk bathing cap, and a pair of fur gloves purchased for the voyage may have been discarded in the last minute attempts to reduce weight prior to the flight.[48]

After dressing, Jeffries called on Captain Campbell and his wife, then went to the launch site, "where I found my little hervik [*sic*] Captain, and the balloon half filled."[49] Blanchard was dressed more simply than his companion in a greatcoat and flannel trousers which were fastened to his shoes.

They released another small Montgolfier at 11:30. A half hour later Ralph Lane was invited to launch the small gas balloon retrieved after the Rhedarium flight. Both balloons "went very well, and took a very good direction for us."[50]

Soon thereafter the car was tugged into position beneath the expanding envelope. Blanchard, perched on the load ring, began the task of tying the car to the hoop, being careful to insure that the load was evenly distributed. The scientific instruments and the miscellaneous items of equipment to be used on the flight were then loaded on board while twelve workmen held the balloon in check. Eight 10-pound bags of ballast, a bottle of brandy, two silk flags, one

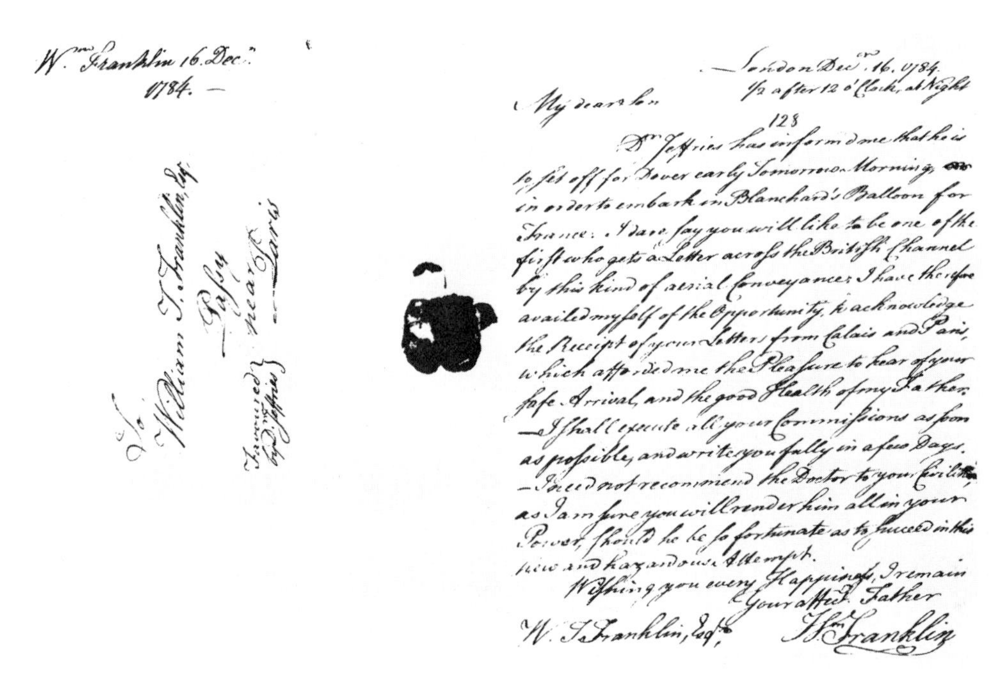

Wm. Franklin 16. Decr. 1784. —

To William T. Franklin, Esqr. Passy near Paris

Favoured by Dr. Jeffries

London Decr. 16. 1784.
½ after 12 o'Clock, at Night

My dear Son

128

Dr. Jeffries has informed me that he is to set off for Dover early Tomorrow Morning, in order to embark in Blanchard's Balloon for France. I dare say you will like to be one of the first who gets a Letter across the British Channel by this kind of aerial Conveyance. I have therefore availed myself of the Opportunity to acknowledge the Receipt of your Letters from Calais and Paris, which afforded me the Pleasure to hear of your safe Arrival, and the good Health of my Father. — I shall execute all your Commissions as soon as possible, and write you fully in a few Days. — I need not recommend the Doctor to your Civilities, as I am sure you will render him all in your Power, should he be so fortunate as to succeed in this new and hazardous Attempt.

Wishing you every Happiness, I remain
Your affectd. Father
Wm. Franklin

W. T. Franklin, Esqr.

The first airmail letter, from William Franklin to his father, Benjamin, via John Jeffries

French, one English, a pair of cork life jackets, a few "biskits," a compass, several inflated bladders, maps, a French dictionary, two packages of blank cards, a pencil, writing paper, a wafer of sealing wax, a watch, a barometer, a hydrometer, and a thermometer were among the odd items to be stored in the car. Of particular note were the several letters contained in one of the inflated bladders. The most important of these was a letter from Jeffries's friend William Franklin to his father, Benjamin.

With Jeffries seated at the head of the car and Blanchard at the rear, the balloon was carefully walked to the edge of the cliff where the silken wings and moulinet were attached. Then, at 1:07, Blanchard ordered the release.

"We rose slowly and majestically from the Cliff," Jeffries reported, "which being at the time of our ascent from it almost covered with a beautiful assembly from the city, neighboring towns and villages, with carriages, horses, etc., together with the extensive beach of Dover, crowded with a great concourse of people, with numbers of boats . . . assembled near the shore, under the Cliffs, afforded us, at our first rising from there, a most beautiful and picturesque view indeed."[51]

An English print illustrating the flight of Blanchard and Jeffries across the channel

The Channel was so smooth that it appeared "like a fine sheet of glass," and the day was so clear that the balloon would at all times remain visible from both the French and the English coasts.

By 1:30 the slowly rotating balloon had begun to swell. Blanchard untied the twin appendix tubes at the bottom of the envelope and allowed them to vent the expanding hydrogen over the side of the car. A few minutes later, alarmed at the sudden and rapid descent of the balloon, the two aeronauts quickly retied the tubes, dropped ballast, and fastened the inflated bladders to the load ring. They then tied twin makeshift rope slings to the ring to serve as seats "to favour our beaver like retreat upwards, in case we were forced down into the water."[52]

These experiments were only temporarily effective. By 2:00 P.M. they were forced to drop their remaining ballast and a parcel of pamphlets in order to rise. Fifteen minutes later, now near the halfway mark, the rest of the pamphlets went over the side.

The "most enchanting and alluring view of the French coast, from Blackness to Cape Blanez to Callais [*sic*] and on to Gravelines," was not clouded by the

A crude print of Blanchard and his balloon

sight of a balloon that "did not appear to be three fourths distended with gaze [*sic*]." Everything not tied down was being jettisoned by 2:30. "Biskits," apples, the wings, the moulinet, the drapery that hung around the car, even the single bottle of brandy went over the side. Jeffries was fascinated to note that a long jet of steam and a "hissing noise" were emitted by the bottle during its descent. "And, when it struck the water, we very sensibly (the instant before we heard the sound) felt the force of the shock on our car."[53]

Observers in Calais, who had been watching the progress of the flight since word had been received at 1:30 P.M. that "a blackish body" had appeared in the air over Dover, were now seriously warned that the balloon was not only sinking dangerously close to the waves but was being blown out of the straits toward the North Sea.[54]

Aboard the balloon, now three-quarters of the way across, the situation was rapidly degenerating. The grapples and ropes were gone now, followed by the aeronaut's coats and trousers. Clad in little more than their cork life jackets, they were, Jeffries later remarked, "merry as grigs to think how we should splatter in the water."[55] Then, four or five miles from the coast, just as they were preparing to climb aloft into the slings to await the inevitable, the balloon began a slow, steady climb. They crossed the French coast midway between Cape Blanez and Blackness, still rising toward their peak altitude.

Benumbed with cold, Blanchard and Jeffries were experiencing the most critical moment of the voyage. Having cast off their life jackets and the rope slings, they were entering a final, rapid descent with no more ballast to reduce their speed. Understandably, Jeffries "felt the necessity of casting away something to alter our course":[56] "Happily it almost instantly occurred to me, that probably we might be able to supply it from within ourselves from the recollection that we had drank much at breakfast, and not having had any evacuation; and from the severe cold, little or no perspiration had taken place, that probably an extra quantity had been secreted by the kidneys, which we might now avail ourselves of by discharging."[57]

Blanchard retrieved two of the flotation bladders from the load ring, which they then filled with what Jeffries estimated as five to six pounds of urine. The bladders were tossed over the side just as the balloon was sweeping down toward the trees: "It so altered our course, that, instead of being forced hard against them . . . we passed along near them in such a manner, as enabled me to catch hold of the topmost branches of one of them, and thereby arrest the further progress of the Balloon."[58]

Jeffries struggled to hold the balloon in the branches, while Blanchard frantically valved gas. Having brought the craft under some measure of control,

NACELLE DU
BALLON de BLANCHARD
QUI le PREMIER
a TRAVERSÉ la MANCHE en 1785

The balloon basket in which Blanchard and Jeffries braved the Channel, preserved in France until destroyed during World War II

the two men then maneuvered it from branch to branch until they reached an open area into which they could descend. Twenty-eight minutes after Jeffries had first caught the top of the tree, Blanchard gave a final tug to the valve line, allowing the gas to escape "with a very loud rushing noise." A little before 4:00 P.M. they finally returned to the earth, as Jeffries proudly noted, in "the *Forest of Guines*, not far from *Ardres*, and near the spot celebrated for the famous interview between Henry the Eighth, King of England, and Francis the First, King of France."[59]

Safely on the ground again, their first tasks were to "empty the envelope, stay warm and recover the use of our limbs, which were much cramped and stiffened from the cold, and the situation we had been confined to in the Car."[60] Fortunately for the shivering pair, they were quickly discovered by a search party armed with extra clothing. The group proceeded to the nearby residence of Viscount Desandrouin for refreshments, then it was on to Calais, where they arrived after midnight.

While Blanchard and Jeffries realized that their flight would attract a great deal of attention, they could not have foreseen the extent to which they would become certified public heroes. The process was clearly in motion the day after the flight when they were officially greeted by the governor, commandant, mayor, chief engineer, and chief justice of Calais. The city then treated them to a banquet attended by "all the officers, Mayors and Aldermen of the City, the King's Procurer-General, and all titled and principal people of the place and neighborhood."[61] Most of the honors, including the freedom of the city and a key in a gold box, went to Blanchard, a French citizen. In Calais, as later at Paris and Versailles, Jeffries received repeated apologies that he, an Englishman, could not be tendered the same honors.

The reception in Calais was only the beginning of a triumphal procession that ended at Versailles on January 12. Blanchard, the debt-ridden veteran of only five ascents, and Jeffries, a man without a country, were instantly acclaimed as heroes of the age by the most splendid court in Europe. They met Count de Vergennes, the architect of French support for American independence, and Madame La Duchesse de Polignac, who greeted them as she was emerging from her toilet, "like a Venus in white muslin," as Jeffries was delighted to note.[62] Of course the highlights of their sojourn in the capital were the formal and informal meetings with Louis XVI and Marie Antoinette. True to form, Jeffries remarked on the king's dress (black velvet and *cordon bleu,* on one occasion, "very plain, with brown leather spatterdashes," on another).[63] One of his fondest memories was of the evening when he "often caught the Queen's lovely eyes" on him at dinner.[64]

The doctor's life in Paris was a round of trysts, dinners, and theater parties with "lovely, lovely women."[65] He was everywhere cheered and applauded. When the crowd at the Italian Comedy discovered that he was in the audience, his name was "echoed from the pit, and universal and repeated claps of applause succeeded, to which we endeavored to return our compliments."[66] At the opera there were "loud claps and shouts of applause, *three* times repeated, before the curtain drew up, and repeated again when the opera was over."[67] At scientific meetings he was greeted by the leading savants of France with "repeated shouts of applause and clapping of hands, encore and encore."[68] He dined with the comte de Buffon, who honored him "with many and great compliments . . . which from such a man are indeed more than compliments."[69]

These first heady days in Paris and Versailles also marked the beginning of Jeffries's personal reconciliation with the leaders of the new United States. On January 14 he paid his first call on Benjamin and William Temple Franklin to deliver the note from William Franklin, one of the few treasured items that had not been jettisoned in the Channel. The patriot branch of the family was as taken with Jeffries as their loyalist father/son in London had been. He dined repeatedly with the Franklins over the next few weeks, joined their theater parties, and was gradually introduced to their broad circle of French and American friends. Jeffries, who only six weeks earlier had reacted violently to the suggestion that he had carried an American flag into the air, now found himself dining with the likes of "the celebrated and brave Commodore Paul Jones, from whom I received many compliments on my enterprise."[70]

Even the Adams family was impressed. John Adams was familiar with Jeffries as a result of their joint involvement in the Boston Massacre trials. The aeronaut met the rest of the family at a dinner hosted by Franklin on the evening of February 14. The company was select, including John Paul Jones, John, Abigail, and Nabby Adams, the Marquis de Lafayette and his wife, Lord Mount Morris, and the two Franklins. Nabby Adams, who was seated next to Jeffries, reported that even in this exalted company he was "the man of the day."[71]

The dinner marked the beginning of an important relationship for the Adams family. Following their transfer to the Court of St. James, John Jeffries would become their family physician and would attend the birth of his young admirer Nabby Adams's first child.

Franklin and Adams were able to offer Jeffries some immediate assistance. Franklin interceded with the duke of Dorset in order to obtain a government pension for the expatriate American. William Temple asked his father to write to Dr. Blagdon that the Franklins would be most pleased to see Jeffries elected a member of the Royal Society, "free of all expense."[72]

When Jeffries passed through Dover on March 2, he was given the freedom of the city and the honorary title Baron of the Cinque Ports. Finally, at 4:15 P.M. on March 5, he arrived home again, "thank God, at Margaret St., Cavendish Square, London."[73]

It was not yet apparent that the epic cross-Channel flight had been the doctor's last balloon voyage. While still in Paris, he had spent a great deal of time with Pilâtre de Rozier. Together the two men visited the workshop where a large, mixed hot air and gas balloon with which de Rozier hoped to fly the Channel was being constructed.

When de Rozier visited London that spring, Jeffries met with him several times and attended one of the chemical demonstrations for which the Frenchman was so well known. On this occasion, de Rozier inhaled hydrogen, then lit the gas as he blew it out through a small glass tube. The resulting "very strange and fierce flame" would melt sheet lead or glass. Unfortunately, "immediately, after the first experiment, Le Sieur de Rozier became pale . . . his eyes staring and twitching and he fell down in a swoon."[74]

Jeffries repeatedly offered to support de Rozier's Channel venture in return for permission to accompany him. At one point in early June it appeared that de Rozier had agreed to the proposition, but, for reasons that remain unclear, it was Pierre Romain, not John Jeffries, who died with de Rozier in the disastrous explosion of the large balloon over Boulogne at the very outset of his Channel flight on June 15, 1785.

Through the spring of 1785 Jeffries attended the London ascents staged by Lunardi, Blanchard, and others. His relationship with Blanchard was rapidly approaching the breaking point however. Jeffries was unable to understand why his "amiable little Captain," who was not only making money from his London flights but was also the recipient of financial gifts from the king of France, could not redeem his original £100 note.

Nor had their other disagreements, which had led to the serious split just before the Channel flight, been resolved. John Sheldon, still an ardent Blanchard supporter, threatened to call Jeffries out should the doctor publish his allegations relating to the presence of Mrs. Sheldon in the party at Dover. Jeffries dealt with Sheldon's threat lightheartedly, offering to meet him in the balloon of his choice, but the situation was now far too explosive for humor to be of much value in easing the tension.[75]

The clash of two enormous egos lay at the center of the controversy. Each man was convinced that the other was attempting to capture the major share of the credit for their joint achievement. Jeffries, with his patrician condescension, and Blanchard, the feisty egalitarian, were now irreconcilable and would remain bitter enemies for the rest of their lives.

That spring both men were marshaling support, calling on influential friends whose assistance might be of some value in damaging the other man's reputation. Angry letters passed between the one-time partners and their backers.

The "paper war" reached a feverish pitch following the publication in 1786 by John Jeffries of *A Narrative of Two Aerial Voyages of Doctor Jeffries with Mons. Blanchard*. Blanchard regarded the pamphlet as a condemnation and rushed to the newspapers with his side of the story. His complaints began with the title page, which suggested that Jeffries had been the prime mover. Moreover, Jeffries had dedicated his account to the duke of Dorset, in spite of his knowledge that Blanchard had promised the honor of the dedication to the baron de Breteuil.

But this was only the beginning. The entire pamphlet was "filled with lies," Blanchard fumed. Even the meteorological data with which Sir Joseph Banks had been so pleased was fabricated, he claimed, for the all-important thermometers had been shattered when the car struck a chimney on take-off. At any rate, "the Doctor was utterly incapable of making any [observations]," Blanchard

FRANÇOIS PILATRE DE ROZIER.
President of the Museum established at Paris in 1781 under the Patronage of Monsieur and Madame; Inspector of the Cabinet of Physick, Chymistry & Natural History of Monsieur; Secretary of the Cabinets of Madame, Pensioner of the King; Member of several National and Foreign Academies & an honorary Member of the Thornville or Balloon Club of London.
From an Original Picture in the Possession of Col. Thornton (being the only Portrait he would ever permit to be Painted) by whose desire it is Engraved, to perpetuate the memory of that great Man.
François Pilatre de Rozier the First Aeronaut was born at Metz on the 30th March 1756. — In 1782 he performed the Experiment described in the Picture of inspiring & expiring inflamable air before the Royal Family at Paris & repeated the same Experiment in London on May 27th 1785 before the Members of the Balloon Club. — On the 14th of June 1785, M. Pilatre de Rozier, accompanied by M. de Romain, ascended with his Balloon from Boulogne with an intention to cross the Channel to England; at an elevation of 3600 feet the inflamable air took fire and exploded the Balloon, which descended with such an accelerated velocity as to crush the unfortunate Adventurer. —
Published by W. Faden, Charing-Cross, Jany 9. 1786.

Pilâtre de Rozier, one of the first two human beings to fly and the first man to die in the air

reported to *Lloyd's Evening Post*, "which his companion ascribes to his ignorance, to his being petrified with fear, and to his frequent recourse to the brandy bottle during their voyage."

In short, Blanchard claimed that Jeffries had been "a mere bag of ballast" who had paid the French aeronaut "for immortalizing his name." He called attention to an "English" scheme to seize him before the flight so that full honors would go to Jeffries. The lead belt, he continued, had never existed. According to the reporter, Blanchard seemed "almost literally driven to his wits end by this attempt to rob him of his laurels."[76]

And so one of the great adventures in the history of flight ended not in triumph but in bitterness and recrimination. The hatred would fester for another decade before a final resolution was reached in an American courtroom.

Ironically, both Jeffries and Blanchard continued to profit from their fame. Blanchard, now the most celebrated aeronaut in Europe, resumed a career during which he would make thirty-nine more ascents between the Channel crossing in 1785 and the beginning of an American tour in 1793.

Jeffries also prospered. His medical practice quickly expanded to include some of the great names of the realm. His marriage to Hannah Hunt in 1787 was also a success, producing ten children by 1804. When Jeffries returned for a visit to Boston following his father's death in 1789, he discovered that, contrary to his own expectations, it was possible for at least one loyalist to live comfortably in the United States once again. The Jeffries family returned to Boston for good that year, their past political sins forgiven.

By the time of his death at the age of seventy-four on September 16, 1819, Jeffries was universally revered as one of Boston's great men. Oliver Wendell Holmes would remember him not only as a kind and good-natured general practitioner but as "a leader of medical opinion in America."[77] But, for the children of early nineteenth-century Boston, John Jeffries was a heroic figure for quite a different reason. He had, after all, been "up in the balloon."

Chapter 4  Blanchard in America

NEWS OF THE AERIAL CONQUEST OF THE ENGLISH CHANNEL, COUPLED with reports of the success of Peter Carnes and the rapid progress of ballooning in Europe, continued to fuel American interest in flight. Fresh reports of proposed manned balloons circulated in a number of American cities. In February 1785, for example, the *New Jersey Gazette* reported that a large balloon measuring 30 feet in diameter, "to which will be affixed a Gallery or boat," was being built at an academy in New York.[1] According to the *Gazette of the State of Georgia*, the $500 cost of the venture was to be covered by subscriptions.[2] The unnamed fellow behind this project, who was identified only as "a man who for some time past has amused the vulgar in different ways," finally announced that he would make a flight on July 4, 1785. When the anonymous showman failed to produce a balloon on the appointed day, the sheriff served him with a court summons and offered protection from the angry mob of disappointed spectators.[3]

Less ambitious but far more successful unmanned balloons were soon being flown in all parts of the United States. The age of flight arrived in Charleston, South Carolina, on the afternoon of February 7, 1785, when a "young gentleman" launched what may have been the first hydrogen balloon in the nation. Released from the city green, the aerostat rose "gradually into the air" until it was out of sight and flew some distance beyond the city limits into neighboring St. John's parish—certainly a spectacular performance if the little craft was a hot air balloon. A second balloon launched from a residence on Queen Street the following Saturday landed on St. John's Island.[4]

In April 1785 a hot air balloon 5 feet tall and 3½ feet in diameter launched from Boston landed twelve minutes later on the north slope of Breed's Hill in Charlestown. A few days later the same balloon climbed "very high" above the Burying Ground Point in Charlestown and landed ten minutes later near Battery Point, three-quarters of a mile from the take-off point.[5]

These two flights inspired other amateur mechanics in the Boston area to try their hand at balloon building. One of these balloons, "either not being properly shaped or balanced or not being let off by a chymical genius, produced a faux pas." While the news report is not specific, we can presume that the "faux pas" was a minor fire.[6]

Mr. Buffelot, a New York showman, launched a splendid red-white-and-blue specimen accented with thirteen silver stars at Philadelphia on September 28, 1786. Buffelot inflated his craft at six that evening before "a respectable and numerous company." It rose to an altitude of 3,000 feet, in spite of high winds, and landed 6 miles away on the New Jersey shore thirty-five minutes after launch.[7]

American academies were also becoming interested in aerostation. Ezra Stiles, the president of Yale College, had read news accounts of the de Rozier-d'Arlandes ascent in December 1783. The following July 14 Stiles noted in his diary that he and Josiah Meigs, Yale's senior tutor, were constructing a hot air balloon some 5 feet high and 3½ feet in diameter from fifty sheets (two quires) of glued paper.

This craft was apparently never flown, but on April 25, 1785, Meigs did launch a second balloon. It was an extraordinarily simple device, a paper cylinder 8 feet high and 8 feet in diameter, capped by a paper dome that added 3 more feet to its height. Stiles noted in his diary that the mouth of the balloon was nearly one foot in diameter, but his accompanying sketch shows a straight-sided aerostat with the mouth as large as the maximum diameter of the craft.

The balloon was first flown to an altitude of 300 feet on the evening of April 24, 1785. The next day Meigs inflated it in three minutes over a fire of "kindling shavings" laid "on the square before the college." The meticulous Stiles recorded that it was released at 12:33 and thirty seconds and "passed about fifty or sixty rods at an altitude nearly equal the Top of the Brick Steeple and settled on Mr. Pierpont's house and rolled off to the Ground at 12:35."[8] The flight had been short, with the balloon passing through the air parallel to Church Street and flying over three churches. "Its passage was lofty and majestic." A third balloon "conflagrated and burnt out aloft in the Air" on May 3, 1785.[9]

Only a week later, on May 10, a fourth balloon, the most spectacular to date, was launched from Chapel Street in New Haven. An 11-foot sphere, it was decorated with the figure of a flying angel holding a trumpet in one hand and a U.S. flag in the other. A 7-foot ribbon attached to the balloon carried the motto "Nil Jutentatum nostri liquere" in Latin, Greek, Hebrew, Chaldee, Arabic, French, German, and, added Stiles, "I think in English."[10]

Unfortunately this balloon also "took fire in the ascent, and being converted into a Pyramid of flame at its greatest height, exhibited a grand and pleasing object to the spectators, who had only to regret that the same spectacle was not shown in the night."[11]

The students and faculty of the College of William and Mary were also flying balloons. Thomas Jefferson, an alumnus of the college, had introduced the subject in a letter to James McClurg, a science professor there, in the spring of 1784. As Rev. James Madison, cousin of the future president and himself president of William and Mary, remarked that April, Jefferson's letter "has put us on attempting an Aerial Voyage." As yet, however, the local balloon enthusiasts had not been successful. Madison had been experimenting with the smoke of various materials. "I find the Air from Straw much more inflammable than any I ever collected from Wood etc.," he remarked to Jefferson. "But [I] have not been able to observe the extreme levity which must be necessary for the Purposes of a Balloon." Madison had apparently been burning these materials in a gun barrel, perhaps misunderstanding the current reports from France that hydrogen could be produced by heating iron filings in a metal tube. Clearly, these experiments had been frustrating. "We do not understand the Method used in France for collecting such vast quantities of it [hydrogen]," he complained to Jefferson. "If you should have been informed relative to this, I should be most [happy] for your Communication."[12]

By the early spring of 1786, the "Balloon Club" at William and Mary, still under Madison's direction, had finally succeeded in producing balloons. Madison informed Jefferson that the group had flown a number of small aerostats filled with "inflammable air." The gas was produced by heating pit coal in an iron tube. In addition, large hot air balloons up to 20 feet in diameter were also flown at William and Mary.[13]

Madison was pleased with the result of his effort, particularly as he was unaware of other American balloon activity. He seems, in fact, to have been quite out of touch with the latest developments in European and American aeronautics. Still, he foresaw the possibility of a great future for the balloon:

> *Who would have supposed a few years past that . . . the bold aeronaut should dare to attempt excursions in so rare a Medium, and even be able to direct his course nearer to the Wind than the best Sailing Vessels. It is probable that these aerostatic machines will in time be applied to other purposes than a mere Philosoph. Experiment, tho in that respect alone they are certainly very valuable. Yet I have seen no result of observations made by them, relating to several matters for which they seem particularly adopted. Such as the rate of decrease in the density of the atmosphere at different Elevations, also the Rate in which its Temperature varies, Meteors in general, Propagations of Sounds, Descent of bodies, etc., are all proper*

subject of Investigation, and which no doubt will be investigated as those Machines are more perfected.[14]

Ballooning was to remain a tradition at William and Mary into the following century. In May 1801 John Skelton Watson, a William and Mary undergraduate, informed his brother that "the Spirit for Balloons has been in a rage amongst us."[15] Only a month before he had "been engaged in the construction of an Air balloon."[16] Shortly thereafter, Watson and a friend attempted to send their 6-foot aerostat up from the Court House Green between the Duke of Gloucester and Nicholson streets in Williamsburg. As the wind was blowing "rather fresh," they set fire to the small container of "Turpentine Spirits" hanging below the mouth of the balloon in the lee of the courthouse wall. It rose along the wall and caught in the eve, where the small device hung in flames for some time. The "young philosophers" were mortified to discover that their failure had become something of a joke around the town. Anxious to redeem themselves, they inducted several fellow students into their "balloon club" and tried and failed once again at the same spot with an 8-foot aerostat.

"Our credit," Watson remarked, "we even thought the credit of our college, was now at stake."[17] Allied with an even larger number of students, they constructed their masterpiece, a 20-foot balloon ornamented with sixteen blue stars. This time the balloon brazier was filled with spirits of wine ("more heat, less flame"). The "numerous concourse" that gathered to witness the flight gave a "general shout" as the craft rose into the air. "I never saw so great and so universal [a] delight as it gave to the spectators. It bore a striking resemblance to the moon in partial eclipse."[18]

Balloons had also appeared in New York by the summer of 1786. On July 7 of that year the first small aerostat was sent aloft from 50 Bowery Lane. But James Deeker is the man who deserves credit for introducing New Yorkers to balloons in a serious fashion.

The haphazard approach to orthography adopted by so many eighteenth-century American newspapers often leads to difficulties in identifying personal names. The man who was probably James Deeker is variously identified as James Decker, Joseph Deeker, and Joseph Decker. In view of the fact that he unquestionably identified himself as English, however, and claimed to have made a series of flights beginning with an ascent from Bristol on April 9, 1785, this can only have been the same James Deeker who supervised the inflation of Blanchard's balloon prior to the Channel flight.[19]

Certainly Deeker was no novice. Following the ascent from Dover, he had flown his own balloon from Quantrell's Gardens in Norwich on June 1 and June

22, 1785. Both ascents were a bit chancy, involving ripped envelopes and narrow escapes from disaster aloft.

Following the Bristol and Norwich ascents, no further word is heard from Deeker in England, but on June 10, 1789, he placed his first advertisement in a New York newspaper. Deeker proposed to construct a new manned balloon 25 feet in diameter with which he would ascend from the city. Subscription books to fund the project were opened and subscribers would be able to observe the balloon under construction in the Exhibition Room at No. 14 William Street. In addition, visitors would have an opportunity to see Deeker's "speaking image," a moving wax automaton actuated by air pressure. The words spoken by the "image" were provided by a hidden colleague who answered questions from the audience in a "proper and delicate" tone.[20]

Deeker launched a large, 24-foot balloon from Fort George, the rectangular fortification at the foot of Manhattan Island, on August 7. This first balloon to be launched from the Battery rose to the cheers of a "large concourse of spectators" and flew south, then west, on a twenty-minute flight that carried it 9 miles to a landing in the Harlem River.[21]

The *New York Journal and Weekly Register* carried a letter from one correspondent who had been "delighted to observe the variety of emotions which the passing of the balloon . . . occasioned among the several classes of people." He took particular glee in reporting that one family had run screaming to their cellar "for fear the balloon would break their skulls."[22]

On August 11, Deeker inserted in New York newspapers a note of thanks to those who had attended the ascent. He also announced that he would launch a 30-foot balloon from a lot near Mr. Seaman's racetrack at the corner of Eagle, Suffolk, and Cellar streets on August 15. This flight was made at 6:20 P.M. on the appointed day. The balloon flew 15 miles in twenty minutes, landing on a farm in Flushing. A small hole about the size of a dollar had been burned in the envelope, which was retrieved by a Mr. Prince, a New Yorker who had already flown two small balloons of his own.[23]

Early in September Deeker announced that he was close to finishing work on a gigantic, 100-foot balloon with which he would make a flight from Seaman's lot on September 23. Although several papers mention the 100-foot diameter, it is doubtful that Deeker could have constructed so large an envelope in so short a time. When Ezra Stiles visited Deeker's exhibition rooms on September 22, he reported a 30-foot balloon on display but made no mention of the larger aerostat.[24]

The *New York Packet* reported that fully two-thirds of the population of the city was on hand for the final Deeker launch. Another newspaper offered the perfect encapsulation of the scene at Seaman's lot that day: "The day arrives—crowds of spectators fly to the theatre of action—our hero partly inflates his aerial vehicle—the upper retainer is loosed—and, alas, the gas fails!—the balloon falls!—the fire communicates!—and the expectations of thousands ascend in fumo!"[25]

Deeker blamed the disaster on the press of the crowd and the wind. Others were not so kind. He was accused of designing a balloon that was nothing more than "a bubble." One correspondent went so far as to suggest that at least Deeker had kept his promise of leaving the city right after the event, but not before having dipped "into the purses of the generous and disappointed spectators."[26]

Deeker's failure did little to dampen the "balloon rage" in New York. A play entitled *The Air Balloon* was featured at the American Theatre Company's playhouse the month following the disaster. Taverns like the Balloon-House on Broadway traded on the popularity of aeronauts. Unmanned balloons continued to rise above the New York rooftops. Occasionally the newspapers took note, as in November 1790, when two especially large balloons were released from Brooklyn Heights and disappeared over the North River toward Philadelphia.

Three years later, in June 1793, James West, an actor at the John Street Theatre, noting that "Aerial Excursions" had "become a topic of much conversation in America," was "happy to have an opportunity of exhibiting a Balloon to the Ladies and Gentlemen of New York."[27] West sold tickets that would admit patrons to the theater as well as to an exhibition and launch of a 12-foot and a 36-foot balloon. West, who claimed to have flown similar balloons for George III and his family, promised to launch his "beautifully varigated" aerostats later that week from the College Green. Apparently he did.[28]

For over a decade, mechanically inclined Americans could count on being able to draw a modest crowd of paying spectators to witness the launch of an unmanned balloon. The fact remained, however, that for nine years following the limited success of Peter Carnes no human being had ventured aloft into American skies. The United States could boast a sizable number of balloon enthusiasts whose appetites had been whetted but not satisfied.

By late 1792 Jean Pierre Blanchard, now clearly established as the leading European aeronaut with a grand total of forty-four ascents to his credit, was looking for richer pickings. As usual, he was facing financial disaster. Following his forty-fourth ascent, from Lübeck on July 3, 1792, he had been arrested and imprisoned at Kufstein, in the Tyrol, by Austrian officials who feared that he was spreading the radical doctrines of the French Revolution, as, in fact, he probably was.

Immediately after his release from prison, he turned his attention to the

United States, where he hoped to reap the profits that had thus far eluded him in Europe. After obtaining some acid and other supplies from London, he sailed from Hamburg aboard the brig *Ceres* on September 30, 1792.

Blanchard landed at Philadelphia on December 9. It is not at all clear whether his wife, Victorie, and their son and three daughters accompanied him on the *Ceres*. If not, they certainly arrived in the United States soon thereafter, for news reports of Blanchard's activity over the next three years mention the presence of his family.

As in the past, Blanchard had little trouble generating publicity. All Philadelphia was quickly aware that, as the *Federal Gazette* remarked, "The celebrated Mr. Blanchard . . . has arrived among us."[29]

Establishing himself at 9 North Eighth Street, he was advertising copies of his publications for sale within days of his arrival. By December 22, Mr. Roset, Blanchard's secretary, was ensconced in Oeller's Hotel selling tickets for the ascent at $5.00 a head. Roset's books were to remain open only until December 25, at which time the date and place of the flight would be announced and tickets distributed to subscribers.

Blanchard made every effort to boost ticket sales. There would be only one flight, he explained, for he had just enough acid (4,200 pounds) for a single inflation, and no more was available in the city. Responding to suggestions offered by newspaper correspondents, he agreed to keep the subscription books open until January 8 so that country people would have an opportunity to purchase seats. Blanchard also decided to offer cut-rate $2.00 tickets to a special section located behind the regular $5.00 seats.[30]

With ticket sales underway and the site chosen, Blanchard announced that he would begin his aerial journey promptly at 10:00 A.M. on January 9, 1793, weather permitting. Excitement mounted as the day approached. Blanchard was inundated with advice and questions. He responded with a no to those who wished to accompany him. He similarly discouraged those who hoped to chase the balloon on horseback. "If the day is calm," he noted, "there will be full time to leave the prison court . . . as . . . I will ascend perpendicularly; but if the wind blows, permit me, gentlemen, to advise you not to attempt to keep up with me, especially in a country so intersected with rivers, and so covered with woods."[31] The publication of this response from Blanchard was, of course, a direct challenge to the horsemen of Philadelphia.

Blanchard's arrival in the city also provoked much political controversy and debate. The French Revolution was a touchstone dividing the numerous American political factions. Republican leaders such as Philip Freneau, editor of the *National Gazette*, showered praise on Blanchard as a symbol of gallant French

Jean Pierre Blanchard at the time of his American tour

Republicanism. These were the men who invited the aeronaut to a supper at Oeller's Hotel on January 3, where fifteen toasts were offered by "the friends of equality and the French Revolution . . . to commemorate the glorious successes of the Republican armies over Austria and Prussia."[32]

For Freneau and company, Blanchard and his balloon symbolized French achievement. "What cannot her courage and genius attain?" he queried.[33] It was a theme that would find its way into Republican poetry as well:

If the English should venture
to sea with their fleet,
A host of balloons in a
trice they shall meet,

The French from the Zenith
their wits shall display,
And pounce on these sea-dogs
and bear them away.[34]

Another Philadelphia editor was even more explicit in portraying Blanchard in terms of the promise of the French Revolution: "Great Blanchard! as you wing your way toward the heavens, there announce to all the planets of the universe, that Frenchmen have conquered their interior enemies, and that their Exterior Foes have been repulsed by their intrepidity; Dart through Olympus and tell to the Gods that Frenchmen have been victorious. Implore the aid of Mars, that the Arms of France may crush the ambitious designs of tyrants FOR EVER."[35]

For some, the balloon represented virtue itself. One writer for Freneau's paper remarked: "It would seem worthwhile to determine by a few experiments whether the human race, if generated in the higher regions of the atmosphere, would not partake of a more exalted nature, more inclined to virtue and less to evil."[36]

Freneau agreed with these sentiments:

By science taught, on silken wings,
Beyond our groveling race you rise,
And soaring from terrestrial things,
Explore a passage to the skies—
O, could I thus exalted sail,
And rise with you beyond the jail.[37]

The Federalist opponents of the Francophile Republicans took a very different view of the French Revolution, Blanchard, and ballooning in general. Blanchard was, according to one such critic, "intruding into regions where he had no business."[38] Another New England Federalist accused the Frenchman of giving a "cheap performance."[39]

Even Vice-Pres. John Adams remarked with some alarm that "Blanchard . . . is to set all the world upon a broad stare at his balloon," and wished that a political opponent would "make it an interlude" and travel back to Europe with the Frenchman.[40]

Wednesday, January 9, 1793, dawned hazy and pleasant in Philadelphia. It had been the most extraordinarily mild winter in the memory of the oldest inhabitants. The temperature on the day of Blanchard's ascent would climb toward a relatively balmy 50 degrees, so that the thousands of spectators who flocked to the neighborhood of the Walnut Street Prison would be able to hang out open windows, clamber onto roofs, or stretch out on the grass in comfort.

Blanchard's choice of the jail as a launch site had caused some initial concern. At the time of the flight the building housed thirty-eight criminals, twenty-eight men and ten women, whose crimes ranged from burglary, forgery, and housebreaking to highway robbery and fencing stolen goods. Still, the yard was large enough to accommodate 4,800 ticket holders. There was even a separate entrance through which ladies could be admitted to the yard without passing through the prison. Moreover, the prison was centrally located just across the square from the State House. The enclosed yard would save Blanchard the expense of constructing a fence, and, most important, the city fathers had offered the use of the site free of charge.[41]

Cannons positioned at Potter's Field began to fire at six that morning. By 9:00 A.M. the early haze had burned away and the enormous crowd was in place. Even Blanchard was impressed by the number of spectators: "Accustomed as I have long been, to the pompous scenes of numerous assemblies, yet I could not help being surprised and astonished when . . . I turned my eyes toward the immense number of people, which covered the open places, the roofs of houses, the steeples, the streets and the roads."[42]

Dr. Benjamin Rush, an associate of Foulke and Morgan, reported that the city was so crowded with visitors from New York, Baltimore, "and other distant parts" that hotel rooms were simply unavailable, and the theaters were so crowded that hundreds were turned away from evening performances. The small towns surrounding Philadelphia were virtually deserted on the day of the ascent. Samuel Mickle of Woodbury, New Jersey, noted that "all of Woodbury, almost, was going to see it."[43]

Benjamin Franklin was not in the crowd, having been buried three years before in the Christ Church Cemetery at Fifth and Arch streets. His grandson, Benjamin Franklin Bache, who had shared the Franklins' quarters in Passy while attending boarding school with John Quincy Adams and Peter Jay Munro, was present however. Bache, now a Philadelphia newspaper editor, had been one of several local science enthusiasts who assisted Blanchard in preparing for the flight. Others in the group included Caspar Wistar, Dr. Glentworth, Dr. Nassy, and Peter Legaux.[44]

Blanchard appeared at 9:00 A.M. dressed in a "plain blue suit" topped off by a cocked hat with white feathers. The inflation began soon after his arrival and went very quickly. The balloon itself, a veteran of five prior ascents, was constructed of heavily varnished yellow silk. The basket was painted blue and covered with spangles. Blanchard was still trying to row himself about the sky with the oars attached to the car. In addition, he would load the basket with nine 10-pound bags of ballast, a French edition of his pamphlet on the Sheldon flight, the usual inflatable bladder for letters, a compass, "philosophical instruments," a small bottle of brandy, a few biscuits, two cork life jackets, and a double-sided silk flag, one side sporting the American ensign, the other the French tricolor. Blanchard carried a few other odds and ends on his person. The day before the flight, for example, Benjamin Rush had given him a pulse glass so that the aeronaut could keep a record of his own heartbeat during the ascent.[45]

By 9:30 A.M. Blanchard's balloon was beginning to swell with hydrogen. The sun had burned away the clouds, "until they appeared no more than cobwebs on the irradiated atmosphere."[46] George Washington came forward at 10:00 A.M. to bid the aeronaut *bon voyage*. The president, who received Blanchard sometime before, now presented him with a "passport" that would guarantee his safe conduct:

TO ALL WHOM THESE PRESENTS SHALL COME

The bearer hereof, Mr. *Blanchard a citizen of France, proposing to ascend in a balloon from the city of Philadelphia, at 10 o'clock,* A.M. *this day, to pass in such place as circumstances may render most convenient THESE are therefore to recommend to all citizens of the United States, and others, that in his passage, descent, return or journeying elsewhere, they oppose no hindrance or molestation to the said* Mr. *Blanchard; and, that on the contrary, they receive and aid him with that humanity and good will which may render honour to their country, and justice to an individual so distinguished by his efforts to establish and advance an art, in order to make it useful to mankind in general.*[47]

The artillery offered a final salute at 10:09 as Blanchard climbed into the basket with his companion, a small black dog. After carefully dumping a bit of ballast, he ordered Nassy and Legaux to remove their hands from the basket. The first aerial voyage in America had begun.

The take-off, Benjamin Rush noted, was "truly sublime." On his first appearance above the wall there was a universal cry of "Oh! Oh! Good Voyage."[48] Flourishing his hat and waving the flag, Blanchard rose straight up from the prison yard, then began to move toward the southeast. Gen. John Steele, comptroller of the United States Treasury, was stunned by the sight. "Seeing the man waving a flag at an immense height from the ground, was the most interesting sight that ever I beheld, and tho I had no acquaintance with him, I could not help trembling for his safety."[49]

By 10:30 the aeronaut was hard at work conducting the simple experiments he had been asked to perform. B. F. Bache supplied the barometer used on the flight, while Dr. Caspar Wistar had provided six sealed bottles "containing diverse liquors," which were now exposed to the atmosphere and resealed. Approaching his maximum altitude of 5,812 feet, he attempted to use Dr. Rush's quarter-minute timing glass to measure his pulse, but finally gave up and resorted to his own pocket watch. Finally, he brought out the lodestone presented to him by Dr. Glentworth and discovered that it would not lift as much weight at altitude as it had on the ground.[50] Blanchard then packed his instruments safely away and snacked on "a morsel of biscuit and a glass of wine" while he played with the dog and enjoyed the scenery. He also sniffed a bottle of ether placed in the car by Dr. Nassy and Mr. Legaux, "which refreshed me very much."[51]

Unwilling to fly too far into the country, he began to vent gas, hunting for a spot to land. After several abortive attempts he was on the ground once again at 10:56 in Gloucester County, New Jersey, just a mile east of Woodbury. Blanchard had covered a distance of some 15 miles in roughly forty-five minutes.

The aeronaut, whose English was a bit shaky, had some difficulty communicating with the local citizens who made their way to the scene, some of whom came armed. The remains of the wine and Washington's passport saved the day. Samuel Mickle noted, however, that the balloon remained "ye subject in almost every quarter" of Deptford Township for weeks to come.[52]

Blanchard and his equipment were immediately carted back to Cooper's Ferry on the Delaware River. At 6:30 that evening the exhilarated aeronaut delivered his personal account of the exploit to the president of the United States.

As had so often been the case in his career, Blanchard's Philadelphia ascent was a personal triumph but a financial disaster. He estimated his total expenses at $500. The gate receipts had amounted to only $405. Far from growing rich

An American representation of Blanchard's Philadelphia balloon

in the new world, he had not broken even.

Blanchard may have been poor, but he was famous. Benjamin Rush was only one of many who could talk of little but the gallant Frenchman: "For some days past the conversation of our city has turned wholly upon Mr. Blanchard's late Aerial Voyage. It was truly a sublime sight. Every faculty of the mind was seized, expanded and captivated by it. 40,000 people concentrating their eyes and thoughts at the *same* instant, upon the *same* object, and all deriving nearly the *same* degree of pleasure from it."[53]

Rush was charmed by Blanchard's description of his harrowing adventures aloft and became fully converted to ballooning. Flight, in Rush's mind, was a religious imperative. "The first command to subdue the earth," he remarked, "like every other divine command, *must* be fulfilled. The earth certainly includes water and air as well as dry land. The first and last of these have long ago yielded to the dominion of man. It remains for him only to render the air subservient to his will."[54] The doctor went so far as to echo the sentiments of the French physicians who had spoken of "balloon cures" a decade earlier: "I have destined them [balloons] to be employed as remedies where a sudden and extensive change in the body is required."[55]

Undaunted by the poor financial return of his first flight in America, Blanchard was determined to remain in Philadelphia and recoup his losses. Joseph Ravara, the consul general of Genoa, mounted a fund-raising campaign designed to reimburse the aeronaut for his efforts. When this project failed, Gov. Thomas Mifflin of Pennsylvania offered Blanchard free use of a lot on Chestnut Street between Seventh and Eighth, just above the "New Theatre." Blanchard constructed a large building on this site that would serve as his base of operations for months to come.

The new "aerostatical laboratory" housed the balloon and car used on the flight of January 9 as well as additional mechanical curiosities that Blanchard hoped would lure paying customers through his doors. Beginning in April 1793 he exhibited a wheeled automaton known as the "Curious Carriage." Blanchard would drive the carriage around the interior of his laboratory at specified times each day to the delight of visitors. The vehicle moved "with such rapidity as the conductor wishes to give it," and featured a mechanical eagle flapping its wings, which "seems to give it motion."[56]

In the spring of 1793 Blanchard advertised an upcoming ascent to be made with Joseph Ravara. This time, however, he was determined that the flight would not be made until sufficient cash to cover expenses and provide a modest profit was in hand. Local newspapers carried his plea not to "dishonor this great capital, the Athens of America . . . where every species of genius ought to meet its reward."[57]

Business was in full swing by May. The aerostatical laboratory was open every day. A twenty-five cent ticket would enable the visitor to inspect the balloon to be used in the upcoming ascent as well as a number of smaller aerostats. The admission fee rose to fifty cents at 5:30 P.M., when the Curious Carriage was demonstrated. Guests were asked not to bring dogs "accustomed to the chase," for experience had taught Blanchard that such animals were liable to attack the carriage and "break the Eagle, whose wings flap."[58] Visitors could also purchase copies of the aeronaut's account of his recent flight, *Journal of My Forty-Fifth Ascent*, and could add their names to the subscription list for tickets to the next flight.

But luck was running against Blanchard. Subscriptions trickled in, but before he was satisfied with the amount, his balloon was severely damaged by vandals. He then turned to flying small balloons of the sort Philadelphians had seen many times before, but with a fascinating twist. Animals were sent aloft, automatically released at altitude by a fuse, and recovered by means of a parachute, or "falling screen."

The first of these animal drops took place on June 6, 1793, when a dog, a cat, and a squirrel were safely returned to earth. The event, according to the *General Advertiser*, was witnessed by "few paying, but many nonpaying spectators."[59]

Another parachute flight was scheduled to be made from the Chestnut Street laboratory in mid-June. Blanchard was now reduced to pleading that "those persons who are acquainted with the expenses of the artist will honor him with their company *inside* the said place."[60] As the animal flights continued into the early summer, newspapers such as the *Federal Gazette* took up Blanchard's cause: "All are now subscribers to the distinct merit of that enterprising philosopher Mr. Blanchard. There appears to be a prevailing disposition to compensate him. We hope all will step forward before it is too late."[61]

Apparently the pleas for support had some effect. The parachute flights of June 29 and July 3, 1793, drew good crowds both "within and without the enclosure." The indoor demonstrations of the Curious Carriage continued throughout the summer as well.[62]

The late summer of 1793 brought extraordinary difficulties for everyone in Philadelphia. On Monday, August 19, John Foulke, Benjamin Rush, and Dr. Hugh Hodge met at the home of Peter Le Maigre, a balloon enthusiast who had been one of the leading subscribers to Dr. Morgan's project. Le Maigre's wife, Catherine, was desperately ill with a disease that was all too familiar to the three physicians—yellow fever. The official death toll would reach 4,044 persons over the next four months, but the mere recitation of the number of dead does not suggest the horror that gripped Philadelphia. Business ground to a halt as thousands deserted the city, only to discover that Philadelphians were unwelcome in many towns and villages whose inhabitants feared the spread of the plague. The federal government came to a halt and ceased to function as officials from the president to the lowliest clerk sought refuge outside the city.

Those Philadelphians that remained in town were certainly unwilling to gather in public places. As a result, like other showmen, the proprietor of the aerostatical laboratory found attendance at his establishment dwindling to nothing.[63] In a desperate attempt to recover his losses, Blanchard announced the opening of yet another subscription to support his long-awaited forty-sixth ascent in 1794. Once again, the attempt was a failure. Now, after more than a year of unrelieved discouragement in Philadelphia, Blanchard decided to move on.

He sailed for Charleston, South Carolina, aboard the coasting schooner *Success* in the fall of 1794. The *General Advertiser*, at least, was sorry to see him go and advised the citizens of Charleston that "Blanchard will gratify [them] . . . if they encourage him."[64]

Blanchard's activity in Charleston was a repetition of his experience in Philadelphia. He exhibited the "Philadelphia balloon," inflated with air, and promised to make his next ascent from the city, provided money was forthcoming. It was not, and Blanchard once more offered tickets of admission to his now standard "aerostatical experiments."[65] Small animals were sent aloft in a car shaped like a "Liberty Cap." When the balloon reached a "considerable height," an explosion was heard, and "there immediately springs from the cap a beautiful parachute of silk," which carried the animals back to earth.[66]

The citizens of Charleston were pleased by Blanchard's "facility and punctuality." As in Philadelphia, his appearance in the city created serious interest in ballooning. The "alert managers" of the French Theatre were quick to capitalize on this interest, featuring performances of the play *Harlequin Balloonist* six times in 1794. This was of little help to Blanchard however.[67]

Charleston having proved as unsatisfactory as Philadelphia, Blanchard transferred his operations to Boston in the summer of 1795. Scattered news articles and a handful of surviving letters from the aeronaut describe a scheme to raise $3,000 to fund several ascents. But Bostonians were as tight-fisted as the people of Charleston and Philadelphia, and the plan failed.[68]

Disaster visited the Frenchman in yet another form while he was in Boston. At the very moment when Blanchard was exhausting the last of his savings, John Jeffries brought suit for the repayment of his ten-year-old loan. When the two men had parted company in 1785, Jeffries had been angered by Blanchard's failure to repay the £100 debt and by his cavalier attitude toward the expenses of the venture. Only the doctor's patrician sense of noblesse oblige had saved Blanchard from an immediate lawsuit.[69]

The intervening decade had done little to ease the American's injured sensibilities. Jeffries reopened the case. Soon after Blanchard's arrival in the United States, he sent his first note demanding repayment to Blanchard in Philadelphia on January 30, 1793. Several letters had passed between the old partners when, on February 24, Jeffries retained a lawyer to bring suit in Pennsylvania courts. Blanchard was enraged at the suit and sent Jeffries an "almost scandalous and ungrateful letter full of perjury and lies."[70]

Blanchard had struggled with a bruised ego for ten years. He had always believed that Jeffries had attempted to steal credit for the Channel flight. Now, with the threat of a lawsuit at hand, all his venom burst to the surface.

When the Curious Carriage went on display in the aerostatical laboratory that April, it featured paintings on either side of the driver's compartment representing Blanchard's peculiar vision of the flight from Dover to Calais. On one side, Jeffries was shown "surrounded with bottles for his provision," while the heroic Frenchman struggled alone to operate the scientific instruments. On the opposite side, both men were portrayed "nearly naked." Jeffries was emptying a bottle of brandy to forget the danger, "whilst his captain is calculating and working."[71]

Jeffries became aware of the libelous paintings on May 14, when he received a package from Philadelphia containing an unsigned note in French and news articles describing the carriage. He immediately called Blanchard's performance to the attention of Francis Ingraham, his Philadelphia lawyer. At the same time he was putting all the bills relating to the Channel venture in order and soliciting letters from gentlemen in England and in France who could testify to his perfectly sober condition at the conclusion of both flights.

The suit was long and involved, but the ultimate judgment was in the American's favor in the amount of $370. The money meant very little to Jeffries, who had reestablished himself as one of Boston's leading physicians. For Blanchard, however, whose American tour had been nothing short of a financial disaster, the sum was enormous. In the spring of 1796 he moved from Boston to New York for one final attempt to achieve an American triumph.

Blanchard was no stranger to New Yorkers. As early as October 1793, a waxwork statue of the French aeronaut had been displayed along with images of Benjamin Franklin, Alexander Hamilton, John Hancock, and "several beautiful young ladies," in the Tammany Museum on Broad Street, just a block up from Battery Park.[72] Gardiner Baker, the proprietor of the museum, was now very much interested in playing a role in Blanchard's planned New York ascent.

Baker, one of New York's pioneer showmen, was described by a friend as a "snub-nosed, pock-pitted, bandy legged, fussy, good-natured little body, full of zeal and bustle in his vocation."[73] His museum, the city's first, was a wonder. In addition to the wax figures, it was packed with animals, both alive and stuffed, Indian artifacts, small monuments commemorating great events in American history, and such modern mechanical marvels as a guillotine and an air gun. In 1794 he expanded his operation to include the city's first zoo, an outdoor menagerie that drew repeated complaints from neighbors who objected to a large collection of noisy, smelly animals in the middle of the city.[74]

Baker approached Blanchard in June 1796, offering to provide a home for the aeronaut's family, to pay all expenses, and to handle all business arrangements for the ascent in exchange for a seat in the car. Baker would not even accept a cut of the profit. All monies beyond the expenses of the operation would go to Blanchard.

The Frenchman welcomed the opportunity to turn the business operation over to Baker so that he could concentrate on lecturing and preparing the balloon for flight. Baker forged ahead, distributing subscription books to coffeehouses throughout the city. In an effort to boost subscriptions, he also wrote and published a short history of ballooning, complete with a prospectus of the upcoming flight.

Early in July 1796 Baker informed New Yorkers that he had collected only 1,000 of the 3,000 subscriptions required to fund the ascent and complained that "the fatigue of waiting on the citizens at this warm season is very great."[75] He was careful to explain the special problems of funding a balloon ascent:

> *Experience has proved, that of all subscriptions which have been opened in great cities of Europe, those relative to Aerostation have been attended with the least success. The reason of this is so evident as to need no explanation. If a subscription be opened for a fine theatrical Piece, people in general add their names to the list with little or no hesitation; whilst at the same time, while aerial ascensions attract many more admirers than a comedy, in so much that people have passed from one kingdom to another to be witnesses of this interesting spectacle; yet often times, out of more than four or five hundred thousand spectators, there have not been more than one thousand subscribers.*"[76]

Blanchard, Baker noted, had faced special problems during his American tour: "Although the public has never compensated Mr. Blanchard for his labours in any other way than by their applause, he has the satisfaction to [know] that

Kings, Princes, Sovereigns, and Senates, have always indemnified him for his labours of this kind in Europe. This, however, . . . has not been the case in America."[77]

At some point that summer, Baker had invested in the construction of a special building, measuring 36 feet by 38 feet, to house Blanchard's balloon. The building was open each day from 4:00 to 6:00 P.M. so that ticket holders could pay repeated visits to watch as the envelope was repaired and inflated with air, the net laid in place, and the car attached.

In spite of all Baker's efforts, subscriptions were still slow in coming. The ascent, which had been scheduled for August, was cancelled, with the promise that the flight would be made as soon as the goal of $3,000 in ticket sales had been achieved.

The end came at 5:00 P.M. on the afternoon of September 14 when a tornado ripped through the center of New York. The "Balloon House," which stood directly in the path of the storm, was completely destroyed. Portions of the foundation were carried more than 20 yards, while light boards were blown 300 feet. Large corner beams bolted to an adjoining inn snapped like match sticks. Water, fragments of glass, and wood splinters covered the site.[78]

The balloon had burst with a "tremendous roar" as the cyclone swept through the building. The three workmen who had been inside the structure at that moment escaped with minor cuts and bruises. Blanchard's sixteen-year-old son, who had been working on the roof, was much less fortunate. Thrown violently to the ground, he lingered a few days, then died.[79]

Relations between Blanchard and Baker, which were already strained, broke down completely in the wake of the September 14 tragedy. Blanchard openly attacked Baker, holding him directly responsible for the cancellation of the originally scheduled flight as a result of negligent fund raising. Moreover, he charged, the showman had wasted what little money was available on a useless pamphlet, too many expensive newspaper advertisements, and an improperly constructed "Balloon House." He claimed that Baker had demanded to be "absolute master" of the project, reducing the hero of forty-five ascents to the role of "aerial Coachman."[80]

Blanchard called attention to what he regarded as Baker's callous cruelty following the "fatal accident." The showman, he charged, had forced the Blanchards to live in abject poverty. "You know very well," he remarked, "the unhappy situation to which I was reduced by the ingratitude of Dr. Jeffries in Boston by his inhuman and unjust suit of law."[81]

Blanchard complained that he was so "entirely ruined" that he could not afford $10 for his son's burial. "Happily," he continued, "there are some sensible and good natured hearts; and it is to these that I owe my present support."[82]

The Frenchman asked Baker to release the $164.35 remaining from the subscriptions already collected so that all old debts could be cleared and Blanchard would be free to open a second subscription to fund construction of a new balloon. The original subscribers, he reasoned, would not object to this procedure, for they had at least been able to see the balloon in the shed before the storm.

Baker was quite understandably enraged. "When you could receive from my hands some of my own and the public's money, you come to see me in a brotherly manner," he fumed. "But because I will not give you more money you have deserted me and sent a stranger to talk to me."[83] Blanchard was a man of "ridiculous disposition," a "pretended philosopher." "I believe," Baker continued, "that if the perspiration that has parted from me to serve this man was collected in a vessel it would be nearly enough to drown him."[84]

Baker pointed out that while he did have the $164.35 on hand, this would not cover $400 in outstanding bills. In addition, the storm had destroyed some borrowed lumber that would have to be replaced. Nor did Baker believe that the 1,200 ticket holders would be satisfied with simply having seen the balloon on the ground.

In fact, the aborted ascent would cost Baker a great deal of money and goodwill. His promise to redeem all the balloon tickets for admissions to the Tammany Museum did not satisfy subscribers like Henry Hays, a cobbler who brought a $200 lawsuit over the ticket issue in May 1797. A $200 suit brought by Ephraim Hart for back rent on the house provided for the Blanchards proved far more serious.[85]

Certainly Baker's reputation had suffered to some extent. One angry New Yorker offered his acid thoughts to the editor of the *New York Argus*:

> *Since the attempts which Mr. Baker has hitherto made to astonish and please his fellow citizens have failed, I have thought of two expedients . . . First employ some indefatigable artist to make him a large kite, and in order to gratify the curiosity of his fellow citizens more effectually, and to relieve his distressed friend Mr. Blanchard and his family from starvation, the kite's to be kept in some convenient place and to be shown at six pence for each person, but to* subscribers *gratis.*
>
> *As soon as $100 shall be subscribed—one half to Mr. Blanchard to relieve his present necessities, the KITE shall be immediately prepared for aerial flight . . . a basket will be at the end of the tail of the kite to hold* one *person only—Mr. Baker.*

Should this attempt fail by the kites being destroyed in a tornado, the other plan is to have Mr. Baker blown full of inflammatory air; this with the assistance of his head, which may be supposed to be specifically much lighter than atmospheric air, will cause him to perform with ease his aerial voyage. And to make the prospect still more grand, and to compensate for all the former disappointments, it is intended to have four small kites ascend with him, which will be a scene, not only grand and astonishing, but such a one as was never but once *before exhibited to mankind.*[86]

Blanchard completely abandoned Baker and struck out on his own with new sponsorship. Establishing himself back in Ball's Alley, Blanchard repeated the small-scale animal parachute drops of the kind he had made in Philadelphia. One of his balloons was destroyed by rocks and snowballs thrown by vandals early in 1797. He was back in business by March 4, however, with a new balloon, decorated in red, white, and blue, which he hoped would be impervious to this sort of attack. The *New York Gazette* reported that the ascent of this balloon "was beautiful," but the parachute "did not prove strong enough." One of Blanchard's anonymous "quadrupeds" was killed by the shock of landing.[87]

Blanchard's disastrous American tour came to an end in May, 1797, as the *New York Diary* reported:

Blanchard has at last taken his flight—not in the air, as he had proposed, but on dry land. He made his exit with his family, about the first instant; whither we cannot tell. All hopes of an ascension, however, are at an end, and this circumstance is a proof that his conduct toward Mr. *Baker lost him the confidence of the citizens. In* Mr. *Baker's last publication,* Mr. *Blanchard is accounted to a man unworthy of confidence and that, should the citizens patronize him, deceit and ill treatment would be the certain consequences. The event has completely realized* Mr. *Baker's predictions.*[88]

Blanchard finally made his long-awaited forty-sixth ascent, not in the United States, but at Rouen on August 12, 1798. Back in France once again, the aeronaut was able to improve his situation. Following the death of Victorie Lebrun, his first wife, he married Marie-Madeleine-Sophie Armant, an eighteen-year-old admirer, in 1798. His grand total of ascents rose to fifty-nine by the time of his death at age fifty-six on March 7, 1809. The new Mme Blanchard subsequently carried on the family tradition, becoming the best-known woman aeronaut prior to her death in a fall from a balloon on July 6, 1819.

Difficult, abrasive, and possessed of a monumental ego, Blanchard's personality had blocked his route to success time after time. There can be little doubt that both John Jeffries and Gardiner Baker had been more than justified in turning on the Frenchman, who was surely the poorest of partners. In spite of his deficiencies, however, Jean Pierre Blanchard had been a man of enormous courage and perseverance. He brought the balloon to America at a time when no native of the United States was willing to follow in the footsteps of Peter Carnes. He made the first free flight in the United States and thrilled thousands of Americans with his lectures and parachute demonstrations. In return, he suffered the loss of his son, public humiliation, financial ruin, and the destruction of his balloons and equipment. Seen in these terms, his behavior and attitude are not difficult to understand.

Chapter 5 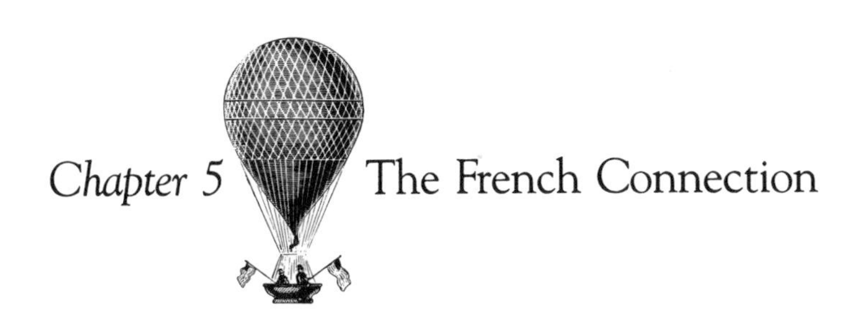 The French Connection

JEAN PIERRE BLANCHARD WAS THE GREATEST OF THE FIRST GENERATION aeronauts to operate in the United States, but Peter Carnes, who was never to rise more than ten feet from the ground, had been a much more astute businessman. Carnes knew when to quit.

For all the interest its inhabitants had taken in the subject of flight, eighteenth-century America posed insuperable obstacles for the professional aeronaut. Population centers large enough to furnish a reasonable number of paying spectators were few and far between. Transportation costs were high in a land of enormous distances. Knowledgeable assistants and such essential supplies as sulphuric acid were simply unavailable even in the largest American cities. Not until the mid-1820s would population increases, internal improvements, and the establishment of pleasure gardens, circuses, and other entertainment centers in major cities make it possible for an itinerant aeronaut to earn a living in the United States.

During the early years of the nineteenth century, Americans remained content with unmanned balloon ascents. A small aerostat was relatively inexpensive to construct and could be quickly inflated and flown by anyone interested in making a few easy dollars. The anonymous "transient person" who launched a balloon some 10 feet high and 8 feet in diameter from Springfield, Massachusetts, on September 30, 1796, was typical. A Boston newspaper commented that the flight was conducted in "a noble and majestic manner."[1] Small balloons were also an important feature of the early American theatrical scene. As early as the 1790s, the New York actor James West, it will be recalled, employed balloon launches to lure record crowds to his playhouse.[2]

By 1800 New York impresarios were commissioning large inflatable figures that could be exhibited and then released from yards and alleys in the theater district. In July of that year one "ingenious showman" exhibited the "Great Mustapha," an "aerostatic machine in the shape of A Giant, thirty feet high—

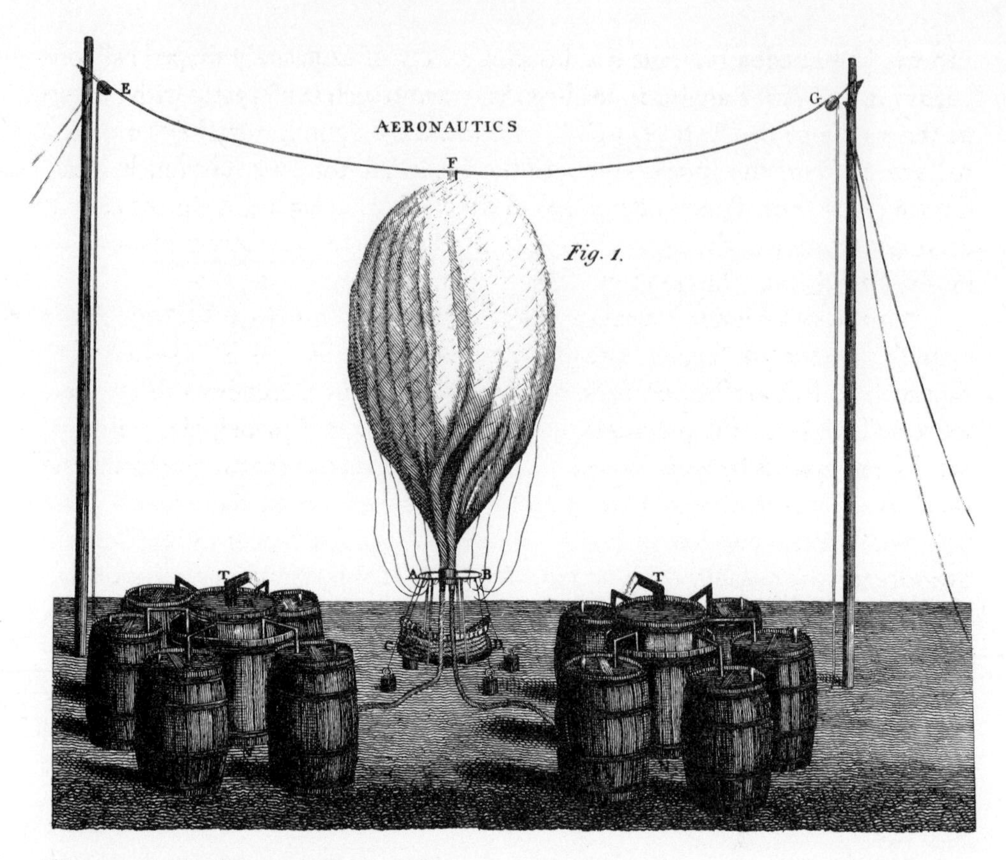

The process of inflating a hydrogen balloon

dressed in Turkish habit." "The Great Mustapha," the predecessor of the enormous tethered figures that would one day delight millions of holiday parade watchers, was finally released at the end of July 1800.[3] Another transparent colossal figure was released from New York's Vauxhall Gardens at 4:00 P.M. on June 14, 1824. This balloon was constructed of goldbeater's skin in the shape of "A *Grecian Hero, of Scio,* clad in Oriental Armor."[4]

Similar figures were favorites with visitors to Castle Garden. A large elephant was inflated at the old fort on the Battery on several occasions. But the most spectacular of the Castle Garden balloons was an "Ancient Knight in Armour . . . astride the American Horse Eclipse." Constructed at a cost of $400, horse and rider took to the skies on August 16, 1824.[5]

It should be noted that these balloon figures were not an American innovation. The Entslen brothers had flown a variety of exquisitely shaped balloons beginning in 1785. Ranging from allegorical figures such as a Pegasus with Perseus in the saddle to the "Air Nymph," an aerostatic lady in a formal gown with a balloon coiffeur, the Entslen figures were launched from several French cities. Occasionally these figures offered social or political comment, as in the case of a flying caricature of the hypnotist and scientific quack Franz Mesmer constructed by the Frenchman Thomond in 1785.

New York's Mount Vernon Garden and Adonis Hotel on William Street offered a course of "expert lectures" on aerostation by one Richard Croslie, capped by a balloon launch on October 27, 1800.[6] In December of that year, John Bull Ricketts, the proprietor of Ricketts' Circus in Philadelphia, presented Mr. Blount and his balloon. Blount's craft was advertised as sporting "a handsome car," so we can assume that he made tethered flights from the circus amphitheater. Ricketts pleaded with the citizens who owned homes on the hillside overlooking the amphitheater not to sell tickets as they had for previous attractions at the circus.[7]

In July 1805 Mr. Pinchbeck offered the citizens of Boston an "Aerial Excursion in which the Public will not be disappointed." His large unmanned aerostat, an "Orbicullar," as he called it, contained 1,100 yards of silk and had been constructed at a cost of $4,000. It seems more than likely that the reputed size and cost of Pinchbeck's aerostat were more thoroughly inflated than his balloon ever was.[8]

Similar ascents staged by theater or pleasure garden owners and private entrepreneurs continued well into the mid-nineteenth century. On July 14, 1816, Shadrach Turnbull launched a large aerostat some 53 feet in diameter from South Christian Street in Suffield, Connecticut.[9] Mr. Pelubit, "a French gentleman," typified the early balloonists who supplemented their incomes by staging ascents in New England towns and villages. On November 4, 1819, Pelubit sent "an illuminated balloon" up from the State House Square in Hartford, Connecticut. An "ingenious and industrious man with a large family," his success on this occasion was so encouraging that he immediately announced that more balloons would be released the following week, one of them carrying a "cage containing some live animals." Pelubit eschewed ticket sales to the events. Rather, six gentlemen were appointed to pass through the crowd following a successful ascent "to collect what may be given . . . to remunerate him for his trouble and expenses."[10] Pelubit was only one of many balloon promoters operating in the United States. The citizens of Wilmington, Delaware, were treated to the ascent of a large aerostat apparently similar to Pelubit's on August 23, 1817.

Home-built balloons were also common in the United States during the first two decades of the new century. Notices such as that which appeared in *Nile's Weekly Register*, for August 31, 1816, dotted the nation's newspapers. In this case, students of the "College of Georgetown" (District of Columbia), were advised that the gas balloon they had launched the month before had been found at Cover Point, just a few miles above the mouth of the Patuxent River on Chesapeake Bay. After a flight of almost 50 miles, the little balloon had been attacked and destroyed by frightened locals.[11] Students at the "University in Baltimore" fared even better. The small balloon they launched on the evening of May 30, 1820, was found the next day in Chester, Pennsylvania, having flown 90 miles.[12]

By the 1830s, children's craft books issued by American publishers often included instructions for making balloons. The author of *Parlour Magic* (Philadelphia, 1834) advised his young readers that "one of the simplest and most beautiful experiments in aerostation is to take a turkey's maw, or stomach, properly prepared, and to fill it either with hydrogen or carburetted hydrogen produced from coal."[13] Budding aeronauts were advised to obtain the turkey maw from an optician. The author, anxious to steer his young readers away from corrosive sulphuric acid, even provided a detailed description of a safer method of producing hydrogen in the home by heating iron wire in a gun barrel.[14] The author recommended that young experimenters launch their balloons "off a high staircase, and observe it ascend to the cupola or light, where it will remain near the highest point til the escape of the gas allows it to descend."[15]

Most children's "how-to-do-it" books of the period recommended construction of much simpler hot air balloons however: "Make a balloon by pasting together gores of bank post paper; paste the lower end round a slender hoop, from which proceed several wires terminating in a kind of basket, sufficiently strong to support a sponge dipped in spirits of wine. When the spirit is set on fire, its combustion will produce a much greater degree of heat than a regular flame: and by thus rarifying the air within the balloon, will enable it to rise with great rapidity to considerable height."[16] Mass circulation magazines such as *Scientific American* also provided detailed instructions for manufacturing small balloons of "fine, thin, close textured tissue."[17]

Authors who encouraged American children to build and fly hot air balloons earned little praise from their fellow citizens, most of whom agreed with the editor of *Atkinson's Saturday Evening Post* who, in September 1838, decried the dangerous "practice of inflating balloons and setting them afloat with lamps, or vessels of burning alcohol attached to them."[18]

The *New York Evening Post* complained that "the city is endangered almost every night, by boys sending up paper balloons in almost every direction. After ascending in the air some distance, they take fire and down they tumble on to the roofs of houses and stables. . . . Is there no remedy for this alarming practice?"[19]

A Philadelphia editor agreed that small fire balloons, the charred remains of which could be seen in many city streets, represented a definite public hazard. He cited the recent case of a fire balloon launched by Charles Benclon of Camden, New Jersey, which had crossed the Schuylkill at Kingesing and descended on a ferry separating the farms of James Maloney and Evan Thomas, Jr. The resulting alcohol spill started a fire that was extinguished only after "great exertions."[20]

In this case, the editor admitted, a catastrophe had been averted, "but," he asked, "what should prevent them from descending, at night, upon the roof of a bridge, a barn, or house, and setting it on fire?" He painted a picture of whole families "destroyed in their beds" and of "vast tracts of valuable timber" burned to the ground.[21]

Other aspects of American aeronautical enthusiasm during the first two decades of the century are a bit more puzzling. As in the case of a New York advertisement of 1806 touting the ascent of "Mr. Charmock's Naval Balloon," one is often left with unanswered questions regarding the appearance, even the basic nature, of the aerostat in question.[22]

The "Patent Foederal Balloons," or "Vertical Aerial Coaches," are even more puzzling. The story begins with one Moses McFarland who, on October 28, 1799, was granted the first patent awarded by the U.S. government for an aeronautical device.[23] The following year John Graham announced that he had purchased the exclusive right to exhibit "Archimedeal Phaetons, Vertical Aerial Coaches, or, Patent Foederal BALLOONS" in Hartford, Connecticut.[24] The fact that Graham had "obtained a deed of the exclusive right" to exhibit "Patent Foederal BALLOONS" suggests that the device in question was McFarland's.

Graham's announcement offers some intriguing clues which indicate that the "Federal Balloons" (to modernize the spelling) may not have been balloons at all. The machine was, he noted: "Strengthened and supported in all its parts—rendered perfectly safe and secure, and elegantly repainted and decorated—system and regularity established, so that voyagers may be treated with slow and steady or more rapid movements as they may order; so that persons of a timid cast will enter with assurance and be much delighted; others may progress 500 yards per minute."

Graham remarked that ladies had ridden in safety and advertised that "persons out of health who contracted with the late proprietor and others whose health is thereby invigorated, may be accommodated *gratis*, when the coaches are not thronged."[25]

It is safe to assume that there was no balloon in Hartford, Connecticut, in the spring of 1800 capable of carrying a party of eight into the air. This fact, coupled with the mention of "boxes" and the adjustable speed of the "machine," suggests that Graham was advertising some sort of an amusement ride.

Late in June 1800 Graham transferred his operations to a new site on Hartford's South Green. He continued to extol the virtues of his enterprise. "Invalitudinarians may regain their health by a sudden change of air and atmosphere, and a sudden revulsion of the blood and humors, and . . . healthy persons may experience pleasure and delight," he noted.

> *The superior elevation of the present position, will afford additional prospects of the delightsome surrounding country,—Strict attention will be paid to the voyagers—and their orders respecting rapid or easy movements, strictly obeyed,—Those who prefer either extreme will please to arrange themselves in classes of 8 persons, and enter at a time, to avoid the contradictory orders, which are sometimes given, this may be easily effected by entering the boxes thus marked, and all be satisfied. The proprietor finding the throng too great for the Machines erected in this town, would dispose of a right in any of the capital towns in the county, a few in the vicinity of this city excepted.*[26]

Graham was apparently successful in selling the rights to exhibit the machine in other localities, for in July 1800 one Phineas Parker advertised an exhibition of the "Patent Federal Balloons" at New York's Vauxhall Garden where visitors would enjoy "a rich variety of landscapes equal to any in the world, and alternate views of the waters of the East and North [Hudson] Rivers, of the City of New York and the neighboring villages."[27]

By September 1800 the "Patent Federal Balloons" on Hartford's South Green were under the new management of Azariah Hancock, who pronounced that the "Balloons" were "stronger than ever they have been before."[28]

While Americans contented themselves with the vicarious thrills provided by aeronautical amusement rides and the ascent of small aerostats, European aeronauts worked to realize the potential that Franklin and others had predicted for the balloon. The French Corps d'Aérostiers, formed in 1794, demonstrated the military utility of observation balloons at the battles of Maubeuge and Fleurus. Scientists pondered the atmospheric data provided by Joseph Louis Gay-Lussac, who reached an altitude of 23,000 feet in 1804. The distance from Calais to Dover seemed insignificant when compared to the aerial journeys of up to 246 miles that had been made by 1807.

A balloon-borne parachute of the sort employed by Louis Charles Guille

European aerial showmen had also been busy developing new stunts with which to attract ever larger crowds. André Jacques Garnerin pioneered the most daring and spectacular of these innovations when he made the first manned parachute drop from a balloon on October 22, 1797. Garnerin's original parachute was refined by other aeronauts who added parachuting to their repertoire during the following decade.

Louis Charles Guille brought the manned parachute to the United States in 1819, twenty-six years after his countryman Jean Pierre Blanchard had ascended from the yard of the Philadelphia jail. Guille, a veteran of nine flights and seven parachute drops in France, had first come to public notice on November 13, 1814, when he attempted to fly a large dirigible balloon from the Champ de

Mars. The craft stood 90 feet tall and measured 75 feet in length. The operator's car was suspended far beneath a large elliptical gasbag that could be tilted nose up or down by means of a system of lines and pulleys operated by the aeronaut. The balloon-glider did not fly, but it did attract a great deal of attention and marked Guille as an important aeronautical innovator.

Guille's American tour began with an ascent from New York's Vauxhall Garden on August 2, 1819. It was not a good day for flying. High winds had sent rain clouds scudding across the sky during the morning and early afternoon. Guille, determined not to disappoint his first substantial American crowd, rushed to complete the inflation in spite of the weather. By 6:00 P.M. the envelope was full, and the aeronaut had finished rigging the parachute, car, and release lines to the balloon netting.[29]

A gust of wind swept through the Vauxhall Garden just as Guille began to rise, sending the balloon into a tangle of branches. When the craft tore free, Guille abandoned his usual flag-waving response to the cheering crowd in order to clear the broken branches from the parachute lines and release mechanism. Anxious to make the drop in full view of the spectators, Guille was forced to release at a lower altitude than that to which he was accustomed in order to avoid entering the clouds. He dropped 300 feet before the parachute opened, then, caught in the high winds, he was blown across the East River to a landing in Bushwick, Long Island, some 4 miles from his take-off point. The abandoned balloon continued to a landing 30 miles away in a meadow at Fort Neck, South Oyster Bay. A collection taken up in New York coffeehouses furnished $200 with which Guille could pay the ransom demanded by the Long Islanders for the return of his balloon.[30]

Guille and his wife moved to Philadelphia soon after the New York ascent. Arriving late in August, they found that two rival aeronautical teams had preceded them to the city. Guille, the most experienced of the three balloonists, wisely decided to remain in the background while his amateur rivals attempted to fly.

David B. Lee, the American contender, was the first to try his luck, on September 23, 1819. We know very little about Lee, other than the fact that he was possessed of a long-standing desire to fly. In a letter written to Thomas Jefferson in 1822, Lee remarked that "the art of navigating the atmosphere has had my constant study with little exception since childhood. But being branded for a fool by every person to whom I revealed my project, I kept it almost an entire secret till early in the year 1819."[31]

Lee actually had two projects in mind. His great dream was to construct a mechanical heavier-than-air flying machine. Realizing that funding this project

One of Guille's American descents must have appeared very much as depicted in this view of a London parachute exhibition.

would be next to impossible, he set as a second and more reasonable goal the development of a navigable balloon.[32] "Not being one of fortune's favorites," Lee "applied to several individuals for assistance, most of whom were so prejudiced against the very idea of flying that they would not even give me a hearing."[33] He was finally able to obtain financial support for the construction of his balloon "with the necessary apparatus to propel and steer it from a man in Camden, New Jersey."[34] Philadelphia newspapers reported that Lee had two associates, Mr. Bulkley and Mr. Pomeroy. Presumably, one of these men was the financial backer, the other a mechanic.

The best surviving account of Lee's Camden Balloon, as it was called, describes the aerostat as resembling "a Turrapin [turtle] wanting its head, legs, etc." It was, a reporter for the Philadelphia *Union* remarked, "elliptical shaped, but not exactly elliptical."[35]

Thirty thousand persons swarmed across the Delaware River to witness the inaugural flight of Lee's balloon in Camden on September 2, 1819. Delays during the inflation angered the spectators, some of whom slipped through the fence and slashed the leather pipes leading from the gas generator to the balloon. Undeterred, Lee shifted operations to Philadelphia, where he made a second trial on September 7. On this occasion the balloon left the ground, struck a tree, and came back down. When Lee left the basket to investigate the damage, the craft escaped and was apparently lost.

Late in September Lee announced that he intended "to ascend first in a balloon of common construction, and afterwards to carry into operation his principles for navigating airships." As Lee himself admitted, however, "I was doomed to a more luckless fate."[36] Financial support collapsed following the twin failures of September 1819, and his proposed flight did not materialize.

A Frenchman, M. Michel, and his assistant, a Mr. Stanislaus, composed the second team of hapless aeronauts attempting to ascend from Philadelphia that fall. Their flight was scheduled to be made from the Vauxhall Garden on September 8. Michel planned to climb to an altitude of 2,000 feet, then cut his car loose and parachute to earth. Michel and Stanislaus could scarcely have chosen a less propitious moment for their flight. David Lee's twin failures of the previous week had convinced many citizens that "all aeronautic endeavors were so many impositions upon the public."[37]

High winds on September 8 created serious problems during the inflation. The twenty leaky hogshead casks that comprised the gas generator could scarcely fill the suspended envelope against the pressure of the wind. By late afternoon the breeze had begun to fall, and the ascent was rescheduled for 6:15 P.M. The balloon, which had a capacity of 1,000 cubic feet, was three-quarters full and straining at the tethers by six o'clock.

At this point the spectators began to lose control. They could see the balloon bobbing about above the fence surrounding the Garden, obviously ready to go, and could not understand the continued delays. The melee began when a small boy who had crawled over the top of the fence was struck and injured by one of the attendants. The word spread through the crowd that the child had been killed. Those closest to the fence crashed through the barricade, attacking the balloon with rocks and bottles. Acid spread over the ground as the gas generators were broken open with axes. Liquor was "liberated" from the refreshment bars that dotted the area inside the fence.

The Vauxhall Garden pavillion and theater were next to go. The mob first destroyed the stage and scenery and then stole the dresses and other wardrobe items before setting fire to the building. As there were no fireplugs in the vicinity, the conflagration spread over the entire area.

The losses were considerable. The $800 that Michel and Stanislaus had gathered in ticket sales was stolen. The value of the equipment and buildings destroyed was estimated at $2,500. In addition, Mr. Lure, the Vauxhall vocalist, lost his piano and the band engaged to entertain the crowd prior to the flight "generally lost their instruments and music."[38]

Some leading Philadelphia newspapers ignored the riot. Others, like the *United States Gazette*, expressed their shock in no uncertain terms. "A mobbing spirit has not been characteristic of Philadelphia," noted the editor, "and it is with regret that we publish that such a disagreeable riot has taken place."[39]

All this was too much for Guille. Philadelphia was obviously not in the mood for another ascent, particularly from Guille, whom Michel had attempted to portray as his partner. Wisely, Guille retreated to New York, where he scheduled a second ascent from the Vauxhall Garden on October 21. The poor quality of the hydrogen produced on this occasion prevented the Frenchman from ascending. Fearing the reaction of the crowd, however, he released his balloon on a free flight. Thirty to forty thousand spectators watched as the balloon, parachute, and empty wicker car rose rapidly toward the low-hanging clouds. The balloon was seen flying very high over New Haven, Connecticut, late that afternoon and again at Colchester about sunset. It landed at 6:15 that evening on the Johnson farm at Bozrah, near Norwich, Connecticut.[40]

As the balloon approached Bozrah in the twilight, the townspeople seemed uniformly convinced that it was preternatural, but "there was a difference of opinion; whether it would prove *diabolical* or otherwise of a *pacific* or *troublesome disposition*."[41] One poor fellow "ran about half distracted, uttering many incoherent expressions—amongst which were that the Devil and Hell were approaching and that he distinctly saw the Devil's legs and cloven foot."[42]

As the local citizens were uncertain how to empty the balloon, it was chained to an ox cart until Guille could claim it. Colonel Johnson, the owner of the orchard where the craft had come to earth, originally planned to charge Guille $50 for the return of the balloon. This was reduced to $25 when Johnson learned that the aeronaut was "a stranger and unfortunate."[43]

Guille made his next ascent from Jersey City at noon on November 20. The aeronaut cut his parachute loose at an altitude of only 500 feet, slicing his thigh in the process. He fell 200 feet before the parachute opened and landed in the middle of a salt marsh less than 40 rods from the take-off site. Aeronaut and car were quickly retrieved "and borne back to the place of departure amidst the huzzas and shouts of the multitude."[44]

Guille's reputation was spreading throughout the country. Newspapers in cities as far removed as Cincinnati and Savannah reported his activity and expressed the hope that the aeronaut would see fit to visit them. Anna Boyd McHenry of Baltimore was one of thousands who wished "to take a flight with Mr. Guille."[45]

Mrs. McHenry was part of an "immense crowd" that gathered to witness Guille's first Baltimore ascent in April 1820. The aeronaut had originally sought permission to stage his flight "in the yard of the Penitentiary, in that of the Jail, or the Court-House, the City Hospital, and the Alms-House." When officials refused to allow him to use any of these city facilities, Guille built his protective fence in "the wagon-yard of one of our Western Taverns."[46]

Guille remained in Baltimore through the spring and early summer of 1820, waiting for another opportunity to ascend. He inflated his balloon at least once during this period but was unable to launch because of a rent discovered in the envelope.[47]

Guille placed notices in local newspapers expressing his desire to make an ascent so as not "to leave an unfavorable impression on the public mind." He pointed out that his stay in Baltimore had cost him $600, while he had collected only $400 in ticket subscriptions to his two abortive inflations. City newspaper editors took up his cause, calling attention to the aeronaut's courage, his success in Paris and New York, and the fact that Baltimore theater admissions averaged $500 to $800 a week. In spite of these entreaties, Guille returned to Philadelphia without having flown in Baltimore.[48]

The aeronaut scheduled his first Philadelphia ascent in July 1820. The balloon was blown into some trees, tangling a sandbag intended to upend and empty the envelope after the parachute was cut loose. When Guille stepped from the basket to clear the rigging, the balloon escaped from the ground handlers and rose out of sight. The spectators remained calm, but the sheriff, perhaps remembering the Vauxhall riot of 1819, personally escorted Guille and his wife through the crowd to his own carriage.[49]

Guille did ascend from Philadelphia's Vauxhall Garden on October 14, 1820. Launching at 4:00 P.M., he climbed almost immediately into a cloud bank hanging less than a thousand feet overhead. As he was unwilling to commit himself to the parachute without a clear view of the ground, the aeronaut had little choice but to remain with the balloon.

It was to be the most dangerous of ascents; Guille's balloon was intended to do nothing more than carry a parachute to altitude. It had neither the valve nor the ballast required for a long-distance aerial voyage. Guille remained in the air for ninety minutes, finally coming to earth on the farm of George Woolsey 8 miles northwest of Trenton, New Jersey. His descent was so catastrophically rapid that the parachute rigged between the gasbag and basket blew open and eased the shock of landing. Once on the ground, he was dragged through a cornfield "about the length of Market Street," coming to a halt only when the balloon struck a grove of trees.[50]

On May 21, 1821, the *New York Post* reported that the aeronaut was once more "in the city" and planning an ascent from Joseph Delacroix's Vauxhall Garden on Fourth Avenue, south of Astor Place, on May 15.[51] Guille's wife would first make a tethered flight, after which an inflated human figure would be released. Guille would then make the "grand ascent."[52]

High winds and problems with the balloon forced postponements until June 7. When the launch did take place, "the violence of the winds" led Mrs. Guille to cancel her flight, while the aeronaut rose only 50 feet into the air and skimmed over the rooftops of Manhattan for less than a mile before landing near Love Lane.[53]

During one of these 1821 Vauxhall ascents, perhaps that of June 7, Guille not only came close to losing his life but also became the object of the first American lawsuit resulting from aeronautical activity. Soon after the lift-off, his foot became tangled in a dangling line. While struggling to free himself, he fell partly out of his small basket, grabbing the valve line as he went. The crowd that had gathered to witness the flight, realizing that Guille was in serious peril, chased the balloon through the city streets. He descended into a small garden owned by a gentleman named Swan and was then dragged through the neat rows of beets and potatoes for a distance of 30 feet. His calls for help brought the spectators swarming over Swan's fence to the rescue. In the process the garden was destroyed. The outraged Swan brought suit for $90 in damages to his small crop. Originally heard in Justices Court, the case eventually reached the New

York State Supreme Court where, in January 1822, Justice Charles J. Spenser awarded damages to Swan, arguing that Guille's cries for assistance had drawn the crowd onto Swan's premises.[54]

Almost seven decades later, the case was commemorated in the best aerostatic tradition by a poetic student of legal history:

> *For parachutes the courts care little,*
> *Balloonatics no rights enjoy;*
> *But they will not abate a little*
> *'Gainst those who garden-shoots destroy.*[55]

Guille announced that he would make a parachute ascent at Boston's Washington Garden at 4:00 P.M. on September 3, 1821. His name disappears from American newspapers soon thereafter.[56]

The American tours of Blanchard and Guille were remarkably similar. While Guille had made more flights than his predecessor, he had been no more successful financially. Both had lost important lawsuits.

Eugène Robertson, the third foreign aeronaut to tour the United States, was undaunted by the experience of the two Frenchmen. Robertson was an extraordinary fellow, a second generation member of one of the great European ballooning families.

His father, Etienne, had been born to wealthy parents in Liège in 1763. He made the first of the great European scientific ascents on July 18, 1803. Flying from Hamburg with a German music teacher named Lhoest, he claimed to have reached an altitude of 23,526 feet. The following year he published a satyrical design for an enormous balloon to be known as *La Minerva.* With a crew of sixty, outhouses, kitchens, and facilities for farm animals, the craft would remain in the air for six months at a time.[57]

For Etienne Robertson, as for his sons, Dimitri and Eugène, ballooning was only one facet of a successful show business career. The elder Robertson used his scientific training to mount spectacular demonstrations aimed at debunking the charlatans of the age. His ability to create such startling audiovisual effects made the elder Robertson a millionaire by the time of his death in 1837. It also enabled his son to conduct an American tour so successful that it would lead to the establishment, at long last, of a firm tradition of aerial showmanship in the United States.

Eugène Robertson had made his first flight with his father at the age of five. After studying physics with J. A. C. Charles and serving an apprenticeship in ballooning and showmanship with his father, Eugène made his own solo flight at Lisbon on May 14, 1819. This was followed by a second ascent on December 12, 1819, a flight from Pôrto on June 25, 1820, and a third launch from Lisbon on April 8, 1822.[58]

Eugène Robertson's American tour was inspired by French news reports of the emotional outpouring that greeted citizen Marie Joseph du Motier, the Marquis de Lafayette, during his final triumphal visit to the United States. The Congress and Pres. James Monroe had invited Lafayette to tour the nation at government expense as part of the celebration of the fiftieth anniversary of the Battle of Bunker Hill.

Robertson was determined "to carry into the middle of this American celebration, a balloon, the fruit of French genius, the infant whose birth certificate had been signed by Benjamin Franklin."[59] With financial assistance from his family, Robertson took passage on a merchant vessel bound for the United States. He arrived too late to ascend in Boston on June 17, 1825, the date of the Bunker Hill celebration, but he did contract with the manager of New York's Castle Garden to ascend on July 9, during Lafayette's visit to the city.

Robertson made a wise decision in choosing to fly from Castle Garden. Opened in 1824, it was the newest, most elaborate, and best known of the public pleasure gardens that would play such an important role in the early history of ballooning in the United States.

Some of these gardens, notably Niblo's, in New York, were closely patterned after such European originals as London's Vauxhall and Ranlegh Gardens or the Tivoli in Paris. They featured winding tree-shaded pathways, imported statuary, and secluded summer houses where refreshments could be purchased. Others, like Philadelphia's Vauxhall, the site of the disastrous balloon riot of 1819, were primarily theaters surrounded by open courtyards where pyrotechnic displays and other outdoor entertainments could be staged. New York's Vauxhall, founded by Joseph Delacroix in 1796, offered everything from waxwork displays and arc-lighted transparent paintings to public dancing. The Mapleton Garden in New York was typical of those relatively simple establishments that were little more than open-air taverns.

Castle Garden was the epitome of the American pleasure garden. Situated on a small island just west of the Battery on the southern tip of Manhattan, the structure had originally been constructed as one of a series of harbor fortifications. Following a long period of negotiation, Congress turned Castle Clinton, as it was then known, over to the city of New York in March 1820. Four years later the Common Council announced that the fort had been leased for a five-year period to three businessmen who petitioned for a ten-year lease, pointing out that they planned to transform the castle into an amusement center for the citizens.

Castle Garden opened on July 3, 1824, just in time to greet Lafayette and eight thousand New Yorkers with a long pyrotechnic display and the release of several unmanned balloons. This pleasure garden and similar resorts in New York, Philadelphia, Baltimore, and other American cities combined functions of the municipal stadium, urban park, nightclub, theater, beer garden, amusement park, and fairground. Gaudy spectacles and mass exhibitions that could be enjoyed by thousands of spectators simultaneously were their stock-in-trade. As the editor of the *New York Sun* noted with some distaste, the American appetite for this sort of entertainment was voracious, and none too discerning. "So numerous of late have become the amusements throughout our city, the greater part of which are of the most trifling nature, and so filled are our public papers with puffs magnifying their importance and thus gulling the public, that the scene has become absolutely disgusting," he noted.[60]

As we have seen, the proprietors of various pleasure gardens had been sponsoring the launch of unmanned balloons since the late eighteenth century and had even provided a launch site for unsuccessful aeronauts such as Lee and Michel. Whenever possible, Guille had launched from a pleasure garden as well. But it remained for Eugène Robertson and the proprietors of Castle Garden to exploit fully the potential of the manned balloon ascent as a public spectacle capable of attracting a mass audience.[61]

The weather on July 9, 1824, was perfect for Robertson's first ascent from Castle Garden. Inflation began on schedule at 3:00 P.M. One hour later a spectator noted that the partially filled balloon "exhibited the appearance of a young elephant, without his proboscis, and quivered about on the piece of canvas on which it was placed, as if it were just beginning to become instinct with life."[62]

The aerostat, which Robertson had brought from France, was a gorgeous thing. It was relatively small, only 21 feet in diameter, with a capacity of 5,875 cubic feet. The blue envelope was decorated with a silver band that carried a star for each state in the Union. The hemisphere above the band featured a large American eagle, complete with lightning bolts in one claw, an olive branch in the other, and the motto "E Pluribus Unum" in its beak. Two long ribbons, one bearing the name of Washington and the other Lafayette, were attached to the car. Robertson was as stylishly attired for the occasion as his balloon. He wore a "species of French uniform," a blue coat with embroidered collar and cuffs and a hat trimmed with lace.[63]

Nine thousand spectators had gathered inside the castle walls by 6:00 P.M., while thousands more waited outside the gates. "The immense number of spectators out on this occasion was truly astonishing," reported the *New York Evening Post*:

> *From the roof of the saloon in the garden the Battery presented one of the finest and most diversified spectacles which the eye could behold, and the whole space was almost filled, and presented an immense block of human beings. The harbor round the garden was literally crowded with small boats full of persons, and the steamboat* James Kent, *stationed a short distance from the western extremity of the garden, was literally crowded and hung around with spectators. The house-tops and balconies were also filled, and the wharves and the shipping in the neighborhood, were in the very same state. There could not be fewer than thirty or forty thousand persons out on this occasion.*[64]

Lafayette, who had dined in Jersey City that evening, arrived aboard the steamboat *American Star* at 7:00 P.M. With the guest of honor on the scene, Robertson topped off his casks with fresh acid in order to hurry the inflation. The aeronaut then attached the car to the dangling lines of the net and climbed aboard.

Robertson and his balloon were walked around the garden and brought to a halt in front of the canopy that sheltered Lafayette. The general stepped forward to bid the aeronaut *bon voyage*. Robertson now discovered, much to his embarrassment, that the balloon was leaking hydrogen and refused to lift over the wall. An assistant, M. Fontaine, who had flown in France, recharged the generator and added yet more gas to the balloon.

At 7:30 Robertson finally climbed over the imposing wall of the garden, barely missing one of the lamp posts with its open flame. The balloon was still heavy, forcing the aeronaut to drop his flag and grappling hook in an effort to reduce weight.

The balloon gained altitude slowly, passing very close to the spire of Trinity Church as it moved straight up Manhattan, over the Bronx, and on to a landing 2 miles north of Newtown, Connecticut. Robertson was back in New York accepting the congratulations of Lafayette only four hours after take-off.[65]

The aeronaut's American tour was off to a solid start. For once, a manned balloon ascent had been a financial success. Robertson had cleared 6,000 francs, and the proprietors of Castle Garden had made a tidy profit as well. Certainly, New Yorkers were pleased. As one editor remarked: "Mr. Robertson appears to be the master of his business—he attended personally to all the arrangements, and entered the car with perfect composure. Our citizens have never had so good an opportunity to witness an ascent, and the public were perfectly satisfied with the performance."[66]

Robertson's second ascent from Castle Garden on September 5, 1825, was as successful as his first. Eight thousand paying customers wedged themselves into the garden, while another 50,000 gathered in the neighborhood to watch as the balloon rose above the walls. On this occasion the aeronaut flew over the Hudson to a landing in a New Jersey wood some 5 miles from Hoboken.[67]

Robertson did not make another flight until the fall of 1826. He apparently remained in New York during the intervening months, however, for local newspapers reported that he was offering theater audiences the "optical illusions and phantasmagoria" pioneered by his father.[68]

The *New York American* began advertising Robertson's third ascent the following July. Originally scheduled for August 10, 1826, part of the proceeds of the flight were to have gone to an indigent local artist. When the weather forced a postponement of the launch, Robertson began to rethink his plans for the first ascent of 1826. The aeronaut announced that the flight, now rescheduled for September 20, would be a night ascent featuring an exhibition of fireworks launched from the balloon.

Robertson's New York friends tried to dissuade him, calling attention to the unhappy fate of Mme Blanchard. During a pyrotechnic ascent on July 19, 1819, the balloon which Blanchard's widow was flying caught fire in midair. Just as it appeared she would reach the ground safely, Mme Blanchard toppled from her car, struck a roof, and fell to her death in the street.

Robertson promised to take every precaution, including suspending the fireworks a full 200 feet beneath the balloon. As announced, the aeronaut made the first night flight in American history on September 20, 1826. The usual large crowd gathered for a performance that began with the launch of 6-foot and 8-foot pilot balloons and small parachute drops. The aeronaut took to the air precisely at 9:00 P.M., offering an aerial spectacle visible throughout the city, after which he landed at "Flatbush, on Long Island."[69]

Robertson's fourth ascent, on October 10, 1826, featured yet another first for the New York audience. The aeronaut had constructed a new balloon for this flight. The central envelope, with a capacity of 16,000 cubic feet, was surrounded by four "attendant balloons," each 10 feet in diameter. This new aerostat was the first in America large enough to carry two people aloft. Robertson had arranged for the second person to be a woman. Several small balloons were launched prior to the main event, including one that dropped a kitten back to earth via parachute.

The aeronaut and his fair companion, whose name has been lost to history, went aloft just as the sun was setting at 5:40. Soon after the launch, spectators

Eugène Robertson flies from Castle Garden, October 10, 1826.

noted that "the little flotilla of balloons was extremely agitated." One of the four small balloons was so overinflated that it burst "with a shock and a report."[70]

Robertson had prepared several "special experiments" to be conducted aloft. One that seemed particularly appropriate involved observing the "effect" produced by imbibing champagne at high altitudes. While most of the champagne blew out when the bottle was uncorked, the few drops that remained "were particularly lively and stimulating."[71] Thus fortified, the pair proceeded to shout through a speaking trumpet in order to study the echoes reflected back to the balloon.

Robertson descended near Elizabeth Town, New Jersey, at 7:00 P.M. and turned his companion over to Mr. and Mrs. Brown, the startled owners of the farm on which they had landed. He then reascended alone to conduct further experiments with an electrical machine, a pair of evacuated Magdeburg hemispheres, and other items of scientific apparatus.

The aeronaut claimed to have reached an altitude of 21,000 feet, where "respiration was laborious and painful—faculties were blunted—the cold was insufferable, especially in the hands . . . which felt like iron, producing torpidity and cramps."[72] Robertson made his final descent that evening at Westfield, New

Jersey, 8 miles from Elizabeth Town.

Anxious to see more of the United States and to exhibit his balloon in an area that had yet to witness a manned ascent, Robertson departed New York for New Orleans soon after his October flight. He arrived in December with letters of introduction from friends in Paris and New York. Robertson's arrival in New Orleans was perfectly timed. The wealthy residents who deserted the city for upland plantations in order to escape the ravages of yellow fever during the spring, summer, and fall were now back in town and anxious to add their names to the list of subscribers supporting Robertson's flight.

With cash in hand, the aeronaut proceeded to select as his launch site the large, open courtyard of the old Ursuline convent. The ascent, he announced, would be made on Washington's birthday, February 22, 1827. The editor of the *New Orleans Argus* noted the emotions of the spectators on the day of the flight. "Their spirits were exalted," he remarked, "all of their hearts were stirred, and tears ran from the eyes of many persons."[73]

As Robertson rose to 600 feet, he saluted the applauding crowd and noticed that an unbroken line of carriages still extended along all the roads leading to the convent. Realizing that he was being blown toward the sea, the aeronaut landed only twenty minutes after take-off; having flown a distance of some 7 miles. His plans to descend at the edge of a cypress grove were thwarted by a sudden gust of wind that carried him into the center of the swamp. Robertson's initial fear that he had fallen into a savage area where he would wander undiscovered until he died either of starvation or from a venomous snakebite was relieved when he heard a group of slaves splashing to the rescue. The men carried Robertson and his equipment to the edge of the forest where M. C. Blanche was waiting with his carriage to return the aeronaut to a hero's welcome in New Orleans.

Robertson made a second ascent from the city on April 29, 1827. On this occasion he was accompanied by Mme Oliver, a young dancer ("jeune prêtresse de Therpsychore"). The two aerial travelers passed over the city and descended, once again, into a cypress forest. That evening Mme Oliver received a standing ovation at the theater, while Robertson dined with the governor of Louisiana. The aeronaut had intended to conclude his tour of the south with an ascent at Mexico City, but the early onset of the yellow fever season, coupled with a bad case of intestinal flu, forced him to change his plans and return to New York.[74]

Still determined to visit Latin America, the aeronaut made his next ascent in Havana on April 19, 1828. Robertson planned a second flight from the same city with Virginia Marette, a young woman visiting the city from New Orleans, but chose instead to return to New York with Miss Marette when it proved difficult to obtain sufficient acid before the beginning of the yellow fever season.[75]

Robertson's seventh ascent was very nearly his last. Staged at New York's Castle Garden on September 26, 1828, the flight attracted the standard crowd of fifteen to twenty thousand spectators. The aeronaut, having experienced difficulty in clearing the castle wall on previous occasions, inflated his balloon in the far corner of the open parade ground. He ascended to shouts of the crowd and flew straight into a 100-foot flagpole that he had tried to have removed from the top of the castle wall the day before.

The envelope was so completely torn that the gas escaped in a rush, overturning the car and leaving Robertson, his foot tangled in a line, dangling head down from the top of the pole. One of the spectators milling around the base of the flagstaff attempted to shinny to the rescue but was unable to climb to within less than 15 feet of the trapped aeronaut. Finally, after a harrowing fifteen-minute struggle, Robertson succeeded in pulling himself upright by means of the lines used to raise and lower the flag. He freed his foot and slid down the pole "in his exhausted state almost as fast as if falling, and was caught by his friends below."[76]

In one sense, Robertson realized, the accident offered new opportunities. New Yorkers had become familiar with his activity and were, perhaps, a bit jaded. The incident on September 18 not only called attention to the aeronaut but underscored the danger inherent in every ascent. In an effort to take full advantage of these favorable circumstances, Robertson scheduled his eighth and final American ascent for October 28, only ten days after the accident. As he hoped, a crowd nearly as large as that which had gathered for his first New York flight crowded onto the Battery at the appointed time. Virginia Marette, who was to accompany him on the trip, emerged from Robertson's quarters beneath the parapet at 5:00 P.M. An admiring reporter for a New York newspaper was obviously much impressed by the young lady, "a pretty looking woman, just turned twenty, dressed in white under pink crepe. Her head was handsomely attired, with pink crepe bows and puffs—but without plumes."[77]

Once Robertson and Miss Marette were in the car, the balloon was walked around the compound and released to float over the city to the northeast. As darkness was falling, the flight lasted only a few minutes. The aeronauts landed in the Hudson between Bushwick Creek and Williamsburg, some 200 yards from shore, but were promptly rescued by an observant boatman.

Robertson pronounced this flight "one of the most pleasant he has ever made." Members of a delegation of Winnebago Indians visiting the city were not so certain. When asked about his reaction to the flight, one of the Indians replied that he thought "nothing of it." Another remarked that he thought the flight of "no use," while a third could only conclude, "Americans foolish!"[78]

Robertson returned to France late in 1828. After four years in Europe, however, the aeronaut was back in New York for a second American tour. He began a flight from Castle Garden on July 4, 1834. Accompanied by a gentleman named Isaac Edge, the aeronaut disappeared into a cloud bank soon after launch, leading the editor of the *New York Sun* to wonder if Robertson had "encamped in the sky."[79] In fact, the pair landed on the Schuyler farm, near Newtown, Connecticut, at 9:30 that night.

The aeronaut made two more New York flights, on July 18 and September 24. At some point during this period Robertson was contacted by Don Antonio Panot, who offered the aeronaut 10,000 pesos for a flight in Mexico City. Robertson accepted and made his first Mexican ascent there, from the Plaza de Toros in San Pablo at 11:00 A.M. on February 12, 1835. The president of the Republic of Mexico was in the audience as Robertson took off, flew toward the southeast, and landed two hours and 22 leagues later in a grove of trees near the village of Chalma, at the base of the Zempaola Mountains.

A second flight scheduled for March 22, 1835, was postponed until April 5, then April 30, and finally cancelled because of adverse weather. The Mexican press did not take these disappointments kindly. One editor provided some poetic advice to the aeronaut:

> *Prepare another short season like*
> *the one you have just had,*
> *For you have been well paid,*
> *to be a failure.*[80]

Robertson finally made his second Mexican ascent on September 13, 1835. He waved the Mexican flag and dropped leaflets of his poetry to the applauding crowd as he drifted over Mexico City for less than fifteen minutes, then landed at Balbuena.

The new president of Mexico, General Santa Anna, was present for Robertson's third Mexican flight less than a month later on October 11. Once again Robertson chose to venture aloft with a young woman. His companion's mother had laid down certain strictures however. There would be no champagne or speaking trumpets on this occasion. In order to obtain the consent of the family, the aeronaut promised to remain at a low altitude and to land as soon after take-off as practical, so that the girl's mother would be present when the voyagers stepped from the car.

The ascent went exactly as ordered. Thirty minutes after the launch from

Mexico City, the pair were back on the ground in the Culebritas Valley. The flight drew particular attention since it was the first time a Mexican national had flown.

The details of Robertson's final three years in Mexico are a bit hazy. He made at least one more ascent from Veracruz during this period. Ironically, Eugène Robertson, who had fled New Orleans and Havana a decade earlier in order to escape the ravages of yellow fever, died of the disease in Veracruz in 1838.

European domination of American aeronautics came to an end with Robertson's death. During his first tour of the United States, Robertson himself had planted the seeds of an American aerostatic tradition. By the time of his death, those seeds had long since born fruit.

Chapter 6 The First American Aeronauts

EUGENE ROBERTSON'S CONTRIBUTIONS TO AMERICAN AERONAUTICS amounted to far more than simply having made a greater number of flights in the western hemisphere than any of his predecessors. By demonstrating that an aerial showman could make money in the United States, he inspired the first American aeronauts to enter the field.

Mme Johnson, the first woman to fly in the United States, began her career less than four months after Robertson's initial New York ascent. Very little is known of her background. Even her first name is uncertain. The *New York American* reported that she had been associated with Blanchard in Europe. If the mysterious Mme Johnson was a citizen of the United States, then she was the first American to make a solo flight.[1]

Following Robertson's lead, Mme Johnson ascended from Castle Garden at 5:00 P.M. on October 24, 1825. As the *National Gazette* noted, this flight had been "long promised and twice deferred."[2] The aeronaut dropped a great deal of ballast at take-off, so that she rose much higher "than any balloon we have ever before seen," then moved southeast, "nearly opposite to the direction Mr. Robertson took on his second ascent."[3] The "fair voyager" was dunked in the waters of a salt marsh near Flatlands, Long Island, on landing.

By 1828 Mme Johnson was centering her activity at Niblo's Garden in Manhattan, where she caused a near riot on July 23, 1828. Some 20,000 spectators had gathered for an ascent that was canceled because of problems with the gas generator. After several unsuccessful attempts to rise out of the garden, Mme Johnson released her balloon on a flight with no one aboard, but the crowd was unsatisfied. "Many riotous persons outside [i.e., non-ticket holders], on learning that no person went up with the balloon, made an effort to get into the Garden, but were fortunately repulsed by the officers, not, however, until they had bruised several of them."[4]

Following a successful ascent from Camden in October 1828, Mme Johnson

Charles Ferson Durant, "The American Aeronaut"

risked the possibility of a second riot when she refused to fly from the Liberty Inn Garden in Philadelphia's Polar Lane on October 24. She informed the "thousands of both sexes and all ages" who had assembled for the flight that she "would not make the attempt—inasmuch as the amount of money received at the gate would not pay half the expenses of inflating the balloon."[5]

Mme Johnson's pleas for support must have been heeded on this occasion. Three days later, on October 27, she did make the ascent, flying 4 miles "at a considerable height" to a landing at New Town, near Germantown, thirteen minutes after take-off.[6]

Mme Johnson was not the only American to be inspired by Eugène Robertson. Charles Ferson Durant, a nineteen-year-old resident of Jersey City, had been among the gaping thousands who came to witness Robertson's first American ascent. Durant, a native New Yorker, was born in a house on James Street on September 19, 1805. While he was not, as he later claimed, the first American to fly, he was certainly the most significant of the first generation American aeronauts.

As the *New York Evening Journal* observed at the time of the aeronaut's death on March 2, 1873, "Mr. Durant was a wide-awake, go-ahead American. He was a man of remarkable talent, if not genius; and he was continually engaged in a novel scheme about something or other. He seemed to take aptly to everything he laid his hand to."[7]

Durant was certainly a man of many parts. His business interests were diverse. He was, at various points in his career, a dealer in fish and oysters, an exhibition balloonist, a candy store owner, a gentleman farmer, a partner in the New York printing and lithography firm of James Nairns and Company, and the guiding spirit behind the Kepler Court, "a utopian scheme" for a shopping and business mall in Jersey City. Durant's activity in local politics ranged from "pugnacious encounters" with the Jersey City political machine to an unsuccessful candidacy for mayor on the People's Party ticket in 1849.

Durant earned minor fame as an amateur scientist as well. While his formal education ended with grammar school, he had a serious interest in mathematics, botany, chemistry, and astronomy. He spent many years collecting, classifying, and studying the seaweed and other plants native to New York harbor, and, in 1851, wrote *Algae and Corallines of the Bay and Harbor of New York*, a quarto volume that contained pressed specimens of some of the two-hundred-odd species described in the text.[8]

Durant, an amateur ichthyologist, kept seven aquarium tanks in his home and exchanged specimens and information with Harvard's Louis Agassiz. His interest in chemistry led him to buy out a small bankrupt apothecary shop in order to furnish his own laboratory. He was among the first to raise silkworms in the United States and was the recipient of several medals honoring the terrestrial globes and lightweight barometers that he constructed. Durant was bitterly opposed to the pseudo-scientific fads of the day. An outspoken critic of the famous Fox sisters, "Spirit Rappers" of Hydesville, New York, he was once threatened with lynching by spiritualist sympathizers in Providence, Rhode Island.[9]

Charles Ferson Durant, the businessman, civic leader, naturalist, and tilter at windmills, was certainly an interesting and admirable fellow. But it was Durant the aeronaut that captured the imagination of the public. Americans were, of course, familiar with the exploits of Blanchard, Guille, Robertson, and Mme Johnson, but none of these balloonists had been accorded the publicity and adulation that greeted Durant. As Peter Parley, the prolific author of educational and moral tales for the children of mid-nineteenth-century America, commented, "No one [of the contemporary aeronauts], we believe, has been more distinguished."[10]

Durant's aeronautical career began on July 9, 1825, when he watched Eugène Robertson rise gracefully above the walls of Castle Garden. Durant must have impressed Robertson as well. When the two men met sometime after the ascent, the aeronaut invited his young admirer to accompany him on the voyage to Havana in 1828.

Durant was then dispatched to France for additional instruction in the fine points of ballooning under the tutelage of Etienne Robertson. According to Durant family tradition, Charles enjoyed his months in France. He visited museums and galleries and even obtained the recipe for some particularly delicious candies that he would produce in his own confectionary shop after his return to New York.

Durant made his one and only European ascent on August 3, 1829. Eugène Robertson, now back in France, was his check pilot on this occasion. Many years later Durant provided his wife, Elizabeth, with a stirring account of his first aerial adventure. She, in turn, passed the story on to her daughter, Emma Durant. By 1932, when Emma related the tale to Paul Edward Garber of the Smithsonian Institution, it had taken on a distinctly legendary flavor: "While far above the earth and sailing onwards, something broke on the outer surface of the balloon, . . . well up toward the top. It must be attended to or disaster might follow! With a jack knife held in his mouth, the nimble and fearless young pupil climbed far up the broken cord which he soon made perfect again, while Robertson waited in amazement, fearful for the life of his pupil. But the pupil thought nothing of the feat and had felt no fear."[11]

After two years with the Robertsons in France, Durant made his American debut at Castle Garden on September 9, 1830. The proprietors had spared no expense in publicizing the event. Newspaper advertisements and handbills distributed throughout the city lauded Durant as the "First Native Citizen of the United States that ever Attempted an Ascension in a Balloon," and called attention to the fact that he was a "Pupil of the Celebrated Eugene Robertson."[12]

The members of the Common Council of New York were invited to attend the ascent as a body. Alderman Stevens "thought it improper for the Corporation to go as a body to witness any exhibition where the loss of life might ensue."[13] Alderman Palmer agreed, fearing that they might next be asked to attend exhibitions of "rope dancers or dancing bears." These two gentlemen were overruled however. The council, following the lead of Alderman Strong, judged the objections "ridiculous" and voted to attend.

They were joined by 20,000 other New Yorkers. One reporter noted that "for hours before the ascent the streets and avenues leading to the site of the exhibition were literally lined with the beautiful, the gay, the sober and the serious, the fashionable and the plain, the old and the young."[14] The *New York Evening Post* reported that the Battery was "literally covered with an enormous multitude of every age, sex, condition, and colour," all of them drawn to the scene "to see a man risk his neck for their amusement and for their money." Actually, the number of spectators risking their money for a seat in the castle was "a rather moderate three thousand."[15]

Three cannons were fired at 1:30 P.M. to mark the opening of the ticket booths and the start of inflation. Durant required two hours to fill his French-built balloon with 10,000 feet of hydrogen. At 3:30 P.M. the spectators were treated to a selection of "National Airs" played by the band as they watched the release of two small balloons and enjoyed the sight of the large balloon straining at the tethers.

The balloon was constructed of green silk and featured an "elegantly ornamented" portrait of Gov. De Witt Clinton of New York. A reporter for the *New York Sunday Mail* rhapsodized over "the rays of the autumnal sun that glittered on the gilded car and gave it the appearance of a pagoda." "Ribbands, wreathes of flowers, and luxuriant festoons of silk, beautifully variegated," were draped over the basket.[16]

Durant climbed into the car a few minutes after 5:00 P.M. and was walked around the court, scattering copies of his poem, "The First American Aeronaut's Address," to the cheering crowd:

Good bye to you—people of Earth,
I am soaring to regions above you.

But much that I know of your earth,
Will ever endure me to love you.
Perhaps I may touch at the moon,
To give your respects, as I pass, Sirs.
To learn if the spheres are in tune,
Or if they are lighted with gas, sirs.[17]
[. . . and so on for five more verses]

He finally cast off at 5:20, offering the spectators a final thrill as he barely edged over the wall. Durant remained in the air for two hours, releasing two carrier pigeons as he moved across the bay and over Staten Island to a landing at South Amboy, New Jersey. The aeronaut spent the evening at Arnold's Hotel in Perth Amboy, where he regaled leading citizens with the story of his flight. Solomon Andrews, a future mayor of Perth Amboy, who would himself make a considerable mark in the history of American aeronautics, was one of those in attendance that evening.[18]

Anxious to take full advantage of Durant's new fame, Mr. Marsh, the Castle Garden booking agent, immediately scheduled a second ascent for the afternoon of September 22. Once again, leading New Yorkers, including the acting governor and members of the Court of Errors, joined 20,000 fellow citizens on the Battery. Newspapers were pleased to report that the garden had been well filled.

Durant began inflation at 2:00 P.M., but the discovery of two ruptures in the envelope delayed the launch until 6:00. As soon as a rope dangling from the car was disentangled from a lamp post on the castle wall, the aeronaut moved rapidly to the northwest, traveling toward Hoboken and Weehawken until he was "but a speck to the eye." He caught the attention of travelers on a turnpike one mile south of Hackensack at 6:45 P.M., dropped a line, and was towed into town. Coming to rest on the City Common, Durant was overwhelmed by volunteers who sought to assist him in deflating and packing the balloon. The aeronaut was the guest of honor at every stop on the trip back to New York. He was a particular favorite with the "numerous company of ladies" who asked him to tea. Charles Ferson Durant's rise to fame had begun.

The next ascent, on August 23, 1831, was preceded by a wave of very effective advertisements. "The eagles will be surprised this afternoon, should the weather be suitable, by a visit from Mr. Durant," noted the *New York Journal of Commerce.*[19] The *Commercial Advertiser* agreed, advising readers that "the really curious will want to be inside the Castle to see the whole thing, not just a transient glimpse." Prospective spectators were urged to "pay for the trouble, experience, and danger" of the aeronaut.[20]

A Durant poster

The *National Gazette and Literary Register* described the scene at Castle Garden on launch day:

> *At five o'clock the balloon was filled, and the car, beautifully ornamented with flowers, was attached to the cords. The most perfect calmness pervaded the atmosphere; the national banner hung lazily from the flag staff, and the streamers from the masts of the vessels lying motionless in the waters . . . There were, perhaps, four or five thousand spectators within the walls of the Garden and at least thirty thousand people on the Battery. All things being ready, the intrepid aeronaut stepped into the car at twelve minutes past five. The balloon having been allowed to ascend so as to raise the car to a height of about twenty feet, Mr. Durant bowed to the admiring throng, and cut the cord which bound him to earth, himself.*[21]

With scarcely a breath of wind blowing, Durant rose straight up. At an altitude of 2,000 feet, he parachuted a rabbit to a landing in the Bay, where the "pretty voyager" was quickly rescued by waiting boatmen.

Durant, having showered the spectators with copies of his latest poem, dropped ballast and rose to an altitude of 3,455 feet. Twenty-one minutes after launch, the aeronaut had scarcely moved from his position directly over the castle. He valved gas, dropped close to the surface of Buttermilk Channel, and caught the attention of the crew of a boat which towed him back to shore. The towline was then passed from hand to hand until, at 6:10 P.M., Durant returned to earth on the spot from which he had taken off fifty-eight minutes before.[22]

The flight, which cost Durant and the owners of Castle Garden some $1,000, brought a return of only $1,200. The aeronaut was satisfied with his small profit, remarking to a *New York Standard* reporter that he was "happy to make 30,000 people happy."[23]

"People again poured forth by ten thousands upon the Battery" to witness Durant's next "adventurous flight" on September 10, 1831. In all likelihood, it was the largest crowd yet gathered for a balloon ascent in an American city and, as the *Mercantile Advertiser* observed, "one of the handsomest scenes of the kind ever witnessed."[24] A traveler approaching the Battery aboard the steamboat *Ben Franklin* described the extraordinary spectacle. "The justly celebrated promenade [Castle Garden] presented nothing to the eye but human heads and faces," he marveled. The Hudson, "the noble river itself," was crowded "for an extent of a mile in Radius, taking Castle Garden for a center, [with] nothing but steamboats, smacks, ferries, pilot boats, row boats, and, in fact, all other kind of boats that ever floated on the water of this Western Babel."[25]

Durant rose "rapidly and majestically" at 5:15 P.M., "soaring off to the west, in the direction of Jersey City."[26] He passed so low over New Durham and Hackensack that he was able to converse with persons on the ground. Durant had hoped to fly as far as Paterson, but contrary winds and the approach of darkness forced a descent at the edge of a wood some 500 yards from the center of Paramus and 22 miles from Castle Garden.

Durant's enormous success and growing popularity were a bit puzzling to the editor of the *New York Evening Post*, who complained that all the morning newspapers had been "thrown into great raptures" in describing the latest flight. They found the English language "too poor to express their extacies [*sic*]," he remarked. For his part, the editor could not understand why Durant merited any more praise "than is due him who turns a somersault over half a dozen horses in the circus, who dances on a slack rope, or walks on the ceiling with his head downward, or performs any other like feat for money."[27] "Actually," he continued, "circus riders, rope-dancers, sword swallowers, and other showmen and jugglers" were more interesting, particularly in view of the fact that Durant seemed unwilling to fly very far.[28]

The editor of the *Evening Post* certainly found few sympathizers in the columns of the other city newspapers. The *Standard* "marveled that one of our contemporaries should be so opposed to aerostation and aeronautics."[29] The editor of the *Evening Journal* regarded the "vile, ill-tempered paragraphs" as "stodgy," "conservative," and "aristocratic." It was, he remarked, just this sort of "pure feudalism" that had hampered Franklin's early experiments in electricity and Gov. De Witt Clinton's efforts to construct the Erie Canal.[30]

Durant could afford to ignore the controversy. His profits from the fourth ascent had been large enough to cover the purchase of a house at Paulus Hook, establish his candy story in partnership with a Mr. Kenney at No. 20 Water Street, and finance work on a new balloon designed to support two people and "six hundred pounds of machinery for experiments in navigating the air." In June 1832 the bright silk gores were dipped in a special varnish containing turpentine, linseed oil, and gum-elastic. The fabric was then spread over poles and chains to dry overnight. When Durant checked the envelope, he discovered that spontaneous combustion had reduced it to smoldering cinders. He displayed the remains in his shop, but the flying season of 1832 was a total loss.[31]

The aeronaut was back with a third balloon, "entirely of his own construction," in the spring of 1833. Six hundred yards of Levantine silk had been fashioned into an envelope 47 feet in diameter and 28 feet tall. Thirty-seven major seams had been double sewn with a backstitch. The finished balloon had a disposable lift of 200 to 300 pounds.[32]

Durant scheduled his fifth flight from Castle Garden on June 1, 1833. In spite of low, scudding clouds and intermittent rain, the garden was "one sweet round of confusion" shortly after the gates opened at 1:30.

A reporter for the *Traveler* found himself at a loss for words in attempting to describe the scene. "Beauxes and Belles were tripping it along, bedizened out in their best bibs and tuckers, and all the boys of the town, indeed apparently of every other town in the world, collected in one vast swarm, screaming and leaping about like—like—alas!"[33] He marveled at the profusion of "light eyes" and "sunny smiles" and watched "boys laughing, girls gabbing, and dandies humming love diddies" as they enjoyed the music of Mr. Marsh's band. Small balloons were released at 3:00 and 4:00 P.M., while spectators were examining a large tethered balloon that was supposed to resemble a dolphin but was described as looking more like a cross between a whale and a flying fish.[34]

After all the excitement of a record crowd in the castle, the ascent was a bit anticlimactic. Durant zipped straight up and out of sight into the low-hanging clouds twenty-five seconds after release. Those in attendance, who were accustomed to enjoying the sight of the aeronaut suspended in the sky long after take-off, were disappointed. Durant, however, was delighted with his view of "the rich downy softness and white fleecy accumulation of clouds, piled in waves as far as the eye could see," all of this beneath the "clear cerulean expanse" of the sky.[35]

Durant made his next ascent less than two weeks later, on June 14, to celebrate the joint visit of Pres. Andrew Jackson and Black Hawk, the famed Sauk and Fox chief who, only a year before, had led his people in an uprising against settlers in Iowa and Illinois. It was an exciting time in the city. "Our streets," noted the *New York Gazette*, "throng with assemblages of citizens anxious to see the President and exchange salutations with the memorable."[36]

In addition to the usual entertainments of the city, which included "grog shops invitingly open on the corners of the streets, billiard tables, cock-fights, and games of chance in half-concealed alleys," the city fathers had planned a series of public displays, parades, and receptions in honor of their distinguished guests.[37]

Durant's ascent was certainly the best attended of the festivities. The crowd within Castle Garden was so large on this occasion that the aeronaut realized a profit after expenses of $12,000. The journalists of the city were fascinated by the mass of humanity streaming onto the Battery to witness the flight. Mr. Brooks, editor of the Portland (Me.) *Gazette*, "rambled around through the crowd, enjoying the sights, sounds, and even the odors of New York, the kitchen of American cities."[38]

The *New York Gazette* focused attention on the plight of a forty-year-old woman, dressed in a black silk dress, a white crepe shawl, a straw bonnet, and carrying a chocolate-colored parasol, who had vanished from Castle Garden on the day of the flight. The disconsolate family waited for news at 41 Rutgers Street.[39]

Another reporter discovered "a modern Cinderella" who lost her slipper in the press of the enormous crowd. Several weeks later she discovered the shoe in the coat pocket of the young man who was escorting her to a dance. He had found it lying on the old parade ground of the garden on the day of the flight and, unaware of its owner, had absentmindedly stuck it in his pocket.[40]

President Jackson greeted the crowd from the roof of the saloon at 5:15. Having spent an active day, he explained to Durant that he was too tired to remain for the ascent. The proceedings were further delayed while Jackson was escorted from the garden. The short bridge connecting the castle to Battery Park collapsed just after the president had crossed over it, dumping a dozen pedestrians into the channel. Fortunately, all of them were fished out with nothing more than minor injuries.[41]

Durant finally ordered the release of his balloon at 5:55 P.M. After being blown against the wall of the saloon and bouncing precipitously across the roof, he asked to be pulled back to earth for a second start, which went much more smoothly. The short flight came to an end with a landing on a Long Island racetrack. He returned to the garden to receive the congratulations of his friends.

Black Hawk, who had watched the beginning of the flight from a ship at anchor off the Battery, "expressed astonishment and surprise at the scene and thought Mr. Durant must be a brave man." His son—"Tommy Hawk, as the wags have it"—was indifferent, being much more impressed by a young actress whom he had seen on the stage.[42]

Durant took advantage of his new fame by scheduling his first ascents outside the New York area, but he intended to be very careful in selecting launch sites in other cities. He was willing to discuss the possibility of a flight with the proprietor of any amphitheater seating 5,000 persons or more. Alternatively, he offered to fly from any city willing to offer a suitable cash guarantee large enough to cover the cost of constructing a temporary enclosure.

On June 19 the aeronaut was in Philadelphia to open negotiations with Mr. Jaggers, the owner of an entertainment center on Broad Street. While Jaggers raised the funds with which to build a fence and additional seating for his establishment, Durant moved on to Baltimore, where he was very much interested in arranging a flight from Observatory Garden, overlooking the harbor on Federal Hill. Once again, however, he was unable to reach an immediate agreement.[43]

Other cities were also anxious to see Durant fly. The *Boston Gazette* published an open letter to the aeronaut. "Bostonians are marvelous lovers of the wonderful; they will stand on tip-toe to look at a horse, mackeral, or walk a mile to see a conjurer bite off his own nose." The people of Boston, the letter continued, "are tired of chasing small balloons" and would turn out in record numbers for a manned ascent.[44] The aeronaut replied that he would love "to see Boston from on high"—once he received a guarantee of a suitable launch site.[45]

Like his colleague in Boston, the editor of the Albany *Microscope* complained that the people of Albany had "frequently seen small balloons, [but had] never beheld such a vehicle transporting a human being." He predicted that if Durant would agree to "survey" Albany and Troy from the air, citizens would stream through the ticket booth in unprecedented numbers.[46]

Albany had one great advantage in the person of Mr. Meeks, the proprietor of a pleasure garden on the corner of Swan and Lafayette streets. While the seating capacity of the establishment was not up to Durant's standards, Meeks promised to begin an immediate program of expansion and refurbishment. Satisfied, the aeronaut agreed to fly from Albany on August 8, 1833.[47]

By August 1, when Durant returned to Albany with his balloon, twenty demijohns of sulphuric acid, and a dozen casks of iron shavings, the town was already buzzing with excitement. The Hudson River Line had arranged for the steamboat *Ohio* to make special passenger runs to small towns and villages along the river. Albany businessmen, led by Mr. Crittendon, "mein host at the Eagle Tavern," had raised additional funds to assist Meeks in expanding his amphitheater to seat 6,000. Philip Hooker, a local architect hired to conduct a safety inspection of Meeks's new tiers of seats, commented that while the benches were "neither plained [*sic*] nor cushioned, for strength they are ample."[48]

Durant did his part to build excitement. For some weeks, articles had been appearing in eastern newspapers discussing the aeronaut's plans for a long-distance aerial voyage making use of high-velocity winds to be found at high altitudes. Durant claimed to have attained airspeeds of up to 60 MPH when accidentally caught in such a wind on the occasion of his second ascent. Now he suggested that the Albany ascent would afford an excellent opportunity to test his theory. If conditions permitted, he announced, he might choose to continue his flights from upstate New York to Philadelphia or even Baltimore.[49]

As predicted, "a countless throng" crowded into Meeks's new amphitheater on August 8, 1833. "Nothing could exceed the harmony which pervaded the garden," remarked one proud observer. "Not a whisper of impatience, or a look of displeasure was heard or seen."[50]

Durant ordered his release at 5:05 P.M. Far from flying to Philadelphia or

Baltimore, the aeronaut returned to earth at 6:47 P.M. on the farm of Peter Slingerhand, a half mile from the town of New Scotland and only twelve miles from Albany. Everyone was pleased with the outcome of the ascent. Durant's share of the gate receipts had totaled $1,300. The citizens of Albany were so impressed by what they had seen that a town meeting immediately after the flight resulted in pledges of $400 in support of a second ascent.[51]

The people of New Scotland were so enthusiastic that they voted to rename their town Durant. This was going too far, even for the aeronaut's supporters. The Baltimore *Journal of Commerce* hoped that, accustomed as he was to "inflation," Durant would not become "puffed up" as a result of this new honor. An Albany resident remarked that as much as he admired the aeronaut's courage and daring, he believed that it would be much more fitting to rename the town after "some revolutionary worthy or distinguished civilian whose achievements have conferred more solid advantages on the site."[52]

The year 1833 marked the peak of Durant's career. On September 5 he announced that he would fly from the Observatory Garden on Baltimore's Federal Hill. Mr. Stanley, the proprietor of the garden, squeezed 6,000 extra seats into his arena to prepare for the ascent, which was scheduled for September 26.

The scene in Baltimore that day was a familiar one to Durant. Baltimore Street and the other avenues leading to Federal Hill were jammed with pedestrians and carriages for two hours prior to the launch. The *Baltimore American* noted that the city seemed deserted by 5:00 P.M. "The whole population had gone out to see the wonder."[53] It was a record turnout, with as many as 50,000 spectators present. This was of little comfort to the aeronaut, however, for the garden itself was only "tolerably filled" with perhaps 2,000 people. A general fear that the grounds would be dangerously overcrowded was apparently responsible for the slack attendance.

It was fortunate that there were not more spectators in the garden for the ascent. As the balloon was cut loose at 5:27, everyone in the amphitheater surged toward a section of makeshift seats constructed over the roof of the saloon in order to get a last look at the departing aeronaut. The structure collapsed under the sudden load, dropping the frightened but uninjured people into the bar room.[54]

Durant trailed more excitement in his wake as he moved 13 miles across country to a landing in Bel Air, Maryland, at 6:44. Washerwomen, field hands, carriages, horses, and farm animals scattered as the aeronaut passed low overhead. One elderly woman, who had been confined to bed for weeks, became so excited that she rushed from the house "and actually clambered over a fence with as much alacrity as a young girl."[55] A crowd of two hundred people who had followed the balloon during its final few minutes in the air greeted Durant as he climbed from the car. He distributed the flowers that decorated his car to a bevy of ladies besieging him with offers of tea and sampled a "very superior old wine" provided by a local admirer.[56]

The citizens of Baltimore, much impressed by Durant's performance and embarrassed by the poor financial return which he had received, took immediate steps to remedy the situation. At an open meeting held at the City Exchange on the evening of September 30, a "numerous and respectable" assembly created a committee charged with raising $5,000 to sponsor a second ascent. Within a week, subcommittees had sprung up in each of the city's twelve wards to conduct a door-to-door ticket sales campaign.[57]

It was an extraordinary effort. For days Baltimore newspapers carried a continuous flow of reports announcing meetings of the ward committees and the progress of the fund drive. Durant's praises were sung in a flood of amateur poetry that recalled the early days of ballooning in America:

Get on thou fearless mariner!
And leave all meaner things,
Unto earth's multitudes below,
Thou greater be than Kings!
Thou manly tenant of the air,
God speed thy gallant bark,
And bring thee safe to earth again,
As Noah in the Ark![58]

According to the *Baltimore Visitor*, Durant was "the pet of the city."[59] The aeronaut moved quickly to take full advantage of the situation. With the money pouring in from zealous ticket sellers, he constructed a large temporary enclosure with a sixteen-foot fence on the grounds of the Observatory Garden. He was careful to emphasize that the new seats had been carefully reinforced. Even "ladies of the most delicate and nervous constitution need not entertain the slightest apprehension."[60]

Five thousand spectators who had paid one dollar a head filled the amphitheater on the afternoon of October 14. When a gun sounded at 2:00 P.M., the crowd pushed to one side of the enclosure where the envelope had been spread over a sheet of canvas. The hydrogen was produced in twenty casks. Leather pipes connected the head of each cask to one of two large wooden reservoirs where the gas was cooled and cleaned of impurities by bubbling through water. The aeronaut had purchased the sulphuric acid for both of his Baltimore ascents from Ellicott's Chemical Laboratory. He also brought several hundred pounds of newly cut nails to insure pure hydrogen.

Durant went aloft at 4:43 P.M. He flew east, passing over Chesapeake Bay on his way to the Eastern Shore. Perhaps realizing the difficulties he might encounter on an overland journey back to Baltimore, he dropped anchor next to the steamboat *Independence* while still a mile from shore. Captain Pearce ordered his crew to haul Durant down to the deck, and by 7:00 P.M. the aeronaut was enjoying a pleasant dinner as the ship steamed back toward Baltimore.[61]

Durant made three final ascents in Boston during the following year. The first of these was nearly his last. The aeronaut went aloft from a special enclosure built at the foot of Boston Common at 5:50 P.M. on July 31, 1834. Passing over the open water at Nahant, Durant hoped to reach a landfall at Cape Ann but began a rapid descent long before he reached the cape. Skipping four or five times over the surface of the water, the balloon finally came to a stop with the basket and its occupant completely submerged. Durant struggled to the surface and floated about, buoyed up by a gum-elastic life jacket, until he was rescued by Captain Spaulding of the schooner *Minor*.[62]

Having repaired his balloon and suffered one postponement because of bad weather, Durant was in the air once again on August 25, 1834, flying from the enclosure on the Common to a landing at Mount Auburn forty minutes later.[63] Durant made his final ascent on September 13, 1834. The flight, which had been rescheduled three times because of rain and high winds, was a relatively long one, covering 17 miles from Boston to Lincoln, Massachusetts, in two hours and twenty minutes.[64]

Charles Ferson Durant brought his aeronautical career to an end following this third Boston ascent, apparently as a result of a promise made to his new bride.[65] In all, Durant had made only thirteen flights, twelve of them in the United States. When compared to the record of many of the aeronauts who followed, this seems a very small number. But Durant's impact cannot be measured by simply counting the number of times he had ventured aloft or by calling attention to the fact that his flights were usually of short duration.

Charles Ferson Durant was the first of the great nineteenth-century American balloonists. He had made minor contributions to balloon technology, most notably by developing a barometer specially suited to the needs of aeronautics. Far more important, Durant fixed the image of the daring aeronaut in the minds of the American public and inspired an entire generation of native balloonists. His short career marked the beginning of a golden age in the history of American aerostation. Long after he had made his last ascent, at a time when other men and women had completed longer and more spectacular flights, Durant would still be regarded as the nation's premier aeronaut.

The seeds planted by Durant first took root in Baltimore. Founded in 1729, the city had a reputation as a fun-loving town, boasting a wide variety of theaters, parks, pleasure gardens, monuments, and racecourses. Notices calling attention to the latest theater attractions filled the city newspapers. Advertising posters were plastered on walls and fences around town: "Bills of all descriptions and sizes are posted at the corners and in public places, and it is quite common to behold a knot of gawkers reading them with the intentest interest, ever and anon counting the scanty store of coppers in their pockets to ascertain whether they can afford a frolic or not."[66]

The competition for the limited entertainment dollar was fierce. In late September 1833, while Durant was preparing for his first Baltimore ascent, the city was also hosting a French opera company, "patronized by many who know as little about the music as they do about the language of the monsears." A group of Italian performers, "the fragments of a quixotic humbugging expedition to this country," were performing in the city at the same time. The Front Street Theatre featured an English touring company "sustained by Mrs. Austin and Mr. Walton, with a shocking long list of miserable chorus singers." The museum and the Menagerie and Equestrian exhibition were drawing record crowds as well.[67] Baltimore pleasure gardens were crowded with citizens out for a good time. The "grand gala" fireworks exhibition at the Columbian Garden was so successful that Mr. Wagner of the rival Observatory Garden was forced to counter with a working volcano constructed on Federal Hill.[68]

The spectacle-loving citizens of Baltimore who frequented the pleasure gardens and theaters were anxious for another opportunity to see a human being carried into the sky. In the wake of Durant's twin ascents, one young daredevil after another announced his or her intention of joining the New Yorker in the air.[69] By the end of 1834, the city would be recognized as the undisputed capital of American aeronautics. As early as July 1834, the editor of the *Evening Star* had observed with pride that his city was "as fruitful of aeronauts as of monuments and riots."[70] "The aeronautic spirit is up," concluded the *Patriot.*[71]

James Mills, "a practical mechanic of Baltimore," was first on the scene, flying from Mr. Stanley's Observatory Garden on April 3, 1834. His homemade balloon stood 43 feet tall and was 25 feet 3 inches in diameter. The ascent, which *Nile's Register* regarded as "more splendid than either of Mr. Durant's," lasted seventy minutes as Mills covered 16 miles to a landing at Bodkin, Maryland.[72]

The aeronaut made his second flight from the Fairmount Garden on May 1, 1834. It was a "perfect success," a "beautiful and imposing spectacle," in spite of low-hanging clouds and drizzle. Having passed 14 miles across Chesapeake Bay from North Point into Kent County, Maryland, Mills prepared to descend, only to be caught in a windstorm and carried back across the bay to a landing near North Point at 7:10. He had flown a total distance of almost 50 miles in three hours, reaching an estimated altitude of 2¾ miles at one point. The editor of the *New York American* was delighted to note that young Mills, "self taught, and dependent alone on his own unaided efforts, may fairly take rank with the most successful aeronauts of the age."[73]

On May 26, 1834, Mills announced that, before his third ascent that afternoon, "such ladies and gentlemen as desire it, will be permitted, free of charge, to ascend to some distance from the earth, the balloon being detained by a cord." Mills was rained out on this occasion, but only a month later, on June 26, he was in the air once again, flying from a lot on the corner of Spruce and Broad streets in Philadelphia.[74]

The "beautifully navigated balloon" lifted off at 4:30 P.M. Ninety minutes later, having covered a distance of 40 miles, Mills attempted to land near Hanover Furnace in Burlington County, New Jersey. Dropping anchor in a salt marsh, he encountered unexpected high ground winds that shredded one side of his car. Mills was dragged through the swamp, "sometimes on the ground, sometimes forty or 50 feet up." Forced to jump 8 feet to earth, Mills incurred a bruised face and arm and a twisted ankle as he watched $2,000 worth of balloon and equipment "sail away for the coast of Spain or Portugal," as the observant diarist Samuel Brecks noted.[75]

Concerned Philadelphians established a fund to compensate Mills for his loss, but he remained grounded for the remainder of the summer while constructing a new balloon.

Balloon enthusiasts in Baltimore were already turning their attention to other local aeronauts. Nicholas Ash, an artist who was described as "a very deserving and respectable young Man," attempted his first ascent from the Observatory Garden on June 2, 1834. Ash's balloon, the largest yet seen in the city, featured silk gores dyed six different colors and "arranged in a style which reflects much credit upon the taste and judgment of the artist."[76]

The aeronaut's hopes were dashed by a fresh northwest wind that prevented full inflation of the balloon. After two hours of struggle, Ash removed the envelope from the netting and released it on a free flight in an effort to calm the disgruntled spectators. The *Chronicle* noted that the balloon looked like "an umbrella, or parachute, and took various other forms," as it drifted 2 miles before falling safely to earth near Fort McHenry.[77]

Fearing that his failure to fly on this occasion might end his aeronautical career before it had properly begun, Ash rushed into print with assurances that he would attempt a second launch on July 7. This time he was successful, rising out of Fairmount Garden at 3:45 P.M. After striking a neighboring chimney, he climbed to an altitude of three-quarters of a mile. Only fifteen minutes into the flight, while still over the center of the city, the balloon began to rotate, twisting the valve lines and confusing the aeronaut, who inadvertently opened the valve. The balloon fell precipitously onto the roof of a house near the corner of Pace and Lexington streets and was then blown into the gable of a nearby market house before depositing Ash safely, if unceremoniously, onto Lexington Street.[78]

Nicholas Ash became the first man to ascend from the District of Columbia on July 30, 1834. Andrew M. Land, the proprietor of the Analoston Garden on Mason's Island (now Theodore Roosevelt Island), opposite Georgetown, had engaged the veteran of one short and nearly catastrophic flight to introduce the citizens of the nation's capital to the wonders of aerostation. Special ferries carried spectators from the Georgetown docks to the garden. Those who preferred could pay a small fee to watch the proceedings in comfort from the upper stories of the Old Stone Warehouse, 300 yards away on the Georgetown shore.[79]

The *National Intelligencer* reported that "nearly all" the residents of Georgetown, Washington, and Alexandria were on hand for the flight. Ash, fearing "the impatience of part of the spectators," cast off before his balloon was completely filled with its 12,000-cubic-foot ration of hydrogen. Nevertheless, he remained in the air for two hours, moving in and out of the clouds before descending on the Travers farm some 4 miles from Georgetown at 8:00 P.M. The flight had been a complete success. Ash lost little time in advertising a second ascent, which was postponed when he fell ill a day or two after the first flight.[80]

The number of Baltimore residents anxious to fly now seemed to be growing daily. William A. Woodall, Jr., announced that he would make his first ascent from Fairmount Garden on July 16, 1834, accompanied by "a young lady of the city."[81] All went well on the day of the flight until the basket was attached to the gasbag at 5:30 P.M. A sharp tug on the netting ripped a huge hole in the envelope, allowing the hydrogen to escape in a rush and dropping the crumpled fabric on the heads of the ground crew. The "respectable audience" inside the enclosure seemed content to have witnessed the accident, but the "crowd of disorderly persons" outside the gate, exasperated because they could not enjoy "a gratification for which they had not paid," rushed against the enclosure forcing Mr. Stanley, the proprietor, to call out the city police.[82]

While Woodall rushed to repair his balloon for a second try, yet another potential candidate for aeronautical fame was emerging in Baltimore. An anonymous lady advertised that she would launch a small balloon manned by a "cupid effigy doll" from Fairmount Garden on July 28, 1834. While her aerostat was too small to carry an adult aloft, she offered to give tethered rides to any brave children in the audience. The profits of this first exhibition would go toward the construction of a full-scale balloon in which she would make a flight. Her advertisement closed with a poetic plea for support:

> *To assist aspiring Genius, in a*
> *female of this land,*
> *Let not the arts and sciences, to*
> *man be all confined,*
> *Ye citizens of Baltimore, now prove your liberal mind.*[83]

The anonymous lady aeronaut performed as promised, but Mr. Stanley was far from pleased. It seems that a Mr. Hamblin had approached him three weeks earlier requesting that a Baltimore widow be allowed to fly from his establishment. Stanley agreed to the proposition but had been mortified to discover that the aeronaut was nothing more than a "big wax doll" and that the leaky balloon was "only capable of carrying up a small child, and that with improper risk."[84]

Hugh Frazier Parker was far more successful than his anonymous female rival. At 4:00 P.M. on July 5, 1834, Parker piloted the balloon *Washington*, "a new, large and tricolored" aerostat out of the enclosure on Federal Hill. Half an hour later he returned to earth 5 miles beyond the city limits of Baltimore. Parker's flight was an agreeable surprise to those who had been disappointed by several recent failures.[85]

Parker, "a worthy and respectable Baltimore mechanic," was back in the air on August 6, 1834. Waving a flag and sprinkling copies of a printed address over the heads of the cheering crowd, Parker lifted off with a twelve-year-old girl passenger who "gave manifest and gratifying proof that she was quite collected and free from apprehension." A half hour later the pair landed in the water off Fort McHenry, where they were rescued by a waiting boatman, having suffered nothing more than damp shoes. The balloon was towed back to shore, at which point Parker returned to the basket and was "floated through the streets, and finally to Fairmount, the original point of departure."[86]

George Elliott was the last of the Baltimore balloonists to make his appearance in the summer of 1834. Having witnessed a great many ascents, the residents of Baltimore were delighted to discover that they would have an opportunity to see balloons "hunting in couples on July 28, 1834."[87] Elliott, a newcomer flying a 15,000-cubic-foot balloon named the *Lafayette*, planned to go aloft from Federal Hill at the same time William A. Woodall attempted to launch his own repaired aerostat. When the great day arrived, however, the would-be aeronauts discovered that they were unable to produce enough hydrogen to inflate both balloons. All the gas was transferred to the as yet untried *Lafayette*. Elliott invited Woodall, who had already failed to ascend once, to redeem himself by making the flight alone. The gratified Woodall began his first flight at 6:00 P.M., landing safely one hour and thirty-five minutes later.[88]

Elliott made his own first flight two weeks later on August 12, 1834, accompanied by a six-year-old boy, John Quincy Adams Redman. They were abruptly dunked into the Patapsco River later that afternoon when the balloon's valve stuck open.[89]

It had been an extraordinary summer in Baltimore. As usual, the city newspapers were divided over the meaning of what they had witnessed: "We have not, thank God, either the cholera or the yellow fever among us, but we have been troubled for a month or two past with a disease decidedly contagious in nature—balloon fever."[90]

The enthusiasm was unmistakable, but there were serious questions to be answered as well. One editor feared that "there may yet be some breaking of necks." Others questioned the sanity of the parents who had allowed the two children to go aloft. The *Providence Journal* was bored with the whole thing: "We have never seen any practical good result from the voyages of the 'aeronauts,' as they call themselves. They all 'gracefully ascend,' and just as 'gracefully descend,' and there is an end to the matter. It is time for some of the adventurers to make a grand discovery, or the glorious discovery of the glorious art of 'aerostation' will come to be considered another of the humbugs of the day."[91]

Such dismal pronouncements did little to dampen American interest in the balloon. The railroad and steamboat were working a revolution in American transportation. Who could say that the balloon would not extend this revolution into an entirely new dimension?

> *The latest mania among our people is ballooning. Every few days a glittering globe is seen suspended in the heavens, or slowly floating over the town, freighted with adventurous citizens who tempt the skies in search of new sensations. . . . In view of these things, there would seem to be danger that young America is about to revolutionize the rail and note the car a slow coach. Who is sure that we are not on the eve of very great changes in our locomotion, when instead of the pony phaeton, Smith will invite Maria . . . to take an airing, or in lieu of his fast pair, Tompkins will propose to Jane to drop him in his Hydrogen Parachute.*[92]

Of course the balloon continued to inspire weak newspaper humor. At the height of the balloon madness that summer of 1834, Baltimore papers carried an advertisement for yet another ascent: "Miss Felunia Mouser begs leave to inform the citizens of Baltimore that she has completed her splendid Mouse-Skin balloon, which she has called the 'Whittington,' in honour of the illustrious patron of her ancestor. Miss Mouser planned to go aloft at the sign of the 'Puss-in-Boots,' but warned 'Aunt Tabby' and other of her female friends of weak nerves to stay away from the ascent, as the excitement might prove over-powering."[93]

By the fall of 1834, the small cadre of aeronauts who had made their debut in Baltimore that summer were demonstrating their newly acquired aeronautical prowess in other parts of the country.

James Mills, who had suffered the mishap in New Jersey, remained the best-known and most experienced member of the group. Having constructed a new balloon with funds provided by the Philadelphia subscription, he was back in the air on September 16, 1834, going aloft from a lot on Broad Street and landing 16 miles away near the Lancaster Turnpike. The flight was, as the *Philadelphia Evening Star* opined, "all that himself or his warmest friends could have desired."[94] On October 6, 1834, Mills flew from Camden to Morristown, New Jersey. The aeronaut was particularly pleased by the large number of paying spectators gathered for this ascent.

His next flight was from Lancaster, Pennsylvania, on November 1, 1834. Rising above the clouds, Mills enjoyed the sight of "bright sunshine, with a vast field of white opake [*sic*] vapors below me convoluted into singular forms, and presenting a variety of elevations and cavernous depressions."[95] When he dropped back down through the overcast two hours after take-off, he was surprised to discover that he had flown all the way to Oakton, Maryland.

With the return of good flying weather the following spring, Mills ascended from Fairmount Garden accompanied by a Miss Phillips on May 4, 1835. The pair landed safely in Peach Bottom Township, York County, near the banks of the Susquehanna River, at 6:20 P.M. Mills was especially proud of Miss Phillips's *sangfroid*. She had, he noted, remained "perfectly calm and *highly* delighted; busying herself in throwing out feathers to ascertain when the balloon rose or fell; she always reported her observations to me."[96]

Mills made his ninth and final ascent from York, Pennsylvania, on July 25, 1835. He crossed the Susquehanna to a landing not far from an inclined plane railroad near Columbia. Mills had covered approximately 11 miles in a little more than an hour.[97]

Mills, who had advertised a tenth ascent to be made from York on August 22, was found dead in his hotel room in York on August 16, 1835. Baltimore newspapers reported that he had burst a blood vessel, but in view of the fact that he had been varnishing a section of his balloon in the confined room, it seems more likely that he was asphyxiated.[98]

Other Baltimore aeronauts were enjoying less success than Mills. George Elliott occasioned another balloon riot when he failed to ascend as advertised from Camden, New Jersey, on August 25, 1834. The launch site was a poor one. The 2,000 paying spectators that squeezed into the enclosure found themselves in an area so dusty they had to hold handkerchiefs over their mouths and noses. Security was so lax that non-ticket-holding rowdies broke through the fence. The ensuing noise, hustle, bustle, and confusion made it very difficult for Elliott to supervise the inflation.

The aeronaut had advertised that he would fly with a lady passenger dressed as a bride. When the partially inflated aerostat refused to leave the ground, Elliott attempted a solo flight. Still unable to go aloft, he attempted to add more hydrogen to the envelope, but the balloon still refused to rise. The spectators, "disgusted at the whole scene, pulled down the balloon and very gently, under the circumstances, cut it to pieces."[99]

Elliott promised to make a second "gratuitous ascent" in Camden as soon as he had completed work on the new balloon. Baltimore newspapers advised against such a flight. "In our opinion, the greatest part of the assemblage do not deserve such a favour."[100]

Elliott apparently agreed, for as soon as his new aerostat was finished, he was off to Charleston, South Carolina, where he flew on October 22, 1834.[101] The weather was so perfect that Elliott's balloon remained in view until it appeared as "a speck scarcely visible in the great conclave above. The golden rays of the late afternoon sun playing on the blue gas bag created the most magnificent phenomena we have ever witnessed."[102]

The aeronaut landed in the water when the valve stuck open. He was forced to climb onto the load ring, where he sat bobbing about for a half hour until rescued by Captain Albert of the pilot boat *Cora*.

Elliott made his second flight from Charleston on November 3, 1834. After untangling his balloon from the trees into which he had flown at take-off, he rose out of the amphitheater and flew out of sight to the southwest, finally making a difficult landing in high winds on the Hugh Wilson plantation some twenty miles from Charleston. Uncertain what to make of the sight of a great billowing globe descending from the sky, black slaves watched as the aeronaut bounced over the top of a group of pine trees before being dragged across the open ground. As the blacks helped carry the folded aerostat to the nearest road, they were already singing an extemporaneous song about "Massa Elliott and his

Balloon."[103]

George Elliott's next ascent was made from New Orleans on December 28, 1834. Things went wrong from the outset. Several persons were injured when violent winds carried the balloon through a row of spectators benches. Elliott was then driven against the chimney of a neighboring arsenal, breaking his thigh and dumping so many bricks into the car that the balloon was unable to rise very far. Careening against one house after another, he was blown toward the waterfront. His head, hands, and face were severely lacerated by broken glass from the many windows he had smashed. The balloon finally came to rest when the basket caught in the rigging of a ship. Elliott was rescued, "blinded by blood from wounds over his eyes, the broken thigh bone protruding through his drawers and pantaloons."[104]

Physicians cut four inches of bone from Elliott's leg, and feared that amputation might be necessary. The aeronaut seems to have survived, however, and may, in fact, have been the same Mr. Elliott who flew from Baltimore to Havre de Grace, Maryland, on November 18, 1835. In the fall of 1854, during a scheduled ascent from Richmond, Virginia, a Mr. Elliott allowed a young man named Carrier to make a short tethered ascent prior to the aeronaut's departure. Carrier simply cut loose and "sailed rapidly off towards the sky." The Richmond *Dispatch* noted that Elliott was so overcome with disappointment that he fainted.[105]

The fall of 1834 was indeed a difficult season for aeronauts. Hugh Frazier Parker was called a "quack aeronaut" when he failed to ascend as advertised in the Portsmouth-Norfolk, Virginia, area in September 1834. Parker had sent two apprentice aeronauts, Wallace and Stanley, to fulfill the engagement, while he remained in Baltimore nursing a sick family member. Wallace and Stanley bungled the flight, barely edging out of the infirmary yard at Briggs Point, then bobbing over the fence and hovering a few yards over a nearby creek. They were unceremoniously dumped into the stream, forced to choose, as one local observer noted, between drowning and being mobbed by the dissatisfied crowd. Parker was outraged by the ineptitude of his substitutes and denied any connection with the fiasco.[106]

The citizens of Frederick, Maryland, were similarly disappointed two days running that fall. On September 12 a Mr. Simpson, who had been identified as William A. Woodall's partner earlier that year, rose only 250 feet into the air and flew less than a mile in a very small balloon. The next day, September 13, Woodall was unable to fly at all.[107]

Things were no better in Chambersburg, Pennsylvania, when, on October 15, 1834, a Mr. Miller, the designer of "a winged balloon," failed to leave the ground. The citizens of Baltimore also suffered frequent disappointment. The newspapers were dotted with advertisements from would-be aeronauts such as Joseph Ames and W.A. Swift, with no indication that such flights ever took place. Baltimore editors expressed the fear that the number of unsuccessful "aerial aspirants" pouring forth from their town would soon give the city a bad name.[108]

The string of unrelieved failures was soon reversed however. Hugh Frazier Parker gave the citizens of Fredericksburg, Virginia, an opportunity to see a human being in flight when he ascended just after sunset on October 15. He remained in the air for two hours and claimed to have reached an altitude of 2 miles before descending on the Limbrick farm between the Potomac and Rappahannock rivers.

Parker may have made a second ascent from Fredericksburg. It is certain that he was in Washington, D.C., for a flight on Saturday, October 8, 1834. While the details are cloudy, Parker apparently made the Washington ascent as a favor to Nicholas Ash. Ash had fallen seriously ill following his first flight from Mason's Island. At the time of his "severe indisposition," the aeronaut had already scheduled a second ascent in Washington and another in Philadelphia. There is no record of his having made either flight, or any others, for that matter, but, as Parker's reports of his ascent on October 8 are addressed to the "patrons of Mr. J.J. Ash," we may assume that Parker was fulfilling an engagement for his sick friend.[109]

The flight was a triumph. Parker went aloft from an enclosure constructed on the square north of President's House. Casting off at 4:55, he was blown southeast, certain "that man never beheld a more beautiful sight than the view of Pennsylvania Avenue, leading, as it does, to the two splendid houses which obstruct its way."[110] Passing over the White House, Parker noted that Pres. Andrew Jackson was observing his progress from the terrace. The aeronaut saluted Jackson by "waving over him the Stars and Stripes which he has so nobly defended," then dropped a guinea pig to earth aboard a parachute.

Parker passed over Alexandria at 5:00 P.M., remarking on the enormous crowd gathered on the docks to cheer him on. He finally landed near Blackstone's Island at the mouth of the Potomac, some 110 miles from his point of departure. Like Elliott, Parker was severely dashed about at the end of his anchor line while attempting to convince the watching locals that he was not a supernatural being dropped from the sky.[111]

Parker did not fly again until the late summer of 1835, when he traveled to York, Pennsylvania, to fulfill an engagement scheduled by James Mills prior to his death. The local citizens became so infuriated when Parker failed to go aloft on two occasions that the shaken aeronaut was forced to accept an offer of protective custody from the local sheriff. It is not too surprising to discover

An American balloon ascent, circa 1835

the name of Hugh Parker missing from the list of active balloonists soon thereafter.[112]

William Woodall, who had experienced so many failures and so few successes during the previous months, traveled to Savannah for an ascent on January 27, 1835. True to form, he allowed his balloon to escape the netting while the car was being attached. The envelope, which blew out to sea, represented a $700 loss and the end of Woodall's checkered aeronautical career.[113]

The old figures fading from the scene were quickly replaced by new faces, men and women who had been inspired and trained by their scarcely more experienced predecessors. It seems likely, for example, that Samuel Wallace (or Wallis), who advertised a proposed night ascent late in 1834, was the same Wallace who had attempted to fill in for Hugh Parker at Norfolk that September. If so, Wallace was still plagued by misfortune, for his promised night flight failed to materialize.[114]

Zebulon Mitchell was a far more successful second generation Baltimore aeronaut. He donated a portion of the proceeds from his first ascent, on March 23, 1835, to benefit the "poor seamstresses of Baltimore."[115]

"The Balloon Ascension," by Nicolino Calyo, shows a balloon ascension circa 1835 from Baltimore's Fairmount Garden. Photo courtesy of the Maryland Historical Society, Baltimore, Maryland.

In late May 1835 Mitchell "happily effected an ascent" from the Old Council Chamber Hill in Richmond, Virginia. The aeronaut called attention to the educational nature of his exhibition, particularly for those "who wish to obtain a knowledge of Chemistry." A "very respectable number of citizens" paid to witness the launch. An even larger group of non-paying citizens pulled down the protective fence in order to obtain a better view of the proceedings. Fearing a riot, Mitchell ordered the balloon towed out of the enclosure and onto the brow of a nearby hill. Doffing his hat, waving his flag, and tossing printed verses to the boisterous crowd, he ascended "in the approved style on such occasions."[116]

Zebulon Mitchell also trained apprentice aeronauts. Jane Warren, a Baltimore matron, was his best-known pupil, if ten minutes of instruction can be said to have qualified her as a pupil. Mitchell had planned to make a solo ascent from Fairmount Garden on September 14, 1837. Mrs. Warren, who had never met Mitchell, stepped forward just before the flight was to begin and asked that she be allowed to make the ascent in his place. Word that a woman was about to fly spread rapidly through the crowd. Mitchell, grasping the temper of the spectators, recognized that he had little choice but to accede to the woman's request. After some quick words of warning from Mitchell, Mrs. Warren set off on a two-hour flight that included a circuit of the city and a trip across Chesapeake Bay to a landing in Kent County, on the Eastern Shore of Maryland.

After some extended instruction from Mitchell, Jane Warren made her second ascent from Fairmount Garden on September 28, 1837. Baltimore editors, clearly entranced by the sudden appearance of a new heroine, urged their readers to patronize the ascent, "for the aeronaut is a lady, and she must be rewarded for her enterprise."[117]

Mrs. Warren continued to ascend from Baltimore at least through the fall of 1837. Her most unpleasant experience occurred early that October when she was dunked into Chesapeake Bay near Poole's Island, between the mouth of the Bush River and Burden Point. Her struggles to escape the mud and water were completely ignored by the crew of a passing schooner. Finally rescued by a party of surly fishermen, Mrs. Warren was treated "most inhospitably." Following her return to Baltimore, she broadcast her complaints regarding the "brutes in britches" whom she had encountered on her trip.[118]

The great age of Baltimore ballooning had come to an end by 1837. James Mills, George Elliott, Hugh Frazier Parker, William Woodall, Nicholas Ash, Zebulon Mitchell, Samuel Wallace, and Jane Warren—the successful and the unsuccessful, the genuine aeronauts and the quacks—had passed into history.

They were an extraordinary group, these pioneers, as notable for their failures as for their successes. Their generally dismal record is not so difficult to understand. As John Wise noted, "the most of them were not philosophically acquainted with the subject. For this reason did so many of them fail in the simple process of a balloon ascension."[119] Unlike Guille, Robertson, and Durant, the Baltimore balloonists did not have the advantage of apprenticeship with an experienced aeronaut. Viewed in these terms, it is surprising that men and women like James Mills, Nicholas Ash, George Elliott, and Jane Warren were able to establish themselves as professional balloonists.

Chapter 7 A Golden Age Begins

THE MANNED BALLOON ARRIVED IN "THE QUEEN CITY OF THE WEST" with considerable fanfare in November 1834. At the time, Cincinnati was not yet half a century old, having been founded in 1786, two years after Peter Carnes had sent young Edward Warren up from Baltimore. Yet the city was far from the raw frontier town that many residents of older, East Coast cities imagined it to be.

The census of 1830 had identified Cincinnati as the sixth largest metropolis in the United States and had also underscored its growing importance as a major American industrial center. It was "Porkopolis," the nation's leading meat processor, where half a million hogs were slaughtered in a good year. Westerners ate the resulting ham and bacon with greens cooked in Cincinnati lard, washed it down with Cincinnati beer and whiskey, and cleaned up the mess with soap from the vats of thirty Cincinnati reducing plants. They dressed in clothes manufactured by a garment industry that employed 950 factory hands and 9,000 additional pieceworkers who operated from their homes. By mid-century Cincinnati foundries would be turning out a thousand stoves each day to heat homes furnished with the products of the city's burgeoning woodworking industry.

The river was the source of Cincinnati's phenomenal prosperity. The east end of the waterfront was the industrial heart of the city, where some thirty steamboats were coming off the stocks annually. Some of the finest practical engineers and mechanics in the world found employment aboard the steamboats or in the machine shops and foundries supplying high-pressure steam engines to the boatyards.[1]

If the attention of most Cincinnatians was focused on the river and the canals that carried the city's commerce, there were at least a few who dreamed of traveling through the air. As early as 1815 a local showman named Gaston had flown large, unmanned balloons as a part of local Fourth of July celebrations. Cincinnati newspapers had also given considerable coverage to aeronautical events in Europe and the eastern United States. A serious attempt to lure Guille to the city in 1819 had failed.

Thomas Kirkby arrived in Cincinnati with his balloon in the late summer of 1834. Newspaper accounts noted that he had come from Baltimore, so we may assume that he had witnessed at least some of the recent flights in that city.

Immediately after his arrival in the Queen City, Kirkby began to prepare for his first ascent by ordering the construction of a large amphitheater capable of seating 4,000 to 5,000 persons. The admission charge of fifty cents would permit the spectator to observe the inflation of the balloon, witness the release of small trial balloons to test the direction and velocity of the wind, and enjoy the fine "Band of Music" provided by Kirkby for the "entertainment of the spectators." Preparations were complete by November 20, and the aeronaut announced his intention of taking to the air on November 27. The gates were to open at noon and the ascent was to take place at 3:00 P.M.

By noon on the appointed day, the apparatus used to inflate the balloon was in place. A number of casks, connected by leather pipes, surrounded the two uprights that supported the limp envelope. The balloon had a 10,000-cubic-foot capacity, which meant that two and a half hours would be required to inflate it fully. As afternoon passed into evening, however, it became apparent that the balloon was not filling properly. By dusk, Kirkby, forced to admit defeat, distributed "checks" that would admit the bearer to a second attempt. The problem lay in the generating apparatus, not the balloon. Laboring through the night, he attempted to seal the casks so that the precious gas would not leak out of the pipes.

On the morning of November 28, Kirkby offered a public apology and promised that a second attempt would be made at two o'clock that afternoon, "at which time the public may rest assured it [the ascension] will positively take place." That afternoon, as those who had paid the day before filed into the amphitheater, a crowd began to gather outside where they could have a fine view of the proceedings once the balloon was launched. When it became apparent that the gas generator was still not functioning as it should, Kirkby again distributed checks to the paid audience and promised a successful ascent in a few days. The crowd outside was unwilling to allow a second failure to pass so lightly however. Some had been standing in the cold since noon without benefit of the band music inside. Realizing that the aeronaut was about to give up for a second time, they refused to allow anyone to leave the amphitheater. Although the crowd was inspired by a "determination to level every thing connected with it [the balloon]," Kirkby was able to convince them of the folly of mob action, and they dispersed.

The next morning the mob collected outside the amphitheater again, threatening to destroy the balloon and the apparatus. The timely arrival of Mayor Samuel W. Davies and a squad of nineteen officers saved the aeronaut and his equipment from a fate already familiar to unsuccessful balloonists. Mayor Davies assured the crowd that an ascent would take place and that on the occasion either he or John J. Wright, a prominent auctioneer, would take a place in the car with Kirkby. The mob, feeling that this demonstration of faith was sufficient guarantee, broke up. Nevertheless, the nineteen men stood guard all that night.[2] On November 29, in a considerably less confident tone, Kirkby offered a public apology.[3]

A week later the aeronaut "respectfully informed" the citizens that the fault had indeed been in the generator and that the ascent would take place in the middle of the coming week. Kirkby had solicited the aid of a number of "scientific gentlemen," including Drs. Slack, Flagg, and Riddle of the medical college. With their help he built an entirely new generator and announced December 15 as the date of the next attempt. A cannon was to be fired on the half hour from nine o'clock to three o'clock on the day of the ascent. Every precaution was taken to avoid a repetition of the fiasco of November 28. The public was assured that "city officials will be present to preserve order." To avoid confusion, persons holding checks were to redeem them at the office of Esq. Harrison before coming to the amphitheater.[4]

By three o'clock on the afternoon of December 15, "the largest [crowd] that we have ever seen collected in this city" had gathered to witness the spectacle. It seemed for a time that Kirkby would again disappoint the spectators, for the balloon was still far from fully inflated when the scheduled ascent was to take place. The crowd became restless and began to "talk about using the poor man up, because he was unable to peril his life for their entertainment." A small striped balloon was released to quiet the mob, but it soon became apparent that if the ascent were not made shortly, real trouble might ensue.

Fortunately, it was not long before Kirkby's balloon began to swell. The aeronaut climbed into the car and called for the restraining lines to be released. To the amazement of the audience, and quite possibly to the performer's as well, the balloon rose slowly out of the amphitheater. The editor of the *Cincinnati Chronicle and Literary Gazette* described the reactions of the crowd: "'Drizzle me if he ain't off,' muttered a disappointed rioter who thought he had not been fairly used inasmuch as he came there on purpose to have a row—'Well,' said a grey-headed son of the soil at our elbow, 'I've seed a mighty chance o' things in my day, but nothin quite so pokerish curus as that.'" Rising steadily now, Kirkby waved the Stars and Stripes and accepted the shouts of encouragement offered by the crowd below. The balloon resembled a "brilliant star" shining in the late afternoon sun as it disappeared to the east.[5]

The wind, which was blowing from the southwest at take-off, had shifted to the west, carrying Kirkby over the bend in the Ohio River. Approaching Columbia, east of Cincinnati, he had already reached his maximum altitude of 2½ miles. From this height, the village appeared as "a confused mass of buildings with no discernable outline." In order to fight the cold, Kirkby put on his overcoat and resorted to "a draft of generous cordial which I had in the car as a companion." "It was indescribably beautiful. The regularity of its plat, the bright light cast upon it by the setting sun, covering the roofs with apparently a tissue of silver, contrasted with the black lines which marked the streets running north and south and the sombre shades of those laid out east and west—the landscape of the country surrounding it drawn out in miniature, dotted by the cheerful hand of industry with innumerable farms;—the beautiful Ohio appearing like a silver cord carelessly thrown upon the picture."

In spite of this graphic description of the sights, Kirkby was in no position to enjoy the scenery. Although the balloon was no longer rising, it was spinning "in a constant whirl," and the aeronaut became airsick. As he passed between Milford and Batavia, he began to lose altitude rapidly. Realizing that the flight could not continue much longer, Kirkby prepared to descend in Clermont County. As he approached a large swamp three miles from Williamsburg, he brought his epic voyage to a successful conclusion in the top of a tree on the farm of one Samuel Riley. The farmer and a number of his neighbors arrived soon after and were able to extricate the balloonist and his equipment from the branches. The flight had covered a distance of about 31 miles in slightly less than an hour.[6]

In the wake of this successful ascent, Kirkby was referred to as a man of science and his voyage described as a "beautiful and sublime spectacle." He was, however, in financial trouble. The expense of the balloon and equipment, the construction of the amphitheater, and the acid and iron scraps, as well as the necessity of readmitting those disappointed by the abortive attempts and the cost of an entirely new generator, had taken what small profit he might have expected. Now several hundred dollars in debt, Kirkby welcomed the opportunity to recoup his losses with a second ascent.[7]

He announced that, weather permitting, the next flight would be made on Christmas Day 1834. As on the previous occasion, a cannon would be fired at half-hour intervals to inform the public that the ascent would take place. It was hoped that the balloon would develop sufficient lift to permit Dr. Riddle to accompany Kirkby so that he could conduct scientific tests in the upper atmosphere. Unfortunately, the sky was overcast on the 25th and 26th, but the 27th dawned cold and clear. The extreme cold kept attendance to a minimum, and preparations for the ascent proceeded without incident. Dr. Riddle was disappointed however; in spite of the trouble-free inflation, the balloon refused to leave the ground with both men aboard. Kirkby then decided to go alone. Free of the restraining lines, the aerostat made a rapid vertical ascent, describing a half circle over the city at an altitude of three-quarters of a mile before disappearing behind a range of hills to the east. Kirkby remained in view for thirty-seven minutes. He brought the balloon to rest in a soft, plowed field 2 miles from Milford after a flight of 13 miles.[8]

Although he had now completed two successful flights, the aeronaut remained "poorly remunerated for his trouble, expense and risk." His second Cincinnati ascent was also his last in the state. While he is reported to have made a flight in Louisville on March 7, 1835, no further mention of him is to be found in Cincinnati newspapers.[9]

The achievements of Thomas Kirkby were soon overshadowed by the spectacular flight of Richard Clayton from Cincinnati to Monroe County, Virginia, in April 1835. Compared to this flight, which set a world distance record for free balloons, Kirkby's ascents seemed insignificant and his pioneering efforts were soon forgotten.[10]

Clayton, a native of England, was born in 1811. Immigrating to the United States, he established himself as a watchmaker on the corner of Sycamore and Second streets in Cincinnati in the early 1830s. Immediately following Kirkby's flights, Clayton announced that he would attempt a voyage "of unusual length" on April 8, 1835. He obtained permission to use Kirkby's amphitheater for the inflation and launching of his *Star of the West*, "the largest and most splendid silk Balloon in the United States, and the first ever constructed West of the Mountains." Clayton devoted far more time to preparations than had Kirkby and was able to avoid the delays that had plagued the earlier aeronaut. The balloon, of "his own manufacture and made entirely in this city," had been carefully constructed of 4,500 square feet of the finest oiled silk, varnished to make it as airtight as possible. Fully inflated, the *Star of the West* stood almost 50 feet high, with a gas capacity of 18,000 cubic feet capable of producing 1,000 pounds of lifting power. The car was a splendid creation of wicker covered with watertight canvas and draped with blue silk, decorated with pink and gilt tassels and a pink silk lining. In an effort to draw larger crowds, Clayton advertised that from an altitude of one mile he would "let down a Parachute, 125 square feet in surface, containing a living animal, which will descend with safety to earth."[11]

Shortly before five o'clock on the afternoon of April 8, Clayton ordered the restraining lines released, and the *Star of the West* rose slowly out of the amphitheater as the aeronaut waved a small American flag to the crowd below. Beneath the balloon car was a small basket containing a dog, and as the balloon ascended the animal's loosely trailing parachute could be clearly seen. While still over the city, Clayton untied the lines holding the parachute and allowed the dog to fall free for a few moments until the canopy billowed open. The dog was retrieved just east of Main Street by two men who had followed his descent, and was borne in triumph back to the amphitheater. The animal's owner was later offered a large sum of money for him but declined, commenting that he felt the dog was popular enough to run for vice president.[12]

Clayton remained aloft all night, traveling 350 miles from Cincinnati to Monroe County, Virginia. The area where he landed in a tree was later renamed Clayton, Virginia (later West Virginia), in his honor. The aerial voyage, which set a world distance record, was carried as a major news item all over the globe, and the man who had elevated the Queen City to new fame was lionized by its citizens. No task now seemed too difficult for this ingenious mechanic. The recent appearance of Edgar Allan Poe's classic tale "The Unparalleled Adventure of One Hans Pfaall," in which a luckless Dutchman journeys to the moon in a balloon constructed of old newspapers, coupled with the publication of the *New York Sun*'s "Moon Hoax" articles, prompted the editor of Cincinnati's *Western Monthly Magazine* to suggest that Richard Clayton attempt a unique missionary expedition:

> *If . . .* Mr. Clayton *will undertake to prosecute a voyage to the newly discovered regions in the moon, and open a correspondence between its amiable inhabitants and those of our own globe, we will back him against a world of infidels. . . . He may easily engage the services of a few schoolmasters and schoolmistresses, from a neighboring part of our continent, who, for the purpose of doing good, would make sacrifices, in order to teach our language to the benighted people of that strange land, . . . instruct the poor lunatic heathen how to erect splendid dwellings, and magnificent churches, and furnish them with the best lawyers, doctors, schoolmasters, and divines. . . . Excellent books might be prepared for the instruction of the lunarians [and] their language doubtless needs to be corrected.*[13]

Another Cincinnati writer remarked that while a French and an English aeronaut had traveled higher, experienced greater danger, and performed more meaningful scientific experiments, "the voyage of Mr. Clayton . . . in a practical point of view—the only aspect in which the subject should be considered—[is]

of greater result than either of them."[14]

On the wave of the enormous public response to his flight, Clayton announced that he would next attempt a voyage to the Atlantic Coast. On May 13, 1835, the "prince of aeronauts" stepped into the *Star of the West* with a small bag of mail for delivery in the East. The basket, overloaded with ballast, refused to rise swiftly enough. Before Clayton could release the extra sandbags, the craft careened into a building and deposited him on the roof.

Public donations enabled the aeronaut to construct a second balloon, and at 5:45 P.M. on July 4, 1835, he launched his craft for the East Coast once again, carrying a packet of letters entrusted to him by the Cincinnati postmaster. By ten o'clock he had flown straight into the teeth of a storm. "Water poured down upon me," he recalled later, "the blankets and sheets, ballast bags, mail bag, and everything on board were wet."[15]

Clayton rose above the storm into weather which, "though drier, was much more unpleasant."[16] He struggled to read his thermometer and barometer with the aid of a phosphorus lamp as the temperature plummeted: "The silk had become hard as wood, icicles hung from the valve rope; I stood in wet clothes upon blankets and sheets that were stiff with ice, and if ever I felt cold in my life it was at this time."[17]

Ill, soaked, and frozen to the bone, Clayton struggled to remain awake, stamping his feet and beating his arms and legs. By 1:00 A.M., everything was going overboard as the gas cooled and the balloon sank lower. Having cast out all the sand ballast, six bottles, "and their contents and my possessions," the anchor and lines, blankets, sheets, his greatcoat, and instruments, the aeronaut knew that if he could remain aloft until sunrise the balloon would rise once again in the heat of the day.

But his efforts were to no avail. The balloon slowly descended straight down into the top of a grove of trees. Exhausted, Clayton secured his craft to the upper branches and "laid myself down in my cold disagreeable berth, without any great coat, blankets, or drink, to warm me, and with nothing to gaze upon but darkness, or to listen to but frightful noises in the woods."[18]

The following morning he was able to maneuver the balloon from branch to branch, finally reaching an area where he could descend twenty feet to the ground on a rope. After a mile walk through the woods he reached the home of a Mr. Bryant, who informed him that he had come to earth 18 miles from Chillicothe in Peeble County, Ohio. *The Star of the West* remained aloft, "ornamenting the trees," until an area of woods was cleared the next day and it could be safely retrieved. Clayton deposited the mail, which seems to have been the first ever carried aboard an American balloon, in the post office at nearby

Waverly.

Late in 1835 the "prince of aeronauts" embarked on an extended tour of the South. Christmas 1835 found Clayton preparing for an ascension from New Orleans. This was to be an experimental flight, the first time that an American balloon would be carried aloft by city-illuminating gas.

The bright, yellowish light of the gas jet was already inaugurating a new era in urban life. Jan Baptista van Helmont had obtained a flammable gas from burning charcoal as early as the seventeenth century. By the 1790s William Murdock, one of James Watt's engineers, had lit factories and public displays with the gas produced by reducing coal to coke in a sealed iron retort. It remained for Frederick Albert Windsor, a German showman living in London, to popularize gas lighting and to establish the London Gas Light and Coke Company during the early years of the nineteenth century.

Gas lighting was demonstrated in Philadelphia and in Newport, Rhode Island, by 1812. Within four years both Baltimore and New York had installed experimental gas plants. In 1827 the "Great White Way" was a reality, with gas street lights stretching from the Battery to Grand Street.

Illuminating gas was a boon to aeronauts as well. The English balloonist Charles Green had first flown from London with coal gas in 1821. Green realized that the product of Windsor's Westminster and London Gas Light and Coke Company, while heavier than hydrogen, would radically speed inflation, would do far less damage to the envelope than the older acidic home-brewed gas, and would penetrate the balloon fabric much more slowly. City gas was also six times less expensive than hydrogen and had a higher specific heat than either air or hydrogen, so that it would react more slowly to changes in temperature. In other words, a balloon filled with coal gas should fly farther than one filled with hydrogen.[19]

It seems likely that James H. Caldwell was responsible for Clayton's experimental flight with illuminating gas. An actor, theatrical producer, and entrepreneur, Caldwell had established the first American theater in New Orleans. He was an early convert to gas lighting and had imported the necessary retorts and fixtures required to illuminate his theater. Early in the spring of 1833, Caldwell persuaded the Louisiana legislature to grant him the exclusive right to produce gas in New Orleans. His New Orleans Gas Light and Banking Company was in full operation by 1835. Not satisfied with having illuminated New Orleans, Caldwell had led the fight to introduce gas lighting to Memphis and Cincinnati as well. It seems likely that he met Clayton during a visit to Cincinnati in 1835 and convinced him to follow Green's lead in flying with city gas.

Clayton's flight was a complete success. He traveled 25 miles from New

Orleans to the island of Petit Coquille, near Fort Pike, where he "came near to being devoured by Gallinippers, who swarm there in immense quantities."[20]

On March 2, 1836, he watched his friends "dwindle to Lilliputian size" as he made his eighth "Grand Ascension" from Natchez, Mississippi. On this occasion he flew 15 miles to a landing at Lebanon Springs.[21]

Clayton continued to make news with several Cincinnati ascents from April through July 1837. When the *New York Gazette* advised him to retire from ballooning "and go to work at some *honest* employment," local newspapers were quick to leap to his defense.[22] "We will inform our friend on the *Gazette*," one Cincinnati editor remarked, "that Mr. Clayton is one of our honest and industrious mechanics, having carried on the watch making business in this city for many years. His first attempt at ballooning was in Cincinnati; but neither that, or any of his subsequent adventures, have been at the expense of his business."[23]

He flew from Louisville's Woodland Garden on July 31, 1837. Clayton was angered to discover that the thousands of spectators ringing the garden prevented paying customers from entering. After only forty-five minutes in the air, he descended on the Churchill farm, 4 miles beyond the city limits. After a restful night he reascended the next morning and was carried back over Louisville to a landing near Bardstown.[24]

The Louisville flight reinforced Clayton's confidence in the *Star of the West* ("which is the best balloon that was ever constructed in the United States") and convinced him that it was quite possible to remain in the air for up to a week at a time and travel thousands of miles.[25]

The aeronaut next flew from Allegheny, Pennsylvania, on August 31, 1837. Clayton had constructed a 3,000-seat amphitheater for the event and announced that two tons of sulphuric acid, 3,000 pounds of "iron turnings," and ten tons of water would be required for the inflation.[26] Roughly an hour after his 5:00 P.M. launch, Clayton was observed by passengers on a canal boat near Johnstown. The group, which included "several gentlemen of taste and science," was "electrified as they watched the balloon rise above a thunderstorm." "The wind and flackes [*sic*] of lightning were playing around him, illuminating his path through the air."[27] One member of the canal boat party remarked that he "would willingly forsake this tedious earth to be with him a while in a purer world midst earth and sky."[28]

As the balloon passed over the Pennsylvania Canal, the "emotions" of the aeronaut's "bosom" would have surprised the rhapsodic observers below: "The cold was now intense, and the water which was condensing from the gas and running down the neck of the balloon froze into circles, and hung around the silk and valve cord. This was a severe and disagreeable situation for me, who had been sick for nearly three weeks, fed upon gruel and tea, and had only the day before stole from my sick chamber, contrary to the wish of my physician."[29]

Clayton, noting the rapid approach of the thunderstorm, made a safe landing on Laurel Ridge, some 8 miles from Johnstown. Before he could set off in search of a place to spend the night, however, the wind carried the tethered balloon through trees and over rocks "until she struck with dreadful violence against the strong limb of a tree." Clayton dangled fifteen feet from the ground in his basket:

> *The storm continued to increase, the rain pouring down in torrents, night came upon me, and no one came to my assistance. I now had to make my bed for the night in my little car, which is only five feet long and two feet wide; this mode of spending the night I should not have disliked had it been fine weather, but the rain poured down, and perfect darkness existed. Occasionally a flash of lightning would show me my dreary situation, and to heighten my pleasure, a clap of thunder would ring through my ears. Here I lay with my clothes wet through, and every thing around me saturated with rain, anxiously wishing for morning to arrive.*[30]

When finally rescued the next morning, Clayton discovered that some "ignorant and superstitious residents" of Laurel Ridge had observed his landing the night before. Taking the balloon for some monster riding in the teeth of the approaching storm, they had not come to his assistance. The aeronaut, nevertheless, made a second flight from Allegheny on September 13, landing in Fayette County, some 4 miles from Brownsville, Pennsylvania.[31]

Disaster struck Clayton on the occasion of an ascent from Louisville the following spring. While inflating his balloon at 2:30 on the afternoon of April 9, 1838, escaping hydrogen was ignited by a spectator's cigar. The explosion injured five persons, including Clayton. The noise, remarked the Louisville *Gazette*, was "like the discharge of artillery. . . . The balloon was instantly burned, and the fire was communicated to the wooden portion of the cistern containing the water and other ingredients from which the gas was manufactured. We were present at the scene about an hour afterwards, when fire was still issuing in flashes from all parts of the surface of the water with a sound like the faint rumbling of distant thunder."[32] Clayton's loss was estimated at $1,500. The aeronaut's injuries cannot have been too serious, for he was back in the air on June 13, flying from Cincinnati.

Richard Clayton continued to fly for a few more years. On the evening of July 31, 1839, he made another ascent from the Pittsburgh area. As late as July

1842, he made the first balloon flight in Columbus, Ohio. As had so often been the case, the experience was something less than pleasant for Clayton:

> *A great quantity of water poured down upon me from the neck of the balloon. This water was taken into the balloon in the form of vapour when the gas was generating and afterwards, when it was exposed to extreme cold, condensed and fell in copious showers of rain upon me. Being drenched with water . . . I felt extremely chilly and rather sick at the stomach; the sickness was occasioned partly by inhaling a goodly quantity of hydrogen gas. A teaspoonful or two of brandy and a little excellent cake, prepared by a fair friend of mine, restored me to my proper feelings.*[33]

As late as 1849 advertisements continued to identify Clayton's watch shop as the "Balloon Store." By the 1850s the reference was dropped from advertisements. Clayton's stellar career as an exhibition balloonist was over. His name disappears completely from the Cincinnati directories after 1859.[34]

A host of other aeronauts had followed Richard Clayton into the sky during the decade of the 1830s. Some of these men and women had short and relatively undistinguished aeronautical careers. Mr. S. Hobart, for example, made a flight from Lynchburg, Virginia, in September 1835, which newspapers aptly characterized as a "perilous and foolish voyage."[35]

Hobart flew directly into clouds and was out of sight for most of the flight. The aeronaut faced a myriad of dangers, both real and imagined. At one point he feared a collision with "two brilliant meteors." Both disappeared "without coming into contact, very much to Mr. H's relief, who feared the ignition of the gas."[36] Hobart should not be taken to task too severely for his scientific ignorance. So experienced an aeronaut as John Wise also feared being struck by a meteor or comet.

Having recovered from his close encounter with the meteor, Hobart flew into a gale, "the balloon and car being whirled to an immense height." He claimed to find breathing difficult and "feared the balloon would burst, and that some of his blood vessels would be ruptured."[37] Hobart yanked wildly at the valve line and descended precipitously into the top of a pine tree. The aeronaut was "hurled from the car with considerable force, and was severely jarred by the fall, breaking his valuable barometer, and damaging some of his other instruments."[38]

Flights like Hobart's led some American newsmen to suggest that "the ballooning mania will be arrested one of these days by the dreadful sacrifice of human life." Hobart was not to sacrifice his life however. He continued to fly during the next decade. Perhaps his best-known ascent of the period took place on October 9, 1841, when, flying from St. Louis with a young woman passenger, he was forced to climb onto the load ring to free a tangled valve line.[39]

A number of unsavory rogues tried their hand at ballooning prior to the Civil War. Henry Munro, for example, variously described as a "famous aeronaut" and "the hero of a recent ascent from Brooklyn, New York," was arrested in August 1837 for the theft of a $130 watch from Mr. George C. Howe of Chathorn Street.[40]

But if some of the emerging aeronauts were inept bunglers or scoundrels, others were airmen of extraordinary skill and exemplary courage. These were the men who would dominate aerostation in the United States for the next half century. Among their number were to be found some of the greatest airmen in American history.

William Paullin was one of these. He was born in Philadelphia on April 3, 1812, but very little is known of his early years. According to family tradition, he began work on his first balloon at twenty-one and made his first tentative flight in August 1833.[41]

Philadelphia newspapers reported "another" Paullin ascent in August 1834. "We seldom remember such an outpouring of the people," remarked one newsman. Paullin, like Clayton, dropped a dog to earth by parachute before landing 2 miles from Gloucester, New Jersey.[42]

An aborted ascent from Camden, New Jersey, in early June 1836 apparently created serious problems for Paullin. When he did fly on June 21, from Philadelphia, the crowd "manifested such symptoms of dissatisfaction that . . . Mr. Paullin thought it prudent to leave his patrons at a very early hour." The spectators trailed the aeronaut up Market Street "and occasionally noticed him by a volley of hisses."[43]

On July 27, 1837, Paullin went aloft from the Philadelphia Gas Works near the Schuylkill River. It was an experimental flight, "a private balloon ascension," as the *National Intelligencer* styled it. Paullin was interested in testing the lifting quality of city-illuminating gas, apparently unaware of Clayton's similar experiment in New Orleans nineteen months before.[44]

One editor was bemused to consider that balloons had become so common that an aeronaut like Paullin could actually wish to avoid publicity. The *National Intelligencer* reported that all Philadelphia had been surprised at the spectacle "of a Balloon sailing majestically over the city with an individual passenger in the car appended to it."[45] There were those who supposed "it to be Mr. Clayton, the adventurous aeronaut, come all the way from Cincinnati."[46] Philadelphia newspapers were quick to resolve the mystery the next day, noting that Paullin, "after passing somewhere between the planets Saturn and Venus," had returned safely to earth near Mount Hope, New Jersey.[47]

Paullin made an announced ascent from Philadelphia that September. Once again he inflated his balloon with "common gas such as we burn for light, obtained by an attachment to the main pipes of the gas company on Filbert Street."[48]

Hundreds of spectators entered the enclosure to watch Paullin tap the gas main. "In general, however," noted a visiting Boston reporter, "curiosity was gratified from the streets, the windows, the roofs, the trees, and each corner and vantage, into which or upon which, the human frame could be squeezed."[49] Paullin landed safely near League Island after an uneventful flight of fifty-five minutes.

The aeronaut continued to gain experience and build his reputation over the next several years. Paullin made one of his best-publicized ascents of this period when he and John Wise staged a dual launch from the courtyard of the Pennsylvania Farmer's Hotel in Philadelphia on July 4, 1840.

John Wise was born in Lancaster, Pennsylvania, on February 24, 1808. His father, William Weiss, married Margaret Trey on August 14, 1796. John Wise was the fourth of eight children, each of whom was educated in both German and English. As was customary in third generation Lancaster County immigrant families, young John anglicized his patronym. Rather than adopting the direct translation "White," he chose the phonetic pronunciation "Wise." Even so, old neighbors like Mathias Zahm would occasionally slip and speak of balloon ascents by John Weiss.[50]

Wise was educated in local schools. He was apprenticed to a cabinetmaker from age sixteen to twenty-one. As late as 1835 he was employed as a piano maker. Wise later claimed that he had first become interested in aeronautics when, at age fourteen, he read an article on ballooning in a German-language newspaper. He decided to construct a balloon while living in Philadelphia in the spring of 1835. In his own words, "When I first conceived the idea of making a balloon, I had never seen an ascension with one, nor had I any practical knowledge of its construction."[51] We can only assume that he was absent from Lancaster at the time of James Mills's ascent the previous year.

Wise began his aeronautical career by making a careful study of the atmosphere and contemporary balloon technology. His initial intent was simply to taste the pleasures of flight for himself without selling tickets or soliciting financial assistance. Wise therefore took care to keep expenses at a minimum.

The balloon would not be constructed of silk but of much less expensive domestic sheeting muslin. He eschewed costly dopes and varnishes in favor of eight pounds of birdlime suspended in linseed oil. After two coats of this home-brewed varnish, the fabric was cut into gores, sewn, and given a third coat to seal the needle holes. The finished balloon was 28 feet in diameter, had a gas capacity of 13,000 cubic feet, and weighed 186 pounds. Wise persuaded a local woman who normally produced fish nets to weave his network from ordinary cotton seine twine. His wicker basket measured only 2½ feet in diameter and the bottom was less than 4 feet deep.[52]

Having expended all his limited resources in the construction and outfitting of his balloon, Wise was forced to sell tickets to his first ascent in order to cover the cost of inflation. Fully aware of the long history of riots associated with unsuccessful ascents and with the admonition of his friends that he was "going right in amongst the butchers, . . . a very rough determined class of people," ringing in his ears, Wise scheduled his first flight for April 30, 1835.[53]

Stormy weather led him to postpone his aerial debut until May 2. An inexperienced ground crew literally threw Wise and his half-inflated balloon aloft from the empty lot on the corner of Ninth and Green streets at 4:00 P.M. "with considerable projected force."[54] Wise spent the next few moments bouncing off the neighboring chimneys and wildly jettisoning two bags of sand ballast. He came to earth again less than 400 yards from his starting point. Handing his coat, boots, and instruments to the bystanders, he bellowed a command to those who were holding the edge of the car and was off a second time on a flight that ended near Haddonfield, New Jersey, 9 miles from Philadelphia.[55]

Wise attempted a second ascent on May 18, 1835. While his back was turned, a barrel of sulphuric acid was spilled on the envelope, and the area around the valve was slit with a knife. Wise could only conclude that the damage "could not have been the work of accident or carelessness, but of design—the work of some malicious person unknown."[56]

July 4, 1835, found Wise venturing aloft from Lebanon County, Pennsylvania. His problems began almost immediately as the dangling appendix, through which expanding gas could escape, became doubled up between the envelope and the load ring. Unable to reach the valve line, which was now trapped inside the balloon, Wise watched in horror as the balloon swelled tight against the network.

While standing on the rim of his car attempting to cut a hole in the bottom of the balloon, Wise heard "a report like that of suddenly bursting a paper bag."[57] The tin neck pipe had punctured the envelope and cut through two of the netting cords, allowing the bag to bulge alarmingly through the gap. Wise was now making a "tolerably rapid descent" toward the village of Womelsdorf, Pennsylvania, where a county militia unit drilling on the green greeted the aeronaut

John Wise

with a volley of musketry. Tossing some papers overboard, Wise rose once again to avoid the milling crowd. He finally made a safe landing 4 miles from Reading.

Following two additional flights from Reading in August, Wise had enough money to construct a silk balloon. He scheduled one final launch of the old muslin balloon from Lancaster on October 1, 1835. Forced to rely on his usual ground crew of inept volunteers, the aeronaut was dashed against the eaves of a two-story house and spilled onto the roof. He watched, dazed, as his balloon plunged into "a chasm of dense black clouds." "Thus," he recalled, "ended the experiments with a machine that had given me much more trouble than reputation as a skillful aeronaut."[58]

Wise named his new silk balloon the *Meteor*. Completed in the spring of 1836, the *Meteor* was 24 feet in diameter, pear-shaped, and constructed of white sarsenet silk. Each of the gores was treated with a standard varnish of gum-elastic and turpentine. During its construction Wise discovered, as had Durant, that the spontaneous combustion of treated balloon fabric was a serious problem. A full dozen of the newly cut, doped gores of the *Meteor*, rolled and drying in a loft, were reduced to "a putrescent mass, emitting heat and smoke, and were rendered useless."[59]

The citizens of Lancaster, Pennsylvania, established a fund to pay for a free public ascent of the *Meteor* on May 7, 1836. "The day was extremely disagreeable," Wise remembered, "a succession of showers ushered in the morning's dawn, intervals of most undesirable brevity left little reason to induce the belief that an attempt would be hazarded—much less that, if made, it would be successful."[60]

In spite of the weather, Wise launched from the town common shortly after 5:00 P.M., having been preceded into the cloudy sky by several small balloons, including one shaped like "the Flying Dutchman."[61] Once in the air, Wise had a relatively uneventful flight, amusing himself by timing echoes, observing the clouds, and asking directions of those he passed on the ground before landing that evening between Bel Air and Port Deposit, Maryland.

With the *Meteor* stacked out in the yard of the Stump farm, Wise began the slow process of emptying the envelope and stuffing it into the basket. The vented hydrogen, hanging close to the earth in the humid air, was ignited by a lantern, "making a report like a pack of artillery."[62] Wise was thrown ten feet through the air. The blast set fire to the clothes of some bystanders. Others received severe burns to their exposed hands and faces.

Agonizing screams continued throughout the night as the burn victims received what little medical assistance could be provided. Wise expressed immediate concern for a group of blacks that had been standing near the balloon, then he lapsed into shock. "An agonizing pain . . . concentrated itself in my hands and face. I felt as though the very heart's blood was oozing through the skin, the watery fluid of the system oozed out in profuse drops."[63] Blind and in excruciating pain, he was moved to Port Deposit, then to Philadelphia: "By dint of blood-letting, wholesome diet, and the constant application of cooling cataplasms, I was out in ten days afterwards with a new skin on my hands and face, determined to make a new balloon, feeling satisfied in my own mind that all my sufferings were overpaid by the experience that I had gained in the adventure.[64]"

In fact, Wise was not all that anxious to return to the air. In the immediate aftermath of the horror of May 7, he began work on a 25-foot balloon dubbed the *Experiment*. Once again the varnish failed to dry properly. Wise, who was living at various spots in Lebanon and Dauphin counties, Pennsylvania, at the time, struggled to salvage the envelope but was finally forced to release it unmanned. "Pecuniarily bankrupt in the business, and almost so in reputation as an aeronaut," Wise retired to Philadelphia, where he found employment as a scientific-instrument maker.[65]

He found it impossible, however, to forget the thrills he had experienced aloft or to ignore his own visions of the future of aerial transportation. The death of the English experimenter Robert Cocking while testing an experimental parachute sparked Wise's return to aeronautics. He was convinced that Cocking's parachute represented a genuine advance and that the tragedy could be ascribed to faulty workmanship and a poor choice of materials.

In the fall of 1837 Wise reviewed a letter from Mr. George Diehl of Lebanon County announcing that he had found the old *Experiment* envelope and would be happy to return it. Wise seized the opportunity to fly once more and, at the same time, to make a trial of the Cocking parachute.

He repaired the old balloon and enlarged it by adding a new equator section so that he could fly the balloon when it was filled with heavier carburetted hydrogen from the Philadelphia Gas Works. He made the first ascent with the rebuilt balloon from the corner of Filbert and Broad streets on September 18, 1837. Determined to prove that the Cocking parachute would work if sturdily constructed, he dropped a dog via a regular parachute and a cat aboard one of the small inverted Cocking parachutes. Wise noted that the hastily repaired seams of his balloon were opening and began his own descent into Chestnut Street as soon as he saw that the animals were safe. Swept against a three-story building, he dropped a rope to the crowd in the street, asking them to hold the balloon while he climbed through a window.[66]

His next ascent, in October 1837, was a private venture sponsored by a group of wealthy Philadelphians in honor of another visit by Black Hawk, Keokuk, and their retinue to the East Coast. This flight was particularly puzzling to Keokuk, who had not been present for the earlier Durant ascent. The Indian was curious about the speaking trumpet. Was it to be used in talking to the sky? Convinced that the "pale-faced *medicine man*" was not a humbug, Keokuk stepped back and allowed Wise to cast off on a voyage that carried him to a sawmill some forty miles from the starting point.[67]

Wise "gratified" the citizens of Easton, Pennsylvania, in the spring of 1838. The old balloon was now covered with patches and had become "fragile" as a result of the aging varnish. The aeronaut discovered that the envelope would no longer hold gas on the first trial. His second attempt, on May 26, 1838, was more successful, ending at Schooley's Mountain, 20 miles from Easton. When the balloon was pulled from its box for a third ascent that July, it was found to be so hot that it was immediately thrown in a horse trough to prevent spontaneous combustion.[68]

Exasperated, Wise now began a series of experiments with different sorts of varnish. He determined that linseed oil alone served as an acceptable treatment for balloon fabric. Stuck in Easton, the aeronaut began construction of a cambric muslin balloon, reasoning that this material was easier to obtain, would be simpler to handle, and required fewer stitches than silk. Moreover, the absorbent muslin could be soaked in the linseed oil rather than painted, a much more time-consuming process.[69] The balloon was finished in record time. Twenty-four feet in diameter, it differed from professionally made balloons in that the top of the fabric surrounding the valve, where the strain was greatest, was not doubled.

Wise cast off from Easton in his hastily built balloon at 2:00 P.M. on August 11, 1838. As had so often been the case, a severe thunderstorm was brewing as Wise took off. He released two more animals in parachutes (one normal, one a Cocking model), then prepared to conduct another experiment. Still fascinated by parachutes, he had designed this new craft with a special line which when pulled would immediately empty the gas from the envelope. If all went according to plan, the bottom of the bag would gather in the upper hemisphere, forming a parachute.[70]

Considering the thunderstorm brewing beneath him, Wise had already decided to forego the experiment, when the expanding envelope pulled the cord taut and, in Wise's words, "the balloon exploded":[71]

> *Although my confidence in the success of the contrivance never for a moment forsook me, I must admit that it was a moment of awful suspense. The gas rushed from the rupture with tempestuous noise, and in less than ten seconds*

BALLOON ASCENSION.

The cut herewith represents the late ascension of Mr. John Wise, at Philadelphia, in his balloon Hercules. It was a grand affair, and was sketched for us on the spot. The cost of the balloon and rigging was $2600. It was manufactured of prepared silk. Its size is immense, and said to be the largest ever made in this country. It is capable of containing 41,000 cubic feet of gas. At five minutes past six o'clock on the day of ascension, about 37,000 cubic feet had been obtained, when Mr. Wise, not wishing to weary the patience of his friends, disconnected the tube from the balloon, and prepared for a departure from *terra firma*. At a quarter past six o'clock, a topical ascension was made. The *voyageurs* were Mr. Wise, his wife and son, Miss E. Denton, and W. R. Stockton of Spring Garden. The balloon rose gracefully to the height of over one hundred feet, and remained stationary for a few minutes. It was then drawn down by means of a windlass to which the end of the rope was affixed. At half-past six o'clock, the rope was cut, and the balloon, with the same persons, shot upwards, and continued to rise to a great height, perpendicularly. It afterwards took a north easterly direction, and was perceptible to view for nearly an hour. The audience within the enclosure was entirely orderly, and expressed the greatest approbation of the skill and success of the æronaut. A band of music was provided for the occasion, and contributed to fill up the tedium incident to the filling up the huge vessel. The streets in the vicinity of the yard were filled with persons, a large proportion of whom were females, accompanied by children. As the cause of the delay was not communicated to the "outsiders," much dissatisfaction was felt by them—but those who stood their ground, despite of the piercing rays of the hot sun, when the ascension did take place, went on their ways rejoicing, from having viewed such a perfect and interesting balloon ascension. The æronaut and his company landed safely about five miles north-east of Camden. Our Philadelphia friends have generally paid much attention to the subject of aerial navigation, and the Alleghany Mountains were crossed in this manner as early as the summer of 1837. It is said that Lucifer himself is the "Prince of the Air," but we shall not be at all surprised to see his dominions invaded by some enterprising Yankee in a profitable style of travel. In Pennsylvania, the Dutchman and the Yankee seem to be all the time trying their strength in honorable rivalry. Dr. Franklin's paper-kite led to the discovery of some very important first principles of science which have since benefited the whole world. Therefore we say to our scientific ballooning friends, who really desire the cultivation of utility,—Go on and prosper! Or, let them take the motto of New York, and cry out,—"Excelsior!" Our humble endeavor will be to aid in the publicity or illustration of all such flights of true genius. In discoveries or inventions of this nature, mankind are benefited not merely in one branch of science, but by the analogical theories and practical applications which arise from the general diffusion of its use. No reasonable mind can doubt that all the elements may be made subservient to our will. To man has already been given the means of locomotion upon the earth and upon the water; why not the air itself, so far as animal life can exist in a rarefied atmosphere, be now at our control? The time is close at hand when we should be as likely to think of living in a vacuum as not to travel in air.

AN AUTHOR'S WEAKNESS.

Shall I confess a weakness? The only set-off I know to these rebuffs and mortifications, is sometimes in an accidental notice or involuntary mark of distinction from a stranger. I feel the force of Horace's *digito monstrari*—I like to be pointed out in the street, or to hear people ask in Mr. Powell's court, *which is Mr. H——?* This is to me a pleasing extension of one's personal identity. Your name so represented leaves an echo like music on the ear; it stirs the blood like the sound of a trumpet. It shows that other people are curious to see you; that they think of you, and feel an interest in you without your knowing it. This is a bolster to lean upon; a living to your poor, shivering, threadbare opinion of yourself. You want some such cordial to exhausted spirits, and relief to the dreariness of abstract speculation. You are something; and, from occupying a place in the thoughts of others, think less contemptuously of yourself. You are the better able to run the gauntlet of prejudice and vulgar abuse. It is pleasant in this way to have your opinion quoted against yourself, and your own sayings repeated to you as good things. I was once talking with an intelligent man in the pit, and criticizing Mr. Knight's performance of Filch. "Ah!" he said, "little Simmons was the fellow to play that character." He added, "There was a most excellent remark made upon his acting it, in the Examiner (I think it was)—*That he looked as if he had the gallows in one eye and a pretty girl in the other.*" I said nothing, but was in remarkably good humor the rest of the evening. I have seldom been in a company where fives playing has been talked of, but some one has asked, in the course of it, "Pray, did any one ever see an account of one Cavanagh, that appeared sometime back in most of the papers? Is it known who wrote it?" These are trying moments. I had a triumph over a person, whose name I will not mention, on the following occasion. I happened to be saying something about Burke, and was expressing my opinion of his talents in no measured terms, when this gentleman interrupted me by saying, he thought, for his part, that Burke had been greatly over-rated, and then added, in a careless way, "Pray, did you read a character of him in the last number of the——?" "I wrote it!"—I could not resist the antithesis, but was afterwards ashamed of my momentary petulance. Yet no one, that I find, ever spares me.—*William Hazlitt.*

BALLOON ASCENSION, WITH A CAR CONTAINING FIVE PERSONS, AT PHILADELPHIA.

John Wise ascends from Philadelphia with four passengers.

> *not a particle of hydrogen remained in it. The descent at first was rapid, and accompanied by a fearful moaning noise, caused by the air rushing through the networks, and the gas escaping above. In another moment I felt a slight shock. Looking up, . . . I discovered that the balloon was canting over, being nicely doubled in, the lower half in the upper.*[72]

Wise breathed an enormous sigh of relief as the contraption oscillated safely back to earth. Following a normal ascent from Allentown, Pennsylvania, on September 8, 1838, he was off to Philadelphia for another demonstration of his exploding balloon on October 1.[73]

Wise made several more ascents from Allentown during the flying season of 1839. Late that year he allowed Wellington Dunlop, "a young man of Berks [Bucks] County," to make a solo ascent in his balloon. "Mr. Dunlop," Wise remarked, "is under the impression that his trip was a greater benefit to his health than all the medicine he has taken for ten years past."[74]

William Paullin and John Wise, now two of the nation's most experienced aeronauts, arranged their dual ascent from the Pennsylvania Farmer's Hotel in Philadelphia on July 4, 1840. Wise had sold his old cambric balloon, the veteran of a dozen ascents, to an amateur. He was now the proud possessor of a new silk aerostat, 23 feet in diameter. Discovering on the day of the flight that his new balloon would not lift his weight when inflated with the relatively heavy city gas, Wise discarded his basket and made the flight standing on a board.[75]

Wise cast off first, followed by Paullin a few seconds later. The two men conversed easily as they jockeyed to see which of them could climb the highest. Paullin won the contest and the two balloons gradually moved apart to make safe landings.[76] Wise and Paullin were to remain friends and associates for many years. In the immediate aftermath of their dual ascent, however, their paths diverged.

Paullin sailed for Valparaiso, Chile, in September 1841. Over the next six years he would fly in several South American nations. One of his most harrowing experiences came during a flight over the St. Jago volcano. He was nearly suffocated by the rising fumes, fearing all the while that the heat and ascending embers would ignite the hydrogen.[77]

Paullin arrived in Lima, Peru, in the wake of a revolution. Peruvians had seen a number of small balloons at this point. Don Manuel Florez, an Argentine, had already flown a hot air balloon in Lima. Paullin, however, would be the first to fly what the Lima newspapers regarded as "a very modern" hydrogen balloon.

He made his first ascent from the Plaza de Achao on Sunday, June 19,

1842. As had so often been the case, he released small balloons, a fish and a snake, an hour before the flight. Six hundred people gathered in the plaza, with an additional 12,000 to 15,000 outside the fence.

Paullin climbed into his car ("the size of a child's cradle") at 4:45, waved his kerchief, and gave the order for release. Three minutes later he entered a cloud bank, then reappeared. Lima newspapers were filled with the comments of spectators who wished they could share his "magnificent view" and who were "envious of his solitude." High in the blue sky, remarked one journalist, "one's soul could imagine that it has passed the confines of the sepulcher."[78]

Exciting as it may have been, the flight netted only 280 pesos for Paullin, not enough to cover his expenses. The aeronaut's Peruvian sponsors were anxious for a second flight, on condition that it be effected in some more lucrative fashion.[79] Volunteers were appointed to comb the city for subscriptions. The effort was a success, and the repeat flight was scheduled for July 17, 1842. Two more small balloons were launched at 1:45. A large "Gulliver" figure, "as in the land of the Lilliputs," *Gulliver*, 11 feet tall, had been flown and recovered at several of the aeronaut's American ascents.[80] Paullin himself went aloft at 2:45. This time he added a new twist for his South American audience, "parachuting a dog from high altitude."[81] The flight was short, only fifteen minutes in length.

In later years Paullin claimed to have flown in many South American countries, as well as in Cuba, Haiti, Puerto Rico, and Mexico. It is certain that he was back in the United States by 1847.

Wise and Paullin, of course, were not the only aeronauts barnstorming the Americas during these years. Louis Anselm Lauriat was their most active rival for public favor. Born at Maria Calante on the island of Guadeloupe, French West Indies, on September 13, 1785, Lauriat immigrated to Salem, Massachusetts, in 1806, where he supported himself as a teacher of French. One of his pupils noted her impressions of Lauriat in her diary. "He is a very pleasant man, and his conversation interesting," she remarked, adding that "he possesses a degree of diffidence unusual in Frenchmen." Not everyone would have agreed with this assessment; a Boston reporter, writing in 1853, remarked that Lauriat "talked incessantly upon a variety of topics."[82]

By 1822 Lauriat and his bride, Sarah Dennis Lauriat, were living in Boston. Eleven children arrived in rapid succession as Lauriat worked as a goldbeater, chemist, and assayer.

Unlike most of the others who took up ballooning during the decade of the 1830s, Lauriat was no longer a young man. Fifty years old at the time of his first ascent, Lauriat, nevertheless, remained active until 1848, by which time he had completed nearly fifty flights.

According to family tradition, Lauriat was present for Durant's two Boston ascents in 1834. Inspired to try his own hand at the game, he made his first ascent on July 17, 1835. Boston reporters were careful to note that his balloon was silk and "not one of gold beater's leaf," as might have been expected considering Lauriat's profession.[83]

Jeremiah Goodwin, who was visiting Boston from his home in Maine, recorded Lauriat's second ascent, made as part of the city's Fourth of July celebration:

> *The display in Boston to-day has been very imposing and brilliant—The military parades, musick, cannonading, bell ringings; the procession of the Unions of All Trades, with their banners, work-shops, ship, etc., all in operation while in procession; the ascension of a dozen small balloons from the foot of the Common, and then the elegant ascension of the large Balloon with Mr. Lauriat and his daughter; the tremendous host of persons on the Common, 40,000 or 50,000, and to close, the display of fireworks;—were altogether brilliant sights, worth seeing.*[84]

Late in July 1835 Lauriat made yet another ascent, from Providence, Rhode Island. It was, the aeronaut reported to a friend in the city, "a most beautiful excursion among the clouds."[85] He remained in the air for an hour and twenty-five minutes and landed only 19 miles from Boston.

Lauriat continued to make flights at an unprecedented rate. He was in New York for a flight from Castle Garden on August 20, 1835, landing on the farm of Judge Terhune after only thirty minutes in the air.[86]

His son accompanied him on an ascent from Troy, New York, on September 9, 1835. On this occasion the two aeronauts traveled 20 miles to Berlin, New York. On September 23 he was back in New York for a flight with his daughter Aurelia. The balloon was apparently heavy, for Lauriat was forced to make a precipitous landing on the rocks near the Battery shore. While Miss Lauriat was placed aboard a rescue boat, her father was towed back to Castle Garden for a solo ascent. His daughter returned to the residence of Mr. Marsh, the lessee of the garden, "where she received the congratulations of thousands at her providential escape from dangers which seemed to threaten her life."[87] Lauriat returned to the garden at 1:00 A.M., having landed safely at North Hempstead, Long Island.

Another ascent in Rochester that September brought tragedy to a group of more than one hundred spectators perched on the roof of a workshop. Inevitably, the building collapsed "with a tremendous crash, made horribly terrific by the fearful screams of those upon it."[88] This accident led some newsmen to question the wisdom of balloon ascents. "They are always attended with more or less mischief," noted one observer, "and never do any good."[89]

Lauriat was back in the air the next spring, flying from Boston on June 17, 1836. "The Amphitheater was pretty well filled," observed the *Daily Times*, "but the millions—the great bulk of the anxious and admiring spectators—did not deem it worth while to enter the Narrow pass."[90] All those present, inside or outside the gate, were soaked by a sudden downpour.

Lauriat continued to fly throughout the remainder of the summer and fall. He launched from Lowell, Massachusetts, on July 4, 1836, admiring the "most superb view of the city of Lowell and adjacent towns. The variegated verdure and richly cultivated surface of the earth, between here and Newburyport, presented at once to the eyes a scene of astonishing beauty and grandeur—looking like a vast garden interspersed here and there with little running brooks and small spots of water, for so did the Merrimack and numberless lakes and ponds appear at that elevation."[91]

Lauriat moved on to New York late that summer, flying from Castle Garden on August 28 and October 13, 1836. So it went for the next several years, a round of ascents from cities all over the Northeast. He flew from Boston on August 3, 1837; Salem, Massachusetts, on July 4, 1838; and Concord, New Hampshire, on September 21, 1838. On November 9, 1838, he was unable to ascend from Portsmouth, New Hampshire, and sent his ten-year-old son up alone.[92]

Lauriat ran into serious difficulties during an ascension from Chelsea, Massachusetts, on June 17, 1839. Attempting to land in a high wind, he was dragged over hills, through trees, across the surf line, and out to sea. He clung to his torn netting as the balloon was carried 30 miles toward Cape Ann. Finally rescued by Capt. John Pierce of Wellfleet, his balloon escaped through a gaping hole in the net "and went off on the wings of the wind with greater rapidity than a steam-engine, and was a total loss: it cost one thousand dollars."[93]

While Clayton concentrated on introducing the West and the South to the balloon and Wise and Paullin restricted their activities to the Middle Atlantic states, Lauriat was firmly establishing himself as the premier aeronaut of the Northeast.[94] He flew from the Colonnade Garden, Brooklyn, on October 9, 1840, and advertised an ascension from Boston on August 25, 1842, as his forty-fifth.[95]

Like Paullin, Lauriat also became something of an internationalist. He was, for example, the first aeronaut to fly in Canada. Charles Ferson Durant had toured Montreal during the summer of 1836. At that time Mr. Guilbault, the proprietor of a local botanical garden, had offered the aeronaut $500 to ascend from the city. Durant's refusal led the disappointed editor of the *Montreal Gazette* to comment that the "aerial navigator was more lofty in his demands than could be afforded from the support of a *Montreal* audience."[96]

It remained for Louis Lauriat to introduce Canadians to the excitement of a manned balloon ascent in the summer of 1840. The aeronaut traveled north to St. John, New Brunswick, aboard the steamer *North American*, which had begun making weekly runs to Boston the previous year. Lauriat, billed as a "Professor of Chemistry and Aerostatic Exhibitions," originally announced that he would ascend from the local Barrack Ground on July 23, 1840, then rescheduled the flight for August 10.[97]

Flying a balloon that he had named the *Star of the East*, Lauriat traveled over Loch Lomond, where 500 black Americans had settled following the War of 1812, to a landing 21 miles from St. John. The number of paying spectators was so obviously short of what Lauriat required to break even that the admiring Canadians passed a hat to collect an additional $100.[98]

Lauriat also traveled the path to Mexico that had been blazed by Eugène Robertson.[99] He opened negotiations for a Mexican ascent with two promoters, Wright and Lapham, in 1840. In June 1841 Lauriat informed his son Gustavus that he was negotiating a Peruvian tour with a Senor Palacios. The aeronaut hoped to sign a one-year contract with a $10,000 guarantee, but neither the Mexican nor the South American venture materialized.[100]

Lauriat was still active in the summer of 1842. That July, having completed an ascent from Castle Garden in New York, he was very nearly incinerated while waiting for the boat scheduled to return him to Boston to meet an exhibition date.

> *We had last night a miraculous escape from a dreadful explosion and conflagration. A rocket fir'd from the garden took a downward course and struck over the building containing a large quantity of gun powder and ready made fire works, igniting the center part where only a small quantity of the combustible was located, the explosion was instantaneous—and terrific; but fortunately did not communicate to the next room, and if it had, many lives, balloons, etc. would have been extinct. I was at the time taking a bath, swimming within 20 feet of the building. I made out to find my clothes for the bath house was envelop'd with dense smoke and when I reached the saloon where the Balloon was, the danger was over.*[101]

In 1848 as Lauriat's aeronautical career was coming to a close, *Boston Notions* offered an interesting summary of his thirteen years in the air, complete with a somewhat exaggerated catalogue of the dangers to be encountered aloft:

> *Mr. Lewis* [sic] A. *Lauriat says, that he has ascended in his Balloon 48 times from various places between the British Provinces and Mexico, and the highest altitude he ever attained, was at 24,500 feet, as measured by the Barometer and Revolving Index; that being 3,000 feet beyond the upper clouds: there the thermometer ranged from 12 to 15 degrees below freezing point, and at that elevation the air was so rarefied as to cause the gas in the balloon to expand nearly a third more in capacity than it was on leaving the earth: and the difficulty of breathing was such as to cause three times respiration to one below: his pulse before starting being at 70; rose to 110 a minute; causing small blood-vessels to swell and straining for vent, producing great pain in the forehead; at last streams of blood from his nose gave relief to his head; still, owing to the extreme lightness of the atmosphere and constant and free evaporation through the pores of his body created incessant thirst that water would be constantly desirable.*[102]

In 1849 Louis Lauriat, now fully retired from ballooning, was off on yet another great adventure, this time in company with thousands of his countrymen. He sailed for California, making the long voyage around the Horn and experiencing the obligatory storm while making the passage through the Strait of Magellan. He went ashore in Lima, Peru, finally visiting the city where he had planned to fly seven years before.

Once in California, the former aeronaut failed to strike it rich in the goldfields but was able to establish himself as an assayer and small businessman. Lauriat revisited Boston in the early 1850s but returned to Sacramento, where he died in August 1858. An obituary in the *Sacramento Union* provided an admirable summary of Lauriat's strengths and weaknesses:

> *Lewis* [sic] *Lauriat, an old resident of this city, an accomplished chemist, and, in his palmy days, a leading aeronaut, died at the County Hospital, about midnight on Monday night, in his ninety-second year. According to the record at the Hospital, he was a native of the West Indies. Since his residence in Sacramento, until a year or two past, he pursued the business of an assayer of metals; but within the latter period, became addicted to the use of deleterious beverages, which undoubtedly precipitated his decease. It is reported that, at an early date, he was elected Professor of Chemistry in the College of Marseilles, France. Some time prior to his decease, he informed us that he had made some sixty ascensions in balloons, and we remember distinctly that he was the first person we ever saw ascend. The ascent was made from Castle Garden, New York. The remains will*

be interred at nine o'clock this morning, from the County Hospital. Deceased was a member of the Masonic fraternity, many members of which recently exhibited some solicitude for his proper care and maintenance.[103]

One of Louis Anselm Lauriat's dreams would long outlive him. As early as June 1838, the *Boston Evening Transcript* reported that "Mr. Lauriat . . . [is] preparing to cross the Atlantic in a balloon."[104] Many others during the next century and a half would share Lauriat's desire to fly the Atlantic in a balloon. No one, however, was to pursue the elusive goal of the Atlantic with more enthusiasm than John Wise.

Wise had continued to fly from various Pennsylvania towns following his dual ascent with Paullin. In August 1840 he made one flight from Lancaster and two from Chambersburg. By the spring of 1843, he had added Danville, Lewistown, Wilkes-Barre, Bellefonte, York, Gettysburg, and Carlisle, Pennsylvania, to his growing list of launch sites. At this point he had made roughly fifty flights and was already generally recognized as the nation's most experienced aeronaut.[105]

Unlike many other balloonists, Wise had a serious interest in atmospheric science and nursed an honest desire to demonstrate the potential of the balloon to improve the human condition. For Wise, the "atmospheric shell" was "full of unexplored philosophy." It was, he believed, "the great laboratory of earthly life that gives us being."[106]

Wise genuinely loved the sensations of flight. The sky was a new world to be explored, full of new sights, strange phenomena, and a wealth of experiences that had been closed to human beings since the beginning of time: "The varied phenomena of inter-aerial conditions are not yet primarily explored. This ocean of life and light, covering the whole surface of the globe, accessible to a depth of five or six miles without let or hindrance, subject to all the variations that heat and cold can impart to matter, and giving us a means of transition from one part of the world to another, from the Orient to the Occident, and from Africa's sunny heat to polar icy cold, is worthy of a higher order of experience than mere spectacular sight-seeing in acrobatic performances."[107]

The dangers feared by the aeronaut were, Wise reckoned, much overstated. At any rate, all pioneers had to expect some hardship as the price exacted for opening new lands for the benefit of mankind:

"Ah!" says the inexperienced, "but how about the danger, the fear that must overtake one when suspended between heaven and earth, hanging as it were by cobwebs? It makes one shudder to look up at a person so situated." That is very

> *true, but the shuddering is all to those below, while those above are entirely freed from it as soon as the connecting cord is cut. As they sail upward, a delight bordering on ecstasy takes the place of fear; and this may be claimed as one of the strongest arguments that the air, as well as the water, is intended by kind and generous Providence to come under the dominion of man. It will become the stepping-stone to a higher civilization, and the Society for the Advancement of Social Science should not ignore it in their deliberations.*[108]

The notion of crossing the Atlantic via balloon as a means of focusing public attention on the potential utility of aeronautics first occurred to Wise during the winter of 1842-1843. The aeronaut had long ago become convinced of the existence of great currents in the sky, perpetually blowing in one direction as though they were rivers of air. "Like the sea, composed of a system of rivers traversing its great body throughout its length, breadth, and depth, the atmosphere is filled with a system of tides and currents as regular and periodic as the blood in its courses through the venous and arterial system of the human body."[109] "The ancients," he noted, had been "well acquainted with this order of nature, or they could not have stated it so tersely as it stands written in the first chapter of Ecclesiastes, sixth and seventh verses: 'The wind goeth toward the south, and turneth about into the north; it whirleth about continually; and the wind returneth again according to his circuits.'"[110]

During the winter of 1842 Wise discussed with scientific friends in Philadelphia the possibility of traveling to Europe in the clutches of a current of air flowing constantly to the east. Convinced that such a voyage was possible, he informed friends in Lancaster that he would visit them during the course of a flight that would start at Carlisle, Pennsylvania, on May 3, 1834. He would, he announced, be carried to Lancaster "via the *atomospherical current that always blows from west to east* in the higher regions of the air."[111] True to his promise, Wise landed on the outskirts of the city at 4:45 P.M. on the afternoon of May 3, having flown the 54 miles from Carlisle in less than two hours. As John W. Farney of the Lancaster *Intelligencer and Journal* noted, "He had *directed* his chariot with admirable generalship."[112]

Now there was to be no holding Wise back. He announced his plans for the ocean flight in his friend Farney's newspaper in June 1843.[113] The announcement was republished throughout the United States and Europe, exciting a varied response. There were those who viewed the project as "the effervescence of a disordered intellect." Others believed that "Mr. Wise may yet be destined to *soar above* the fame of such common men as Robert Fulton and Oliver Evans."[114] Wise even received an offer from two U.S. Navy midshipmen who volunteered to serve as navigators on the voyage.

The one thing Wise did not receive was the money with which to fund a transatlantic voyage. That being the case, the aeronaut submitted a petition to the United States Congress on December 20, 1843, requesting $15,000 to cover the costs of the flight.[115]

This was not the first time Congress had been petitioned to fund an aeronautical project. On March 25, 1822, Congressman Milnor of Pennsylvania introduced a petition on behalf of one James Bennett, a Philadelphia mathematician. The document declared that Bennett "is the inventor of a machine by which a man can fly through . . . the earth's air, can soar to any height, steer in any direction, start from any place, and alight without risk of injury."[116] Bennett sought a forty-year monopoly "to the right of steering flying machines through that portion of the earth's atmosphere which passes over the United States, or so far as their jurisdiction may extend."[117]

The petition was referred to a select committee whose members included Milnor and Representatives Golden, Stevenson, Fuller, and Poinsett. This was the first committee in congressional history to deal with an aeronautical matter.

No details of Bennett's machine are available, but there was apparently some question as to appropriate credit for the device. On April 1, 1822, Representative Keyes introduced a second petition, this one from David B. Lee, the would-be Philadelphia balloonist whose failure to ascend had led to the Vauxhall Riot of 1819. The Keyes petition, which was presented to the existing subcommittee, argued that Lee, not Bennett, was the real inventor of the flying machine in question. Both petitions were tabled and forgotten.

Twenty-one years later, John Wise's document suffered the same fate. As Wise noted, "this petition was received, read, and referred to the Committee on Naval Affairs where it slept."[118]

But Wise was not a man to be discouraged easily. The transatlantic flight would remain, as he remarked, "the dream of my lifetime."[119] Wise discussed the project at length in his first book, *A System of Aeronautics*, published in 1850. He assaulted the halls of Congress with a second petition for funding in 1851.

With the assistance of a friend, the aeronaut was successful in interesting a number of influential congressmen in the project. Two of these men, Sen. Stephen Douglas of Illinois and Representative Maclean of Maryland, introduced the new petition in the Senate and the House.

Wise requested a congressional appropriation of $20,000, which he would use to construct "a balloon of such material as will retain hydrogen gas for weeks and months, of a diameter of one hundred feet, which will give it an elevatory power of sixteen tons."[120] Wise would then make "numerous" ascents in the Washington area in order to demonstrate the military potential of the observation balloon. Having put his balloon on public display, he would make one long-distance flight from St. Louis to the East Coast.

The Atlantic flight, the final step in his program, would begin from New York. He planned to take six or eight "attendants," all of whom would share the cramped quarters of a lifeboat that Wise planned to convert into a balloon car.[121]

Senator Douglas presented the Wise petition as requested, commenting that the aeronaut was "a man of great intelligence and scientific attainments." Senator Bright, chairman of the Committee on Roads and Canals, to which Douglas suggested referring the bill, was far from enthusiastic. Bright's refusal to accept the petition drew gales of laughter from the senators. "I think," he remarked, "it is the province of this committee to look to roads and canals on *terra firma*; but when it comes to navigating the air, the prayer does not find a friend in me."[122]

The Foreign Affairs Committee likewise declined to consider the petition. It was finally sent to the Committee on Naval Affairs where it died a quiet death. As Wise remarked: "I speedily gave up all hope of obtaining assistance from Congress."[123]

Chapter 8 The Golden Decade

AN EXTRA EDITION OF THE NEW YORK SUN FOR APRIL 13, 1844, INFORMED readers that John Wise's "great dream" had been realized much sooner than anyone had supposed possible. At 2:00 P.M. on Tuesday, April 9, Sir Everard Bringhurst, Monck Mason, Robert Holland, Harrison Ainsworth, William Samuel Henson, and a Mr. Osborne, all names well known to English aeronautical enthusiasts, had landed the "steering balloon" *Victoria* near Fort Moultrie, at Charleston, South Carolina.

The headlines trumpeted the fact that "the Atlantic has been actually crossed in a Balloon!" "The great problem is at length solved," continued the article. "The air, as well as the earth and the ocean, has been subdued by science, and will become a convenient highway for mankind."

No great difficulties had been involved. The crossing had been made in complete safety with the "machine" under the complete control of the crew at all times. All of this had been accomplished "in the inconceivably brief period of seventy-five hours from shore to shore!" The account of the flight was full of convincing technical details and laced with exciting passages extracted from logs kept by Mason and Ainsworth.

The first doubts about the authenticity of the account began to appear as early as Monday, April 15. Even the *New York Sun* was quickly forced to admit the awful truth. "The mails from the South last Saturday night not having brought confirmation of the arrival of the balloon from England, we are inclined to believe the intelligence is erroneous."

It was more than erroneous. The story was a complete fabrication foisted upon the *Sun* by Edgar Allan Poe, a young writer who had arrived in New York from Philadelphia in a driving rainstorm on April 6. Broke, with a sick wife to care for, Poe had concocted the tale and sold it to the editor of the *Sun* as a straight news story.[1] Poe's balloon hoax became a legend in journalistic circles. Like so many successful hoaxes, the account was convincing not only because of the skill with which it had been contrived but because it met public expectations.

Aeronautics was a subject very much on the public mind. Clayton and Lauriat had both talked of flying the Atlantic. Wise had petitioned Congress for assistance in mounting just such a project. Charles Green, the English aeronaut who was an associate of many of the individuals mentioned in Poe's story, had recently published a pamphlet outlining his own plans for a transatlantic balloon voyage. In short, the readers of American newspapers were fully prepared to accept the news that a balloon had crossed the Atlantic.

The decade and a half that followed Poe's balloon hoax represents the high-water mark in the history of American aerostation. A legion of itinerant aeronauts, veterans and newcomers alike, crisscrossed the nation during these years, flying in cities from Maine to California. Universally honored as "Professor This" or "Madame That," their exploits were featured on the front pages of great newspapers and the new illustrated magazines such as *Harper's* and *Leslie's*. They were a breed apart, the first generation of footloose, barnstorming aerial showmen.

John Wise was to remain the best-known member of the group for the next forty years. He was a genuine professional now. After 1847 he no longer listed his occupation as instrument maker. One Lancaster neighbor recalled that he seemed to have no "regular employment" other than that provided by the balloon kept tethered behind the fence that surrounded a vacant lot near his home.[2]

Now fully dependent on the balloon for his livelihood, Wise was forced to expand his area of operation. In 1847 he flew from Utica (June 7) and Syracuse (July 17), before making two ascents from Buffalo (July 31—his sixty-second flight—and August 6). The season ended with ascents from Rochester (August 14) and Oswego.[3]

By 1851 Wise was allying himself with traveling tent shows touring the Midwest. That summer, for example, he flew for John M. Kinney, the proprietor of Kinney's Mammoth Pavillion, a Columbus-based show that visited towns and villages all over Ohio. In 1852 Wise flew twice from Columbus (July 4, October 2), as well as from Zanesville (October 2) and Cincinnati (October 9, his 127th flight).

He returned to the Kinney troupe the following year. An ascent from Portsmouth, Ohio, in the face of a thunderstorm (June 3, 1852) was followed by flights from Chillicothe (June 10); Circleville (June 17); Lancaster, Ohio (June 24); Mansfield (July 17); Minerva (July 31); Akron (August 7); and Cleveland (September 15–16).[4]

Having milked the Pennsylvania and Ohio crowds, Wise moved on to New

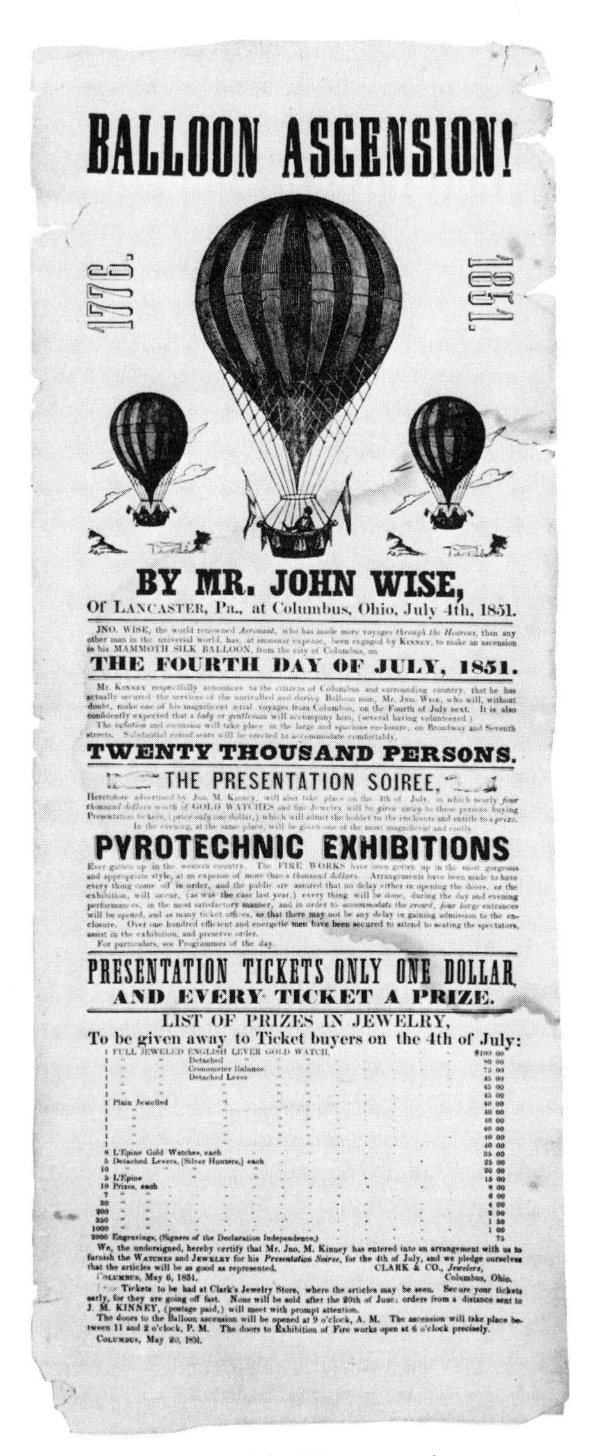

Poster announcing a John Wise ascent from Columbus, Ohio, July 4, 1851

England, where he concentrated his activity in the years 1855–1858. Throughout this period he continued to make occasional flights closer to home in the Lancaster and Philadelphia areas.[5]

Wise had, of necessity, become an experienced balloon maker. He moved rapidly through a series of envelopes he had made himself. The *Comet*, *Vesperus*, *Rough and Ready*, *Ulysses*, *Old America*, *Young America*, *Jupiter*, and *Gannymede* were but a few of the craft he built and flew during these years.

Wise not only sold new balloons to other aeronauts but frequently foisted off his own veteran craft on newcomers to the field. *Vesperus*, for example, having been repeatedly ripped, torn, and several times exploded, was sold to a South Carolina amateur.

In the fall of 1854, Wise sold "an old balloon, considerably worn," to Miss Lucretia Bradley of Easton, Pennsylvania, for $100. He noted that Miss Bradley was "a woman of more than ordinary qualifications." "Active and elastic as a deer," she was "possessed of a spirit of determination that knew no discouragement in anything she undertook." The *Eastonian*, a local newspaper, commented that she was "a brave, enthusiastic, and accomplished Yankee girl, and doesn't want to be anything else."[6]

Wise was not eager to sell the balloon to Miss Bradley. It was small, much patched, and brittle with age. The network was "irregular," and the appendix, or neck pipe, was far too small to allow sufficient gas to escape in the event of a rapid ascent. To all of this the intrepid woman remarked that if the balloon was strong enough for Wise, it was strong enough for her.

Miss Bradley made her one and only ascent with the craft in January 1855. The balloon shot into the air at an alarming rate. She valved gas continuously, while additional hydrogen roared out through the appendix, but the envelope continued to swell. The balloon finally exploded "like the discharge of a cannon." Looking up, Miss Bradley saw that the entire lower half of the gasbag was "all shattered."[7] Plummeting toward the earth, the novice aeronaut was saved by the fact that the fabric gathered in the upper hemisphere to form a parachute. If we are to believe Miss Bradley's account, we must admit that she was indeed a paragon of valor. She claimed to have spent the few minutes of her descent "in singing a song of praise to the Creator for the sublime beauty and grandeur that surrounded me in this my first aerial adventure."[8]

Lucretia Bradley ordered a new Irish linen balloon from Wise and constructed an apparatus to generate hydrogen by passing steam over hot iron. In spite of her high hopes, she was unable to drum up public support and retired from aeronautics after her single, spectacular ascent.

Throughout his career John Wise held what was, for the period, a remarkably high opinion of women aeronauts. "Woman," he would remark, "when really determined, seems to be more daring than Man."[9] He applauded the sentiments of his daughter-in-law, Louisa Wise, who made her first flight from Lancaster on September 18, 1869. "I am not an advocate of woman's rights in the modern acceptation of that term," she had commented, "but have, nevertheless, a notion that a lady might take a ride through the ethereal regions of space without sinning against the proprieties of her sex, or in the least infringing upon the good order of 'a time for all things.'"[10]

John Wise would always regard Lucretia Bradley as "one of the most heroic women in America."[11] Miss Bradley was, however, far from the only woman to venture aloft during these years. Both Wise and William Paullin were of assistance in enabling Mme Delon to fly from Philadelphia on June 25, 1856. Very little is known of Mme Delon before she began her preparations in a lot at the corner of Callowhill and Seventh streets that morning. Wise and Paullin supervised the inflation and had apparently instructed the aeronaut as well. Mme Delon flew to the northeast, crossing the Delaware and passing over the villages of Richmond, Ararringo, Beverly, and Burlington, before landing a mile and a half north of Tacony at 5:20 P.M.[12]

Women would continue to make ascents during the years prior to the outbreak of the Civil War in April 1861. The unidentified lady who made a "daring exhibition" at a Cambridge, Ohio, fair in October 1858 and the mysterious Miss W., who ascended twice from New Orleans in 1858, were typical.[13]

None of these women were yet prepared to join their peripatetic brothers in touring the summer circuit. All too often these feminine ventures aloft were viewed with jocularity. When Mrs. E. W. Davis flew with Prof. Samuel Wilson from Tuscumbia, Alabama, aboard the smoke balloon *American Eagle* on July 10, 1858, for example, the Nashville *Gazette* remarked that "by and by, the ladies will ascend by simply filling their crinolines with smoke for that is all they lack of being angels."[14]

Of course Wise did not restrict his teaching activities to women. His own son Charles E. Wise was his most successful pupil. Charley Wise, who made his first flight from Shannondale Springs, Virginia, on September 3, 1853, at the age of seventeen, would follow his father's footsteps into the life of a professional aeronaut.[15]

Wellington Dunlop, a young Bucks County man, had been the first graduate of the John Wise school of aeronautics. Dunlop had gone aloft from Kutztown, Pennsylvania, in the summer of 1839 after but a few minutes of instruction from the master. Another pupil, Charles Henry Brown, "a young man of much intelligence and a great enthusiast in aeronautic matters," later moved to England

and kept his old instructor current on the state of the aeronautic arts in Great Britain and on the Continent. Brown later immigrated to Australia where he made some of the first ascents on that continent.[16]

Wise also made an occasional practice of offering tethered ascents to spectators. Following an ascent from Danville, Pennsylvania, in the spring of 1852 he allowed a "rustic-looking young country girl" to ascend 200 feet into the air, "much to the satisfaction of herself and the amusement of the bystanders." Wise then sent up a trumpeter from the crowd to blow a few blasts, "which acted like magic in bringing people across the Susquehanna Bridge." The aeronaut was suddenly overwhelmed as spectators crowded to enter the car at a quarter a head. Before the day was over, Wise had cleared an additional $80.[17]

While Wise was touring the land, thrilling the crowds, training other aeronauts, giving rides, building balloons for himself and others, and generally learning to earn his living as an aeronaut, he continued to gain expertise. There were few technical improvements to be made to the gas balloon as originally constructed by J. A. C. Charles and the Robert brothers, but Wise was quick to take advantage of those that did appear. He was one of the first American balloonists to make use of Charles Green's invention of the dragline. This was a long rope attached to the basket which helped to stabilize altitude. When the balloon dropped close to the ground, the end of the rope dragged over the earth, reducing the weight carried by the craft and halting its descent. When the balloon rose, the total weight of the line was carried, and the ascent was arrested without valving gas.

Wise introduced a few innovations of his own. We have noted the origins of the rip panel in his "exploding" balloon of 1838. When fully developed, the rip panel was a lightly stitched section of fabric in the upper hemisphere of the envelope, the top of which was attached to a line passing through the appendix to the basket. When landing in a high wind, the aeronaut could tug the line to open the panel, allowing all the hydrogen to escape at once so that the billowing envelope would not fill like a sail and drag the balloon over the countryside.

Wise was also among the first to notice the impact of solar heating on a dark-colored or black balloon. Flying from Lewistown, Pennsylvania, in April 1842, he commented in his log: "I have at present use of a black balloon, which creates a superior atmosphere around itself in the cold upper regions of the air, from the radiating superiority of that color over a lighter one."[18]

The aeronaut certainly retained his interest in science throughout the decade of the 1850s. Meteorology, or the "geology of the atmosphere," as he preferred to call it, was his particular interest. While Wise remained a bit of a crank in other scientific areas (for example, attempting to conduct physical

Mme Delon rises above the heads of a Philadelphia crowd, June 25, 1856.

experiments to prove the "hollow earth" theory of John Cleves Symmes), he was a master at reading the clouds and the winds. Moreover, he possessed an excellent working knowledge of contemporary atmospheric theory. He quite accurately characterized the ocean of air as a heat engine: "That great centrifugal furnace of our equator, with its seething band of heated and volatized vapor, that is pumped up as it were by a tremendous irrigating engine, as if made expressly to send heat and moisture to the uttermost parts of the earth, is of itself a grand and interesting study before which the heaving fires of Vesuvius are comparatively tame."[19]

Wise had begun conducting simple meteorological experiments for Profs. J. K. Mitchell and William Espy, both pioneer American students of the atmosphere, as early as October 1837. Joseph Henry, however, the first secretary of the Smithsonian Institution, was the aeronaut's primary tutor in the theory and practice of atmospheric science. Wise was so impressed by Henry's ability to blend the many conflicting bits of information into a weather forecast, that he nicknamed him "Old Probability."[20]

Wise had little doubt about the importance of understanding the mechanism of weather. "The very laws of life," he would remark, "are more dependent upon what is overhead than upon that which is underneath our feet." Nor did he doubt the importance of the balloon in exploring the mysteries of climate. "When we want to find out what abounds in the wilderness," he delighted in pointing out, "we must go into it."[21]

William Paullin may have been less interested in science than John Wise, but he was no less active following his return from his Latin American tour. Like Wise, Paullin concentrated his activity in the Midwest after 1850. On October 23, 1852, for example, he flew from Zanesville, Ohio. Having wagered that he would land on the same spot from which he had taken off, the aeronaut hired one Elijah Ross, a local gunsmith, to follow his balloon and tow him back to Zanesville.[22]

Paullin returned to Ohio the following year, opening the season with a flight from Sandusky on June 23. While preparing for a Columbus ascent that September, he encountered Silas M. Brooks, a wandering musician who was destined to become Paullin's best-known trainee and one of the great aeronauts of the nineteenth century.

A native of Plymouth, Connecticut, born on December 19, 1824, Silas was the son of Willard and Maria Markham Brooks. Young Silas, his sister, and his brother George, who would also become an aeronaut, grew up in Burlington where their father worked as a cabinetmaker.[23]

Brooks was living in Forestville, New York, in the summer of 1848 when he was approached by Phineas T. Barnum, the renowned New York showman, whose attention had been drawn to the young man's native ingenuity and mechanical skill. Barnum had recently come up with a new scheme to bilk the public. Always an astute judge of the temper and taste of the time, Barnum recognized an opportunity to turn a handsome profit from the contemporary enthusiasm for the occult, mystic religion, and the hidden mysteries of the past. He proposed to field a "Druidish Band" to commemorate the arcane achievements of the Celtic priests who were reputed to have built everything from Stonehenge to the strange earthen mounds that dotted the landscape of the American Midwest. The band would tour the nation attired in the most gorgeous and costly floor-length belted robes "of Black Silk Velvet, trimmed with heavy Gold and Silver lace." The robes were to be decorated with badges, jewelry, and cabalistic symbols. The uniform was to be topped off with a "Druidish hat," a black hood completely covering the head and shoulders. Each member of the band was to wear sandles, a grey wig that would reach to his shoulders, and a long fake beard.[24]

Barnum's band could scarcely sally forth armed with trumpets, trombones, and saxophones. Appropriately "Druidish" instruments would be required. When a New York musician initially chosen to design and build the instruments gave up the task in disgust, Barnum's manager recommended that they turn to Silas Brooks.

With $75 in hand, Silas drew his brother George into the project, collected a supply of cattle horns from a tannery, and went to work. As Brooks recalled in later years, "We had the hardest time imaginable to get a horn that would make a noise, while a tune seemed out of the question." Within a month the brothers had produced a "Druid Horn," a suitably arcane concoction of sheet brass, iron, leather, and highly polished horn.

Delighted, Barnum ordered five additional horns, hired three German musicians to play first, second, and tenor Druid horn, and appointed Silas Brooks to play the "Base Druid Horn." After an out-of-town tryout in Collinsville, Connecticut, the "Druidish Band" opened at Barnum's American Museum on Christmas Day 1849, playing "the most wild and secret music ever offered to the public."[25]

The band was an enormous success, but Barnum became bored with the project and sold his shares to Brooks and another band member, G. W. Farnsworth. Farnsworth, in turn, sold out to a theatrical promoter named Brown.

In the spring of 1853, Brooks and Brown, having expanded the band to a tent show including one hundred animals and eighty employees, began a tour of the West. They encountered William Paullin while playing in Columbus,

Ohio, and immediately hired the aeronaut to accompany the troupe at the considerable salary of $100 a week. Paullin's first ascent as part of the show was in Canton, Ohio. Brooks was delighted. "The show went so well that we speedily made money," he recalled.[26]

Brooks made his first flight when Paullin was taken ill in Memphis later that season. By the time of Paullin's ninety-fourth flight, from Hartford, Connecticut, on July 4, 1854, Brooks had abandoned the circus and was working as the aeronaut's junior partner. On this occasion they made a dual ascension, launching three minutes apart from South Meadows and landing a mile from one another near Manchester, New Hampshire.[27]

By 1855 the two aeronauts had parted company. Brooks was opening new ground in the area west of the Ohio Valley. A reporter describing Brooks's career in a 1906 obituary claimed that the aeronaut had made the first flight in Michigan during this period. He was certainly the first to fly in Illinois and Iowa.

Brooks made his first ascension from Chicago on July 4, 1855. At the time, the aeronaut was flying a balloon known as the *Eclipse*, which was owned by S. D. Ledgar, who promoted Brooks's appearance. The launch, which began from a lot at the corner of Randolph and Peoria streets, was a disaster from the outset. Taking off in a high wind, Brooks found himself moving rapidly toward Lake Michigan. He valved gas and descended, flying "a mile a minute" straight into a telegraph line bordering the Michigan Southern Railroad tracks. Brooks fell from the basket as the balloon "mounted into the heavens like a freed bird, and sailed off into the eastern sky."[28]

This was only the first occasion in which Lake Michigan would place an aeronaut in danger. In years to come, some of the nation's finest balloonists, including the great John Wise, would meet death in the cold waters of the lake.

Contemporary news accounts ridiculed Brooks, pointing out that this was the third balloon he had lost in the past few weeks. The *Eclipse* was eventually recovered, but only after Brooks had relocated in Rockford, Illinois, where he was constructing a balloon of his own, the *Comet*. Ledgar retained the services of O. K. Harrison, a local man who was planning to continue flying the *Eclipse*.[29]

Brooks made one of his first ascensions with the *Comet* from Rockford on August 14, 1855. Over the next three years he would fly from a dozen other Illinois towns, including Joliet, Peoria, and Ottawa.[30] Brooks arranged the first flight in Iowa on October 9, 1856. Advertisements touted the event as "the greatest novelty that had ever visited" the city of Muscatine, where the third Iowa State Fair was in progress. On the day of the flight, Brooks lectured to the crowd gathered inside the twelve-foot canvas fence that protected the launch area. He announced that his young assistant, John Leonard, native of Burlington,

Connecticut, would go aloft. Leonard had already made several ascents under Brooks's supervision. On this occasion he flew 15 miles to a landing 3 miles north of Wilton in Cedar County.[31]

At least one other attempt was made to fly, from Keokuk, on October 27, 1856. It was, according to local news reports, something of a disappointment: "The large crowd to witness the ascension of the balloon yesterday afternoon, seems to have inspired it with their spirit of impatience, and in one of those fretful freaks . . . it [the balloon] rolled and tumbled and escaped away on the wind into the blue ethereal, leaving the Aeronaut looking for all the world as if he had 'let something go.' The only regret seems to have been that the poor fellow didn't go with his balloon."[32]

Silas Brooks moved to St. Louis after the 1856 season, temporarily retiring from aeronautics to accept a position as manager of a St. Louis museum. His brother George joined him in St. Louis, where city directories listed his occupation as "dealer in liquors" during this period.

After Brooks's temporary retirement, John Leonard struck out on his own, returning to Connecticut where he made a total of seventeen ascents during the summer of 1857. His best-known flight of this season was made from Hartford on September 25 with a young woman passenger. The Hartford *Courant* noted that there were, "without exaggeration," 25,000 people on the ground to witness this flight, which was staged on historic Charter Oak Hill. Leonard's companion, Miss Jane Wright, an eighteen-year-old woman "of rather pre-possessing appearance," from New Burlington, "appeared to be strong-nerved, and manifested no apprehension." Leonard's mother, on the other hand, "seemed to be anxious for her son's safety, but he answered her motherly entreaties by words of cheer."

The balloon was relatively small for carrying two persons. Made of four hundred yards of oiled cotton, it was designed to fly with 15,000 cubic feet of city gas. The pair cast off at 2:00 P.M., with Leonard hanging out of the car waving a kerchief. They remained in the air for one hour and forty minutes, landing near Bloomfield.[33]

Even before Brooks, Leonard, and others began flying in the upper Midwest, the balloon had skipped across the continent to make its appearance in the California goldfields. A Mr. Kelley staged the first balloon ascension west of the Rockies in Oakland, California, on August 28, 1852. Three ferries were kept busy steaming across San Francisco Bay that day, their decks crowded with passengers on each trip. Kelley's balloon, a silk specimen some 40 feet in diameter, was inflated in a small lot on Third Street, near Broadway.

As the crowd grew restless during the long inflation, Kelley decided to go aloft at 3:30 P.M. with the balloon only half-inflated. A reporter for the San

Francisco *Herald* described the scene: "A dense crowd collected round the balloon, and the aerial voyager took his seat in the car, which was released from confinement, and away went the balloon, not into the upper air, but along the street in a southeasterly direction, banging Mr. Kelley against the ground in anything but an agreeable manner, and knocking up quite a dust until it was captured."[34]

A second attempt to launch the balloon with a smaller pilot aboard also failed. In desperation, Kelley removed the car from the balloon to reduce the weight. A light board was tied across the load ring and an even smaller spectator was chosen to make a third attempt, but the balloon still refused to rise.

Fortunately for Kelley, the crowd was good-humored. Rather than destroying the balloon, they began to pester the aeronaut to allow individual spectators to sit on the board and bounce a few yards into the air. One of these volunteers, a sixteen-year-old boy named Joseph Gates, who had come to Oakland that morning to sell oranges, got more of a ride than he had bargained for. When Gates, the lightest aspirant to try the balloon, was tossed into the air, he continued to rise. He appeared to panic momentarily as Kelley shouted up instructions for descending, but a series of catcalls from friends in the crowd apparently persuaded Gates to continue his voyage. Worried spectators watched the balloon disappear in the distance, believing, as one reporter noted, that no "hope of the safety of the youth can be entertained."

These fears were put to rest the next day when Gates returned to San Francisco to report that he had "enjoyed himself first rate." The boy aeronaut had flown for a considerable distance, enjoying the view before he decided to descend. The valve line, apparently tangled in the confusion of the network, would not function and finally broke in Gates's hand. The young aeronaut had then clasped a knife in his teeth and "shinned up" the balloon to a point eight or ten feet above the ring, cut a hole in the fabric, and returned to straddle the board as the balloon descended to safety. He had landed in an area known as the Suisun Plains, walked five miles to the safety of a cabin, and returned to San Francisco by boat the next day.[35]

By January 1857 Prof. Samuel Wilson had made a number of short but very successful ascensions in the San Francisco area. Wilson complained that a lack of funds had prevented him from returning east to obtain a larger balloon, but the Bay area newspaper *Echo of the Pacific* seemed to be satisfied with his performances to date, commenting that he had "worked audaciously" to please his fellow citizens with flights and popular lectures on aeronautics.[36]

Now, Wilson announced, he was moving south of the border, having heard that "the Spanish people have shown themselves eager to watch balloons." He

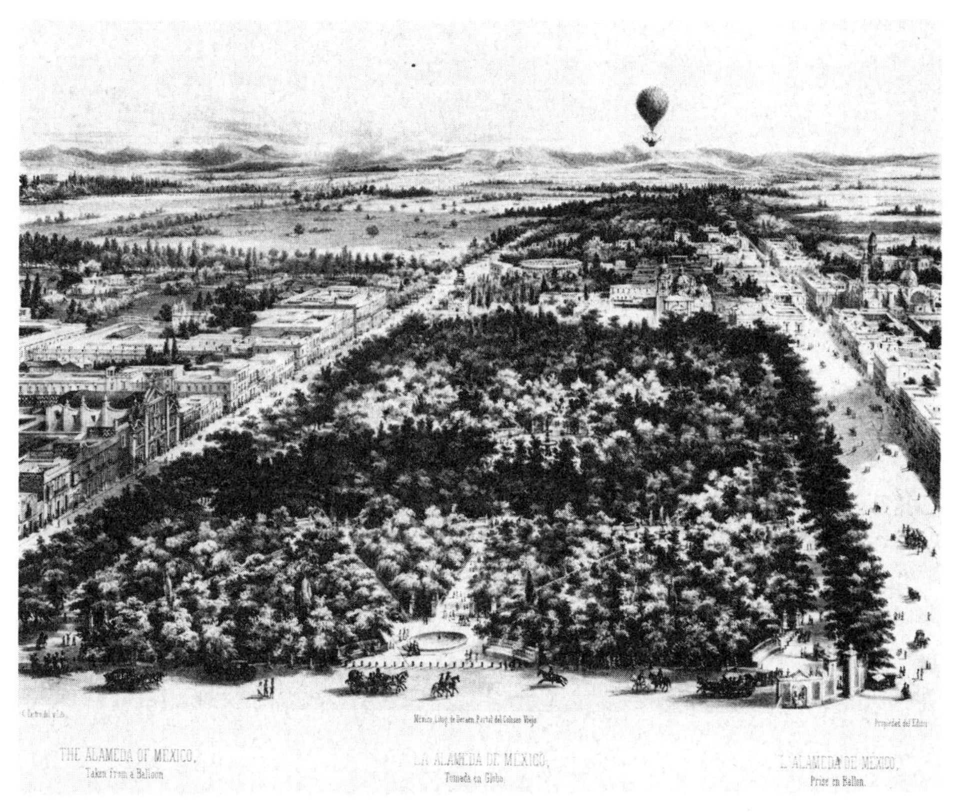

"The Alameda of Mexico," by Jules Arnout, circa 1846, shows a balloon ascent.

made his first Mexican ascent from Guadalajara on March 29, 1857. This was followed by a second ascent in the Alisco region on April 5. Two additional flights from the Plaza de Torros del Pases Nuevo in Mexico City on June 14 and June 21 brought Wilson to the brink of financial disaster.

The presence and support of Gen. Don Ignacio Comonfort, president of the Mexican Republic, attracted a large crowd for a third flight in the city on July 5. On this occasion Wilson flew with a young lady in the basket of his balloon *Monteczuma*. A week later, on July 12, the aeronaut allowed a young Mexican friend, Antonio Vazquez, to ascend with a young woman disguised as "America" to protect her reputation from scandalmongers. Virtually the entire Mexican government attended Wilson's next ascent. Newspapers complained that the aeronaut had admitted the official party at no charge, while "enthusiasts without influence had to pay."

Wilson's final flight in Latin America came that September in honor of Mexican Independence Day. Soon thereafter the aeronaut "disappeared from Mexico." A local reporter suggested that he had left the country in such a hurry that it seemed "as if the clouds had carried him off." Actually, Samuel Wilson had simply returned to the United States, where he was to remain touring the Midwest for the two years remaining before the outbreak of the Civil War.[37]

Of course, aeronautical activity was not restricted to the Mississippi Valley and the Far West during the mid 1850s. A new generation of balloonists was making its mark in the New England and Middle Atlantic states. Some, like Mary S. Rangard, who flew from Springfield to New Salem, Massachusetts, on July 4, 1855, had short careers and attracted little notice.[38] Others, like Joshua Pusey, proved a bit more durable.

Pusey had begun to fly as early as September 21, 1850, when he traveled through a snowstorm while voyaging from Reading to Haddington, Pennsylvania. "What was strange to him, and will perhaps be so to everybody," reported the Hartford *Daily Courant*, "was the fact that the snow flakes ascended."[39]

Pusey made "a most successful ascension" in May 1851, flying from a lot at the corner of Callowhill and Seventh streets in Philadelphia to Moorestown in Burlington County, New Jersey. By July 1856 he was sufficiently well known to be booked as the featured aeronaut at Mme Tournaire's Circus in Cincinnati. Local newspapers reported that most of the delegates attending the Democratic National Convention preferred viewing Pusey's ascensions to participating in the interminable political debates.[40]

Flying in a high wind from Eastern, Massachusetts, on July 21, 1857, Pusey crossed the Taunton River and descended into the top of a tree. "The balloon was by this time cutting up all sorts of antics, rolling and surging, and dashing the car about in a manner anything but comfortable and assuring to the occupant," he reported to the local press. "Tearing away from this tree, it struck the earth, rebounded about forty feet, and dashed onward like a maddened steed, the anchor out and dangling in the air, tearing away stone walls, fences, rooting up trees, upturning immense stones and everything in its course." Pusey was finally saved by Eunice Parrish, a fourteen-year-old girl who courageously seized a dangling line and made it fast to a tree. Boarding a train to return home to Philadelphia, the aeronaut characterized his experience by noting that "there is much fun going up, but often very little coming down."[41]

Indeed, Pusey seems to have experienced repeated difficulties in "coming down." On the occasion of a flight from West Chester, Pennsylvania, he was blown over the Delaware River and across the mouth of the Schuylkill to a rough landing ten miles from Camden, New Jersey. The aeronaut was then dragged "through a pretty extensive piece of woods" before his progress was halted by two alert Jersey farmers. No sooner had the stunned Pusey stepped from his basket than the balloon *Louisa* broke free. The tattered remains were found by the crew of the steamer *Boston*, which was bound from New York to Philadelphia.[42]

Samuel Archer King and James Allen were the two most significant American aeronauts to emerge during the early 1850s. King was a native of Tirricum, Pennsylvania, born on April 9, 1828. As a young boy he was intrigued by accounts of John Wise's performances. During the summer of 1851, he was presented with a book in which the English aeronaut Monck Mason described a long balloon voyage from London to Weillburg, Germany.[43]

King was thus inspired to begin work on his first balloon, which was complete and ready for a flight from the Fairmount Water Works in Philadelphia on September 25, 1851. The balloon had a capacity of 12,500 cubic feet and was only half-inflated when the aeronaut was informed that the city could supply no more gas. Forced to decide between postponing his first ascension or flying with a heavy balloon, he forged ahead.

It was a ragged start. Bouncing across the wires of the Callowhill Street Bridge, he dipped twice into the Schuylkill River, scrambling onto the load ring as the basket scraped up and over a dam and was dragged through a grove of trees before coming to a halt in a cornfield.

Perhaps feeling the necessity to climb back into the saddle as soon as possible, King was back in the air the next day, remaining in the air for several hours on a flight that undoubtedly rebuilt his confidence. This was the first of 454 ascensions the aeronaut would complete prior to his death on November 4, 1914.[44]

Over the next ten years, King centered his activity in the Northeast, making ascensions in Pennsylvania, New Jersey, Delaware, Rhode Island, Connecticut, Massachusetts, and New York. These years were full of the hazards and delights so familiar to the aeronauts of the period.

Following a series of ascents from Wilmington, Delaware, he flew from Wilkes-Barre, Pennsylvania, on June 16, 1856. Caught in a high wind, he was driven into the trees where his balloon was destroyed, dropping the aeronaut forty feet to earth. Fortunately, he fell through branches that broke his fall. Though King was knocked unconscious, he escaped serious injury.[45]

By the time of a scheduled Philadelphia ascension on October 8, 1856, King had acquired a partner, James Allen, who was to gain fame as "The New England Aeronaut." Born in Barrington, Bristol County, Rhode Island, on September 11, 1824, he was the ninth child of Sylvester Allen, a ship's captain.

James grew up in Providence, where he worked in a cotton mill and as a farm laborer, seaman, printer, and jeweler. Ill health forced him to move to Wilmington, Delaware, where in 1855 he met Samuel Archer King.[46] After a period of furious reading in the aeronautical literature, Allen began an apprenticeship with King which would mature into an enormously successful and mutually profitable four-year partnership.

By May 16, 1857, Allen was making solo flights in the 25,000-cubic-foot *Zephyrus* under King's supervision. The two men were now flying singly, in pairs, and simultaneously launching in separate balloons. A dual ascension from Providence on July 4, 1857, was typical of their performance during this period.[47]

Allen launched by himself at 4:45 P.M., followed by King, "the somewhat celebrated and daring aeronaut," who carried two passengers. After a one-hour flight, the two balloons had become separated. King and Allen, in a demonstration of their virtuosity, maneuvered into wind currents that carried them so close to one another that "if this course had been continued, the balloons would have met." Allowing themselves to separate once again, they landed safely a mile apart.[48]

And so it went at Worcester (July 28, 1857), New Haven (August 11), Lowell (July 6, 1858), Manchester (July 27), Norwich, Providence, Paterson, Charlestown, and Boston, as the two aeronauts skipped across the Northeast.[49]

King and Allen gained a great deal of experience during these two early years and had a great deal of fun as well. Decades later they would fondly recall several coincidences that had enabled them to astound spectators. One of King's favorites had occurred while he was preparing for a flight from Wilmington, Delaware, in the summer of 1855. King had been approached by a gentleman named Richardson who wished to shake his hand. During the course of their conversation, Richardson invited the aeronaut to pay him a visit if he was ever in the neighborhood. After a flight of several hours, King threw out his anchor and waited for the assistance of a man who was running toward him. "I eyed him sharply," recalled King, "and recognized my friend Richardson. Our conversation of the morning instantly recurred to me. Here is a good joke, thought I, and as he reached me I said to him—'Well, here I am, Mr. Richardson, according to promise.'"[50]

But there was a more serious side to their enterprise as well. Like John Wise, King and Allen became serious students of atmospheric science. Their simple observations on wind speeds and currents, clouds, variations in temperature and pressure, and other weather phenomena proved useful not only to the aeronautical fraternity but to budding meteorologists as well.

The two aeronauts were also responsible for introducing at least one bit of

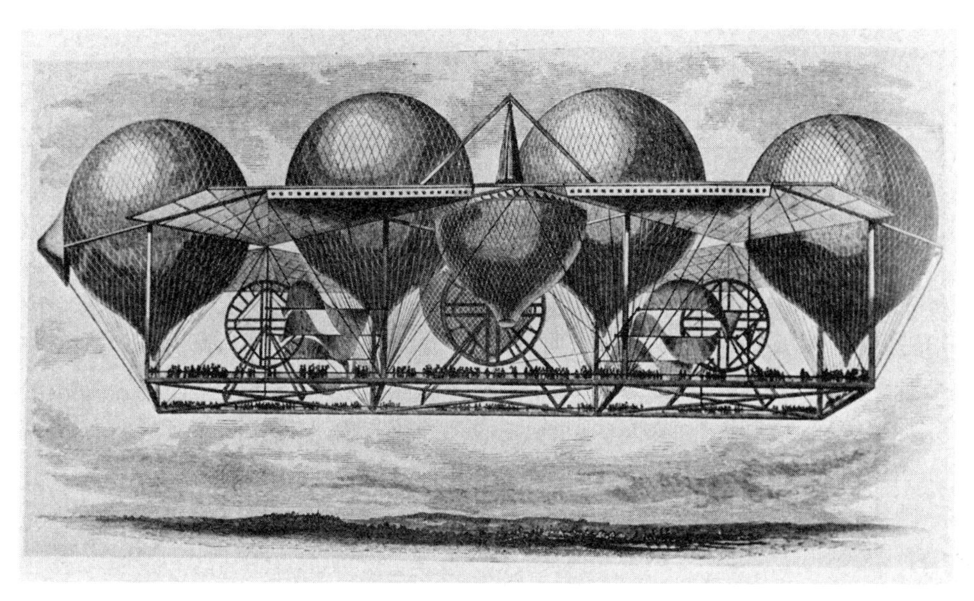

Pétin's airship

new balloon technology to the United States. Flying from New Haven on August 15, 1857, they had experimented with the dragline, an altitude control device recently invented by the English aeronaut Charles Green. Wise, Paullin, Brooks, and other American balloonists were quick to follow the lead of King and Allen in adopting the device.[51]

Choosing to fly in an area famed for its long, harsh winters, King and Allen were forced to seek occasional employment during the off-season. King, for example, worked as a professional photographer, a chemist, and a dresser for the great tragedian Edwin Forrest.

The native American aeronauts, men like Wise, Paullin, Brooks, Leonard, Wilson, Pusey, King, and Allen, who were touring the nation during the 1850s were joined by a new generation of European colleagues. Ernest Pétin was one of the first to arrive. He was a curious character with striking features: a high, domed forehead, long, blond locks, and piercing eyes. Born in Amiens on March 12, 1812, he was a politically active haberdasher whose plans for aerial navigation had come to official attention following the revolution of 1848.[52]

Pétin envisioned a giant airship supported by four large balloons enclosed in an enormous wooden framework. Inclined planes, revolving sidewheel propellers, and other strange appurtenances were to be used in directing the motion of the craft. Pétin's project was widely discussed in both the technical and the popular press, capturing the attention and support of a number of prominent Frenchmen, including the writer Théophile Gautier.

Pétin was able to fund construction of three large red balloons that would, he hoped, eventually be incorporated into the airship. He made his first balloon flight on August 3, 1851, under the tutelage of M. Poitevin, who had earned his reputation making ascents while seated on the back of a horse. The aeronaut's high hopes were dashed, however, when French officials refused to encourage the airship project because of fears that, if successful, it might destroy existing customs arrangements. Unwilling to accept defeat, Pétin decided to undertake a fund-raising tour in America.

Pétin sailed for the United States aboard the three-masted schooner *Emperor* in January 1852. By April of that year, he was describing his plans for the airship in a series of lectures delivered through an interpreter at New York's Broadway Tabernacle. The aeronaut was off to a slow start. The crowds that gathered to hear his talks and gaze at his paintings and models were small.[53] Pétin was soon writing articles outlining his ideas for American newspapers and promising an eventual Atlantic flight in his airship.[54]

The aeronaut was forced to cancel his plans for an ascent on horseback from New York for unknown reasons. His first attempted flight from Bridgeport, Connecticut, on July 5, 1852, was scarcely more successful. The balloon was cut to pieces when it flew into telegraph lines immediately after take-off. The aeronaut suffered little more than a bruised ego as a result of his twelve-foot fall to the ground.[55]

Pétin was back in the air over Bridgeport in the second of his three "airship balloons" only ten days later on July 15, 1852. The aerostat was 70 feet in diameter and 100 feet tall, with a gas capacity of 35,000 cubic feet. It was decorated with French and American flags and featured a "boat" 20 feet long, large enough to carry the aeronaut and two assistants, Gustave Reynauld, a French mechanic who had accompanied Pétin to America, and Mr. Wood, a Bridgeport admirer.

The three men ascended "like an arrow . . . in magnificent style." "Far beneath us, shining like molten silver, lay Long Island Sound, dotted with vessels, which appeared like specks on its bosom, while Long Island appeared in the distance, and far beyond, the broad expanse of ocean." The party landed at River Head, Long Island, after a 50-mile flight.[56]

Pétin scheduled his next flight from Bridgeport on September 6, 1852. Now claiming that his balloon was the largest ever flown in the United States, he went aloft with Reynauld, J. W. Dufour, an interpreter, and Mr. Seech, a schoolteacher from East Bridgeport. Blown out to sea, the four men struggled for two hours before rescue arrived in the shape of a lifesaving crew from Bridgehampton. The $3,500 balloon was a total loss.[57]

Undaunted, Pétin went aloft from Springfield, Massachusetts, on September 30, 1852. It was, the Springfield *Republican* noted, "a magnificent ascension": "In the foreground, arose the huge monster of a balloon—sleek and smooth, blood-red—caught in a net, forty men pulling at his whiskers to keep him down. In the enclosure, upon the level ground, grouped around the car or boat, stood a large number of persons and beyond the enclosure in the streets, on the fences, on houses, on trees—not only near, but, in some instances, at the distance of half a mile, were disposed the uncounted thousands."[58]

Filling the giant balloon had presented a very real problem for Pétin. The aeronaut was able to produce only 20,000 cubic feet of hydrogen, which was stored in large metal tanks awaiting transfer to the balloon. An additional 50,000 cubic feet of gas was provided by the Springfield Gas Company over the period of a day and a half, to avoid creating a gas shortage in the city. Messrs. Ingersoll and Allen and several officials of the U.S. Armory had been influential in arranging the ascent and convincing the gas company to cooperate with Pétin, who, speaking only French, had some difficulty dealing with American civic officials.

Pétin and Reynauld cast off at 4:15 P.M. They dropped flowers out of the basket as they passed up and over the wall, and passed over the Armory grounds on their way to a landing in East Windsor ten minutes after sunset.[59]

Pétin made his next ascension from Lafayette Square in New Orleans on December 25, 1852. Flying over Lake Pontchartrain with three companions, Pétin discovered that his balloon had developed a catastrophic leak. The travelers received a dunking in the cold water but were quickly rescued by the steamer *Alabama.*[60]

While waiting for a new balloon to arrive from storage in New York, Pétin lectured on his airship scheme at the Louisiana College on Dauphin Street. In spite of his poor performance to date, he was named "Man of the Century" by the *Deutsche Zeitung*, a local German-language newspaper, in April 1853. Pétin was not destined to make another flight from New Orleans. When his new balloon finally arrived, it was damaged by vandals who broke into his workshop. Discouraged and close to bankruptcy, the aeronaut decided to try his luck in Mexico.[61]

Pétin made the mistake of launching his Mexican tour with the usual series of talks outlining his hopes for aerial navigation, admitting that he had been unable to interest French or American investors. As early as February 1854, before Pétin had even scheduled his first ascension, Mexican newspapers were already referring to him as a "liar, cheat, and poseur of the first order." Nor did Pétin's actual performance inspire much confidence. His record of cancellations due to improper inflation and bad weather was unparalleled. Gaston Tissandier, the great French historian of ballooning, notes that, true to form, Pétin "died miserably" in his native France in 1878.

The experience of Eugène Godard and his wife stood in direct contrast to that of their countryman Ernest Pétin. As Charles Dollfus has noted, "The dominant name in aerostation during the second half of the nineteenth century is that of Godard."[62] Eugène Godard was the head of a family that included his father, Pierre Edmé, uncle Abel (known as Fanfan in the family), Jules, sister Eugénie, and cousin Fanny, plus many of their wives, husbands, sons, and daughters. Among them they made thousands of ascents in Europe, North America, Asia, and Africa. They absolutely dominated European aeronautics for over sixty years.

Eugène Godard had set all of this in motion. He was born in Clichy, near Paris, on August 27, 1827. The son of a master mason, he suffered a difficult and impoverished childhood. Godard, who was almost entirely self-educated, had won an architectural prize and was working as an architectural draftsman when he made his first balloon flight aboard a paper Montgolfier at Lille in 1847.[63]

Over the next seven years, Godard was to amass an enormous body of experience in the air. If we are to believe the publicity flyers the aeronaut distributed after arriving in the United States, the poor boy from Clichy had transformed himself into "Mons. Eugène Godard, the celebrated French Aeronaut, Member of the Academy of Arts and Sciences of Paris, Chief Aeronaut of Austrian Armies . . . , who has crossed the Alps and Carpathian Mountains and English Channel, and ascended to a height of 22,300 feet above the earth, traveled in a Single Voyage to the distance of 450 Miles at the rate of 90 Miles in 55 minutes."[64] Much of this, including the speeds and altitudes attained and his claims of service as "Chief" aeronaut with the Austrian forces besieging Milan in 1848, is public relations hyperbole. The fact remains, however, that Eugène Godard spent a great deal of his time in the air between 1847 and 1853. After 1850 he became a regular performer at the Hippodrome, a Paris pleasure spa, where he made regular flights, including many with passengers. He had also begun a lifelong series of experiments with very large hot air and gas balloons during these years.

The American press followed Godard's early career with some interest. In November 1851 *Scientific American* described a romantic moonlight journey across

M. EUGÈNE GODARD
D'après une photographie de M. Panajou, de Bordeaux.

Eugène Godard

Paris made by Godard and half a dozen companions.[65] A year later the same journal reported that Godard seemed "bent upon rendering . . . [his] profession every day more and more perilous":

> *During the whole of the past year the ascensions from the Hippodrome have been made with gymnasts suspended beneath the car, executing their terrible exercises during the passage of the balloon to the clouds. The last experience was the reverse of this. It consisted of the descent of a parachute from an enormous altitude, with M. Godard hanging below it. He turned somersets and performed all sorts of rigadoons in the air, from the time when the cord was cut till it was time to look out that he touched the ground with his feet. The experiment was successfully and gracefully performed. This is the most wonderful feat of lofty tumbling ever performed by mortal man. It takes the French to do these things in grand style.*[66]

The date of Godard's arrival in the United States is uncertain. As he was awarded a Silver Medal for Improvements in Aerial Ascension at the New York Exhibition of Industry of all Nations in 1853, it seems quite possible that he was in the country at that time.

The aeronaut had arrived in New York with his wife and five balloons. They were, he hastened to inform booking agents, "the most beautiful balloons in America." The smallest, the *Niagara*, was a one-man model with a capacity of 12,000 cubic feet. The *Zephyr* and the *L'Eole* were aerostats of 17,000- and 24,000-cubic-foot capacity respectively. The 36,000-cubic-foot *Leviathan* could carry "five or six persons, or a horse." His largest envelope, the *Transatlantic Telegraph*, had a capacity of 106,000 cubic feet and could carry as many as twenty passengers.[67]

Godard announced himself ready to "make ascensions in every style yet known, either scientific or eccentric." His repertoire of "eccentric" flights ranged from the performance of stunts in midair on a trapeze bar to Mme Godard's much-publicized ascensions on horseback. The aeronaut was quick to point out that his balloons were designed to be flown with city gas so that disappointments due to problems encountered during inflation would be held to a minimum. Godard offered advice to a fund-raising committee seeking to sponsor a flight and, moreover, could double as a magician and traveled with "one of the most beautiful and rich Cabinets of Magic so that he could give Magical Representations" to "help defray expenses and make handsome profits besides."[68] Like Robertson, the Godards would succeed because they combined extraordinary skill, courage, and daring in the air with superb showmanship.

The record of early Godard ascensions in the United States is sketchy. They were reported to have flown the balloon *Know-Nothing* from the New York Hippodrome on October 21, 1854. Presumably they had simply renamed one of their original balloons, but it is interesting to note that these French Catholic visitors to America would name their craft in honor of the virulently anti-papal and nationalist political party.[69]

In an advertising circular designed to interest "Their Honors, the Mayors of Cities, the Presidents of State and County Fairs" and others that might be interested in an appearance by the aeronauts, Godard also cited a series of tethered passenger flights made at Buffalo, New York. The pair had then transported their fully inflated balloon over thirty miles to Lockport, where they "arrived the next day and made a magnificent ascension with the same gas." It is not clear whether or not these ascensions were made before or after the Hippodrome flight.[70]

The Godards spent a portion of the winter of 1854–1855 in the South. On January 1, 1855, Eugène flew 190 miles from New Orleans to East Felicia, Louisiana, in three and a half hours.[71] The couple may have flown from Cuba during this period as well.

They worked their way up the Mississippi Valley that spring and summer, pausing to fly in St. Louis before moving on to Cincinnati where they planned to make "Grand and Novel Balloon Voyages." The first ascent was to be staged from the city lot near the corner of Ninth and Plum streets on September 22. Admission to the canvas-enclosed launching area was to be fifty cents, half-price for children and servants.[72]

After a two-day delay, the 32,600-cubic-foot aerostat *Ville de Paris* was ready to depart. This balloon did not appear on the list of the envelopes that the Godards originally brought to the United States. Either they had ordered a replacement craft from France or had purchased a new one in this country.

J. C. Belman, the self-proclaimed "balloon editor" of the Cincinnati *Daily Gazette*, had been invited to accompany Eugène Godard and his wife on their first ascent from the city. After performing his stunts on the trapeze, Godard climbed into the basket by means of a rope ladder, released more ballast, and quickly rushed into the skies. The "Metropolis of the West" now appeared as "a Lilliputian villa, built of toy houses by children on the parlor floor."[73] The trio caught a current of air blowing to the northeast and passed over Cumminsville as the moon rose. At an altitude of 4,000 feet, they flew over Spring Grove Cemetery, "where dots of white embedded in green foliage marked the reposing places of the remains of mortals." Rising to 9,000 feet, they viewed a Cincinnati, Hamilton, and Dayton train approaching the city, moving like "a snail upon a thread." The Miami and Erie Canal resembled a "chord of silver adorning a rich green velvet." The Godards and Belman enjoyed an alfresco repast of turkey, chicken, duck, bread, and cakes, with wine and cordials. Passing over Mount Pleasant, they were greeted with cheers by farmers returning from the fields. Eleven miles west of this village, they again dropped ballast and rose to 15,700 feet, at which point they caught a current that blew them toward Glendale, Ohio. Maintaining an altitude of more than 15,000 feet, Godard filled a champagne bottle with water and tossed it out of the basket. Belman claimed that he distinctly heard the bottle break three minutes and twenty-five seconds later. By 9:00 P.M. they had reached Hamilton, Ohio, where they conversed with the inhabitants through a speaking horn. Several hundred people pursued the balloon to the landing area three miles from the city. Godard and Belman removed the seats and extra ballast from the basket and attached long ropes to the balloon. They then reascended to 65 feet and were pulled by the spectators into Hamilton, where they spent the night.[74]

Monsieur and Madame Godard hoped to draw even larger crowds of paying spectators to their second Cincinnati ascent by reducing the admission fee to twenty-five cents and by promising a "Grand Aerostatic Festival" featuring their large balloon *America*. This 95,000-cubic-foot aerostat was to be inflated with illuminating gas equal to that consumed by the entire city over a three-day period. The large balloon car was constructed in the shape of a building, labeled "Godard's Hotel." The Godards sold tickets for balloon rides, promising their passengers a sumptuous meal served at 12,000 feet in the balloon's salon, a small room decorated with historical and mythological paintings.

"All the world, at least so much of it as be located for some miles within the vicinity of Cincinnati," appeared at the city lot on October 1, 1855, to witness the ascent of this "monstrous craft." Taking advantage of the reduced admission fee, an estimated 3,000 persons paid for seats in the canvas enclosure, while countless others gathered in the surrounding streets. So crowded were conditions within the launching area that one of the two hastily constructed tiers of spectator seats collapsed, resulting in a number of minor injuries and a hasty exit by the occupants of the second tier. By 6:00 P.M. the injured had been removed and preparations for the ascent continued. Passengers on this flight included Eugène Godard, J. C. Belman, Col. William Latham, general agent for the Great Miami Railway, George Hoel, a river pilot, and William G. Crippen, a *Daily Times* reporter.

Moving to the northeast after take-off, the *America* passed over the suburb of Mount Auburn and the nearby towns of Sharon, Lockland, and Reading, Ohio. Godard ignored the danger of an approaching storm and attempted to climb above the clouds, reaching an altitude of 17,450 feet:

Here sensations of chilliness came over us, and the atmosphere being damp, our party experienced very unpleasant feelings. Overcoats, shawls, wines, cordials, etc. were not sufficient to keep us from shaking, and for a time, the "chills" had control of us. The storm was terrible, the lightning leaped from cloud to cloud accompanied by peals of thunder. The gloom was fearful. Never had I beheld such warring of the elements. . . . The wind which drove the clouds onward so rapidly, also hurried us on at the rate of seventy miles an hour.[75]

Godard instructed his passengers to crouch in the bottom of the basket, while he pulled the valve cord to descend. The balloon's anchor, caught momentarily in the branches of a tree, broke loose, placing the voyagers at the mercy of the winds a second time:

Suddenly we felt our car rushing over the tops of trees, crashing and tearing the limbs as the balloon was driven along. Mons. G. gave us the valve rope, and mounting the side of the car, he ordered us to hold fast. In another moment we landed in a corn field, and by the force of the wind we were dragged and bumped along the ground, a distance of half a mile, now through a fence, then striking a stump or tree, or whirling through the corn stalks at a fearful velocity; our heads rapped each other, and not infrequently we saw stars all around.[76]

Godard was hurled to the ground as the basket struck a large tree; Latham and Hoel were tossed out one on top the other; Belman and Crippen remained with *America* which came to rest near a large, dead tree. Godard, Latham, Crippen, and Belman escaped with a variety of cuts and abrasions, while Hoel fared less well with three broken ribs. Latham and Belman made their way to the nearby home of George E. Smith, who dispatched a rescue party. Smith informed them that they had landed near Caesar's Creek, 3 miles south of Waynesville and 51 miles from Cincinnati.[77]

The Godards spent the 1856 season in the Northeast, with Eugène performing his usual aerial gymnastics. They began a series of ascents on horseback with a flight from Manchester, New Hampshire, on July 4, 1856. They went aloft at 8:20 P.M., after waiting most of the day for the wind to die down. It must have been an extraordinary sight: Godard standing on the back of the horse dangling beneath the balloon while Mme Godard, alone in the basket, acknowledged the "immense cheering" of a crowd that covered "acres and acres" of the surrounding countryside. The horse, according to a report in the Manchester *Mirror*, was "hanging his head low down, with eyes intently fixed upon the earth, without struggling a particle."[78] After a twelve-mile circuit, they anchored the balloon in the top of a grove of trees at 8:55 P.M. The horse seemed quite content to graze on the tree tops until the entire ensemble could be maneuvered to the ground. The citizens of Manchester were delighted. The only complaint heard as the crowd dispersed that evening was that "the horse did not carry as good a head and tail as was shown on the bills."[79]

In spite of the total loss of the balloon *America* from an accident during an inflation in Boston on August 3, the 1856 season was a profitable one for the Godards. The receipts for a single Boston ascent on July 21 had totaled $3,000.

When the pair arrived in Montreal that August, the first order of business was to advertise for seamstresses to work on a new balloon. The *Canada*, which was exhibited at the Bonsecours Market soon after construction, was quickly pressed into service for flights on September 8, 15, and 22, carrying three passengers aloft on each occasion.[80]

The Godards returned to the United States that fall and closed their 1856 season with a series of flights from Philadelphia. Most of these ascents were made with the larger balloons, so that passengers could be carried at $50 a head. The well-known local daguerreotypist W. L. German went aloft on October 25. Several weeks later, on November 12, Godard made another typical ascent with William Van Osten, Edward Bills, E. Watson, and T. M. Coleman. The aeronaut spiced the occasion with low-level, high-speed passes over the treetops and a touch-and-go landing before bringing the four-hour flight to a close 18 miles from the take-off point.[81]

The Godards toured the Midwest once again in 1857. On October 29 Eugène and his wife carried P. W. Huntington and R. H. "Rocky" Thompson aloft from the Capital City Premium Fair and Exhibition Grounds in Columbus, Ohio. Godard performed his usual acrobatics on the trapeze bar prior to landing near Reynoldsburg, Ohio.[82]

The two aeronauts had signed on as a stellar attraction with Search's New Orleans Minstrel Show in Columbus and were advertising an upcoming ascent on horseback for late September and early October. By November 9 poor weather was still preventing a flight. The Godards were in such dire straits that thirty-six local citizens gave a benefit dinner at which $89.50 was raised so that the aeronauts could return to the East. In spite of these short-term cash flow problems, however, the Godards were prospering. They could look forward to an 1858 season that was full of profitable bookings and at least one exciting special event. Flying in Philadelphia early in 1858, Godard had made the acquaintance of John Steiner, a local aeronaut with considerable experience. The two men had agreed to stage a special balloon race in Cincinnati in 1858. With several months to publicize the event, they could hope to attract a record crowd.[83]

The 1858 season seemed to hold a great deal of promise for other American aeronauts as well. They were riding the crest of a wave of balloon enthusiasm. Their names and faces were familiar, their exploits legend. Like Godard and Steiner, many American aeronauts were fully prepared to take advantage of this situation by attempting the most spectacular flights of their careers. As a result, the final two years of the decade would be filled with stirring aeronautical events. The stage was set for a short but unprecedented period of aerial excitement. From 1857 to 1860 the attention of the American public would be focused on the balloon as never before.

Chapter 9 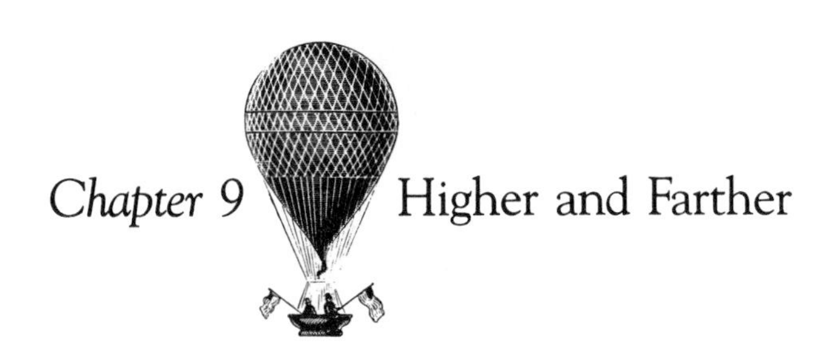 Higher and Farther

AS THE GODARDS BEGAN THEIR TREK WEST IN THE SPRING OF 1858, THEY sensed a new spirit abroad in the land. The editor of the Philadelphia *Ledger* also took note of the enthusiasm for the balloon that was sweeping through the Midwest:

> *In many of the towns of the interior, ascensions of balloons are taking place, lives are lost, and new men, after not more than one or two ascents with others, set up on their own account. . . . Thousands are moved to the attempt because of the descriptions of the newspaper press. . . . We doubt if for mere amusement any such ascensions should be countenanced. But in nine cases out of ten this is not mere amusement with those who ascend. Each one hopes to make some discovery, or to acquire that knowledge by which he can hereafter make a long and more important voyage.*[1]

The Godards arrived in Cleveland, Ohio, on June 30, 1858. Eugène Godard advertised that he would make his four hundredth ascent from the Public Square on July 5. The story dominated the newspapers for days before and after the event. There was coverage when the city council voted its thanks to the Cleveland Gas Light Company "for their liberal donation of gas for the balloon ascension," while Godard regaled reporters "with interesting facts connected with ballooning to which we have heretofore been strangers."[2] The flight went as planned but created some ill will for the Godards because of a controversy as to whether the representative of the Cleveland *Herald* or the *Review* should have been allowed to accompany the aeronauts on their voyage.

Undaunted, the Godards traveled as far west as Iowa that summer, having flown in St. Louis and Springfield and been forced to cancel a Cairo ascent because of bad weather. They were back in Cincinnati that October, arriving just as the publicity surrounding their upcoming balloon race with John Steiner was approaching its peak.

Steiner was an experienced aeronaut who had made his fortieth ascension the year before. A native of one of the German states, he had apparently immigrated to the United States in 1853. Judging from the outlandish spelling and grammar of his Civil War dispatches, he had not yet mastered the English language.

Steiner had first come to national attention the year before when he very nearly drowned in Lake Erie. By 1858 he had added his name to the growing roll of those who hoped to fly the Atlantic. When he launched from Erie, Pennsylvania, on a June afternoon in 1857, however, his goal was to become the first man to fly across the lake to Canada.

As Steiner reported, "The weather could not possibly have been more unfavorable than it was at the time I had determined upon to start."[3] A storm was obviously brewing, with fierce winds blowing out over Lake Erie.

Steiner cast off with some trepidation. He felt a "dreary sense of loneliness" as he watched "man and his numerous works . . . receding rapidly away . . . apparently going down, down—its inhabitants appearing like little black pins on a cushion." Reaching an estimated altitude of 3 miles and moving rapidly toward Canada in the face of the storm, Steiner found himself "renewed" by the "beautiful look-out." Flying in and out of the clouds, he could see the lake almost from end to end and was able to count thirty-eight ships plowing across the surface of the water. But Steiner lost his short-lived sense of well-being as he reentered the clouds:

> *Imagine my feelings when I realized the fact that the clouds upon which I had often gazed in my childhood, and fancied to be the very sky itself, glided along beneath me. Oh! What a scene was transpiring around me! As I cast my eyes about I saw huge masses of vapor sailing towards me, like mountains enveloped in mist, or wreathed into all manner of shapes and appearing like gigantic phantoms. Every moment the surrounding masses of clouds were illuminated by flashes of lightning, succeeded by terrible crashes of thunder, in the very midst of which I seemed to be floating, and my excited imagination led me to fancy that I would feel my frail car quiver at every shock.*[4]

Just as Steiner was approaching the Canadian shore below Long Point, the wind reversed, blowing the balloon back down the lake toward Buffalo. With the approach of dusk, Steiner realized that he could no longer hope to make a landfall in the safety of daylight. Moreover, he noted, "After my ramble through the clouds I felt the cravings of nature, and therefore had no desire to spend a

MR. STEINER'S BALLOON SUSPENDED OVER LAKE ERIE.

John Steiner over Lake Erie, June 1857

supperless night floating through space."[5]

Glancing down, the aeronaut spied the wake of the ship *Mary Stewart*, which was making her way up the lake. He immediately valved gas, hoping to land near the ship, but miscalculated and touched down three miles upwind. Seven minutes later Steiner bounded past the steamer, twenty yards astern. The balloon had struck the water repeatedly, each time rebounding twenty feet or so into the air. Steiner's anchor was dragging through the water, but it did little to halt his progress as he went "rushing down the Lake at railroad speed." Even when a small boat crew from the steamer made fast to the anchor line, the balloon continued to sail down the lake.

Steiner had little choice but to leap from his balloon and swim for the boat. He returned to Buffalo aboard the *Mary Stewart*, then returned home to Philadelphia before continuing to Cleveland, every step of the way bemoaning the loss of his $500 balloon. The aeronaut had every reason to feel disconsolate. Newspapers reported that this was the third expensive balloon he had lost. The tattered remnants of the craft were eventually discovered about one hundred miles in the interior of Canada and returned to Steiner in exchange for a $100 reward.

Steiner found it difficult to reject his penchant for over-water flying. He launched from Toronto in August 1859 and flew 160 miles, almost the entire length of Lake Ontario, to a landing 8 miles south of Oswego, New York.[6] In the fall of 1858, however, both Steiner and Godard were absorbed in preparations for the "Great Balloon Race."

Early that fall Cincinnati newspapers began to carry advance notices of the race, which was scheduled for October 18, 1858. To stimulate spectator interest, reports were circulated that the two aeronauts were bitter rivals and that Godard had challenged Steiner to determine once and for all who was the finest aeronaut in the nation.[7] To further insure maximum attendance, nationalistic sentiment was injected into the contest with the announcement that Steiner would fly the Stars and Stripes from his basket while Godard would display his native tricolor. Excitement mounted as the day of the race approached. According to one reporter, side betting was animated, "for a race of this kind is something new under the sun."[8] Both participants received inquiries from individuals wishing to join in the race. Steiner was approached by a young woman who offered to join him as a passenger, but who refused to pay for the privilege. Another young Cincinnatian who wanted to go along was refused when he demanded $100 plus return travel expenses in the event of a failure.[9] While Steiner ultimately entered the race alone, Godard announced that he would be joined by George Hoel, the river pilot who had accompanied him on the disastrous flight of October 1, 1858.

Monsieur Godard arrived in Cincinnati a few days ahead of his rival and made a number of ascents to publicize the forthcoming race. His balloon, the *Leviathan*, was the larger, boasting a 36,000-cubic-foot gasbag. Steiner's 30,000-cubic-foot *Pride of the West* was constructed of the finest Irish linen, which made it appear almost transparent in flight. The egg-shaped gasbag could develop 1,000 pounds of lift; the wicker car, four feet high and trimmed with red velvet and gilt lace, carried 600 pounds of ballast.[10]

Shortly before the great balloon race, J. C. Belman, the intrepid reporter who had accompanied Godard on the 1855 flights, announced that he too would enter the competition. He bragged that he would not only beat his rivals but would far outdistance them, as he planned to journey to the North Pole and return "by way of the equator."[11]

October 18, 1858, dawned clear and balmy, with a slight wind blowing

STEINER JUMPING FROM HIS BALLOON INTO LAKE ERIE.

Steiner's leap, June 1857

from the southwest. A crowd estimated at 30,000 to 40,000, each paying the twenty-five-cent admission fee, had gathered in the amphitheater to witness the inflation and launch. Many more, however, preferred to watch from the side-walks. Pickpockets and thieves took full advantage of the situation, reaping a rich harvest of watches and wallets.[12]

By three o'clock that afternoon the area was "one dense mass of human beings. Wagons, carriages, carts, and vehicles of every description, were drawn up along the curb stones of the streets bounding the square. The streets were completely blockaded for a considerable distance in every direction. The fences were taken possession of, and the trees were waving with young and old America in the branches. . . . To have attempted a count, would have been like enumerating the leaves of the forest."[13]

Belman's balloon, the *Niagara*, smallest of the three, was fully inflated and ready for launching by the announced time of 4:00 P.M. Waving an American flag, he moved slowly to the northeast at an altitude of about one mile. Nearing Carthage, Ohio, his craft fell to a half mile and was caught in an air current that spun it round and round. The aeronaut dropped ballast and rose out of this current, but was forced to land soon after on the farm of Andrew Riddle, one and a half miles from Glendale, far short of his announced goal of the North Pole.[14]

In the amphitheater, both Steiner and Godard required more time than expected to fill their balloons, and the crowd outside grew restless. Spectators broke through the line established to keep them at a distance, creating such a furor that Steiner was unable to hear the instructions of Mayor N. W. Thomas, head of the committee supervising the race. Perhaps mindful of the experience of other aeronauts caught by a menacingly disappointed crowd, Steiner ordered an immediate release. Some inexperienced members of the ground crew handling the restraining lines held on too long, and the *Pride of the West* began its climb by careening through a refreshment tent and scraping over the steeple of a nearby church. The aeronaut dropped ballast immediately and was soon out of danger, with no serious damage to the balloon or its equipment. Once safely in the air, he dropped advertising leaflets and waved his flag while awaiting the ascent of Monsieur Godard. The Frenchman seated himself on the hoop above the car and, with Hoel in the basket, made a safer, if less eventful, ascent. In a few minutes the *Pride of the West* and the *Leviathan* were within shouting distance of each other. Godard then uncorked a bottle of wine and toasted "the Great Republic" and "the greatest Aeronaut in America, Professor Steiner." Steiner in turn drew a bottle of fine Catawba from his provisions and returned a toast to "the justly celebrated French Aeronaut, Monsieur Godard." Hoel offered the traditional tribute to "sweethearts and wives."

The two balloons, now at 15,000 feet, were approaching each other so rapidly that a collision seemed unavoidable. Godard released more ballast but was unable to gain altitude rapidly enough and struck the *Pride of the West* midway up the gasbag. Godard and Hoel were then able to push the two balloons apart, and they rose swiftly to an altitude of 14,000 feet. Both aerostats were now traveling toward the northeast at about forty miles per hour. Steiner released ballast and rose, only to see Godard drop to 500 feet. He assumed that his rival was preparing to land but soon discovered that Godard was only buzzing farmers at work in a field. He later remarked that when Godard descended, "the country people . . . shouted like a hundred steam whistles." After forty-five minutes in the air, the *Pride of the West* and the *Leviathan* lost sight of each other.

Passing over Dayton, Ohio, at 6:30 P.M., Steiner dropped handbills printed with greetings, which were blown far beyond the city limits. At 6,000 feet the aeronaut became quite chilly as the sun set, and prepared to settle down for the night. He stowed his provisions and let down a 3,000-foot dragline that was tied to his wrist. The balloon would lose altitude as the gas contracted in the cool night air, and as it sank toward the ground more of the rope would drag, thus lightening the load and arresting the balloon's fall. If an obstacle were encountered, the tug of the rope on the aeronaut's wrist would awaken him. Ready for his night aloft, Steiner wrapped himself in a shawl and fell asleep. At about 10:30 he was shaken to a rude awakening as his craft catapulted into a tree, tangling the aeronaut in the ropes and throwing his provisions and ballast to the ground. The sudden loss of weight caused the *Pride of the West* to gain altitude rapidly, forcing Steiner to valve gas in order to bring the balloon under control. Having stabilized his ascent, he glimpsed Lake Erie gleaming in the distance, and, in view of the loss of his food and ballast, Steiner decided to land immediately rather than attempt a crossing. He put down on the farm of A. G. Townsend, just outside the corporation limits of Sandusky, Ohio. Seeing no one about, he deflated the balloon and fell asleep in the field. The next morning, with Townsend's assistance, Steiner packed his equipment, placed his aeronautical gear on an Adams Express car, then boarded a train for Cincinnati.

To Steiner's astonishment, Godard and Hoel boarded the same train about twenty miles south of Sandusky. The two had enjoyed a less eventful trip in the *Leviathan*, descending in Huron County at about ten o'clock the previous evening. Since the race was to be decided in favor of the man who covered the greatest distance, John Steiner, who had traveled 230 miles, was declared the winner. While most Cincinnati reports praised the courage and skill of the aeronauts and prophesied a brilliant future for the art of ballooning, editorials in other sections of the state were not so glowing. The Cleveland *Herald* declared the race a "hum-bug" and wondered what possible good could result from such a venture.[15]

Exactly one month before the great balloon race in Cincinnati, Silas Brooks and Samuel Wilson had made their own contribution to the aeronautical excitement that was building in the Midwest. Brooks, while retaining his position with the St. Louis museum, had hit the aeronautical trail once again in the summer of 1858. Now an employee of the Ericson Hydrogen Balloon Company, he made at least three Illinois ascents that year: Jacksonville on July 3, Springfield on July 5, and Alton on July 10. Brooks must have faced the usual financial problems, for we find the editor of the *Illinois State Journal* suggesting that a collection be taken up among the spectators outside the enclosure to help defray the costs of the Springfield ascent.[16]

George Brooks was back in the air as well. In the fall of 1857 he had published a harrowing account of a flight from the St. Louis area during which he had encountered high winds, snow flurries, and violent thunder and lightning.[17]

Silas Brooks had encountered Samuel Wilson while touring Illinois late in the summer of 1858. Wilson had returned to the United States from his abbreviated Mexican tour in 1857, and worked his way through the South, making ascents as he went. We have noted his flight with Mrs. E. W. Davis at Tuscumbia, Alabama, on July 10, 1858.

When Brooks fell ill prior to a scheduled ascension from Centralia, Iowa, on October 23, 1858, Wilson offered to stand in for him. The flight covered 20 miles and ended with Wilson descending safely into the top of a tree on a farm owned by one Benjamin Harvey.

Harvey was delighted with his surprise visitor and supervised the crew of hastily assembled volunteers who worked to extricate the balloon. With this accomplished, Wilson fell into a conversation with the bystanders, while Harvey climbed into the basket, hoping to make a tethered ascent. He quickly realized that he was too heavy for the partially inflated balloon and placed his three children in the car. The craft was still too heavy, and Harvey instructed his oldest daughter to climb out, leaving eight-year-old Martha Ann and three-year-old David Isham by themselves. The stage was now set for one of the great balloon melodramas of the century.

The inexperienced ground crew, startled to find the balloon rising at last, lost their hold on the tether. The dangling grapnel hook tore through a rail fence as the balloon climbed out of the yard. The distraught parents listened as the plaintive cries of "pull me down, father" grew fainter. Horsemen were immediately dispatched to alert the countryside, while a party of men and boys did their best to follow the drifting balloon.

The news of the runaway balloon created a sensation in Centralia. Brooks was roused from his sickbed to assure the gathering crowd that the balloon would descend on its own within two or three hours, probably somewhere within a 30-mile radius of the take-off point. The aeronaut sent his assurances to the parents that all would be well, remarking that the only real danger might occur if Martha Ann stepped out of the car first after landing, accidentally allowing her younger brother to rise again when free of her weight. Local citizens were much less confident, expressing fear that the children might freeze to death in the air or starve to death on the ground.

At three the following morning, Mr. Ignatio Atchison stepped out on his porch to observe the much-publicized passage of Donati's comet. As his eyes grew accustomed to the darkness he was startled to note "an immense spectre rising from the top of a tree twenty yards away." He roused his family and approached the mysterious object, finally coming close enough to hear a faint cry from the top of the tree. "Come here and let us down. We are almost frozen." The lost had been found.[18]

Atchison immediately sent for help and began cutting away the branches that imprisoned the children in the balloon. News of the rescue was announced in Centralia churches on Sunday "amid ecstasies of joy." When the children returned the next day, they were greeted by cannon salutes "and a general jubilee." William B. Matthews, the local daguerreotype "artist," captured the image of the young hero and heroine on a silvered plate. Dressed in their Sunday best, Martha Ann and David Isham Harvey are seen seated in a rosewood armchair much finer than anything to be found in their own prairie home, their faces a mixture of curiosity and pride.[19]

The fall of 1858 was the season for runaway balloons. The same newspapers that exalted in the safe rescue of the Harvey children were also reporting the continued search for the remains of a promising young aeronaut named Ira Thurston who had last been seen on the morning of August 16, 1858, desperately clinging to the remnants of a balloon that was slowly drifting out of sight over Lake Erie.

Thurston, who gave his hometown as Lima, New York, had first come to public notice in September 1850 when he advertised an ascension from the Provincial Exhibition being sponsored just over the border in Niagara, Canada. On this occasion Thurston had disappointed the Canadian patrons who had stood "so many hours . . . in the burning sun," only to have the aeronaut finally admit that he would be unable to fly. The St. Catharine's *Journal* noted that "the people conducted themselves admirably under the disappointment" but advised Thurston that "all possible contingencies, except those arising from wind and weather, should be provided against by any man calling together such a multitude of people."[20]

Thurston had persisted however. On October 6, 1851, he had made the first flight from Cleveland, Ohio. Going aloft in the balloon *Buffalo* from the foot of Erie Street, the aeronaut flew to a safe landing in East Cleveland. Over the next seven years Thurston would continue to fly from towns in New York, Michigan, and Ohio. After moving to Adrian, Michigan, Thurston began flying with a partner, W. D. Bannister.[21]

Local merchants in the Adrian-Monroe, Michigan, area were anxious to encourage the efforts of the two aeronauts, who had been "instrumental in putting hundreds of dollars into the hands of the businessmen of this city, and

Ignatio Atchison discovers the Harvey children, October 24, 1858.

Martha Ann and David Isham Harvey after the rescue

it will be a shame if they are allowed to suffer pecuniarily by making ascensions here."[22]

The pair usually flew together in Thurston's balloon *Adrian*. In the best show business tradition, they advertised the craft as being constructed of the best India silk, 126 feet in circumference, and capable of holding the "enormous amount" of 240,000 gallons. This was so much more impressive than giving the diameter or the cubic footage.

Shortly before nine o'clock on the morning of August 16, 1858, Thurston and Bannister climbed aboard the *Adrian* for what promised to be nothing more than a normal aerial voyage. Thurston was dressed in a pair of checked trousers, boots, a snuff-colored coat, and a top hat. The aeronaut had accented his outfit with a black silk kerchief tied around his neck. A pair of gloves were stuffed in his coat pockets, along with a silver watch, a jack knife, $1.36 in a buckskin purse, a letter to a Philadelphia address quoting the price for a new balloon, and a few calling cards. It was a gorgeous day. The whole world seemed spread out before them as they rose into the air.

> *The whole end of Lake Erie was studded with picturesque islands, an occasional vessel the only sign of life on the vast expanse of water. Nearer, the city of Monroe [Michigan] buried in a rich grove, which only afforded an occasional sight of the buildings. Further to the south, lay Toledo and Maumee City; just underneath and behind was the city of Adrian, and the neighboring towns of Manchester, Tecumseh, and Quinton. The Michigan Southern Railroad and its branches were distinctly traced out, and the trains could be seen traveling along them. Still following around, the eyes rested on Detroit, its tall spires and endless mass of building, and the beautiful river which flows so silently by it. Farmers working in the fields; characters moving along the highways; boats sailing upon the waters; the hustle and stir of city life, with the quiet and retirement of the country and prairies, rivers and lakes.*[23]

After forty-five minutes in the air, Thurston and Bannister landed in Riga Township, Lenawee County, 17 miles from Adrian. The aeronauts remained in the car for half an hour after touchdown, chatting with spectators while slowly valving gas to empty the balloon. Finally Thurston suggested that they climb from the car. They deputized the bystanders to hold down the balloon while they detached the basket, removed the netting, and overturned the gasbag to allow the hydrogen to escape more rapidly through the open appendix. Bannister watched as his companion crawled inside the netting and clambered to the top of the balloon, which was being held low to the ground. As the fabric billowed around him, Thurston sat down with the 13-inch wooden valve held between his knees. Holding the valve flappers open with his hands, he allowed his own weight to force the gas out of the envelope.[24]

Thurston was following a standard, if dangerous, procedure for rapid deflation. The catch came with the next step. The aeronaut ordered all hands to let go the lines. Under the best of circumstances, the free envelope would rise a short distance into the air and, because of the aeronaut's weight, flip over to allow the gas to escape through the appendix. So experienced an aeronaut as John Wise had occasionally resorted to this procedure until one day when, caught in the netting, he was sent flailing over a grove of trees "like an eel in a net." When the balloon opened, he suffered a severe fall.[25]

Thurston was to be less fortunate. Warning the bystanders that they "might see another ascension now," he ordered them to release the lines.[26] The envelope rose and flipped as expected, but the loose lines had wound themselves around the appendix so that the gas could not escape. When last seen, Thurston was winging his way toward Lake Erie seated precariously on the frail valve disc, his arms wrapped around the balloon. One of the ground crew, J. Westerman, grabbed a line in an attempt to pull the bag back down, but wisely chose to let go when the rising balloon lifted him ten feet in the air. Bannister and the spectators watched helplessly as Thurston and the balloon dwindled to a speck, then disappeared.

Before Bannister had returned to Adrian with the car and netting, reports had already begun to filter in. Don Munger, a conductor on the Great Western Railroad, claimed that Thurston had landed safely and was on his way home. His story proved false, but another, more creditable, witness claimed to have seen the aeronaut as he passed over the lakeshore. The black kerchief was still around his neck, she noted, and he kept raising and lowering his legs as though they were tired or cramped. A boy living near Baptiste Creek, Canada, on the shores of Lake St. Clair, had been startled by a whistling noise in the sky some time later. He looked up just in time to see something fall into a nearby wood.[27]

Several days later the balloon was found near Baptiste Creek. The valve had been ripped from the envelope, and there was no sign of Thurston. When the remains of the envelope were exhibited in Adrian, schoolchildren petitioned for release from class to file past the torn fabric. "They were accommodated," noted a local newsman, "and were much more demonstrative than the older visitors."[28]

On September 25 a group of Monroe, Michigan, residents found a body "very much decayed," resting under two feet of water at the edge of a marsh. When it was determined that decomposition was much too far advanced for this to be Thurston, the corpse was allowed to remain where it lay.

Thurston's disappearance was now a matter of interest across the nation. William Henry Hoag, a fifteen-year-old boy scouring the woods four miles east of Sylvania, Michigan, for his father's sheep, finally found Thurston's remains the next March. The bones had been cleaned of flesh but were identified on the basis of the clothing and the contents of the pockets.

Thurston was not the first American to die in a balloon accident. Timothy

Winchester had disappeared over Lake Erie during a flight from Norwalk, Ohio, in the fall of 1855. Winchester was a newcomer to aeronautics, having made his first flight from Milan to Hudson, Ohio, earlier that fall. Prior to the ascent from Norwalk he had announced that he would go "higher and farther than any aeronaut had ever dared to think of going."[29] We will never know whether Winchester achieved his goal or not. He was never seen again.

While the attention of the entire nation was riveted on the life and death adventures being played out in the skies of the Midwest during the summer and fall of 1858, John Wise, Samuel Archer King, James Allen, and others were enthralling audiences in the Northeast. Nor was the South neglected.

For years well-known aeronauts had made quick sweeps through Dixie, flying in major cities all over the region. Ascents such as that offered by a Mr. Lehmann from the Algiers section of New Orleans in the fall of 1846 had become fairly common. Flying 23 miles in just over half an hour, Lehmann descended near a fire tended by plantation slaves. His balloon caught fire and burned.

Not until the 1850s, however, did the South develop an indigenous group of aeronauts who confined their activities to the region. Alexander J. B. DeMorat was one of the first aeronauts to make repeated ascensions in the South. DeMorat claimed that he was a graduate of St. Cyr and had gained his initial balloon experience as chief engineer of aeronauts for the Fourth Army Division during the Second Republic. Applying for a position in the Union Army Balloon Corps in 1862, he claimed to have made 185 ascensions.[30]

At the time of DeMorat's second ascent from Congo Square (the Place d'Armes), New Orleans, on December 31, 1856, local newspapers remarked that the aeronaut had only 60 ascents to his credit. We may assume, then, that if DeMorat was honest with both newspaper interviewers and the U.S. Army, he completed some 125 flights in America between 1857 and August 1862.[31]

DeMorat occasionally ventured out of the South, as in the spring of 1857 when he traveled to Canada and made a number of flights in the northeastern United States. In June 1857 he flew from Newark, New Jersey, with two young men. They passed over New York City, thrilling to a sight "beautiful in the extreme," before proceeding toward Long Island Sound and on to a landing at Clinton, Connecticut, two and a half hours after take-off.[32]

But it was in New Orleans that DeMorat enjoyed his greatest popularity. Between 1856 and 1859 he became the city's reigning aeronaut. As on December 31, 1856, when he flew from Congo Square to the Rimbaud plantation with James J. Lane of the New Orleans *Picayune* and H. Paine, a local hotel agent, DeMorat invariably drew "a numberless crowd."[33]

Like so many other successful aeronauts, he was a superb showman. Prior

THE AERONAUT THURSTON CARRIED OFF ON THE BALLOON.

Ira Thurston's ride, August 16, 1858

to an ascent in January 1857, he staged a well-publicized footrace between a Mohawk and an Iroquois Indian chief. This was to be an exciting day for DeMorat as well. He flew 35 miles to a landing in a swamp where he thrashed about until rescued by a party of hunters who stumbled on him the next day.[34]

DeMorat performed his most outlandish stunt on February 8, 1858. Since the early years of the nineteenth century, European aeronauts had been ascending seated on the back of various animals. Like the Godards, most balloonists had chosen to go aloft on horseback. A few had been more imaginative. The French aeronaut Margat, for example, had gone aloft on June 5, 1817, seated on a gorgeous white stag named Coco.

DeMorat and a colleague, S. S. Smith, added a peculiarly American touch to this penchant for animal ascents when they launched from Congo Square astride two young alligators. Providentially, their flight was brief, ending in a flower garden at the corner of Felicity and Camp streets. The editor of the *Daily Picayune*, apparently uncertain what to make of this, could only comment that while DeMorat and Smith had "been high often," he was certain that "the alligators . . . were never so high before." We are blessed with a print of Margat and Coco. Would that we had a similar illustration of DeMorat and Smith on their mounts.[35]

DeMorat was also quick to take advantage of interest in "balloon racing." On February 15, 1858, months before Godard and Steiner paired off in Cincinnati, DeMorat and a visiting English aeronaut, Richard Wells, staged their own aerial competition. DeMorat flew the hydrogen aerostat with which he had made so many ascents. His name, A. J. DeMorat, was inscribed in large block letters across the face of the gasbag. Wells went aloft with "a fancy cotton concern" inflated with alcohol vapor. DeMorat's balloon was the favorite with the betting fraternity "as it presented the most approved appearance in the ribbed rotundity of its silky sides."[36]

Wells's balloon took an early lead. As one editor commented, it seemed "natural" for the "alcohol" to "get high." But when "anybody . . . gets high under alcohol, he, she, or it, is likely to have a speedy fall." So it was with Wells. His balloon stopped short at 1,000 feet. DeMorat hurtled past him "and sailed off majestically in the blue ether."[37] As the alcohol began to condense in Wells's balloon, the craft descended rapidly toward the street. Thousands of people, "animated by an amiable desire to be 'in at the death,' ran frantically toward the spot which promised to be the scene of the catastrophe."[38]

Fortunately, Wells landed safely on the roof of a house attached to the rear of the Bank of Louisiana. The aeronaut clung to the roof while the balloon tumbled into a courtyard crowded with spectators. One man watching from a neighboring roof became so excited that he slipped and tumbled into the street. Severely injured, he was carried to an apothecary shop, trailed by many of the crowd, who assumed that Wells was the victim of the accident. The aeronaut was finally rescued, while DeMorat continued his flight to a safe landing near the Belleville Iron Works. "So," remarked the New Orleans *Picayune*, "terminated the first balloon-race of the season."[39]

DeMorat was to suffer financially as a result of the race. Mr. Rawlings sued the city for $500 in damages caused as the crowd following Wells's balloon had swarmed through his garden. As a result, the Board of Aldermen refused to consider DeMorat's request for permission to make a series of eight flights from Tivoli Square.[40]

But aeronautical activity continued in New Orleans. Wells barely escaped disaster that March when his hot air balloon caught on the eaves of a building. Over the next three months, DeMorat, Wells, S. S. Smith, and a gentleman named Norble continued to fly from the city. It was DeMorat's turn to be rescued from the Mississippi following an ascent from the Pavillion Gardens that May.[41]

While DeMorat, Wells, and company were entertaining citizens of New Orleans, Charles E. Cevor and his companions were performing a similar function for the residents of Savannah, Georgia. In view of John Wise's positive inability to spell the names of even his close friends correctly, it seems quite possible that "Mr. Crever," a pupil to whom he sold his balloon *Comet* in 1843, was the same Charles Cevor who was to become one of the best-known of the southern aeronauts.[42]

Cevor had placed his balloon *Montpelior* on exhibit at the Armory Hall in Savannah in February 1860. A "nearly disastrous" ascent by Richard Wells that March led the editor of the *Daily Morning News* to hope that sufficient money would be pledged to tempt Cevor to give the citizens a proper show. In fact, Cevor did ascend from the Armory Hall Yard with his friend Mr. Dalton on March 8. Climbing to an altitude of over 10,000 feet, they were caught in a wind blowing out to sea. Working frantically, Cevor ordered Dalton onto the rim of the basket as he valved gas as rapidly as possible. Eighteen minutes after take-off, they struck the water of Calabaya Sound. The balloon billowed like a sail, towing the two airmen into the Atlantic. Cevor's anchor finally caught, holding the two men, who remained submerged to the chest. George A. Savage and a party of slaves finally came to the rescue, but the balloon was lost.

Cevor and Dalton returned to Savannah on March 12. The city was proud of its new heroes and hailed the ascension as "one of the great events of the season."[43] The *Montpelior* was finally found near the Sewanee River in Florida, but Cevor had already raised the funds to construct a new balloon, the *Forest City*. The envelope, which was sewn in a large storeroom owned by Mr. E. W. Baker, was an object of interest to the citizens of Savannah even before the first flight was scheduled from the Barracks Yard on the Monday after June 21.[44]

For all its excitement, the 1858 flying season had been little more than

The narrow escape of Cevor and Dalton, March 8, 1860.

preparation for the next year. Over a distance of almost a century and a quarter, 1859 stands out as something of an *annus mirabilis* in the history of American aeronautics. It was a year dominated by record flights and during which many new aeronauts appeared, and everyone seemed bent on flying across the continent or ballooning over the Atlantic. Great long-distance aerial voyages, well-publicized crashes, and the search for lost balloonists dominated the news. It was a spectacular culmination to an era that had begun when Peter Carnes had first sent Edward Warren aloft from Howard Park in Baltimore.

For the citizens of Dayton, Ohio, the year 1859 opened with the appearance of "Professor" Wilson, who offered lectures on ballooning and natural science as well as lantern slide talks on his "world travels." As Dayton newspapers fail to provide the professor's first name, we cannot be certain whether or not the aeronaut in question was Samuel Wilson.

The mysterious Mr. Wilson claimed that the Rothschild family had invested $15,000,000 in his aeronautical experiments, but all he could produce in the way of a balloon was a small hot air aerostat constructed of muslin coated with glue and ochre, to which a clothes basket was attached with rope to serve as car. He refused to risk his life in this makeshift contraption, but he was more than willing to accept applications from volunteers like Jacob Sellers, an employee of Frank Welty's Ice Cream Parlor. Sellers's tethered ascent came to a premature end when the large holes in the balloon, which had been tied off with twine, opened up several hundred feet in the air.

It is not too surprising to learn that Wilson was forced to search for a new aeronaut. A Dayton teenager named Brown made the second ascent on May 10, 1859. The inexperienced ground crew mishandled, then dropped, the tether lines, sending the balloon skidding over the city in wild gyrations. The terrified Brown spent most of his harrowing trip standing up in the tiny basket with his arms clasping the balloon.

The latest performance was too much for the citizens of Dayton. Now openly critical of his wild claims of aerostatic experience, they invited him to leave town by the most expeditious route.[45]

To the north of Dayton, yet another aeronaut, B. F. H. Lynn, the editor of the *Erie Dispatch*, was attempting to cross Lake Erie. Forced down in the lake following a Cleveland ascent on July 4, Lynn and a companion bobbed about in the water until rescued by a passing steamer.[46]

The attention of the entire nation focused on St. Louis that June as a party of aeronauts and adventurers led by John Wise prepared to launch themselves on the first stage of what they hoped would culminate in a much longer over-water flight. Wise was finally about to attempt his long-contemplated and oft-postponed Atlantic crossing.

The episode began, not with Wise himself, but with a twenty-nine-year-old "practical seaman" from Troy, New York, John La Mountain. La Mountain had received a better than average education in ballooning during a "novitiate" of half a dozen ascents with Wise. He was an apt pupil. Wise believed that his student's experience as a seaman "made him proficient in the management of sailing paraphernalia, and . . . in the prognostication of weather."[47]

In fact, La Mountain was a much more remarkable man than Wise was willing to admit. Thin and ascetic, a pair of clear, penetrating eyes and a strong chin masked by an unruly spade beard dominated his sharp features. An admirer would later characterize him as "a brave man, who probably does not know what fear is."[48] La Mountain's considerable reserves of courage and resourcefulness would be tested to the limits during the summer and fall of 1859.

Following his apprenticeship, La Mountain had purchased a balloon from Wise and made a few ascents on his own. He had inherited more than skill in the air from his mentor however. The elusive Atlantic had already begun to fascinate him.

Late in October 1858, he placed an advertisement in the Troy *Whig* calling for investors willing to finance the construction of a $30,000 balloon capable of flying the ocean. The most promising response to the notice came from Mr. O. A. Gager of Bennington, Vermont. La Mountain took his balloon to Bennington and offered Mr. Gager a ride. A deal was struck soon thereafter. Gager, "being better supplied with worldly goods, agreed to supply the necessary money, whilst Mr. La Mountain was to supply his time and mechanical skill."[49]

The aeronaut returned to Troy, where he worked for six months to construct the balloon, which was to become famous as the *Atlantic*, in rented quarters on the fairgrounds at nearby Lansingburg, New York. La Mountain put in twelve-hour days supervising a group of local seamstresses laboring to cut and sew the 2,200 yards of Chinese silk. The finished product had a circumference of 180 feet and employed an estimated six miles of cordage in the network. When the balloon was fully inflated and both the large wicker basket and the specially constructed boat slung beneath it were attached, the entire aggregation stood 120 feet high.[50]

The boat, constructed by a New York shipbuilder, was 16 feet long and 4½ feet wide, complete with oars, oarlocks, and hand-operated "propeller wheels" on each side to propel the craft if it should be forced down at sea. Capable of housing six men on a 500-mile open sea voyage, it weighed only 118 pounds. The *Atlantic* itself had a capacity of 120,000 cubic feet of gas and could lift a total weight of 25,000 pounds. The finished vehicle was estimated to have cost $30,000.[51]

With the balloon complete, La Mountain and Gager drew John Wise and two additional investors, a Mr. Johnson and a Mr. Gilbert, into the scheme. Together they formed the Trans-Atlantic Balloon Corporation. Wise was named director-in-chief, La Mountain the aeronaut, and Gager the scientific observer.

In reporting on the construction of the *Atlantic*, St. Louis newspapers had remarked that the proposed voyage was "one of the grandest experiments ever projected in this country."[52] Wise and La Mountain favored the notion of a shakedown cruise from some inland city to the coast, so an offer from St. Louis city officials to provide free city gas if the flight were to begin from their city was particularly welcome.

The aeronauts arrived in St. Louis with their balloon in late June 1859. Contrary to expectation, they refused to wait for the Fourth of July. Their epic voyage was scheduled to begin on July 1. By one o'clock that afternoon, a huge sheet of canvas had been spread over the ground at Washington Square. A large pipe installed between a gasometer at the Seventh Street depot and the launch site had conducted over 50,000 cubic feet of gas to the balloon by 6:00 P.M.

John La Mountain

Silas Brooks, who proposed to lead the voyagers across the Mississippi, then inflated his much smaller, one-man balloon, the *Comet*.

A reporter for the St. Louis *Herald* peeked into the car of the *Atlantic*, noting that a United States Express Company agent had already placed a ten-pound bag of letters and papers aboard. The contents were to be forwarded to their destination when the balloon landed. He noted:

> *Besides the business like package, the cargo consisted of nine hundred pounds of sand in bags, a large quantity of cold chickens, tongue, potted meats, sandwiches, etc., numerous dark-colored, long-necked vessels containing champagne, sherry, sparkling catawba, claret, madeira, brandy, and port, a plentiful supply of over-coats, shawls, blankets, and fur gloves, a couple or three carpet bags chockfull of what is called a "change"; a pail of iced lemonade and a bucket of water, a compass, a barometer, thermometer, and chart, bundles of the principal St. Louis newspapers; [advertising] cards of candidates for clerkships in several of the Courts, tumblers, knives, and perhaps other articles which have escaped me.*[53]

Mr. Baker, who was responsible for maintaining order, had his hands full.

A scuffle ensued when "a thirsty individual" attempted to steal the wine from the car. Once this fellow was turned over to the authorities, Baker struggled to clear the milling spectators from the area around the balloon. Wise supervised one hundred "strong armed men," two on each line, as the aerostat was maneuvered to the area where the boat and basket were loaded and ready for attachment.[54]

The sandbags hanging from the netting were removed one by one as the lines were made fast. With all preparations complete, La Mountain, Gager, and William Hyde, a reporter for the St. Louis *Republican*, whom the aeronauts had agreed to take aboard as supercargo, climbed into the boat. Wise clambered up into the willow basket rigged fourteen feet above the boat. Standing on the rim of the car, the director-in-chief introduced his companions to the multitude. Mr. Gager, he remarked, had furnished the capital, Mr. La Mountain the skill, and now he expected to perform his part and come in for a third of the glory.[55] Brooks launched his balloon once the speeches were complete. Moments later Wise ordered the *Atlantic* released.

Rising majestically above the city, Wise was particularly impressed by the sounds emanating from below. "With the clatter and clang of its multifarious workshops, and the heterogeneous noises of a great commercial emporium, it gave out sounds more like a pandemonium than that of a great civilized choir of music. At greater heights these sounds were modulated into cadences."[56]

As Wise and Gager worked to equalize the pressure on the thirty-six ropes from which the car and boat were suspended, La Mountain prepared the balloon for an all-night voyage. They had passed over the Mississippi and were well into the prairie country that lay to the east when they saw Brooks land at dusk.

At about the same time, all the aeronauts were startled to note that their balloon had become incandescent. Wise marveled that the envelope seemed "illuminated" with an extraordinary brightness. "So powerful was this," added La Mountain, "that every line of the netting, every fold of the silk, every cord and wrinkle, were as plainly visible as if illuminated by torches." The light was so bright that Wise was able to read his watch throughout the night. La Mountain was probably correct in identifying the phenomenon as St. Elmo's fire.[57]

As the temperature dropped, John La Mountain took over the night shift, while Wise, Gager, and Hyde settled down for the night. The balloon continued to climb until, by 11:30, it was fully distended. Fearing the excess gas escaping through the appendix might have overcome Wise, who was isolated in the upper basket, La Mountain asked Gager to climb up the rope to investigate. Wise was nearly comatose, breathing "spasmodically," and had to be shaken back to consciousness.

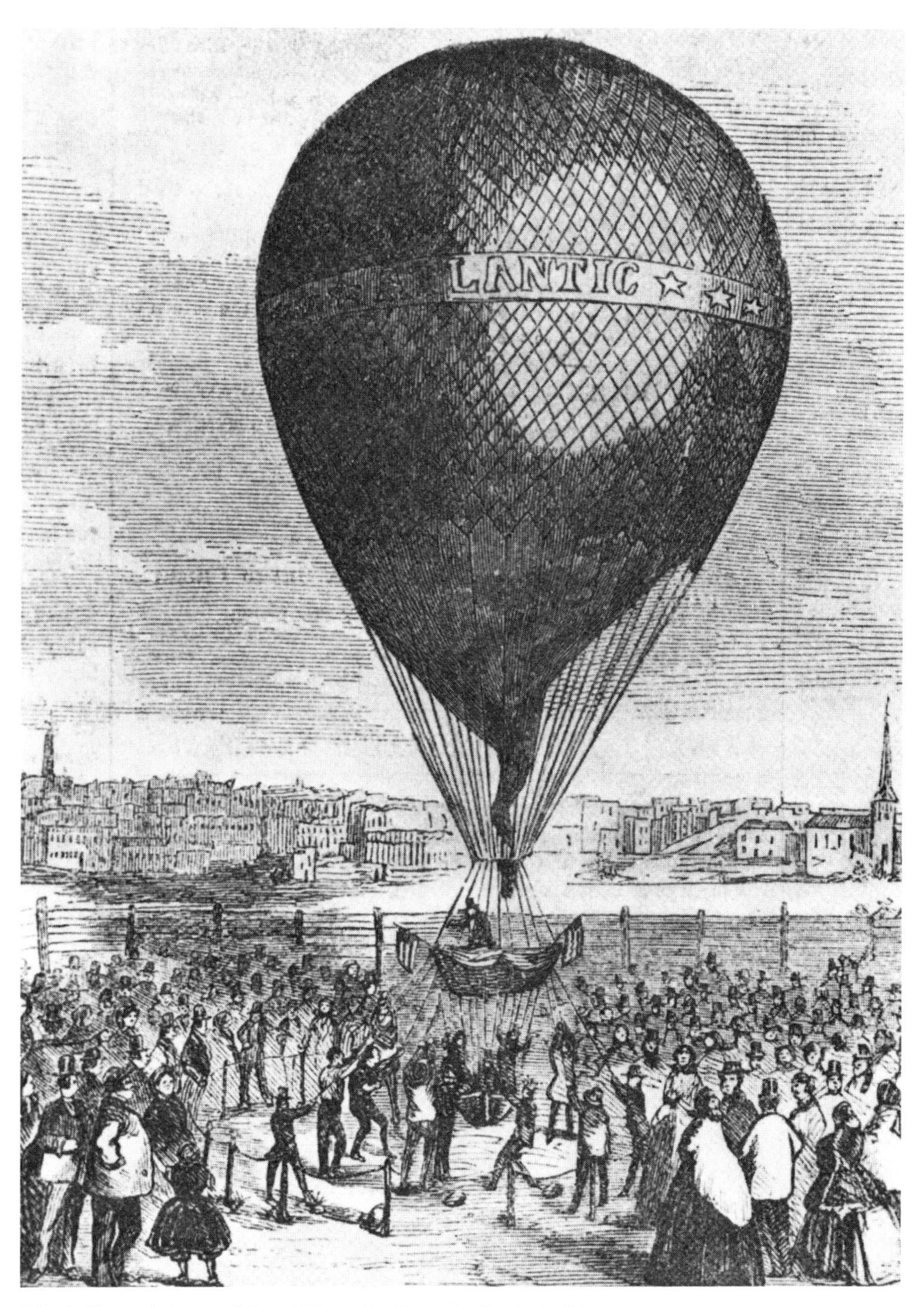

The balloon Atlantic *lifts off from St. Louis, July 1, 1859.*

The excitement was over by 1:00 A.M. and everyone but La Mountain had gone back to a fitful sleep. The entire party was up again within three hours, passing the time until dawn by hallooing to the ground and attempting to cheer La Mountain, who, Wise remarked, "had got a little out of humor" as a result of his uncertainty about their location.

At sunrise Lake Erie was clearly visible in the distance. A quick check of their maps indicated that the city they had passed a short time before must have been Fort Wayne, Indiana. They were correct. A resident of Fort Wayne had seen the balloon passing 6 miles north of town at 4:00 A.M.

Wise accepted La Mountain's advice that they valve gas, descending within 500 feet of the surface in the hope that they could catch a current of wind blowing toward Buffalo. They were moving rapidly now, passing seven steamers as the *Atlantic* was carried toward the Canadian shore. They missed Buffalo, passing by Lockport, New York, on the left. Lake Erie had been conquered at last.

The decision about their ultimate destination remained. Shouting between the boat and basket, the group decided to try for Rochester, where Gager and Hyde would be landed, enabling Wise and La Mountain to continue to Boston or Portland. Their problems began shortly thereafter, as they were suddenly caught in what Wise described as "a terrific gale." The winds roared like "a host of Niagaras," he recalled, "and the surface of the earth was filled with clouds of dust." Unable to land with such high ground winds, the aeronauts now faced the prospect of moving into a storm brewing over Lake Ontario. Hyde, inexperienced and uncertain, could only shout up to Wise, "This is an exciting time, Professor, what should we do?" "Trust to providence and all our energies," came the reply.[58]

La Mountain fully realized the gravity of their situation. "Above," he noted, "the clouds were as black as ink; around, the winds were howling as if alive with demons; and below, the waters, capped with foam, and lashed by contending air currents, swept up in swells fifteen feet high, that ran in every conceivable direction."[59] "O! how terribly it was foaming, moaning and howling," agreed Wise.[60]

Remarking, "I guess we are gone," Hyde struggled to join Gager and Wise in the basket, while La Mountain remained below to dump ballast. The balloon struck the water so violently that La Mountain had to leap into the rigging as three planks on one side of the lifeboat were crushed. The canvas outer cover of the boat prevented the water from entering and dragging the balloon under, but the situation was obviously desperate.[61]

The spirit of cooperation was rapidly breaking down as well. Wise and La

The Atlantic *returns to earth in Henderson, Jefferson County, N.Y.*

Mountain shouted to one another over the roar of the gale, unable to agree on the best course of action. La Mountain was appalled to learn that Wise proposed swamping the balloon in the lake, "in other words to leave us at the mercy of the waves fifteen feet high, to swim forty or forty-five miles to shore!"[62]

Instead, La Mountain called for his companions to toss everything that might be thrown overboard down into the boat. He hoped that by dropping these items when they approached the surface again, and by cutting up the boat itself, he could keep the balloon aloft until they reached shore.

At 1:10 P.M. La Mountain finally caught sight of land. They swept ashore one hour and seventeen minutes later. La Mountain scrambled aloft with the others as the balloon began to tear through the trees behind the beach: "On we went, the balloon surging, heaving, and literally mowing its way, sweeping off the tops of the branches, tearing up trees, swinging sometimes almost vertically, and cleaving a clear path through the woods by our course. At last, just as we were about despairing, the balloon caught in the last tree in the woods—a monstrous elm—the silk gave way, swung backward and forward at a tremendous rate several times, then dropped at successive stages fifty feet down the branches, and we were safe."[63] As Wise noted with a sigh of relief, "We had come to the land, or rather the tree, of Mr. T. O. Whitney, town of Henderson, Jefferson County, New York."[64]

It had been a voyage of epic proportions. The four men had traveled an estimated 809 miles in twenty hours, forty minutes, establishing a world distance record that would stand until September 1910. But there was little sense of triumph in the wake of the flight. Refusing to recognize that there was more than enough credit for everyone, the aeronauts immediately fell to bickering in the press.

In the immediate aftermath of the flight, articles began to appear under each man's by-line. The account credited to Wise was generous to a fault. La Mountain was described as a "hero" who had repeatedly risked his own safety to assist his fellows. In a letter to the editor of the *New York Tribune*, however, explaining why the balloonists had not flown directly to New York, Wise had been a bit less charitable, blaming his companions for the problems encountered at the end of their voyage:

> *The reason why it was not done is this: Some of our party did not provide themselves with extra clothing. Immediately after leaving St. Louis, I took the balloon to an altitude at which she was making due east. In this current we sailed until some of my companions shivered with the cold, so that the balloon quivered. Mr. La Mountain had taken no extra clothing, and the other two were not fully provided for the change of temperature. I had on two undershirts, woolen drawers, cloth coat, Cassimere pants, and over these I had two woolen blankets; but the expostulations of my companions to come down into a more congenial temperature could not be unheeded. I admonished them, however, of our advertisement to sail for New York; but in response was told that if we got into the State, the program will be fulfilled. I also told them that the lower current would take us on the lakes, as it was coming from the southwest; but to this it was answered that we could cross the lakes if we had ballast enough when we got to them.*[65]

Wise also expressed his doubts about the efficacy of the "propeller wheels" that La Mountain had designed for the boat. In addition, he announced his desire to raise $6,000 with which to construct a new balloon to challenge the Atlantic.

La Mountain immediately countered with his own letter to the *Tribune*. "Mr. Wise says that we could have reached New York had I not neglected to provide myself with extra clothing!" he fumed. "Had he been less sleepy and more observant during the trip, he would have made no such statement." La Mountain claimed that Wise "did not once" remind us of our promises to reach New York, "and he knows it." Wise, he continued, intimated that he, La Mountain, planned to use the "fans" on the boat for propulsion aloft. On the contrary, there had never been any intention of using the propellers for anything other than propelling the boat if they should be forced down in the water.[66] The depth of La Mountain's feelings was exposed in his final paragraph:

> *From first to last, I have been robbed of just credit, slandered, ridiculed and placed in a false position by this man, whom I allowed to accompany me. He has conveyed everywhere the idea that I was a fellow of some pluck, but having no scientific knowledge of ballooning, and that his wisdom barely compensated for my blunders. I go to Watertown today for the wreck of my balloon, which will at once be rebuilt. In October, I shall cross the Atlantic. If, meanwhile, Mr. Wise considers the matter of sufficient importance to test our relative capacities—scientifically considered in a trial trip from San Francisco to the Atlantic seaboard with balloons of equal size—he knows my address.*[67]

In his own published account of their adventure, La Mountain lost few opportunities to needle his old instructor. He called attention to the "lesson" Wise had learned when he fell asleep in the cloud of escaping hydrogen, and portrayed Wise as being close to panic at various points, screaming orders such as: "For god's sake, La Mountain, throw overboard anything you can lay your fingers on."[68]

Gager was uncertain about the reason for La Mountain's outburst, but he felt compelled to rush to the defense of John Wise. La Mountain was simply mistaken, he informed the press. He was particularly upset that both the *Troy [New York] Times* and *Frank Leslie's Illustrated* had carried a false account of the voyage under his name. Wise, he noted, had been far more than a "mere passenger." Like himself and John La Mountain, Wise had been a full partner in the enterprise and still owned a one-third share in the remains of the *Atlantic.*[69]

Privately, Gager was writing to Wise, informing him that he had not seen La Mountain since the three men parted company in Adams, New York, after the flight. Gager had learned, however, that a Mr. Demers, of the *Troy Times*, had been encouraging La Mountain to believe that Wise was attempting to steal the honors, and had, in fact, himself written several of the articles attributed to La Mountain.[70]

Certain of Gager's support, Wise wrote once again to the *New York Tribune* in an attempt to clarify his position. "I think, with Mr. Gager," he began, "that we ought to have no controversy about this matter, but be thankful that our lives were saved." He felt compelled, nevertheless, to refute some of the charges leveled against him. Wise pointed out that he was a valid partner and had invested time and money in the project. A written contract with Gager gave Wise the title director-in-chief. Moreover, he could produce letters from La Mountain dated as early as January 1859, begging for advice and offering Wise the opportunity to superintend the construction of the *Atlantic*, "for I know you can do it better than I."[71]

In a similar letter to *Harper's Weekly*, Wise summed up his position in no uncertain terms: "Permit me to say I have no ill feeling against Mr. La Mountain. I verily believe that no trouble would have arisen with him had he not placed himself in the hands of an indiscreet relation connected with the *Troy Times*. This person is attempting to build up a reputation for Mr. La Mountain by destroying mine, which I have been more than twenty years in acquiring.[72]

La Mountain was determined to continue working toward an Atlantic flight on his own. On July 4 he was back in Watertown, New York, having taken possession of the *Atlantic*, and promising another trial flight from Chicago once he had completed repairs.

But La Mountain did not go to Chicago. He remained in the Watertown area into the fall, completing repairs on the *Atlantic*. In spite of the bitter feelings that continued to divide them, Gager was apparently allowing him free use of the balloon, which had been reduced a full one-third in size during the repairs.

Time was pressing if La Mountain hoped to be the first to fly the Atlantic. John Wise was no longer his only rival. T. S. C. Lowe, a relative newcomer who had made his first public ascent only the year before, was hard at work on the *City of New York*, an enormous balloon specifically designed to conquer the Atlantic.

La Mountain, still attempting to raise cash for another distance flight as practice for the proposed ocean crossing, was now eager to make his first test flight with the rebuilt *Atlantic*. John A. Haddock, the editor of the Watertown *Reformer*, had returned from a trip to Labrador soon after the conclusion of the St. Louis flight. Recognizing that the people of the area were "in a state of considerable excitement upon the subject of aerial navigation," Haddock offered to fund the second flight of the *Atlantic*. He was expecting to produce an exciting article and "not to be absent from home more than 10 or 12 hours at the longest, and to have a good time."[73]

La Mountain originally scheduled his ascension with Haddock for September 20, 1859, but a severe storm forced a postponement until September 22. At 6:25 that evening the aeronauts shook hands all around and stepped into their car to cries of "God Bless" and "Happy Voyage." La Mountain called "All aboard," and the *Atlantic* literally sprang into the air and moved rapidly off to the northeast. Fifteen minutes later they were seen passing over a village 15 miles from Watertown. Residents of the town of Fawler in St. Lawrence County, 39 miles from the take-off point, saw them not long after. The *Atlantic* was last spotted over the town of Pitcairn, 11 miles farther on, "going due east with great velocity." Then they were gone, vanished as completely as had the Harvey children or Ira Thurston.

Speculation was rife. A search party from Watertown set out for a likely wilderness area. La Mountain's booking agent, A. J. Morrison, offered a $1,000 reward for "the discovery of the bodies alive, and $500 if dead."[74]

Over the next week, there were reports that the aeronauts had been found safe and sound in one place or another. Each of these reports eventually proved false. Finally, on October 22, 1859, word came that La Mountain and Haddock had landed in the Canadian wilderness, near Filliman's Creek, some 150 miles north of Ottawa and 300 miles from Watertown.[75]

The two aeronauts had an extraordinary tale to tell. It was a story that would quickly spread through the pages of both American and Canadian newspapers. The aeronauts had encountered problems from the very outset of their voyage. They climbed away from Watertown so rapidly that Haddock immediately noticed a severe ringing in his ears. The temperature dropped noticeably. Haddock's feet were cold, and he noted that the damp sandbags had frozen solid. La Mountain, who had been ill for several days before the flight, was shivering. Haddock wrapped his companion in a shawl and spread blankets over their knees and feet. The "abominable smell" of the gas belching through the appendix added to Haddock's misery. "The result," La Mountain noted, "was that we were soon considerably lightened."[76]

Swathed in overcoats and blankets, their hands protected by extra gloves that Haddock had stuck in his pocket before take-off, the aeronauts curled up in the basket and prepared to spend the night aloft. It was far from quiet. Their rest was disturbed by barking dogs, locomotive whistles, and the sound of rushing water. By 8:20 the friendly lights of towns were no longer visible, and La Mountain, fearing that they had flown into a wilderness area, decided to descend. Haddock grabbed the top of an approaching spruce and La Mountain lashed them to the tree with the dragline. Unwilling to climb down from their perch in the dark, they rolled themselves in their blankets to wait for morning.

A heavy rain began at dawn. The men remained in the basket, hoping that the sun would appear and heat the gas so that they could reascend and survey the countryside. When the rain continued, they jettisoned everything in the basket and rose high enough to enable them to obtain an unobstructed view of wilderness extending in every direction. Worse, from La Mountain's point of view, this was a spruce forest: "It was spruce! A very messenger of evil tidings. No spruce grew in the New York wilderness, that I knew, and the hardy tree was native only of colder climates. We must therefore be over Canada. If this was so we were above the great wilderness. This I knew was almost unbounded—its only limit being the Arctic Circle."[77]

Immediately valving gas, La Mountain set the balloon down in a small clearing: "After jumping out, I knew that it was necessary to abandon the balloon. . . . There was no alternative, the work of the *Atlantic* was done."[78]

The situation was clearly very serious. La Mountain had little idea where they were, and neither of the two possibilities that came to mind offered much consolation. They might have descended in John Brown's Tract, a 4,000-acre wilderness of spruce, pine, and hemlock along the northeastern border of New York. Alternatively, they could have overflown the border and reached the Canadian wilderness. In either case, their best hope of reaching civilization lay in moving southeast.

After walking less than a mile, they reached a small stream flowing to the west. Here they found an abandoned camp and a half-barrel of rotten pork bearing the stamp "Montreal." Sure now that they were in Canada, Haddock and La Mountain continued downstream all day Friday, September 23, eventually crossing the creek on a floating log, then striking a blazed path that led to a deserted lumber road. Following the road to its terminus, they saw a ramshackle shanty on the opposite bank of a small river.

La Mountain lashed some fallen timbers into a makeshift raft and poled himself across the stream. Haddock, who was much heavier, pulled the raft back and attempted the crossing, only to be dunked when the frail craft collapsed in midstream. Hungry and soaked to the skin, the exhausted men tried and failed to build a fire by rubbing two sticks together. They were finally reduced to spending a dismal, cheerless night beneath a pile of straw in the shack. "I will not attempt to describe our thoughts as we lay there," remarked Haddock. "Home, children, wife, parents, friends, with their sad and anxious faces, rose up reproachfully before us as we tried to sleep."[79]

By daylight they noted that the roof of the shack was constructed of split, hollow logs. La Mountain bound these together with twine, creating a second raft on which to travel downstream. After moving a short distance around a bend, they came to a spot where a fallen pine completely blocked the stream, forcing them to disassemble the raft and rebuild it again on the other side. While so engaged, they heard a series of gunshots in the vicinity. When their shouts brought no response, they continued downstream. The two men would later discover that they had been less than three miles from a large farm carved out of the wood to feed the Bilmour and Company lumbermen working in the area.

La Mountain and Haddock pushed on all day Saturday, September 24, "weary, exhausted, almost broken down with effort and yet seeing nothing to indicate any probability of relief from our sore distresses." They remained on the raft until ten o'clock that night, soaked once again by a cold drizzle, finally poling ashore to huddle, "muskrat fashion," beneath a clay ledge. Unable to sleep, they were on the move again by midnight but were forced to seek shelter a second time when the rain began to pour steadily.[80]

Daylight found them back on the river, "dripping with water; chilled to the very marrow of our bones; pale and hollow-eyes, and with those terrible sensations of ringing in the head, dryness of the lips, and parching of the throat that precede starvation."[81]

They reached some rapids soon after sunup on Sunday and decided to abandon their raft. After traveling less than half a mile through thick underbrush, they realized that "salvation depended on going back and getting it down the creek." Stick by stick, they sent the raft through the rapids, wading up to the waist in the frigid stream to dislodge the occasional snagged timber. Back on the raft, they spent the rest of the day exploring the shore of a large lake for an outlet or for some sign of civilization.[82]

Both men were now close to collapse. For the past three days they had not had food except for two tiny white frogs that they had eaten raw ("Not hind quarters alone; we were not dainty; fore-quarters, heads, bones, and all. I never tasted a sweeter morsel in my life"); a single raw oyster that Haddock insisted La Mountain take; and a few handfuls of high bush cranberries ("whose acid properties and bitter taste probably did us more harm than good").[83]

Haddock noted that their clothes were now in tatters: "My pantaloons were slit up both legs, and the waistbands nearly gone. My boots were mere wrecks, and our mighty wrestlings in the rapids had torn the skin from ankles and hands. La Mountain's hat had disappeared; the first day out he had thrown away his woolen drawers and stockings, as they dragged him down by the weight of the water they absorbed."[84]

La Mountain, who had been ill and had not eaten for some time prior to take-off, was in particularly bad shape. He seemed to be loosing the sight in his right eye, and, as Haddock noted, "his face was shriveled so that he looked like an old man."[85]

Pushing their raft into the mouth of a small stream the next morning, they heard two more gunshots. Soon thereafter, Haddock spied a whisp of smoke rising above the forest. They quickly poled to shore, where they found a birchbark canoe overturned on the bank. While La Mountain guarded the canoe (and prepared to gobble down the carcass of a duck that he found secreted beneath the boat), Haddock hurried up the trail to a shanty, where he found Angus Cameron and a party of Indian and French Canadian lumbermen. They were saved.

The aeronauts were returned to civilization in easy stages. On their way out of the wilderness, the party paused at the remains of the balloon long enough for La Mountain to retrieve the valve, while Haddock cut the letters "TIC" from the gasbag as a souvenir. The trip back to Watertown via Ottawa, Ogdensburg, Potsdam, Antwerp, and Evan's Mills was nothing short of a triumphal procession. The aeronauts were given special railroad cars, and there were speeches and celebrations at every stop. The superintendent of the Potsdam and Watertown Railroad gave special permission for the train to make five-minute stops at every station en route "in order to allow the people to satisfy their curiosity and express their congratulations at the fortunate recovery of the aeronauts."[86]

The climax came in Watertown when thousands flocked to the station to greet the returning heroes. Bonfires blazed in the square, skyrockets streaked overhead, and cannon "belched forth the immoderate joy of the people over the safe return of their favorites."

The appearance of the aeronauts was something of a shock to family and friends. They had traveled from Canada in borrowed clothes and a three-weeks growth of beard, so that they looked "decidedly rustic." After a few minutes with their families, La Mountain and Haddock appeared at Washington Hall, where Haddock spoke for three-quarters of an hour, moving his audience to tears.[87] John La Mountain continued to make local ascents until the outbreak of the Civil War, but any immediate hope of flying the ocean had died with the loss of the *Atlantic* in the Canadian wilds.

John Wise continued his career as an aerial barnstormer as well. He had returned to St. Louis immediately following the first *Atlantic* flight, where he made his 231st flight on July 30, 1859.

On August 16, 1859, Wise and his son Charles flew their balloon *Jupiter* from Lafayette, Indiana. The inflated balloon was accidentally released while being walked from the local gasworks to the city square. Aware that the three daguerreotype "artists" waiting in the square would be disappointed if they were unable to photograph the balloon before take-off, Wise rose to 3,000 feet, then valved gas to descend at the edge of town. When the balloon had been walked back into the square, Charles Wise made a half-hour flight that was cut short by a leaky valve.

With repairs completed the following day, John Wise made a third ascension on August 17. This was a flight of some significance, for several days earlier, on August 15, Postmaster Thomas Wood of Lafayette had announced that "all persons who wish to send letters to their friends in the East by balloon today must deliver them at the post office previous to 12 p.m. as the Jupiter mail closes at that hour."

By August 17 a grand total of 23 pamphlets and 123 letters had been marked for delivery "via the balloon Jupiter." This was not the first time mail had been carried aboard a balloon. Jeffries had carried letters from London to Paris; Clayton had carried mail from Cincinnati; and Wise, La Mountain, Gager, and Hyde had transported the express mailbag aboard the *Atlantic*. However, this was apparently the first time that stamped letters, which had passed through a post office, had been flown.

It was, at best, a haphazard means of transporting the mail. La Mountain had dropped the express bag from the car of the *Atlantic* during the mad rush to lighten the balloon. The mail was found six miles west of Oswego, New York, on July 4 and forwarded to Adams, where it was postmarked and redispatched.

The mail that Wise carried from Lafayette aboard the *Jupiter* was delivered in an even more bizarre fashion. Approaching Crawfordsville, Indiana, and short of ballast, Wise fashioned a parachute from a nine-foot muslin sheet and dropped the locked mailbag to earth. It was later recovered and sent east by a Colonel Reed, agent for a local railroad with a U.S. mail contract.[88]

Throughout the fall of 1859 and the spring of 1860, the team of John Wise and son centered its activities in the New York area. There John Wise's name was closely linked with one of the aeronautical tragedies of the period. Augustus Connor, a Wise pupil who had purchased the balloon *Venus* from his instructor, made an ascension from Manhattan's Palace Garden. The upper seams of the old balloon had given way during inflation, forcing Connor to attempt makeshift repairs on the scene.

Brushing aside the advice of friends, the novice aeronaut stepped into the basket and ordered his release. Just as the *Venus* was rising above the surrounding buildings, it was blown against the glass windows of a dance studio. Connor was dashed against the window, while the balloon rose a short distance and caught on an eave. It burst and fell into an adjoining yard.

Connor was found lying unconscious on the roof of the concert hall. His cuts and bruises did not appear serious, but he died of internal injuries at eleven that evening without regaining consciousness.[89]

Wise was accused of negligence both in training the unfortunate young man and in selling him a balloon that had clearly not been airworthy. Wise responded

CONNOR SHAKING HANDS WITH HIS WIFE.

Augustus Connor lifts off from New York.

The unhappy end of Mr. Connor, as freely interpreted by a Leslie's *illustrator*

by noting that Connor had been a headstrong pupil who rejected his instructor's advice. Moreover, Connor had not employed the equipment supplied by Wise, choosing to fly with his own inadequate network rather than the new, strong cordage Wise had supplied with the *Venus*. Wise was, in his own mind, entirely blameless.[90]

In spite of his continued activity, Wise, like La Mountain, was unable to raise sufficient funds with which to mount a new attempt to cross the ocean.

In the fall of 1860, a new aeronautical star was rising. Thaddeus Sobieski Constantine Lowe, a relative novice, was the only balloonist in the United States still in a position to challenge the Atlantic. If John Wise was the most successful of the nineteenth-century American aeronauts, T. S. C. Lowe was rapidly becoming the best known. Six feet tall and broad-shouldered, he cut a handsome figure with his sweeping handlebar mustache and clear, penetrating eyes.

Lowe, a native of Jefferson Mills, Coos County, New Hampshire, was born on August 20, 1832. While he had nothing more than the normal grammar school education of the period, he took an early interest in botany, geology, and chemistry. "My fondness for science found expression in many ways," he would recall with great pride in later years. "I saved every dollar I made—read all the scientific books I could obtain—courted the society of men who knew something—and tried in every way to store my mind with knowledge that would be of service to me along the line of scientific investigation."[91]

His youthful experiments in aeronautics began at age sixteen when he sent cats, colored lanterns, and American flags aloft aboard large kites. Lowe launched his "scientific" career two years later when, while convalescing from an illness with relatives in the White Mountains, he attended a lecture on chemistry offered by a traveling performer. The young man served so ably as a volunteer demonstration assistant on this occasion that the "Professor" hired him to stay on in this capacity for the remainder of the lecture tour.

By age twenty Lowe had purchased his own portable laboratory and embarked on the tour circuit for himself. "My lectures were simple," he remarked, "but they contained scientific truths to a marked degree."[92]

Lowe met his wife, Leontine Augustine Gachon, the daughter of an officer of Louis-Philippe's Royal Guard who had fled Paris following the Revolution of 1848, during one of these lectures. His first balloons, small hydrogen aerostats employed in his demonstrations, were also constructed during this period.

In his unpublished autobiography, "My Balloons in Peace and War," Lowe claimed to have acquired his first balloon in 1856 and to have made a number of short, unpublicized ascensions with it. He first came to public attention as the result of a series of flights in Ottawa in the spring of 1858. A scheduled flight on May 24 in honor of Queen Victoria's birthday was aborted, but he did fly from the city in celebration of the laying of the first Atlantic cable. Lowe was back in Portland, Maine, on July 4, 1858, releasing a large number of small hydrogen balloons from his basket in flight. A few of these aerostats were later recovered 500 to 600 miles at sea. This gave him pause. "If the balloons had been larger," he reasoned, "they would doubtless have crossed the Atlantic. The uncertainty of the working of the Atlantic Cable also suggested a balloon to reach Europe in the upper current, and a number of businessmen offered to interest themselves in the project. So my dream of a vast airship, capable of transporting troops, freight, and large numbers of passengers rapidly took shape."[93]

Lowe continued to fly for the next year, saving his money and attempting to interest others who might be willing to invest in the construction of a balloon suitable for challenging the Atlantic.

Lowe was also making serious contact with other aeronauts during this period. He had contacted John Wise in the spring of 1859 while plans for the

Thaddeus Sobieski Constantine Lowe

Atlantic voyage from St. Louis were still coalescing. While Lowe's letters to Wise have not survived, the Pennsylvanian's response makes it quite clear that the two men were discussing a partnership that would include joint flights, the possibility of an Atlantic crossing, and a balloon tour of Japan.

Wise informed Lowe of his problems with La Mountain, complaining that he was expected to work on the *Atlantic* project for "nothing," if, he remarked, La Mountain would not pay his expenses from current proceeds and allow Wise to make fund-raising flights on his own. "So you see," he informed Lowe, "there is nothing in the way for you and I to go into any plans, so long as my acts in the matter will not be prejudicial to Mr. Gager, or his scheme, for as honor among men is and ought to be observed." Clearly the seeds of Wise's friendship for Gager and his distrust of La Mountain had already been sown.

Wise also expressed some doubts about the quality of the *Atlantic* balloon to Lowe. Having seen the first samples of the silk La Mountain was using in its construction, he remarked that "it cannot last long," as the material would "not wear good."[94]

As Wise became more heavily involved in the *Atlantic* project, Lowe continued to develop his own plans for the Atlantic crossing. He began work on the great balloon that was to become known as the *City of New York* in mid-July 1859. The construction centered in a Hoboken, New Jersey, lot, proceeded quickly, and drew an enormous amount of publicity. Lowe himself claimed to have completed the craft in ninety days, and judging from contemporary press reports, this must have been very close to the truth.

The *City of New York* had a diameter of 130 feet, stood 200 feet from the valve to the bottom of the dangling lifeboat, weighed 3½ tons, and had a gas capacity of 725,000 cubic feet. Lowe calculated that the monster would have an aggregate lift of 22½ tons when inflated with hydrogen. Inflation with coal gas would cut the lift in half.

The envelope was constructed of 600 yards of oiled, twilled cloth sealed with three coats of varnish. Seventeen sewing machines had been used to stitch the enormous gores of fabric. The rattan basket was 20 feet in diameter with walls 4 feet high. A circular seat ran around the interior wall. Pegs and shelves were provided for instruments and "roomy coffers" for food and extra clothing. O. A. Gager, who had financed construction of the *Atlantic*, was interested in Lowe's project as well and provided a special lime stove for heating and cooking. Canvas walls and ceiling, complete with windows, would enclose the aeronauts in the basket.

A small hole in the floor of the car led to a rope ladder that provided access to the lifeboat hanging below. Thirty feet long with a seven-foot beam, the boat

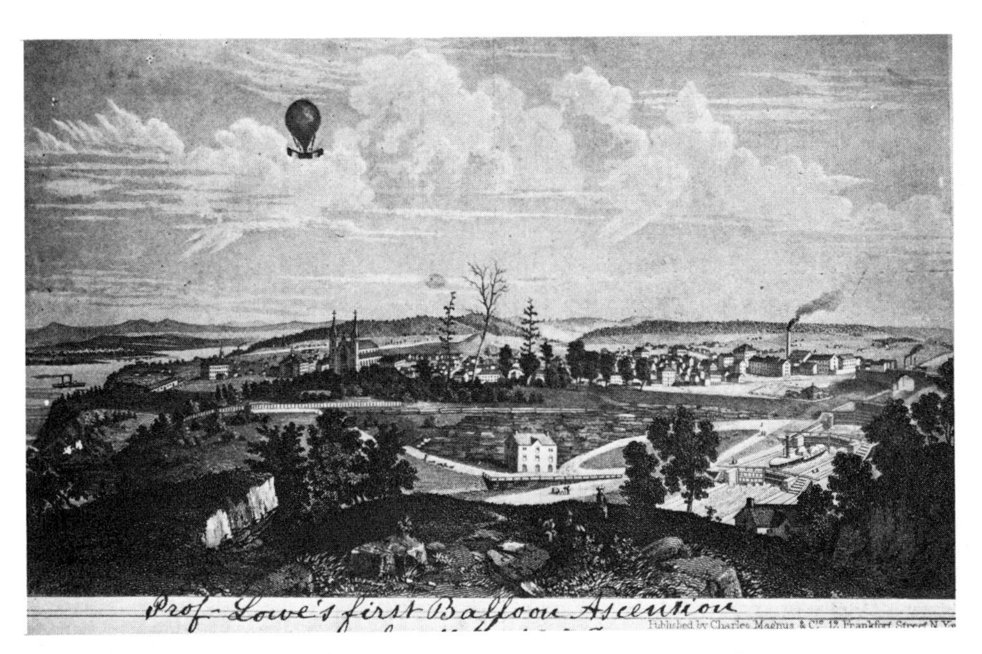

T. S. C. Lowe rises above Ottawa, Canada, July 17, 1858.

was powered by a four-horsepower Ericsson caloric engine. The craft was designed to carry sufficient water and provisions to feed a crew of ten for six months. An India rubber cover provided protection from the elements in the event they were forced down in the water. Floats, extra line, experimental "propeller wheels" for controlling the altitude of the balloon, and other items were also carried aboard the boat.[95]

By late September 1859 Lowe had even begun to address the problem of crew selection for the voyage. Two or three "scientific seamen" would be included, as well as "a scientific landsman, a member of Congress from New England, whose name is yet a secret, a New York editor, and one or two others."

Lowe's *City of New York* was not, of course, the only large balloon in America. The *Atlantic*, at 120,000 cubic feet, was really not in the same class, but Charles C. Coe was to unveil an even larger balloon in 1860. Coe and a companion, Joseph S. Cottman, had made an exciting ascension from Rome, New York, on September 29, 1859. Flying before a crowd of 10,000 spectators gathered at the local fairground, they had climbed to an altitude of over 10,000 feet when the valve jammed. The balloon burst with an explosion "about equal to the discharge of a musket." Two enormous rents allowed all the gas to escape.

The aeronauts dropped for over a mile "at a most terrible speed." Providentially, the torn fabric gathered in the upper half of the netting, slowing their fall. The balloon caught in the top of some tall trees, dangling the thoroughly frightened men seventy feet in the air. They were able to climb down the trees to safety.

Charles Coe and Q. L. Andrews brought what they claimed was the largest balloon in the world to Cleveland, Ohio, in September 1860. If the news reports were accurate, their aerostat was indeed a monster. It was said to stand 208 feet tall and 118 feet in diameter with a cubic capacity of 1,736,000 feet of hydrogen. This would have made it larger than either the *City of New York* or Nadar's French balloon *Le Géant*.

Coe and Andrews claimed to be on their way to St. Louis, where, like the *Atlantic* voyagers, they would begin a test flight in preparation for an ocean crossing. The two aeronauts had brought a small balloon that they hoped to fly in Cleveland. The small aerostat was destroyed on the ground in a storm. The fate of the large balloon is unknown.[96]

In the fall of 1859, however, public attention was focused on T. S. C. Lowe and the *City of New York*. Lowe recalled with pride that "everything about the ship aroused the greatest curiosity and interest; men, women and children haunted the construction hall, and looked upon the whole affair as something of the nature of a fairy tale."[97] Newspaper comment was widespread and mixed. But everyone, those that laughed and those that applauded, looked forward to the moment of departure.

One mark of the extraordinary public interest in Lowe's project is found in the number of letters he received from those who wished to accompany him on the voyage. Ferdinand Gross, a resident of Pittsburg who had flown with Eugène Godard in 1857, posed a typical query as to "the practical conditions for passengers and what chance is there for the undersigned?" Vernon Henry Vaughn, a senior at Chapel Hill and editor of the University of North Carolina yearbook, was also anxious to go. Lowe would probably have looked much more favorably upon the application of Sam Jackson, a ship's carpenter who had served as a carpenter's mate on the U.S. frigate *Niagara* during the laying of the transatlantic cable. Volney Anderson of Albany also begged for a seat in the car, while William Hendricks, a New Yorker, volunteered "advice on science for the voyage." Another petitioner, from Dayton, Ohio, admitted that he was "poor as Job's Turkey," but offered to provide Lowe with a foolproof propulsion system for aerial navigation.[98]

Lowe moved his balloon to the Crystal Palace at Forty-Second Street and Sixth Avenue to begin inflation on September 30, 1859. For days, spectators gathered to see the enormous aerostat swell to full size. But the visitors to the

The huge gas meter through which New York City gas was distributed to Lowe's balloon

Crystal Palace were disappointed. The envelope was stretched over a bed of straw scattered near the reservoir. A large tent had been pitched to protect the fabric, gas lines ran from Forty-Second Street, and workmen scurried around the area; but the balloon itself bore little resemblance to the magnificent craft portrayed on the posters plastered all over the city. As late as November 11, the huge balloon was continuing to draw enormous crowds, but, as one disappointed reporter noted, "the aerial ship began to gradually increase its bulk—spreading out laterally, but retaining much the same vertical height as on Tuesday." By

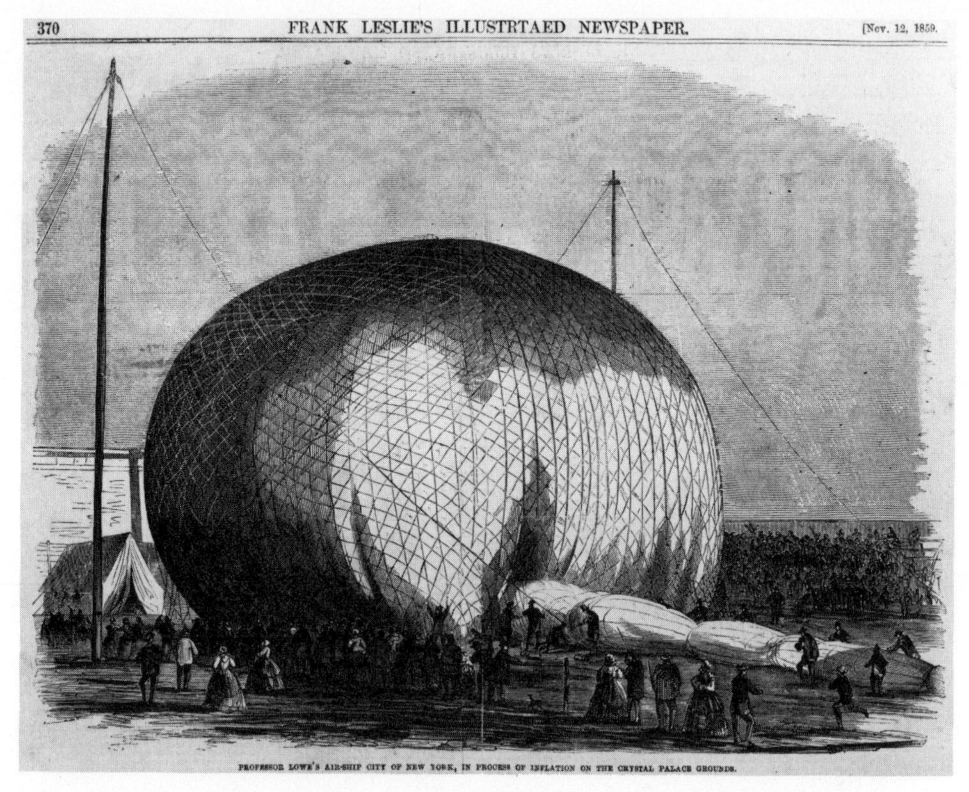
370
FRANK LESLIE'S ILLUSTRTAED NEWSPAPER.
[Nov. 12, 1859.
PROFESSOR LOWE'S AIR-SHIP CITY OF NEW YORK, IN PROCESS OF INFLATION ON THE CRYSTAL PALACE GROUNDS.

The City of New York *being inflated on the grounds of the Crystal Palace, New York*

November 16 Lowe was forced to offer his apologies for the slow progress of the inflation. The city was not equal to the task of supplying sufficient gas. The trip across the Atlantic, which had been scheduled to begin the following day, would have to be postponed. Lowe promised the citizens, however, that he still hoped to touch the shore of Great Britain with his balloon before the enormous English iron ship *Great Eastern* reached the United States. The editors of the *New York Tribune* were less certain. The balloon, now renamed the *Great Western*, was, they noted, less than one-third full after two weeks of inflation.[99]

In an effort to stave off criticism, Lowe flew from the Crystal Palace in a small balloon on the afternoon of November 17. Wrapped in a "rough, heavy fur coat" and a "curious fur hat," surrounded by the cords supporting the basket, Lowe was said to resemble "a Russian bear seen through the bars of a menagerie cage."[100]

A contemporary cartoon pokes fun at two marvels of the age, the Great Eastern *steamship and Lowe's* Great Western *balloon.*

Lowe was soon forced to recognize that at this rate the *Great Western* would never be inflated. The gas was now escaping from the envelope faster than the gasworks could pump it in without blacking out the city. "I was on the horns of a dilemma," he recalled, "when I received an invitation to take my airship to Philadelphia." This invitation was extended by a group of Philadelphia enthusiasts headed by John C. Cresson, president of both the Franklin Institute and the Philadelphia Gas Works. The Pennsylvanians also offered to pay all transportation expenses.[101]

Lowe accepted the offer but realized that the season was now too far advanced to attempt a crossing before spring. He placed the large balloon in winter storage and took his "fleet" of small aerostats to Charleston, South Carolina, "in which location and delightful climate I spent the winter in ascending and studying various air currents."[102]

PROFESSOR LOWE'S BALLOON.

We give this week an engraving of Professor Lowe's air-ship, City of New York, as it will appear when the inflation is completed. In our last number will be found a complete description of the balloon and its appurtenances, unnecessary here to repeat. The professor announces that he will depart on his great trans-Atlantic voyage some day this week. The enterprise is a bold one, and we hope some practical good may come of it.

FRANK LESLIE'S ILLUSTRATED NEWSPAPER. 391

Nov. 19, 1859.

CITY OF NEW YORK

PROFESSOR LOWE'S MAMMOTH BALLOON, CITY OF NEW YORK, AS SHE WILL APPEAR WHEN FULLY INFLATED.—SEE PAGE 394.

The City of New York, *as Lowe hoped she would appear on her first flight*

The Great Western, *during inflation in Philadelphia*

The aeronaut was back in Philadelphia early the next spring, anxious to begin the Atlantic flight. His sponsors were just as anxious to insure that all contingencies had been considered. They asked that Lowe first make a series of flights from the city in his smaller balloons. One of his first opportunities came when he and William Paullin were invited to entertain a Japanese embassy visiting Philadelphia. Less than ten years after Commodore Perry had opened Japan to the West, the Japanese had sent a delegation of nobles and distinguished citizens to establish reciprocal diplomatic relations.

As Lowe noted, "The fete day was in the nature of a reunion, as Commodore Perry and all of his officers available were in attendance, as well as the elite, science and beauty of Philadelphia."[103] The aeronaut was incorrect in assuming that Commodore Perry himself was present on this occasion; the famous naval officer had died two years before. Lowe exhibited the *Great Western* and, with Paullin, whom he insisted on referring to as his "assistant," flew two smaller

Lowe's Constitution, *flown during the celebration of the arrival of the first Japanese embassy in Philadelphia. The illustration appeared in a Japanese children's book.*

balloons.

On July 4, 1860, Lowe made two ascents in one day. As the *Philadelphia Inquirer* noted, "The proceeds will be to complete his transatlantic equipment."[104] In fact, the *Great Western* had already been in the air on at least one occasion. In late June 1860 Lowe went aloft with Garrick Mallery, the associate editor of the *Philadelphia Inquirer*. In large measure, Lowe planned the flight to silence the criticism of "self-constituted professors and atmospheric voyagers, to the effect that it was madness to risk the untried experiment, and, as to our friends, we were ridiculed, threatened, cajoled, bribed, wept over by turns."

Lowe, overjoyed to be in the air aboard his gigantic balloon almost a year after it had been constructed, leaped onto the hoop. "Here at last is *The Great Western* afloat," he proclaimed, "after all the prophecies against her, and half a million witnesses to the fact!"[105]

With the safe landing at Medford, New Jersey, Lowe must have believed himself ready to begin the Atlantic flight. He had already failed to make good one promise. The *Great Eastern* docked in New York on June 28, 1860, the day of the *Great Western*'s first flight.

Lowe maintained his contact with John Wise and other balloonists, including William Paullin, throughout the summer and fall of 1860. Wise, who had been offered an opportunity to attempt the crossing with Lowe the previous fall, remained very much interested in Lowe's progress. Writing to him on September 9, 1860, Wise remarked:

> *Since yesterday your present position was presented to me in such a light that it put me really under a slight conviction that you had some serious intentions to make the attempt in good faith. Now if I am right permit one to say that your plan and machinery is [sic] too incongruous—and the odds therefore are decidedly against you, both as to making a fair start, as well as in the successful termination of the trip if you should ever get started. These were the reasons for my refusal of your invitation to accompany you when you had prepared the outfit for last fall in New York.*
>
> *From a conversation with Mr. Paullin two weeks ago I learned that your balloon was somewhat dilapidated. He represented to me it was entirely unfit to be used with some of the ponderous contrivances in the detail of the machinery. I wish you to understand that it was not Mr. Paullin who [said] these things. He gave me* no opinion *about the matter. He only gave me a description of the condition of the Balloon and the result of the trial trip.*[106]

In spite of his doubts about the airworthiness of the *Great Western*, Wise still recognized that it was the only balloon capable of crossing the Atlantic that season. He suggested that Lowe might "increase the probability of success" by making a few "alterations of some parts of your machinery." Wise suggested that, should Lowe see fit to spend $500 in making improvements, he would volunteer his services as "scientific director" for the trip. If Lowe preferred, Wise would be prepared to remain in the background and assist "in getting you fairly started."[107]

Wise's offer came too late. On September 8, 1860, the day before he wrote to Lowe, the *Great Western* burst during inflation. Paullin's observations about the condition of the great aerostat had been correct. A full year of incomplete inflation had taken its toll on the balloon.

Lowe was not deterred however. He returned once more to Cresson and his other backers with a request for funds with which to construct a replacement balloon. The group did not rule out such a possibility, but they thought it prudent to ask the nation's unofficial chief scientist, Joseph Henry, secretary of the Smithsonian Institution, to rule on the existence of the high altitude currents that Lowe believed would carry him to Europe.

Henry, who, as noted earlier, was an acquaintance of John Wise, offered a cautious endorsement, tempered with advice:

> *It has been fully established by continuous observations collected at this institution for ten years, from every part of the United States, that, as a general rule, all the meteorological phenomena advance from west to east, and that the higher clouds always move eastwardly. We are, therefore, from abundant information, as from theoretical considerations, enabled to state with some confidence that on a given day, whatever may be the direction of the wind at the surface of the earth, a balloon elevated sufficiently high, would be carried easterly by the prevailing current in the upper or rather middle region of the atmosphere.*
>
> *I do not hesitate, therefore, to say, that provided a balloon can be constructed of sufficient size, and of sufficient impermeability to gas, in order that it may maintain a high elevation for a sufficient length of time, it would be wafted across the Atlantic. I would not, however, advise that the first experiment of this character be made across the ocean, but that the feasibility of the project should be thoroughly tested, and experience accumulated by voyage over the interior of our continent.*[108]

Lowe was so encouraged that he immediately set out for Washington to seek Henry's council and support. Their meeting was, in the aeronaut's words, "one of the highlights of my life." Henry reinforced his initial suggestions that a long test flight begun in some inland city would build public confidence in Lowe's ability to cross the ocean.

Accepting Henry's advice, Lowe packed one of his larger balloons, the *Enterprise*, and traveled to Cincinnati, the city he had chosen for this trial ascension. Once in Cincinnati, Lowe became acquainted with Murat Halstead, editor of the Cincinnati *Daily Commercial*, who publicized the aeronaut's local flights and introduced him to other wealthy Cincinnatians who offered support and the opportunity to raise needed cash through a series of lectures on aeronautics. A lot adjacent to the Commercial Hospital, at Twelfth between Plum Street and Central Avenue, was selected as the launching site for the *Enterprise*. The inflation was begun on the night of April 19, 1861, only seven days after Confederate gunners opened fire on Fort Sumter. By 3:30 the next morning all preparations had been completed. After a fresh stack of *Commercials* had been tossed aboard, Lowe gave the order to release the restraining lines, and the *Enterprise* rose rapidly into the sky.

The *Enterprise* traveled northeast and north until it had reached an altitude of 7,000 feet, at which point it was blown slowly to the east. The "atmosphere was quite frosty" and the temperature fell from 45 degrees to 15 degrees above zero as Lowe continued his journey, now heading southeast. As the sun rose at 5:05, the *Enterprise* was at an altitude of 8,000 feet on the Kentucky side of the Ohio River. By 7:00 A.M. the sun had heated the gas, carrying the balloon up to 11,000 feet. By mid-morning Lowe had passed over the northern range of the Cumberland Mountains, remaining in a current blowing to the southeast.

> *Being desirous of ascertaining with more certainty my exact whereabouts, I let off gas, and gradually descended to within a short distance of the earth, with the hope of seeing someone to inquire of. . . . Seeing some persons at work in a field, I descended near to them, and asked, "What State is this?" The men, without answering, looked in all directions but upwards. . . . I again sang out at the top of my voice, when the reply came, "Virginia". . . . I then asked what country and threw out some sand to clear the tops of some tall trees. This struck the ground with a spatter and caused them to look up, and instead of answering the question a yell of horror arose from them, and if the fleetness of foot is any indication of fright they must have been terribly frightened.*[109]

Soon after this encounter, Lowe crossed the Alleghenies and continued south along the front of the Blue Ridge Mountains, searching for a low spot where he might cross to the sea with safety. Unable to find such a spot, the aeronaut dropped sixty pounds of ballast until he rose over the mountains: "Here, the thermometer fell to 10 degrees below zero; the water, fruit, and other things froze, and it required all the clothing and blankets I had to keep me warm. . . .

I had cleared the mountains whose tops were covered with snow, and was rapidly moving to the east. It was now 12 o'clock, and I could distinguish the blue ocean in the eastern horizon. . . . Knowing that the coast in that direction was an uninhabited swamp, and being desirous of landing near a railroad, I concluded to descend and look for a good place."[110]

Lowe initially descended in Spartansburg Township, on the boundary of North and South Carolina. Landing near a group of field workers, he asked for assistance but discovered that the men had no intention of giving aid to the pilot of such a "hellish contrivance that had frightened them so." Lowe describes the incident: "They would not believe that I had sailed from the State of Ohio that morning, and informed me, that they would be very thankful if I would leave, and ordered the negroes to let go of the ropes they were holding. . . . I threw out a bag of sand and commenced to ascend, [and] at that moment one of the bystanders seeing the bag of sand fall sang out, 'Hello stranger, come back, I reccon [*sic*] you have lost your baggage.'"[111]

Lowe drifted twenty miles to the east at an altitude of 7,000 feet before attempting a second landing. During this time he heard "many discharges," which he took to be muskets firing at him, and as he began to descend he prepared to signal those on the ground that he was friendly.

> *When within a half mile of the earth I heard loud cries of terror, and saw people running in all directions; but I was determined to land for good this time, let come what would, and in five minutes more the anchor took a firm hold in a short scrub oak, and the car gently touched the ground. Thus fast, the globe gently swinging to and fro, presented a very lifelike appearance. I soon noticed some heads peeping around the corner of a log hut that stood near by. . . . I called to them to come and assist me, at which they took no notice until I threatened to cut loose and run over them, after which two white boys, three old ladies and three negroes, in a body ventured within twenty feet of me. At that moment a gust of wind caused the balloon to swing over near to the ground, and a general stampede took place, which caused me to abandon all hope of getting any assistance.*[112]

Led by a "stalwart-looking young woman, six-feet high and well proportioned," the group ventured up to the edge of Lowe's basket. The woman informed Lowe that he had frightened a number of old people who, believing that the day of judgment had arrived, were inside the log hut praying. She also told him that the group of men that had fired shots at the balloon were expected to arrive momentarily. By the time Lowe had packed the gasbag and equipment into sacks, a band of "rough-looking fellows" arrived, threatening to destroy the "devil that could travel through the air." Allaying his fears, however, the young woman assured him there was no danger, "for all the men then in the neighborhood were cowards as all the brave ones had gone to the wars."

A guard of nine armed men loaded the aeronaut's balloon onto a wagon and escorted him to Unionville, South Carolina, intending to place him in the town jail until they received instructions from state officials. The jailer was uncooperative, however, and "positively refused to allow any such animal as they described to come into the building." Lowe was then taken to a hotel, where he found "persons of fine education," who recognized him and promised that he would be conveyed safely back to Cincinnati.

The next morning Lowe began his return trip but was apprehended in Columbia, the capital of South Carolina, by Southern sympathizers who feared he might be bearing dispatches to the Union North. "This brought together a number of learned and scientific gentlemen, who at once knew me by reputation, and saw my position, and I was immediately released, and furnished with a passport by the Mayor of Columbia. From this time until I reached Cincinnati, no more impediments were placed in my way."[113]

After Lowe arrived back in Cincinnati on April 26, 1861, he concluded from his experience that it would be "an easy matter to cross the Atlantic in less than two days." Still, he decided to abandon his transatlantic flight plans and to offer his services instead to President Lincoln as a reconnaissance balloonist. Murat Halstead agreed to recommend Lowe to Secretary of the Treasury Salmon P. Chase. Realizing that an extended stay in Washington might be necessary to convince the War Department of the value of balloons as observation platforms, Lowe advertised that he would make a second Cincinnati ascent in the *Enterprise* to raise funds for his visit to the nation's capital.

On the afternoon of May 8, 1861, Professor Lowe, with Junius Brown of the Cincinnati *Daily Press*, W. W. Ware, and Jacob C. Freno of Philadelphia, boarded the *Enterprise*. A short flight was made to the village of Bethel in Clermont County, where Freno and Ware left the party to return to Cincinnati. Lowe and Brown ascended again, apparently hoping for another long voyage in order to persuade the officials of the War Department of Lowe's ability and experience. The long flight was certainly achieved, for the pair landed in Hamilton, Ontario, having been blown far to the north by unexpected high winds. Lowe took advantage of the situation by making a series of successful ascents in Hamilton in honor of Queen Victoria's forty-second birthday. Following these ascents, Lowe returned home to Philadelphia to prepare for demonstration flights in Washington.

As noted earlier, Confederate artillery had opened fire on Fort Sumter at 4:30 A.M. on April 12, 1861, exactly one week before T. S. C. Lowe took off from Cincinnati. With war imminent, Lowe was not the only American balloonist anxious to volunteer his services to the cause. John Wise, John La Mountain, James Allen, and William Paullin would all fly for the Union armies. Others, like Alexander J. B. DeMorat, would be rejected. Charles Cevor and Richard Wells, DeMorat's opponent in the great New Orleans balloon race, would both fly beneath the Stars and Bars of the Confederacy.

Exhibition ballooning would remain a popular form of mass entertainment during and after the war. Many veterans, including Wise, Allen, Paullin, Silas Brooks, and Samuel Archer King, would continue their aeronautical careers. A new generation of aeronauts would appear on the scene to thrill the public with ever more daring feats. The challenge of the Atlantic would continue to tantalize balloonists.

The fact remains, however, that the sight of a free balloon wafting across the sky would never again elicit quite the same excitement as it had during the early years of the century. The "Golden Age" of American ballooning had come to an end.

Chapter 10 To Navigate the Air

FERDINAND VON ZEPPELIN ARRIVED IN ST. PAUL, MINNESOTA, ON AUGUST 17, 1863. Only twenty-five years old, his distinctive features masked by a bushy crop of "Burnside" whiskers, a beard and mustache, the young graf, or count, was on the final leg of a long journey that took him through a number of American cities.[1]

As a first lieutenant in the army of Württemberg, he had convinced both his father and his king that he would profit by joining the ranks of the foreign military observers accompanying the Army of the Potomac. Zeppelin traveled to America via Ostend, London, and Liverpool, arriving in New York aboard the Cunard steamer *Australasia* on May 6, 1863. The count enjoyed the sights of Philadelphia, then made his way south to Washington, D.C., where he registered at the Willard Hotel sometime prior to May 21.

Alexandre Davidow, second secretary to the Russian embassy, and Rudolf Schlieden, the Hanseatic ambassador to the United States, were particularly helpful in introducing Zeppelin to American officials. Schlieden was especially impressed by the young man whom he regarded as "an extraordinarily nice little man, with a range of interests unusual in a cavalry officer."[2]

Zeppelin was soon visiting such luminaries as Secretary of the Treasury Salmon Portland Chase, and had obtained an interview with President Abraham Lincoln. With the appropriate military passes and a letter of introduction to Gen. Joseph Hooker in hand, the count and his black servant Louis proceeded to the headquarters of the Army of the Potomac at Falmouth, Virginia.

Zeppelin quickly made friends with a German-speaking Swedish officer and other members of Hooker's staff. He found little difficulty in circumventing rules forbidding foreign military observers from taking part in combat operations. On the evening of June 20, 1863, he joined a Union cavalry detachment carrying critical dispatches to Gen. Alfred Pleasonton. The following day, Zeppelin narrowly escaped capture by Confederate troops while he was performing reconnaissance during a skirmish at Ashby Gap. He made his way back to Fairfax Court House and apparently took part in some of the fighting around Fredericksburg. He must have left the army soon thereafter, however, for he was not at Gettysburg on July 1–3.

Zeppelin returned to New York and witnessed the draft riots of July 13–15, then traveled to Buffalo and on to Detroit. As he told his father, his plan in traveling to the West was "to get actual impressions of the cultural history of this part of the world."[3]

The young nobleman was intrigued by towns such as Cleveland, Toledo, Buffalo, and Detroit, which seemed to have sprung suddenly to full growth. "All," he informed his family, "are cities brought into existence as if by magic during the very recent past to satisfy the demands of traffic on the great waterways. . . . These places seem to have sprouted overnight."[4]

In the company of Alexandre Davidow and another Russian friend who met him in Cleveland, Zeppelin booked passage on the lake steamer *Traveller* for a "grand excursion to the romantic regions of Lake Superior." The five-day boat trip was something of an idyll. Zeppelin found American beer very "heavy stuff" and enjoyed flirting with a party of American girls on board. Many years later the aging count would recall that the "beautiful American girls" had been "as anxious to get acquainted with me as I was with them. They finally broke the ice by flipping apple seeds into my face, and then we had a jolly talk."[5]

Writing from Superior, Wisconsin, on July 30, Zeppelin informed his family that he was about to strike out for the wilderness with his Russian friends and two Indian guides. The little party was on the trail for eighteen days, hiking overland, then canoeing down the St. Croix River to Crow Wing, "a small town of about two hundred people who trade with the Indians." From there they traveled by stage to St. Anthony, then proceeded to St. Paul in a hired carriage.[6]

The trip was tame by American standards, but an exhilarating experience for the urbane aristocrat. "I regret that I cannot give you a description of our fortnight in the wilderness," he wrote his father. "The impressions are too many, the impressions too varied! . . . There is a feeling you have when you know you are all alone in the midst of primitive, unspoiled nature. You can laugh, cry, shout, throw yourself in the rushing current of a river; you can set fire to the woods, and no human soul knows about it. You are alone with the creator in his magnificent temple."[7]

He described the hardships of the march, which included difficult canoe portages, thirst as the result of an unusual drought, and hunger when they ran low on ammunition. Zeppelin noted that his companions had come to regard "water rat" [muskrat] as "a good dish." He also described the Indians of the region: "Above all, there are the Indians themselves—their ways and customs; their wild calls and songs; the strange war stories; their duelings; their skill (one of our guides shot a duck in flight at sixty paces with one shot and one bullet); and what is more there is the [Indian] language which can create the word 'Schingku baba' for eggs."[8]

As Zeppelin and his Russian companions stepped from their carriage and prepared to register at St. Paul's International Hotel, they may have noticed a hastily constructed board fence enclosing a corner lot across the street. John Steiner, the victor of the great Cincinnati balloon race, was now a veteran with a full year's service as one of the most active, successful, and disgruntled aeronauts in the Union Army. Having resigned in December 1862 as the result of a pay dispute and a general feeling that his talents were not fully appreciated, Steiner was now completing a series of ascensions in the upper Mississippi Valley.

The aeronaut had advertised that he would make both free flights and "Army," or tethered, ascents with passengers during his time in St. Paul. The first flight, scheduled for August 17, the day of Count Zeppelin's arrival, was canceled because of bad weather. When the weather cleared on August 20, Steiner faced problems with the local gasworks, which could supply only 36,000 cubic feet of gas for the 41,000-foot balloon. The aeronaut was unable to offer passenger rides to any but the lightest citizens of St. Paul. With Sen. Alexander Ramsey aboard, for example, Steiner could not rise above the housetops. Ramsey's ten-year-old daughter Marion, however, was able to make a full ascent.

The exasperated aeronaut finally resorted to allowing several adults to make solo ascents to the limit of the tether. A reporter for the St. Paul *Pioneer* noted that "the ascensionists" took responsibility for their own safety after Steiner had provided "all the necessary information as to the conduct of the balloon."[9]

Count Zeppelin was one of those that flew in Steiner's balloon. "I have made the acquaintance of the famous aeronaut Prof. Steiner, who has invented a new kind of balloon suitable for military reconnaissance," he remarked to his father, spicing his comments with a bit of hyperbole.[10] It is clear that the young officer realized the military potential of the balloon. His description of a tethered flight with Steiner on August 19 sparkles with enthusiasm:

> *Just now I ascended with Prof. Steiner, the famous aeronaut, to an altitude of six or seven hundred feet. The ground is exceptionally fitted for demonstrating the importance of the balloon in military reconnaissance. The Mississippi with St. Paul, and westward of the latter, a mile from the river, lies a ridge of hills running parallel with it. This forms a very good defensive position against an aggressor marching up through the valley. There is no tower, no elevation high enough to*

> *study the distribution of the defender's troops on the gentle, open slope behind his battle line. From the high position of the balloon these could be completely surveyed. Should one want to harass with artillery fire the troops deployed in reserve on the other slope the battery could be informed by telegraph where their projectiles hit. The above technique has at times been used with great success by this country's armies. No method is better suited to viewing quickly the terrain of an unknown, enemy occupied territory.*[11]

Zeppelin went on to describe a free flight Steiner had made at Edward's Ferry, Maryland, passing completely over the Confederate lines while observing the enemy through binoculars. The problem, he noted, resulted from the susceptibility of the balloon to high winds and from the inability to navigate the craft. Steiner, he believed, had a solution to these difficulties. He planned to construct a very small balloon, just large enough to carry two observers aloft. "He has given it a very long, thin shape," Zeppelin continued. "Furthermore, he has added a strong rudder and in that way the balloon is hindered less by the wind and it will reach its destination more smoothly and more surely."[12] The aeronaut planned to return east and construct his new balloon following his current tour. He would then travel to Paris to demonstrate his invention before an applauding world.

Steiner's rosy vision was, of course, nothing more than a pipe dream, but it was one that Ferdinand von Zeppelin would never forget. Over half a century later, a hero to his countrymen, an ogre to its enemies, the aging count would recall: "While I was above St. Paul I had my first idea of aerial navigation strongly impressed upon me and it was there that the first idea of my Zeppelins came to me."[13] It was only fitting that Zeppelin's first thoughts of aerial navigation should have been inspired by the activity of an American aeronaut. The subject had long been of interest to citizens of the United States.

There were two possible approaches to the construction of a flying machine that would truly operate under the control of the aeronaut. One possibility was to follow the example of the birds and design a winged, heavier-than-air craft.[14] Most nineteenth-century Americans regarded the notion of heavier-than-air flight as the province of fools, dreamers, and eccentrics. The possibility of a navigable balloon seemed much more reasonable and intriguing. The great problem of flight itself had been solved, after all. Nothing seemed to remain but to devise a means of propelling a balloon against the wind and to provide some method of enabling the aeronaut to control its motion.[15] The emerging professional aeronauts, the men and women closest to the subject, realized just how difficult these problems were. As a result, the many schemes developed for dirigible balloons in the United States after 1820 were the product of inventors who had little prior experience in aeronautics. The aeronauts themselves watched the proceedings with interest and offered occasional comment, advice, or support, but were seldom directly involved in the effort.[16]

In 1784 Jean Baptiste Marie Meusnier, a brilliant French military engineer, had been the first to consider the possibility of a navigable balloon. Meusnier, an early associate of the Montgolfiers, had been detailed to determine the altitude and rate of climb of early French balloons. Having provided a theoretical base for aerostation, Meusnier turned his attention to the problem of navigability. Eschewing the attempt to control a round, free balloon by means of rudders, oars, wings, or propellers, he designed an elliptical gasbag complete with an internal balloonet. This was a smaller balloon placed inside the gasbag. It could be inflated with air as hydrogen was valved, thus maintaining internal pressure and the rigidity of the outer envelope.[17]

Over the next several decades a number of other figures, notably Sir George Cayley, borrowing Meusnier's notion of the streamlined, spindle-shaped envelope, continued to produce airship designs, none of which was ever constructed.

Edmond Charles Genêt brought the dirigible airship to America. The first minister plenipotentiary to the United States appointed by the revolutionary French government, Citizen Genêt arrived in Charleston, South Carolina, on April 8, 1783. He immediately fitted out four privateers to sail against the English, then traveled to Philadelphia via Camden, Salisbury, Richmond, and Baltimore, preaching the doctrine of "Liberté, Egalité, Fraternité" every step of the way.

Genêt, soon ensconced among Thomas Jefferson's coterie of Republican friends, became the center of bitter political controversy in a young nation struggling to remain aloof from the war raging in Europe. The French minister connived to seize Florida and Louisiana from Spain, encouraged revolution in Canada, and planned to send the French fleet anchored in New York against the British West Indies. By August 1793 George Washington felt compelled to request that the French government recall Genêt.

The minister, however, did not return to France. When his replacement arrived from Paris in February 1794, he married the daughter of the governor of New York and settled on a farm in Rensselaer County. Genêt remained an active and controversial force in American life, offering his thoughts on a wide range of subjects from scientific agriculture to the industrial growth of the United States.

In 1825 Genêt published a slim volume that he believed would overturn Newtonian science and replace the steam engine with the balloon as a servant of mankind. Entitled *Memorial on the Upward Forces of Fluids and Their Applicability to Several Arts, Sciences, and Public Improvements*, the book outlined the way in which balloons, pulleys, levers, and other simple machines could be combined to raise and lower canal boats on an inclined plane, recover fouled anchors, posts, snags, "and other heavy bodies" from underwater, and lift boats "stranded or grounded on shoals, bars, or alluvials."[18] In addition, Genêt designed "hydrostatic cranes" to be used in lifting boats and a "hydronaut," a side-wheel vessel driven through the water by "hydrostatic" balloons rising and falling through a fluid.

Genêt's masterpiece, however, was his "Aeronaut" or "Aerostatic Vessel." The flat-bottomed, domed, elongated gasbag had been modeled after three distinct fish, "viz: the trunkfish of South America, which may be seen at Mr. Scudder's Museum in the city of New York; the blue lump fish and rostrated daree. . . . From the first, I have taken the large flat belly and high dorsal ridge; and from the other the various articulations which suited the best my purpose."[19]

The hydrogen-filled gasbag was to be 152 feet long, 46 feet wide, and 54 feet high. It would contain 1,023,000 cubic feet of gas and lift 73,462 pounds. A horizontal wheel 66 feet in circumference and twin side-wheels, each 20 feet in diameter, would provide forward motion. These were to be driven by two horses. Extendable wings and a rudder were also provided.

Genêt estimated that the finished airship would have an empty weight of 10,000 pounds. The three attendants, various pieces of chemical apparatus, horses, and provisions would add an additional 3,400 pounds, leaving 60,062 pounds for passengers and freight. The cost, he believed, would not exceed $10,000, including a $1,000 fund "for mistakes and contingencies."[20]

Controversy swirled around Genêt's scheme for aerial navigation. Felix Pascalis, president of the American branch of the Linnaean Society of Paris, was his most vocal defender and admirer. Thomas Jones, professor of mechanics at the Franklin Institute and editor of the *Franklin Journal*, was his most bitter critic.

Jones ridiculed Genêt's attempts to update Newtonian science and "render obsolete" the steam engine. He noted that the Frenchman had "contrived . . . as we take it, a *perpetual motion*," and suggested that government officials who had granted a patent on Genêt's airship should have enforced their own rules and refused the application "until Mr. Genêt *made it go*."[21]

Professor Benjamin Silliman of Yale, editor of the *American Journal of Science and the Arts*, provided the most reasonable appraisal of Genêt's proposal and the controversy as a whole. Silliman realized that, while Genêt's aerostatic schemes were, in large measure, an extension of much older plans for achieving perpetual

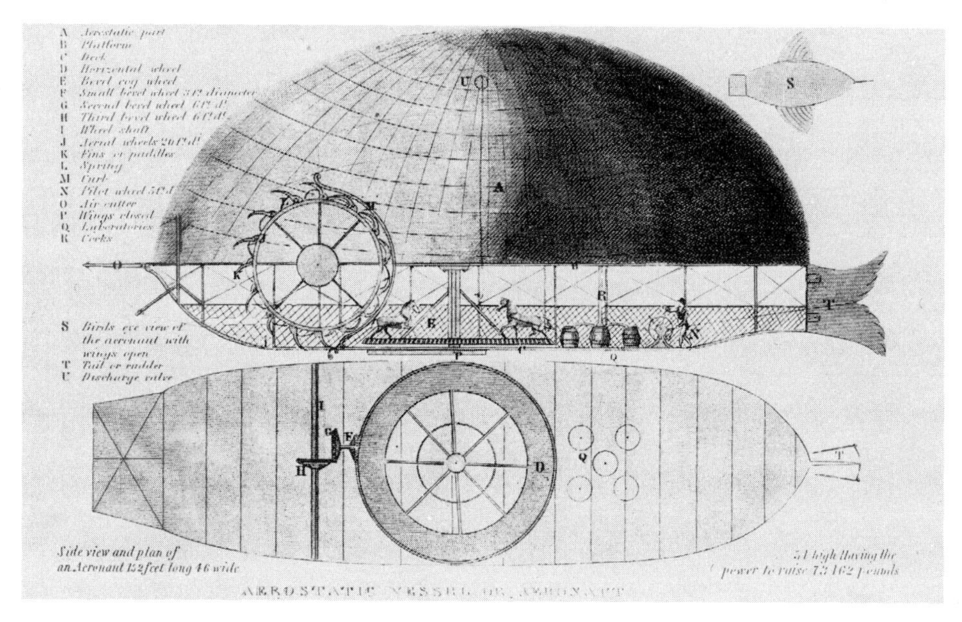

E. C. Genêt's *"Aerostatic Vessel, or Aeronaut"*

motion, his basic airship design had merit. He realized that it could never fly as proposed. It was far too fragile and, with its unique two-horsepower propulsion system and churning paddle wheels, woefully underpowered. In spite of these deficiencies, Silliman argued that Genêt deserved a hearing, "although he may utter some things new and strange."[22] To ridicule the notion of aerial navigation without a full discussion would, he believed, be to risk destroying an idea that might someday bring enormous advantage to mankind.

Throughout the nineteenth century the dream of aerial navigation would loom large in the American consciousness. The ingenious Yankee, the poor boy who rose to power and prominence by devising the better mousetrap, was emerging as a genuine folk hero. Technologists like Samuel Colt, Eli Whitney, Robert Fulton, Charles Goodyear, Samuel F. B. Morse, and, ultimately, Thomas Edison, Alexander Graham Bell, and the Wright brothers became the standard exemplars for American youth. These men represented the success of traditional values such as hard work, perseverance, dedication, and know-how. Most important, they achieved their success by applying these virtues in a new technological environment. They seemed the perfect heroes for a new age in which Americans were coming to regard their nation as an inexhaustible fount of technological innovation.

For a generation that had seen the telegraph reduce communication time from days and weeks to minutes and seconds and the railroad and steamship link nations and continents, the possibility of an airship did not seem so fantastic. Leading American scientists, men such as Joseph Henry and Alexander Dallas Bache, agreed with Benjamin Silliman's belief that serious schemes for aerial navigation deserved full consideration.[23]

Hopeful inventors kept a steady stream of patent applications, covering a wide range of airship designs, flowing to Washington throughout the century. The United States Patent Office approved numerous applications during the 1800s. They came from every corner of the growing nation:

> *I contemplate an oblong balloon, whose capacity should be but a few pounds less in ascensive powers, when provided with a vertical and horizontal set of rotary inclined planes, than the burden it is to carry. Then, but a small portion of the power of a man to the horizontal planes would cause its ascension, and a proportionate power to the vertical ones would cause its forward progress. The steering would be effected exactly in the manner of a ship, bird, or fish. I have, also, in my head the application of a well-known, very common principle of power, little inferior to the steam, which has but trifling weight, and requiring neither fuel nor fire, which could as certainly be brought into action, as that of human or any other force.*[24]

This letter echoed a dream shared by thousands of amateur mechanics from Maine to California.

American scientific and mechanical journals also provided a forum where flying machine schemes could be discussed at length. *Scientific American*, which was to become the most popular of the technical periodicals aimed at a general audience (and which was founded by Rufus Porter, one of the best-known American airship pioneers), kept its readers current on the latest developments in the field and allowed would-be aerial voyagers to present their own plans for the conquest of the skies.

Viewed from the perspective of an age that regards supersonic passenger flights and voyages to the moon and planets as commonplace, it is all too easy to regard this activity as little more than empty and eccentric dreaming. It is certainly true that most of those who submitted patent drawings, wrote letters to the editor, and published plans for airships had little serious hope of realizing their dreams. A few more serious experimenters were to emerge during the four decades following 1830 however.

These were men who, unlike Genêt, had every intention of carrying their plans beyond the blueprint stage. Models were built and flown. Learned scientific committees offered their endorsements for several airship designs. Companies formed to construct full-scale, man-carrying flying machines attracted relatively large amounts of money from serious investors. The subject of aerial navigation was taken very seriously indeed.

John H. Pennington was one of the first American airship promoters to attract wide notice. Pennington, a Baltimore piano maker and tuner, had been intrigued by the notion of flight since childhood. In his own words, he shared his ideas with the world only "after years of careful application to his trade."

He published his first pamphlet, *Aerostation, or Steam Aerial Navigation*, in Baltimore in 1838. The booklet described a "Steam-kite, or inclined plane, for navigating the air." The craft was a heavier-than-air device featuring two steam-driven paddle wheels on either side of a flat, elliptical plane. A steel engine fired by turpentine, alcohol, and sulphur would be housed in the crew compartment beneath an overhead disc-wing measuring 125 feet wide by 375 feet long. Pennington estimated that the finished "steam-kite" would weigh 950 pounds with a full load of fuel.[25]

Pennington's original plan for a heavier-than-air machine attracted little interest. There was widespread recognition of the fact that such an aircraft represented a complete leap into the unknown and had no chance of success.

During a lecture trip to Wheeling, Virginia (now West Virginia), Pennington encountered John Wise, who was planning an ascension in the city. Wise advised Pennington to abandon any hope of heavier-than-air flight and to substitute a large balloon for the inclined plane. Discouraged by the lack of response to his plan, Pennington adopted the aeronaut's suggestion.[26]

The inventor had completed plans for a new machine by 1838 and deposited a model and a patent application in Washington, D.C. He petitioned Congress for assistance in constructing his machine and provided a complete description of the revised scheme in a new pamphlet, *A System of Aerostation, or Steam Aerial Navigation*, published in 1842.

Like Genêt, Cayley, and so many of his predecessors, Pennington initially planned to construct a streamlined balloon patterned after an aquatic animal. In this case, the balloon was dubbed the "Steam Dolphin Balloon." As before, the craft would be driven through the air by "bevelled wheels powered by a steel steam engine fueled by turpentine or alcohol."[27]

John Wise complained that Pennington probably did not deserve a patent, for "the application of beveled wheels to the propulsion of aerial machinery."

Nor, of course, was the streamlined balloon an innovation. Jean Baptiste Meusnier, Sir George Cayley, and Citizen Genêt had all suggested the use of similarly shaped envelopes.

Wise overlooked, however, the most significant of Pennington's contributions to airship technology. All airships described previously had relied on the pressure of the gas to maintain the rigidity of the envelope. Pennington suggested the addition of a hollow "spinal frame," or keel, to be constructed of white oak or hickory. This frame would serve as both a longitudinal stiffener and a point of attachment for the crew compartment. Pennington had taken a major step forward by proposing a rudimentary semirigid airship.

Persistence and a determination to carry his plan to a successful conclusion were major factors separating Pennington from the aeronautical dreamers that preceded him. When his initial attempt to interest Congress in his airship plans failed in 1842, he began touring the country, exhibiting his static models and drawings in an effort to subscribe the money required to carry the work forward. In Wheeling, for example, a traveling band known as the "Harmonium" staged a benefit performance "to assist him in his endeavors to demonstrate the practicability of his plan." *Scientific American* applauded Pennington's success in "calling attention to the subject."[28]

In January 1847 the same journal reported that Pennington, "who has been trying to fly for two or three years," was experimenting with a gunpowder engine to power his flying machine. "We believe it is as likely to make him fly as anything," the editor remarked.[29]

By 1850 Pennington had formed a corporation to finance the construction of a 230-foot airship with "proportionate beam as per a water craft." The craft, which was to be powered by a ¼–½ horsepower engine, would cost $10,000. The firm of Pennington and Company offered 200 shares of stock at $50 a share in the spring of 1850. Individuals were limited to holdings of five shares.

Pennington's initial fund-raising efforts failed to produce the required cash. Visiting Washington, D.C., in the fall of 1853, however, he informed his friend Joseph Gates that his hopes were far from dead. "I wish to inform the public that . . . J. H. Pennington and Company is still alive in law," he remarked. Moreover, he was proud to announce that the firm was in "no way committed to any power either public or private."[30] Pennington had come to the capital hoping to interest federal and state officials in investing in the venture, and he was also soliciting "moneyed aid" from individuals.[31]

Pennington outlined the sales pitch he planned to direct toward Congress in a letter to Mr. Mehan, the librarian of Congress. "The government," he believed, "should watch the progress of Aerial Navigation with a 'Jealous Eye'

. . . There is no standing still now, for if we are not a-going ahead some other nation will go for us: therefore Rob us of that National Glory that would accrew [*sic*] from such an achievement as Aerial Navigation."[32]

Warfare, he argued, was "prolix and harmless" compared to the revolution that would come with the airship. "In the purposes of war 10 men could do the destruction of 10,000 in the same given time." Mail delivery would also be enormously improved.[33]

Pennington hoped that the federal government would purchase two-thirds of his stock. This money could go toward the construction of a "Ship House" where the craft would be constructed and housed. The structure would be 350 feet long and "proportioned after the fashion of one of our National Ship Houses at Washington Navy Yard." Both ends of the building would "be made to open or close at pleasure to meet the emergency of Arrival or Departure."[34]

The other one-third of the money required would come from a public sale of stock and would be used to construct the airship. Pennington had apparently been giving more thought to the design of the machine, for he commented to Mehan that "Bamboo and Rattan should form very important items in the construction of the Aerostatic fraim [*sic*] or . . . Sceleton."[35]

The inventor warned Mehan that if Congress failed to finance the project, he would not hesitate to sell his plans and experience to the first European government to offer $50,000. Pennington's apparent haste to begin work on a full-scale balloon is not difficult to understand. He was no longer alone in the field.

On September 24, 1852, Henry Giffard, a French experimenter, had made the world's first successful airship journey. Giffard had traveled 17 miles at a speed of 6 MPH aboard an airship powered by a three-horsepower steam engine. His gasbag was a standard elongated envelope with the operator's car and power plant suspended forty feet below.

Giffard's achievement received little attention in the United States however. Pennington was far more concerned about his American rivals, particularly Rufus Porter. He believed that Porter and other "Neophytes" were attempting to make his ideas their own, but he hastened to assure potential investors that his airship design was fully protected by a "fence at Law." Moreover, he believed that Porter, in particular, was doomed to failure for "it would take more money to build a suitable shiphouse than Mr. P. has received or can receive under the present auspices.[36]

In fact, Rufus Porter was a much more formidable rival than Pennington was willing to admit. The Baltimore inventor was never able to begin work on a full-scale machine. Porter, however, a man blessed with absolute self-confidence and a genius for self-advertisement, was to come much closer to success.

A native of West Boxford, Massachusetts, born on May 1, 1792, Porter personified the popular image of the ingenious Yankee. He had enjoyed an idyllic New England boyhood. Like so many other boys of his time and place, Porter was absolutely fascinated by machinery. At age six he built a working cider press. Within three years, miniature waterwheels driving sawmills, trip hammers, "and a variety of complicated mechanical movements" dotted the small streams that ran across the Porter farm. As he grew older, Porter was given the run of his father's workshop where, on a homemade lathe, he labored to produce "a variety of curious things, mostly original, although I was generally required to work at least 12 hours a day on the farm in summer and was accustomed to work at chopping and hauling pine timber in the forests in the winter."[37]

In addition to all this, Porter managed to squeeze in what was for the period a fair education. This included two quarters at the Freyburg Academy, Freyburg, Maine, and the Dartmouth College preparatory school. He would always regard his informal observation of mills and machinery, however, as the most important phase of his education. On the basis of these observations "and by a habit of analogical reasoning, I obtained a general knowledge of all the principles of the mechanical powers, hydraulics, hydrostatics, inertia, momentum, reaction, projection, frictional and resistance, &c."[38]

Before he was twenty, Porter had worked as an ornamental sign, house, and ship painter. He had also "acquired some knowledge of marine discipline, while working as a sailor during several weeks employment aboard 'armed vessels.'" His musical talent led him to enlist "in regular service at a fort," where he operated a fife and drum school. He spent part of the year 1813 teaching school, and soon thereafter offered his first invention, "a large, horizontal wind mill," to the public.[39]

Over the next half century Rufus Porter would pursue a bewildering variety of careers. He was a poet, dancing master, shoemaker, machinist, printer, and journalist. Art and invention were the two great themes binding his life however. Each summer for most of his active years he would take to the road as an itinerant painter. Since his "discovery" by art historians in the 1940s, Porter has come to be recognized as one of the most active and distinguished folk artists of the early nineteenth century. His stark and simple portraits of the men and women of New England remain prized family heirlooms. His landscape murals, dominated by drooping, funereal willows and hills rolling away toward the sea, still decorate a number of New England homes and taverns. Several have been salvaged from demolished buildings and preserved in museums.

Porter was much better known to his contemporaries as an inventor than

Rufus Porter

as an artist. He developed a staggering number of mechanical ideas, contributing a "rotary plow" and a "field engine," for "harrowing, sowing and rolling at the same time," to agriculture. His "car for moving houses and other ponderous bodies" was to be propelled by a single horse walking on a treadmill geared to the front wheels. In later years he decried the cruel use to which horses had been put on Porter treadmills designed to thresh, saw, and raise water.

Transportation was always one of his primary interests. He sketched a "horse power boat" propelled by one of his treadmills and a three-wheeled steam carriage weighing 600 pounds which, he claimed, could negotiate "common roads" at a speed of 10 MPH.

He was extraordinarily proud of the "steam farmer," which could "ascend or descend hills with safety and facility, and will plow, harrow, sow, reap, mow, cart hay, thresh grain, shell corn, saw wood, make fence or go to market." His minor inventions ranged from floating docks to "a new method for rowing boats," hot air ventilating systems, fog warning devices for ships, life preservers, fire alarms, the revolving cylinder for firearms, trip hammers, ice making machinery, chain stitch sewing machines, a clothes dryer, a double-handled rake, and a walking stick that unfolded into a one-legged seat.[40]

In addition to his own inventions, Porter labored to establish communal workshops where poor young mechanics could live and work while perfecting their ideas. He established the first of these Inventor's Institutes in Boston. While residing in that city, Porter himself worked on experimental canal boats, an alarm clock that would "ring an occasional alarm and at the same time light a lamp," and "some twenty other different new inventions."[41]

Several years later Porter established a similar institute in Hoboken, New Jersey. Founded in 1847, the New Inventors Institute, like its Boston predecessor, was a self-contained campus complete with machine shops, factories, and an engine house, all designed to "promote genius and enterprise by encouraging and aiding that useful class of men . . . inventors." Established as a joint-stock company, the facility would provide a site where "practical mechanics" could experiment, perfect, and produce finished inventions. The project demonstrated Porter's recognition of the importance of cooperation and communication in fostering technological progress.[42]

Rufus Porter exercised his most profound influence on American industry through his activities as a publicist and journalist. His publications covered a spectrum of topics, including dancing manuals, instruction books for aspiring artists, and religious tracts, but he concentrated on the production of a series of journals designed to preach the gospel of technology.

Porter launched his first paper, the *New York Mechanic*, in 1841. The first weekly newspaper in the United States to specialize in reporting the latest developments in technology, the *Mechanic* reached a circulation of 3,000 to 4,000 during Porter's tenure as editor.

Porter founded his most ambitious and influential publication, *Scientific American*, in 1845. The new journal extended the tradition of technical journalism begun with the *New York Mechanic* to a national audience. It was well illustrated and aimed at keeping mechanically inclined Americans abreast of the latest developments in the field. Each issue featured several long articles offering a detailed treatment of some recent innovation. The inside pages offered pungent editorial comments on inventors and new machines. Readers were never left in doubt about where the outspoken editor stood on any issue.

Porter regarded his journal as more than a means of bringing new ideas to public notice however. He sought to use *Scientific American* as a means of organizing and assisting inventors. He advertised a "general patent agency" which, he boasted, could "furnish to enterprising young men better facilities for making money than can be found elsewhere in the United States."[43]

Porter edited the *Scientific American* for only a year, but it was to evolve through the rest of the century along a pattern he had established. The journal was Porter's most important and lasting contribution to American technology. Roger Burlingame, the pioneer historian of the role of the machine in American history, has noted its influential educational and inspirational role: "Perhaps no other single factor has been a greater boon to old-school inventors than the *Scientific American* with its up-to-the-minute lists of patents, its lucid illustrations and diagrams, its sometimes over-enthusiastic articles about innovations, some of which came to naught but most of which stimulated some amateur mechanic or incipient engineer."[44]

A fire in the offices of *Scientific American* forced the uninsured Porter to sell the journal to Alfred Ely Beach and Orson D. Munn for $800 in 1847. Undeterred, he immediately launched a new paper, the *Scientific Mechanic*, that same year.

By 1847, however, Porter was already focusing most of his attention on the problem of aerial navigation. Flight had interested the inventor since his childhood. "At the age of seven years," he recalled, "my mind was much occupied with the subject of aerial locomotion." Like most small boys, he "watched the motions of the hawks, pigeons, and humming birds, but more especially those of the barn swallows." He conducted the usual experiments, using a pair of broad shingles as wings, and was delighted to find that "I would leap farther from the top of a log by aid of the wings than I would do without them."[45]

Over the next decade Porter continued to ponder the possibility of human flight. He designed an engine "to be operated by the force of the gases liberated by a slow combustion of gunpowder composition," calculated the relative strength of materials that might be used in constructing a flying machine, and studied "the motions of wings or fans."

> *I could not, in theory, strain up the power to a sufficient degree of intensity to surmount the obstacles. I had duly investigated the feasibility of training and harnessing a team of one hundred hawks or buzzards to sustain an aerial car, but as yet knew nothing of hydrogen gas, nor any other material capable of atmospheric buoyancy, except smoke or rarified air. I observed the apparent buoyancy of the thistle-down, often experimented with them, and endeavored to discover the principle by which they ascended in the air. . . . I had no books, nor any scientific person to consult; but had become acquainted with most of the general laws of natural philosophy by observation.*[46]

None of this was of much help until Porter witnessed his first flight of a small balloon in Cambridge, Massachusetts, probably in 1819. Within a year he had, in his own words, "invented the main principles of the aeroport." While the dimensions and details of the Porter "aeroport" would vary over the next four decades, the basic design remained the same. The balloon would be constructed in the shape of a "revoloidal" spindle (i.e., a cigar shape pointed at both ends). An underslung car would carry crew and passengers.

Porter remarked that "the most prominent and main object which I had immediately in review when I invented the . . . Aeroport was the liberation of Napoleon from St. Helena, where he was then imprisoned."[47] His own firm belief in the power of technology to improve the lot of mankind soon furnished a series of more reasonable justifications for undertaking the work:

> *Of the many advantages which are expected to be derived from the use of the aeroport, may be mentioned a thorough explanation of hitherto unknown parts of the world. . . . The facilities for communicating intelligence in a more substantial form than can be done by telegraph; daily newspapers will be circulated from the principal cities, throughout the length and breadth of the continent; intelligence from South America and from Europe will be as fresh in the United States as was the news from this metropolis ten years ago. Fresh fruits from the south will be common articles in the northern markets much earlier than the ordinary season for them, while the south will enjoy the advantages of the northern forests and fisheries; and people who suffer in health from the severity of the southern summers or northern winters, may with facility accommodate themselves with perpetual*

> *spring. The social habits of mankind will be promoted, and many truly vicious and deleterious as well as ridiculous customs which now prevail, will give way to those more rational, reasonable, and salutary; and common justice and equity, which have been so long discarded by popular governments, will be gradually advanced, and systems of laws and customs, more in accordance with common sense and common comfort, will prevail throughout the world.*[48]

Porter's utopian vision of a world peacefully restructured by the flying machine contrasted sharply with the more bellicose perspective of John H. Pennington noted earlier. On a more mundane level, Porter was certain that his aeroport would revolutionize world commerce. He compared the relatively moderate cost of a very large airship of up to 1,000 feet in length to that of a large merchant ship. His aeroport, he argued, could carry 1,160 tons of "ponderous merchandise" much more rapidly and efficiently than a ship. It could be made of double canvas for strength and carry a car with fifty apartments for passengers:

> *Such an aerial transport would not be expected to run against a head wind, but take advantage of the aerial currents, whereby, in some cases, it would run 50 or 60 miles per hour. In case of strong head winds, the craft would come to moorings, if over land; and if over water, it would* lay to, *by means of a large plate, made of plank, and which, having ropes attached to its four corners, and a bar of iron attached to one edge, would be dropped into the water, and assuming a vertical position, would effectively hold the aeroport from drifting with any considerable motion to leeward. Such aeroports will be more in demand, however, and preferable for overland service, when they can run by daylight, and come to moorings at night.*[49]

Having conceived the basics of his airship as early as 1820, Porter did little to realize his dream until 1833. In that year, he constructed his first model airship in Bristol, Connecticut. The dimensions of the little craft are unknown, but we can assume that, like his other early models, it was propelled by a clockwork mechanism. Porter's only comment was that he could "not succeed in making it operate."[50]

The inventor provided the first descriptions of the aeroport in two letters published in the American edition of the *Mechanic's Magazine* in November 1834 and January 1835. He suggested two different airships. The first, a "tethered observatory," was an unpowered craft some 125 feet long and 25 feet in diameter. With 30,000 cubic feet of hydrogen in the envelope, the balloon would have a disposable lift of 700 pounds. A 1,000-foot-long jointed pipe connecting the airship to a gas generator would constantly replace leaking hydrogen, and a fixed tail would keep it pointing into the wind. Such a "tethered observatory" would enable a city watchman to spot fires, criminal activity, or approaching ships. In addition, citizens purchasing a ticket could obtain a bird's-eye view of their town.[51]

Porter referred to his second design as the "travelling balloon." Designed as a genuine navigable airship, this craft would be 500 feet long and 50 feet in diameter. Like the "tethered observatory" and all subsequent Porter designs, the "travelling balloon" was a "revoloidal spindle." The inventor had apparently conducted a series of simple tests demonstrating that such a cigar shape "whose length is equal to five times its diameter, sustains, when pointed to the wind, but one twenty-fifth part of the force that is sustained by a cylinder of the same diameter."

A ten-horsepower steam engine with charcoal-fired boilers would power the four Archimedes' screws that drove the ship through the air at 50 MPH. Hydrogen would be produced on board to keep the balloon fully inflated. Eight wooden rods, or battens, were to be stitched at equal intervals around the interior of the gasbag as stiffeners. The "Saloon" for the crew and 100 passengers would be suspended from the battens. A four-vaned tail controlled from the saloon would be attached where the eight poles met at the tail of the gasbag. A long exhaust pipe led from the engine to a point beneath the tail so that stray sparks would not ignite the gas.[52]

Porter's "tethered observatory" disappeared from his subsequent plans, but his thoughts on the "travelling balloon" continued to mature over the next decade. The revised craft made its next appearance in Porter's *New York Mechanic* in 1841. It was clear that the inventor had given serious thought to the practical problems that would be encountered in the construction of the machine. Another small flying model, presumably powered by a clockwork mechanism, had been an important part of this learning process.

The basic dimensions of the ideal "travelling balloon" remained unchanged in the 1841 article. On this occasion, however, the inventor provided a comparison of lift and costs for 300-, 125-, and 75-foot aeroports.[53] Porter provided additional details on the propulsion system as well. He suggested the use of small copper tubes in the boiler to insure both speed in firing the engine and safer operation. The required power had been reduced from ten to four horsepower, with a concomitant reduction in estimated airspeed. The Archimedes' screws of 1834 had also been replaced by propellers with standard blades.

When the aeroport reemerged in the pages of *Scientific American* on September 18, 1845, Porter had effected still more changes in design. The craft had now been fixed at 350 feet in length and 35 feet in diameter, and the number of internal battens had been increased from eight to sixteen.[54]

Porter had also conducted his first engine experiments during this period. He had built a two-horsepower steam engine, and planned a ten- to thirty-horsepower model for the large aeroport. Such a power plant, linked to "two spiral fan wheels 16 feet in diameter," would, he estimated, drive the aeroport at a speed of 100 MPH.

Like John H. Pennington, Porter faced the dual problem of differentiating his plans for aerial navigation from those of others and justifying his own scheme. As the mid-century mark approached, a variety of airship designs emerged in the American press. The same issue of the *Mechanic's Magazine* that had published Porter's first airship article in 1834 had also carried a letter and drawing from a Michigan resident identified only as B.G.N. They portrayed a craft that was distinguishable from those of Pennington and Porter only in detail.[55]

Porter was very much concerned about his rivals. Although he was no longer editing *Scientific American* in October 1847, he certainly agreed with the new editor's assessment of John H. Pennington: "Mr. Pennington . . . not having the fear of common sense before his eyes," remarked the journal, "has published proposals for the formation of an Aerial steamship navigation company. If the vessel proves as light as the vision of the projector, they will need an extra amount of ballast."[56] Porter chose to focus his personal wrath on another contender for aeronautical honors, Muzio Muzzi, an Italian visitor to the United States, whose model airships were attracting much favorable attention in scientific and engineering circles.

Muzzi, the son of Prof. Luigi Muzzi of Bologna, had developed a craft quite unlike the classic oblong dirigible balloon that had evolved from Meusnier's drawings. Instead, Muzzi suggested a great disc-shaped, lenticular balloon that would be flown upright and edge-on into the wind. Later variants of the Muzzi design would include large revolving propeller wheels on either side of the gasbag. The models actually flown in Europe and America, however, were unpropelled and featured only two rectangular planes mounted horizontally beneath and on either side of the balloon and a large triangular rudder at the rear.[57]

Muzzi's notion was quite simple. The horizontal planes, which could be moved to various angles of attack, would serve as inclined planes which by their resistance to the air would translate the rise or fall of the balloon into forward motion.

Muzzi had first demonstrated models of his flying machine at an Italian scientific meeting convened in the palace of the Chevalier brothers in Pisa on October 10, 1839. The action of the planes was illustrated by a model operating

An Italian print of Muzio Muzzi's flying machine

in a water tank. A hydrogen-filled model was also flown in the air. Muzzi received "the warmest congratulations and praise" from those in attendance. The Chevalier brothers were sufficiently impressed to place a plaque commemorating the exhibition on the wall of the villa.

The earlier performance was repeated in Florence on November 14, 1839, this time in the presence of Gabinetto Fisico, his Imperial Royal Highness the Duke of Tuscany, who requested one of the small craft for the ducal museum. Muzzi continued his demonstrations in Europe, flying his models at Leghorn on May 30, 1841, and applying for patents in France (May 12, 1842; no. 6,837) and the United States (October 16, 1844; no. 3,799).

Unable to raise the money to build a full-scale machine in Europe, Muzzi came to the United States in 1844. His initial reception in America was quite as warm as it had been in Europe. On December 7, 1844, the inventor offered his standard demonstrations to a distinguished group of American scientists and engineers gathered in the chapel of the University of the City of New York. James Renwick and John W. Draper, professors of chemistry and natural philosophy at Columbia College and New York University, respectively; James Tallmage, president of the host college; John Ericsson, already recognized as one of the world's leading mechanical engineers; and James J. Mapes, a local professor of mechanics and a well-known consulting engineer, were among those present for Muzzi's American debut. The scientists and technicians were impressed, but Muzzi's project ultimately floundered in the United States, primarily, it would seem, as a result of Rufus Porter's opposition.[58]

Porter's widely circulated review of Muzzi's performance in New York was merciless. "We doubt," he began, "whether any event or circumstance has transpired within the present century, which has served to develop the ignorance and gullibility of the citizens of New York, to so great an extent, as that of the introduction of Signor Muzzi's contrivance for aerial locomotion." Porter viewed "with mingled sensations of regret and vexation" the applause Muzzi received from "the most popular city papers . . . and several of our professors of science, and men of reputed scientific attainments."[59]

Porter hastened to point out the weaknesses in Muzzi's scheme for propelling the craft. The roller coaster motion required to move the machine forward would quickly exhaust both gas and ballast, he argued. Moreover, it would almost certainly be ineffective in propelling the craft in a modest breeze.

While Porter was drawing attention to the deficiencies of his rivals, he was continuing to extol the virtues of his own aeroport in the pages of the newly founded *Scientific American*. The front-page overview of the "travelling balloon" published in the issue of September 18, 1845, was followed by more detailed articles covering air resistance and other specialized topics.

Porter assured his readers that his plans to construct an aeroport were moving forward, although he admitted that money remained a major problem. He had hoped to fund his aeronautical work with the profits from *Scientific American*, but "disappointments with regard to our anticipated circulation" and the loss of his own investment in the venture as the result of a disastrous fire made this impossible. Moreover, Porter feared that "a rank and bitter prejudice against the project" might cause a further drop in circulation.[60]

Nevertheless, Porter persevered. In March 1846 he announced that arrangements had been made to begin work on an aerial ship that would be placed in operation later that spring. In order to encourage investment in the project on the part of *Scientific American* readers, the editor promised that he would accept one dollar of any stock purchase in payment for a year's subscription to the magazine.

There was, as yet, little substance to Porter's optimistic talk of airships and stock issues. Having established his third journal, the *Scientific Mechanic*, in 1847,

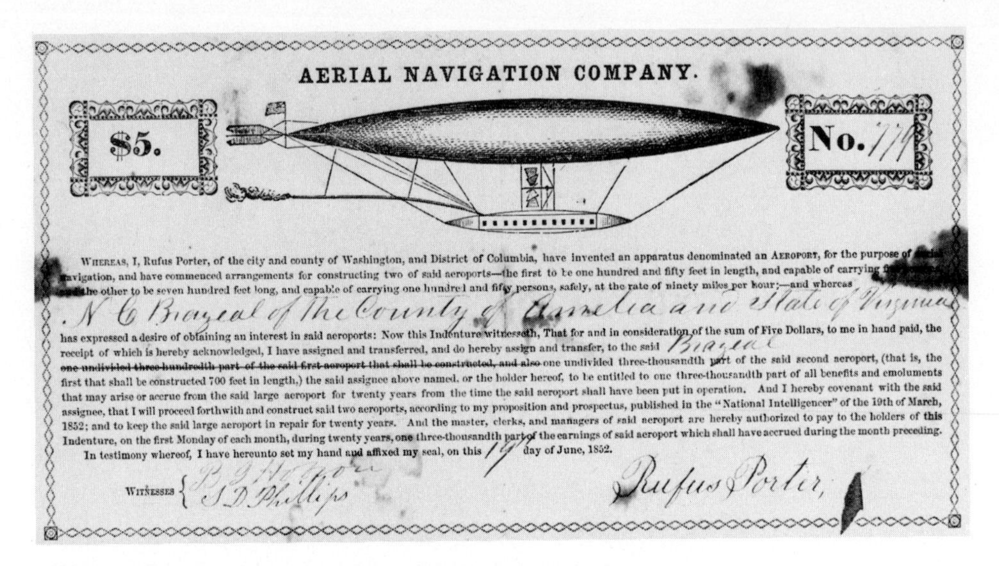

AERIAL NAVIGATION COMPANY.

$5.

No. [illegible]

WHEREAS, I, Rufus Porter, of the city and county of Washington, and District of Columbia, have invented an apparatus denominated an AEROPORT, for the purpose of [illegible] navigation, and have commenced arrangements for constructing two of said aeroports—the first to be one hundred and fifty feet in length, and capable of carrying [illegible] the other to be seven hundred feet long, and capable of carrying one hundred and fifty persons, safely, at the rate of ninety miles per hour;—and whereas

N C Brazeal of the County of Amelia and State of Virginia

has expressed a desire of obtaining an interest in said aeroports: Now this Indenture witnesseth, That for and in consideration of the sum of Five Dollars, to me in hand paid, the receipt of which is hereby acknowledged, I have assigned and transferred, and do hereby assign and transfer, to the said Brazeal ~~one undivided three-hundredth part of the said first aeroport that shall be constructed, and also~~ one undivided three-thousandth part of the said second aeroport, (that is, the first that shall be constructed 700 feet in length,) the said assignee above named, or the holder hereof, to be entitled to one three-thousandth part of all benefits and emoluments that may arise or accrue from the said large aeroport for twenty years from the time the said aeroport shall have been put in operation. And I hereby covenant with the said assignee, that I will proceed forthwith and construct said two aeroports, according to my proposition and prospectus, published in the "National Intelligencer" of the 19th of March, 1852; and to keep the said large aeroport in repair for twenty years. And the master, clerks, and managers of said aeroport are hereby authorized to pay to the holders of this Indenture, on the first Monday of each month, during twenty years, one three-thousandth part of the earnings of said aeroport which shall have accrued during the month preceding.

In testimony whereof, I have hereunto set my hand and affixed my seal, on this 19 day of June, 1852.

WITNESSES { B. J. H[illegible]
S. D. Phillips

Rufus Porter,

A stock certificate for Rufus Porter's Aerial Navigation Company

he returned to the practice of featuring the "travelling balloon" on the front page in the hope of exciting the interest and financial support of his readers.[61]

But 1847 was to be a year of change and activity for Porter, during which he built and flew his first two successful airship models. Other than their weight, about two pounds, no details are available as to the size of these craft. Porter did note that, like the unsuccessful model constructed in Bristol, Connecticut, thirteen years before, they were propelled by clockwork mechanisms. While the first model was being demonstrated by trusted friends in New York, Porter traveled to Boston with a second "larger and improved model," which he flew in Temple Hall.

The inventor apparently continued to offer indoor model flights accompanied by "semi-public" discussions of his airship proposals in both cities for the next three years. He also acquired a partner, a Mr. Robjohn, during this period. It is tempting to suppose that an influx of cash from Robjohn made possible the construction of the initial models.[62]

Public reaction to these first practical demonstrations of the aeroport was mixed. Porter certainly had little difficulty attracting large crowds of applauding spectators, and most newspapers offered enthusiastic support as well. The editor of *Scientific American* attended a lecture offered at the New York Tabernacle in 1849. He was delighted with the amusement, although something went wrong

SCIENTIFIC MECHANIC

INVENTORS' ADVOCATE, PATENT OFFICE REPORTER, AND EXPOSITOR OF ARTS AND TRADES.

VOL. I.] NEW-YORK AND WASHINGTON, D.C., SATURDAY, DECEMBER 25, 1847. [NO. 16.

THE SCIENTIFIC MECHANIC,
PUBLISHED EVERY SATURDAY BY
RUFUS PORTER,
At 128 Fulton st., (Sun Buildings,) N.Y.

TERMS:
One Dollar a-year in advance.
THREE CENTS a Copy delivered.

Each number of this paper is furnished with several Original Engravings, in illustration of New Inventions, Scientific Principles, and Curious Arts. The editor being practically as well as theoretically acquainted with the sciences of Mechanics, Hydraulics, Pneumatics, Chemistry, and Electro-magnetism, and being a practical master of fifteen different Arts and Trades; and having introduced upwards of fifty new inventions, he is confident of the ability to furnish a paper more useful to mechanics and amateurs in Science, than any other periodical of the same price in the world.

In this paper will be found early notices of the progress of Invention, Enterprise and Improvements in the United States, and particularly those in the process and facilities of the mechanic arts and trades.

CORSETS.

When I was down in Boston town,
A month ago or more,
I saw a very sing'lar thing
I never saw before.

'Twas hanging in a window case,
Upon a string a-straddle—
Looked something like an hour-glass,
And something like a saddle.

I asked of several citizens,
Who chanced to be at hand,
"What was it?" but their gibberish
I could'nt understand.

One fellow called it "a restraint
On certain parties placed,
Like a decree in chancery,
To stay the tenant's waste!"

Another—just the queerest chap
Of any in the swarm—
Said "'twan't the glass of fashion, but
It was the mould of form."

nother said "twas a machine
A lady used to rig her—
To bring her form and life unto
The very smallest figure."

At last a little girl came out,
And—think of my amaze!
She asked me "if I wouldn't please
To buy a pair of stays?"

Of course I'd heard of "stays" before,
But strike me deaf and dumb!
If ever I, until this hour,
Suspected "them was um."

Well—isn't it exceeding strange
That any maid or wife,
Just for a "little taper" should
Put out the "lamp of life?"

I know that lunatics must have
Straight jackets put about 'em—
But women in their wits should make
A shift to do—without 'em!

Play upon Words.

A bat about a farmer's room,
Not long ago I knew
To *fly*. He caught a *fly*, and then
Flew up the chimney *flue*.
But such a *scene* was never *seen*,
(I am quite sure of that,)
As when with sticks all hands *essaid*.
To hit the *bat* a *bat*.

Lost Time.

I threw a bauble to the sea,
A billow caught it hastily;
Another billow quickly came,
Successfully the prize to claim;
From wave to wave, unchecked, it passed,
Till tossed upon the strand at last.
Thus glide unto the unknown shore
Those golden moments we deplore;
Those moments which not thrown away,
Might win for us eternal day.

REMARKABLE WEATHER.—The Boston and Providence papers report that dandelions, violets and water cresses have within a few days past, flourished in the open air, and men and boys have been bathing at the beaches. The Winter thus far has been more favorable to the poor, than to the coal dealers.

THE TRAVELLING BALLOON.

ÆRIAL NAVIGATION.—The practicability of travelling rapidly and safely through the air, has been already established, as far as theory can establish a point without actual experiment; and the most important principles on which success in this, mainly depends, have been already thus established. The specific gravity of hydrogen gas is less than that of atmospheric air, by something more than one ounce per cubic foot; and consequently a cubic foot of this gas being enclosed, has a buoyant power of one ounce in atmospheric air. A hollow globe, five feet in diameter, may be made of oiled silk, of less than one pound weight; yet this globe, being filled with hydrogen gas, will possess a buoyancy in atmospheric air of more than four pounds. This sufficiently illustrates the first principles of ballooning; but as it is plain that a spherical balloon cannot be propelled with any considerable velocity on account of atmospheric resistance we have adopted the eliptic spindle form, which will encounter—as will be proved in our next number,—less than 1-2000th part as much resistance as a globe of equal diameter. A balloon of this figure, 350 feet in length and 35 in diameter, being inflated with hydrogen, will have acquired a buoyancy equal to 12,000 pounds. It will require in its construction 16 long rods, extending the entire length, and estimated to weigh 1000 pounds. 2,200 yards of strong linen cloth which, when varnished, will weigh 1000 pounds, thus, leaving a balance of buoyancy of 10,000 pounds. An eliptic, or revoloidal spindle-shaped saloon is attached to the balloon, being suspended immediately under it by cords or wires. It is covered with cloth, excepting 100 feet of the central part, which has a permanent floor, and is ceiled with thin boards. The weight of this part of the apparatus is estimated at 3,000 lbs. We have already constructed and put in operation a steam engine and boiler capable of working two horse-powers, but weighing only 200 lbs., from which it is estimated that an engine and boiler, of a size to work two horse-powers, may be made within the weight of 1000 lbs. One or two spiral fan-wheels of 16 feet in diameter, together with the requisite machinery for communicating the power of the fan-wheel; and the requisite ropes and rigging connecting the saloon to the balloon &c., may all be comprised within the weight of 1000 lbs.; thus leaving a ballance of buoyancy of 5,000 lbs, for passengers, baggage, &c. A rudder consisting of two broad fans intersecting each other at right angles in the centre, is attached to the stern end of the balloon, by a ball and socket, or universal joint. Four arms project in opposite directions from the rudder near its connecting joint; and from the ends of these, small lines extend to the interior of the saloon, and by means of which the direction of the balloon is completely governed, both horizontally and vertically. With regard to the velocity of this ærial ship, it is susceptible of demonstration, and will be shown in a future number, that ten horse-powers is sufficient to balance the atmospheric resistance at a velocity of 100 miles per hour: but if it were otherwise, there is sufficient buoyancy to admit of engines capable of producing 30 horse powers. This balloon being once inflated, is intended to be kept constantly afloat, being moored at about 100 feet distant from the earth, when not in use. When passengers are to be received or landed, the balloon will descend to within a few feet of the ground, and a light-seated car, will descend with passengers, to receive them. When passengers are received, a corresponding weight of stone blocks will be deposited, and *vice versa*. The saloon will be furnished with an apparatus for the production of hydrogen, sufficient to supply whatever quantity may escape by leakage. The interior of the balloon will contain an arrangement of lines and pulleys, by which the circumference may be contracted sufficient to cause the balloon to descend whenever occasion requires. These lines, which are attached to the rods, and by which the rods are drawn towards the centre, are operated by a small iron rod or wire, which passes from the interior, down through a small stuffing box, to the saloon. By means of this rod and system, the vertical movement of the balloon will be principally governed. With regard to the safety of this mode of travelling, we think it can be readily made to appear that there will be less danger in travelling over land in ærial vessels, than there now ordinarily exists in travelling by either sailing vessels or steamboats. In the first place, the balloon will be less liable to accident, than either steam vessels or railroad cars, on account of being less exposed; resting on the elastic atmosphere, and floating with the current, except what forward motion is given it by the propelling machinery, it cannot be affected by gales or squalls, like a stationary object, and its motions must be comparatively gentle. In the second place, the balloon will never soar so high but that in case of accident, or any derangement of the machinery, it may safely descend to the earth, in ten minutes: an idea much more consolatory to the passengers, than that of being five hundred miles from land, and surrounded by a violently raging element, to which, in case of either fire or leakage, the marine passenger must commit himself. A third consideration is, that the balloon will be furnished with an improved parachute for each passenger, and of which each may avail himself in less than one minute, in case of any extraordinary emergency, and thus descend safely to terra firma, much easier than he could paddle himself to shore, on a cotton bale, even from the middle of Long Island Sound. The steam engine by which the balloon is propelled, will be very small, and the boiler being constructed of small copper tubes, there can be no possibility of damage by explosion: and no accident can possibly happen to the balloon, that would cause a sudden descent, so as to prevent the passengers from having sufficient time to avail themselves of the parachutes. Moreover, their baggage would all come to land in due time. We shall give further and more minute particulars on this subject, with sectional illustrations, and mathematical demonstrations: but these must be deferred for future numbers.

SHARP SHOOTING.—The editor of the Alabama Flag recently fired the following squib at Prentice:

"Why is the editor of the Louisville Journal's career through life like a celebrated tune? Because it is the Rogues' March.

Prentice, who is a "dead shot," loaded his piece, and taking good aim, let the Flag man have it right in the teeth after the following manner, to wit:

"Why will the editor of the Flag, at the close of his career through life be like a tune of Paganini's? Because he will be *"executed on one string."*

There is exhibited at the Exchange Reading Room, Philadelphia, a well polished ox horn, four feet nine inches long, and sixteen inches in circumference. It was brought from Rio Janeiro.

Mr. Robert Reed, of Scotland, has adopted the practice of boxing up and burying his bees in the earth in the Autumn, and finds them come out bright and healthy in the Spring.

A simple mixture of rye meal and unslacked lime, is recommended as a destructive food for rats. They immediately seek water, which, by slacking the lime, will destroy them.

The Westminister (Md.) Caroltonian gives the following description of the mode of making Iron wire fences in that vicinity; "The posts are about one half the ordinary size, planted firmly at the distance of ten feet apart with nine strands of wire drawn tightly through a half-inch auger hole and tightly plugged at each hole; the wire is of the size of that used for the handle of the Yankee bucket, and to combine them more firm, wire of a lighter description is wound through the middle, which prevents the hogs from separating them and creeping through.

There are now in operation in the United States, thirty one lines of telegraph with nearly 4000 miles of wire. Fourteen other lines, in progress, or contemplation, will make an aggregate of 8000 miles.

Dr. J. Patten of Donellsville Ohio, has lost his life by accidentally taking Strychmia instead of Morphine. Doctors should know better than take such drugs themselves.

"Boy, why don't you go to school?"—"Bekase, sir, daddy is afeard if I larn anything now, I shan't have nothing to larn when I go to the catemy."

THE PHANTOM BALLOON.—Mr. Gale, the æronant, witnessed a phenomenon during his ascent from Rosherville, which is said to be unprecedented in the annals of balloons.—Having become becalmed and perfectly stationary, for a full quarter of an hour, he was preparing to descend, when he suddenly beheld a perfect "counterfeit presentment" of the machine depicted on the clouds to the southward. This appearance continued for about four minutes, when the balloon descended to a lower region and the image gradually faded away.

THE UNEMPLOYED.—There are nearly 12,000 factory hands in Manchester, Eng., without the means of employment. In Ashton-under-Lyne, 8,377 hands have struck work, while the unemplyed in that town and Lancaster, Preston, and Bolton amount to 4,679.

"Will you give me that ring on your, finger," said a village dandy, to a girl, 'for it resembles my love for you—it has no end.' 'Excuse me, sir,' was the reply, 'I choose to keep it; it is emblemetic of mine for you, it has no begining.

A coquette is defined as one who wants to engage the men without engaging herself.

LIST OF PATENTS
ISSUED FROM THE UNITED STATES PATENT OFFICE,
For the week ending Dec. 4, 1847.

To Curtis Wooster, of Philadelphia, Penn, for improvement in guages used in finishing spiral Hand Rails. Patented Dec. 4,

To Elias Hall, of Persia, New York, for improvement in self-acting Cheese Presses. Patented Dec. 4, 1847.

To George Pollock, of New York City, for improvement in Registers for Furnaces. Patented Dec. 4, 1847.

To Richard Albert Tilghman, of Philadelphia, Penn., for Improvement in decomposing alkaline and other salts. Patented Dec. 4, 1847.—Ante-dated Feb 1, 1847.

To Richard Albert Tilghman, of Philadelphia Penn., for improvement in decomposing P a h-Felspar, for obtaining certain salts.—Patented Dec. 4, 1847. Ante-dated Feb. 1, 1847.

To Reuben Cornitt, of Georgetown, Missouri, for improvement in connecting rods, pitmans, &c. Patented Dec. 4, 1847.

To Nathan Sawyer, of Baltimore, for improvement in Brick Presses. Patented Dec. 4, 1847.

To Harvey B. Nash, of Kingsbury, New York, for improvement in Bedstead Fastenings. Patented Dec. 4, 1847.

To Joseph Tuers, of Jersey City, N. J., for improvement in Boats for sailing. Patented Dec. 4, 1847.

DESIGN.

To D. F. Goodhue, and Charles Guild, of *Cincinnati, Ohio, for Design for Stove Plates.* Patented Dec. 4, 1847.

The Queen of England.

The income of the Queen of England is one million of pounds sterling—$5,000,000. From this £200,000 is taken to pay what is called the "Household," which consists of the great officers of state, such as the Lord Chamberlain, &c. The queen has seven or eight palaces, but only four are used by her—Buckingham Palace, Windsor Castle, Claremont, and Osborn House. The others are St James', Hampton Court, Newmarket and Kensington. In these, apartments are given to the widows and daughters of persons who have deserved well of the country. When the queen goes in state, the procession consists of the Life Guards in half armour; three of the royal carriages with four horses each, having the great officers of state in them; then the corps of Yeoman of the guard dressed in the costume of the time of King Henry the Eighth, with long halberts in their hands; after these are forty footmen in liveries of red, literally covered with gold; then the marshal-porters, also gaudily dressed. The state carriage is drawn by eight cream colored horses, in scarlet morocco leather harness thickly studded with gold. The queen is generally accompanied by Prince Albert, the principal lady of the bed Chamber, and the Duke of Wellington carrying the sword of state. She wears upon this occasion a coronet of brilliants, and diamonds of immense value upon her dress, which is of the richest materials, and over her shoulders she has a crimson velvet mantle, lined with ermine, and richly embroidered with gold.

Prince Albert has from the British nation, independent of the Queen an income of £30,000, $150,000, and holds places under the government to the value of 40,000 pounds, or making in the whole $350,000 a year. The queen has 320 domestic servants, 110 of whom wear the royal livery of scarlet and gold. At a royal banquet the whole dinner service is of gold—the queen sits in the centre of the table, and the lords in waiting take the head and foot.

Thanksgiving Commodities.

It has been a matter of curious remark that the appointment of thanksgi g in Maine was made so unusually early; but we have recently learned that pumpkins attained an enormous ize in that State the past season. In Sangerville, especially, one has been produced which weighs one hundred and fifty seven pounds, and many others nearly as large.

Enterprise in Maine.

The bridge recently erected between Wiscasset and Davis' Island and Edgecomb, (Me.) is now passable for carriages. It is about 4000 feet long. Another long bridge is to be built beyond the present one, and then the distance to Damariscotta will be shortened about four miles. The whole road will be opened to th public during the next season.

Porter's "Travelling Balloon" in an 1847 incarnation

with the mechanism, and wished Porter and Robjohn "all success."[63]

The editor of the Boston *Bee* concurred. "Mr. Porter's 'flying machine' did all that it promised on Wednesday evening," he informed his readers. "It rose above the audience and went round the hall, exactly as he said it would, and the spectators gave three cheers for the successful experiment."[64]

The *New York True Sun* offered a ringing endorsement following one of the final demonstrations in the city in March 1850:

> *The Aerial Steamer Model was again tried at the Merchant's Exchange yesterday afternoon, and with brilliant success. It described a circle of the rotunda eleven times in succession, following its rudder like a thing instinct with life. With its description of each circle, burst after burst of applause arose from the excited throng, and followed it throughout its journey. At the close of the performance, three loud cheers were given for the steamer, and the auditors quitted the rotunda with every manifestation of pleasure and delight.*[65]

The *New York Evening Post* added its voice to the admiring chorus in 1849:

> *The exhibition at the Tabernacle last evening, of a spindle shaped balloon, floating in the air and moving around the chandelier, propelled by a chronometer spring acting upon fans, was very amusing. If Messrs. Porter and Robjohn can with a steel spring propel their balloon a couple of rods, it is very natural to infer that steam will propel it any given distance. It is supposed by some that a balloon moved by steam, and the principles illustrated at the lecture last evening, can be made to move at a rate of one hundred miles per hour. One of the prerequisites to secure this velocity will be to have the bag of wind entrusted to some one more cautious than even the wise and valorous Ulysses.*[66]

This sort of approval was far from unanimous however. Porter complained that at the time of his original New York demonstration in 1847 he "encountered much opposition from parties who were interested in marine navigation, railroads, etc."[67] One "renowned practical balloonist" suggested that a "balloon (or float) in the form of a revoloidal spindle could never be rendered sufficiently buoyant to support its own weight." Another critic argued that no material, including "boiler iron," could withstand the atmospheric resistance to be encountered at speeds of 100 MPH.

Porter was also bitter about the reaction of newspapers such as the *New York Sun*, the *Pathfinder*, and the *Globe*, "which published articles of derision without reasons." Others, notably the *New York Tribune* and the *Farmer and*

Mechanic, seemed ever ready "to laud and commend every fallacious project that was introduced in preference to my plan."[68] The views of the Philadelphia *Bulletin* were typical of this group:

> *Though every man of sense is, or ought to be, aware of the impossibility of steering a balloon, or any other aerial machine, yet it seems there has been found, in New York, a fellow who was knave or fool enough to advertise for exhibiting a flying machine at the tabernacle; and that here were found Dogberries sufficient to fill that huge building. We have heard nothing more ridiculous, since a theatre was once filled, on the other side of the Atlantic, to see a man get into a pewter pot. It would seem as if the gullibility of human nature kept pace with the wit of knaves, and that nothing could be proposed for an exhibition too preposterous to find believers. In this very case, the thing proposed was impossible. A ship is steered in the water, because the action of the wind on the sails, and of the hull in the water, can be brought to counteract with each other by means of rudder. Now a flying machine operates in one element, and hence can never be steered. Yet, as in the analogous instances of perpetual motion, there will be found dolts to believe in it, we suppose, to the end of time. Alas! poor humanity.*[69]

In spite of this mixed reception, events occurring 3,000 miles away would bring fresh opportunities for national publicity which, in turn, would breathe new life and fresh cash into the venture. Porter was about to offer "the best route to California."

Chapter 11 Aeroports, Aereons, and Avitors

ON THE MORNING OF JANUARY 24, 1848, JAMES MARSHALL, A TRANSplanted New Jersey carpenter who was supervising the construction of a sawmill in a bend of the South Fork of the American River in California's Coloma Valley, spotted a bit of glitter in a millrace. "My eye was caught by something shining in the bottom of the ditch," he later recalled. "I reached my hand down and picked it up; it made my heart thump, for I was certain it was gold. The piece was about a half the size and shape of a pea. Then I saw another."[1]

News of Marshall's discovery spread down the California coast that spring and summer. By July the rough adobe towns of southern and central California were half deserted as their residents swarmed north toward Sacramento.[2]

In a letter to Secretary of State James Buchanan written in late June 1848, Thomas Larkin, a wealthy Monterey merchant, gave fair warning of what the rest of the nation might expect during the next several years. "If our countrymen in California will forsake employment as clerks, mechanics, and workmen at from two to six dollars a day, how many more of the same class in the Atlantic states earning much less will leave for this country under such prospects?"[3]

By the late summer and early fall, news of the discovery of gold in California was beginning to appear in eastern newspapers. The effect was quite as extraordinary as Larkin had suggested. As the *New York Herald* reported, "Poets, philosophers, lawyers, brokers, bankers, merchants, farmers, clergy men—all are feeling the impulse and are preparing to go and dig for gold and swell the number of adventurers to the new El Dorado."[4]

The thousands of adventurers who sold what they owned in order to outfit themselves for the goldfields faced one overwhelming difficulty. California lay 3,000 miles to the west of the Atlantic Coast, on the far side of an empty, almost unknown expanse that previously had been categorized as the Great American Desert.

Inland gold seekers swarmed to St. Louis and St. Joseph where they joined what formerly had been a trickle of overland immigrants. For the residents of eastern cities, a sea voyage around the Horn seemed simpler and quicker. The route had, after all, been traveled for many years by ships engaged in the California hide trade. There were also those who braved disease and tropical jungles in an attempt to cut short the four-to-six-month, 18,000-mile voyage around the Horn by booking passage to the Chagres River, making their way across the Isthmus of Panama, and traveling the relatively short distance up the Pacific coast to San Francisco.

Rufus Porter was quick to take advantage of the need for a fast, safe means of traveling to California. Having fully demonstrated, he believed, the feasibility of his aeroports through the model flights begun in 1847, he rushed into print with a pamphlet aimed at travelers bound for California.

Entitled *Aerial Navigation: The Practicality of Travelling Pleasantly and Safely from New York to California in Three Days Fully Demonstrated, With a Full Description of a Perfect Aerial Locomotive With Estimates of Capacity, Speed, and Cost of Construction*, the little booklet featured the standard cut on the back cover. The bold script set over the airship informed readers that here at last was the "best route to California."[5]

The aeroport described in the 1849 pamphlet was a much more ambitious craft than those that had been outlined earlier. The revoloidal spindle had grown to 800 feet in length, 50 feet in diameter, and 838,000 cubic feet in capacity. Twenty-four spruce rods, each 1½ inches in diameter, would serve as battens.[6] Porter was now suggesting that these rods be united in such way as "to form the skeleton or frame of [the] spindle."[7]

Porter intended to cover his framework with cloth coated with gum-elastic (india rubber). The saloon, 180 feet long, 10 feet square at the center, and tapering to a point on each end, would be suspended 60 feet beneath the bag by flattened steel wires.

> *This saloon is made of painted cloth supported by a very light framework, and about 80 feet of the central part is furnished with windows, and a floor of thin boards sustained by four rows of vertical wires extending upward to the float [balloon]. The saloon is furnished with seats, which may be readily transformed to beds for those who may have occasion to sleep on board. In the center of the saloon is an apartment 6 feet by 12, in which are adjusted six light tubular boilers of two horse-powers each, and two steam engines, the power of which is applied to two fan wheels or propellers mounted between the float and the saloon.*[8]

Porter suggested that the engine and boiler be suspended independently from the framework of the envelope. An elevator, four feet wide and eight feet long, furnished with seats and railings, would be located near the boiler room. Operated by a windlass connected to the engine, the elevator would move passengers and freight to and from the tethered airship.

Porter had also devised a new method of controlling the buoyancy of the airship. Interior cords would connect each of the 24 battens. The operator in the saloon could draw these battens together to compress the gas and descend, allow the gas to expand in order to rise, or compress either end to trim the craft. Two propellers, each with eight blades and a 20-foot diameter, would revolve at 200 RPM to drive the craft through the air at a top speed of 100 MPH.

The inventor also included a detailed bill of materials in the pamphlet. He estimated that one aeroport would contain: 20,000 feet of spruce rods; 8,000 yards of vulcanized cloth; 800 feet of ½-inch boards for the saloon flooring; 300 yards of painted cloth for the saloon walls; 12,000 feet of cast steel wire; 6 boilers; and 2 engines. The finished aeroport, fully fueled and ready for loading passengers and freight, would weigh 14,000 pounds.

Porter also provided estimates for smaller airships. Another steam-powered model, 200 feet long and 25 feet in diameter, would weigh 2,360 pounds and cost $1,700. A third and even smaller aeroport, 120 feet in length, would be powered by a hand crank and cost only $500.

Porter took great care to explain the safety features built into his aeroports. There was, he argued, "less danger of any accidents to the float [balloon], than to the hull of a marine sailing vessel." Like the great airships of the early twentieth century, the "float" would contain several "apartments," or gas cells, so that if the bag were punctured, all the hydrogen would not escape. The invention was planned to fly at moderate altitudes of less than 1,000 feet. Moreover, each passenger would be furnished a neatly packed parachute 8 feet in diameter. A copper wire could be tossed to the ground to protect the craft from lightning.

Porter waxed eloquent in describing the pleasures of an aeroport voyage:

> *Let our gentle readers imagine themselves to be visiting the pleasant and excellent literary establishment on the summit of* Mt. *Holyoke on a sunny morning in the balmy month of June, and gently descending thence toward the verdant plains which border the meandering* Connecticut, *and then at an elevation of only 8 or 10 feet from the ground sailing moderately over the rich fields of broom and grain; and over the flower spangled fields of grass, waving to the western breeze, and conversing by the way with the merry farmers, as they follow their recreative avocation of hay-making; then ascending with accelerated velocity to the altitude*

> *of refreshing temperature and returning to New York to dine. Or suppose yourselves leisurely cruizing* [sic] *along by the steep and rugged sides of the Rocky Mountains, and laughing at the astonished countenance of the harmless grizzly bear, or at the agility of the frightened antelope; and then descending to the extensive prairies to watch the prancing of the wild horses, or the furious rushing of hordes of Buffaloes.*[9]

Porter emphasized that, while it was quite possible for an aeroport to cross the Atlantic, it was primarily intended to facilitate immigration to California. In addition to 100 passengers and their baggage, each ship would carry fuel and provisions for a voyage of forty-eight hours, which, the optimistic inventor predicted, "will be several hours longer than the time required for the passage, under ordinary circumstances."[10]

Readers were informed that the firm of R. Porter & Company, with offices at 128 Fulton Street in New York, was making "rapid progress in the construction of an Aerial Transport, for the express purpose of carrying passengers between New York and California." The craft was scheduled to begin service on April 1, 1849. The books were open for 300 passengers willing to put $50 down toward passage to the goldfields during the coming season. The full price for board and passage was to be $200. The round trip would require seven days. Porter claimed that almost 200 deposits had already been received.

The publication of the pamphlet created immediate interest, though not all of it was favorable. The very possibility of flying to California inspired a series of classic comic prints that did little to build confidence in the venture but did publicize Porter's aeroport.[11]

The fact that their ideas were not regarded seriously in all quarters seems to have had little impact on Porter and Robjohn, who forged ahead with a building program. In September 1849, *Scientific American* reported that Robjohn had completed a canvas gasbag 240 feet long and 50 feet in diameter. Two five-horsepower oscillating steam engines were on hand. The car was under construction as well.[12]

On May 4, 1850, the same journal reported that a tornado had swept through Hoboken, "sadly damaging" the airship. Porter and Robjohn seem to have reached a parting of the ways following the storm. Robjohn apparently chose to rebuild the damaged craft. In September 1851 a reporter for *Gleason's Pictorial Drawing Room Companion* visited Robjohn's establishment "on the plain west of Hoboken village." Inside a large enclosure measuring 290 by 275 feet, he found the old aeroport was almost completely rebuilt.

The new car was a particularly impressive structure, 64 feet long, 6 feet wide and 6 feet 4 inches tall, just as described in Porter's pamphlets. It consisted of canvas walls stretched over a board frame, with light board flooring. Glass windows and wooden doors were provided. The lower half of the forward cockpit section was completely glassed to improve pilot visibility.[13]

Pictorial reports noted that the two steam engines now developed twelve horsepower, rather than the five suggested by *Scientific American*. The engines, which had a 20-inch stroke, were described as being "very perfect" and a "curiosity." Robjohn could move the pistons through a complete cycle by blowing through a valve, "carrying a driving wheel of five feet in diameter."[14]

The rudder was also regarded as "worthy of minute examination." Robjohn remarked that the craft had cost $5,500 to date and was within several hundred dollars of completion. The aeroport was protected by a high fence, but those interested in a close view of the work were invited to contact Robjohn at his office, 166 Bowery Street in New York.[15]

In spite of the fact that the Robjohn craft was obviously very close to completion in the fall of 1851, there is no indication that it was ever finally assembled or tested. Nor is there any mention of Porter in this final article, suggesting that the two men were no longer working together.

Rufus Porter had moved to Washington, D.C., in 1850, where he supplemented his income by working as a consultant to inventors visiting the Patent Office. He also published his second aeronautical pamphlet that year. Entitled *An Aerial Steamer, or Flying Ship*, the booklet was essentially a repetition of his earlier articles and pamphlets. As usual, the dimensions of the aeroport had been altered.[16]

This time the length of the craft was reduced to 700 feet. Two ten-horsepower engines would drive two eight-bladed propellers. As before, Porter emphasized the safety of his machine. There was no mention of California in the new pamphlet. The inventor was now far more interested in attracting investors to underwrite the construction of a new airship.

Porter also prepared a memorial for presentation to both houses of Congress in 1850. The inventor requested a $5,000 appropriation "to extend his experiments, and apply his invention on a scale of practical utility."[17] He was careful to couch his appeal in national terms, calling attention to "the severe sufferings and dangers to which thousands of citizens are annually exposed in journeying between the eastern and western sections of the United States."[18] He extolled the potential of the aeroport for carrying the mail and pointed to the importance of similar congressional support in establishing the utility of the telegraph.

Porter's effort, however, was to little avail. "I applied to five different members before one could be found who would consent to introduce the memorial," he recalled. "Subsequently, neither of the committees to whom it was referred deemed the subject of sufficient importance to give it any attention, nor permit me to introduce any illustrations thereof."[19]

Porter tried yet another method of fund raising in 1851, sending letters soliciting investment "to some of our first capitalists, and men in high office."[20] He outlined his scheme in the letter, noting his success with the models, and described the support received thus far from "many scientific mechanics and engineers."[21] The inventor promised to construct an aeroport capable of carrying 150 passengers for $15,000, adding that he was, for the moment, aiming much lower. He planned to begin work on a "Pioneer" aeroport designed to carry only five passengers. Porter invited the recipients to become "the philanthropists of the world" by investing in the program. In return, he promised to provide either an annual guarantee of $15,000 or a complete airship in the sponsor's name once the program was in full operation. There were no takers.

Finally, in March 1852, Porter issued his first prospectus for an open stock issue. Three hundred shares were offered at $5.00 per share. The resulting $1,500 would be used to construct the small *Pioneer* aeroport. Each stockholder was entitled to a one three-hundredth share in the *Pioneer*, a one three-thousandth part of the first large aeroport, and all profits to be derived from that craft for a period of twenty years. He estimated that the purchase of a single share would guarantee a $10 to $20 return each week for the entire period.

Porter's two recent pamphlets, news stories describing the flights of his models, even the satirical prints, had prepared the ground for the creation of the joint stock company. His stock circulars drew an immediate and favorable response.

By May 18, 1852, 500 citizens, many from areas as distant as Indiana, Illinois, Mississippi, and Wisconsin, had subscribed to the $1,500 required for the construction of the five-man *Pioneer* aeroport. The number of subscribers climbed to 700 that July, five months after Porter had begun issuing shares.

In spite of this success, financial problems remained. Most of the sales were single purchases of the minimum $5 stock certificate. In addition, a large number of subscribers that had accepted certificates and promised to pay on demand found themselves, in Porter's words, "disappointed pecuniarily, and could not command the money so readily as they had anticipated."[22] Porter thought it wise to carry these individuals on credit, hoping to obtain the money from them in the future. This action certainly helped make friends for the program, but it did little to ease Porter's immediate need for cash to support the construction of the *Pioneer*.

The inventor would continue his efforts to expand his base of investors through 1853. He reduced the business of running a joint-stock company to a fine art. Nicely printed form letters notified subscribers when their stock payments fell due. Printed placards and posters describing the aeroport and inviting inquiry were distributed throughout the nation. Twenty issues of a newsletter, the *Aerial Reporter*, were printed to keep investors informed and interested in the progress of the firm. Porter also distributed news articles and reprints of favorable press notices describing the flights of the model aeroports to newspapers across the country.[23]

In the end, however, the $400 Porter had personally invested in the scheme and the contributions of a few very large subscribers, notably William Markoe, formed the financial backbone of Rufus Porter and Company.

A native of Philadelphia, Markoe had become interested in flight when, as a boy of seventeen, he attended one of William Paullin's ascents. Befriending the aeronaut, Markoe assisted at subsequent launches and made one flight with Paullin from Camden, New Jersey. Many years later Markoe reminded Paullin of this flight: "With what zeal I used to hold the cords during inflation and with what a gratifying boyish sense of self importance I used to take part in the management of your ascensions and 'order about the gaping boobies' who wanted to help but were rather afraid of the big thing; and when I recollect our ascent together and the whacking bump and the upset we got when we struck, I feel quite like a boy again."[24]

Several years after his experience with Paullin, William Markoe, now a divinity student, attended one of Porter's model flights in New York. Mildly interested, he purchased a copy of *Aerial Navigation*. Ordained in 1849, Markoe and his family moved to Delafield, Wisconsin. The young minister's thoughts continued to return to Porter's aeroport however. In response to a letter from Markoe inquiring about the status of the project, Porter described his move to Washington and the establishment of the company, but complained that "his great difficulty was a lack of funds."[25] Markoe, a relatively wealthy man, immediately purchased a $1,000 block of stock and later persuaded relatives to invest an additional $1,550 in the venture.[26]

Porter began work on the *Pioneer* aeroport on May 18, 1852. He bought lumber for the flooring and battens in Washington and set a local mill to work cutting the material to size. The inventor then traveled to New York to purchase the required fabric. He also contracted with two New York machine shops for parts of the boilers.

Back in Washington by May 22, he contracted with three local machine shops for the engine and for additional metalwork. The finished lumber, fabric, and a special varnishing machine were moved to a construction site two miles from the Capitol, where a large tent, measuring 24 by 170 feet and standing 22 feet tall, had been erected at the work site.

Porter was employing five full- and two part-time workers on the site by midsummer 1852. The foreman was a Massachusetts farmer turned carpenter and piano maker who had assisted Porter with the earlier model demonstrations. The inventor regarded him as a man with "a full share of native genius and energy." An English machinist, "an expert workman at various mechanical branches," a young Virginia steam engineer, and a cabinetmaker rounded out the day shift. "A steady, old fashioned Virginian, half farmer and half carpenter," filled in as needed during the day and served as night watchman. Two other young men and a boy also found occasional employment at the aeroport work site.[27]

The work progressed slowly from the outset. As early as July 17, 1852, Porter was offering his apologies for running over the budget and behind schedule. "Most of the work being entirely new and diverse from any ordinary branch of business," he explained, "all the workmen employed must be expected to progress but moderately until they acquire expertness by practice."[28]

Porter complained of the complexity of the enterprise and admitted that much time was lost "in consequence of my inability to superintend personally five or six different branches in progress at as many different places, at the same time."[29] In his original prospectus the inventor had estimated that six weeks would be required to produce the first airship. By the end of July he was forced to admit that he could "not now venture to guess at the number of days that may yet be required to put this first machine in operation."[30]

Problems continued to mount through late summer. The original order for spruce lumber was delayed when the planing mill broke down. Some of the machined metal parts proved unsatisfactory and were returned to the shop. The Georgetown machinist commissioned to produce the engine and boilers forced lengthy delays, pleading the imminent arrival of skilled workmen from Harper's Ferry. When the specialists finally appeared on the scene, they proved "most of the time unfit for business through intemperance."

Porter's inexperience as a supervisor created difficulties and bottlenecks as well. "In my anxiety to hasten forward the work, I have constantly had more hands employed than could work to advantage; and in the peculiar nature of the business it often occurs that those employed on some parts have to wait for the accomplishment of other parts in which the first cannot assist."[31]

The *Pioneer* slowly took shape, in spite of the difficulties. By the end of August 1852, both of the ten-horsepower engines were on hand. Work on the saloon was approaching completion as well. Composed of 140 pieces of spruce, the floor of the car was 20 feet long and six feet wide. It weighed only 20 pounds but would support, Porter estimated, forty persons.

Constructing the great envelope, with its complex system of battens tacked to the fabric, was a particularly difficult learning experience for Porter and his crew.[32] When two inner partitions of the gasbag proved too heavy, they were removed and lighter material was substituted. Inept workmanship forced the crew to rebuild the frame of the saloon after the windows and doors had been installed. New "T"-shaped battens had to be installed in the gasbag. Two hundred bracing wires had to be adjusted time and time again. Test inflations of the gasbag uncovered defects in the hydrogen generators. This, in turn, necessitated "a forced march to New York and another to Baltimore" to procure $1,600 worth of additional zinc and sulphuric acid.[33]

Porter was at his wit's end. "If you are impatient, he remarked to stockholders on November 6, "you may be sure that the trial of my patience is vastly more severe than that of yours."[34]

Heavy autumn rains and the early onset of winter proved even more severe trials to Porter's patience, pocketbook, and peace of mind. A swarm of visitors appeared at the work site on Thanksgiving Day. One of these "Thanksgiving rowdies" slit the envelope. A severe rainstorm the next day increased the rent and dumped gallons of water into the gasbag. When Porter visited his tent the next morning, he discovered that these pools of water had frozen the fabric to the ground. Although the gasbag was thawed and dried, it was apparent that the frail, patched canvas could not stand another inflation. By mid-February Porter realized that he would have to construct another envelope in the spring.

The loss of the expensive gasbag was only the beginning of a disastrous winter. Porter moved the engines and boilers into a rented shop but was forced to purchase a new tent to protect the nearly finished saloon and other components that were too large to move inside. The winds of February and March made short work of the tent and wooden car. When spring arrived, only the long spruce battens were usable.[35] It was a time of personal tragedy for Porter as well. A married son living in Washington contracted a lingering illness and died, leaving a second family totally dependent on the elder Porter for support.

The completion and exhibition of his largest flying model was the only bright spot in an otherwise dismal season. Designed to encourage a second round of stock sales that would finance a new season's work on the *Pioneer*, the new airship was 22 feet long and 4 feet in diameter. It was also the first of Porter's models to be powered by a small steam engine. The oiled silk balloon was built around a skeleton of twelve ⅛-inch battens.

The saloon, seven feet long, hung three feet beneath the envelope. Porter had applied a number of detailed touches, such as painting the faces of happy passengers looking out the windows, which added to the air of verisimilitude as the model flew around a crowded hall.

The model was first flown at Carusi's Hall, a collection of dancing, meeting, and assembly rooms at the corner of Eleventh and C streets, N.W. On the evening of March 25, 1853, Carusi's was festooned with flags and bunting. Porter opened the evening with an "interesting lecture." The crowd rose to its feet as the large model was walked into the hall and released to hover over the heads of the spectators. When the steam valve was opened and the position of the rudder set, the craft "moved gracefully around the room to the great delight of the spectators."[36]

The performances at Carusi's Hall continued into mid-April. The *National Intelligencer* reported that "the assembled spectators manifested much excitement of admiration and gratification."[37] The pupils of several local schools were admitted to a special Friday afternoon showing of the model. Members of the U.S. Senate attended a showing on the evening of April 12. The next day the *Washington Evening Star* reported that the performance was "highly satisfactory, and elicited frequent applause from the excited audience. . . . Never prior to the introduction of Mr. Porter's model aeroport, has anything appeared upon which creeping humanity could base a rational anticipation of the long-desired art."[38]

There were those who regarded Porter's demonstrations as an "optical illusion, or some peculiar affliction of the imagination." The editor of the *Star* would have none of this: "There is the tangible fact before them—a real, mechanically constructed steamship, with its wheels, engine, and cargo, floating in the air, and occasionally shooting forward in direct-lines or circles, according to the dictates of its engine and helm."[39]

His success with the new model enabled Porter to present a confident front in the spring 1853 issue of the *Aerial Reporter*. He grandly asserted that "the prospect is much better than it was at the commencement, inasmuch as I have one year's experience in the business." The engines, boilers, and other metal items were still sound, and he hoped to construct a "new and improved float" in half the time required to build the original.[40]

The project was now crippled by financial difficulties however. Porter did scrape together sufficient cash with which to begin work on the new gasbag in 1853, but construction proceeded at a much reduced level. He was forced to give up his large and expensive rented lot and transfer operations to a vacant lot—near his home—which city officials offered rent-free. At least his problems with the "eight or a dozen" workmen employed the previous summer had been solved. He was now reduced to a work force of five men. Even so, Porter was unable to complete work on the second envelope that season. It was obvious that the drive and energy had gone out of Rufus Porter and Company.[41]

Porter continued to urge existing stockholders to increase their investment and renewed his efforts to raise funds, particularly in the Washington area. By late summer 1853 even Porter had begun to lose heart. In a letter thanking William Markoe for yet another stock purchase, the inventor expressed his own rising doubts:

> *Oh! William Markoe,—All my successive, combinations of misfortunes, disappointments and adversities, even with the addition of distressing sickness in my family, have not yet moistened my eyes as did your last letter.*
>
> *Now, I am firmly determined that I will never employ a dollar of this money from you, until I can command enough to give me assurance of being able to complete the work, without suspension: but will hold it to be returned to you, if the work fails of completion.*[42]

And fail it did. In the final issue of the *Aerial Reporter*, issued on April 27, 1854, Porter was, as usual, full of apologies. He was finally forced to admit that the work on the *Pioneer* could not resume without more cash. He planned to construct a 35-foot model capable of "running against the wind" out-of-doors. If exhibition flights with this model refilled his coffers, he would complete the man-carrying aeroport.[43]

In fact, Rufus Porter's aeronautical career had come to an end. Friends and supporters continued to hope that Porter would make a new start. Finally, however, even faithful William Markoe was forced to admit that Porter was finished: "Again and again I reviewed accounts from him of the progress of the work; again and again he wrote that everything was almost ready; that he was getting ready to inflate; that in about a month or week he expected it to be afloat; but invariably it would follow that some unexpected circumstance had knocked the whole concern on the head until the next season when very much the same series of events would be gone through again."[44]

Markoe did not abandon his old friend. He invited him to join the Markoe family in Wisconsin. Although this offer was refused, the two men continued to correspond. As late as 1857, Porter was still assuring his "Dear Brother Markoe" that he was on the verge of putting together another $1,500 for a fresh start on a large aeroport.

Markoe, now corresponding with Paullin, Wise, and other professional balloonists, became the first to fly in his adopted state of Minnesota. On September 22, 1857, he went aloft from St. Paul aboard the *Minnesota*, a large balloon of his own construction. Accompanied by several friends, he flew 45 miles in an hour and a half.[45]

This was much more of a flight than Rufus Porter was ever to make. Twice, once with Robjohn and once through the joint-stock venture, Porter had actually constructed man-carrying airships and been within weeks, perhaps even days, of flying them. On both occasions he was defeated by inadequate financing and his own lack of managerial skill.

In view of his ultimate failure and the extravagance of his claims, it is all too easy to dismiss Rufus Porter as a crank and a dreamer. This is unfair. Porter conceived at least the rudiments of several features that would later be incorporated into the giant, rigid airships. He saw the need for an internal structure to hold the shape of a hydrogen-filled envelope and realized the advantages to be gained by partitioning the gasbag. He constructed three large flying models and inspired considerable interest in the possibility of aerial navigation in all parts of the United States.

In purely practical terms, it is only fair to note that Rufus Porter came very close to success, though not, it is true, the success of which he dreamed. His aeroports would not have crossed the continent in two days nor braved the North Atlantic winds. If Porter had actually completed either of his full-scale craft, however, it is quite possible that he would have become the first man to demonstrate the ability to navigate the air at low speeds in a calm. The opportunity to write his name in history as the constructor of the first frail airship was within his grasp.

Instead, that honor went to Henry Giffard, the talented French engineer who, on September 24, 1852, at a time when Porter was racing to complete the *Pioneer* aeroport, propelled a gasbag over Paris at a speed of 6 MPH with a three-horsepower steam engine. Rufus Porter was aware of Giffard's effort, but he regarded the news of the flight as little more than an unsubstantiated rumor.

The fact that most Americans had ridiculed Porter's promise of an intercontinental airship did not mean that they rejected the basic notion of a navigable balloon. Many would have agreed with the sentiments of John Seymour, of Lynne, Ohio, expressed in a letter to the editor of the *New York Sun*:

> *I have recently learned by an extract from your paper that a steering air balloon with wings has recently been invented. This I know will be regarded by most people as the production of a visionary brain. That, however, does not render it certain that this is the real character of the invention. Were I to say that I believe the regular mail would within fifty years be transported in winged air balloons, I would be regarded with contempt. But is such an event more improbable than*

steamboats or steam carriages were fifty years ago?[46]

Seymour also expressed his vision of the future of aerial navigation in a letter to Rufus Porter. "I have no very strong confidence that Aerial Navigation will soon be made a safe and economical method of transportation for men or things," he admitted. "But for discovery, scientific investigation, the conveyance of special messages, and perhaps for some of the most important mails, I think it can be, and will be, made valuable."[47]

For Seymour and other Americans interested in aerial navigation, there was little need to look overseas for news of aeronautical progress. The years following the collapse of Rufus Porter and Company were filled with American attempts to develop navigable balloons.

Some of these schemes, like that of the Massachusetts inventor Ira Smith, remained unknown outside of aeronautical circles. Most aeronautical thinkers, however, attempted to generate all the publicity they could.

Like Porter, a number of these experimenters attempted to interest the United States Congress in their schemes. E. D. Tippet, "an old and respectable teacher and inventor" from Washington, D.C., submitted his proposal in 1853. Tippet noted that his plan would be used to perfect a "Magnificent Aerostatic Machine." He was extraordinarily secretive for a man who hoped to persuade practical politicians to support his plan. The propulsion system he "profoundly keeps to himself," although he did assure anyone who would listen that his was "the only plan which will ever answer the purpose."[48]

Having failed to interest Congress, Tippet traveled to New York where he hoped to form a company to construct his craft. Failing in this and finding himself "without means for his proper support," Tippet unveiled his plans to the editor of *Scientific American.* His balloon featured a "condensing reservoir" so that gas could be withdrawn from the envelope and condensed when the aeronaut wanted to descend and reintroduced in order to rise. Two propellers were to drive the craft forward. The judgment of *Scientific American* that "Mr. Tippet's plan is the best we have yet seen" was not shared by many knowledgeable observers.[49]

Isham Walker, a resident of central Tennessee, submitted his petition to Congress on November 4, 1854. Walker, "a good practical millwright and Engineer with fifteen years experience in this and the northern states," planned an airship 100 feet long, 16 feet high, and 10 feet wide. The machine was to be built of sheet copper and wire. It would feature two gas cells, each 30 feet in diameter, placed at either end of the craft, while a third 80-foot long "reservoir" would extend along the top of the machine. Four "air engines" at the rear would drive twin 15-foot propellers at 400 RPM. The cabin would be 40 feet long, 8 feet tall, and 10 feet wide. Walker noted that his ship would cleave the air at 500 MPH.

Walker's supporters called attention to "the untold importance" of the invention, "not only to the post office department of the Republic, but to the whole Civilized world." None of this seems to have had much impact on Congress, where the petition disappeared without a trace.[50]

The desire to navigate the air had spread to the American South as well. A Mr. Davidson, of Mobile, Alabama, became the object of considerable ridicule when he suggested that Sir John Franklin, the leader of a party of English explorers who had disappeared into the arctic mists in 1847, might be found and rescued by means of his "balloon locomotive."[51] "Science has its revenges," he remarked in a letter to the editor of a Mobile newspaper, "and sooner or later they will come upon those who ridicule the idea of practical aerial navigation." In defense of Davidson, the editor of *Scientific American* noted that "the steamboat, the locomotive, and the magnetic telegraph . . . have undergone and triumphed over the doubts and sneers of men lacking the genius to comprehend them, and the generosity to give them a trial."[52]

Franklin Kelsey, of Middletown, Connecticut, suggested a very different means of aerial locomotion. He planned to string an endless overhead wire looped over pulleys on telegraph poles. The line would be moved over the pulleys by a steam engine, tugging tethered balloons laden with passengers and freight along with it.[53]

A Massachusetts man, Capt. John Taggert, conducted some tests with a more traditional "steering balloon" during the decade of the 1850s. Taggert's work had first come to public notice late in 1848 when he unveiled a picture and model of his proposed craft in Boston. President Everett and Professor Threadwell of Harvard College, as well as Mr. Pook, a local naval architect, had all "expressed favorable opinion of the project."

Taggert claimed to have invested $15,000 of his own in the venture and hoped to raise that much again by subscription. He promised a first flight by July 4, 1849. The "little captain" did venture aloft from Lowell, Massachusetts, at 4:00 P.M. on July 4, 1850. He remained in the air for an hour and a half in a "pear-shaped" balloon "with his flying machine attached."

Taggert flew past Dorset, Tewksbury, Haverhill, Reading, Andover, Ipswich, Georgetown, Lawrence, Danvers, and Salem. After flying out to sea a short distance, he started back toward Lowell, where he planned to land at his starting point. He was forced to land at Middletown because of the breakdown of his apparatus. Local newsmen were convinced that Taggert could have completed his voyage had it not been for the accident.

Later that summer, Captain Taggert brought his "propeller balloon" to New York. He barely escaped death when his craft was dashed against a wall, dumping the car and aeronaut on the ground, while the balloon escaped over Long Island Sound. Captain Taggert wisely chose to withdraw from the field.[54]

Solomon Andrews was much more persistent. Like Rufus Porter, Andrews fit the classic mold of the American inventor-mechanic. A native of Herkimer, New York, born on February 15, 1806, Andrews spent most of his life in Perth Amboy, New Jersey, where his father, Josiah Bishop Andrews, served as minister of the First Presbyterian Church, health officer of the port, local physician, and president of the Middlesex County Medical Association. Solomon followed in his father's footsteps, graduating from the Rutgers Medical School and serving as a druggist, physician, health officer, and three-time mayor of Perth Amboy.[55]

Andrews built his national reputation, and his fortune, as an inventor however. He developed tricycles, sewing machines, nutcrackers, tobacco filters, barrel-making machines, gas lamps, a kitchen range, the system that enabled moving trains to snatch mailbags from a station, and a prison lock that remained in use for many years.

At age twenty-six Andrews developed and demonstrated a special combination lock. To dramatize his faith in the device, the inventor locked $1,000 in an iron chest chained to a lamp post at the corner of Broad and Wall streets in New York. None of the city's famed picklocks was able to retrieve the cash. Unlike Porter, Andrews was successful in marketing and profiting from his inventions. The combination lock alone would bring some $30,000 to its inventor.

For over twenty years Andrews poured the profits from his other inventions into airship research and construction. He had been bitten by the flying bug when, as a boy of seventeen, he found himself daydreaming while listening to one of his father's sermons. "Looking out of a window at the soaring of an eagle in his winding way through the air, I caught as with an electric shock, the key to the whole system of aerial flight," he recalled forty years later.[56] Andrews regarded this moment of youthful vision as the turning point of his life: "From that moment my aim of life was fixed. The study of medicine, and of the sciences generally, were influenced by the one idea. . . . The acquirement of various trades, and skill in workmanship, was determined by the resolution to construct a flying machine. And my whole life has been spent to obtain by assiduous labour and attention to business the pecuniary means for its accomplishment."[57]

Andrews came to share Muzio Muzzi's belief that gravity was a sufficient motive power for an airship. "In existing methods of transportation," he announced, "the chief difficulty is to produce motion." Given velocity, "it is easy to give direction and control." The problem with earlier aeronautical experimenters, he believed, stemmed from their concentration on power. They had, he noted, forgotten that "motion is the natural condition of the balloon, needing only direction and control."[58]

From the outset, Andrews planned to link several standard cigar-shaped balloons to form an inclined plane, and then to swap gas or ballast for forward motion. He performed a number of experiments, "the most important and successful of which was the flight of a paper kite, not attached by any string, but free to move in any direction."[59] Andrews claimed that on one occasion two witnesses had seen his kite move over 100 feet against a strong wind.

Andrews's first attempt to construct an airship grew out of his work in founding an Inventor's Institute in Perth Amboy in 1847. This Institute, in which Rufus Porter was also involved, was established in a series of abandoned buildings, and was intended to support young inventors and promote their efforts.

The level of Porter's involvement in the project is not clear, but Andrews was clearly a central figure in the Institute. Within a year it had become apparent that Andrews regarded the enterprise as a perfect means of organizing and funding an airship program. He proposed retaining one of the 250-by-290-foot lots controlled by the Institute and selling the ninety remaining lots to raise $15,000 with which to construct a suitable hangar and begin work on an airship.[60]

In a pamphlet issued to interest potential lot purchasers, Andrews announced that the airship hangar would measure 100 feet long, 40 feet wide, and 32 feet high. The method of operating the "Aerial Car" was to remain a secret until the machine made its first flight in the spring or summer of 1849.

While Andrews offered no details of his machine, he took some pains to justify his plans: "This will be called a visionary scheme by many who are inclined to doubt the feasibility of any mechanical invention until it is completed. And others may conclude that the inventor is a monomaniac. The same was said of Watt, Arkwright, Whitney, Fulton, and Morse. It would be a matter of regret were there no more such."[61]

That summer Andrews was placing ads for "two active working men to be employed in the construction of the Aerial Car." He was also searching for "90 persons in the City of New York or in the United States who, having the means, have also the faith enough in the possibility of Aerial Navigation, to induce them to invest the sum of $100 each, in the experiment, and that too upon a valuable equivalent in real estate. . . . In short, are there any besides poor inventors who believe Aerial Navigation feasible?"[62]

Certainly Andrews found little support in the editorial offices of *Scientific American.* As early as July 1848 the journal had commented that both the "Aerial

Solomon Andrews

Car" and the Inventor's Institute were "hum bugs."[63] Ultimately, Andrews was unable to fund this initial venture. His dream of aerial navigation was to lie fallow for fifteen years, while the inventor grew more prosperous.

Enlisting in the Union Army in 1862, Andrews served as a surgeon with the Army of the Potomac. Having observed what he regarded as "the vain attempts to reconnoitre the enemy by means of captive balloons,"[64] he was convinced that he could more effectively serve the cause by developing a navigable balloon. He resigned his commission and asked Secretary of War Edward M. Stanton to provide $5,000 with which to construct the machine. In spite of Andrews's guarantee "to sail five or ten miles into Secessia and back or no pay," his request was refused.[65]

Andrews then retired to Perth Amboy where, in October 1862, he began construction of the first "Aereon," as he was now styling his craft. He apparently drew most of the $10,000 cost of the vehicle from his own pocket. Built in the old British Barracks in Perth Amboy, which also housed the Inventor's Institute, the Aereon was moved from its makeshift hangar onto the town common for the first time in June 1862.

As a reporter for the *New York Herald* noted, the machine consisted of three "seegar shaped" balloons, each a "varnished cylinder of linen" 80 feet long and 13 feet in diameter. Like Porter, Andrews had recognized the necessity of subdividing the balloon into separate gas cells. John Wise produced 21 cambric bags to the inventor's specifications and was on hand in Perth Amboy to supervise their installation.[66]

A large net was spread over the top of the balloons. The wicker operator's car dangled 16 feet beneath the gasbags, supported by 120 cords. The aeronaut moved weights to the front or rear along a line to force the nose of the Aereon up or down. A small rudder, 17 feet square, was mounted on the rear of the central cylinder.

Andrews himself made the first ascent on June 1, 1862. Ellis C. Wite, a local architect, and Hamilton Fonda, a foreman at the local Hobbs Lock factory which produced Andrews's invention, were responsible for the ground handling operation. S. V. R. Patterson, cashier of the Perth Amboy City Bank, was among the astonished spectators, as was John Wise himself.

It was a fairly short flight made in a breeze stiff enough to keep the 30-foot banner attached to the balloon blowing completely unfurled. Andrews climbed to 200 feet, traveling with the wind, then, to the astonishment of the spectators, he turned the craft in a half circle and flew against the wind to land near the spot from which he had taken off.[67]

The Aereon was back a month later minus the Wise gas cells, which could

The Aereon

no longer be inflated. Andrews found his large balloons to be gas-tight however. The aeronaut also dispensed with the sliding weights, having discovered that he could force a climb or descent simply by walking to the front or rear of the car.[68]

When Andrews made his third flight that August, even local doubters were forced to admit that his system seemed to work for short flights. Andrews now felt sufficiently confident to schedule a "grand experiment" for September 4, 1862.[69]

The day was an enormous success. Andrews began with his usual short ascents, rising no higher than 1,000 feet, then quickly coming about and swooping down to a landing. The spectators were impressed by both the maneuverability and speed of the Aereon. As a *New York Herald* reporter commented, the capabilities of the machine "were demonstrated beyond all possibility of doubt."[70]

For his big demonstration of the day, the inventor ordered the basket removed and tied the rudder at a sharp angle, then released the craft on a final free flight. The inventor estimated that the Aereon spiraled toward the clouds at a speed of 120 MPH. It began by describing roughly twenty circles a mile and a half in circumference. Entering the overcast, the balloon could be seen to send the clouds into swirling motion. It was the last anyone ever saw of Aereon number one.[71]

With the experience and publicity of one airship behind him, Andrews was at last able to obtain a hearing from federal officials. Following an interview with President Lincoln and the presentation of letters of endorsement from leading citizens of Perth Amboy, the inventor was invited to demonstrate the principles of the Aereon. In 1864 Andrews presented yet another petition to Congress, then flew a four-foot model of his craft before members of the military committees of both the House and Senate in the basement of the Capitol Building.[72]

The petition, lecture, and demonstration so impressed Robert C. Schenck of Ohio, Chairman of the House Committee on Military Affairs, that he persuaded Secretary of War Stanton to name a special scientific commission to investigate Andrews's claims. Joseph Henry, secretary of the Smithsonian Institution, Alexander Dallas Bache, superintendent of the U.S. Coast Survey, and J. C. Woodruff, a major in the Corps of Engineers, agreed to serve on the panel.

In July 1864 Andrews gave a second demonstration of his model, the gas cells of which had been prepared in Paris. Flying the little craft for the special commission in the Great Hall of the Smithsonian Institution, Andrews apparently made a favorable impression on the distinguished group.[73]

In opening their report to Stanton, Henry, Bache, and Woodruff called attention to the fact that tethered observation balloons had not been as successful as Union officials had at first supposed they would be. Andrews, they noted, "well known to the government as one of the most ingenious and successful inventors of this country," offered an alternative. At first sight, they admitted, his notion seemed chimerical, but the demonstration had convinced them that "it is not impossible that he can really perform what he has asserted he can do."[74]

It was scarcely a ringing endorsement. From experiments on so small a scale, the commissioners were not prepared to vouch for the value of the invention of Dr. Andrews.

In spite of a final recommendation that the War Department fund a second experimental man-carrying Aereon, Stanton chose to ignore Andrews. The inventor was granted a patent for the Aereon, then waited for months for the judgment from Washington. Finally, in March 1865, Andrews received a letter from Congressman Schenck:

> *Nothing was done in regard to the aerial ship. I found, after various discussions, that I stood alone in my Committee as a friend and favorer of the project, willing to make or recommend an appropriation for testing the invention in actual practice. Some of the Committee made distinct and positive opposition; others were willing*

> *that I should make a report, but on my own responsibility, and without their support in the House of Representatives, and all but myself were incredulous.*
>
> *Under these circumstances I thought it most advisable not to bring the subject forward for what promised to be a hopeless effort against doubt and prejudice.*[75]

As the Civil War drew to a close, Andrews attempted to interest investors in the Aereon as a peacetime carrier of goods and passengers. On May 10, 1865, he published his first pamphlet, *The Art of Flying*. The booklet called for the formation of an Aerial Navigation Company, a joint-stock venture to underwrite the construction of a new Aereon. Twelve thousand dollars, he estimated, would be sufficient to construct a large airship of "durable materials."[76]

When the requisite funds were not forthcoming in three months, Andrews determined to forge ahead on his own. He planned to salvage the varnished pongee silk from the *Union*, the *Intrepid*, and two other veteran Union Army balloons he had purchased in the spring of 1864. The material would be used to build a single Aereon of newer design. He would fly this balloon over New York, then exhibit it and invest the profits in a major airship-building program.

Andrews changed his mind once again in October 1865. Deciding that "those subscribers who have shown liberality and enthusiasm" should be allowed to make a profit (and bear a portion of the initial expense), he founded a small joint-stock company. Thirty-four investors subscribed a total of $6,000 in the Aerial Navigation Company. Several stock purchases were quite large, ranging from $1,000 to $1,500. The majority of the subscribers limited their investments to $50 or $100. Andrews valued his own contribution, two of the old Army balloons, his patent, and his expertise, at an additional $2,000, bringing the total capitalization to $8,000.

The bylaws adopted by the shareholders left little doubt as to who was in charge. Andrews was to be president and permanent director of the firm "as long as he wants." He nominated both the corporate officers and the board of directors. Shareholders signed articles that bound them not to divulge information about the Aereon either to the government or the "New York press" without Andrews's approval. When stock certificates valued at $18,000 were eventually issued, half were allocated to Andrews. Moreover, half of all company profits up to $100,000 would be donated to the president's favorite charity, the Inventor's Institute.[77]

Andrews called the first stockholders meeting of the Aerial Navigation Company to order on October 16, 1865. New Yorker G. Waldo Hill, a $500 investor, was the "permanent director's" choice as president pro tempore, while C. J. Hopkins, who represented the largest single block of investors in the Erie–

Blooming Grove, Pennsylvania, area, was elected secretary pro tempore.

The second stockholders meeting on October 25 unanimously approved the final slate of officers. Andrews was named president, George W. Trow was elected vice president, and C. M. Plumb, secretary. Significantly, Emmett Dinsmore, the largest single investor ($1,000) and the head of a family that had contributed a total of $2,200 to the venture, was Andrews's candidate for treasurer. The same group dominated the seven-man board of directors (Andrews, Trow, Dinsmore, Hill, Hopkins, and Cyrus J. Poole).

With the preliminary organizational tasks out of the way and a constitution and set of bylaws adopted, the Aerial Navigation Company filed papers of incorporation with the secretary of state in Albany on November 27, 1865. They were now fully prepared to proceed, as the bylaws indicated, "to construct without unnecessary delay and with the money in the Treasury one Aereon or airship and to make one trip from New York to Philadelphia and back." By January 1866 open meetings of all stockholders had been replaced by a single annual meeting. Day-to-day business was now the province of monthly board of directors meetings.

Progress was rapid. The firm leased a lot at the corner of Green and Houston streets in New York for $350.00 a month. An additional $149.32 was expended to grade the area and $575.00 to fence it. A pamphlet, *The Aereon, or Flying Ship*, was published, and tickets of admission were sold to those that desired to watch the ship under construction.

The Aereon itself was first inflated in April 1866. The new craft bore little resemblance to the original Aereon of 1862. Andrews had fashioned a single envelope 86 feet long and 42 feet in diameter from two of his war surplus balloons. The gasbag was no longer described as cylindrical or "seegar" shaped. As one New York observer remarked, "It resembled a long lemon terminating in a sharp point at either end."[78]

A long, wide, leather belt passed along the longitudinal axis over the top and bottom of the balloon. This strap could be loosened or drawn tight by cords in the car. In this way Andrews hoped to be able to compress the gas and descend without the need to valve hydrogen. The car, "shaped like a cradle," measured 8 feet in length and 2 feet in width. It was supported by the net that passed over the top of the balloon.[79]

The Aerial Navigation Company was entering a period of financial crises as the balloon was approaching completion. Treasurer Dinsmore reported an outstanding debt of $3,100 at the fifth board of directors meeting on May 18, 1866. There was little choice but to forge ahead, attempting to convince stockholders to increase their investment by fifty percent and rushing to make a first flight.

The great day arrived on May 27, 1866. Andrews had decreed that Trow, Plumb, and Hill would accompany him on the inaugural voyage. The process of filling the balloon, which had begun at 9:00 A.M., was complete by three o'clock that afternoon. There were few visitors inside the fence to witness the launch an hour later. Andrews had been announcing for several weeks that the flight would be made but he had not publicized the precise date, apparently in the belief that by surprising New Yorkers with the sight of an airship cruising overhead he would attract a flood of eager new subscribers.

The flight proceeded smoothly following a shaky take-off. The *New York Herald* offered a breathless description of the scene. "Lifting herself out of the enclosure, the wind caught the bow of the peerless ship, as if to caress her with its first blessing, before the main part had fairly risen from the premises."[80]

Things looked a bit less idyllic from the car. George Trow recalled that the ropes controlling the rudder had become tangled as they rose out of the enclosure. The Aereon drifted to an altitude of 2,000 feet before Andrews regained control.

The aeronauts then discovered that they were unable to force the nose down far enough to descend against the freshening wind. After circling and cruising over the city for half an hour, they landed in a wooded area near Astoria, Long Island. Andrews was displeased with the performance of the craft and informed reporters that he planned to add an extra 12 feet to the length of the car in order to enable the crew to move farther to the front or rear for improved pitch control. The rudder was also to be enlarged.[81]

If Andrews was unhappy with the flight, the citizens of New York were delighted. Throngs of gawkers streamed out of the Union Club and the Eclectic, while others craned their heads out of windows and crowded the roofs. Pick-pockets had a field day while prosperous New Yorkers scrambled to obtain one of the souvenir cards Andrews scattered from the basket.[82]

The excitement did not result in the fresh investment the inventor had envisioned. The demonstration had been very expensive. In the wake of the first test flight, the members of the Aerial Navigation Company found themselves deeper in debt. Treasurer Dinsmore's plea that each stockholder increase his shares by fifty percent was to no avail.

Andrews was able to pay for one more inflation, on June 5. This time all New York was ready and waiting. An enormous crowd gathered outside the enclosure at the corner of Green and Houston at 5:00 P.M. as Andrews and Plumb climbed into the car. The third occupant of the basket, a reporter for the *World*, had to give up his seat when the balloon refused to rise.

With the preliminaries complete, *Aereon II* rose straight up, then began circling as Andrews struggled to free the rudder which had jammed once again. Gaining control at last, he ordered Plumb to the rear of the car and began to climb into the wind. Spectators could clearly see the sand ballast being blown back and away from the rapidly moving balloon. Little more than a speck, the craft finally disappeared into the clouds, over Blackwell's Island. Half an hour after take-off, the voyagers made a safe landing near Oyster Bay, Long Island.

As before, the watching crowd was elated. The entire city seemed to have turned out for the flight. As the disappointed reporter noted, "the fair sex seemed to become oblivious of the presence of the fashionable swells, and in their abstraction ran their sun shades in their eyes; while the gentlemen, equally absent-minded, played havoc with the hoops and other various appendages of the perambulating milliners frames."[83] Passengers on the Brooklyn Ferry marveled at the way in which the airship rose and fell, undulating its way against the wind.

The flight had been most impressive, but it failed to attract desperately needed cash. Hopelessly in debt, facing the need to refurbish the envelope, Andrews, like Pennington and Porter before him, had been defeated by inadequate financing. Solomon Andrews had made his last flight.[84]

The ultimate failure of men with the talent and preparation for the task of constructing an airship was widely recognized, but it did not deter Frederick Marriott. Marriott's airship experiments were the continuation of his lifelong pursuit of the dream of flight.

Born in Enfield, England, on July 16, 1805, Marriott, as a young man, worked in India as an employee of the British East India Company for several years, then returned home to settle into a post with the Bank of England. He was gradually drawn into publishing, investing his wife's modest fortune in a series of journals. The best known of these was an illustrated paper, the *Weekly Chronicle*, which developed into the enormously successful *Illustrated London News*.

Selling his interest in the *Weekly Chronicle* to a partner in 1842, Marriott had joined William Samuel Henson and John Stringfellow in establishing the Aerial Steam Carriage Company. Building on the pioneering work of Sir George Cayley, Henson and Stringfellow had designed a large monoplane with twin pusher propellers that would be driven by a steam power plant. Marriott played a major role in planning the widely circulated engravings designed to interest investors in the venture. These prints showed an Aerial Steam Carriage in full flight over places ranging from Hyde Park to the pyramids and the Taj Mahal. The pictures did much to fix the image of the aeroplane (a word Marriott himself coined) in the public mind as a craft with fixed wings, a fuselage, and a tail.[85]

Convinced that they would never raise sufficient cash with which to construct a full-scale Aerial Steam Carriage, Marriott sold his stock to Henson and Stringfellow in 1842. After another few years as a magazine editor, he set sail for California, in the words of a friend, "to seek his fortune and further the interests of aerial navigation."[86]

After crossing the Isthmus of Panama and being shipwrecked on the voyage, Marriott reached California and established himself as a San Francisco banker. In 1856, now a wealthy man, he returned to journalism, founding one of the most popular West Coast papers of the era, the *San Francisco Newsletter*. Marriott filled the columns of the *Newsletter* with brilliant prose sketches like those offered by a young, redheaded reporter friend, Mark Twain. The editor also delighted in taking on the quacks and corrupt politicians that flourished in the wide-open city by the bay.[87]

In the midst of all his financial and journalistic endeavors, Frederick Marriott had not forgotten the flying machine. Recognizing the need for closer communication with the East Coast, he founded the California Aerial Steam Navigation Company in the mid-1860s. By 1868 work had begun on the "Avitor," a large, winged, flying airship that was under construction in a San Jose barn.[88]

Contemporary accounts of the craft disagree as to its length (28, 30, 37, even 95 feet). There was some agreement on its height (11 feet) and width (8 feet). A light framework of wood and cane ran around the lower half of the spindle-shaped balloon, bracing the envelope and supporting the engine, propellers, wings, and tail.[89]

The wings, white fabric stretched over a wire frame, were triangular surfaces, five feet square, attached to the forward half of the frame. The triangular cruciform tail was attached at the rear of the frame.

The steam engine and boiler were tiny jewels, together measuring only a foot long by four inches wide and six inches tall. A spirit lamp brought water to a boil, driving the piston with a two-inch head through a three-inch stroke. Twin two-bladed propellers were placed on each side of the gasbag.

The Avitor made its first flight at a San Jose racetrack, Shell Mound Park, on July 2, 1869. With the rudder set to carry the craft in a great circle, the Avitor was released on a tethered ascent. Lines ran from the nose and tail, so that ground handlers, moving at a fast dog trot, could keep the airship in check.

The propellers carried the Avitor twice around the field when one of the lines gave a tug to the rudder, sending the machine racing in a straight line for a quarter of a mile. Finally brought under control, it was tugged to the ground. The first outing had been a great success.[90]

A second trial conducted on July 4, 1869, was less successful. Gusty winds

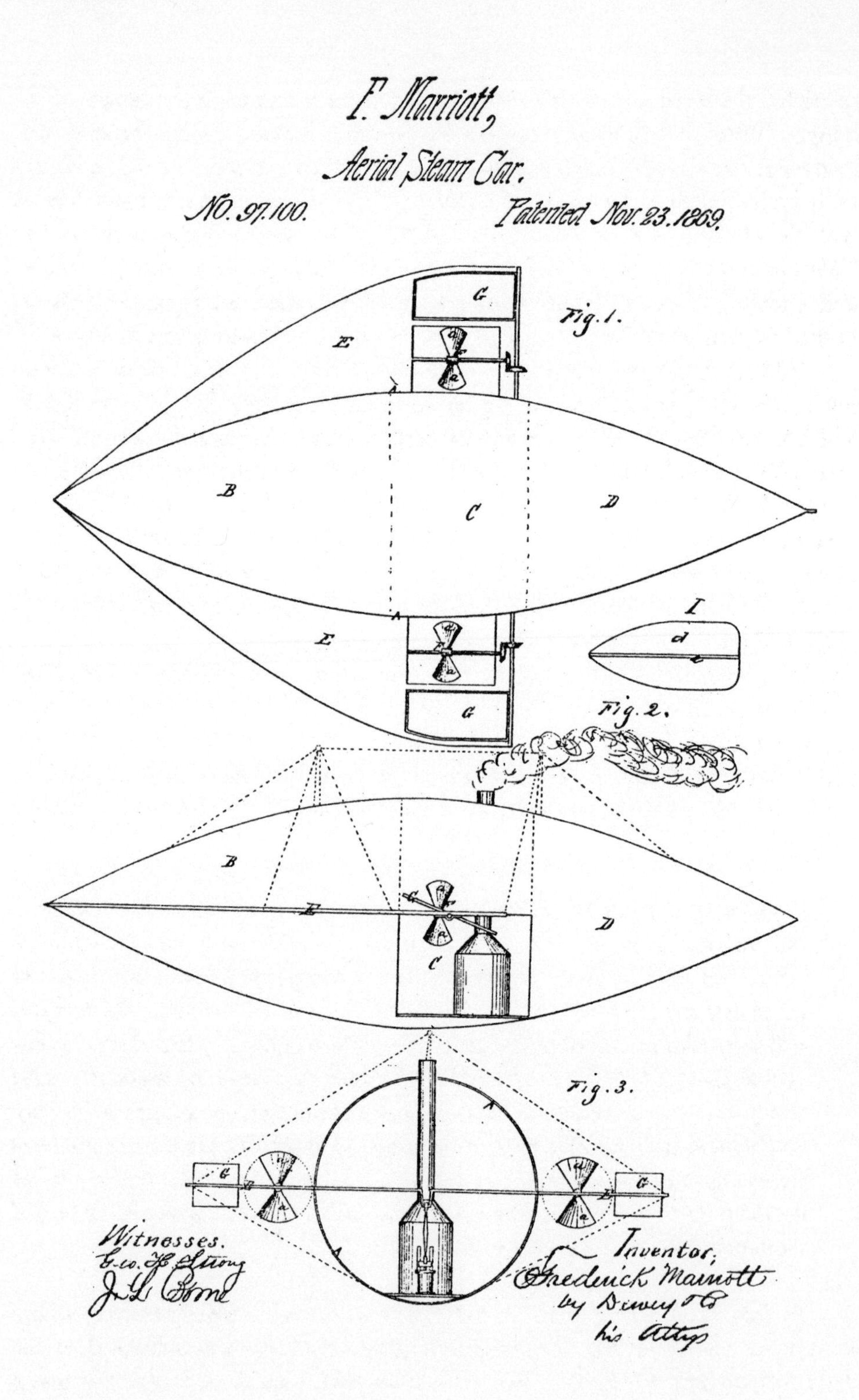

Patent drawings for Frederick Marriott's "Aerial Steam Car," or "Avitor"

restricted the performance to several short trips back and forth the length of the hangar. When it was noted that the balloon was leaking badly, the craft was retrieved, resewn, filled, and released once again. "The guests," noted an admittedly prejudiced reporter for the *Newsletter*, "cheered loud and many fairly danced with delight and success." The editor of the *Alta California* was less charitable: "We are compelled to say that the performance looked something of a failure, and a feeling of regret, rather than of ridicule pervaded the company present, at the lowering of the hopes that had been raised by its over sanguine inventor."[91]

Whatever the attitude of those actually present, the world press was very much interested in the craft being test flown at a remote San Jose racetrack. The London *Dispatch* called attention to "the wonderful California flying machine," which "will soon wing its way across the Rocky Mountains from San Francisco to New York."[92] The *Overland Mail* boasted that "the first extensive flight of a vessel for aerial navigation is accomplished, and the grand problem of atmospheric locomotion is realized."[93] Even the prestigious *Journal of the Aeronautical Society of Great Britain* congratulated Marriott on his success in managing a "practical test" of a balloon partially sustained by wings.[94]

Like virtually all his predecessors in the business of aerial navigation, Marriott found it difficult to hold his imagination in check. "Next in point of speed to the telegraph," he informed his readers, the Avitor would open "communication between distant points, bearing its men and messages through the air, while the railroad drags its heavy burdens of freight."[95] Furthermore, Marriott believed:

> *No savages in war paint shall interrupt its passage over and across our continent. No malaria, or hostile tribes nor desert sands shall prevent the exploration of Africa, no want of water, the examination of central Australia, nor ice floes or ice mountains the search for the Northwest Passage. No underground railways will be needed to accommodate the crowded thoroughfares of Broadway, no circuitous passage to find a narrow isthmus between continents, no waiting for trade winds; no necessity of lying becalmed under tropical suns; no extortions for huge corporations who monopolize the great routes of travel. No tax for crossing New Jersey; no states under tribute to railway companies. Man rises superior to his accidents when for his inventive genius he ceases to crawl upon the earth and masters the realms of the upper air.*[96]

By July 31, 1869, the Avitor had been transported to San Francisco, where it went on exhibit at the Mechanics Pavillion. Admission was charged to the daily indoor flights. Marriott had already begun to speak of building a much larger Avitor 150 feet long.

It was not to be. The original craft was destroyed in a fire. The funds for a new machine were not forthcoming. Marriott made a final effort to revive his Avitor project in the late 1870s, then turned his attention to heavier-than-air flight. The years prior to his death on December 16, 1884, were spent in designing and unsuccessfully attempting to patent an "Aeroplane Steam Carriage."

The work of Frederick Marriott brought to a close an era in the history of American attempts to navigate the sky that had begun with the arrival of Edmond Charles Genêt in the United States a century before. The first powered, man-carrying airship had already been flown in America at the time of Marriott's death. A small, one-man balloon powered by a handcrank, it was a far cry from the giant 100 MPH, transcontinental craft envisioned by Pennington, Porter, Andrews, and Marriott. Nor did the little airship see service in war or commerce. Like the free balloon, the first American airship was introduced by showmen.

Chapter 12 The Civil War Aloft: Origins

AT 4:30 ON THE MORNING OF APRIL 12, 1861, A GROUP OF OFFICERS STANDing on the walls of Fort Sumter, South Carolina, heard a dull thud and saw a flash of light burst among the bonfires flickering on the Charleston shore. A mortar shell, its path traced by a burning red fuse, arched across the sky to burst above the fort. Thirty-four hours and 4,000 shells later, Maj. Robert Anderson ordered the Stars and Stripes lowered. The Civil War had begun.

President Lincoln's call for 75,000 state enlistees on April 15 met with an overwhelming response. Men rushed to enlist in every corner of the Union. Martial music was heard throughout the land. So many streets were draped in red, white, and blue that the *Detroit Free Press* feared bunting might become a scarce commodity. While young men flocked to the colors, women organized Florence Nightingale Associations, sewed regimental flags, and lined the parade routes as their swains marched off to war.

State governors had little difficulty meeting their quota of troops for the Union line. Gov. William Dennison of Ohio faced a much more common problem. Required to raise thirteen regiments, he was forced to apologize that "owing to an unavoidable confusion in the first hurry and enthusiasm . . . of our people . . . I can hardly stop short of twenty regiments."[1]

Gov. William Sprague of Rhode Island was anxious to insure that the single regiment required of his state would be the first to report for national service. On April 17 he proudly informed Secretary of War Simon Cameron that the First Regiment, Rhode Island Detached State Militia, accompanied by the Providence Marine Corps of Artillery, under the command of Col. Ambrose E. Burnside, was fully organized, equipped, and ready to march. One member of the Marine Corps of Artillery drew special attention in the local press. As the *Providence Post* reported on April 19, "ALLEN, our New England Aeronaut, has promptly volunteered his services, with balloons for reconnoitering purposes, and leaves us tonight for the seat of operations. He possesses the qualities that a man needs in the professions; good temper, courage, quickness of perception, and that order of mind which enables him to act in an emergency. Success to Allen."[2] James Allen, "the New England Aeronaut," was well on his way to becoming the first American military balloonist.

Civil War ballooning is the one aspect of the history of American aerostation that has received a great deal of attention over the years.[3] Historians tracing the rise of American air power invariably open their accounts with a discussion of the Civil War aloft. The emphasis is usually on the romance and excitement of this pioneer effort. T. S. C. Lowe is portrayed ascending amid shot and shell, while the genteel and patriotic ladies of Richmond sacrifice their petticoats to provide the precious silk for a Confederate balloon. This was the stuff of legends.

The reality was a great deal more complex than the legends suggest. It is certainly true that much was accomplished. Aeronauts on both sides of the Mason-Dixon line made dozens of tethered ascensions and free flights in order to reconnoiter enemy positions. They served as artillery spotters and conducted successful experiments with both telegraphic and visual communication systems. With the aid of portable gas generators mounted on wagons and boats, they attempted to demonstrate the value of their services by carrying on their activity in the heat of battle.

Ultimately, however, the attempt to demonstrate the military utility of the balloon must be judged a failure. While a few commanders, notably Benjamin Butler and Fitz-John Porter, were eager to take advantage of aerial reconnaissance, others were openly hostile. The aeronauts themselves were divided by feuds and animosities rooted in the prewar years. The end came in 1863 when the Union Balloon Corps was disbanded, at the very moment when two armies were blindly feeling their way toward a meeting at Gettysburg. It was a sad conclusion to an experiment that seemed to hold much promise at the outbreak of hostilities.

American aeronauts had long been anxious to demonstrate the potential of the balloon in making military observations. They could cite a long list of precedents, beginning with the formation of the French Corps d'Aérostiers. Organized in April 1794, the corps made its first observations at Charleroi and Fleurus in June of that year and remained on active service with the revolutionary armies until disbanded in March 1799.

Jean Margat had made at least one ascent during the Algerian campaign of 1830. Louis and Eugène Godard continued the tradition of French leadership in military aeronautics with a series of ascensions during the Italian campaign of 1859.

Other nations had also experimented with war balloons. The Russian Army sent observers aloft during the siege of Sebastopol. Both the Danes and the Milanese had scattered propaganda leaflets from balloons, and Austrian troops sent balloon-borne bombs aloft during the Siege of Venice in 1848.[4]

The U.S. Army turned its attention to the balloon for the first time during the Second Seminole War. The Seminoles, a splinter group of Alabama and Georgia Creek Indians, had resettled in the Florida swamps during the eighteenth century. Fiercely independent, they sided with the British during the War of 1812, encouraged and harbored runaway slaves, and conducted sporadic raids against American settlements in southern Georgia.

An undeclared war raged between the U.S. Army and the Seminoles from 1835 until 1842. Ten thousand regulars, 30,000 state militia volunteers, and $30 million from the federal treasury were poured into the conflict. The toll was appalling. Over 1,500 soldiers and countless Indians and settlers lost their lives. In the end, the Seminole leader Osceola was captured while attending a negotiating conference under a flag of truce, and 3,800 pathetic, half-starved Seminoles surrendered. Some 300 Indians remained in the swamps, somehow managing to eke out a living while hiding from state and federal troops.

The Seminole War was one of the longest and most bitterly contested of all Indian-White conflicts. It was a guerrilla war for which the U.S. Army was neither prepared nor equipped. The frustration of pursuing a mysterious and elusive enemy through a trackless wilderness was compounded by the complaints of politicians and citizens who were unable to understand why the army could not bring the war to a speedy and satisfactory conclusion.

Perhaps observation balloons could succeed where more traditional means had failed. That, at least, was the belief of Frederick E. Beaseley, a disgruntled "patriot" who, "in common with all true lovers of this country," had been "affected with extreme pain in witnessing the progress of events in our Florida war."

Bemoaning the fact that "the Republick has exerted her utmost strength and spent millions of her treasury" in an unsuccessful attempt to subdue a band of "savages," Beaseley offered his solution in a letter to Secretary of War Joel Poinsett on October 12, 1840. "As the great difficulty in this war with Savages has arisen from the peculiar nature of the country, and the facility with which in their hummocks and hiding places they can elude our troops and render themselves inaccessible to all approach and discovery," Beaseley suggested the use of balloons combined with a telegraphic network. Two to four aerostats could be launched to overfly the Everglades when the winds were right. Cavalry patrols would protect the launch sites and follow the balloons in order to be on hand for the landing. The aeronauts' reports would then be carried to a nearby telegraph station to alert the troops as to the enemy's position. Tethered observation flights could supplement the riskier free ascensions in calm weather.[5]

Secretary Poinsett thanked Beaseley for his suggestion and informed him that a number of other citizens had submitted similar plans which were under consideration. These were not empty words. The War Department was giving very serious thought to the possibility of employing balloons in Florida.

A month before Beaseley's letter arrived, the secretary had received a similar proposal from Col. John H. Sherbourne. Sherbourne's thoughts carried special weight, for he had been a member of an 1837 Seminole peace delegation and was thoroughly familiar with the local terrain. The colonel suggested making a series of tethered night ascents with instruments to observe the Indian campfires so that their ground position could be triangulated.

Poinsett was intrigued, particularly when Sherbourne reported the following month that he had contacted the veteran aeronaut Charles Ferson Durant. Durant offered to sell the army one small balloon and all necessary equipment for $600. He could construct an even larger aerostat for $900.

Sherbourne had also sought the advice of former Secretary of War Benjamin Butler (not to be confused with the Civil War general of the same name). Butler, he reported, had remarked that the balloon idea was "the best that the Secretary can adopt in the present state of things in Florida." Colonel Sherbourne offered to take the whole matter in hand. He would solve the remaining technical problems, provide the equipment, and superintend operations in Florida.

Poinsett "applauded" Sherbourne's zeal but warned that it would be impossible to allocate the funds immediately. Late dispatches suggested that the Seminoles might soon be defeated, and, at any rate, he could take no action without consulting Warren K. Armistead, then serving as commanding general in Florida. Armistead rejected the plan, refusing even to consider the possibility of using a balloon.

Sherbourne refused to shelve his scheme. He resubmitted the idea late in 1841 as the war continued unabated. On this occasion, the adjutant general sought the opinion of Gen. Edmund P. Gaines, commanding the Western Department, who was no stranger to the problems plaguing the troops in Florida. Gaines offered a qualified and halfhearted endorsement. The balloon could prove valuable in the Everglades and might even be useful to Indian fighters in the West, he agreed. Once the Indians realized what was afoot, however, they would make every effort to destroy the balloon. Failing this, they would, in all probability, devise camouflage and other means of deceiving the aerial observers.[6] With so little enthusiasm on the part of field commanders, Poinsett chose not to proceed.

It was left for John Wise to reopen the issue of the military balloon during the Mexican War. Wise suggested a plan that he believed would "render the capture of the Castle of San Juan de Ulua as feasible and easy as the launching of a frigate."[7] The aeronaut proposed to construct a very large balloon, 100 feet in diameter, the basket of which would be loaded "with percussion bomb shells and torpedoes." The balloon could then fly over the fort, far beyond the range of Mexican artillery, while the aeronauts dropped their deadly missiles onto the helpless troops.

Wise's proposal, which he originally published in a Lancaster newspaper, attracted much attention and comment. The editor of the Philadelphia *Public Ledger* noted: "The public have been amused by the many comments upon Mr. Wise's plan of taking San Juan de Alloa by balloons. This new method of besieging a fortress has been discussed in every vein of seriousness, wit or contumely, as the idea seemed feasible, funny or absurd to various minds."[8] A former governor of Kentucky was heard to remark that while the idea was the product of a man of genius, "I think it will be a very troublesome matter to enlist volunteers for that service."

Wise formally submitted his plan to the War Department on December 10, 1846. His proposal was rejected out of hand.[9]

When President Lincoln issued his call for volunteers to march against the Southern Confederacy, Wise responded with enthusiasm and alacrity, but he refused to offer his services as a balloonist. Rather, the fifty-three-year-old veteran aeronaut informed Gov. Andrew G. Curtis of Pennsylvania that he was "forming a company here in [Lancaster] of picked men, now nearly full and well drilled in the rifle manual,"[10] who would serve for three to five years.

While John Wise was busily drilling his troops and waiting for the governor's orders, James Allen was making his first trial flights as a military aeronaut. Allen had emerged as the central figure in northeastern aeronautical circles during the years prior to the Civil War. Other members of the group included Allen's mentor and partner, Samuel Archer King, his brother Ezra Allen, and independent balloonists such as William H. Helme, a dentist in the Allens' hometown of Providence, Rhode Island.[11]

Between 1857 and 1861 Allen and King had continued to crisscross the Northeast, delighting the citizens of Providence, Worcester, New Haven, Lowell, Manchester, Taunton, Paterson, and other towns. Their activity was universally applauded by spectators, newspaper reporters, and the civic officials that sponsored the flights.

In addition to making their normal public ascents, King, Allen, and Helme were involved in a number of important balloon experiments that seemed to have military implications. In the summer of 1860 Helme had persuaded J. William Black, a leading Boston photographer, to attempt the first aerial photography in the United States. Ascending with Helme over Providence, Black exposed two initial plates. The results proved unsatisfactory since the collodion cracked, damaging both plates. Nevertheless, Helme reported that "the buildings in one [photograph] were sharply defined, while the other was blurred by motion."[12] Several months later, in October 1860, Black produced the first genuinely successful aerial photos in the U.S. from the basket of Samuel Archer King's *Queen of the Air* tethered over Boston.

Other American aeronauts were also interested in the possibility of aerial photography. John Wise had arranged to carry the New York "artist" W. J. Kunns and his camera aboard the *Atlantic* in 1859, but inclement weather forced the cancellation of the flight.

These early experiments sparked an extended debate about the military utility of aerial photography and were a major factor in stimulating War Department interest in aeronautics. Charles Seeley, a member of the American Photographic Society, called that group's attention to the subject in the summer of 1860. Seeley's suggestion that the society offer its advisory services in the area of balloon photography to Secretary of War Simon Cameron was enthusiastically endorsed by John W. Draper, a prominent American scientist then serving as president of the organization. *Scientific American* took Cameron to task for his failure to respond to Draper's subsequent offer.[13]

The cry for army support for aerial photography grew more strident after the fall of Fort Sumter. Seeley's thoughts, which were published in the *American Journal of Photography* in June 1861, were echoed by some of the nation's leading newspapers. Readers of the *New York Tribune* and the *Philadelphia Ledger* were soon calling for aerial "daguerreotypes" to be taken of enemy troop concentrations.[14]

Alexander Dallas Bache, a friend of Draper's and the head of the U.S. Coast Survey, was instrumental in opening an effective channel into the War Department. He brought J. W. Black's photographs of Boston to the attention of his uncle, Maj. Hartman Bache, head of the U.S. Army's Topographical Engineers, and urged that some attention be paid to this new application of photographic technology. Major Bache apparently agreed with his distinguished nephew, for he wrote to Black's partner, a Mr. Batchelder, to obtain a detailed account of the procedure for making aerial images. Hartman Bache was to play a central role in directing the early balloon activities of the Army of the Potomac. But in view of the initial importance of aerial photography in War Department thinking, it is ironic to note that not a single photograph would be taken from a balloon during the course of the Civil War.[15]

Thus, as James Allen made his way south toward Washington in late April 1861, the stage had been set for the U.S. Army's first venture into aeronautics. There was, however, very little agreement on what steps should be taken to incorporate observation balloons into the structure of the army. As Allen had already discovered, confusion reigned.

Allen had initially offered his services directly to federal authorities but had received no reply. As the *Providence Daily Post* informed its readers, "The brief time allowed for negotiations with the proper department at Washington before the departure of Governor Sprague with the first half of our regiment, rendered it impossible to complete any arrangements."[16]

Rather than asking Allen to remain behind while officials in Washington tried to decide where to assign him, Sprague enrolled the aeronaut as a member of Ambrose E. Burnside's Marine Corps of Artillery. William H. Helme was also enlisted as a private in the unit and was apparently assigned to Allen's section.

May 1861 was a difficult month for Allen. Ordered by Governor Sprague to travel separately and join the Rhode Island troops in Washington, he loaded two balloons, a portable gas generator, 5,000 feet of tether rope, and "an ingeniously contrived windlass" aboard a train bound for Philadelphia.[17]

The City of Brotherly Love bristled with activity as thousands of troops from all of the Northeast sought to reach their duty stations. Allen was particularly disturbed by the harsh discipline meted out to raw recruits. "I witnessed one poor fellow drop," he noted in a letter to his brother Ezra, which was later published in the *Providence Daily Post*. "He was shot for disobeying orders . . . I was within three yards of him. Six shots were fired at him."[18]

Finally, on the evening of April 23, Allen made contact with Colonel Burnside and the Marine Artillery, which had been shunted to Easton, Pennsylvania, until the roads to Washington were secured. Presumably he joined the regiment soon thereafter and remained with the unit until it reached Washington late in May.

Allen contacted the War Department once he arrived in the capital. Contemporary news reports noting the activities of the aeronaut during the first week of June 1861 indicate that he had been appointed an official "aeronautical engineer." Maj. Albert J. Myer, the chief signal officer, was detailed to supervise Allen's preliminary ascents.[19]

The aeronaut inflated the larger of his two balloons at a gas outlet on the corner of Massachusetts Avenue and Third Street on June 9. He then supervised the Rhode Island troops who towed the aerostat to the regimental camp at

Caton's Farm, a mile north of the Capitol Building.

Allen made his first official flight with the balloon later that day. Standing 50 feet tall and measuring 35 feet in diameter, the balloon attracted a large crowd as Allen ascended to the end of his tether and obtained his first bird's-eye view of the city.[20]

The First Rhode Island Regiment broke camp on June 10 with orders to travel to Chambersburg, Pennsylvania, via Baltimore. There they would join a column commanded by Maj. Gen. Robert Patterson that was marching to secure the federal arsenal at Harper's Ferry, Virginia. Washington newspapers reported that Allen and his balloon would accompany the troops and "will be used to reconnoitre the force and position of the enemy at Harper's Ferry."[21] It seems likely that Allen did travel with Burnside's regiment, but he did not make any observation flights, perhaps as a result of the failure of his gas generating equipment, which would later prove difficult to operate under field conditions.

Maj. Hartman Bache was giving serious thought to the role of aeronautics on the battlefield during this same period. On June 10, 1861, while Allen was presumably packing his balloon for the train ride to Chambersburg, Bache wired John Wise in Philadelphia. The major inquired about the salary of an aeronaut and the price of a balloon capable of lifting 500 pounds.

When he failed to receive an answer the next day, the major wired a second inquiry concerning the aeronaut's whereabouts to federal officials in Philadelphia. He discovered that Wise was thought to be at home in Lancaster (still drilling his troops), but that William Paullin was present in Philadelphia and might be able to answer Bache's questions.[22]

Convinced that John Wise was the man for the job, Bache finally reached the aeronaut on June 12. Wise lost no time in wiring a price of $300 for the balloon, adding that his own services would be "gratis, in the cause of the Union."[23] After consulting with Capt. A. W. Whipple and Lt. Henry L. Abbot, both topographical officers who had seen Allen's balloon and who were familiar with T. S. C. Lowe's work, Bache wired Wise on June 26, requesting a second estimate on a sturdier silk balloon with a gas capacity of 20,000 cubic feet. The aeronaut replied that such a craft, with all necessary equipment, would cost $850 and would be constructed in two weeks. Bache ordered Wise to proceed and bring the new balloon to Washington as quickly as possible.[24]

The number of aeronauts anxious to enlist their services in the Union cause was now growing rapidly. John La Mountain was scarcely two weeks behind James Allen in applying for an army position. He had canvassed the citizens of his hometown of Troy, New York, obtaining the signatures of thirty-three leading editors, physicians, businessmen, and local politicians on a petition attesting to the fact that La Mountain "has the reputation in this community of being a highminded and excellent citizen . . . and a successful and skillful aeronaut."[25] The petitioners declared their "full confidence in the skill of Mr. John La Mountain, as one of the most scientific balloonists of the country, and being familiar with his capacity, [we] have no hesitation in recommending him . . . and . . . in assuring . . . that he is a gentleman of thorough practical information, upright, honest, and indefatigable, and one who may be relied on to perform all he promises, when performance is possible."[26]

La Mountain sent copies of the document to both Secretary of War Cameron and the New York State military commissioners. He appended a long discourse justifying the military utility of the balloon on the basis of historical examples drawn from his considerable knowledge of the observation balloons that had been employed by European armies.[27]

A week later La Mountain sent a second letter to federal authorities, adding that he could, in addition to his own balloon and experience, provide "a portable apparatus of my own invention and used solely by myself with which I can manufacture gas for inflation from the decomposition of water, at a barely nominal rate of expense." His generator "with balloons and all appurtenances necessary could be carried in a single wagon in the train of an army and the balloon filled in a few hours notice wherever there is coke or charcoal and water."[28]

After waiting a month without receiving any reply, La Mountain was surprised by a letter from Maj. Gen. Benjamin F. Butler. Butler, commander of the Department of Virginia, with headquarters at Fort Monroe, offered to hire the aeronaut as a civilian balloonist, promising to assist La Mountain in obtaining steady employment with the army if his services proved valuable. Butler was convinced that La Mountain and his balloons could not "fail to prove of much benefit."[29]

The aeronaut had originally promised to start south for Fort Monroe by mid-June, but financial difficulties in Troy forced repeated postponements. Fortunately, O. A. Gager, who had financed the *Atlantic* trip of 1859, came to the rescue once again. Visiting Fort Monroe, Gager was informed of La Mountain's position by General Butler. He hurried back to Troy and provided La Mountain with enough cash to reach Washington with his equipment. While in the capital, the aeronaut received encouragement from Secretary of War Cameron, who approved La Mountain's service and stated, "If General Butler, or General McDowell, or any of the other Generals should want a balloon attached to their camps, all that they would have to do would be to order whatever they might want."[30]

Finally arriving at Fort Monroe in late June, La Mountain passed Cameron's comments on to Butler and admitted that his vaunted portable gas generator was not suitable for service in the field. He would, therefore, require a total of $480 in supplies before his balloon could be inflated. Butler, taken aback, agreed, but only if La Mountain would forego his salary until he had demonstrated his worth.[31]

La Mountain's comment to General Butler made it clear that the first of the feuds that would eventually mar the history of the balloon corps was already under way. During his stay in Washington, he had discovered that Thaddeus Sobieski Constantine Lowe was also in the capital. Lowe had made a very impressive series of demonstration flights during which he carried a telegraph key aloft and offered instant reports of his observations to the president and other high-ranking officials. La Mountain now informed General Butler that Lowe's performance had been little more than a publicity stunt. "I have a simple method by which I can convey intelligence from above, without any expense whatever," he insisted.[32]

T. S. C. Lowe may have been a latecomer, offering his services after Allen, Wise, and La Mountain were already involved, but he quickly made up for lost time. Lowe, it will be recalled, had almost become the first casualty of the Civil War.

His 900-mile flight from Cincinnati, which was intended to demonstrate his ability to cross the Atlantic, had come to an end in a field near Unionville, South Carolina, just after one o'clock on the afternoon of April 20, 1861. The welcoming committee of belligerent backwoodsmen was more than a little daunting: "The people among whom I had fallen are called clay-eaters, because of their often repulsive and disgusting dietary habits, and in appearance they were wild and ferocious brigands, with long, dirty, unkempt hair and long beards, mostly sandy red in color, reaching to their short rotund stomachs, wearing slouch hats, and with unclean faces and hands, and ragged and tattered clothes. [They were] mounted on shaggy horses, each man with a shotgun over his shoulder."[33] Jailed several times as a Yankee spy as he made his way north, Lowe escaped each time only when one of the better-read local citizens recognized his name.

Returning to Cincinnati on April 26, 1861, Lowe persuaded his friend and sponsor, Murat Halstead, editor of the *Cincinnati Commercial* and a leading figure in state politics, to write Ohioan Salmon Portland Chase, secretary of the treasury, urging that Lowe be appointed to command a Union Army balloon corps. Halstead complied and was pleased to report that Chase had responded favorably.[34]

In his letter to Halstead, the secretary noted that he had spoken to Cameron

T. S. C. Lowe

"and other distinguished officials" on Lowe's behalf. There was, he continued, much interest in ballooning, but some differences of opinion existed concerning the choice of a chief aeronaut.[35]

Murat Halstead's advice that Lowe proceed at once to Washington in an effort to press his case was seconded by Joseph Henry. The secretary of the Smithsonian Institution, whom Lowe had met while planning the transatlantic venture, wrote to the aeronaut on May 28 urging that he come to Washington at once.[36]

Lowe chose to take his time. He made a public ascent with the *Enterprise* in Cincinnati on May 8, then traveled to Hamilton, Ontario, where he flew on May 24 and May 29 in celebration of Queen Victoria's forty-third birthday.

He returned to Philadelphia around June 1, finding the summons from Joseph Henry and an urgent cable from Murat Halstead, who had already arrived in the capital. "Can't you prepare the balloon and come here?"[37]

Lowe arrived in Washington on June 5. He called on Joseph Henry, who immediately arranged a meeting with Simon Cameron, then accompanied Lowe to a conference with President Lincoln on the evening of June 11.

With $250 from War Department coffers, Lowe forged ahead with preparations for a demonstration flight. Launching the *Enterprise* from the grounds of the Columbian Armory (now the site of the National Air and Space Museum) on June 18, Lowe rose to an altitude of 500 feet. He was accompanied by George McDowell of the American Telegraph Company and Herbert C. Robinson, a local telegrapher, who would send Lowe's message to the world from a telegraph key in the basket. Powered by batteries in the War Department, Lowe's key connected to a telegraph office in Alexandria and also to the White House. His message told far more about the aeronaut's public relations sense than it did about Confederate troops across the Potomac:

> *To the President of the United States*
> *Sir:*
>
> *This point of observation commands an area nearly 50 miles in diameter. The city, with its girdle of encampments, presents a superb scene. I have pleasure in sending you this first dispatch ever telegraphed from an aerial station, and in acknowledging indebtedness for your encouragement for the opportunity of demonstrating the availability of the science of aeronautics in the military service of the country.*
>
> *T. S. C. Lowe*[38]

It was a masterstroke. By telegraphing messages to the president, the War Department, Gen. Winfield Scott, Alexandria, and Philadelphia, Lowe had enormously strengthened his own position and focused attention on his own candidacy for the yet-to-be-established office of chief aeronaut.

Once the telegraphic demonstration was complete, Lowe ordered the balloon hauled closer to the ground. With the three men still in the car, the *Enterprise* was towed to the Executive Mansion, where President Lincoln inspected it from a second-story window.

With the *Enterprise* safely staked out on the South Lawn, Lowe spent the night at the White House. The aeronaut discussed the military potential of the balloon with President Lincoln far into the evening:

> *The President was intensely interested in my outline of the proposed Aeronautic Corps and after the departure of his secretaries and assistant, we discussed the possibilities of the service and the details of operation. He was especially interested*

> *in my plan for directing the fire of artillery on an enemy that the gunners themselves could not see. We talked til late into the night, and then retired, he wearied with the cares of State, and I almost too excited to sleep, so enthused was I at the prospect of being directed to form a new branch of the military service.*[39]

Lowe was back in the air the next morning, demonstrating his capabilities once again for the president, members of the cabinet, and military authorities, including Captain Whipple, Bache's subordinate at the Topographical Engineers.

Small wonder that John La Mountain was disgruntled. Lowe was the man of the hour. Newspapers that had given scant attention to Allen's activity less than two weeks before now trumpeted Lowe's success to the world. La Mountain must have been particularly chagrined to discover that most of these enthusiastic accounts focused on Lowe's use of the telegraph. The editor of the *Boston Transcript* predicted that the balloon-telegraph combination would enable a Union general to remain "accurately informed of everything that may be going on within a long day's march of his position in any direction."[40] The *New York Herald* spoke of "War as a Science—the Important Combination of the Balloon and Telegraph," while the Washington *Evening Star* reported that "the aeronautical telegraphic enterprise in Washington has proved, so far, to be a complete success."[41] A third newsman prophesied that Lowe and his "telegraphical balloon" would "help to equalize the advantages gained to the Confederacy and lost to our side by the perpetual reports of their spies and the blabbering of super-serviceable newspaper reporters."[42]

Joseph Henry, who had been instrumental in arranging the demonstration flights, remained Lowe's most powerful ally in Washington. At the time of his first meeting with Lowe on June 6, Secretary Cameron had asked Henry for a full report on the tests he had agreed to fund. The scientist delivered the document to the War Department on June 21.

Henry regarded the experiment as a complete success and expressed every confidence in Lowe's ability to perform the duties of a military aeronaut. Lowe's balloon was capable of inflation with either city gas or hydrogen generated in the field. It would hold a charge for several days and, once inflated, could be "towed by a few men along an ordinary road, or over the fields, in ordinary calm weather, from the places where it is filled, to another 20 or more miles distant."[43]

When allowed to ascend the length of its tether, the balloon commanded a view of the country for 20 miles around. Moreover, Henry commented, when the winds were right, the aeronaut could make a free flight over enemy lines, then rise to a current of wind blowing in the other direction and return to safety. Henry closed with a clear recommendation: "From all the facts I have observed, and the information I have gathered, I am sure that important information may be obtained in regard to the topography of the country, and to the position and movements of an enemy, by means of the balloon, and that Mr. Lowe is well qualified to render service in this way by the balloon now in his possession."[44]

Thanks to Whipple and Bache, Lowe soon had an opportunity to demonstrate his capability to function under battlefield conditions. Brig. Gen. Irvin McDowell, commanding the Department of Northeastern Virginia, had recently received reports suggesting that 20,000 Confederates had gathered in the area between Fairfax Court House and Manassas. McDowell ordered Whipple, then serving as topographical engineer in the area, to arrange a reconnaissance by T. S. C. Lowe.

On the afternoon of June 22, 1861, Lowe refilled the *Enterprise* from a Washington Gas Company main and, with the assistance of a fifteen-man detachment from the 8th New York Infantry, moved the balloon over the Long Bridge to meet Whipple at Arlington House. Lowe apparently made one short ascent early that evening.

By four o'clock on the morning of June 23, the aeronaut and his balloon were on the road to the observation point suggested by McDowell at Falls Church. Following a short ascent at Baileys Crossroads to determine that there were no Confederate pickets blocking the route, the party proceeded to its destination, where Lowe made a series of ascensions over the next two days.

Several topographical officers, including Whipple, ventured aloft as well. Gen. Daniel Tyler ordered one of his officers to produce a sketch map based on balloon observations, noting the location of distant Confederate campfires.[45]

Lowe's activity made believers of a number of the officers with whom he came in contact. General Tyler informed McDowell that while he had originally doubted the value of aeronautics, "the sketch map had convinced him" that "a balloon may at times greatly assist military movements."[46] Favorable newspaper reports of Lowe's activity in Northern Virginia were also widely circulated.

Back in Washington on June 26, Lowe continued to lobby for an official position as head of army aeronautical activity. Captain Whipple informed Lowe that the Bureau of Topographical Engineers had decided to establish an observation balloon unit. He discussed field operations with Lowe and obtained the aeronaut's estimate for the construction of a military balloon.[47]

Lowe's high hopes were dashed when he paid a visit to Whipple's office on June 27. The officer told Lowe that the first contract for a military balloon had been awarded to John Wise, who had underbid Lowe by some $200. The aeronaut rejected Whipple's suggestion that he might be employed to operate the Wise balloon: "To the latter part of his remarks I replied that I would not be willing

PROFESSOR LOWE MAKING A BALLOON ASCENSION ON A RECONNOITRING EXPEDITION TO VIENNA.—SKETCHED BY OUR SPECIAL ARTIST.—[SEE NEXT PAGE.]

T. S. C. Lowe ascends to observe Confederate activity during the march toward Vienna, Virginia.

to expose my life by using so delicate a machine where the utmost care in construction was required, which should be made by a person in whom I had no confidence. I assured him that I had greater experience in this business than any other aeronaut and that I would guarantee the success of the enterprise if entrusted entirely to my direction."[48]

Lowe freely admitted that his own ego was now involved. "It is difficult for a man to give the details of his own work without conveying the impression of egotism," he remarked, "yet since my whole success lay in my ability to adapt original thought to operation, I speak of it not in a spirit of self-laudation but to be informative."[49]

Lowe obviously had little professional regard for either Wise or La Mountain, whom he considered to be his principal rivals. Wise, he admitted, "had won considerable distinction in his profession." La Mountain "was simply a balloonist. . . . Neither had the least idea of the requirements of military ballooning nor the gift of invention which later made it possible for me to achieve success."[50]

Lowe castigated his two rivals for having "assiduously courted the attention of whatever officer of the Topographical Engineers happened to be in charge of the Aeronautical Corps."[51] This attitude, coming from a man whose friendship with Murat Halstead and Joseph Henry had carried him into the inner sanctums of the War Department and the White House itself, can only be regarded as remarkable.

Lowe's reactions to Bache's purchase of the balloon and services of John Wise was simply to withdraw from the field. Through the month of July 1861, he continued to capture the attention of the nation's press and government officials with a series of tethered ascents from the front lawn of the Smithsonian Institution. During this same period, James Allen and John Wise returned to the capital in an attempt to establish themselves as leaders of the army aeronautical program.

It was a confusing situation. Hartman Bache of the Topographical Engineers was now convinced that a balloon corps should be established, but he refused to structure a formal organization. Rather, Wise, Allen, La Mountain, and Lowe were being played against each another in the hope that the most successful aeronaut would emerge from a sort of informal competition.

On July 8, 1861, Capt. Whipple detailed Lt. Henry L. Abbot to call on the services of James Allen for a reconnaissance of a Confederate position near Washington. Their first attempt to inflate Allen's large balloon in the field on July 9 was plagued by the failure of the aeronaut's gas generator and the clumsiness of an inexperienced ground crew.[52]

Abbot attempted a tethered ascent in the half-filled balloon that day, but the craft bobbed and rocked so severely that nothing could be accomplished. Things were worse when the inept crew assigned on July 8 was replaced by a completely new group of men the following day.

Abbot retained his belief that the balloons "may be useful in the coming campaign," but he added that a formal organization was required. This would include not only an aeronaut but a trained officer-observer and a permanent cadre of troops that could be trained as balloon handlers. Moreover, he continued, the generation of hydrogen under field conditions was so difficult that arrangements should be made to inflate the balloons with city gas and move them by wagon to the observation point.[53]

Abbot's recommendations were accepted. Allen's two balloons were ordered back to Alexandria, where they would be prepared to accompany Gen. Daniel Tyler's advance division as part of the Union army's attack on the Confederate troops gathering around Manassas, Virginia.

The smaller of Allen's two balloons burst during inflation on the morning of July 14. The large balloon in which all flights had been made to date was turned over to sixty gaudily dressed men from the 11th New York Zouaves who had been assigned to tow the craft to Tyler's headquarters at Falls Church, Virginia. The party had traveled only a short distance when a sudden gust dashed James Allen's last hope to obtain the position of chief aeronaut. Lt. Henry Abbot described the scene: "We carried along nearly to the point where our branch road diverged when suddenly a furious gust occurred. The detail, struggling and shouting, was slowly pulled toward the river in spite of their efforts, until the balloon in one of its stately plunges struck a telegraph pole. There was a puff of gas, and our work was ended."[54]

In later years James Allen would attempt to cast a more favorable light on these events by insisting that the disaster had occurred a week later, at the time of the Battle of Bull Run. Abbot's recollections and contemporary news accounts leave little doubt that the two balloons were lost in Alexandria and en route to Falls Church on the morning of July 14.

John Wise lost the new balloon he had constructed for Major Bache under similar circumstances. As requested, Wise had added some significant details to better adapt the balloon to military purposes. With a 20,000-cubic-foot capacity, it was constructed of a double thickness of the best raw silk. The netting and rigging were designed to meet the special needs of observers and mapmakers. The willow and cane basket featured an iron floor to protect the occupants from sharpshooters below.

Wise completed work on the new balloon on July 16, 1863, only two days after the destruction of James Allen's two aerostats. John and Charles Wise reported to Hartman Bache on July 18.

Uncertain as to the status of the Wise balloon and aware that Allen could no longer be of service, Bache had asked Lowe to join Irvin McDowell's troops moving toward Centreville and Manassas on July 17. Now with Wise on the scene, Lowe had to step aside. Engaged in filling the *Enterprise* at a Washington gas main near the Columbian Armory, Lowe was ordered to halt the operation so that Wise could inflate his balloon. Bache and Maj. Albert J. Myer, chief signal officer, had arranged for Wise and his balloon to be transported to the front by a wagon, accompanied by a squad of twenty picked men.[55]

The group moved through Georgetown and over the Aqueduct Bridge into Northern Virginia in the predawn hours of July 21. After passing through Fairfax Court House on the way toward Bull Run, the party encountered more difficult terrain. The trees became so thick that Wise had to allow the balloon to rise above the branches while the handling crew struggled to maneuver it from the ground.[56]

By noon the noise of battle could be heard drifting in from Manassas.

Hurrying toward the action, Major Myer now took command. As they moved rapidly between the trees, the balloon became wedged in the branches. During the struggle to free the craft, the bag was badly torn, forcing John and Charles Wise to return to Washington from the very edge of the battlefield.[57]

Repairs were completed by July 23, and Wise was once again prepared to attempt observation flights. The disaster at Bull Run created a general fear that Confederate troops might immediately advance toward the capital. T. S. C. Lowe was already making ascensions from Fort Corcoran, a large earthen fortification constructed opposite the Georgetown Heights to protect the Virginia end of the Aqueduct Bridge.[58]

At Captain Whipple's insistence, Wise moved his balloon over Georgetown's Aqueduct Bridge and toward Arlington House early on the morning of July 26. In spite of the fact that the aeronaut had ordered his handling crew to weight the basket with their knapsacks and weapons, a stiff breeze brushed the tether ropes against telegraph lines, which cut the balloon free. The craft fell to earth near Arlington House, so badly damaged that it was no longer of use.[59]

Captain Whipple, who had now been twice disappointed by the Wise balloon, addressed the aeronauts in a manner that "was not of the most polite and delicate character." The straitlaced Wise not only objected to Whipple's use of profanity in the presence of his son Charles but remarked that "with regard to the Bull Run Campaign, the balloon part of the disastrous affair was just about as good as the fighting part."[60]

Wise was not yet ready to withdraw from the field. Convinced that recent experience had demonstrated the problems inherent in moving inflated balloons, he submitted plans for a portable field generator that would decompose water to form hydrogen. When cost estimates for producing the generator ($7,000) proved ten times higher than the cost of a balloon, the Bureau of Topographical Engineers rejected his scheme. Wise returned to Lancaster where, by October 1861, he was serving as an officer with Company F, 9th Cavalry, Pennsylvania Volunteers. Retiring several months later for reasons of health, John Wise resumed his career as an aerial showman.[61]

T. S. C. Lowe was scarcely more successful during the difficult days of July 1861. He had reinflated the *Enterprise* and started for the front as soon as he heard of Wise's failure to reach Bull Run. Lowe established himself at Fort Corcoran in Arlington, where he made a number of tethered flights. Then, desperate to demonstrate his abilities by obtaining information on the Confederate troops reportedly sweeping toward Washington, he decided to make a free flight.

The aeronaut planned to pass over Confederate lines and then locate a current that would carry him back to safety. He launched his balloon shortly after daybreak on July 24, and as he had hoped, a west wind swept him over outlying Confederate pickets. He then rose and recrossed Union lines, only to draw fire from Union soldiers. Not having equipped himself with an American flag, and fearing that the soldiers might shoot first and ask questions later, the aeronaut was forced to continue his voyage to a landing on the Mason plantation, five and a half miles from Alexandria, in the no-man's-land beyond Union lines.

Lowe's wife, witnessing the friendly fire directed at her husband, had rushed to the scene and enlisted the services of men from the 31st New York Volunteers in locating the aeronaut. The troops could see Lowe from a distance but were unable to move into an area thick with Confederate skirmishers. That evening Leontine Gachon Lowe, disguised as a farm woman, led a horse-drawn covered wagon to the spot where her husband had concealed his balloon. With the equipment packed and Lowe hidden in the wagon, she then proceeded back down the road to the safety of Alexandria.[62]

Captain Whipple ordered Lowe to take to the field again on July 29. On this occasion Lowe was forced to empty his balloon en route to the observation point in order to avoid being caught in a storm.[63]

Whipple had had enough. He had honestly attempted to assist the efforts of three different aeronauts before, during, and after the Battle of Bull Run. Each time he had been disappointed. There was little doubt in his mind that the balloon was too delicate and temperamental an instrument to operate under field conditions. He expressed this opinion in a telegram to Woodruff and Bache on August 1, remarking, erroneously, that "Lowe's balloon filled at our expense has burst. I wish to have nothing more to do with it."[64]

Ultimately, Whipple was to be overruled by higher authority. Lowe had spent a second evening with President Lincoln in the White House on July 25. After listening to a description of the chaos that reigned as three aeronauts tried to operate independently within the loose control of Whipple and Bache, Lincoln asked Lowe to explain the situation to Lt. Gen. Winfield Scott, the senior U.S. military commander. The aging general, who had not the slightest interest in military ballooning, refused to see Lowe. Finally, on July 27, President Lincoln escorted the aeronaut to Scott's office. Explaining that Lowe was his "personal friend" and the man "who is organizing an Aeronautic Corps for the Army, and is to be its Chief," Lincoln instructed his commanding general to facilitate the aeronaut's work in every way possible.[65]

When Captain Whipple had suggested abandoning the balloon experiments, he was unaware of the president's interest in the program and the steps that Scott had already taken to comply with Lincoln's instructions. In addition,

Whipple had received a letter from Joseph Henry asking him to reconsider. The secretary explained that Lowe's failure to ascend on July 29 had been due to a storm. Henry closed his plea with a strong affirmation of his continued belief in Lowe's ability:

> *Mr. Lowe came to this city with the implied understanding that, if the experiments he exhibited were successful, he would be employed. He has labored under great disadvantages, and has been obliged to do all that he has done, without money. From the first he has said that the balloon he now has was not sufficiently strong to bear the pressure of a hard wind, although it might be used with success in favorable situations and in perfectly calm weather. I hope that you will not give up the experiments and that you will be enabled, even with this balloon, to do enough to prove the importance of this method of observation, and to warrant the construction of a balloon better adapted to the purpose.*[66]

Whipple dutifully forwarded the letter to Bache's office, where it joined others from American scientific leaders such as Dr. Cresson of the Franklin Institute, who affirmed that Lowe was "unsurpassed in the requisite qualifications for military aeronautics."[67]

Such ringing endorsements, coupled with President Lincoln's clear desire to continue the balloon program, left Hartman Bache little choice but to reverse his subordinate's decision.

Lowe appeared at Bache's office on August 2, 1861, prepared to plead for another chance. Instead, Bache informed the aeronaut that Whipple had been ordered to contract with Lowe for a new military balloon. Moreover, the aeronaut would receive $5 a day during construction and $10 once he entered service as an official government aeronaut. Naturally, the Topographical Engineers would pay all the expenses related to the construction of the balloon and the acquisition of the necessary supplies and equipment.[68]

Lowe proceeded to Philadelphia to construct the balloon, a 25,000-cubic-foot aerostat to be named the *Union*. The aeronaut was back in Washington with the balloon by August 28, when he was ordered to inflate the *Union* at the Columbian Armory and report to Fort Corcoran.[69]

Confederate troops had invested Mason's, Clark's, Munson's, and Upton's Hills opposite Washington. Fairfax Court House and the roads leading toward Arlington and Alexandria were also in Rebel hands. Lowe's task was to observe Confederate troop movements in these areas. Moving between the fort, Chain Bridge, and Baileys Crossroads, he flew on twenty-three of thirty-four consecutive days during August and September 1861. Lowe's observations proved both accu-

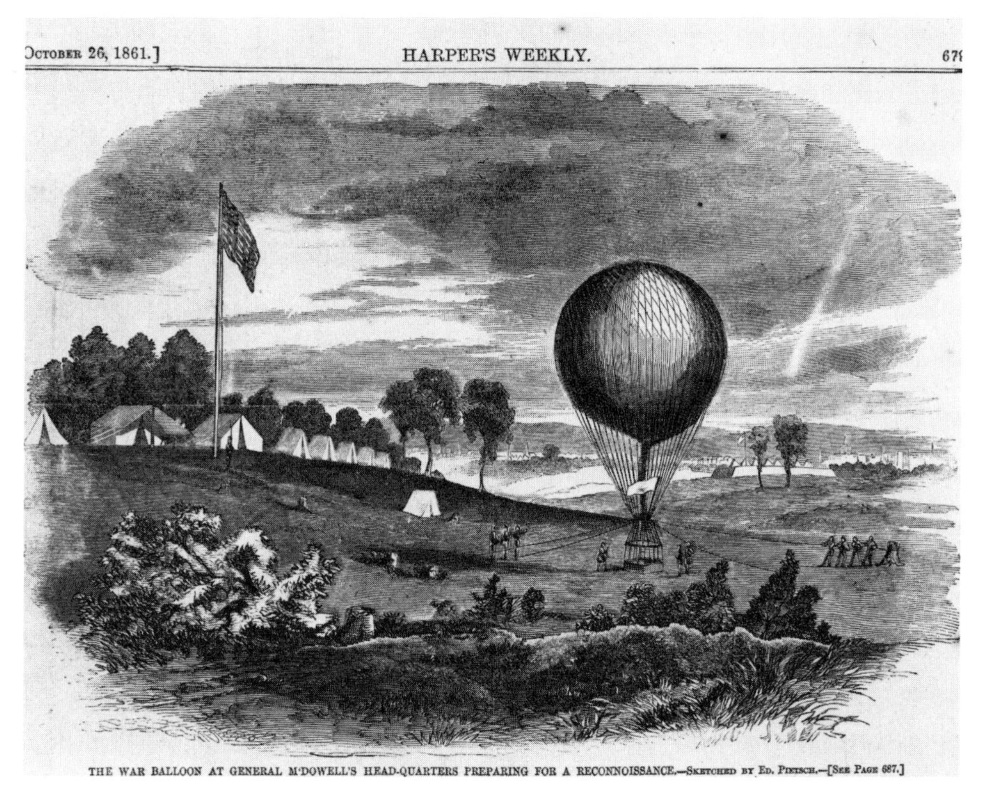
OCTOBER 26, 1861.]
HARPER'S WEEKLY.
THE WAR BALLOON AT GENERAL M'DOWELL'S HEAD-QUARTERS PREPARING FOR A RECONNOISSANCE.—SKETCHED BY ED. PIETSCH.—[SEE PAGE 687.]

Preparations for an ascension from General McDowell's headquarters in the fall of 1861

rate and informative. He was also able to establish procedures and explore techniques that would form the basis for the later operations of the Balloon Corps. A permanent ground crew was appointed and trained. In order to maintain the schedule of daylight ascensions, Lowe moved the balloon back to Washington for reinflation at night. When recharging was not required, Lowe made a number of night ascensions to count Confederate campfires.[70]

Army officials were suitably impressed by the aeronaut's performance. On September 5, Gens. Irvin McDowell and Fitz-John Porter went aloft together. Two days later, on September 7, Gen. George McClellan made a two-hour flight, followed by a second several weeks later. Other generals who ascended during this period included John H. Martindale, W. F. Smith, and Samuel P. Heintzelman. Each of these officers had been thoroughly impressed. As Fitz-

John Porter remarked to Lowe on September 9, "You are of value now."[71]

Lowe added artillery spotting and adjustment to his repertoire of skills late in September. Working in concert with a battery under the command of Gen. William F. Smith, Lowe used both telegraphic and flag semaphore signals to direct fire on Confederate positions in and around Falls Church. Capt. Frederick F. E. Beaumont, an English observer, commented that the artillery fire would have been completely ineffective without Lowe and his balloon.[72]

The Confederate reaction to Lowe's activity also indicated that the balloon was a weapon of some consequence. Gen. Pierre Gustave Toutant Beauregard resorted to black-outs, the dispersal of troops, and camouflage in order to disguise his strength from T. S. C. Lowe. Beauregard also ordered false campfires to be lit and dummy artillery batteries established. Black logs or long stove pipes "of different caliber" were set in place to deceive aerial observers. Such "Quaker Gun" emplacements did occasionally puzzle Lowe during September. He also commented on a greater than usual number of late afternoon campfires which suddenly appeared when he made an ascent.[73]

Gen. Fitz-John Porter also noted the care that Confederate troops took to mask their activity from aerial observation. During his ascents from Fort Corcoran in September, Porter observed evidence of recent work on a fortification. He watched for two hours without seeing a single worker. He then ordered the balloon hauled down out of sight. After a half hour on the ground, he suddenly bobbed back up just in time to see a squad of Confederate workmen scurrying for cover.[74]

The great southern artillery officer Edward P. Alexander took particular delight in directing the fire of his batteries at the Union balloon. In a letter written to his father during this period, Alexander gloated that "we sent a rifle shell so near old Lowe and his balloon that he came down as fast as gravity could bring him."[75]

Although few details are available, it seems likely that Lowe's success inspired the first Confederate aeronautical activity in the fall of 1861. Reports of southern observation balloons had appeared in the northern press as early as July 23-24, but these seem to have been nothing more than confused accounts by those who had actually seen Lowe's balloon. The earliest southern balloon was a small unmanned signal craft sent aloft from the positions ringing Washington in July 1861. The reports of manned observation balloons hovering over Confederate positions near Fairfax Court House as early as June 18 can probably be discounted as confused sightings of Lowe's balloon.

It is probable, however, that Confederate commanders P. G. T. Beauregard and Joseph E. Johnston did have a balloon at their disposal by the end of the summer. On August 22 Johnston suggested to Beauregard that "the balloon may be useful. . . . Let us send for it; we can surely use it advantageously."[76]

The origin of this first southern balloon remains a mystery. Reports reaching northern newspapers suggested that "several aeronauts had offered their services to the Southern Government as early as May, a considerable time previous to the offers made to the Federal Government by Professor Lowe and others."[77] Southern newspapers contain no mention of any aeronauts in Confederate service during these early months of the war, but one officer on Beauregard's staff noted that when the general was unable to obtain a balloon through official channels in Richmond, he contracted for one through "a private source." This balloon was apparently poorly constructed, for the officer in question, Beauregard's inspector general, remarked that it was not successful. It seems likely that the mysterious balloon did make at least one ascension. On September 4, 1861, a manned observation balloon was seen rising from the Confederate fortifications on Munson's Hill, a time and place at which it could have been confused with Lowe's craft.[78]

By the end of September, Lowe's activity had created a core of enthusiastic support for aeronautics among the general officers of the Army of the Potomac. Fitz-John Porter was a particularly influential partisan. On September 16, at Porter's request, Lowe had submitted a plan for the organization of a regular Balloon Corps. In final form, Lowe's plan called for the construction of four balloons, two of 20,000-cubic-foot capacity and two 30,000-cubic-foot models, together with portable gas generators to permit field operations.

Porter forwarded Lowe's recommendations to General McClellan, who had also received a request from Division Commander Gen. William F. "Baldy" Smith for his own balloon unit. McClellan acted quickly. On September 25,1861, Lowe received word from Q. M. Gen. Montgomery C. Meigs that, at McClellan's request, the aeronaut was authorized to proceed with the construction of the four balloons "together with such inflating apparatus as may be necessary for them and the one now in use." Meigs would pay all expenses up to $1,500 each for the large balloons and $1,200 each for the smaller ones. Soon after, Lowe was commissioned to order two more small balloons. Thus, by the spring of 1862, Lowe's Balloon Corps held title to seven observation balloons in addition to the original *Enterprise*.[79]

The new balloons, including the *Union*, which went into service in August 1861, were constructed in Philadelphia. Each aerostat featured an envelope of doubled pongee silk. The mahogany valves were fitted with brass hinges and rubber springs. The finished bags were treated with Lowe's special varnish compound of benzine and a drying agent suspended in linseed oil. The netting was woven linen.

In addition to the standard load ring supporting the weight of the basket, a second spreader ring was provided on most of the balloons. Each balloon was also equipped with three 5,000-foot manila tethers and pulley blocks to be used in raising and lowering the craft.

The seven balloons were as follows:

Balloon	Crew	Capacity
Union	5 men	32,000 cu. ft.
Intrepid	5 men	32,000 cu. ft.
Constitution	3 men	25,000 cu. ft.
United States	3 men	25,000 cu. ft.
Washington	2 men	20,000 cu. ft.
Eagle	1 man	15,000 cu. ft.
Excelsior	1 man	15,000 cu. ft.[80]

Three other balloons, *Pioneer*, *North America*, and *General Scott*, have also been identified by later historians as having been employed by the Balloon Corps. If so, they were certainly not included in Lowe's regular inventory.

In addition to the balloons, Lowe constructed twelve portable gas generators, six of which were in service by the spring of 1862. Designed for use with a standard army wagon body, each apparatus consisted of an eleven-foot tank standing five feet high. Internal shelves were provided to enable the operator to distribute the iron filings evenly. Sulphuric acid was added through a funnel on top, while a gastight grating was provided at the rear for removing the residue. The resulting hydrogen was fed through a limewater washer and cooler to remove the acid and cool the gas. A hand-operated pump was used to fill the balloon.

On the march, a four-balloon company required seven wagons in addition to those for the generator, washer, and cooler. Additional wagons carried the iron and glass carboys of sulphuric acid. If the balloons were to be operated with a telegraph apparatus, several more wagons were added to carry five miles of insulated wire and additional tools and equipment.

The wagon crew consisted of an aeronaut (Lowe or one of his assistants), a captain, and fifty noncommissioned officers and enlisted men to serve as wagoners and ground crew. When Lowe's system was in operation, such a crew could have the aeronaut in the air within three hours of the time the wagon train pulled to a halt.

With work on the balloons and generators under way in the fall of 1861,

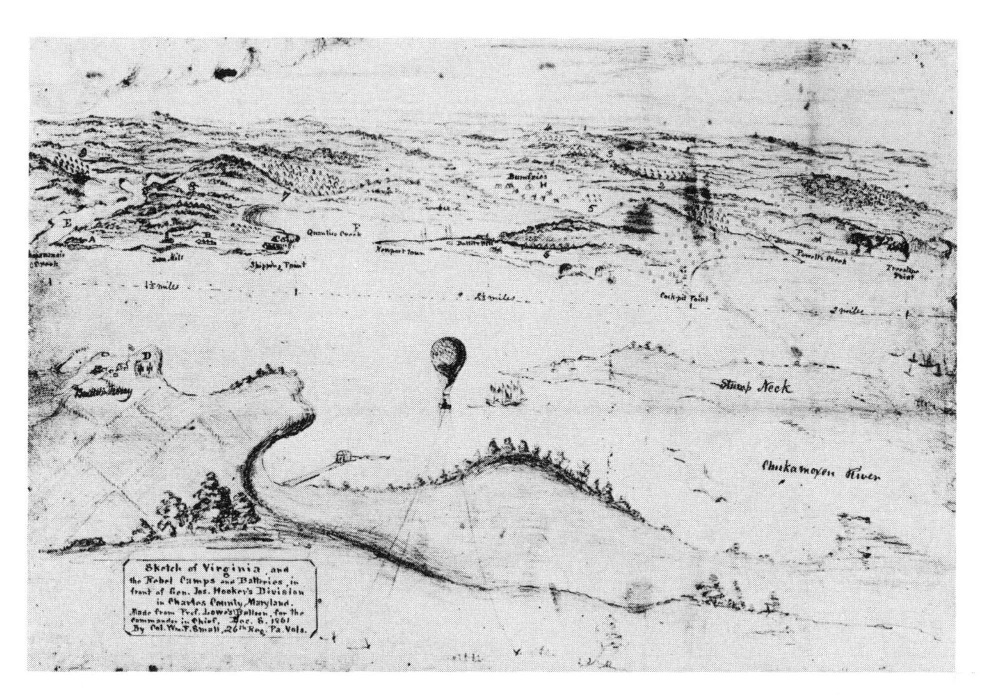

A sketch made by Col. William Small, 26th Regiment, Pa. Volunteers, following an ascent in a Lowe balloon from Charles County, Maryland, on December 8, 1861

Lowe faced the problem of hiring a staff of aeronauts. William Paullin, one of the nation's most experienced balloonists and an old friend with whom Lowe had flown for the Japanese embassy in 1859, signed on board on October 11,1861. Paullin was dispatched in response to Gen. William F. Small's request for a balloonist. He later served with Gen. Joseph Hooker's division at Budd's Ferry on the Lower Potomac.

Paullin was notably successful in pleasing the officers of Hooker's command. General Small was particularly enthusiastic, remarking that Paullin was "entitled to much credit for the skill and zeal displayed in conducting the ascensions."[81] Hooker himself ascended with Paullin and became a convert to the balloon.

In spite of this success, Lowe was forced to dismiss Paullin from the service late in January 1862 when the aeronaut insisted on continuing to pursue his career as an ambrotypist on the side.[82]

John B. Starkweather was another of Lowe's early recruits. Apparently a Boston friend of the aeronaut, Starkweather signed on as a balloonist in mid-

November 1861 and served with Brig. Gen. Thomas W. Sherman's command at Port Royal, South Carolina, before resigning in the summer of 1862.[83]

Ebenezer Locke Mason, who joined the Balloon Corps in December 1861, had followed a variety of careers before the war. Hailing from Troy, New York, he seems to have been acquainted with John La Mountain. His early correspondence with Lowe leaves little doubt that these two men had also been associated before the war. Mason listed himself as a harness maker in Philadelphia city directories, but in a letter to Lowe informing the aeronaut of his availability for service in the Balloon Corps Mason noted that until recently he had been employed as a press agent for "Wyman the Wizard," an Albany magician. Mason's first duty was to supervise construction of the last two Union balloons, *Eagle* and *Excelsior*.[84]

As an aeronaut, Mason was assigned to various points on the Potomac during the winter and spring of 1861 and 1862. When pay for the aeronautical corps proved slow in coming, Mason refused to fly. Lowe, therefore, dismissed him from the corps in spring 1862.[85]

Ebenezer Seaver, the only other aeronaut to join Mason's strike, was dismissed at the same time. Recruited by Lowe in November 1861, Seaver apparently had only limited prewar ballooning experience. Assigned to Fitz-John Porter's divisions, Lowe's old area of operation, Seaver replaced Paullin at Budd's Ferry and participated in the Richmond campaign early in 1862.[86]

In addition to William Paullin, Lowe was eventually able to recruit a number of other very experienced aeronauts. James Allen, having failed to establish himself as head of the army aeronautical program, signed on with Lowe on March 23, 1862. His brother Ezra Allen's service with the Balloon Corps began that January. John Steiner had joined Lowe's operation in December 1861.[87]

Jacob C. Freno, a Philadelphia lawyer with a taste for aeronautics, was recruited in January 1862. Freno had made a number of flights with Lowe and, apparently, Paullin. He had served as Lowe's assistant during the Cincinnati flights of 1861, then enlisted in the 66th Pennsylvania Infantry. Drummed out of the service for cowardice, gambling, and a variety of other crimes, Freno created severe problems for the Balloon Corps. After several months of good behavior, he suffered a lapse, and Lowe was finally forced to dismiss Freno "for repeated absence without leave, for expressing disloyal sentiments, opening a faro bank for the purpose of gambling, and for the demoralizing effect which he had upon subordinates."[88]

Nor was this the end of Lowe's problems with Freno. In March 1863 the disgruntled ex-aeronaut broke into Lowe's Washington storage area and vandalized the balloon *Constitution*. Freno evaded capture by federal officials and

Two members of Lowe's Balloon Corps: left, Ebenezer Seaver; right, John Starkweather (1862).

sent a series of wild charges and accusations against Lowe to the provost marshal general's office.[89]

John R. Dickinson was the ninth and last of Lowe's aeronauts. A mysterious figure, Dickinson's name appears in only one of Lowe's dispatches, when he was reported to be in command of the *Union* at Washington in October 1861.[90]

A number of other aeronauts were rejected for service with the Balloon

Corps. Samuel Archer King, Allen's old friend and partner, was among this group, as were Alexander J. B. DeMorat and Edward La Mountain, John La Mountain's brother.

In addition to the aeronauts, the roster of the Balloon Corps included nonflying personnel. Lowe's father Clovis served as a balloon repairman and assistant to several of the aeronauts. John O'Donnell relieved Lowe of paperwork, while Richard Brown managed the balloon boat *George Washington Parke Custis*, which Lowe employed on the Potomac during the Peninsular Campaign. Civilian telegraphers and teamsters were also employed by the corps.

A variety of officers, noncoms, and enlisted men were detailed to Lowe from 1861 through mid-1863. One of the chief aeronaut's major administrative problems throughout the history of the corps related to the need to maintain a permanent cadre of trained military personnel.

The status of Lowe and his aeronauts within the military establishment was never too clear. Capt. Frederick F. E. Beaumont, the English observer, remarked that Lowe was chief aeronaut, "whose exact rank I could never quite make out, but it was not lower than a captain, or higher than a brigadier. He was a civilian and by profession an aeronaut. He was very highly paid, the same as a brigadier, and as military rank, I believe, in America, is in some way attached to and determined by, the pay received, I fancy Prof. Lowe must have been a brigadier; at any rate he was a very clever man, and indefatigable in carrying out his work."[91]

In fact, Lowe's pay of $10 a day was just a bit more than that of a full colonel. John Steiner, Ebenezer Seaver, and John Starkweather, all of whom Lowe detailed to distant commands, earned $5.75, while Allen, who served as Lowe's assistant, was paid $4.75. Ezra Allen, Mason, Freno, Paullin, Clovis Lowe, and O'Donnell each received $3.75.[92]

Administratively, Lowe's Balloon Corps moved from bureau to bureau like an unwanted child. Originally the responsibility of the Bureau of Topographical Engineers, the aeronauts later came under the supervision of the Quartermaster Corps, the Corps of Engineers, and finally the Signal Corps. During most of this period, Lowe's activity was controlled by a number of field commanders, in addition to the Washington administrators, so that he was often uncertain as to whom he was responsible. As a result, Lowe frequently operated as he saw best. This attitude did little to endear him to military officials attempting to regulate the balloonists.

In the winter of 1861, however, these problems still lay far in the future. With his new balloons on hand and his corps of aeronauts formed, the nation's newly appointed chief balloonist looked forward to inaugurating a new era in

American military history.

At this moment, just as Lowe was struggling to demonstrate the value of the Union Army's investment in the balloon, John La Mountain appeared like a dark cloud on his horizon. La Mountain had remained an independent civilian aeronaut, with no official relationship to any military bureau or to Lowe's organization. His only responsibility was to Gen. Benjamin F. Butler, commander of Fort Monroe, the man who had hired him.

La Mountain had compiled an enviable record under Butler's command. Still flying the old *Atlantic*, constructed of bits and pieces salvaged from the giant aerostat that had carried the aeronaut from St. Louis to Henderson, N.Y., and then into the wilds of Canada, La Mountain had made repeated ascensions to observe Confederate positions beginning on July 31, 1861. As early as August 3, he had made tethered flights from the deck of the steam tug *Fanny*.[93]

While La Mountain failed to warn of the movement of Confederate troops advancing toward Hampton Roads on August 7, he did observe Rebel strength on the James River and noted Confederate shipping in the area. Butler flew with La Mountain, made use of his sketch maps of Southern positions, and listened with interest to his plans for a large balloon that could be used to "shell, burn, or destroy Norfolk or any city near our camps."[94]

La Mountain brought his career at Fort Monroe to a close with a night ascension on August 10. Having used up all his acid and iron, he obtained Butler's permission to return to Troy, New York, to obtain his large balloon *Saratoga* and a specially designed hydrogen generator that operated on the basis of the decomposition of water.

Prior to his departure, La Mountain had submitted an invoice for $1,200 and obtained Butler's signature on his proposal for a balloon to be employed as a bomber. When the aeronaut returned to Fort Monroe on September 12, he discovered that General Butler had been replaced by Gen. John E. Wool. La Mountain was horrified to discover that Butler had neither briefed Wool on the department's aeronautical program nor submitted the balloonist's invoices. Wool immediately dispatched La Mountain to the War Department in Washington with a request for instructions. Secretary Cameron, having recently received Gen. William F. Smith's request for a balloonist, decided to reassign La Mountain to Smith's command.[95]

George McClellan, "Baldy" Smith's superior and a firm supporter of T. S. C. Lowe, saw the necessity of cooperation between the newly appointed chief aeronaut and John La Mountain. McClellan instructed Fitz-John Porter to interview both aeronauts to determine whether they would be willing to cooperate. Porter regarded the meeting, which took place in Lowe's tent at Fort Corcoran, as a success. The three men agreed that while La Mountain would remain an independent employee not subject to Lowe's control, he would cooperate with the aeronauts of the corps if that became necessary. The newcomer would be carried on the rolls of the Bureau of Topographical Engineers at a salary matching Lowe's, $10 per day. He would be assigned, not to "Baldy" Smith, but to William B. Franklin's division at Cloud's Mill, Fairfax County, Virginia.[96]

During the course of this meeting, La Mountain unveiled his plan for a series of free flights over the Confederate lines, using currents of air blowing in different directions at different altitudes to move back and forth across enemy territory. Lowe, as a result of his one unfortunate try at a free flight on June 24, was dubious. Porter, however, saw the value in conducting two aeronautical experiments, the Balloon Corps tethered operations and La Mountain's one-man foray into free-flight observations.

La Mountain made his first trial flight with the balloon *Saratoga* on October 4. The trip was an experiment. The aeronaut made no attempt to cross Southern lines but flew over Washington to a landing near Beltsville, Maryland. La Mountain conducted captive operations over the next ten days, while Lowe was still in Philadelphia arranging for the construction of the new balloons. He made his first free reconnaissance over the Confederate lines near Camp Williams on October 15.[97]

La Mountain was back in the air for a long flight from Cloud's Mill on October 18. Passing over Confederate territory at 1,400 feet, as he later informed a reporter, he "saw distinctly every Confederate position between the [Potomac] river and the Blue Ridge."[98] He described troop concentrations at Fairfax Station, Manassas, and Centreville, saw gun batteries on Aquia Creek, and noted the movement of several trains.

Approaching Union lines, the aeronaut was "disagreeably saluted" by a volley of gunfire from the Union troops of Gen. Louis Blenker's command. Bullets cut sections of the network, pierced the lower part of the envelope in several places and narrowly missed the aeronaut's head. Once on the ground, he was further threatened with bayonets and his balloon was vandalized. In spite of General Blenker's insistence that his troops were innocent, McClellan issued stern orders to protect both of his aeronauts from friendly fire.[99]

La Mountain continued to build his reputation over the next few weeks. He was assigned a permanent detachment of forty men, and his two balloons, *Atlantic* and *Saratoga*, were purchased by the Army.

La Mountain, unlike Lowe, usually flew alone. One exception to this rule came in mid-November, when a reporter accompanied the aeronaut on a voyage over enemy lines. The resulting account is worth reprinting in full as a rare first-person account of a La Mountain free flight:

> *Being at his [La Mountain's] quarter, he kindly invited me to accompany him on a reconnaissance. Having faith with him in that ever-present upper current towards the east, I did not hesitate, though I knew that the lower current towards the west would in a very few minutes carry us far over inside the enemy's lines. Stepping into the car with him, he cut loose, and in a moment, as it were, our great army lay beneath us, a sight well worth a soul to see—brown earth fortifications, white tented encampments, and black lines and squares of solid soldiers in every direction. So enchanted was I with the scene that I well nigh forgot that we were drifting enemyward, until Fairfax Court House lay beneath us, and I had my first sight of the enemy, in the roaming squads of rebel cavalry visible in that vicinity. Soon Centreville and Manassas came in full sight, and there in their bough huts lay the great army of the South. All along they stretched out southeasterly toward the Potomac, on whose banks their batteries were distinctly visible. So plain were they below, their numbers could be noted so carefully that not a regiment could escape the count. . . . I do not have time at present to give you a full description as I would like, of that great scene, a scene that would furnish material for a long letter; enough to say that it was superlatively grand and interesting.*
>
> *The Professor, satisfied with the reconnaissance, as well he might be, after noting down the strength of the forces and their position, discharged ballast and started for that higher current to bring us back. Now I acknowledge I looked anxiously and (I am sure I was excusable) nervously for a backward moment, conscious that to come down where we were was death, or at least the horrors of a Richmond tobacco prison. Up, up we went, but still bearing west and south. I looked at the Professor's face. It was calm and confident, so I felt assured that all was right. That assurance became a settled thing, when in a few moments we commenced passing gently back to the east.* We had struck the Professor's current. *Back, back we went, as though a magnet drew us, until our own glorious stars and stripes floated beneath us, and we came down gradually and smoothly into the encampments of General Franklin's division, where we were surrounded by enthusiastic soldiers who had been excitedly watching the trip.*[100]

La Mountain's period of successful operation was short-lived however. On November 16 the *Saratoga* was lost when inexperienced ground handlers allowed the balloon to escape in a high wind. Aware that he could not continue to make observations using only the ancient *Atlantic* gasbag, the aeronaut began to cast a covetous eye on the new balloons Lowe was storing in Washington until they could be distributed to his new aeronauts serving with other commands.[101]

The truce between La Mountain and Lowe had been shaky at best. T. S. C. Lowe, having overcome the challenge of all other balloonists seeking the office of chief aeronaut, had no intention of knuckling under to La Mountain.

La Mountain's request to General Franklin for one of Lowe's new aerostats was accompanied by charges that the chief aeronaut was hoarding materials required for the full prosecution of the war effort. Franklin passed his aeronaut's message on to McClellan's headquarters with the obvious comment that La Mountain would be of little value without a new balloon.[102]

While McClellan's staff was considering the situation, La Mountain continued to make free flights with the *Atlantic*. His activity drew much attention in the northern press. Southern commanders also seem to have believed that La Mountain's activity posed a greater threat than the tethered flights of the new Balloon Corps. Joseph E. Johnston complained of the "infernal balloon" overflying his troops at Centreville, while James Longstreet urged that steps be taken to prevent the Yankees from taking full advantage "by floating balloons over our heads."[103]

Lowe was alarmed by both the flood of publicity surrounding La Mountain's effort and by the new threat to his balloons. On December 18, 1861, he forwarded clippings that portrayed La Mountain's success in extravagant terms, together with his own complaints about his rival's claims and a direct refusal to turn any of his aircraft over to La Mountain. He particularly objected to an article, "The New Aeronautic Department under Professor La Mountain," which appeared the following month in the *New York Herald*. The article extolled the contribution of the new Balloon Corps in the Bureau of Topographical Engineers. The reporter noted that John La Mountain was the most successful of the Union aeronauts and suggested that General McClellan and Maj. John Macomb of the Topographical Engineers were about to name La Mountain to head the entire effort. Lowe was not even mentioned.[104]

Lowe objected that such articles not only damaged his own reputation but endangered the Balloon Corps by raising unrealistic expectations. In his letters to McClellan and others, including Col. A. V. Colburn of the adjutant general's office, Lowe argued that all aeronautical activity should be placed under his control.[105]

As in the case of his earlier confrontation with John Wise, Lowe's attitude toward La Mountain was a continuation of prewar feuding. La Mountain, whom Wise had regarded as an upstart, in his turn viewed T. S. C. Lowe as a Johnny-come-lately. When his own plans for a transatlantic flight were quashed by circumstances, La Mountain had attacked Lowe and the *City of New York* with ridicule and sarcasm. As we have seen, Wise, who had also ridiculed Lowe, had eventually considered cooperating with his young rival. La Mountain had refused to consider such a step. Their prewar conflicts unresolved, Lowe and La Mountain found it impossible to cooperate for the good of the Union.

As the controversy built toward a climax, McClellan clearly took Lowe's side. His first communication to General Franklin on the subject included an offer to replace La Mountain with one of Lowe's assistants. Franklin responded by commenting that La Mountain "appears to work energetically" and had "done as much and as intelligently as any balloonist could."[106] McClellan, recognizing that Franklin's enthusiasm was lukewarm at best, then informed La Mountain that "all balloons shall be under the superintendence of Mr. Lowe. Upon this basis if you can come to an understanding with Mr. Lowe, it may be of interest to yourself and the service."[107]

La Mountain launched a second effort to capture one of Lowe's balloons in February 1862. Having failed to drum up support among the officers of Franklin's or McClellan's staffs, he now appealed directly to Major Macomb, who had replaced Whipple as the responsible balloon official in the Bureau of Topographical Engineers.

Macomb, ignoring McClellan's preference for Lowe, ordered the chief aeronaut to turn one of his balloons over to La Mountain. When Lowe refused to accept this order, Macomb responded with a stronger note, commenting that "this matter admits of no further delay."[108] La Mountain returned a second time to inform Macomb that "Mr. Lowe refused to recognize authority and would not obey."[109]

Lowe attempted to discuss the issues involved with Macomb but was refused an audience. He then sent Macomb a long letter, explaining that the two balloons in question, the *Intrepid* and a smaller craft, were required for immediate service elsewhere. This was certainly true. Moreover, Lowe called Macomb's attention to the fact that he had explained the situation to La Mountain. Compliance, he added, had simply not been possible. Finally, Lowe attacked his rival in the most direct fashion:

> *[La Mountain was] . . . a man who is known to be unscrupulous, and prompted by jealousy or some other motive, has assailed me without cause through the press and otherwise for several years . . . he has tampered with my men, tending to a demoralization of them, and, in short, has stopped at nothing to injure me . . . so much so that it is impossible for me to have any contact with him, as an equal in my profession, with any degree of self respect. . . .*
>
> *This man La Mountain has told my men that he is my superior, and is considered to be the Commanding General . . . he says that he is paid by Lieu-*

> *tenant Colonel Macomb two hundred dollars more per month than I am paid. . . . I do not think that I should serve this man by giving him possession of my improved balloons and portable gas generator for his examination . . . yet if it is the desire of the General that I should do so, I will most cheerfully comply . . . Without the improvements that I have made in the manufacture and management of balloons, they could be of little service to the Government; add to that my invention of the portable gas generator, which I am using for the benefit of the Government Service. I submit that I should not be interfered with in the management of this matter, at least until I have instructed men in the use of my invention.*[110]

Lowe's arguments were persuasive. On February 19, 1862, McClellan overruled Macomb and ordered that La Mountain be dismissed.[111]

Unknown to Lowe, yet another rival aeronaut was challenging his authority over the balloon program during the fall and winter of 1861. William H. Helme, the Providence, Rhode Island, aeronaut who was reported to have enlisted with James Allen, wrote to General McClellan on October 18, 1861, suggesting the use of hot air balloons for reconnaissance. Helme pointed out that such a large Montgolfier could be inflated in a matter of minutes and would be much easier to transport and maintain. Describing himself as an "amateur aeronaut," Helme offered to construct a demonstration balloon for $500.[112]

McClellan ordered Macomb to approve Helme's project. The aeronaut was to stay within his budget and submit all bills for approval to the proper authority.

The work proceeded quickly. Helme had completed the envelope and attempted his first inflation with an alcohol burner in late November. The crew failed to hoist the balloon high on the support poles on this occasion, causing the bag to be scorched. This necessitated an additional three weeks for repairs.

By December 18 Helme had conducted two successful inflations. Seven to eight gallons of alcohol had been required on each occasion, he informed Macomb, estimating that, with an improved burner, ten gallons of alcohol an hour would be sufficient to inflate and maintain the balloon.

Helme advised Macomb that he would be ready to provide a final demonstration by December 20. There is no indication that such a trial flight was ever made or that any further consideration was given to the purchase of a hot air reconnaissance balloon. Helme, whose activity was apparently not revealed to Lowe, was the least threatening of the rival aeronauts.[113]

The conflict between Lowe, Wise, Allen, La Mountain, and Helme demonstrated one of the most important weaknesses of the early federal balloon program. From the outset, the nation's leading aeronauts had devoted almost as much energy to infighting and jockeying for position as they had to the observation of the enemy. Their unwillingness to mount a cooperative effort toward the achievement of a common goal during the critical early months of the war had done little to build the confidence of military officials in the utility of the balloon. Bache, Whipple, McClellan, Macomb, Franklin, and others had first been confused by the plethora of fiercely independent aeronauts, then exasperated by the amount of time and energy required to resolve their quarrels. It was an inauspicious beginning to the first American experiment in military aeronautics.

Chapter 13 The Civil War Aloft: Operations

BY MARCH 1862 LOWE HAD ACHIEVED HIS GOAL. HE HAD OVERCOME HIS rivals to become the proud commander of a balloon corps that already included seven aircraft, several gas generators, and eight trained aeronauts (Paullin having left the corps in January). If the position of his unit within the administrative structure of the Army of the Potomac remained uncertain, Lowe felt sure that the value of his services would be so self-evident as to enable him to carve an important niche for himself and his men.

The five-month period from November 1861 to March 1862 was a time of consolidation and training for the Balloon Corps. Six portable generators, constructed at the Washington Navy Yard, were delivered to the corps by the end of 1861. Navy Yard carpenters also refitted the *George Washington Parke Custis* as a balloon boat during these months. The *Custis*, originally purchased as a coal barge, had been assigned to Lowe by Secretary of the Navy Gideon Welles. Measuring 122 feet in length and with a 14½-foot beam, the ship was unpowered and had to be operated in conjunction with a steam tug. Lowe had ordered the hull completely decked over, with the exception of a small shack for the aeronaut on the stern. Fitted with one of the portable generators, the barge could be used to move a balloon to any point on the Potomac system.[1]

The *Custis* was first employed in November 1861 when Lowe transferred Paullin and the balloon *Constitution* down the Potomac to Budd's Ferry, Maryland, near Mattawoman Creek. Gen. Daniel Sickles made observation flights at Budd's Ferry, after which the balloon was moved overland to operate in conjunction with Gen. Joseph Hooker's command.[2]

The *Custis* also provided support for the *Intrepid* and the *Union*, which were operated by Lowe and Ebenezer Seaver under Gen. Charles P. Stone's command at Edward's Ferry beginning in mid-December. By this time John Starkweather was en route to join General Sherman's command at Port Royal, South Carolina. Lowe and E. L. Mason continued to operate in the Washington area with the balloon *United States*.[3]

With the balloons operating in different commands, a number of general officers had their first opportunity to explore the possibilities of the balloon. Gen. Joseph Hooker found his early experience with Paullin and Clovis Lowe to be less than satisfactory. Writing to Lowe on November 30, 1861, Hooker stated his disappointment: "The history of the balloon here is one of accidents and failures. I almost despair of being able to turn it to any account at this season of the year. Hitherto we have been prevented from taking satisfactory observations from the windy or smoky state of the atmosphere, and now we have no gas. In the bonds of its present manager I apprehend that we will find it of little or no service."[4]

The situation improved when Lowe provided Paullin with additional supplies of acid and iron so that operations could resume. By early December, Gen. William F. Small was advising General Hooker that "Mr. Paullin, the aeronaut in charge of the balloon, is entitled to much credit for the skill and zeal displayed in conducting the ascensions and seconding my efforts to comply with your directions."[5] A month later, after Seaver had replaced Paullin, even Hooker was making important decisions on the basis of his own observations from the balloon.

John Starkweather faced even more serious problems with the *Washington* at Port Royal, South Carolina. Gen. Thomas W. Sherman had so little interest in the balloon that no flights were made until Sherman was replaced by Gen. Henry J. Benhorn in April 1862.[6]

John Steiner's experience was typical of the problems faced by the aeronauts that Lowe detached for independent service with other commands. McClellan had ordered Lowe to detail one of his aeronauts for service with Gen. Henry W. Halleck, commander of the Department of the Missouri. Reporting to Halleck's chief of staff, Gen. George W. Cullum, at Cairo, Illinois, on February 24, 1862, Steiner found that he was completely ignored. Unable to obtain either supplies or pay, the aeronaut deluged Lowe with pleas for rescue written in his broken English. "I can not git eny assistance here," he complained. "They say they know nothing about my balloon business . . . they even laugh ad me . . . Let me hear from you as soon as possible and give me a paper from Headquarters to show theas blockheads hoo I am."[7] A month after his arrival in Illinois, things had only gotten worse. "All the officers hear are as dum as a set of *asses*," he informed Lowe.[8]

The chief aeronaut discussed Steiner's problem with McClellan, who expedited the German's orders to Halleck. Even this was of little help. Steiner was still complaining after a conference in St. Louis that April: "I came here but I

The balloon Washington *being operated from the balloon boat* George Washington Parke Custis finde that no such papers arrived here. I told the General here how I was situated and ounder wot sircumstands I bin send out here. He loack ad me a moment and then sed I can doo nothing for you and walf off; and such has bin my treedment every where."[9]

Steiner was finally able to get into the air in late March, no thanks to the U.S. Army. Commodore Andrew H. Looe, commanding the naval guns and mortar boats attacking the Confederate fortifications on Island No. 10 in the Mississippi, provided Steiner with a launch platform aboard a flatboat so that he could observe Gen. John Pope's advance and adjust mortar fire against the defenders. Operating from March 25 to March 27, 1862, the aeronaut carried several artillery spotters aloft. In spite of a general consensus that the aeronaut had played an important role in the attack on Island No. 10, the Department of the Missouri continued to ignore Steiner. "I am here like a dog wisout a tail and I dond know ware I will be abel to draw my pay, for no one seams to know eny thing abought this thing," he remarked in yet another appeal to Lowe in March. "I am treed wis contempt and if I had the means to return to Washington I would strait today . . . now that I can git no pay out here."[10]

Henry Halleck refused to consider Steiner's suggestions for the employment of his services. "I am satisfied that General Halleck is no friend to the Aeronautics Core," Steiner informed Lowe. "I could have bin of grade servis at Cornis [Corinth] and explained it to General Halleck at Pittsburgh Landing, but he told me to stay ad Cairo ontill he woulde send for me."[11]

Halleck's attitude toward Steiner was probably the result of a number of factors. McClellan's imposition of the aeronaut's services on the Western Department without consulting Halleck was certainly one of the most important of these.

While John Starkweather and John Steiner were struggling to overcome the indifference of military commanders in the South and West, T. S. C. Lowe and other members of the Balloon Corps were preparing to take part in a great Union offensive against Richmond. General McClellan had originally planned to move four army corps commanded by Generals McDowell, Sumner, Heintzelman, and Keyes south by water to Fort Monroe. Using the fort as a strong point and supply base, he would then push up the peninsula separating the York and James rivers and move toward the Confederate capital.

These plans were altered when President Lincoln, fearing the movement of "Stonewall" Jackson's "foot cavalry" up the Shenandoah Valley to threaten Washington, ordered McDowell's division to remain in the defensive perimeter around the capital. McClellan, thus deprived of one-quarter of his force, faced other problems as well. As Lowe remarked, the spring of 1862 "was a season of terrible storms and incessant rains and never before nor since in my long life have I known a worse one."[12]

In spite of these difficulties, McClellan moved an army of 125,000 men, 21,500 animals, 44 gun batteries, and ambulances, food, ammunition, and other impedimenta 200 miles aboard 400 steamers and sailing vessels.

On March 23, 1862, T. S. C. Lowe and the Balloon Corps were ordered to hold themselves in readiness for transport to Fort Monroe and service in the Peninsular Campaign. This was a distinct change for the aeronauts. Their weeks of static observations from the fortifications surrounding Washington were at an end. Now they were to become a part of a war of movement, serving as the eyes of an advancing army.

Arriving at Fort Monroe with his balloons, generators, and aeronauts aboard the balloon boat *Custis*, Lowe was ordered to move out on April 3 and accompany Gen. Fitz-John Porter's division during the advance up the peninsula.

With its flanks protected by Union gunboats on the James and York, McClellan's troops had advanced as far as Yorktown by April 5. On that day the Confederate Army of the Peninsula under the command of Gen. John B. Magruder was joined by Gen. Joseph E. Johnston and the Army of Northern Virginia. The combined force was arrayed from shore to shore at Yorktown, blocking McClellan's advance.

Lowe arrived in the forward lines with four army wagons, a balloon, and two portable generators shortly after noon on April 5. After taking shelter during an enemy artillery barrage, the chief aeronaut inflated his craft and made his first ascension with one of Porter's staff officers at 5:20. General Porter and Daniel Butterfield went aloft a number of times during the next several days and ordered draftsmen into the air to prepare sketch maps.

Having established the routine of balloon observations in the forward area, Lowe placed James Allen in charge and returned to Fort Monroe to retrieve a second balloon destined for Gen. Erasmus Keyes at Warwick Court House. Returning to Yorktown on the night of April 10, Lowe became confused by the shifting lines and narrowly escaped capture by Confederate pickets.[13]

Back on the road to Yorktown at 6:30 the next morning, he was startled by the descent of a balloon very near him. The aeronaut hurried to the scene in time to see Fitz-John Porter climb from the car. General Porter informed Lowe that he had ridden into the balloon camp at 5:00 A.M. that morning and ordered James Allen to prepare the balloon for a flight. Throwing caution to the winds, Allen attached only a single tether line, rather than the three that were normally employed. As luck would have it, that line had been damaged by an acid spill and broke when General Porter was fifty feet in the air.

Unable to reach the tangled valve cord, Porter noted that he was drifting over the Confederate lines and chose to make the best of a bad situation. "I took good observations, some notes, but mainly instantaneous impressions like a photographic instrument, and had the enemy's position and defences so grafted on my mind that when I descended I was able to give a good sketch of everything." One of the watching soldiers commented that if the general "had been reconnoitering from a secure perch at the tip of the moon, he could not have been more vigilant, and the Confederates probably thought this some Yankee device to peer into their sanctuary in despite of ball or shell."[14]

After completing his hasty observations, Porter climbed into the netting, untangled the valve line, and descended into a current of air blowing back toward the Union camp. Unfortunately, the novice aeronaut had released so much gas that when Lowe first spotted the balloon it was descending so rapidly that the lower half of the envelope had gathered in the upper hemisphere, forming a parachute.

Horsemen, reporters, and soldiers poured toward the landing site from every direction. Porter was safe, but, as Lowe noted, "I found it difficult for a time to restore confidence among the officers as to the safety of this means of observation."[15]

Lowe soon returned to a cycle of regular observations conducted at several points along the Yorktown line. The value of these flights was as apparent to Confederate as to Union commanders. Anxious to deprive the Yankees of their bird's-eye view of Southern positions, Confederate artillerists under the command of E. P. Alexander trained their fieldpieces on the balloons at every opportunity. One war correspondent reported: "It came at length to be our principal amusement in camp to watch the rebels fire at the balloon, as it sailed tranquilly above our picket line, and I have seen many a dollar staked by the 'boys in blue' on the skill of the grey-coated artillerists. It was laughable to watch these bets, and I think I shall not go far astray from the truth, when I say that some very good patriots would have been glad to see the balloon struck, since it would have enabled them to win their wagers."[16]

Writing to Lowe many years later in 1909, Union veteran Lt. Col. W. J. Handy recalled the scene as Confederate gunners fired on Lowe's balloon from behind the red earthen walls of a fortification that had once sheltered the troops of Lord Cornwallis at Yorktown:

> *I was on Grand Guard duty opposite the "Red Fort" in the woods May 2, 1862, when we saw the balloon about to rise, then commenced a heavy cannonading from the Confederate works, shots went over our head, tearing big branches from*

Maj. Gen. Fitz-John Porter

Lowe's portable hydrogen generating system in operation

> *the trees. We were safer inside the line of fire than closer to our own camp. The balloon rose and the firing was soon directed at this Air Target. Shot after shot, shells exploding way up, occasionally the sharp crack of a rifle would be heard when their sharp shooters took a chance shot, kept up the work for half a day. No damage was done, as I remember, except slaughter of fine old trees and great holes in the ground where solid shot struck.*[17]

Another correspondent noted that the balloons never failed to send the Confederates into "paroxysms of rage." In addition to directing rifle and shell fire at the balloon, the frustrated Southerners "cried at it in derision."[18]

Confederate attempts to destroy the balloon were occasionally more damaging to their own troops than to the Yankee aeronaut. On the afternoon of May 2, one of the large Confederate Armstrong guns was elevated to such an extreme angle and so overcharged that it exploded, filling the interior of the "Red Fort" with dust and bits of flying metal.[19]

The Intrepid, *circa May 1862*

Alexander's gunners eventually resorted to digging shallow trenches behind the gun line. When the balloon appeared, the fieldpieces would simply be pushed a foot or so to the rear for additional elevation. Artillerists also learned to ignore a balloon at altitude, aiming their fire at a lower elevation in an attempt to catch the aeronaut during the vulnerable moments of take-off and landing.

None of these expedients proved successful. During the entire course of the war no balloon on either side was ever struck by artillery fire. Still, the balloons, which an officer serving on Jeb Stuart's staff described as hanging "like oranges in the sky," remained irresistible targets. There were a few close calls. During the Battle of Fair Oaks, a battery of rifled guns sent shells whistling through Lowe's rigging.[20]

For the most part, however, shell fire remained far more dangerous to watchers on the ground. Lowe occasionally had to construct shelters for his ground crews. In late April 1862 Porter found it necessary to order Lowe to cease operations during a short period because Confederate artillery fire directed at the balloon was falling among Union troops. Gen. George Stoneman complained of the danger to a corral filled with precious cavalry remounts under similar circumstances. Generals Stoneman, Heintzelman, and McClellan all came under artillery fire while visiting the balloon sites during the Peninsular Campaign.

A New York sergeant described an occasion on which shells aimed at Lowe's balloon fell on a cookshack at Gen. Henry Slocum's headquarters, "scattering camp kettles and cooks, who were just then preparing breakfast."[21] In February 1863 David Hogan of the 13th New Hampshire Infantry was walking sentry duty near Falmouth, Virginia, when a shell intended for T. S. C. Lowe landed in a neighboring cesspool, scattering "about two cartloads of the vile contents for rods around, nearly burying Hogan out of sight." The frightened soldier was rescued unhurt, but, as a messmate noted, "his clothes and appetite are utterly ruined."[22]

Perhaps sabotage would succeed where artillery fire had failed. In March 1862 the Union camp was buzzing with rumors that Confederate spies had been ordered to destroy the balloon. This story may have been more than rumor. An anonymous "ex-Confederate" interviewed by the *Detroit Free Press* in 1886 claimed that he had been one of a party of five men promised $1,000 and a commission as a second lieutenant if they could damage the balloon.[23]

The men traveled separately. The "ex-Confederate," disguised as a lame peddler of tobacco and notions, reached Lowe's camp a full week after he had crossed his own picket line. He discovered that two of his fellow saboteurs had already been captured. So many guards had been assigned to the balloon that he was forced to bide his time.

Still in the camp on the morning of Fitz-John Porter's involuntary free flight, the anonymous spy was one of those on hand when the general stepped from the basket. In the excitement of the moment, he very nearly accomplished his mission:

I had matches, and I had only to destroy the balloon at a flash. I meant to take

Lowe views the Confederate camps behind Yorktown, spring 1862.

> *every risk, but as I drew a match from my pocket, having a filled pipe already in my mouth as an excuse, a big sergeant who stood beside me seized me by the neck, and shouted at me: 'You infernal idiot! Do you want to fire the balloon?' I did, but he had deprived me of the opportunity. Some of the men laughed. Some said I ought to be kicked out of camp, and prudence whispered to me to take myself off while I had the chance. Only two of us out of the five got back to our regiments. What became of the others was a mystery we were never able to clear up.*[24]

T. S. C. Lowe could have enlightened the would-be saboteur as to the fate of his comrades-in-arms. They were shot.

By mid-April 1862 the Confederates had their own makeshift balloon in operation. It was a poor thing, "nothing but a big cotton bag, coated over so as to make it airtight, and intended to be inflated with hot air, as gas was not a thing to be had in those days and in those places."[25]

It seems likely that this was the same balloon that had made brief appearances over the Confederate camps in Northern Virginia the previous winter. Gen. Joseph E. Johnston had brought the craft with him when his troops departed Manassas for the peninsula.

When first spotted hovering above Yorktown on April 13, the balloon caused more amusement than fear in the Union camps. Lowe noted that it had "neither shape nor buoyancy, and predicted that it would burst or fall apart after a week." The Confederate balloon proved far more successful than Lowe supposed.

While the chief aeronaut was probably never aware of the fact, his rival on the other side of the lines was Capt. John Randolph Bryan. Twenty-one years old at the time, Bryan had been serving as aide-de-camp to Gen. John B. Magruder when he volunteered for service aboard the makeshift hot air balloon. General Johnston was more than pleased with his new aeronaut, who was thoroughly familiar with the peninsula south of Yorktown. Armed with a semaphore flag and instructions on the sort of information required, Bryan proceeded to the launch site hidden in a small grove of pines roughly half a mile behind the line.[26]

Arriving on the scene, Captain Bryan found a small party of soldiers clustered around the deflated balloon. Several thousand feet of stout line lay coiled, "sailor fashion," on the ground. Passed around a windlass, the rope was tied directly to the network of cords supporting the "good-sized hamper or basket, in which I was to stand or kneel to make my observations."[27] A generous supply of pine knots, which would be soaked with turpentine to supply the heat for inflation, had been piled in the center of the launch area.

Bryan made his first flight on April 13. He had scarcely risen above the treetops when he noticed an enormous commotion in the Union lines. Artillery officers were gesturing frantically as their gunners worked to elevate their fieldpieces for a shot at the balloon. Signaling for more altitude, Bryan rose out of the narrow danger zone close to the earth. "A wonderful panorama spread out beneath me," he noted. "Chesapeake Bay, the York and James Rivers, Old Point Comfort and Hampton, and the fleets lying in both the York and the James, and the two opposing armies lying facing each other."[28]

The captain quickly sketched a map of the Union positions, noting concentrations of infantry and cavalry and the location of artillery batteries and wagon trains. It was difficult work. This was a hot air balloon, so small it could scarcely lift the aeronaut. The weight of the three lines that Lowe used to keep his balloon from twisting and turning in the air were a luxury the Confederates could not afford. As a result, Bryan spun slowly around at the end of his single line, hurrying to complete his sketch map before the air in his envelope cooled.

Safely back on the ground, Bryan delivered his map to General Johnston. Ordered to transfer the balloon and crew to an area close to Yorktown, the captain made a second ascent a day or two after his first flight. On this occasion he employed a team of six horses to lower the balloon more rapidly.

Several nights later, Bryan was awakened by a courier ordering him to make his first night ascent. General Johnston fearing that McClellan's troops would attack the next morning, asked his aeronaut to discover which approach the blue-clad troops might take. Confederate soldiers, fascinated by the balloon, gathered in considerable numbers to witness this night launch. All went well until Bryan was 200 feet in the air. At that point, one of the bystanders stepped into a coil of rope. Seeing his friend being drawn screaming toward the windlass, another soldier seized an axe and cut the line, releasing poor Bryan on his first free flight.

Blown back and forth between the lines, the aeronaut was relieved to find himself finally coming to earth near the camp of the 2nd Florida Regiment. His relief was short-lived. "They turned en masse, and believing me to be a Yankee spy, followed me on foot, firing at me as fast as they could. In vain I cried to them that I was a good Confederate."[29]

Bryan overflew the enraged Floridians, narrowly escaping a dunking in the York River and landing in an orchard on the Williamsburg side. Requisitioning a horse, he returned to Johnston's headquarters. "The information which I was able to give General Johnston as to the roads upon which the enemy were now moving enabled him to prepare for an attack which was made by them early the next morning just before day. I was among those who awaited their approach, and you will pardon me if I say that it gave me no little satisfaction to aim my rifle at those who had so recently taken a wing shot at me."[30]

While Bryan was struggling to make observations from his shaky aerial platform, Lowe continued his own regular series of ascensions in the face of increasing enemy artillery fire. His observations now had a special importance. After enduring a month of rain and mud in the trenches before Yorktown, McClellan was at last prepared to move against the city. The daily aeronautical reports had proved invaluable in spotting artillery targets for a barrage that was to precede the attack on May 4.

The *Constitution* at Warwick Court House and the *Intrepid* at Yorktown were kept in the air as much as possible during this period. "Baldy" Smith, now commanding the 2nd Division of Keyes's corps, assigned a junior lieutenant, George Armstrong Custer, to part-time duty as an observer attached to the balloon being operated by James Allen at Warwick Court House.

Custer, "Cinnamon" to his friends because of the brand of hair oil with which he coated his flowing locks, was already a flamboyant personality. One member of Smith's staff noted that he dressed "like a circus rider gone mad."[31] This young man was also one of the most daring officers in the Union cavalry. When General McClellan and his corps commanders rode up to the Chickahominy River for the first time, the army chief inquired about the depth of the stream. When it became obvious that no one could answer the question, Custer plunged into the river, rode to the middle, turned, and shouted back, "That's how deep it is, General."[32]

Custer, the superb horseman, was much less certain of himself a thousand feet in the air. Ballooning was, he noted, "a kind of danger that few persons have schooled themselves against, and still fewer possess a liking for."[33] Nor had Fitz-John Porter's inadvertent free flight over the lines done much to build Lieutenant Custer's confidence in the venture. In short, Custer accepted Smith's order with "no little trepidation."[34]

Nevertheless, in late April 1862 Custer, armed with field glasses, pencil, and a notebook, presented himself at the Warwick Court House balloon camp. Asked if he wished to make a solo ascent, Custer was aghast. "My desire, if frankly expressed, would have been not to go up at all; but if I was to go, company was certainly desirable. With an attempt at indifference, I intimated that he might go along."[35]

As the balloon rose into the air, James Allen stood in the basket with his hands on the load ring. Noting that Custer was seated on the bottom of the basket, the aeronaut invited him to stand and enjoy the view. The intrepid cavalryman replied that "my confidence in balloons at that time was not sufficient, however, to justify such a course, so that I remained seated in the bottom of the basket, with a firm hold upon either side."[36] Custer continued the description of his first balloon ascent:

> *I first turned my attention to the manner in which the basket had been constructed. To me it seemed fragile, indeed, and not intended to support a tithe of the weight then imposed upon it. The interstices in the sides and bottom seemed immense, and the further we receded from the earth the larger they seemed to become, until I almost imagined one might tumble through. I interrogated my companion as to whether the basket was actually and certainly safe. He responded affirmatively; at the same time, as if to confirm his assertion, he began jumping up and down to prove the strength of the basket, and no doubt to reassure me. Instead, however, my fears were redoubled, and I expected to see the bottom of the basket giving way, and one or both of us dashed to earth."*[37]

Custer eventually relaxed and began to take notes on the scene spread before his feet.

> *To the right could be seen the York River, following which the eye could rest on Chesapeake Bay. On the left, and at about the same distance, flowed the James River. . . . Between these two extended a most beautiful landscape, and no less interesting than beautiful; it being made the theatre of operations of armies larger and more formidable than had ever confronted each other on this continent before . . . I endeavoured to locate and recognize the different points of interest, as they lay spread out over the surface upon which the eye could rest. The point over which the balloon was held was probably one mile from the nearest point of the enemy's line. In an open country balloons would be invaluable in discovering the location of the enemy's camp and works. Unfortunately, however, the enemy's camps, like our own, were generally pitched in the woods to avoid the intense heat of a summer sun; his earthworks along the Warwick were also concealed by growing timber, so that it would have been necessary for the aeronaut to attain the highest possible altitude and then secure a position directly above the country to be examined. With all the assistance of a good field glass, and watching opportunites when the balloon was not rendered unsteady by the different currents of air, I was enabled to catch glimpses of canvas through openings in the forest, while camps located in the open space were as plainly visible as those of the Army of the Potomac. Here and there the dim outline of an earthwork could be seen more than half concealed by the trees which had been purposely left standing on their front. Guns could be seen mounted and peering sullenly through the embrasures, while men in considerable numbers were standing in and around the entrenchments, often collected in groups, intently observing the balloon, curious, no doubt, to know the character or value of the information its occupants could derive from their elevated post of observation.*[38]

Custer apparently made his first ascent sometime between April 19, when Allen began operating the *Constitution* at Warwick Court House, and April 23, when he moved the balloon three miles to General Smith's headquarters. It is impossible to fix the exact date since neither Allen nor Lowe mentions a flight with the lieutenant.

Custer's initial report must have pleased General Smith, for he was ordered to continue flying with the aeronauts "almost daily," prior to the Confederate evacuation of Yorktown on May 4. In his postwar memoirs Custer claimed to have persuaded his superiors that it would be wise to fly at dawn and dusk in order to count the Confederate campfires, which would be easier to spot through the trees. This claim is patently false, for Lowe had been following this procedure since his early flights in the Washington area.

Custer also claimed credit for the discovery of the evacuation of Yorktown during the night of May 3–4, 1862. He noted that General Smith had ordered him to make two ascents that night, one at dusk, the other at dawn on May 4. Everything appeared normal that evening, but the next morning the novice observer failed to note the usual number of campfires. A closer look after sun-up revealed that the Rebels had withdrawn from the city the night before.[39]

T. S. C. Lowe, who does not mention Custer by name in contemporary dispatches or in his postwar memoir, disagreed with this account. Lowe reported that he had noted nothing at all out of the ordinary during the ascent at Warwick Court House on the evening of May 3. Returning to his own camp behind Union lines at Yorktown, he had been awakened at midnight by an officer from General Heintzelman's staff. The general, noting the red glow of a fire behind the Confederate lines in Yorktown, asked Lowe to determine if Joseph Johnston was withdrawing his troops.

Lowe's predawn ascension on May 4 was inconclusive, but his suspicions were aroused by the absence of the usual number of campfires on the other side of the line. Ascending with Heintzelman at dawn, the aeronaut discovered that Yorktown had indeed been evacuated. Johnston, aware of the impending attack, was withdrawing into a tight defensive ring in front of Richmond.

Union troops had to exercise extreme care in moving through Yorktown, for the departing Confederates had liberally salted the area with mines and explosives rigged with hidden trip wires. Lowe's telegraph operator fell victim to one of these traps when he stepped on a buried torpedo while descending a telegraph pole.

McClellan's troops now resumed their slow push up the peninsula through the incessant spring rains. Lowe and Allen kept pace with the forward elements of the Army of the Potomac. Lowe traveled up the York and Pamunkey rivers aboard the *Custis*. Coming ashore at the White House, he joined Gen. George Stoneman's cavalry command which was spearheading the advance toward Richmond.[40]

By May 20 he was operating from an area near Seven Pines, a small crossroads on the Chickahominy River, only seven miles from the Confederate capital. Stoneman, who had originally predicted that the balloon would accomplish little beyond drawing enemy fire, had come to appreciate the value of aerial reconnaissance during the siege of Yorktown. Between May 21 and 25 Lowe and Stoneman made a number of joint ascensions, observing the Confederate strong points blocking the advance and directing artillery fire to cover the movement of Union troops toward nearby Mechanicsville. Following one of these flights, the general was heard to remark "that he had seen enough to be worth a million dollars to the government."[41]

Lowe took other officers aloft from Seven Pines, as did James Allen, who was then operating his own balloon camp at Warwick Court House. Newspaper reporters and artists were also anxious to obtain an aerial view of Richmond.

The balloon program had been recognized as extraordinarily newsworthy since its inception. As early as June 1861, Frank Leslie, a pioneer in the field of nationally distributed illustrated newspapers, had asked Lowe to take one of his artists on a reconnaissance flight. Now that Lowe could offer a bird's-eye view of the Confederate capital, the number of requests from influential newsmen became even heavier.[42]

One English reporter who went aloft from Seven Pines found more excitement than he had bargained for. "The Confederates fire on the balloon and the first shell passes a little to the left, exploding in a ploughed field. The next, to the right, bursts in mid-air. The third explosion is so close that the pieces of shell seem driven across my face, and my ears quiver with the sound."[43]

By May 22 the number of applicants requesting permission to fly with Lowe was growing to serious proportions. McClellan ordered that henceforth no officer go aloft without his specific approval. Newsmen were completely banned.

His proximity to Richmond enabled Lowe to employ the balloons as a psychological weapon during this period. When aloft, he was readily visible from the city, a constant reminder to the citizens that a Yankee army was encamped on their doorstep. Lowe was always careful to insure that the name of his balloon, *Constitution*, painted in bold letters across the face of the envelope, was facing the Confederate capital.

Richmond newspapers took careful note of Lowe's operations. On May 26 a local newspaper announced: "The enemy are fast making their appearance on the banks of the Chickahominy. Yesterday they had a balloon in the air the whole day, it being witnessed by many of our citizens from the streets and housesteps. They evidently discovered something of importance to them, for at about 4 P.M. a brisk cannonading was heard at Mechanicsville and the Yankees now occupy that place."[44] Richmond papers also described the color "and the various armaments painted on the balloons, as seen through telescopes from the city."[45]

It was a period of intense activity for Lowe and the Balloon Corps. Captain Beaumont, the English observer, described the chief aeronaut's camp. The balloon itself was "snugly ensconced in a hollow on the west bank of the Chickahominy, surrounded by tents and cooking fires forming a small distinct encampment."[46]

A balloon launch scene, Fair Oaks, June 1, 1862

Lowe now had three balloons in operation before Richmond. The *Constitution* and the *Intrepid* were stationed near Seven Pines on the road from Buttom's Bridge to the capital. Clovis Lowe was operating the *Washington* near Stoneman's headquarters at Mechanicsville, while a third balloon was operating in the vicinity of the general headquarters at New Bridge.[47]

By May 29 it was obvious that the Army of the Potomac was in serious difficulty. McClellan and three of his army corps (Porter, Franklin, and Sumner) held a fifteen-mile front on the north bank of the Chickahominy. Keyes and Heintzelman commanded two corps south of the river. The Chickahominy, swollen by the continuing rains, now threatened to split McClellan's army.

Gen. Joseph E. Johnston quickly grasped the opportunity to move against the two corps isolated south of the river. Ascending with the balloon *Washington* at Mechanicsville on May 31, Lowe saw Confederate troops streaming toward Fair Oaks. It was apparent that Johnston was launching his massive attack on the Union troops south of the Chickahominy. With the telegraph lines down, the aeronaut sent a messenger to inform his new commanding officer, Brig. Gen.

The crew prepares to launch a balloon during the Peninsular Campaign.

Andrew A. Humphreys, chief topographical engineer for the Army of the Potomac.[48]

Lowe then rode six miles on horseback to his other balloon camp at the Gaines House, near Seven Pines and Fair Oaks. Discovering that the *Intrepid*, the only balloon in the area large enough to carry both an aeronaut and a telegrapher aloft, was only partially inflated, he made a quick ascension aboard the smaller, one-man *Constitution*. Having confirmed the observations made at Mechanicsville, Lowe immediately descended and "sent off the most important dispatch of any during my whole experience in the military service."[49]

Now that he had twice warned McClellan of Johnston's approach, Lowe faced the problem of inflating the *Intrepid*. With this large balloon he could fly to 1,000 feet with a telegraph operator and instrument through which he could provide immediate reports on the progress of the battle. Realizing that the task of inflating the *Intrepid* with the portable generators would require a precious hour, he immediately ordered a camp tinsmith to transform a teakettle into a funnel, and transferred the hydrogen from the *Constitution* directly into the *Intrepid*.

Back in the air with his telegrapher, Lowe watched the Battle of Fair Oaks

Lowe ventures aloft during the Battle of Fair Oaks.

unfold at his feet. "Of all the battles I have witnessed," he noted many years later, "that of Fair Oaks was the most closely contested and most severe."[50]

Thanks in large measure to Lowe's early warnings of the Confederate advance against Keyes and Heintzelman, McClellan ordered Sumner's corps to cross the Chickahominy in support at any cost. Sumner speeded construction of a makeshift bridge while sending other divisions across a bridge that was already washing away. The confusion among Confederate troops moving into battle enabled Sumner to arrive in time to prevent a rout.

The engagement, fought in the flooded swampland around Fair Oaks and Seven Pines, ended in a draw. Men fought and died in knee-deep water. The wounded suffocated in the mud before the stretcher-bearers arrived.

Lowe and Allen had two busy days in the air on May 31 and June 1, 1862. Lowe believed that he had demonstrated the value of an observation balloon beyond any doubt. General Heintzelman clearly agreed: "From my own observation and experience, with the portable gas generating apparatus and others of

Lowe, standing, with back to viewer, composes a dispatch to General McClellan during the Battle of Fair Oaks.

your inventions, I would consider your balloon indispensable to an army in the field and should I ever be entrusted with such a command would consider my preparations incomplete without one or more balloons."[51]

Maj. Adolphus W. Greely, chief signal officer of the U.S. Army, went even further in an assessment of Civil War ballooning written in 1900. "It may be safely claimed," he noted, "that the Union Army was saved from destruction at the Battle of Fair Oaks . . . by the frequent and accurate reports of Professor Lowe."[52]

While virtually all McClellan's generals recognized the value of Lowe's balloons, some were still not fully convinced that aeronautics had earned a permanent place in the army. Gen. J. G. Barnard's comment was typical: "A balloon apparatus is decidedly desirable to have with an army; but at the same time it is one of the first encumbrances that if obliged to part with anything, I should leave behind."[53] Even Lowe's old friend Fitz-John Porter complained that the experience of Fair Oaks had demonstrated that professional aeronauts were "not of themselves successful observers" because of their lack of military training, and should always be accompanied by an officer familiar with the country.[54]

For the moment, however, Lowe had secured his position in the Army of the Potomac. The aeronauts continued to ascend throughout the month of June as McClellan began moving the bulk of his army into positions south of the Chickahominy. The general had survived Fair Oaks but lost the overwhelming advantage he had enjoyed prior to the battle. The mighty Army of the Potomac would approach no closer to Richmond during this campaign. McClellan was content to settle into his new positions, preparing to conduct a lengthy siege of Richmond.

But he was facing a new and much more aggressive enemy. Joseph E. Johnston, severely wounded at Fair Oaks, had been replaced by Robert E. Lee as commander of the southern troops which he would soon weld into the legendary Army of Northern Virginia. Lee was constructing his own defensive works while planning a means of striking the Union army from an unexpected direction. With most of McClellan's troops concentrated south of the Chickahominy, Lee was curious about the extent of the force under Fitz-John Porter that continued to hold a railhead on the north bank through which all the army's supplies were flowing.

On June 12 the new commander ordered his favorite cavalry officer, the dashing James Ewell Brown Stuart, on a ride completely around the Army of the Potomac. Stuart crossed the Chickahominy to the north, then swung close to the great supply base at White House on the Union army's right flank before passing behind the army and returning to his own lines on June 15. With his suspicions regarding the weakness of McClellan's right flank, Lee was ready to take an enormous risk. He would recall Stonewall Jackson, who was still operating in the Shenandoah Valley, threatening Washington and tying up troops that would otherwise be serving with McClellan. Leaving a weak screening force in the trenches before Richmond, Lee and Jackson would then turn the Union army's right flank and retrace Stuart's movement around McClellan.

Lowe witnessed the beginning of Lee's attack, which came against Porter's position at Gaines' Mill on June 27. E. P. Alexander, the Confederate artillery chief, watched the first of what became known as the Seven Days Battles from a balloon as well. Alexander's balloon was the most colorful and best-known of the aerial craft to see service during the Civil War. Known as the "Silk Dress Balloon," the patchwork envelope was to become a central element of Southern folklore. Unfortunately, there was little more than a grain of truth behind the cherished legend.

The story had its roots in an article published by Gen. James Longstreet in

A fanciful Currier and Ives view of the Battle of Fair Oaks with a Lowe balloon

the *Century Magazine* in 1886:

> *The Federals had been using balloons in examining our positions, and we watched with envious eyes their beautiful observations as they floated high up in the air, and well out of the range of our guns. We longed for the balloons that poverty denied us. A genius arose for the occasion and suggested that we send out and gather together all the silk dresses in the Confederacy and make a balloon. It was done, and soon we had a great patchwork ship of many and varied hues. The balloon was ready for use in the Seven Days Campaign. We had no gas except in Richmond, and it was the custom to inflate the balloon there, tie it securely to an engine, and run it down the York River railroad to any point at which we desired to send it up. One day it was on a steamer down the James when the tide went out and left the vessel and the balloon high and dry on a bar. The Federals gathered it in, and with it the last silk dress in the Confederacy. This capture was the meanest trick of the War and one I have never yet forgiven.*[55]

The truth behind the "Silk Dress Balloon" was more complex, though far less romantic, than Longstreet suggested. The balloon had been constructed by Capt. Langdon Cheeves in Savannah, Georgia.

Cheeves, a signer of the South Carolina Ordinance of Secession, was the son of Langdon Cheeves, Sr., a judge and member of Congress who served as a Speaker of the House and one of the Treaty of Ghent commissioners. Young Langdon had served in the Seminole War, practiced law in Columbia, South Carolina, and become a rice planter with extensive holdings on the Savannah River. With the outbreak of war, Cheeves enlisted as an engineer, overseeing the construction of Confederate dams and fortifications. He was present during the federal movement against Port Royal, South Carolina, and may well have seen the balloon piloted by Lowe's assistant, John Starkweather.

Cheeves constructed his own balloon in Savannah's Chatham Armory during the spring of 1862. Silk was, of course, a scarce commodity in the wartime Confederacy, but the captain was able to buy sufficient quantities of multicolored ladies dress silk in Savannah and Charleston. Purchasing the colorful fabric in forty-foot lengths, he joked to his daughters, "I am buying up all the handsome silk dresses in Savannah, but not for you girls."[56] It seems likely that this lighthearted comment was the basis for the fully embroidered story told by Longstreet.

After coating his gasbag with a solution produced by boiling rubber car springs in naphtha, he packed the balloon and proceeded to Richmond where he placed the craft at the service of President Davis and General Lee. E. P. Alexander was quick to take advantage of this opportunity to observe the Confederate advance against Gaines' Mill on June 27, sending the first warning of Gen. Henry Slocum's crossing of the Chickahominy in support of Fitz-John Porter's beleaguered divisions.

Transferred down the James aboard the armored ship *Teaser* on July 4, 1862, the balloon was captured by the Union ironclad *USS Monitor* and the *USS Maratza*. *Teaser*, aground and attacked by the two superior ships, was abandoned by her crew. Langdon Cheeves, "the father of the Confederate Air Force," died in the Union attack on Charleston in July 1863.[57]

Unknown to James Longstreet and E. P. Alexander, who had immortalized the Cheeves aerostat, a second "silk dress" balloon was constructed in the Charleston-Savannah area in the summer of 1862. Like Cheeves's craft, this balloon was apparently constructed of "variously colored pieces of silk."[58]

This second balloon was the product of Charles Cevor, the well-known prewar balloonist. Joseph Jenkins Cornish III, an authority on Confederate aeronautics, estimates that the Cevor balloon was roughly the same size as Cheeves's aerostat, 7,500 cubic feet.[59] Unlike the other Confederate balloons, Cevor's aerostat was directly funded by army authorities. On August 21, 1862, P. G. T.

Beauregard, commander of the Department of South Carolina and Georgia, authorized the payment of a detailed list of $475.83 worth of materials that went into the construction of the balloon. Operating under Maj. D. B. Harris, Beauregard's chief of engineers, Cevor earned $140 a month (as compared to Lowe's $10 per diem), while his assistant, A. E. Morse, was paid $100 a month.[60]

Soon after the balloon was completed, Cevor and Morse took it to Richmond. By November 21, 1862, they were back in Charleston, where Capt. W. H. Echols of the Office of Chief Engineer of that city was ordered to begin carrying Cevor on his rolls and to pay any bills required to keep the balloon in good repair. Cevor must have been fairly active during these months, for on February 16,1863, he requested permission to open a charge account for the Confederate government with the Charleston gasworks in order to ease the burden of red tape surrounding each inflation.[61]

While the surviving records do not permit us to document the operational career of Cevor and his balloon in any detail, it is clear that the craft was lost during the Union siege of Charleston in the spring or early summer of 1863. According to one account, it was blown loose during a high wind and captured by Union troops. Exhibited in the Patent Office in Washington, the balloon was cut up, and small pieces were distributed as souvenirs. Confederate aerial activity came to an end with the destruction of the Cevor balloon.[62]

With the sky to themselves once again, T. S. C. Lowe, Clovis Lowe, James Allen, and other members of the Balloon Corps continued to operate throughout the Seven Days Battles in front of Richmond. Both Allen and T. S. C. Lowe fell ill, apparently victims of the fevers sweeping through the ranks of Union soldiers who had now served for two months in the Virginia swamps. Allen recovered and took command of the balloons assigned to the expedition. Lowe became so ill by the time the Union army had retreated to Harrison's Landing on the James that he returned to Philadelphia, leaving his father Clovis in charge of the balloons he had been operating.[63] Early in August 1862 Clovis Lowe cooperated with Commodore Wilkes, making balloon observations from 1,000 feet over the James River while being towed by a steamer.[64]

Union troops were able to hold back Lee's advance at Savage Station (June 29), Frayser's Farm (June 30), White Oak Swamp (June 30), and Malvern Hill (July 1). The best that can be said, however, is that McClellan had made good an escape to Harrison's Landing. Robert E. Lee had gambled the Confederate capital and won. The Peninsular Campaign was over. A newspaper reporter viewing the mass of Union troops gathering at Harrison's Landing on July 2 noted that they looked "more dead than alive."[65]

Unwilling to take the offensive again without substantial reinforcements,

McClellan was rapidly losing the confidence of his commander-in-chief. Lincoln traveled to Harrison's Landing for a meeting with McClellan on July 8, 1862. Three days later he announced that Henry W. Halleck, not McClellan, would exercise overall command of the army of the United States. On August 3 Halleck ordered McClellan to begin a final withdrawal from the peninsula.

While the veterans of the Peninsular Campaign were being evacuated, Gen. John Pope was leading a second expedition to capture Richmond. Starting southward along the line of the Orange and Alexandria Railroad, Pope led an army composed of units that had been held back from the fighting on the peninsula to defend Washington. Drawn into battle with Stonewall Jackson at Manassas on August 29–30, 1862, Pope was soundly defeated and fell back to the safety of Washington.

With Virginia free of Union troops at last, Lee immediately moved across the Potomac in an attempt to carry the war to the North. He hoped that such a demonstration of strength might not only draw Maryland into the Confederacy at last but also attract the friendship and open support of France and England.

Reaching Hagerstown, Maryland, in September, Lee dispatched Jackson to capture the federal arsenal at Harper's Ferry while the main body of troops moved north. McClellan, placed in command of what remained of Pope's army, met the Confederates on Antietam Creek, near Sharpsburg, Maryland, on September 17. The result was the bloodiest single day's fighting in the history of the continent. McClellan thus forced Lee to turn back to Virginia, but the cost to both sides was enormous. Thirteen thousand Confederate soldiers fell at Antietam, while the Union army suffered the loss of 12,400 dead and wounded.

While the fighting surged from Second Manassas to Antietam, Lowe and the Balloon Corps remained on the ground and out of the action. The chief aeronaut, thoroughly recovered from the fever that had struck him on the peninsula, was back in Washington on September 5.[66] His immediate problem was to reopen a channel of communication to the military establishment. The administration of the balloon operation had grown much more complex since the days when Hartman Bache and A. W. Whipple had exercised complete control through the Bureau of Topographical Engineers. Prior to First Manassas, Whipple had served as a topographical officer on Irvin McDowell's staff. Thus Lowe's Balloon Corps was administered by the Bureau of Topographical Engineers and directly tied to the tactical leadership of the army as well.

The situation became more complex when George McClellan accepted command of the Army of the Potomac. McClellan, Fitz-John Porter, and others who had taken a serious interest in the balloon were gone, and Lowe soon found himself accepting orders directly from staff officers, with no reference to Whipple.

PROFESSOR LOWE'S BALLOON "EAGLE" IN A STORM.

The balloon Eagle *is buffeted by a storm during the Peninsular Campaign.*

When Whipple asked for a clarification of his own responsibility for the balloon department, Gen. R. B. Marcy, McClellan's chief of staff, ordered Col. John N. Macomb, staff topographical officer, to take over the operation. This was a decision made entirely within the staff of the Army of the Potomac, with no reference to the War Department in Washington.[67]

In the short term, Marcy's decision was helpful to Lowe. He was now free of War Department red tape and serving directly under tactical officers that believed in his project. For the first time his chain of command was clear. Moreover, since Macomb was unfamiliar with the material needs of the Balloon Corps, Lowe was given permission to purchase materials and to position his own balloons and aeronauts, so long as he was able to carry out the orders of McClellan and his staff. It was a period of extraordinary freedom and accomplishment for Lowe.

The aeronaut's civilian status, his lack of familiarity with the procedures and requirements of military accounting, and his willingness to incur debts to insure the quickest possible response to orders created serious difficulties however. On March 31, 1862, responsibility for Lowe's activity was transferred from Macomb to Lt. Col. Rufus Ingalls, chief quartermaster of the Army of the Potomac. Under Ingalls's watchful eye, Lowe was now required to submit all material requests through the Quartermaster's Office. At the same time, Brig. Andrew A. Humphreys, McClellan's chief topographical engineer, assumed direct tactical control of the corps. Humphreys, a man who understood Lowe's problems and needs, was also invaluable in assisting Lowe through the administrative complexities of the Quartermaster's Office. Thus, during the Peninsular Campaign, Lowe had at last arrived at a reasonable modus operandi with the U.S. Army.[68]

All this was lost during the weeks following the withdrawal from Harrison's Landing. Back in Washington with his aeronauts and balloons in early September, Lowe discovered that Humphreys was now commanding a division and would no longer be available as supervisor and intermediary for the Balloon Corps. Uncertain where he should turn, Lowe addressed his requests for travel orders permitting him to rejoin the army to A. V. Colburn, the assistant adjutant general. Colburn advised the aeronaut to remain in Washington until word was received from McClellan about his intentions for the Balloon Corps.[69]

Lowe was kept in suspense for two weeks. Orders from General Marcy finally arrived on September 18, 1862, the day after Antietam. The aeronaut was told to pack his wagon train with the balloons and equipment and proceed at once to Sharpsburg. Arriving at Antietam during the week following the battle, Lowe immediately met with McClellan. The general expressed regret that Lowe had not been present at the battle "as the balloon would have been invaluable to him during that engagement."[70]

Lowe explained his difficulty in receiving orders and obtaining transportation. He then proceeded to explain the nature of the "delays, annoyances, etc. [that] fell to my lot while connected with the army . . . through my being a civilian employee":[71]

> *From the first I had hoped to be allowed to organize the Balloon Corps as a military branch and applied for a commission to command it. But all that 'the powers that be' would do was to grant me the privileges of that rank but not the authority, consequently I was subject to every young and inexperienced lieutenant and captain, who for the time being was put in charge of the Aeronautic Corps. These young fellows had no knowledge whatever of aeronautics and were often a serious hindrance to me rather than a help. But they were not all unintelligent, and all of the Generals under whom I served expressed keen appreciation of my*

work, and this over-balanced the trouble I sometimes had with minor officers.[72]

McClellan was sympathetic to Lowe's problems. "He replied," Lowe recalled, "that he was fully of the opinion that I could not operate my department to its greatest efficiency unless its full command were vested in me and he would recommend it be made a distinct branch of the Army and that I would be given a commission."[73] Lowe was elated. "At last it seemed that my greatest ambition was about to be fulfilled."[74]

The aeronaut made his first observation flight from McClellan's camp on the morning after his interview with the general. He and James and Ezra Allen kept the balloons in operation at Sharpsburg and on Bolivar Heights overlooking Harper's Ferry for almost a month, but there was little of importance to be seen. The Union army remained in place into early November, licking its wounds while Lee made good his escape to the safety of Virginia.[75]

When McClellan did move south of the Potomac, he placed his army in a perfect position to destroy, one at a time, the widely separated wings of the Army of Northern Virginia. But Longstreet remained safe at Culpeper Court House, and Jackson at Winchester. Lee was certain that the cautious McClellan would hesitate to move against either force. He was correct. President Lincoln, thoroughly disillusioned with the commander-in-chief of his principal army, removed McClellan from his command in November, replacing him with Maj. Gen. Ambrose E. Burnside, the Rhode Islander with whom James Allen had originally marched off to war.

Burnside decided to sidestep the two wings of Lee's army. Making a feint toward James Longstreet at Culpeper Court House, he would drive south with the main body of his troops, cross the Rappahannock River at Fredericksburg, and move on to Richmond before the Army of Northern Virginia could move south to protect the capital. Lincoln, who probably would have preferred Burnside to engage and defeat Lee directly, nevertheless approved the new commander's scheme. Burnside moved with alacrity, reaching Falmouth, opposite and a bit upstream from Fredericksburg. Here he paused. The small Confederate force in Fredericksburg had destroyed the bridges over the Rappahannock. Rather than attempting to find the shallows downstream from the town, Burnside chose to wait for the arrival of pontoon bridging equipment. Lee, recognizing the gravity of the situation, rushed his army to Fredericksburg. Burnside's hesitation had set the stage for a frightful sequel to the carnage of Antietam.

McClellan had sent T. S. C. Lowe and his corps of aeronauts back to Washington at the beginning of his march south into Virginia. Recognizing that he would be moving in the shadow of the Blue Ridge Mountains, where the balloons would be of little value, he intended to bring Lowe south once the situation stabilized.

With his favorite general deposed, Lowe was now forced to seek a channel of communication with the new commander. On November 20, 1862, he dispatched a letter to Maj. Gen. J. G. Parke, Burnside's chief of staff. He informed the general of his capabilities and requirements and requested orders.[76] Parke replied on November 24, ordering Lowe to come to Fredericksburg with his men and balloons as soon as possible. The Quartermaster Department in Washington and the Army of the Potomac were ordered to render Lowe all possible assistance on the journey.[77]

Lowe arrived at Burnside's headquarters in late November. James and Ezra Allen arrived soon thereafter, having returned home to Providence while James recuperated from a fever contracted during the Peninsular Campaign. Burnside informed Lowe that he wished to keep the presence of the balloons a secret until the army had effected a crossing of the Rappahannock.

The pontoon bridge equipment finally arrived on November 25, but Burnside continued to hesitate. Federal troops were now ranged along the east bank of the Rappahannock on either side of Burnside's headquarters on Stafford Heights. From this point, Union artillery commanded both the river and the town of Fredericksburg.

In view of this fact, Lee had wisely decided not to wage a battle for the city. Rather, he would contest the crossing of the Rappahannock with a relatively small force of skirmishers, while his main body of troops, Longstreet's veteran corps of 35,000 men, would hold a virtually impregnable position on a low line of hills paralleling the Rappahannock a mile or two behind Fredericksburg.

Burnside hesitated so long that Lee began to fear that the Army of the Potomac might suddenly shift around Fredericksburg and strike elsewhere. In fact, Burnside was simply attempting to discover a means of overcoming Lee's overwhelming tactical advantage. With only the most limited knowledge of the Confederate defenses, and against the advice of his corps commanders, Burnside remained firm in his determination to confront Lee at Fredericksburg. His plan of battle was loose. Sumner would move across two bridges and move directly against the troops in Fredericksburg on the Union right. William Franklin would cross three bridges on the left, downstream from the town, and attack Stonewall Jackson's corps entrenched in the hills ranged behind Fredericksburg. Joseph Hooker, commanding the Union center, would remain in reserve to support either flank by moving his troops over a single bridge.

Burnside's engineers began to push their six pontoon bridges across the Rappahannock in the early morning fog on December 11, 1862. A cold wind blew the recently fallen snow down the shallow river valley, scarcely ideal conditions for the men working in boats or struggling in the freezing water to lash the floating walkways in place. As the sun began to burn the fog away, Confederate sharpshooters hidden in buildings on the opposite shore opened fire on the helpless bridge builders.

Burnside halted the work and directed a heavy artillery barrage against the town. The destruction in Fredericksburg was frightful, but the artillery fire did little to daunt the Confederate skirmishers. When the firing lifted, they emerged from their basement shelters to open up on the engineers once again. In desperation, Burnside then called for volunteers who crossed the Rappahannock in boats. After bitter hand-to-hand, house-to-house fighting, these men established a protective bridgehead. The town was Burnside's, but his battle had scarcely begun.

Union troops poured across the river throughout December 12, spreading along a mile front stretching from the northern edge of Fredericksburg to Hamilton's Crossing far south along the Rappahannock. Burnside finally issued battle orders early on December 13, the morning of the attack. The Union corps commanders were a bit nonplused by what amounted to a broad suggestion from Burnside that they move against the Confederate troops on their respective fronts and attempt to capture the entire range of hills behind Fredericksburg. No specific objectives were given. No tactical activities were suggested.

The attack was doomed from the outset. Lee's artillery controlled the field. The Confederates, particularly those guarding the base of Marye's Heights directly behind Fredericksburg, held an ideal defensive position. Entrenched in a sunken road hidden behind a stout four-foot high stone wall, Longstreet's corps was impregnable.

The attack began on the Union left as George Gordon Meade moved against Johnston's troops at 8:30 A.M. For a moment it appeared that these Union troops might actually attain the crest, enabling Burnside to move down the hills against Lee's flank. By mid-afternoon, however, Confederate counterattacks had driven the Federals back down the slope. The situation was far more dismal on the Union right wing, where four divisions were hurled against what one participant described as "that terrible stone wall" below Marye's Heights. "Nothing," he added, "could advance farther and live."[78]

Burnside eventually ordered Franklin to mount a new push on the left in an effort to relieve the fire being directed against his troops attacking the stone wall, but his corps commanders had had enough. Franklin simply ignored the order, claiming that it had arrived too late.

The Battle of Fredericksburg ground to a halt with a final unsuccessful twilight bayonet charge against the stone wall led by Lowe's old friend and commander, A. A. Humphreys. That evening the Union corps and division commanders had to present a united front in order to dissuade Burnside from resuming the carnage the next morning.

Ambrose Burnside had not approached the battlefield. He remained at his headquarters, the Phillips House on Stafford Heights on the far side of the Rappahannock, throughout December 13. T. S. C. Lowe's primary task was to report the battle to the commander of the Army of the Potomac. Preparing his balloon on December 12, he was in the air almost constantly on the day of the battle. Lowe remained tethered on Stafford Heights throughout the engagement, so close to the general headquarters that no telegraph or signal flags were required. The aeronaut simply shouted his observations down to Burnside's staff.[79] Lowe commented that a great many officers ascended with him, adding that "much valuable information was furnished the Commanding General."[80]

Many years later Gen. Daniel Butterfield informed Capt. William A. Glassford, a U.S. Signal Corps officer writing a history of Civil War balloon operations, that he had been one of those who ventured aloft from Stafford Heights. His flight with Lowe had, he remembered, been a short one, just long enough to give him a quick view of the engagement before leading his own troops over the river and into battle. Although he had not risen high enough to obtain a view of the Confederate reserves behind Marye's Heights, Butterfield noted that "my short ascent in the balloon had given me a view of the topography, ravines, streams, roads, etc., that was of great value in making dispositions and movements of the troops."[81]

Lowe's memory of Butterfield's short ascent was considerably different. Following his departure from the corps in the spring of 1864, the chief aeronaut informed an influential Philadelphia supporter that General Butterfield had been one of his most vociferous opponents in the army. Lowe ascribed this to the fact that during the ascent at Fredericksburg Butterfield had been so frightened by artillery fire that he ordered an immediate descent. Lowe had ignored the general's instructions until he had completed his observations.[82]

Burnside seems to have appreciated the information received from the balloon. He called particular attention to the work of James and Ezra Allen, who were apparently also operating that day, although none of their observation reports have survived. As noted earlier, James Allen had marched off to war with Burnside's Rhode Island regiment in the spring of 1861.[83]

In spite of these favorable comments, it is apparent that the balloon had little impact on the outcome of the battle. William Glassford was the first to point out the vital contributions Lowe might have made, particularly in view of

Burnside's poor intelligence on Lee's troop dispositions. "With the use of two or even three balloons at Fredericksburg," he remarked, "it seems strange that none of the reports available show that any of the aeronauts seemed to have observed the position of the famous stone wall from the protection of which was delivered a fire so deadly as scarcely to be matched in the history of warfare."[84]

Although it is true that the wall was visible from Stafford Heights, Lowe should not be blamed for having missed the troop concentration in the area. Burnside had forbidden him to fly before the morning of December 13. The early fog that obscured the area at the foot of the Heights did not burn away until 10:00 A.M., leaving only two hours to observe the area prior to the first assault on the wall. After noon, Confederate strength along the sunken road was perfectly obvious to everyone but Burnside.

It should also be noted that Lowe was observing the battlefield from Stafford Heights on the opposite bank of the Rappahannock, a considerable distance to the rear. Moreover, the balloon was under some artillery fire that morning; at one point a Confederate shell passed close to the balloon before striking two miles to the rear.[85]

After two days of hesitation in Fredericksburg, Burnside finally withdrew his battered troops to the safety of a winter camp on the Falmouth shore of the Rappahannock. Lowe and the Allen brothers remained with the Army of the Potomac through the winter of 1862–1863. Burnside did not. Maj. Gen. Joseph Hooker was named to succeed him as commander of the Union's principal army.

It was a quiet season for the Balloon Corps. The aeronauts were aloft when weather permitted, attempting to keep an eye on Confederate troop movements across the river. The work was routine, but it did help to build General Hooker's confidence in balloon observations.[86]

Lowe also experimented with new signal technologies during this period. As early as March 1862 the aeronauts had experimented with flag signals for use in areas not served by a telegraph. Similar tests had been conducted during the Peninsular Campaign.[87]

During the two weeks prior to the Battle of Fredericksburg, Lowe had discussed a new signal balloon with General Burnside and obtained approval to order a number of small "caloric" balloons that could be used to carry signal flags or lanterns aloft day or night. The aeronaut was convinced that his small aerostats would prove useful in sending messages between balloon camps or between cavalry patrols and general headquarters.[88]

Ranging from 6 to 20 feet in diameter, the caloric balloons were hot air craft. The envelope for daytime use featured distinctive markings that would be supplemented by a variety of signal banners. Red and green calcium flares would

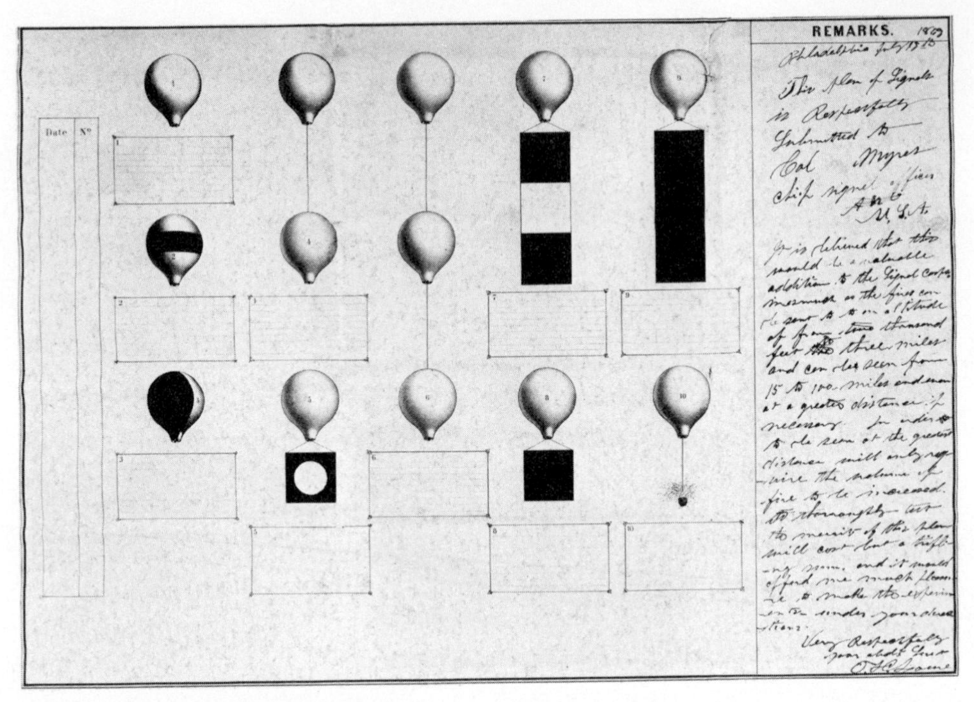

The "caloric" balloon signal system

be employed at night. Lowe estimated that these balloon signals would be visible for distances of up to 40 miles.[89]

Lowe obtained sample caloric balloons but was unable to conduct a field test of the system until after the Battle of Fredericksburg. In March 1863 Lowe complained to General Hooker that his original balloons had been constructed of shoddy materials and were useless. Gen. George Stoneman, the Army of the Potomac's cavalry commander, endorsed the aeronaut's request for the funds with which to construct a new batch of signal balloons, which he believed would be useful to isolated reconnaissance patrols. The request was, nevertheless, firmly denied by Gen. Seth Williams of Hooker's staff.[90]

Lowe continued to press his signal balloon scheme even after his departure from the service. In July 1863 Col. A. J. Myer of the Signal Corps evinced some interest, but not enough to conduct a test. Lowe did cooperate with Stoneman and others in testing the visibility of signal rockets fired from various points in the Union line and observed by the aeronaut.[91]

Hooker's staff also asked Lowe to comment on balloon innovations suggested by others during the weeks of idleness before Fredericksburg. One of these innovations, a new system of balloon inflation devised by a Philadelphian named B. England, particularly upset Lowe, who viewed the proposal as a direct attack on his own system. England claimed that he could cut Lowe's inflation time for a large balloon from fifteen hours to a half hour. Moreover, he promised to reduce drastically the present $400 cost of an inflation and to cut the size of the wagon train required to support a balloon in the field.[92]

Lowe replied that England's claims were false. At present, he remarked, only two and a half hours were required to inflate a balloon, not fifteen. In the spring of 1863 a single inflation cost only $60. England had also overestimated the size of the seven-wagon field train required to support four balloons. Lowe also claimed that England's balloons would be far too light and flimsy for use on a battlefield. Nor did his "new system" of inflation represent any improvement over the methods then in use by the Balloon Corps.[93]

Lowe seemed especially annoyed that two of his old friends and assistants, William Paullin and John Steiner, had endorsed England's plan. In an effort to counter their testimony, Lowe submitted letters from James and Ezra Allen supporting the system currently in use by aeronauts of the Balloon Corps.

James Allen described his own early attempts to operate as a military balloonist prior to Bull Run, admitting that his balloon and equipment had been inadequate for the task at hand. Allen continued by admitting to Lowe that "I can conscientiously say that the government is indebted to you alone for the introduction of this useful branch of public service, and were it not for your improvements in the construction of balloons and the invention of portable gas generators, your untiring perseverance, hard labor, and exposure against great obstacles, Aeronautics could never have been of service to the Army."[94]

Ezra Allen "cordially" concurred with his brother's assessment. Lowe's improvements and his instructions to the aeronauts were the very factors that had made balloons useful in the field.[95]

England's claims were little more than a mild irritation when compared to the far more serious challenges to Lowe's authority over the leadership of the Balloon Corps that appeared in the spring of 1863. Jacob C. Freno, the Philadelphia lawyer-aeronaut whom Lowe had dismissed from the corps in December 1862 for running a faro table and spreading "treasonous rumors," now struck back with a vengeance.[96]

In a sworn statement to the provost marshal of Washington, Freno charged Lowe with a series of crimes. He had, said Freno, placed his father Clovis Lowe on the government payroll at a time when the old man was ill at home in Philadelphia. Freno further charged that Lowe had stolen government property ranging from a saddle to an entire balloon. The chief aeronaut was also accused of a variety of minor infractions, including having requisitioned army rations and forage for himself and influential newsmen when, in fact, he should have paid for these items. He had also used government carriages to take his wife on pleasure trips around the Washington area.[97]

Lowe did his best to refute the charges. The balloon in question was a small experimental craft taken to Philadelphia for repairs. Clovis Lowe had been on duty for fifteen straight months without a sick day. Leontine Lowe had not been in Washington since October 1861.[98]

The aeronaut was certain that his superiors had accepted his explanations. To put a cap on the matter, he accused Jacob Freno of having broken into a Washington warehouse and vandalized the balloon *Constitution* which was under repair at the time. The Allens, Starkweather, John O'Donnell, and others supported Lowe's charges. A warrant was sworn out for Freno, but the Washington district provost marshal's office was unable to locate him.[99]

War Department officials obviously retained lingering doubts about Lowe's financial transactions and the disposition of public property under his control. The poor financial administration of the Balloon Corps had been apparent for some time. Lowe's lack of attention to his responsibilities in this area is illustrated by the minor flap in the Quartermaster's Department when Messrs. E. Pickrell and Company began to complain of an unpaid bill for lumber charged to the aeronautical department. Investigation revealed that Lowe, ignoring all reasonable procedures, had sent the signed invoice approved for payment directly to the Secretary of War.[100]

Lowe's management of his personal finances also left much to be desired. In March 1862 he received a wire from Leontine Lowe. "We are well. Nothing new. You must send money immediately."[101] The long-suffering Mrs. Lowe was still pleading for help in February 1863. "Did not receive money. Need it badly. We are well."[102]

Lowe's inventory procedures were no better than his account books. He was unable to produce receipts for equipment (including whole balloons) that had been removed from service or dispatched to other commands.[103] The aeronaut complained that he was far more concerned with meeting the needs of field commanders than in keeping accurate records for bureaucrats. This attitude created few sympathizers in the War Department. The absence of receipts also made it impossible for Lowe to clear himself fully of Freno's charges that he had misappropriated government property.

The final blow fell on April 7, 1863, when Gen. Seth Williams, Hooker's assistant adjutant general, issued Special Order No. 95, naming Capt. Cyrus B. Comstock to head "the balloon establishment." Comstock, chief engineer for the Army of the Potomac, was quite unlike Humphreys or other officers that had exercised tactical command of the corps. These men had always been willing to issue orders while allowing Lowe to exercise actual control over the aeronauts and equipment.[104]

Comstock fully intended to maintain absolute control over the corps. Some evidence that War Department officials may have been behind the new arrangement is found in a letter from Comstock to Assistant Secretary of War P. H. Waterman detailing his observations on the problems of the Balloon Corps.[105] "On taking charge of the balloon establishment of this Army," he noted with apparent relish, "I found it as I thought, an unsuccessful experiment." Lowe seemed prompted by a stronger sense of his own interests than of those of the government. Comstock judged that the chief aeronaut "has been acting without the knowledge or authority of anyone connected with the Army of which he is an employee."[106] Nor did the new commander have any particular regard for the professional aeronauts. Any "man of intelligence" could learn to manage a balloon in a week, he informed Waterman.[107]

Comstock implemented immediate and far-reaching changes. Clovis Lowe and John O'Donnell were hastily cut from the payroll, leaving only Lowe and the Allens to operate the balloons. Lowe's salary was cut from $10 to $6 a day, reducing him to the status of the Allen brothers. Comstock also demanded an immediate accounting of all public property entrusted to the Balloon Corps.[108]

On April 12, 1863, Lowe submitted a long letter of resignation to Gen. Daniel Butterfield, Hooker's chief of staff. The aeronaut recounted his service to the government and called attention to the financial sacrifices he had made in order to continue the balloon operation. He claimed that he had originally been offered $30 a day, but had himself suggested a reduction to the more reasonable salary of $10 per day:

> *General, I feel aggrieved that my services should not have been better appreciated. As it is, I cannot honorably serve for the sum named by Captain Comstock without first refunding to the Government the excess of that amount which I have been receiving since I have been in the service. This my very limited means will not allow, for it requires the full salary I have received to support myself in the field and my family at home; therefore, out of respect to myself and the duty I owe my family, it will be impossible for me to serve upon any other condition than those with which I entered service.*[109]

To avoid being backed into a corner, Lowe hedged his resignation. In view of increased Confederate activity and the probability of a battle in the near future, the aeronaut would temporarily offer his services free of charge.

If Lowe hoped that his letter would bring about a return to the old policy, he was disappointed. So that the aeronaut would have no doubt where he stood, General Hooker returned the letter of resignation with a mild rebuke for not having sent it through channels with Captain Comstock's approval.[110]

Lowe was now serving without pay, but he was still attempting to exercise his old authority. When the assistant secretary of war sent a query to the Balloon Corps asking if an aeronaut and balloon could be spared for service at Charleston, South Carolina, Lowe replied that James Allen would be made available. Comstock immediately rebuked Lowe for not sending all such messages through him and reversed Lowe's decision. It was clear that Lowe was being given very little room in which to maneuver.[111]

Lowe did have the satisfaction of answering Comstock's technical questions about the manner in which the balloons had been operated in the past. Comstock was rapidly acquiring experience on his own however. He made frequent ascents with the Allens. Flying solo from Falmouth in James Allen's *Washington* on March 23, Comstock remained aloft an hour and a half and ordered himself to be towed three miles down the road. At the same time, Ezra Allen was operating a separate balloon camp at Hooker's headquarters, then at Phillips House on Stafford Heights.[112]

By mid-April the balloons *Washington* and *Eagle* and the aeronauts were being shifted back and forth along the Rappahannock from Banks Ford on the Union right to White Oak Church in the rear and to Fredericksburg itself.[113]

The battle Lowe had predicted was not long in coming. On April 27 Hooker began a movement far down the Rappahannock on the Union right. He planned to cross both the Rappahannock and Rapidan rivers, then move back downstream to fall on the Confederate rear. The maneuver was much more astute than Lee had come to expect of a Union commander. The Army of Northern Virginia, with a vastly superior force approaching its flank, seemed to have little choice but to retreat. Instead, Lee gambled once again. He dispatched Jackson on a most dangerous march directly across the face of Hooker's approaching army, which was then struggling blindly through a dense area of scrub pine and secondary growth vegetation known as the Wilderness. Having accomplished his march, Jackson turned and attacked the Union left flank near a little crossroads known as Chancellorsville on May 2.

Hooker was aware that the Confederate troops had been moving out of Fredericksburg toward his army. Lowe and the Allens were constantly aloft during the preliminaries to Chancellorsville. Lowe, closest to the action, had reported "heavy columns of the enemy's Infantry and Artillery . . . moving up the river accompanied by many Army wagons" on May 1. On May 2, 3, and 4, the days of the fighting at Chancellorsville, he continued to report heavy troop movements to his right.[114]

The Union balloons were fully visible to the Confederates marching toward battle. Early reports of the Chancellorsville action published in the Richmond *Examiner* mentioned "a Yankee balloon aloft at a great height" throughout May 1. There is evidence indicating that Stonewall Jackson took special precautions to screen his movements from Lowe.[115]

General Hooker was very much interested in the balloon observations. Immediately prior to the battle, he requested that flights be made at particular points along the Rappahannock. He seems to have completely misread the meaning of Lowe's early reports, however, assuming that Lee was attempting to withdraw his forces before the Federal attack. Once the battle began, he drew the Army of the Potomac into defensive positions between Chancellorsville and Fredericksburg, handing all of the tactical advantage to Robert E. Lee. The Army of the Potomac, 90,000 strong, was now "surrounded" on three sides by 48,000 soldiers commanded by Lee and J. E. B. Stuart.

John Sedgwick's 6th Union Corps, which had been left behind in Fredericksburg, now stormed the small force of Confederates still holding the position on Marye's Heights. Following this successful engagement, Sedgwick moved up the Rappahannock in an attempt to break through and reinforce Hooker. Failing this, he retreated to Fredericksburg. Hooker, now thoroughly demoralized, also retreated across the Rappahannock to safety.

The balloons flying at Banks Ford and Fredericksburg throughout the engagement had supplied a great deal of useful intelligence. John Sedgwick credited Lowe with having enabled him to attack Marye's Heights with confidence. "The importance of careful balloon reconnaissance and accurate reports therefrom cannot be overestimated," he commented.[116]

The balloon operations during the Battle of Chancellorsville had underscored the lessons of Fair Oaks and Fredericksburg. In the hands of a competent aeronaut and a capable military observer, the observation balloon could provide first-rate intelligence that, as Sedgwick noted, "could not have been procured by any other means."[117]

Yet balloon intelligence was useless unless correctly analyzed and interpreted. As Chancellorsville demonstrated, a field commander could very easily draw the wrong conclusions from accurate data.

Lowe's service with the army came to an end on May 6, 1863. Calling on Comstock in the hope that his recent activity might have caused the captain to reconsider the salary issue, he discovered that nothing had changed.[118] The aeronaut therefore asked to be relieved at the earliest convenient moment, "to which Captain Comstock replied that if I was going I could probably be spared better then than at any other time."[119] T. S. C. Lowe folded his tent and left for Washington.

James and Ezra Allen remained with the army. Having been the first military aeronaut of the Civil War, James Allen was also destined to be the last. After Lowe's departure, the Allen brothers were completely under the control of military authorities with little or no experience in aeronautics. As the Allens informed Lowe, this created a great deal of danger for the aeronauts, who "are often ordered to do things against our better judgment."[120]

The Allens were particularly worried about their rapidly deteriorating equipment. The condition of the balloon *Washington* was especially disturbing. Constructed by Lowe in the early fall of 1861, the balloon had been used by Starkweather at Port Royal, South Carolina. When Allen first inflated the envelope in January 1863, it was "in very bad order."[121] The balloon was full of small holes ranging from 1⁄64 to 1⁄8 of an inch in size. Allen's effort to patch the craft proved difficult, as new holes continued to appear in the weak fabric. The seams, he noted, "are very rotten."[122]

The aeronaut informed army authorities that the aging balloon was unfit for service, but Hooker's staff continued to order it into the air under the worst of conditions. Ezra Allen inflated the *Washington* on May 28 and made several ascensions, one with Lt. H. R. Williams during which the balloon was towed two miles down the Rappahannock. Captain Paine of the Topographical Engineers also made a flight with Allen on May 28.[123]

Hooker then ordered the balloon moved to an area above Falmouth. A strong breeze was blowing on the following morning. In spite of Allen's protests, he was ordered back into the air "without regard to consequences."[124] Flights continued in heavy winds on May 30. "Although the wind was blowing almost to a gale," Ezra Allen was back in the air at 5:30 A.M. on June 1:

> *I at first attained to an altitude of about three hundred feet with five hundred feet of rope out, but I was soon blown down one third of that distance, making me about two hundred feet high, at which time the balloon exploded upon the side exposed to the wind, causing a rent—in a seam only as I afterward ascertained of thirty-six feet in length. The escape of the gas was instantaneous, but with the help of men familiar with the Balloon and their strict attention to my orders, it was brought safely to the Earth.*[125]

In the wake of this near disaster, the Allens pleaded with Gen. G. K.

Warren to provide better care for the balloons. They asked Lowe to use any influence he might have with the secretary of war to obtain funds for the construction of six new balloons. The brothers reaffirmed their desire "to serve their country," but had little hope for the future of army ballooning unless Lowe himself was returned to service.[126] "We are satisfied that if General Hooker had personally known the cause by which you left the Army he would have soon removed it. It was certainly very presumptuous for men without any knowledge of the art to dictate to you, who have spent a lifetime in the study of Aeronautics."[127]

But the Allens continued to carry on alone—and with their old balloons. Northern newspapers like the *New York Herald* had reported Lowe's resignation, "leaving the management of the corps to Mr. Allen . . . who is assisted by his brother. In their hands the silken spheres will continue to furnish a capital observatory from which to reconnoitre the movements of the enemy, and will prove, as they have heretofore, a great auxiliary in whatever operations the army may undertake."[128]

Other news accounts described Ezra Allen's rapid descent in the face of enemy artillery fire late that May, as well as an increasing number of flights by military officers, including a Captain Palere of the Topographical Engineers and a Lieutenant K———, who had been assigned to the corps prior to Lowe's departure.

The Allens were detecting movement in the Confederate lines during the first week in June. Heavy clouds of dust rose into the air as the Army of Northern Virginia embarked on its long march north toward yet another meeting with the Army of the Potomac at the little crossroads town of Gettysburg, Pennsylvania.[129]

Lee had hoped that the Union troops would be slow in pursuing him. They were not. Hooker had abandoned his positions in the Fredericksburg area. Unknown to the Confederates moving up the Shenandoah Valley, they were shadowed by a Union army remaining between them and Washington, D.C.

While there is no firm evidence to mark the exact date of the demise of the Balloon Corps, it seems likely that the Allens were detached from the army as the march north began. They may well have been told to await orders to rejoin the army before the great battle that was sure to come, as had Lowe before Fredericksburg. In fact, the active history of the Balloon Corps had come to an end. The Allens were not recalled in time for Gettysburg.

On August 25, 1863, Providence newspapers reported that Allen had left his hometown to rejoin the Army of the Potomac. One story called attention to the fact that "the utility of the balloon is becoming every day more and more apparent, and is attracting the attention of military and scientific figures in

Europe as well as in this country." The newsman speculated that "at the next session of Congress an Aeronaut Corps will be regularly established as an arm of the military service."[130]

These optimistic hopes for the future of the Balloon Corps were ill-founded. Without the forceful presence of T. S. C. Lowe, the balloons died a quiet death. In retrospect, a number of military officers on both sides of the line found it difficult to understand why aeronautical observations were not continued throughout the war. The comment of E. P. Alexander, the Confederate artillery chief, was typical: "I have never understood," he remarked, "why the enemy abandoned the use of military balloons early in 1863, after having used them extensively up to that time. Even if the observers never saw anything, they would have been worth all they cost for the annoyance and delays they caused us in trying to keep our movement out of their sight."[131]

In fact, as A. W. Greely noted, "It is surprising that such a hybrid organization as the Balloon Corps did such excellent work and held together two years."[132] The reasons for the collapse of the corps in the face of an obvious need for the intelligence it had provided for two years were many and varied. The initial problem had been the inability of leading aeronauts to cooperate at the outset of hostilities. The individual competitive efforts of Wise, Allen, Lowe, and La Mountain were less impressive than if these men had joined hands in a single, well-directed effort. The obvious personal rivalries at work did little to convince capable and concerned officers like Whipple and Bache that the aeronauts would ever be able to work within the confines of their administrative structure.

Once Lowe emerged as the leading figure, he found that the responsible officers in the Topographical Engineers considered that the balloon experiment had already proven unsuccessful. For the rest of the history of the Balloon Corps, the powers that be in the War Department remained, at best, disinterested in the operation. Lowe was never able to build a strong base of support for the corps in any of the government bureaus charged with its administration.

Rather, Lowe concentrated on demonstrating his utility to the field commanders who would make tactical use of the balloon intelligence. His efforts to introduce the balloon in southern and western commands through Starkweather and Steiner and through his final attempt to send James Allen to South Carolina were complete failures. Only within the Army of the Potomac was Lowe genuinely successful in convincing key commanders of the value of the balloon.

There can be little doubt that the one period of unqualified success for the corps had come during the Peninsular Campaign. Here Lowe was working with George McClellan, Fitz-John Porter, and others who believed in the balloon.

With the removal of McClellan after Antietam and Porter's politically inspired dismissal from the army after Second Bull Run, Lowe was required to build a new base of support on the staff. After Fredericksburg the chief aeronaut faced the same problem, exacerbated by the fact that such key officers as Daniel Butterfield were by no means enamored of Lowe's personality or approach. These frequent changes of leadership in the army command, coupled with the absence of serious support for the balloons in Washington, spelled ultimate disaster for the aeronauts.

Had Lowe been able to convince anyone in authority of the need to commission his balloonists, had he been able to carve a permanent niche for the Balloon Corps within the administrative structure of the War Department, the story of the Union aeronautical effort might have had a happier conclusion.

There were, of course, a multitude of contributing factors leading to the death of the Balloon Corps. Lowe's lack of concern with what he regarded as administrative red tape certainly did nothing to build confidence in high places. Nor did his ego and his attempt to circumvent the chain of command by going over the heads of Cyrus Comstock and other responsible officials strengthen his case.

There is, of course, another question to be addressed. Did the balloon contribute to the success of the Army of the Potomac? There can be little doubt that Lowe and his aeronauts did supply useful information of clear military value during the dreary months outside Washington, on the peninsula, at Fredericksburg, and at Chancellorsville. Seldom, however, was this information put to good use. It was Lowe's misfortune to serve with the greatest army in the world at a time when it was commanded by a series of generals ranging from the overcautious to the incompetent. In essence, one can only agree with John Wise's retort to A. W. Whipple's complaint about the incompetence of the balloonists before Bull Run. "The balloon part . . . was about as good as the fighting part."[133]

T. S. C. Lowe had by no means given up his dreams for the Balloon Corps after Chancellorsville. He returned to Washington to prepare a lengthy report on the balloon operations requested by Secretary of War Stanton. True to form, the aeronaut discovered that the two-year history of the corps was contained in a mass of dispatches, telegrams, letters, and reports jammed pell-mell in his luggage. He hired W. J. Rhees, an employee of Joseph Henry's at the Smithsonian with whom he had worked to arrange the original 1861 flights for President Lincoln, to arrange his papers in usable order. The finished report, submitted to the War Department in 1863, was a masterpiece containing the full text of the messages that had given birth and direction to the Balloon Corps, interspersed with Lowe's comments on two years of activity. It was, of course, intended as an extended plea for the institution of a full-scale operation.[134] While Lowe's testimony before the joint congressional Committee on the Conduct of the War has not survived, it is known that the aeronaut did forward the text of many of his telegrams and reports to this body. It can scarcely be imagined that, caught as he was in the midst of his struggle for power with Comstock, he would have allowed this golden opportunity to plead his case to pass by unexploited.

When neither the secretary of war nor the joint committee rose to his bait, Lowe offered to establish an entirely new balloon organization for Col. A. J. Myer of the Signal Corps. Lowe resorted to classic understatement in explaining his availability for service in Charleston, South Carolina, to Maj. Gen. Quincy A. Gilmore. "As the Aeronautic Corps is now situated, my absence, for a time at least, will not interfere with its operation."[135]

When Lowe's own search for new employment failed, his friends attempted to intercede for him. On August 29, 1863, Judge George Harding of Philadelphia wrote to Secretary of War Stanton, explaining that Lowe's problems were the result of Gen. Daniel Butterfield's embarrassment at having panicked under fire while in Lowe's balloon during the Battle of Fredericksburg. Harding also appealed to George Gordon Meade for Lowe's reinstatement.[136]

As late as the spring of 1864, T. S. C. Lowe was still conducting aeronautical demonstrations at the Smithsonian for War and Navy Department commissioners whom he hoped would reintroduce the Balloon Corps. All these attempts to revive military aeronautics failed. As the Civil War ground into its fourth year, there was little enthusiasm for an experiment that many experienced officers and military administrators regarded as having been tried and found wanting.

Lowe must certainly have been aware that he had few friends in the War Department. He was caught in the midst of a losing battle to obtain $3,040.97 in back pay and other expenses incurred during his service with the army. Lowe was now seeking $10 per day for the period April 8 to June 31, 1863. Comstock and the army had taken the aeronaut seriously when he volunteered to remain on duty without pay during this period, but Lowe was having second thoughts. His other charges ranged from $1,500 for a balloon he believed the government had agreed to purchase to miscellaneous travel expenses ($1,558.47) and $142.50, which he had paid to W. J. Rhees of the Smithsonian for assistance in preparing the final report to Secretary Stanton.[137]

Maj. Gen. Edward Canby of the adjutant general's office rejected out of hand all but $280 of Lowe's claim in December 1863. His reasons varied with each item, but official displeasure and animosity toward Lowe were abundantly clear. In answer to the aeronaut's request that the department reimburse him for Rhees's work, Canby remarked that the army had not been responsible for the poor state of Lowe's files and should not be expected to pay for putting them in order.[138]

Nor had the army forgotten Jacob Freno's charges. Canby informed Lowe that the $280 in expenses that were allowed would not be paid to the aeronaut until he could explain a number of apparent irregularities.

> *Neither should the amount be paid until Mr. Lowe can satisfactorily account for a government saddle, which was concealed in a trunk by his father, and conveyed from Harper's Ferry to Philadelphia, October 3, 1862, two silk balloons for which he was responsible in 1862, and a lot of new Government tents, poles, and pins, which together with a new silk balloon belonging to the Government was conveyed by the Steamer* Rotary *from the Peninsula to Philadelphia on account of Mr. Lowe. Explanation should also be made by Mr. Lowe, in relation to forage said to have been drawn by him from the Government and sold to J. C. Jackson about 1st September 1862.*[139]

Lowe did his best to reply to Canby's rejection in April 1864, but the army remained adamant. The aeronaut's claims for travel expenses based on verbal orders were disallowed. If the aeronaut wished reimbursement for the balloon in question, he should present evidence that the government had ordered the craft, orders issued for its use in the field, and proof that the envelope had either been properly disposed of or sent to another command. The forage claims were rejected because bills that sounded suspiciously like these had been paid in 1861.[140]

By mid-July 1864 Lowe was forced to ask Joseph Henry for help in collecting his account. Henry assigned Rhees to the task, and a partial payment was eventually made to Lowe. No serious thought seems to have been given to bringing charges against him for the missing equipment. Neither, on the other hand, was any serious thought given to rehiring Lowe or any other aeronaut.[141] Not until 1893, three decades after T. S. C. Lowe's last military flight, would the U.S. Army count another balloon in its inventory.

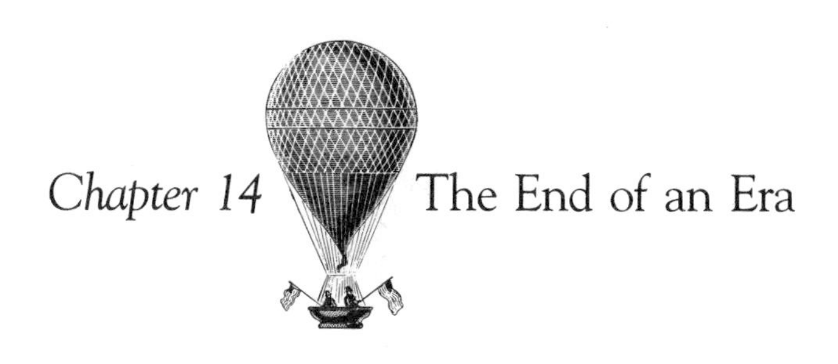

Chapter 14 The End of an Era

THE BALLOON CORPS SLOWLY DISSOLVED DURING THE LATE SPRING AND summer of 1863. A number of former military aeronauts, notably the Allen brothers and John Wise, returned to their prewar careers as exhibition balloonists. For others the Civil War marked the end of involvement in aeronautics.

William Paullin, having been dismissed from the corps for moonlighting as an ambrotypist, turned to photography full time. He died in Philadelphia on December 1, 1871. As one biographical sketch noted, "his intellect was affected for some time before his death."[1]

John La Mountain made a few ascensions following his altercation with Lowe. One of the most terrifying experiences of his career occurred while flying from Bay City, Michigan, in October 1869. La Mountain was operating an old balloon, so leaky that a Mr. Headley, who had been promised a ride, was forced to leave the basket to enable the craft to lift off. The aeronaut then rose into the scudding clouds as though shot from a cannon. He found himself immersed in a mixture of snow and rain that coated balloon, basket, and aeronaut with a layer of ice.

Discovering that the valve was frozen shut, La Mountain yanked the valve line so hard it broke. He climbed onto the load ring and ripped a fifteen-foot tear on each side of the envelope, then returned to the basket, hoping for the best. As the benumbed aeronaut hovered three miles above the earth, he heard the fabric ripping still further, "until, with a crash that sounded like a death knell, the cloth gave way to the pressure, opening a seam on both sides from the bottom to the top." The fabric gathered in the top of the net, slowing La Mountain's descent. In describing his feelings to a newsman, he remembered "distinctly passing through a cloud, and the sensation on regaining sight of the earth. He has an almost distinct recollection of approaching the earth's surface. A dull moaning like the surging of the waves greeted his ears, the flapping of the cloth became louder, and a moment afterward he became unconscious. On regaining his senses, he found himself lying in a wood and his balloon some yards distant."[2]

La Mountain escaped without broken bones or internal injuries, but it was one of his last flights. Only forty-eight years old, he died at home in Lansingburgh, New York, in 1878.

La Mountain's brother Edward, whose application for service as an assistant aeronaut had been refused by military authorities in 1862, came to a more spectacular end. Having established himself as a jeweler in Brooklyn, Michigan, La Mountain continued to accept an occasional exhibition contract. His balloon was a homemade hot air craft with "somewhat of a worn appearance, as if the worse for having been filled with heated air too often." Six lines leading from a twelve-inch wooden block on top of the balloon supported the basket. In the absence of any horizontal lines to keep the envelope from slipping through the crude rigging, the aging balloon was a disaster waiting to happen.

On July 3, 1873, Edward La Mountain was scheduled to fly an exhibition at Ionia, Michigan. The aeronaut arrived in town at the last minute, dressed in the same clothes that he had been wearing for the previous ten days while nursing his desperately ill wife. Anxious to make a quick flight and return home, the aeronaut took fewer precautions than usual. He took off at 2:55 that afternoon, waving his hat to the cheering crowd. Moments later the elation turned to horror as the envelope shifted 90 degrees, spilling heated air and allowing the balloon to slip partially through the vertical ropes.

Plummeting to earth, La Mountain was observed attempting to use his small laundry basket as a parachute. He struck the earth "with a terrible thud that jarred the ground for fifty rods around, and made an indentation in the solid ground eight inches in depth."[3] A local newspaper treated its readers to a description of the body "limp as a rag," with "every bone broken to fragments."[4]

John Steiner disappeared from the scene a bit more quietly. He swung through the upper Midwest after leaving the army, introducing Count von Zeppelin and thousands of others to the wonders of flight. In May 1863 "Captain" Steiner was booked into Philadelphia's Bushnell Garden.

Steiner gradually scaled his activity down, disappearing from the nation's newspapers. By the mid-1870s his name appeared only occasionally, as when he assisted John Wise and his partner W. H. Donaldson in the inflation of the giant *Daily Graphic* balloon. Like Paullin and La Mountain, Steiner's aeronautical career had come to a close.

Lowe was quick to abandon ballooning as well. In an effort to retrieve his financial situation, the chief aeronaut offered the citizens of Philadelphia an opportunity to view their city from the basket of the balloon *Washington*, which

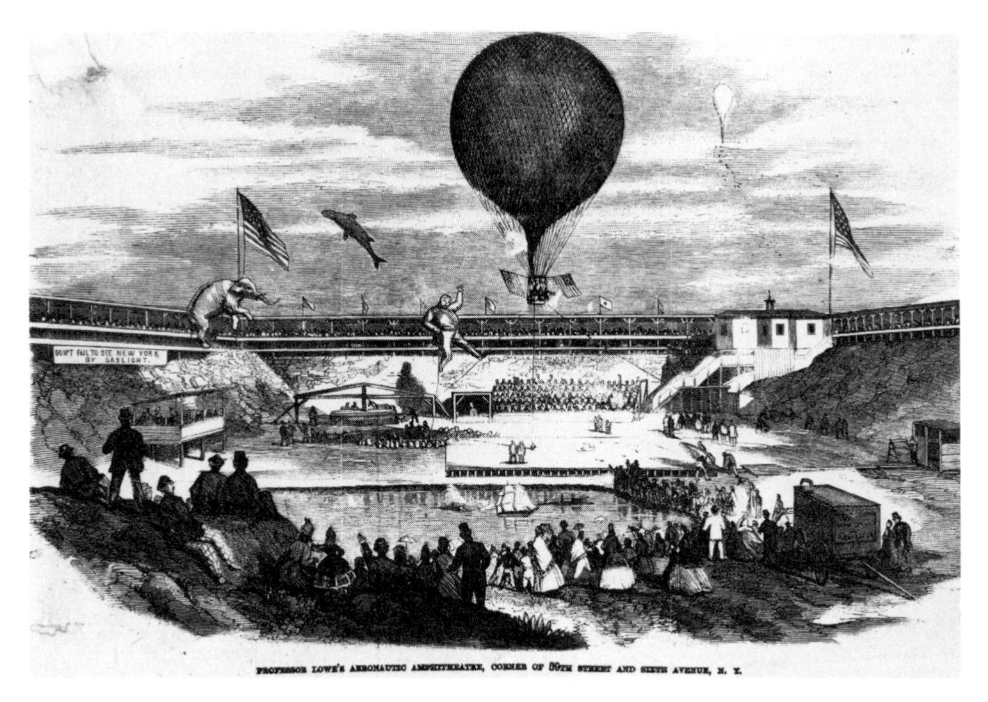

Lowe's "Aeronautic Amphitheatre" at the corner of 59th Street and Sixth Avenue

he had purchased as war surplus. A poster dated October 29, 1864, announced that "15,000 persons have thus ascended in balloons directed by Professor Lowe, *without the least inconvenience or danger.*"[5] In addition to his daily schedule of tethered passenger flights from the corner of Coates and Fifteenth streets and Ridge Avenue, Lowe made free flights at 3:00 P.M. each Wednesday and Saturday, weather permitting.

Lowe transferred his operations to an "Aeronautical Amphitheatre" at the corner of Fifty-ninth Street and Sixth Avenue in New York during 1865–1866. Here, as in Philadelphia, he continued to offer tethered rides and free ascensions. In addition, he added a number of spectacular touches to increase attendance. Lowe reintroduced large figure balloons, sending elephant, fish, and human effigies soaring out of his enclosure. Small-scale naval engagements were staged with model ships in a large artificial pond.

Lowe also expanded his aerial repertoire. Harry Leslie, a local gymnast and trapeze artist, performed stunts on a bar hanging beneath the basket. Lowe even staged the first aerial wedding. On November 8, 1865, Dr. John F. Boynton wed his fiancée at the Fifth Avenue Hotel. The bridal couple and attendants then repaired to Lowe's Amphitheatre for a honeymoon jaunt. The Boyntons had originally planned to be married in the basket, but were unable to persuade Rev. H. W. Beecher to perform the ceremony aloft. They did exchange informal vows aboard the balloon *United States* during the short flight to Mount Vernon, New York.[6]

By 1866 Lowe's interests were rapidly shifting to new areas. In that year he began experimenting with artificial ice-making machinery. Within two years he had outfitted a steamer to carry fresh meat, fruit, and vegetables from Galveston to New York.[7]

When Lowe's refrigeration company failed, he turned to the production of gas and coke in regenerative furnaces. By 1867 the New Lowe Coke Ovens were producing a quality of coke as high as that obtained from Europe. In later years he became well known as the constructor of an inclined railway up Mount Lowe, California.[8]

Although Lowe abandoned aeronautics in 1866, he continued to encourage the Allen brothers, the only Balloon Corps veterans who resumed active long-term careers in ballooning. The Allens apparently began making civilian ascensions as early as November 1863, but their first major public flight was made from Wilmington, Delaware, on July 4, 1864. Together with their partner, a Mr. Horning, the brothers agreed to donate one-third of the proceeds to benefit the families of soldiers.[9]

Following this flight over "the beautifully wooded and romantic valley of the Brandywine," James Allen moved on to New Bedford, Massachusetts, where he and Ezra were back in the air on September 15. They were operating a 33,000-cubic-foot surplus army balloon that stood 60 feet high and measured 38 feet in diameter. In 1865 the Allens began operating two balloons, the *Empyreon* and the *General Grant*. On July 22–23 the brothers made two twin ascents from Lowell, Massachusetts. That August James ascended from Kennebec, Maine, "in fine style."[10]

When the brothers began their 1866 season in their hometown of Providence, Rhode Island, local newspapers proudly boasted that the balloonists had become "the most popular institution in this city."[11] Young James K. Allen, James Allen's eldest son, was occasionally flying with his father and his uncle. James K. would eventually join the family firm as a full-fledged aeronaut. His sister Lizzie would also become a balloonist. The youngest child, Malvern Hill Allen, named after the climactic Seven Days Battle during which his father had drawn particular praise from T. S. C. Lowe, would also fly after 1881.

James and Ezra began the 1866 season with a new balloon, the *Monarch of*

The balloon wedding party, November 8, 1865

the Air. Unveiled at Providence in early June, the craft stood 75 feet tall, measured 38 feet in diameter, and had a gas capacity of 40,000 cubic feet. The very large basket was a completely enclosed and roofed wicker structure, complete with windows. Bedecked with flags and ready for flight, the *Monarch of the Air* was, as one observer commented, "a beautiful sight to behold."

When the new balloon was first flown on July 6, 1866, the passenger list included James Allen, James K. and his cousin Samuel F. Allen, and local citizens O. W. Prince and Thomas Landy. Anxious not to disappoint the crowd, Allen took off in spite of thick fog and an approaching storm. The first attempt to launch the craft, however, was a failure caused by overballasting. "On the second attempt it attained a higher altitude, and amid the wild cheers and stirring music shot away before the gale up the valley, apparently on the line of the Boston Railroad, with terrific velocity."[12]

Teamsters working at the firm of Morrill and Company, manufacturers of printing ink, saw the *Monarch* passing three miles west of Dedham, Massachusetts, at 6:15 that evening. The balloon dropped into a wood, then rose again, and was driven into a grove of high elms. The balloon was "nearly demolished." The passengers, who had been riding on the roof of the car, were scattered through the woods, shaken but not seriously injured.[13]

For ordinary citizens, the wild ride aboard the *Monarch of the Air* would have provided excitement enough. The Allen brothers, however, apparently longed for the days when they had flown with the Army of the Potomac. When T. S. C. Lowe offered them an opportunity to resume the life of military aeronauts, they leapt at the chance.

Lowe had received a flattering letter from Emperor Dom Pedro of Brazil in 1866. For more than a year Brazil and her allies Argentina and Uruguay had been at war with Paraguay. The trouble had begun when Francisco Solano López, the dictator of Paraguay, closed traffic on the Paraguay and Parma rivers to all foreign shipping. Since certain provinces of Brazil and the Argentine Confederation could only be reached by these rivers, the War of the Triple Alliance was inevitable. The war continued for five years (1865–1870) and brought unbelievable hardship to the people of Paraguay.

Hopelessly outnumbered, the despot López continued to feed troops into battle until virtually the entire male population of the nation had been exterminated. When the war finally came to an end with López's death, Paraguay had lost 50,000 square miles of territory. Her prewar population of 1,200,000 was reduced to an estimated 200,000 women and 28,000 men.

Dom Pedro had followed the course of the American Civil War with interest. He had been particularly intrigued by the Balloon Corps. In his letter to

Lowe, the emperor outlined his plan to establish an aeronautic corps for the Brazilian Army. He offered Lowe command of this organization, complete with a commission in the Brazilian Army, a contract for all balloons and equipment, and a handsome salary.[14]

Lowe basked in the glory of the offer but was unwilling to resume the hardships of life as an observation balloonist. Instead, he offered the job to the Allen brothers. "They were," Lowe remarked, "young and still thrilled to the spirit of adventure."[15]

The spirit of adventure was not the only inducement drawing the Allens to Brazil. Dom Pedro offered a monthly salary of $150 to these first two American aerial mercenaries. At the time of their return to the United States in the spring of 1868, the brothers received an additional $10,000 bonus.[16]

James and Ezra Allen arrived in Brazil in the spring of 1867. As they informed their sponsor T. S. C. Lowe, a number of difficulties had to be overcome before they could begin operations.

> *In the first place the iron and acid that we ordered at Rio did not and has not up to this time arrived. We both thought we had taken every precaution to have everything on board the ship that we came down in . . . the acid and iron being particularly mentioned, but upon arriving, found the most important item had not been shipped . . . we learned, however, that it had been put on board another ship that sailed the same day. . . .*
>
> *We however set to work to do something and found that a Frenchman that had been here had iron and acid at Correnties . . . but upon opening the barrels [we] found it to be wrought iron of enormous sizes from 10 penny nails up to 5, 10, and 15 pounds of junk, but nothing daunted, went to work to do what we could.*[17]

The Allens produced sufficient hydrogen to inflate the smaller of their two balloons, while a cooperative Brazilian general procured additional zinc in Montevideo. The brothers also used their time in revarnishing the balloons and preparing their equipment.[18]

By mid-July the brothers had made ten to a dozen tethered ascensions at the front with their small balloon. Ezra noted with ill-concealed satisfaction that these were "the finest ascensions that we ever made and what we cannot say of the United States Army, it is appreciated by all concerned."[19] Certainly the Brazilian commanding general Marquis de Caxias was "more than pleased."[20]

The Allens claimed to have made "hundreds" of ascents in South America, most of them associated with the siege of Humita, a major stronghold on the

Paraguay River. Not surprisingly, the aeronauts informed local reporters that their activity had been a major factor in forcing the evacuation of Humita. Lowe reported that he had received a letter of thanks from Dom Pedro, "who gratefully acknowledged his indebtedness to the American aeronaut, at the same time informing him [Lowe] that by carrying out his methods, the war had been closed after a single battle."[21]

As Harris G. Warren, an American historian of the War of the Triple Alliance, has noted, the Allen brothers were much less pivotal than they would have their friends in Providence believe. Warren claims that crude smoke screens were used to mask operations from the eyes of the aeronauts when heavy artillery fire failed to damage the balloon. The ascents, he claimed, had little effect on the outcome of the siege.[22]

Whatever the nature of their contribution to the Allied victory, the Allens were back in Providence in April 1868 bearing letters from the Marquis de Caxias and other Brazilian officials attesting to their satisfaction with the performance of the aeronauts. The Brazilian minister of war went so far as to remark that James Allen had been "successful beyond all expectation" and stated that he believed him to "be worthy not only of the name of Chief Aeronaut of the Brazilian Army but of the whole world. For his prowess and quick conception I do not think his equal exists."[23] Whatever the reality behind this glowing endorsement, the Allen brothers returned to Providence as genuine heroes.

And heroes they would remain. The Allens continued to fly well into the twentieth century, earning national fame as America's "First Family of Aeronautics."

Not long after their return from South America, James and Ezra dissolved their partnership, each man continuing to fly with his own family. Ezra, although much less active than James, made scattered flights throughout New England. After Mary Frances Penno married Ezra, she also entered the family profession. Mary Allen earned minor fame as an aeronaut during a twenty-year period, finally abandoning aeronautics after suffering a broken leg in a crash several years before her death.[24]

James Allen and his sons James K. and Malvern Hill had completed a grand total of 481 ascensions by July 1891, at which time all three men were still very active. Lizzie and E.T., James's daughter and other son, made occasional flights, as did James K.'s wife and four daughters.

Even the Allens were not purely professional aeronauts. Although they were on the road with their balloons each summer, James continued to work as a printer and James K. as a box maker during the off-season. Long-term contracts with various cities, notably Boston and Providence, formed the basis of their balloon income. Their relationship with the city fathers of Boston, where they were named "official" aeronauts in 1869, 1870, 1874, 1881, and 1884, was typical. They not only provided mass entertainment on the Fourth of July and other public occasions but also offered thousands of Bostonians an opportunity to view the city from the basket of a tethered balloon. "Going up in a balloon has been rendered so safe and pleasant by Professor Allen that an average of a hundred people daily take experimental trips in his *Castle in the Air*, over a third of which number are ladies," commented the *Boston Evening Transcript* on June 21, 1872.[25] Boston was particularly proud of young James K. Allen, who, as one reporter noted, "is no chicken."[26]

While the family remained based in the Boston-Providence area, their activities ranged throughout the Northeast. Bangor, Lowell, Worcester, Norwich, and a dozen other New England towns thrilled to performances by the Allens, as did larger cities like Troy and Paterson in New York and New Jersey. Operating a variety of balloons, including the *Venus*, *Empyreon*, *Castle in the Air*, *Green Mountain*, *Sierra Nevada*, *Comet*, *Veto*, *Glory of the Skies*, and others, most of which were cut and sewn by Allen's wife, the family occasionally extended its activities beyond the Northeast.

In the fall of 1873 James Allen signed a contract to perform tethered ascents at a San Francisco beer garden operated by R. B. Woodward, a former resident of Providence. Allen announced that he would construct a gigantic 200,000-cubic-foot aerostat for use at Woodward's Garden. Woodward and Allen were reported to be considering a transcontinental voyage aboard the large balloon at the close of the season.[27]

When Allen arrived in San Francisco early in 1874, there was no sign of the large balloon. Rather, he brought the *Castle in the Air*, a 70,000-cubic-foot craft that Allen had used for tethered ascensions in Boston since 1872. The aeronaut also brought an assistant, John Starkweather, an old comrade from the Balloon Corps.

Tethered ascents from Woodward's Garden began on the afternoon of February 21, 1874. George Davidson, an official of the U.S. Coast Survey, and reporters from the *San Francisco Call* and the *Examiner* accompanied Allen to 900 feet on the first flight, which tested the operation of the steam-powered winch that held 1,000 feet of cable. Woodward made the second ascension with his son and daughters, while a representative of the Associated Press, a reporter for the *Bulletin*, and Joseph Gruet, a local aeronaut, joined Allen on the third trial. Gruet, a former acrobat and gymnast, had made a number of hot air balloon flights in the area, billing himself under the *nom de l'air* Etienne Buislay.

Throughout the early spring of 1874 Allen and Starkweather "made hay while the sun shone and the wind didn't blow," as one local commentator noted. Multiple ascensions were made on suitable mornings and evenings. Allen's fame spread up and down the coast, and there seems to have been no dearth of citizens anxious for a bird's-eye view of the city.

Allen made at least one free flight that April aboard the balloon *Sierra Nevada*, which may simply have been the *Castle in the Air* renamed. Launching from Woodward's Garden at 4:00 P.M., Allen and Starkweather carried four passengers, including Gruet, on a flight that ended safely on the Overacker Ranch, between Centreville and Niles Station, California, at 6:15 that evening.[28]

Allen seems to have cut his West Coast tour short soon after the *Sierra Nevada* flight. Gruet inherited the position of chief aeronaut at Woodward's Garden. He continued to fly through the summer and fall of 1874, launching fireworks from his balloon and staging aerial races with Mr. Martin, another local aeronaut.

Gruet, or Buislay, as he was usually identified in the press, continued to favor hot air ascensions. He particularly delighted in performing acrobatic stunts on a trapeze bar in place of a basket beneath his balloon. This was an extraordinarily dangerous practice, for it left Gruet with no control over the balloon.

Gruet paid the ultimate penalty for flying under these conditions on October 11,1874. Having performed his stunts on the bar, the acrobat rode the balloon to a safe landing, only to become tangled in some dangling lines. Carried back into the air by a sudden gust and with the air in the balloon cooling rapidly, Gruet fell onto the rocks of Bernal Heights. There he was found unconscious with severe cuts and bruises as well as head and internal injuries. Paralyzed from the waist down, he was carried to his home, where he died several days later.[29]

Following James Allen's short stay in San Francisco, the family's only long-term contracts for flights were with the officials of Dayton, Ohio, where James K. Allen performed each summer from 1876 to 1881. Other members of the family, including James Sr., sometimes accompanied James K. on the Ohio tour, but young Allen remained the favorite with Daytonians.

A *Dayton Journal* reporter commenting on a James K. Allen flight on June 20, 1877, remarked that the aeronaut was "a young man, lithe and muscular" with a luxuriant growth of wavy blond hair and whiskers. On this occasion, his twenty-seventh flight, twenty-six-year-old James K. flew with his wife and daughter. Mrs. Allen was attired as the goddess of liberty, complete with coronet.[30]

In spite of their activity in San Francisco and Dayton, the Allens remained the preeminent "New England Aeronauts." During the course of a joint aeronautical career that spanned six decades, the members of the Allen family concentrated on pleasing the crowds at state and county fairs, Fourth of July celebrations, and other festive local gatherings throughout the northeastern United States.[31]

The Allens were relative purists, members of a community of aeronauts composed primarily of prewar balloonists who continued to operate the expensive gas balloons as opposed to the hot air or smoke Montgolfiers, which were much cheaper to build and required very little expertise to operate. The family eschewed the flashy acrobatics and the spangled, abbreviated costumes preferred by newcomers to the game.

This did not mean that the Allens were not superb showmen. They hired a professional acrobat, Fred Barnet, to perform on a bar beneath their basket. They also staged special ascensions designed to draw maximum public attention to their activity. Following T. S. C. Lowe's example, they performed a grand total of five aerial weddings, two in Dayton, Ohio, and one each in Providence, Lowell, and Manchester. In most of these cases a wedding service performed with the couple standing in the basket was followed by a honeymoon flight in the "Bridal Chariot." Two of the weddings became front-page items in the nationally distributed illustrated newspapers of the era.[32]

The Allens were past masters at the art of garnering headlines with their balloons. "The old sky sailor" James Allen made the front page of newspapers across New England when he arranged an aerial announcement of the score of the 1891 Harvard-Yale game by means of specially designed blue pyrotechnic balloons.[33]

James K. Allen's distribution of hundreds of gift coupons from the basket of the *Boston Post* on Memorial Day 1892 also proved very newsworthy.[34] As one Boston reporter noted: "It rivaled in its way the famous rain of manna. But the *Post*'s cornucopia differed somewhat from the original manna idea, in that it distributed not simply one unseasoned article of diet, of which the ancient wanderers, it will be remembered, grew weary, but orders for an infinite variety of food and drink and wearing apparel, house furnishings—even partly paid for house cigars, livestock, books, stocks and bonds, excursion tickets, even some sets of beautiful false teeth."[35]

The Allens were rightfully proud of the admirable record of safety they compiled during their sixty years aloft. Nevertheless, they experienced more than a few frightening moments. James Allen actually fell from his balloon basket on three separate occasions while flying at very low altitudes. Each time he escaped serious injury. While flying from Rochester, New Hampshire, James K. Allen's balloon burst at an altitude of two miles. As usual, the fragments of the balloon gathered in the upper hemisphere of the net to form a parachute, which returned the aeronaut safely to earth eight minutes after the explosion. As Allen later informed a reporter, "to him the time seemed as many hours."[36]

James K. noted that his most frightening experience occurred during an ascension from Bolton, Vermont. Allen reached an altitude of over 20,000 feet on this occasion.

> *He passed through seven layers of clouds and encountered rain, snow and the brightest sunshine in rapid succession. At the highest part of the voyage his companion fainted, and Mr. Allen had just strength enough at command to pull the valve cord and allow the gas to escape. The feeling that came over him was one of extreme dizziness, coupled with stinging pains in the head and ringing in the ears. The sight presented by the banks of clouds, he said, was marvelously beautiful. So solid did the vapory masses look to the voyagers that they were almost tempted to step over the basket's side and tred [sic] the snowy summits. He said he could have sworn he was on a range of snowcapped mountains, had he not known that his imagination was at work. The disembarkation was effected in Canada, thirty-eight miles from the place of starting.*[37]

The elder James Allen suffered his most serious injuries during a take-off from Haverhill, Massachusetts, on July 12, 1883. His balloon was blown violently against a building, throwing the two passengers into a jumble and resulting in a fractured collarbone and broken rib for the aeronaut.[38]

Malvern Hill Allen suffered his share of stark terror as well. Climbing to an altitude of two miles aboard the balloon *What Cheer* after launching from Providence on July 4, 1889, young Allen found the envelope swelling toward the danger point. Unable to vent sufficient gas through the valve, the aeronaut yanked the rip line in order to parachute back to earth. The crowd, which included Allen's bride, were horrified. "People turned away from the sight with blanched faces," one newsman reported. "When they looked again the balloon had shot below their line of vision. A rush was at once made for East Providence, where it was supposed the aeronaut had fallen . . . he was found, not a mangled corpse, but pale, though smiling. He had landed in the middle of a softbottomed meadow, and this had undoubtedly saved his life."[39]

James K., Malvern Hill and Mary Frances Allen (Ezra's wife) all suffered accidents on July 4, 1891. Mary Allen's twenty-seventh flight had begun from Dexter, Massachusetts, at 4:20 P.M. Caught in a high wind, she descended twenty minutes after take-off and was blown across an open field, breaking her ankle.[40] James K. Allen, flying with four passengers from Boston Common that day, was blown through trees and over stone walls during a similar high-wind landing

James K. Allen escorts an aerial wedding party aloft.

that seriously injured one of his passengers. Malvern Hill Allen took off from New Bedford that afternoon. His balloon was blown through a grove of trees directly into ex-Mayor Wilson's chimney. Allen was thrown from the basket but suffered only minor injuries.[41]

James K. Allen made his most publicized and daring flight on July 4, 1906, at the very end of his long career. Launching from Providence at 12:30 P.M., Allen was carried northwest, circling around Attleboro before being blown south down Narragansett Bay and out to sea. The aeronaut spent the night flying parallel to the Massachusetts coast, passing near both Block Island and Martha's Vineyard. The next morning a fishing dory caught the balloon's dragline, enabling Allen to step into the boat without getting his feet wet. Newspapers hailed Allen's return from the dead, but the aeronaut remarked only that he was "ready for another trip."[42]

In the aftermath of the rescue, James K. even remarked that he would like to attempt an Atlantic crossing if the proper equipment were made available. He would require a 350,000-cubic-foot aerostat for the trip, with a fully provisioned lifeboat. With this on hand, Allen noted, "I'd just as soon sail across the Atlantic in a balloon as ride up Westminster Street in a trolley car."[43]

James K. Allen's comments on a potential Atlantic flight were nothing more than idle remarks for the benefit of the press. During the decades following the Civil War, however, other American aeronauts were taking serious steps toward achieving this old and elusive goal.

John Wise remained at the center of Atlantic enthusiasts. Already age fifty-three at the time of his unsuccessful trial as a military aeronaut, Wise was in semiretirement by the end of the war and made very few flights. By 1870, however, his name was once again appearing in print as the author of a number of articles on ballooning.[44]

Ultimately, Wise was unable to retire from aeronautics. In 1870 he flew the balloon *Vesperus* from Fredericksburg, Virginia, 13 miles to Stafford Court House. He ascended from Chambersburg, Pennsylvania, on July 29 that year. After his landing, Charles Wise made a second flight from the same place with his son John.[45] Having returned to the active life of an aeronaut, Wise was not long in reviving his dream of an ocean crossing. Recognizing that his advancing age represented a serious liability in launching yet another fund-raising effort to support such a flight, Wise began looking for a younger partner. He was particularly intrigued by W. H. Donaldson, a newcomer who had built a spectacular reputation as an aeronaut in a few short years.

Washington Harrison Donaldson was born in Philadelphia on October 10, 1840. A natural athlete, young "Wash" idolized the gymnasts and acrobats that

John Wise in middle age

appeared at local theaters. Donaldson was soon emulating his favorite performers, amusing his friends by balancing himself on top of an eight-foot ladder, walking a homemade tightrope, and exhibiting his skills as a magician and ventriloquist. He launched his stage career in 1857 with a series of appearances at Philadelphia variety theaters. Donaldson varied his act with feats of magic and knife throwing, his brother Augustine serving as a target.

By 1862 he had abandoned the stage to concentrate on developing his abilities as an aerialist. On May 2, 1862, he crossed the Schuylkill River on a 1,200-foot tightrope, concluding his performance with a 90-foot leap into the water. Always searching for new ways to thrill an audience, in 1864 he crossed the Genesee Falls on a high wire 1,800 feet long. In 1869 he became the first performer to ride a velocipede on a tightrope. Between 1857 and 1871 Donaldson made over 1,300 appearances in one or another of his various theatrical guises. He had, by this time, begun to consider the potential of the balloon as an acrobatic platform.[46]

Donaldson's first balloon ascent was made at Reading, Pennsylvania, on August 30, 1871. His balloon, the *Comet*, boasted an 8,000-cubic-foot gasbag, usually filled with illuminating gas drawn from a local main. When the aerostat refused to rise on this occasion, Donaldson, who was completely unfamiliar with the operation of a balloon, threw all of his ballast overboard. Still unable to clear the surrounding buildings, the novice aeronaut soon found himself resting on the roof of a house on Court Street. Donaldson reflected for a moment, then tossed a rope, grappling iron, hat, coat, boots, and provisions from the car. A quarter of a mile in the air, he climbed onto the hoop from which the basket was suspended and practiced gymnastic feats. After an 18-mile journey, the *Comet* began to descend. With no more ballast on board to be jettisoned in an effort to check the speed of his fall, Donaldson struck the ground violently and rebounded into the air "as high as a house." Coming to earth a second time, he was able to grab a rail fence and hold the balloon down while he loaded the basket with stones.

In spite of this near disaster, the aeronaut announced his intention of making a second flight the following week. This time a trapeze bar would be substituted for the normal balloon basket. This second ascent began more auspiciously than the first. The balloon rose straight into the air with the intrepid aeronaut perched on his bar beneath the 26-foot gasbag. He fell backwards, catching his heels on the trapeze, then sprang upright again. He hung by one hand, one foot, and the back of his head, drawing a succession of hearty cheers from the appreciative audience below. During the remainder of his 15-minute performance, he scattered handbills over the neighborhood and continued his gymnastics on the bar.

As Donaldson passed over the city limits of Reading, he began to valve gas in order to descend. Once again his inexperience was to prove costly. A vigorous tug on the valve line caused a rip to appear in the fabric of the gasbag. This enlarged until three-quarters of the balloon was torn. Fortunately, the bag turned

Washington Harrison Donaldson

Donaldson's balloon bursts.

inside out and acted as a parachute to break Donaldson's fall. The aeronaut leaped into the rigging as the damaged aerostat was blown across the roof of a house before ending its journey in the branches of a large oak.

Donaldson continued to make trapeze flights with a new aerostat, the *Gymnast*. This balloon had a 10,000-cubic-foot gas capacity but did little to improve the aeronaut's luck. During the course of a flight from Norfolk, Virginia, on January 18, 1872, the gas valve froze shut. As the balloon continued to rise, the gas expanded, finally bursting the bag at an altitude of one mile. Terrified, Donaldson clung to the balloon rigging as he plunged into a chestnut tree "with the velocity of a catapult." He returned to Norfolk that evening, his tights torn to shreds and "his body bleeding from a hundred gashes."

Four days later Donaldson once again encountered problems with a balky valve. This time he was forced to climb into the rigging and slash the balloon with his knife. The craft came to earth in a cornfield where a strong wind blew it over one thousand feet into a rail fence. Undaunted by his less than impressive performance to date, Donaldson embarked on a midwestern tour in the spring of 1872. Ascending from Columbus, Ohio, on March 28, he struck a smokestack, ripped yet another gasbag, and "dropped rapidly into an adjoining yard."

Flying from Chillicothe, Ohio, a month later, Donaldson's balloon was blown into the wall of a drugstore "with great force." He then smashed into and destroyed two chimneys before rising above the rooftops. At an altitude of one mile, Donaldson found that he was too bruised and exhausted to perform on the trapeze. He descended in a field outside of town, "rebounded three times, striking first a stone wall, then a tree, then a fence, to which at last I succeeded in fastening my trapeze rope." The aeronaut, who by this time must have felt that he had faced virtually every possible danger to be met in the air, encountered yet another difficulty during an ascent from Ironton, Ohio, on May 20. As the gas expanded with increasing altitude, Donaldson was startled by an ominous series of cracking sounds and realized that the netting connecting the trapeze bar to the gasbag was giving way. "This caused me to feel very uneasy and on that account the voyage was devoid of the pleasure to me that usually attends the trip."

The aeronaut continued his tour through small midwestern towns, rebounding off a series of walls, chimneys, and roofs, and coming to rest in an assortment of trees, lakes, and muddy fields. By July 4, 1872, he was in Chicago, a city whose citizens, Donaldson commented, "never feel satisfied until they kill someone or have an explosion or conflagration that will startle the world." Having passed completely over the city, the aeronaut descended safely into Lake Michigan. Unable to deflate his balloon, he was blown through the water until he

TA
Terrible Position.—Netting Breaks and Threatens Donaldson with a Horrible Death.—12th Ascension.—Ironton, May 20, 1872.

Donaldson's mishap over Ironton, Ohio

An incident from Donaldson's 13th ascension

struck a stone pier near the beach.

True to form, by the spring of 1873 Donaldson had become restless repeating his standard balloon trapeze performance. He felt that his growing body of experience in the air had prepared him to attempt a flight that, if successful, would spread his fame around the globe. Like so many balloonists before and since, Donaldson had begun to dream of crossing the Atlantic by air.

Donaldson agreed with John Wise's contention that strong easterly winds to be found at an altitude of 2½ to 3 miles might make it possible to travel the 2,600 miles in less than three days. His plans called for a balloon measuring 80 feet in diameter with a gas capacity of 268,000 cubic feet. Such a craft would require 2,300 square yards of fabric and could lift a gross weight of 9,380 pounds. In addition, Donaldson planned to carry two smaller balloons, each 32 feet in diameter, in which he could store an extra gas supply. An even smaller aerostat would be attached to the gas valve of the main balloon, so that the gas released to allow the large bag to expand with increasing altitude would not be lost. A metal lifeboat suspended beneath the closed balloon basket would contain sails,

The Anchor Rips the Shingles off of a Country School-house.—The Scholars can't stand "Double Thunder."—25th Ascension.

Donaldson tears the shingles off a country schoolhouse.

oars, instruments, and sufficient food and water to sustain a party of three for thirty days. Newspaper accounts of Donaldson's projected voyage attracted a great deal of attention. The aeronaut received a flood of applications from men and women who wished to accompany him.

More important, however, Donaldson's plans had now come to the attention of John Wise. The older balloonist urged Donaldson to submit his scheme to several learned scientific groups whose approval might make it easier to raise the necessary funds. Officials of Philadelphia's Franklin Institute and Joseph Henry, secretary of the Smithsonian Institution, agreed that the transatlantic plan was technically feasible and that Donaldson and Wise had the experience and skill to accomplish the task.

With these testimonials in hand, Wise and Donaldson traveled to Boston, where they hoped to convince municipal authorities to finance the construction of the balloon. Rebuffed, the pair continued to New York for talks with J. H. and C. M. Goodsell, owners of the *Daily Graphic*, who had taken an interest in the venture. On June 27, 1873, Wise and Donaldson concluded a formal agree

Seamstresses at work on the envelope of the Daily Graphic

ment with the publishers.[47]

Wise was not happy with the agreement, which gave the Goodsell brothers absolute control over the venture. He was quite justifiably afraid that the owners of the *Graphic* would milk the flight for all the publicity it was worth. Donaldson, his new partner, seemed all too willing to cooperate in turning what Wise regarded as a scientific expedition into a circus sideshow. Nor would the balloon be ideal. Wise had hoped for silk but settled for muslin in order to stay within the budget. Still, Wise forged ahead, realizing that this venture represented his last opportunity to conquer the ocean.

By mid-July the giant balloon was under construction in the loft of the Domestic Sewing Machine Company at the corner of Broadway and Fourteenth Street. A New York tailor named Fleck laid out the patterns and cut the fabric into gores, while twelve seamstresses working under the supervision of John Wise's niece, Mrs. Ihling, who also made occasional flights, did the sewing. The construction of the balloon consumed 4,316 yards of fabric, bound by an estimated eight miles of stitching.[48]

As each section of the envelope was completed, it was moved to the Brooklyn Navy Yard, where the final assembly would take place. Six painters finished the fabric with 500 gallons of oil, benzene, and varnish, giving the *Daily Graphic* a brownish yellow appearance.

The finished gasbag was the wonder of New York. The Goodsell brothers took full advantage of this curiosity, selling tickets to the assembly area and featuring an endless series of lengthy articles on the front page of the *Graphic*. On August 9, 1873, they began publication of a new paper, the *Balloon Graphic*, solely devoted to the enterprise.

The finished envelope measured 110 feet high and 100 feet in diameter and had an estimated gas capacity of 600,000 cubic feet. A smaller balloon would be attached to the side of the large aerostat. It would be used as a gas reservoir for refilling the main balloon and as an "aerial buoy" that would enable the aeronauts to "ascend the side of the large gasbag to examine the valve, or to repair the netting, if it should become necessary."

The two-story octagonal car was constructed of duck cloth stretched over a frame of ash hoops. Light pine boards served as flooring for both stories. The upper story, fitted with four windows, would house the crew, scientific instruments, navigation equipment, signal flares, and a cage of carrier pigeons that would carry news of the expedition to a waiting world. The lower section housed ballast and provisions.

As in the case of the *Atlantic*, a large cedar lifeboat packed with thirty days' rations dangled beneath the car. Had the complete craft ever been assembled (which it was not), it would have stood 160 feet tall from the keel of the lifeboat to the 3-foot Spanish cedar and brass valve at the crown of the balloon.[49]

The Goodsells had announced that "this voyage is one of scientific inquiry and not for private gain." In spite of their expressed desire to avoid "unnecessary publicity," the brothers bowed to "public pressure" and shifted construction from the Navy Yard to the Capitoline Grounds, a Brooklyn exhibition area favored by circuses and fairs.

The construction site soon resembled a fairground, as thousands of New Yorkers crowded onto the grounds to catch a glimpse of the *Daily Graphic*. Fireworks were exhibited on several evenings and concessionaires moved onto the site. Donaldson cooperated fully, ignoring the work on the large balloon in order to make tethered ascents with his small balloon *Magenta*. The *Graphic* reported that he "had his hands full as all the ladies wanted to go."[50]

There were limits to which this sort of publicity could be stretched. By early

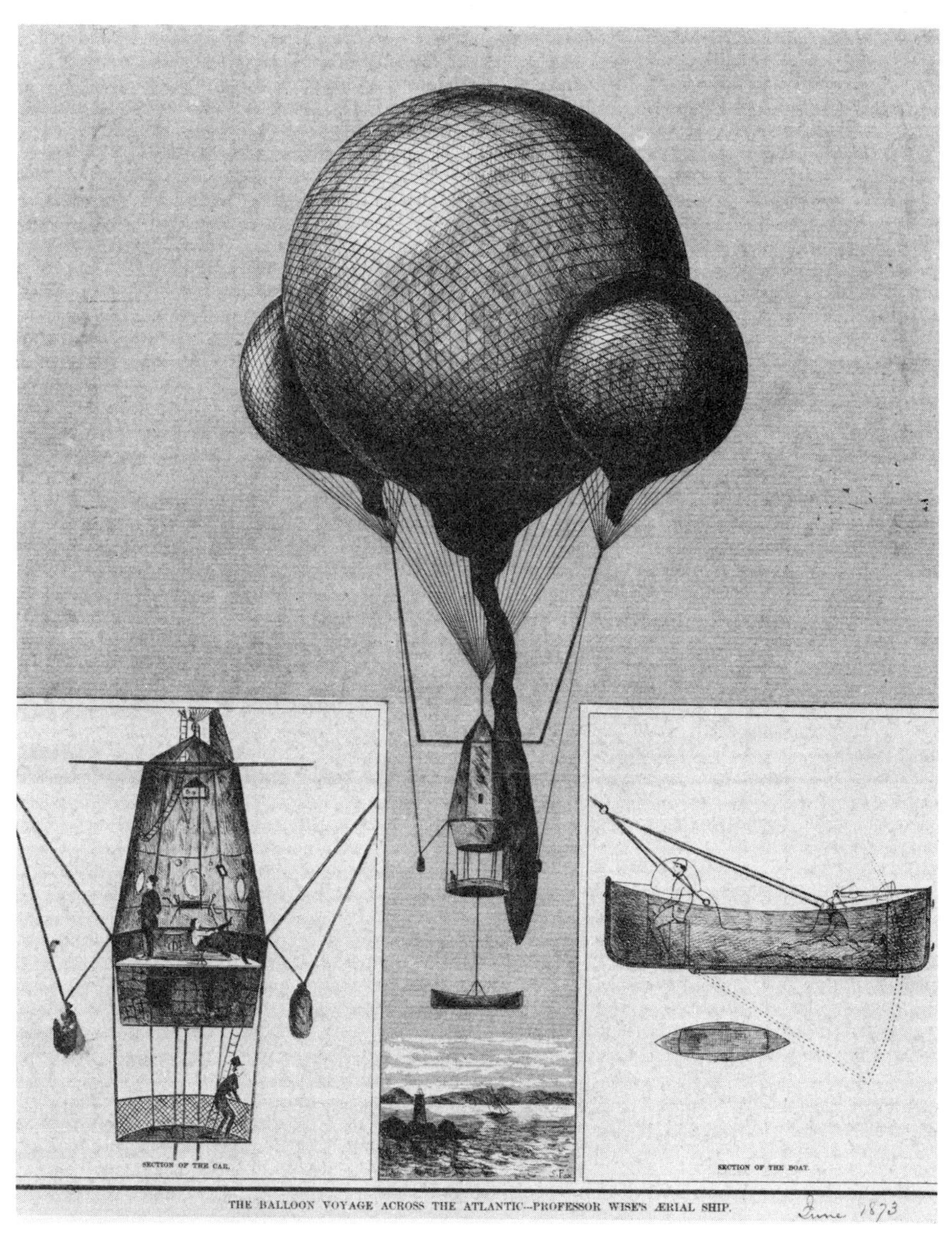

A conjectural sketch of the fully equipped Daily Graphic *in the air*

September 1873 John Wise had lost patience with Donaldson and the Goodsells. Convinced that the balloon was so weak as to be dangerous, he withdrew from the venture. The break was extremely bitter, with the *Daily Graphic* complaining that "Mr. Wise's course from the outset was marked by incapacity, cowardice, and excessive demands for money."[51]

When John and Charles Wise attempted to attend the first inflation of the balloon on September 12, 1873, they were refused admission. Charles responded by talking "loudly and called the whole balloon business a big swindle." Wise added that "the sack was rotten."[52] As a crowd gathered around the two angry men, John Wise spotted one of the Goodsells moving toward them. He shouted at the newsman, accusing the *Graphic* of having defamed his character.

Fortunately, John Steiner was on hand to intervene. Steiner, whom Wise had known for many years, was present to oversee the inflation, a task for which Donaldson was ill-prepared. The German convinced both Wise and Goodsell to discuss their differences in private. Over Charles's protest, Wise agreed and entered the enclosure.

While Wise and Goodsell were settling their differences, Steiner continued the process of inflation. All seemed to be going well when, at about 3:30 P.M., "a pretty large report was heard, accompanied by a rushing sound, and everybody looked toward the balloon, in the top of which a large rent was visible. The escaping gas extended the rent still further, and in less than a minute the balloon had entirely collapsed."[53]

John Wise had been vindicated. While Steiner attempted to protect the reputation of the enterprise as much as he could, it was abundantly clear that he blamed the failure of the balloon on poor materials, improper varnishing, and shoddy workmanship.[54]

In the aftermath of the disaster, Charles Wise reported that the Goodsells had presented his father with $12,000 for the construction of a silk balloon. Donaldson, however, was not so quick to write the original balloon off as a total loss. He convinced the Goodsells that the envelope could be salvaged. Working with Samuel Archer King, James Allen's old mentor, he disassembled the balloon, reduced the size to six gores, and restitched it. The resulting balloon, the *New Graphic*, was exactly half the size of the original.[55]

Their confidence restored, the Goodsells returned to their publicity drive. Donaldson and two representatives of the *Daily Graphic*, Alfred Ford and George Lunt, would serve as primary crew, but the Goodsells hoped to recruit a well-known figure to record the great event firsthand. They were, of course, unsuccessful.[56] The response of author James Parton was typical:

To the Owners of the DailyGraphic*:*

> *I have just received your letter. Good gracious, how flattered I feel. Not because the invitation was addressed to me, but that you ascribe such courage to me. It excites me so and almost drives me crazy to realize that I have been selected to be the official recorder of this expedition and thereby possibly chance to achieve immortal glory in pursuit of my call in that cause. However, I am just now immensely busy, hence forced to decline the great distinction. It was kind of you to think of me. Perhaps another time.*[57]

Donaldson and the two newsmen would make the flight alone, huddled in the lifeboat, the basket having been abandoned due to the reduced lift of the smaller balloon. The long-awaited flight finally began at 9:15 on the morning of October 7, 1873. The *New Graphic* rose rapidly to an altitude of 5,000 feet, then moved toward the east, cheered on by the shouts of the large crowd that had gathered to witness the ascent.

The time passed pleasantly for the three balloonists until about noon, when they encountered heavy winds and rain squalls. As the *New Graphic* passed close to earth near New Canaan, Connecticut, Donaldson and Ford leaped 30 feet to the ground. Lunt was not quick enough and traveled a few additional miles before he was able to jump into the branches of a tree. All three men were safe however.

Samuel King, who had supervised the final inflation of the *New Graphic*, was not surprised that the transatlantic flight had ended so quickly. "Had it been in the charge of a more experienced navigator it would undoubtedly have made a much longer voyage."[58]

Forced to return to the life of a traveling balloonist, Donaldson's first ascent following the *New Graphic* debacle was made from Newark, New Jersey, on October 29, 1873. Descending on Long Island, he was dragged skipping and bumping over the ground, knocking down several people and crashing through a rail fence. The wild ride came to an end only when the balloon struck the side of a house.

Donaldson remained on the ground for the next three months. Then, fully recovered, the aeronaut began a balloon tour of upstate New York during the spring of 1874. Donaldson seems to have profited but little from his three years' experience as a balloonist. Landing in a field near Canandaigua on May 7, he was almost gored by a bull. The next week, after an ascent from Savannah, New York, on May 15, he was pursued by a party of irate apple butter makers who had been covered by the sand ballast carelessly jettisoned by the aeronaut.

Throughout this period Donaldson operated as a barnstormer, alternately taking up passengers in his balloon basket and offering the standard performances

Pouring Sand in the Apple Butter.—48th Ascension.

Donaldson pours sand on the apple butter makers.

up on the trapeze bar. The uncertainties of life as an itinerant balloonist were apparently beginning to worry the aeronaut, however, for he was quick to associate himself with P. T. Barnum's organization when the opportunity presented itself in July 1874. Initially David S. Thomas, who handled publicity arrangements for Barnum's Hippodrome shows in New York, employed Donaldson to make short flights carrying newspaper reporters and others aloft. It was assumed that the grateful newsmen would reciprocate with favorable articles about the show. Thomas was so pleased with the early results of Donaldson's campaign that he provided a large aerostat with a 50,000-cubic-foot capacity. Appropriately dubbed the *P. T. Barnum*, the new balloon was large enough to enable Donaldson to carry parties of four to five passengers on relatively long aerial voyages. Donaldson's first flight with the *Barnum* began on July 24, 1875. Twenty-six hours later the aeronaut and his five passengers, all reporters for major New York papers, came to rest on an upstate farm 400 miles from the point of departure.

Four days later Donaldson made another ascent from the New York Hippodrome with Annie Thomas, the public relations manager's daughter, and three more reporters. On this occasion the group covered 400 miles and four states, landing near Thetford, Vermont, 17½ hours after the start of their flight.

Thomas then ordered Donaldson to transfer his operations to Boston in an effort to spread the good name of Barnum's "new moral show" among the members of the Massachusetts press. While flying from the Boston Hippodrome on August 18, the bag of the *Barnum* was so badly torn that the lifting gas escaped. Once again the remnants of the fabric gathered in the top of the netting so that the passengers were safely, if roughly, parachuted to earth.

Donaldson continued to Philadelphia and Baltimore, where he made a series of ascents in the repaired *Barnum*. As before, his task was to impress local newsmen with a view of their city and the neighboring countryside. On September 25 he reached an altitude of roughly 16,000 feet over the city of Baltimore.

Completing this engagement, Donaldson moved on to Pittsburgh and Cincinnati to publicize Barnum Hippodrome shows in those cities. An ascent from Cincinnati on October 14, 1874, was typical of the flights made during the aeronaut's period of employment with the circus.

Donaldson carried five passengers on this occasion. Included were Ned Ray, a student balloonist, Edwin C. Henderson of the Cincinnati *Commercial*, Louis O'Shaugnessy of the *Enquirer*, James Clegg of the *Gazette*, and Thomas Snelbaker of the German-language *Volksblatt*. The group began their trip at 4:30 P.M., moving up the Mill Creek Valley over the suburbs of Clifton, Walnut Hills, and Spring Grove. Like all balloonists, Donaldson's passengers were particularly impressed by the silence of buoyant flight and the clarity of the sounds that rose from the ground sliding beneath them: "The silence was so intense that our own instinctively low tones were met by sounds of life from below—not only the loud rattle of the passing trains, but even the monotonous croaking of frogs."

During the course of this voyage Donaldson explained the operation of the *Barnum* to the party. He demonstrated the use of the dragline, a 300-foot Manila rope weighing 150 pounds. Normally the rope was kept coiled on the bottom of the basket. When ballast was in short supply, however, the dragline could be pressed into service as an altitude stabilizer. The rope was tossed overboard with one end tied to the load ring that supported the basket. Thus the balloon would not have to lift that portion of the line dragging over the ground. As the lifting gas expanded in the sun, the balloon became more buoyant and began to rise. When this occurred, more and more rope was lifted from the ground, adding to the weight carried by the aerostat. When the gas cooled in the evening, the balloon fell closer to earth, allowing more line to trail on the ground, thus lightening the craft and slowing the descent.

The characteristic sounds made by the dragline could also be used to judge the nature of the countryside over which the balloon was passing during night flights when the ground could not be seen: "When over trees the leaves rustle . . . through cornfields it has the sound of rushing waters; in an orchard there is a little jerk as it leaps from tree to tree, and the rustle of falling fruit; over a rail fence it imitates a buzzsaw; on a board fence it is a fiddle; on a housetop it is a cello; on a barn a double bass; in grass it half whistles and half hisses; in a plowed field it is silent; over a telegraph pole it is an Aeolian harp; in water it first splashes and then is smooth."

Donaldson also allowed the newsmen to inspect the interior of the gasbag by peering through the narrow neck of the balloon. The aeronaut explained that he rarely made use of the large wooden valve at the top of the balloon, preferring to continue a flight until so much gas had been expelled through the neck by expansion that the craft would naturally drop slowly to earth. Donaldson followed this policy during the course of the trip. The aerostat became progressively heavier and periodically approached the ground as the gas cooled after sunset. When this occurred the passengers drew lots to determine which of their number would leave the group at the nearest farmhouse to lighten the load and permit the others to continue their journey.

One of the surprises of the trip was the discovery that Donaldson allowed his passengers to smoke in the basket. He felt that as long as gas was not obviously escaping through the neck of the balloon, there was little danger of fire. The *Barnum* finally came to rest shortly before nine that evening in Morgan Township, Butler County, Ohio. The group had made the 30-mile journey from Cincinnati in less than five hours.

Donaldson's next ascent, his ninety-eighth, was also one of his most spectacular. On the afternoon of October 19, 1874, he went aloft with Miss Mary Elizabeth Walsh, her fiancé, Charles M. Colton, the Rev. Howard B. Jeffries, and two attendants, to conduct the world's first aerial wedding. David Thomas had spent lavish sums to insure that the event would receive maximum coverage in the nation's newspapers. A crowd estimated at 50,000 had gathered at the Hippodrome to witness the ceremony. The great P. T. Barnum and his own bride were on hand to congratulate the bride and groom.

The much-publicized aerial nuptials provided yet another demonstration of Donaldson's consummate skill in capturing headlines. Coupled with his midair gymnastic performances and the abortive ocean crossing, the event helped make him one of the best-known balloonists in America.

He returned for a fourth season with the Barnum show in 1875. On the afternoon of July 15, 1875, Donaldson took the *P.T. Barnum* aloft from the Chicago Hippodrome accompanied by Newton S. Grimwood, a reporter for the Chicago *Evening Journal.* The pair evidently hoped to make a 120-mile flight across Lake Michigan. The *Barnum* was sighted by a schooner 12 miles out at seven that evening, two hours after departure. Crew members later reported that the balloon was flying so close to the water that the basket was occasionally dipped beneath the surface. A small boat was launched to render assistance, but the *Barnum* rose into the air and was blown out of sight before rescuers could reach the scene. Violent rainstorms blew across the lake that night, raising doubts about the safety of the two balloonists. When no sign of the aerostat was found during the following weeks, it was assumed that Washington Harrison Donaldson had met his end at last.

Grimwood's body, still fully clad except for his hat and boots, washed ashore on a Lake Michigan beach one month after the ascent. A rubber life preserver tied around his waist had apparently kept the body afloat. The reporter's personal effects included a sheet of preliminary notes for his story, but these offered no clues on the fate of his comrade.

Over a month later, in late August 1875, a fishing party traveling up the Montreal River discovered a dying man in a fishing hut some 50 miles south of Lake Des Quinzes. The man, who claimed to be Donaldson, was suffering from gangrene as a result of untended breaks in his left arm and leg. Before his death he offered an account of the final ascent. Well over Lake Michigan, a series of wind gusts had forced large quantities of hydrogen from the gasbag. Sinking ever closer to the water, it was apparent that the partially inflated balloon could no

Donaldson launches on his last flight, July 15, 1875.

longer support the weight of two occupants. At this point, according to Donaldson's story, Grimwood drew a gun and attempted to force his companion to jump from the basket. In the ensuing struggle the reporter was thrown overboard to his death. Donaldson was then blown over the desolate Quebec shore of the lake, only to be injured in landing.

Many of Donaldson's friends were unwilling to accept the account of this deathbed story as published in American newspapers. John Wise, for example, produced a short pamphlet based on papers he claimed were found on the body. The aging balloonist argued that Grimwood had voluntarily donned the life jacket and leaped from the basket to save his companion.[59]

Ironically, John Wise was to die under almost identical circumstances only five years after Donaldson. The nation's senior aeronaut had made only four flights between October 1871 and October 1874. On October 17, 1874, he passed an extraordinary landmark, his 450th free ascension. Flying from the roof of a building at the corner of Market and Thirteenth streets in Philadelphia, he carried his old friend Charles Cresson of the Franklin Institute and a Mr. Taylor on a voyage that ended with a safe landing in a cabbage patch on the outskirts of Moorestown, Pennsylvania.

In the fall of 1879 John Wise, now 71, was on a midwestern tour with his grandson, John Wise, Jr. On September 19 John Jr. went aloft from Sterling, Illinois, with the balloon *Pathfinder*, a new craft that had made very few flights. On September 28, 1879, John Wise launched himself from Lindell Park in West St. Louis with a single passenger, George Burr, a young teller with the St. Louis National Bank.[60]

The flight began sooner than planned. Launch had been scheduled for the early evening, but when high winds in the afternoon threatened to damage the fully inflated balloon, Wise decided to ascend at once. The two men scrambled into the basket and took off without throwing sufficient ballast aboard. Before the flight, Wise had remarked that he hoped to rise into the upper atmosphere winds blowing east and travel as far as possible.

Rising into a blustery sky, the two men were carried across the river into Illinois. At 5:50 that afternoon they were seen passing over Alton. The passage of the balloon was also noted by the citizens of Bunker Hill and Carlinville, Illinois, fifty-five miles from the take-off point. Pulling into the little station at Miller, Indiana, thirty-five miles from Chicago on the southern shore of Lake Michigan, an engineer noted the *Pathfinder* moving out over the lake at 11:30 P.M.[61]

Had Wise and Burr remained in the air for another 35 miles, they would have overflown Lake Michigan, recrossing the shoreline somewhere near St. Joseph, Michigan. In fact, they were never seen alive again.

Disregarding the report from Miller, John Wise, Jr., organized search parties that combed the backcountry of Girard and Macoupin counties in Illinois. He found no sign of the missing aeronauts, but refused to give up hope. On October 1, 1873, he wrote his father, who was still in Philadelphia: "Dear Father: I hardly believe he is dead, but I think he is lost in the Michigan pineries. My grandfather had no overcoat, and only a small lunch, I understand that there are plenty of wild berries and so they may be able to live for several weeks. They have a compass and matches to make a fire. The balloon was in good order and we still hope to hear from them."[62]

The answer to the riddle of the aeronaut's disappearance came in the form of a body washed ashore in the dunes near Toleston, Indiana, on the afternoon of October 24, 1879. The remains were not recognizable but bore coat and sleeve buttons marked "G.B." William Burr identified the body as that of his brother George. The body of John Wise was never recovered. The Great Lakes had reasserted their reputation as a death trap for aeronauts.

The loss of John Wise was keenly felt in aeronautical circles. His grand total of 463 ascents during forty-four years of active life as an aeronaut, coupled with his enthusiasm, generosity, and scientific curiosity, clearly mark him as the most distinguished and experienced of American aeronauts.

With the death of John Wise, the ranks of the veteran aeronauts of the Civil War were growing very thin indeed. Only James and Ezra Allen now remained in the air. A number of aeronauts who had begun their careers prior to the war, but who had not served in the Balloon Corps, did remain active however. George and Silas Brooks, for example, were still popular aerial performers.[63]

The brothers had offered their services and equipment to the Union army but had been rebuffed. Undaunted, they announced their determination to "lend a helping hand to put down the rebellion" by raising a picked company of infantry.[64] In spite of such patriotic pronouncements, the "aerial brothers" continued to operate as civilian aeronauts throughout the war.[65] In June and July 1862 they were offering tethered flights to the residents of Hartford at $5 a head. Gentlemen could purchase seats for one of a series of "grand free ascents" for $25 each. "Such a treat," opined the *Daily Courant*, "seldom occurs in a lifetime."[66]

Silas and George Brooks were working with a new partner during this period. E. C. Bassett, a native of West Stockbridge, Massachusetts, had moved to Hartford in 1840. A hairdresser and insurance agent, he also offered lectures on animal magnetism. In addition to his other accomplishments, Bassett was an enthusiastic aeronaut.[67]

As early as 1862 Bassett was operating as a ticket agent for the brothers' performances. On July 4, 1863, he flew the small balloon *Stars and Stripes* at the Hartford Fourth of July celebration, while Silas Brooks had command of the larger *Constitution*. Working together, the three men continued to barnstorm through New England. They flew from New Haven on August 29, 1863, moving on to Middletown (September 2, 1863) and Hartford (September 26, 1864; July 4, 1866).[68]

E. C. Bassett was not only serving as agent and advance man for George and Silas Brooks, he was also their apprentice in aeronautics. By the fall of 1863 he was operating the *Stars and Stripes* on his own. Later that year Bassett relocated to Cleveland, Ohio, where he established himself as an insurance agent. He continued to return to Connecticut each summer, however, to act as the brothers' impresario.

The purchase of a ticket to a Brooks-Bassett show entitled the spectator to far more than just an opportunity to watch a balloon launch. Evening fireworks lit up the sky. Electrical Drummand lights illuminated nearby buildings, including, on one occasion, the dome of the Connecticut State House. Bassett also hired the Denier brothers, trapeze and tightrope artists, to entertain the crowd before the ascent. The aeronauts even resorted to the oldest of aerial stunts, parachuting dogs and cats to earth in order to hold the interest of paying customers.

Once, in early August 1866, George and Silas Brooks operated tethered balloons in association with Messrs. Spencer and Eggleston. Following this engagement in Hartford, the aeronauts' names disappear from local newspapers for over fifteen years. In view of the coverage Connecticut newsmen had given to their activity, it seems likely that the brothers abandoned aeronautics during this period.[69]

Silas Brooks was back in the air with an ascension from Hartford on July 4, 1882. The *Evening Post* described the aeronaut as "a quiet, somewhat weather-beaten man about middle age, and an evident enthusiast in the business despite his quiet way." The flight in the 21,000-cubic-foot balloon *Eagle* was Brooks's 160th ascension.[70]

Brooks continued to make occasional flights over the next fifteen years. He remained a favorite attraction at Hartford Fourth of July celebrations, where he performed in 1883, 1885, 1886, 1887, and 1894. During the last of these annual flights, Brooks became entangled in the branches of a tree soon after take-off. The dazed, semi-conscious aeronaut was rescued with only minor injuries. It was clear that his days as an active balloonist had come to a close.[71]

Bassett died in Connecticut on May 21, 1891, but Silas Brooks lingered on, a well-known figure in Hartford. In 1897 the *Courant* published a lengthy interview with the seventy-three-year-old aeronaut, who reminisced about a long career that had begun with the Druid Band in 1848. Silas Brooks died in the Burlington Poorhouse on April 5, 1906.

But Silas Brooks was not the longest lived of the balloonists who had first taken to the air during the years prior to the Civil War. At the time of Brooks's death, Samuel Archer King, the man that had taught James Allen to fly, was still pursuing an active life as a balloonist.

S. A. King's extraordinary career stretched from the time of his first ascension on September 25, 1851, until 1908, when he retired at age eighty-three. During more than half a century in the air, he made a grand total of 458 ascensions, nearly equaling the record of John Wise. Throughout these years he operated a number of balloons, including: *Queen of the Air* (1857, 33,000 cu. ft.); *Star Spangled Banner* (1862, 50,000 cu. ft.); *General Grant* (1864, 15,000 cu. ft.); *Hyperion* (1869, 65,000 cu. ft.); *Aurora* (1870, 65,000 cu. ft.); *Buffalo*

Samuel Archer King

(1873, 91,000 cu. ft.); *Cloud Nymph* (1874, 91,000 cu. ft.); *King Carnival* (1877, 91,000 cu. ft.); *Pioneer* (1879, tethered, 91,000 cu. ft.); *Atlantic* (1879, tethered, 91,000 cu. ft.); *Ben Franklin* (1908(?), 92,000 cu. ft.).

Having been rejected for service in the Union Balloon Corps, King, like Silas and George Brooks, continued to earn his living as a commercial balloonist. He preceded the Allen family as Boston's favorite balloonist. The aeronaut made an official ascension in the city every Fourth of July between 1859 and 1868.

These were exciting years for King. On July 9, 1862, the aeronaut and four passengers spent an hour bobbing about in Boston Harbor following an emergency landing near Fort Independence. The balloon *Star Spangled Banner* was lost on this occasion.[72]

The July Fourth trip for 1863 was particularly memorable, covering 65 miles from Boston to Deering, New Hampshire. King's 1865 holiday excursion ended before it began when the balloon broke away and exploded during inflation, forcing the aeronaut to ascend in a smaller craft.

King pressed the balloon *Hyperion* into service for a Buffalo ascension on July 5, 1869. The balloon was a 65,000-cubic-foot veteran of six ascents (Boston to New Salem, July 9, 1867; New Salem to Worcester, July 10, 1867; Worcester to North Reading, September 6, 1867; Providence, Rhode Island, to Norton, Massachusetts, October 25, 1867; Norton to Middleborn, October 26, 1867).[73]

Filled with city gas, *Hyperion* was capable of carrying up to a dozen persons on a tethered ascension. For this free flight, however, King had limited the passenger list to four newsmen, one of whom had made fourteen previous flights for the *Boston Journal.*

Casting off at 4:15 P.M., they made a safe crossing of Lake Erie, at one point even blending their voices with those of the crew of a passing tug in the chorus of a popular air of the day, "Up in a Balloon." Having dipped into the water once, the party crossed the shoreline and landed in Eden Township, Pennsylvania, remaining on the ground for fifteen minutes before continuing their journey over Erie and Cattaraugus counties. Looking to the right and left, they could see fireworks rising from village greens and bonfires lit to celebrate the national holiday (a day late, since that year the Fourth had fallen on a Sunday). Rising once again at 8:16 P.M. to survey the terrain for a landing site, King could see nothing but "woods, woods, woods. . . . The distant lights disappeared one by one, and the whole earth seemed wrapped in impenetrable stillness as well as darkness. The stars above us were our only companions. Our overcoats guarded us against the chilliness, which was now that of a frosty November night rather than the mild temperature of a summer evening, as the world's people would feel."[74]

Running dangerously low on ballast, King finally began a blind descent at 11:00 P.M. They crashed into the top of a fifty-foot pine tree, deep in the Allegheny forest. The envelope continued to bob high above the trees, while the basket, its floor now filled with pine needles, was held firmly in the boughs. King was afraid to valve gas and risk a fifty-foot drop, so the men spent the night in the car, singing and talking.

Inspecting the situation the next morning, the aeronaut did open the ripping panel, allowing the party to descend "with a considerable shock," but without injury. After breakfasting on the remains of their provisions, they hiked to the nearest village, Kinzua, eighty miles from Buffalo in Warren County, Pennsylvania. The *Hyperion* had suffered only $300 damage on this, King's most eventful Fourth of July ascent yet.[75]

The *Hyperion* finally met its end on January 1, 1870. King had begun a short tour of the South with a flight from Atlanta, Georgia, on October 19, 1869. He then moved on to Augusta, Georgia, where he was scheduled to fly on January 1. The aeronaut was preparing to cancel this engagement because of a rapidly approaching sleet storm when Dr. Albert Hape, who had made the Atlanta ascent with King, volunteered to try his first solo flight. King inexplicably agreed.

Hape climbed straight into the storm, collecting a layer of ice and snow on the *Hyperion*. The envelope burst, dropping the novice aeronaut precipitously to earth. Hape clung to the side of the basket, which swung like a pendulum throughout the descent. In spite of the usual parachute effect, the impact was enough to knock Hape unconscious. The aeronaut survived, but the balloon "presented a sorry spectacle. Ropes, netting, and fragments of rotten cloth, it lay among the bushes and briars an utter wreck."[76]

For the most part, King confined his activity to the Northeast during the early 1870s. He completed some extraordinarily long aerial voyages during this period. On September 26, 1872, King and a *Boston Journal* reporter named Holden, who had made seventeen previous ascensions with the aeronaut, flew over 500 miles from Plymouth, New Hampshire, to Sayabec, Quebec. Caught in high wind, they were blown over the White Mountains to an early morning landing in a Canadian treetop.[77]

On July 4, 1874, King reunited the veterans of the long *Hyperion* flight of July 5, 1869, for another Buffalo ascension. This time the party flew 400 miles to a safe landing in southern New Jersey.[78] Three months later, on September 5, 1874, King made an eight-hour voyage from Cleveland, Ohio, across Lake Erie to Detroit.

King was famous not only for his long-distance flights with passengers but

King's balloon Buffalo *before a Cleveland ascent*

also for his scientific ascensions. Since the time of John Wise, few American aeronauts had taken any serious interest in science. While European balloonists ventured higher and higher into the upper atmosphere, their American counterparts took little interest in such activity. King, like Wise, was an exception to this rule. Fascinated by meteorology, he made an impressive number of data-gathering flights with pioneer meteorologists of the U.S. Army Signal Corps.

Gen. Albert J. Myer, who as a junior engineering officer during the Civil War had played an important role in the life of the Balloon Corps, served as chief signal officer throughout the 1870s. Myer placed his scientific assistant, Cleveland Abbe, in charge of the U.S. weather service, a new Signal Corps responsibility. Abbe continued and extended Joseph Henry's practice of collecting weather data from a variety of sources. He found that reports of balloon ascensions which were "generally made on the Fourth of July at numerous places in this country . . . helped to elucidate the atmospheric conditions on those days."[79]

Abbe and Myer were quick to take advantage of Samuel King's interest in meteorology. On September 3, 1872, King sent George C. Schaeffer, Jr., a Signal Corps observer, aloft aboard the balloon *Aurora* from Falls Field, Rochester, New York. Schaeffer completed 156 readings with his elaborate instrument package and regarded the flight as an enormous success.[80]

King made a highly publicized flight from Nashville to Gallatin, Tennessee, on June 18, 1877. A. C. Ford, a sergeant attached to the Signal Corps, accompanied King on the flight, as Myer dutifully noted in his annual report. Poor Ford can scarcely have made many observations however. He was crammed into a basket measuring four by seven feet with King and five other passengers.[81]

Samuel King's interest in meteorology and his contacts with the U.S. Signal Corps were major factors shaping his plans for a transatlantic flight. The dream of an ocean crossing had not died with John Wise. Samuel A. King was only one of a number of dreamers who hoped to fly the Atlantic.

Before the *Daily Graphic* failure, two Philadelphia newcomers, Henry C. De Ahna and John A. Light, each offered to undertake the trip. De Ahna, who apparently had no aeronautical experience, nevertheless expressed his willingness to "go ahead."[82]

John A. Light could, at least, claim to have made a number of flights. On August 27, 1873, he took Richard E. Chism, a representative of the *Philadelphia Evening Herald*, on a flight designed to demonstrate the existence of the "eastern current" of which John Wise had said so much. They flew from Philadelphia to a point within two miles of New Jersey's Atlantic coast. Light then proceeded to the heart of the matter, his need for money:

> *I consider we saw enough of the effects of the eastern current to demonstrate the practical ease of a trip across the ocean with proper conveyance and material. If the citizens of Philadelphia will build me a balloon as I wish it built, I will pledge my honor to carry three of us to Europe in it. . . . The scheme is practical, if practical men work it, but the balloon that is to cross the ocean with surety of success must be perfectly tight, and of a capacity of from 600,000 to 800,000 cubic feet. In such a balloon, made of good silk, I will not fear to cross to Europe.*[83]

Clearly, Light hoped to interest the *Evening Herald* in sponsoring his venture just as the Goodsells had supported Donaldson. When this support failed to materialize, John Light faded from the scene.

Light's proposal was at least in the tradition of John Wise. Other would-be transatlantic flyers proposed abandoning the free balloon entirely. F. W. Schroeder, for example, suggested the construction of a unique airship.

A native of Hanover, Schroeder was the son of a Prussian officer. Immigrating to the United States in 1861, he obtained a commission as a lieutenant in a New Jersey regiment. Schroeder became fascinated by aeronautics while assigned to supervise the troops assisting John La Mountain in his early operations with the balloon *Saratoga*.

The German unveiled plans for a cigar-shaped transoceanic airship in the summer of 1878. The bag would measure 96 feet in diameter with a gas capacity of 45,000 cubic feet. A one-half horsepower electric engine would drive a pusher propeller at 1,000 to 1,500 RPM. A pair of enormous flapping, cambered wings mounted between the gasbag and the car (powered by the same overworked one-half horsepower motor) would raise the airship to cruising altitude. Schroeder was convinced that this extraordinary machine would cross the Atlantic in thirty to thirty-six hours. He found few investors willing to support his scheme.[84]

De Ahna, Light, and Schroeder were little more than dreamers who had no serious hope of flying the Atlantic. Such a project would require an intimate knowledge of balloon construction and a great deal of experience in the air. The members of the Allen family, who were certainly qualified on both counts, apparently gave little serious thought to such a flight. Their old friend Samuel Archer King, on the other hand, had given the matter a great deal of thought indeed.

The aeronaut announced his intention of flying the ocean in a letter of April 7, 1879, to the editor of the *New York Herald*. In order to raise sufficient cash to support the venture while at the same time gathering meteorological data on seacoast winds over an extended period, King constructed two large balloons with which to make repeated tethered flights during the summer of

1879.[85]

The two balloons, *Pioneer* and *Atlantic*, were constructed of Irish muslin. Each stood 80 feet tall and measured 65 feet in diameter when inflated with 144,000 cubic feet of gas. The twenty-three gores that composed each bag were cut from a brown paper pattern 102 feet long and 3 feet wide.

The top quarter of each balloon was painted silver "so as to present a dazzling appearance when exposed to the sun."[86] A twelve-foot black and yellow band was painted around the equator of each balloon. The lower half of the bag was painted blue. King noted that the paint scheme served both to seal the fabric and to make it "showy."

Showy they were. The names *Pioneer* and *Atlantic* were painted on the central band of each bag in letters eight feet tall. A Mr. Loeffler, who supplied the fabric, also added a variety of stylish touches to the balloons. The envelope of the *Pioneer* featured a large painting of Christopher Columbus standing on his quarterdeck. A second medallion painting portrayed Boreas, the god of the north wind, and allegorical figures of Europe and America "kept cool by breezes blowing from the nostrils of the windgod." The decorative medallions on the *Atlantic* showed the balloon crossing the ocean and the aeronaut's victory over Neptune.

A Brooklyn rattan factory produced the 250-pound oval baskets. Each was 36 feet in diameter and could carry fifteen passengers. Both balloons were furnished with two separate nets, one of cotton, the other of twine.

King established his "aeronautical observatory" at Manhattan Beach, near Coney Island, New York. The enclosure was a large, fenced area 300 feet east of the Manhattan Beach Hotel. The area was 200 feet in diameter. The top of the fence was capped with a smooth wooden strip that would not abrade the cable in the event that winds moved the balloon outside the structure. Tiered seating was provided, as were a number of small buildings to house offices, reception rooms, and sleeping accommodations for a ground staff.

A small yard extending beyond the fenced enclosure housed the engine, winch, and cable facilities as well as the gas generators, coal bins and the like. King was careful to assure potential customers that his engine, windlass, and pulleys had been constructed by the firm of Otis Brothers, the well-known manufacturers of elevators, and were perfectly safe.

The gas generator resembled a city gas plant. Complex reverberatory furnaces reduced coal to coal gas. Nor was any expense spared in the design of dynamometers for accurately measuring the lift of a balloon aloft, the telephones that connected the aeronaut to his ground crew, and the heavy-duty cable anchorages. All of the arrangements had been submitted to engineering professors at the School of Mines, Columbia College. The total investment, which was supported by a group of New York businessmen, was close to $75,000.

The first of King's two balloons was inflated in late June 1879. By July the aeronaut was making daily passenger ascents to 1,200 feet, offering Gothamites an unexcelled view of their city, while making his own meteorological studies.

From the outset, King hoped that the U.S. Army Signal Corps would see the scientific value of his program and assist in financing the Atlantic flight. It was, therefore, most encouraging to receive a letter from Captain Hogate, General Myer's "sub-chief," requesting the aeronaut's permission to station an officer on permanent duty at Manhattan Beach to assist in gathering weather data. The Signal Corps also provided all the meteorological instruments King was using.

The Manhattan Beach flights brought a great deal of publicity to King. They enabled him to cement his loose relationship to the Signal Corps and provided both a cash return and useful information on winds. The flights also helped to popularize the cause of giant tethered balloons at fairs and summer resorts throughout the remaining years of the nineteenth century. They did not, however, raise enough money to finance the transatlantic expedition. King, disappointed but undaunted, returned to long-distance free flights and a continued drive to obtain Signal Corps support.

The aeronaut's careful study of atmospheric winds had convinced him that John Wise had been wrong. There was no strong westerly current moving across the Atlantic. He was certain, however, that the general trend was from west to east. If a silk version of the *Pioneer* and the *Atlantic* could be constructed so as to remain aloft for weeks at a time, it would eventually reach Europe. In fact, both King and Wise were correct. Wise's river of moving air did exist in the jet stream, but these winds were far higher than the altitudes any nineteenth-century aeronaut attained. King's observations were much more accurate for the altitudes at which a balloon could operate.[87]

King's plans for an Atlantic flight were back in the news during the fall of 1881. The aeronaut had accepted a contract to fly a new balloon, the *Great Northwest*, for a Minneapolis festival, the Great Northwestern Exposition. His flight, which was sponsored by five area newspapers, was scheduled for September 7, 1881.[88]

The local citizens were certainly impressed by the aeronaut's determination and courage. One reporter described him as "a tall, thin man, slightly stooped, eyes as blue as steel, a forehead high—very high, in fact—upper lip clean shaven, but firm cheek and chin and jowl, a very cataract of sun-browned beard dependent. A man who would not impress you greatly at first sight, but who grows upon you with his hearty earnestness and thorough good fellowship." King had announced that his flight would be a very long preliminary to the Atlantic voyage. If possible, he would fly all the way to the East Coast.

Gen. W. B. Hazen, who had replaced the recently deceased Myer as chief signal officer, was anxious to take advantage of King's Minneapolis ascent. One of Hazen's first acts following his promotion was to place Cleveland Abbe in command of a new Scientific and Study Division. Abbe hired a staff of three technical assistants. One of these computers, Winslow Upton, a recent graduate of Brown University, was dispatched to Minneapolis to assist and fly with S. A. King.

Upton would carry a full complement of scientific instruments on the voyage. A series of thermometers, hygrometers, aneroid barometers, and a vertical anemometer were to be mounted on a special shelf in the basket. Thomas Alva Edison, who was reported to have taken a special interest in King's meteorological work, provided a battery-operated light to be used in making instrument readings at night.

In an effort to extend the value of the information gathered on the flight, Hazen and Abbe instructed Signal Corps officers stationed east of the Dakotas and north of Virginia to take a series of standard temperature, pressure, windspeed, direction, and humidity readings during an eighteen-hour period after the take-off of the balloon.

Unfortunately, the promise of the Minneapolis flight was lost. Originally scheduled for September 7, 1881, the ascension was postponed for five days because of heavy weather. Carl Myers's troupe of balloon-borne acrobats entertained those few spectators who did brave the mud and rain, but King wisely resisted the temptation to inflate the *Great Northwest* until conditions improved.

Inflation began on Monday, September 12, 1881. The balloon stood 80 feet tall and 60 feet in diameter, with a gas capacity of 100,000 cubic feet. Early that morning King brought seven large black metal tanks onto the site. The tanks were then loaded with fourteen tons of scrap iron, two carloads of sulphuric acid, and water. After passing through a limewater washer, the hydrogen was carried to the giant balloon through rubberized cloth pipes. A reporter for the *Pioneer Press* was among those who watched as the *Great Northwest* took shape: "Then in the center of the vast field . . . rose a mighty dome, symmetrical, beautiful and wonderful. It rose like Alladin's palace, and its white surface shone in the sun, while all about were standing a wondering throng, alternately watching its rising and graceful form, and the semi-circle of huge black tanks, from which came the unseen force that lifted it."[89]

King cast off at 5:39 P.M., gently bobbing between 3,000 and 4,200 feet as he gingerly dropped ballast to stabilize his altitude. Passing near Minnehaha Falls soon after take-off, the *Great Northwest* dropped into trees which ripped Upton's shelf from the side of the basket seconds after the meteorologist had swept his precious instruments into the relative safety of the car.

Free of the excess weight, the balloon rose rapidly again. Darkness was falling as the party crossed the Mississippi. King, unwilling to remain aloft all night, landed on a farm two miles from the St. Paul city limits. Expecting to continue the voyage on the morning of September 13, the party was grounded by high winds. King watched helplessly as the *Great Northwest* bucked and tossed. Finally, the appendix, which had been tied to conserve hydrogen, tore open.

Volunteers from nearby Fort Snelling struggled to restrain the balloon as the gusts became more and more violent. King's final frantic efforts to valve gas and save the balloon ripped the fabric still further, allowing all the gas to escape and ending any hope of continuing the flight.

A less publicized but much more successful flight began at the foot of Chicago's Randolph Street on October 14, 1881. Flying the *Great Northwest*, now renamed the *A. J. Nutting*, King and Pvt. J. G. Hashagan of the Signal Corps traveled 550 miles during a twenty-hour aerial voyage across Illinois to a landing in a remote Wisconsin swamp.

As Hashagan noted in his report to Signal Corps officials, the trip proved "the feasibility of meteorological observations while in mid-air. With a reasonable amount of precaution and proper packing of instruments, there is no reason why in the majority of ascensions the instruments should not be returned in as perfect order as when taken away."[90] King's association with the Signal Corps continued into the mid-1880s. W. H. Hammon, yet another government observer, accompanied the aeronaut on a series of four flights from Philadelphia's Girard College aboard the *Eagle Eyrie*.[91]

Ultimately, King's efforts to enlist Signal Corps support for a transatlantic flight failed. Like John Wise, King would never fly the ocean. His accomplishments had been extraordinary however. King was, of course, paid to please a crowd of admiring ticket holders and fair-goers. Yet the fact remains that he was less concerned with the trappings of show business than virtually any other professional aeronaut of the nineteenth century. He eschewed the flamboyant costumes and acrobatic stunts employed by contemporaries like Washington Donaldson.

Instead, he focused attention on his prowess with two decades of extraordinarily long flights. King was the most proficient aeronaut active in the United States between 1860 and 1900, but it was his sincere interest in science that set him apart from all other American balloonists. He was the most important scientific balloonist of his generation. King's contribution to the basic understanding of weather systems may have been less spectacular than the high altitude flights of his European contemporaries, but it was ultimately to prove of greater importance.

S. A. King had convinced responsible officials in the Signal Corps that the balloon was an important tool for gathering meteorological data. As a result, this office was to remain the center of U.S. government interest and activity in aeronautics through the First World War.[92] General Hazen and other leaders realized, however, that unmanned balloons could be employed to gather data at considerably less expense than that required to support an aeronaut like King. "It is probable," Hazen reported to Secretary of War Robert T. Lincoln, "that by use of the captive balloon, observations of temperature may be secured without the expense attendant on an extensive voyage, and by use of the smallest kind of free balloon much may be learned in reference to direction and velocity of upper air currents."[93]

Nor were other potential uses of the balloon ignored by Signal Corps planners. As early as 1884 Cleveland Abbe commented: "As the use of the balloon for military purposes has received much attention and is highly appreciated in Europe, it is probable that this may at some time become a duty of the Signal Office."[94]

With both meteorological and military needs in mind, the Signal Corps contracted with S. A. King to produce a balloon manual for army use. The corps also took advantage of the opportunity to send another computer, H. A. Hazen (not closely related to General Hazen), on a flight with James and James K. Allen on June 24, 1886.

When the Weather Bureau was transferred to the Department of Agriculture in 1891, the Signal Corps began to concentrate on signal and observation balloons. By this time Samuel Archer King was no longer flying for the army. He had given up the dream of an Atlantic flight, but he remained an active aeronaut well into the first decade of the new century. His son Frank K. King made occasional ascents as well, but he was never to enter the family business to the extent that the second and third generation Allens did.

S. A. King remained the master of the long-distance aerial voyage. At age seventy-nine his application to participate in the 1907 James Gordon Bennett balloon race was rejected. Anxious to demonstrate that he had not lost his touch, the aeronaut flew six passengers aboard the 100,000-cubic-foot balloon *Ben Franklin*. The party traveled over 500 miles from Philadelphia to Dwight, Massachusetts, with an intermediate overnight stop in Aurora, New Jersey.

King was now a favorite subject for newspaper feature writers who regaled their readers with tales of derring-do provided by the "world's oldest active aeronaut."[95] King's last flight drew particular notice. When the English aviator Claude Grahame-White flew his Farman biplane at Point Breeze near Philadelphia in 1910, Clifford Harmon, president of the Aero Club of Pennsylvania, arranged a flight for the eighty-two-year-old S. A. King. On their first hop, Grahame-White, who was apparently afraid of frightening the old man, simply took off, flew in a straight line and landed. When King announced angrily that he felt cheated, the aviator took him up again for a "real flight." The "venerable professor" informed the press that his hop would remain "one of the most pleasant memories of his life, but not equal to a [ride in a] balloon."[96]

S. A. King, age eighty-six and the veteran of hundreds of forays into the sky, died at home in Philadelphia on November 3, 1914. An active aeronaut for sixty-three years, he was the only member of the pre–Civil War circle of balloonists to bridge the gap to the age of the airplane.[97]

Chapter 15 There's No Business Like Show Business

SAMUEL ARCHER KING AND THE ALLEN FAMILY REPRESENTED A TRADItion of professional gas ballooning that stretched back to the time of J. P. Blanchard, C. F. Durant, and the Baltimore balloonists of the 1830s. These men and women prided themselves on having mastered a difficult and dangerous art. The ability to manipulate a dragline, conserve ballast and gas, and read the subtleties of wind and weather were the marks of a true professional. Experience and sound judgment enabled them to make long flights, to find and use air currents to maintain some control over speed and direction, and to select a proper site and suitable ground conditions for a safe landing.

Before the Civil War, American audiences had been satisfied with these exhibitions of skill, spiced with such occasional show business touches as animal parachute drops, balloon-launched fireworks, and aerial weddings. By the 1870s, however, the complexion of American aeronautics had begun to change. A new breed of aeronaut was emerging. Daredevil acrobats who performed their stunts dangling from trapeze bars slung beneath primitive hot air balloons were thrilling crowds at local fairs and holiday celebrations. After 1882 the acrobats began to give way to the parachutists, for whom the balloon was nothing more than a means of attaining sufficient altitude from which to make a crowd-pleasing jump.

These aerial performers required a great deal of raw courage but a minimum of the aeronautical expertise on which the traditional gas balloonists had prided themselves. The new hot air balloon technology was extraordinarily simple. There were no valves, ballast bags or draglines. The envelopes were nothing more than large cloth bags, usually cotton or muslin.

Unlike the original Montgolfier craft, these nineteenth-century balloons did not even carry a heat source aloft. The envelope was filled with smoke and heated air on the ground. These new craft were familiarly known as smoke balloons, or smokies. The thick smoke that "cured," or sealed, the fabric was essential to the operation. The technique originally developed to inflate and launch a smoke balloon was simple but dangerous. The envelope was held aloft with two long poles while the mouth was placed over an open bonfire.

By the 1880s smoke balloonists like M. L. MacDonald had developed more sophisticated procedures and equipment. MacDonald, who flew under the name Daring Donald, operated a series of typical home-built balloons. Standing roughly 60 feet tall and measuring 28 feet in diameter when inflated, each balloon was composed of forty gores cut from ordinary muslin sheeting. The fabric was sized with a mixture of glue, alum, soda, salt, and whiting, all dissolved in water.

A standard balloon consisted of three sections. The cap, where the internal temperatures would rise dangerously close to the burning point during inflation, was sewn of a double thickness of cloth and came to a peak on top. Uncut strips of fabric 15 feet long and a yard wide made up the central section. In the lower 29 feet each gore tapered from 36 inches at the top to 9 inches at the mouth.

A long cord was sewn into each seam. This felling supported an iron hoop some 8 feet in diameter, usually the felloe of a wagon wheel, which served to keep the mouth open. Four "quarter guys" attached to this iron load ring supported the parachute and the aeronaut. The parachute had a cotton canopy 28 feet in diameter with a 12-inch hole cut in the center. Thirty-two shroud lines led from the edge of the parachute to an 18-inch wooden ring which the aeronaut grasped when the balloon was released.

MacDonald began inflation by digging a trench 18 feet long, 2 feet deep, and 2 feet wide. The trench was roofed with iron sheeting or boards covered with packed dirt. An iron chimney 3 feet tall and 3½ feet in diameter was placed over one end of the tunnel. This was insulated by placing a large wooden barrel around the iron chimney. Earth was packed between the wood and the iron.

The aeronaut then placed a 28-foot pole on either side of the chimney. The envelope was pulled aloft by means of lines attached to rings on either side of the balloon, which ran through pulleys on top of the poles. These ropes supported the bag during inflation. With the envelope suspended on the poles, the mouth of the balloon was placed over the chimney. A wood fire was then set in the far end of the trench. The fire soon created a draught that drew the heat through the tunnel, up the chimney, and into the balloon. An assistant stood inside the balloon with a board to open or close the chimney and a bucket of water to douse sparks.

After fifteen minutes of this, the swollen balloon had already developed so much lift that it had to be held down by men standing on the iron hoop. In order to superheat the air, kerosene was poured on the glowing embers of the fire. A blast of flame so bright it could be seen through the fabric swept down

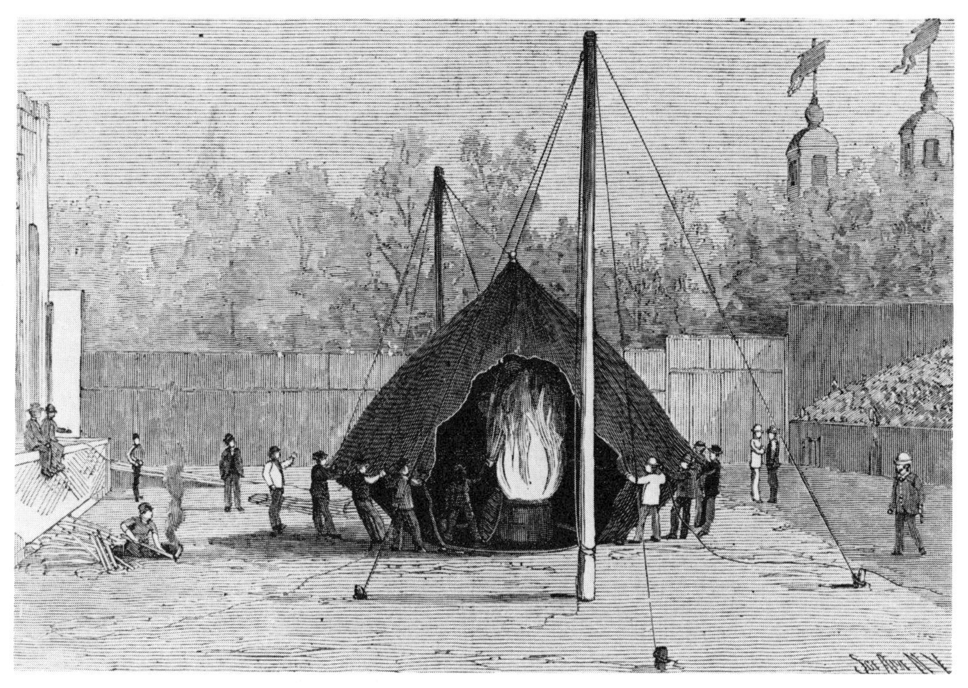

Daring Donald's inflation procedure

the tunnel and into the balloon. This was repeated several times while the aeronaut stood to one side, grasping the wooden ring of the parachute.[1]

When MacDonald ordered the balloon released, he had to run forward so that he would be pulled directly into the air. Because of the size of the envelope (approximately 29,000 cubic feet) and the heat generated, a smoke balloon literally shot into the air. "You were just jerked up," recalls one veteran who "earned his smoke" during the years prior to the First World War. "Blink once and you're at a thousand feet."[2]

A smoke balloon ascension was an impressive sight. The entire arrangement measured 175 feet from the peak of the balloon to the aeronaut dangling beneath the parachute. Once in the air, MacDonald worked his way to a standing position on the load ring, from which he could reach the cutting rope. When he had reached sufficient altitude and was over a suitable landing site, he jerked this line, pulling a pin in a wooden block which allowed the parachute and aeronaut to fall free.

From a normal drop altitude, Daring Donald reached the earth in only thirty seconds, landing with an impact equivalent to that of a jump from six or eight feet. MacDonald's procedure was typical of turn-of-the-century smoke balloonists. Later exponents of the art added a few more touches, such as the use of a quick release safety harness and the use of an oil drum to replace the trench and fire pit. For the most part, however, MacDonald's technique was standard from the end of the Civil War to the end of the smoke balloon era in the 1930s.

The smoke balloonists enjoyed a number of very important advantages over the traditional gas aeronauts. Their balloons were far less expensive to construct. Nor were they required to pay for city gas or to undertake the costly, dangerous, and chancy business of generating hydrogen. Inflation time was negligible compared to the hours required to prepare a gas balloon for flight.

Crowds were especially fond of the new smoke balloons. They were seldom disappointed by an ascension called off because of weather since the balloon could be inflated and the flight conducted in less than half an hour. In many respects, the spectacle was far more interesting. Instead of watching a balloon disappear into the sky, spectators now witnessed the entire ascent. The sight of a human being in flight was tame indeed when compared to a death-defying acrobatic performance conducted hundreds of feet in the air, often culminating in a heart-stopping parachute descent. The aeronaut, once garbed in nothing more exciting than a frock coat and tall beaver hat, was now clad in an abbreviated costume covered with spangles.

But the advantages of the smoke balloon were more than offset by the dangers involved. Once aloft aboard a smokie, the aeronaut was nothing more than a passenger. He or she had absolutely no control over the motion of the balloon. It was impossible to control the rise or fall of the craft, choose a favorable landing site, or ease the shock of impact. At best, the free balloon had always been a captive of the winds, but the smoke balloonist had given up what little control the traditional gas aeronaut had obtained over his vehicle. Raw courage and daring replaced the hard-won survival skills of the aeronaut.

Before 1860 American aeronauts had compiled an enviable safety record. With the exception of poor Augustus Connor, who died as the result of his own inexperience, the few aeronautical deaths in the United States had been a result of drowning. After the Civil War the water hazard remained a serious problem for American balloonists. John Wise and Washington Donaldson both died in the Great Lakes. Prof. G. A. Rogers, a veteran of 118 ascents, was another victim of drowning. Ascending from Boston Common on July 4, 1892, Rogers was forced to descend at sea. One passenger, De Los Goldsmith, swam to the safety of a rescue boat. Rogers and his other passenger, Thomas Lenton, were less fortunate.[3]

Carelessness accounted for a few more gas balloon deaths. John E. Baldwin, for example, died in the air over Greenville, Ohio, when his hydrogen balloon exploded during an aerial fireworks exhibition.[4] Still, in view of the obvious dangers involved, American gas balloonists suffered relatively few fatalities. This can not be said of the smoke balloonists. American newspapers, which had once focused on the skill of aeronauts who made flights halfway across the continent, now carried the grisly details of one aeronautical death after another.

The tally began with the death of T. H. Westbrook, a thirty-four-year-old novice aeronaut who died during an ascension from Sparta in Morrow County, Ohio, on July 4, 1862. He had reached an altitude of only three or four hundred feet when his "rotten" hot air balloon burst. Westbrook struck the ground feet first and died without regaining consciousness. Even at this early date, the penchant of American newspaper reporters for focusing on the ghastly details of the accidents was already apparent. The aeronaut's heels, readers were informed, rammed through the wall of the basket, "sinking to the depth of nearly two inches" into the ground. One of his legs was broken—perhaps both—and the internal injuries to vital parts of his body and brain precluded all hope of recovery. "No human organism could endure such a dreadful, violent concussion."[5]

"Professor" Yard very nearly suffered the same fate during an ascension from Mexico, Missouri, on July 4, 1868. Assisted by another aeronaut named Redmond, Yard followed the classic procedure for inflating his smoke balloon. On this occasion, however, the final coal oil blast set the balloon on fire. Yard was yanked aloft by the flaming envelope before anything could be done. Fortunately, the aeronaut, still in his basket, fell into a grove of oaks from which he was rescued without injury.[6]

"Professor" Wilbur was less fortunate. Preparing to ascend from Orange County, Indiana, in the fall of 1871, Wilbur became tangled in the ropes of his balloon, which was struck by wind gusts prior to launch. George H. Knapp, a local newspaper editor who was to have accompanied the aeronaut, was caught as well but broke free before the balloon rose too high. The horrified crowd watched as Wilbur struggled to climb into the basket, then lost his hold and fell several hundred feet to earth. One reporter was unable to resist describing the way in which the body bored an eight-inch hole in the ground, then rebounded into the air before coming to rest as "a mass of Human Jelly."[7]

Twenty-six-year-old John Short suffered a similar fall at Lafayette, Indiana, on July 4, 1876, flying a very large hot air balloon 95 feet tall and 65 feet in diameter. Rising into the air on a trapeze bar, Short was blown against a building. Although ripped, the balloon continued to rise to 200 feet before falling to earth on Sixth Street, between Main and Ferry. The aeronaut lost his badly fractured

Four harrowing scenes of American men and women aeronauts in various stages of difficulty appear above and on the following three pages.

THE DEATH-PLUNGE FROM THE CLOUDS.—THE TERRIBLE BALLOON ACCIDENT AT PAOLI, IND.—PROFESSOR WILBUR, THE AERONAUT, HURLED TO EARTH FROM THE HEIGHT OF A MILE.—SEE PAGE 2.

Dix-septième année DIMANCHE 16 SEPTEMBRE 1906 Numéro 826

UNE ASCENSION INVOLONTAIRE

Dramatique sauvetage à 300 mètres de hauteur

right leg, but survived.[8]

Reporting the death of J. H. Boley in a smoke balloon in 1873, one local newsman noted: "Those [hot air] balloons are regarded by aeronauts as extremely dangerous, because of their aptitude for catching fire or making too rapid descents when the air in the bag cools too suddenly. None but reckless operators use them."[9] In Boley's case fire was the culprit. Already 300 feet in the air when the flames were discovered, he continued to climb helplessly to 1,200 feet before the fire burned through his trapeze lines, dropping him to his death.

Michael McMann, a De Kalb, Illinois, laborer, also fell victim to fire. Mr. Dennison, a local aeronaut, had advertised an ascent on Thursday, October 19, 1876(?). At the close of the inflation, spectators noted flames working their way up the dry cambric. The surprised ground handlers dropped their lines, allowing the balloon to climb out of the enclosure. McMann, who had been holding one of the lines found himself dangling 300 feet from the ground. Losing his grip, "he fell backward, striking the earth with his back with such force as to produce a concussion heard some distance away. He was utterly crushed."[10]

Frank Hainur of Warren, Ohio, was one of several smoke balloonists who, like T. H. Westbrook and Edward La Mountain, paid the penalty for flying with an aged balloon. Ascending from Glade Springs, Virginia, Hainur "shot skyward like an arrow." Disaster occurred while the aeronaut was hanging by his heels at an altitude of 300 feet.

> *The patched and dilapidated canvas split from bottom to top with a report that was heard miles away. No sooner had the gas escaped than the balloon collapsed and came shooting down as swiftly as it had darted up. The aeronaut saw his situation and as quick as lightning regained his handhold, and began a maneuver to dodge a telegraph wire and post toward which he was falling. This he succeeded in doing, striking the ground with terrible force. . . . All of this was the work of a moment. The crowd was literally paralyzed, women sickening and fainting, and men unable, in their horror, to move. . . . The man was found alive, but dreadfully bruised and mangled. He was calm and cool, and described his feelings as descending, he saw and felt death staring him in the face. He was taken to the Hotel where . . . at 11½ o'clock* P.M. *he paid the penalty of death for his recklessness.*[11]

E. L. Stewart also paid the ultimate penalty for flying with faulty equipment. His hot air balloon burst at an altitude of 2,000 feet over Fayette, Missouri, on Monday, June 18, 1883. Stewart's body fell into a creek where it was found several days later.[12]

San Francisco witnessed several balloon disasters during the 1870s. Joseph Gruet ("Etienne Buislay"), who had flown with James Allen during that aeronaut's short stay in the city, died, it will be recalled, in a hot air balloon accident in 1874.

On October 5, 1879, Professor Colegrove and Charles H. Williams ascended from Woodward's Garden, James Allen's old headquarters. The flight was the culmination of a period of balloon enthusiasm that had begun with Allen's first appearance in 1874. Colegrove himself was a veteran of seven ascensions from Woodward's during the summer of 1879.

The two men took off in "a vicious wind," against the advice of Professor Martin, another aeronaut in the crowd. Passing over Howard Street, the balloon struck a telegraph line which cut several of the ropes tying the basket to the load ring. Colegrove and Williams clung to the basket for a time, then fell to their deaths.[13]

The smoke balloonists, who were already plummeting out of the sky at an alarming rate, began to perform an even more dangerous stunt after Thomas S. Baldwin reintroduced the parachute in 1887. Not since the time of Louis Charles Guille had a human being made a parachute jump in the United States. Suddenly, during the late 1880s and early 1890s, semisuicidal daredevils seemed to be leaping over every small town in America, often with disastrous results.

Samuel Black, who died during an ascent at Beardstown, Illinois, on July 4, 1890, was one of the first victims of the parachute. The flight had proceeded smoothly until the aeronaut cut loose at 2,000 feet. A thin trail of smoke was seen issuing from the canopy as Black fell free. The parachute burst into flame without opening. The aeronaut's body was found a half hour later.[14]

Two aeronauts died at a late-nineteenth-century exposition in Detroit. Edward Cole lost his hold on the parachute ring and fell from an altitude of 60 feet. As one newsman commented, "This whetted the curiosity of the public, which, while being sympathetic, does not object to being furnished with something upon which to bestow the redundance of its sympathy."[15]

John Hogan was the man chosen to receive this "redundance of sympathy." Thirty-four years old, Hogan was one of a family of aeronauts. His elder brother, Edward, had already died in a balloon accident. Professor Bartholomew, who had been employed to supervise the Detroit ascents, persuaded Hogan, who had no aeronautical experience, to replace the unfortunate Cole. The crowd, fully aware of the recent disaster, turned out in full force for Hogan's debut. The event drew "hundreds of people to the exposition who might not have gone to see anything as tame as horse races."[16]

Bartholomew had completed inflation of the large hot air balloon at 5:30

P.M. on August 29. Hogan's problems began immediately after launch. Rather than climbing immediately onto the trapeze bar as ordered by Bartholomew, the aeronaut went aloft dangling by his hands. His arms tiring, Hogan tried unsuccessfully to climb to the bar. Unable either to rest himself on the bar or to transfer his hold from the trapeze to the parachute ring, the aeronaut was finally seen to drop free. The "sympathy" of the crowd was to be exercised once again: "Down he came, wheeling and turning about with arms and legs frantically plunging through the air. Within the first 200 feet he turned over three times. Down he came with terrific force, his black-clothed body showing strongly and clearly in the light. The horror stricken crowd could only watch the awful sight as the body descended."[17]

Professor Hart was severely injured on August 29, 1874. Ascending at a picnic given by H. C. Clybourne in a Wisconsin park, Hart was blown into the trees. His parachute broke free and he fell 50 feet to earth. In spite of a broken leg and internal injuries, Hart was expected to survive.[18] E. E. Craig was severely injured under similar circumstances during an ascension from Topeka, Kansas. A strong wind carried him into a telegraph line, breaking his hold on the parachute ring. He fell 40 feet and struck his head on a buggy wheel.[19]

F. H. Vandergrift, a Cleveland, Ohio, parachute jumper, was drowned following an ascent from Columbus, Georgia. His hot air balloon burst at 2,500 feet. Vandergrift's parachute opened, dropping him into the Chattahoochee River, where he became entangled in the fabric.[20]

The death toll among women parachutists seems to have been particularly high. As early as August 1860, Louise Bates had ascended to an altitude of 10,000 feet over Cincinnati's Coney Island amusement park. Two decades later the park was the site of a tragic accident that claimed the life of Annie Harkness, a young aeronaut from Terre Haute, Indiana. Cutting loose only 500 feet over Coney Island, her parachute failed to open. The local newspaper report illustrates the horrified fascination with the possibility of sudden death that seems to have been such an important element in drawing spectators to parachute exhibitions:

> *Faster and faster descended the parachute. More intense and terrible came the strain upon the madly excited crowd, as horror-stricken it stood waiting for the fearful end. Suddenly there struck upon the ears a dull sickening sound—the end had come. A life had gone out in the midst of pleasure. Quickly a sympathetic crowd gathered around the lifeless, mangled form of the poor girl as she lay in a crushed mass upon the ground. Blood spattered her gaudy clothing and the wreck of the frail parachute that had borne her to her death, and fast running from her broken body, formed little pools about the spot where she lay, and the people*

FELL ONE HUNDRED AND FIFTY FEET.
PLUCKY AERONAUT C. E. SLOCUM'S TERRIBLE TUMBLE FROM HIS BALLOON,
AT ANTWERP, JEFFERSON CO., N. Y., WHICH MAY PROVE FATAL.

C. E. Slocum's "terrible tumble" at Antwerp, Jefferson Co., N.Y.

shuddered, and stood back as they looked.[21]

The crowd may well have stood back, but look it did, and waited for another opportunity to see a human being risk life and limb jumping from the sky.

A Nashua, New Hampshire, newsman had an opportunity to probe the psychology of the jumpers when he interviewed Miss Clara Pattee after she had suffered a broken leg during an 1889 ascent. "Don't you call that jump a rather dangerous thing?" he asked. "Well, I should say so," came the reply. In answer to the question, "But what makes you do it?" she spoke for dozens of her fellow men and women jumpers. "It pays."[22]

The toll would continue to mount so long as it paid. On July 4, 1891, Mrs. Jennie C. Crocker, who flew under the name Nellie Wheeler, simply lost her hold on the parachute ring and plummeted to earth. A thirty-four-year-old divorcée with a young son, Mrs. Crocker was no novice. She had made her first ascent on June 17, 1890, and had enjoyed a successful career prior to the accident.[23]

Other balloon casualties of the period included William Andrews, Edward Clarage, William Dennis, Loretta Deutley, Charles Jones, Louis Tarl, and Clarence Williams. Of course a great many balloonists flew for years without a serious accident. By the 1890s these pure aerial showmen were operating from coast to coast. Many are only names to us: Ira Fiske, George Collard, M. M. Forsman, Leslie Haddock, Professor Montrose, David Thomas.[24]

Most of these late-nineteenth-century balloonists specialized in flights within their own geographic regions. Andrew C. Carlberg, Ford Carpenter, Bob Erlston, Charles Fisher, Eddie Hall, Oscar Hunt, L. G. Mecklem, A. V. Naillen, and P. H. Redmond are among those who seem to have spent most of their time west of the Mississippi. Ruby Duveau, Lewis Nixon, James C. Patton, and S. Zelmo kept the balloon tradition alive in the South. The Midwest, however, remained the center of American aerostation. H. W. Durrel, Professor Drake, Conde Flora, Andy Fowler, Francis Jacobs, John Johnson, Jim McBride, John Murry, Oscar Ruth, Vincent Stone, Professor Strief, I. L. Thompson, W. E. Winterberger, Dorothy Vonda, and Mme Murphy were among the balloonists operating in this area.[25] Most of these performers are now completely forgotten, the record of their exploits buried in the crumbling, yellowed pages of local newspapers.

The number of women who pursued careers in aerial show business is striking. Leona Dare, a former Connecticut circus performer, was typical of the athletic young women who entered aeronautics. Dare introduced her act in Indianapolis in August 1872. Working with a companion, Tommy Hall, she

Ascending from New Bedford, Mass., on July 4, 1868, an aeronaut discovered that two young boys were entangled in his trailing ropes. Cut free (at a much lower altitude than shown here), both boys survived.

performed spins and tricks on twin trapeze bars. She later introduced her unique air daredeviltry to England.[26]

Gradually, aeronautical teams made their appearance. Composed of men and women who specialized in various stunts, these teams operated as touring entertainers. They booked themselves on summer trips through the carnival, church supper, and local celebration and holiday circuit.

The B. McClellan troupe, which began touring in 1891, was typical. For more than a dozen years, this group toured middle America, entertaining the citizens of towns like Piqua, Ohio (July 1891); Greenups, Iowa (October 1892); Greenville, Kentucky (October 1894); Pomeroy, Michigan; and Lewiston, Idaho (July 1902). Well-known aerial performers, including McClellan himself, C. C. Baldwin (no relation to *the* Baldwins), and Miss Sophy Le Claire, flew with the show over the years.

They livened their performances with the standard repertoire of stunts, including fireworks ignited from the balloon, animal descents (the dog was named "Drop"), human parachute jumps, ladder climbs, and trapeze acts. In addition, McClellan added a few touches all his own, for example, the "human cannonball." An aeronaut would ascend inside a long piece of pipe. Once the desired altitude had been obtained, the daredevil would ignite a small powder charge at the mouth of the pipe over his head. Protected from the blast by wadding, he would then drop out of the pipe through the cloud of smoke and parachute to earth, looking for all the world as though he had been shot from a cannon.

The "aerial torpedo" was even more spectacular. The whole balloon would vanish in a puff of smoke while the aeronaut parachuted safely to earth. The triple chute drop, in which three jumpers would drop from a single balloon, was also a popular McClellan specialty.[27]

A few of these troupes, such as that operated by Montz Bozarth, who had first flown at Paola, Kansas, on September 29, 1896, continued to tour well into the twentieth century. The descendants of one of the most popular of the late-nineteenth-century smoke balloon teams, the Allen family, were still operating as late as the 1950s. The founders of this group, Comfort (Curt), Ira, and Martin Allen, were not related to James and Ezra Allen. The three brothers claimed to have become interested in balloons when they witnessed ascents by T. S. C. Lowe during the Civil War. By the mid-1870s the Allen brothers had decided that they should try their hand at the game.[28]

They were smoke balloonists from the outset, and initially confined their activity to the area of the family home in Dansville, New York. Over the next few decades they enlarged their sphere of operations each summer. During the off-season they operated normal businesses. Martin Allen, for example, ran a

A poster for a Leona Dare exhibition

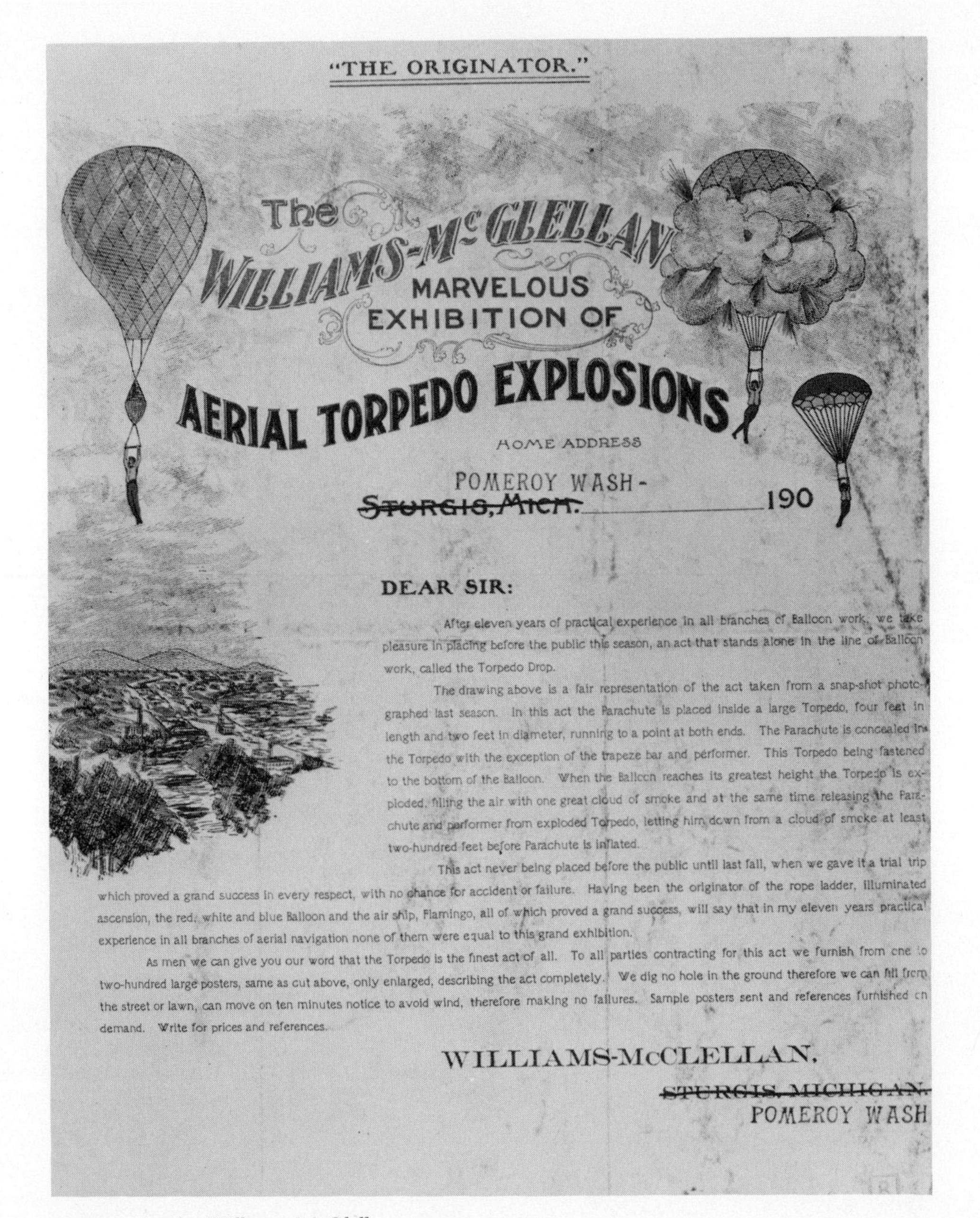

"THE ORIGINATOR."

The WILLIAMS-McCLELLAN MARVELOUS EXHIBITION OF

AERIAL TORPEDO EXPLOSIONS

HOME ADDRESS

POMEROY WASH - ~~Sturgis, Mich.~~ 190

DEAR SIR:

After eleven years of practical experience in all branches of Balloon work, we take pleasure in placing before the public this season, an act that stands alone in the line of Balloon work, called the Torpedo Drop.

The drawing above is a fair representation of the act taken from a snap-shot photographed last season. In this act the Parachute is placed inside a large Torpedo, four feet in length and two feet in diameter, running to a point at both ends. The Parachute is concealed in the Torpedo with the exception of the trapeze bar and performer. This Torpedo being fastened to the bottom of the Balloon. When the Balloon reaches its greatest height the Torpedo is exploded, filling the air with one great cloud of smoke and at the same time releasing the Parachute and performer from exploded Torpedo, letting him down from a cloud of smoke at least two-hundred feet before Parachute is inflated.

This act never being placed before the public until last fall, when we gave it a trial trip which proved a grand success in every respect, with no chance for accident or failure. Having been the originator of the rope ladder, illuminated ascension, the red, white and blue Balloon and the air ship, Flamingo, all of which proved a grand success, will say that in my eleven years practical experience in all branches of aerial navigation none of them were equal to this grand exhibition.

As men we can give you our word that the Torpedo is the finest act of all. To all parties contracting for this act we furnish from one to two-hundred large posters, same as cut above, only enlarged, describing the act completely. We dig no hole in the ground therefore we can fill from the street or lawn, can move on ten minutes notice to avoid wind, therefore making no failures. Sample posters sent and references furnished on demand. Write for prices and references.

WILLIAMS-McCLELLAN.

~~STURGIS, MICHIGAN.~~

POMEROY WASH

A poster for the Williams-McClellan troupe.

jewelry store from 1866 into the 1930s. In addition to their balloon activity, the Allens were the proprietors of various sideshow attractions that accompanied the carnivals with which they worked.[29]

Martin Allen introduced parachuting into the family act in the 1880s. He conducted initial tests with parachutes loaded with sandbags launched from his home on Ossian Street in Dansville. Many years later Irving Burdick, who as a young man assisted Martin in these early experiments, could still recall the day when a failed parachute sent a bag of sand tearing through the roof of a neighboring barn.[30]

Dressed in white silk shirts and black velvet knee pants, the three Allens continued flying and jumping well into the twentieth century. Martin, the last brother to retire, died at the age of ninety. Ira died in 1932 at age eighty-six, while Curt lived to be eighty.[31]

A new generation of Allens took up the family business after 1907, when Curt's son Warren ("Speck," for he was often seen as nothing more than a speck in the sky) made his first ascent. Warren's twin brothers Edward and Edgar also became involved in ballooning at an early age, as did their cousin Stephen, Martin's son.

"Captain Eddie" Allen was clearly the best-known and most successful of the second generation Allens. Captain Eddie made his first jump on September 7, 1912. He "retired" 3,253 jumps later when, aged seventy-eight, he sustained injuries during an ascent from the U.S. National Hot Air Balloon championships at Indianola, Iowa, in 1976.

After six decades, Eddie Allen can still recall the excitement and difficulties of those early days on the road. "People didn't know quite what to think of us in those days . . . we were like men from the moon. I know that back then there was not a hotel keeper in the country who would put sheets on a bed in a room where a known balloonist was going to sleep. They were all convinced we'd steal 'em to patch our balloons."[32]

His four children, Eddie Jr., Gloria, Florence, and Arlene, carried the family tradition into a third generation. The family presented all the time-honored hot air balloon stunts during the depression years. Trapeze ascents, parachute drops, and the human cannonball act remained crowd-pleasing favorites. On June 12, 1937, Eddie Jr., Gloria, and Florence were carried aloft for a triple jump from a large smokie measuring 100 feet tall by 70 feet in diameter. Tragedy struck later that season when Gloria was fatally injured during a triple drop over Blackstone, Virginia. Eddie Jr. and Florence carried on, eventually assisted by their younger sister Arlene, who specialized in the ground-based end of the operation.[33]

Following World War II Captain Eddie once again began flying on his own.

The Flying Allens, circa 1938. Courtesy the Allen family.

In addition to appearances at a variety of fairs and celebrations each summer, he toured the West and Midwest with an auto thrill show. During the 1960s Captain Eddie trained several apprentices in the century-old tradition of smoke balloon parachute jumping. Ultimately, it was a futile effort. The day of the daredevil jump at the county fair had long since passed.[34]

While the smoke balloonists came increasingly to dominate the market for aeronautical thrills, some of the best-known aeronauts in the nation continued to operate gas balloons. A number of newcomers stepped into the vacant shoes of the departing prewar aeronauts.

Carl Myers and his wife Mary, for example, were to earn fame far beyond that achieved by any of the smoke balloonists, most of whom only drew national attention by virtue of a spectacular death. Charles Edgar Myers was born at German Flats, Herkimer, New York, on March 8, 1842. Young Myers obtained a knowledge of the fundamentals of chemistry and electricity while a student at schools in the Mohawk Valley. Unwilling to work on the family farm, he found early employment as a bank clerk and a telegrapher, finally turning his hobby

The Flying Allens during a triple ascent, circa 1938. Courtesy the Allen family.

of photography into a career. In 1867 twenty-five-year-old Carl established himself as a photographer in Hornellsville, New York. As his biographer, Preston Bassett, has pointed out, the photographer's early advertisements indicate he was an ambitious go-getter:

> C.E. MYERS, *'Hornellsville Gallery,'* No. 151 *Main St.*, *Cor.* Canesteo *St. Opposite Park, is the Place to* GET YOUR PICTURE TAKEN . . . *Ambrotypes, Ferrotypes, Tin-Types, or any other* 'TYPES' . . . *Old Pictures Restored, Copied or Enlarged.* . . . STEREOSCOPES *and* ALBUMS *on hand.* . . . *If you.* . . . *do not see just what you want* ASK FOR IT![35]

Myers prospered in Hornellsville, making trips to scenic spots in the Adirondacks to produce popular stereo sets for sale in his establishment. It was here that he met Mary Breed Hawley, seven years his junior, whom he married in November 1871. Within a few years the young couple returned to Mohawk, where Carl became interested in ballooning.

Myers began his long aeronautical career by developing a superior balloon fabric. He boiled linseed oil to the consistency of syrup in a large kettle, added turpentine, then soaked thin cotton sheeting in the kettle. The fabric was then put through a wringer and scraped to remove all excess liquid. Myers claimed that his patented process, repeated three or four times, would completely seal either cotton or silk. The fabric would remain flexible and free of the cracking or gumminess that resulted from other treatments.[36]

Carl and Mary Myers apparently hired a professional aeronaut to fly their first balloons. By 1878, however, Carl had begun to fly on his own. Mary made her first ascension two years later, at Little Falls, New York on July 4, 1880, under the stage name Carlotta, which she would retain for the rest of her career. The crowds clearly preferred to see the pretty Mrs. Myers go aloft, for reasons which the editor of the *Watertown* [N.Y.] *Daily Times* made perfectly clear in describing a flight on July 4, 1882:

> *The spectacle will long be remembered as the finest occurring in the region. The lady was dressed in a jaunty suit of blue flannel trimmed with gold braid, her short skirts revealing neat-fitting garters. A nobby sailor's hat of plaited straw crowned the whole and gave her face a boyish piquancy. She stepped lightly into the frail contrivance which serves Carlotta in lieu of a basket. This consists of a thin wooden platform suspended by hammock twine to the connecting ring of the balloon, and as the "Aerial" gently arose, the entire proportions of her youthful figure could be plainly seen, apparently standing on the very air itself as she waved*

her hat in salute."[37]

For the remainder of their long careers in aeronautics, Carlotta performed most of the exhibition flights, which her husband arranged and supervised. Carl Myers concentrated his attention on experiments with small navigable airships and on the manufacture and sale of balloons.

Carlotta's first ascent from Watertown was also her first solo flight. The day was, she remarked, "a misty, moist, rainy Fourth, dampening all of my expectations of a delightful tour."[38] Carlotta's second ascent, from Sherburne, New York, on August 25, 1880, was uneventful, which is more than can be said for her third foray aloft. Having promised to ascend during a fair at Norwich, Connecticut, on September 9, 1880, she took off in the face of an approaching thunderstorm. After some time spent sailing above "the snow-white mountains of cloud-land," Carlotta descended. She emerged from the clouds to find herself skimming a few feet above a large woods. "Not wishing to alight here I threw over everything dispensable, ballast, ulster, waterproof, rubbers, and lastly two carrier pigeons kept for final messages."[39]

The rain collecting on the envelope had added so much weight to the balloon, however, that the basket continued to bump and skip over the treetops. Finally, trapped in the crown of a large basswood, Carlotta caught the attention of two men and a boy standing on the ground eighty feet below. Refusing an immediate rescue, Carlotta insisted on remaining in the car to supervise the removal of several large branches, allowing her to bring the *Aerial* safely to earth.

Carlotta was now fully launched on an aeronautical career. During the fall of 1880 she made two or three ascents a week, moving through New York, Ohio, Indiana, Michigan, Massachusetts, Connecticut, and Canada.[40] The "monotony" of this long series of flights was broken by occasional "novelty" ascents. Carlotta took particular delight in springing a surprise on the crowd. One of her favorite crowd pleasers was to dress as an old lady and "steal" the balloon, throwing off her shawl and wig once she was in the air.[41] In spite of her interest in collecting meteorological data and conducting aerial experiments for her husband, Carlotta was clearly a master showperson.

The 1881 season was even busier for the Myers. Their first daughter, Elizabeth Aerial, was born that spring. The new mother was back in the air in time for the Fourth of July celebration at Hamilton, New York. After a safe landing she parked the balloon and boarded a train for Utica, where Carl was waiting with another balloon. She made her second ascent of the day at 3:40 that afternoon, then returned home to Mohawk that night.

Carl and Carlotta were fully booked for the rest of the year. Mrs. Myers made the first of her many flights at Saratoga Springs before moving on to St. Paul, Minnesota, where she shared the bill with S. A. King. The pair then moved rapidly through Michigan, Canada, Massachusetts, Connecticut, and New York. After only two years of regular operation, the Myers had already built an enviable reputation for fulfilling their contracts and satisfying the crowds.[42]

Carlotta's balloons were specially designed for quick deflation and ease of packing and transportation.[43] Very small and light, weighing from fifty to sixty pounds each, two of these craft, complete with their collapsible baskets, could be packed in a single trunk. The baskets were especially interesting. Only seven pounds, they consisted of an upper load ring from which a wooden floor was attached by twine netting. Carlotta claimed to be able to exercise some control in landing by drawing one edge of the platform up.

Carl and Carlotta also experimented with a "screw sail" and a "rudder kite" to be used in navigating their balloons. The screw sail was very similar to the moulinets employed by J. P. Blanchard and other early aeronauts. A cloth-covered Archimedes' screw, it measured five feet across the blades and was operated by a hand crank. The rudder kite was a large square of fabric stretched over a wooden frame. It was attached to the edge of the basket by one tip and was controlled by a handle in the car. The rudder kite was designed to be furled and stored in the car to avoid damage in landing.[44]

Carlotta carried one final piece of navigational equipment in her basket. Dubbed the "Flying Dutchman," it was a collapsible sail consisting of five bamboo rods and a universal joint resembling a fishing reel. It was, she believed, "the most effective apparatus that I ever carried aboard the balloon." Nevertheless, she kept it concealed before and after a flight, apparently fearing that other balloonists might laugh at her attempts to sail before the wind.[45]

Carlotta did build an enviable reputation as an aerial navigator. On one occasion she made use of her unerring ability to read changes in the wind to win a bet by flying a predetermined course from the Washington Park baseball field in Brooklyn. Passing over the Brooklyn City Hall, she moved down to the Battery, up the river to the Brooklyn Bridge, on to New York City Hall, and then to a landing in Jersey City. The astounded gentlemen who had laid the bet presented the victorious aeronaut with a special trophy.[46]

In all likelihood, Carlotta also held the unofficial American altitude record. Flying from Franklin, Pennsylvania, with a balloon inflated with natural gas, she reached an altitude of 21,000 feet as measured by an aneroid barometer.[47]

Carlotta's aerial career was short compared to the careers of Wise, Allen, and King. During her ten years in the air, however, she built an enviable safety record. She suffered the usual spate of balloon misfortunes, falling into the tops

The balloon farm

of trees and, on one occasion when flying with seven-year-old Elizabeth Aerial, dropping into the cold waters of York State Lake. But for the most part, Carlotta enjoyed a remarkably trouble-free career.[48] She retired as an exhibition balloonist in 1891, though she continued to serve as an occasional test pilot for her husband's airships.

The Myers moved into new quarters near Frankfort, New York, in 1889. Purchasing a five-acre estate from Mr. Fred Gates, the couple transformed their new home into what soon became internationally known as "the balloon farm." During its prime in the 1890s, the grounds did resemble a balloon farm. A crop of 10 to 20 partially inflated envelopes dotted the lawn, each tagged with its name, the dates and places of prior ascents, and a synopsis of its performance.

The three-story, thirty-room mansion had been transformed into an aeronautical factory. The top floor housed a print shop where Myers prepared his flyers, posters, and an aeronautical newsletter. The upper story also contained a light machine shop, chemical laboratory, carpenter shop, and a loft for laying out fabric. The living quarters were on the second floor, while the ground floor housed offices, the kitchen, dining room, and Myers's superb aeronautical library. Seven bathrooms and a basement storage and shipping room rounded out the mansion.

Each summer the large kettles required for doping cloth were positioned in the yard. Billowing fabric was stretched over lines to dry. Gymnasts who would tour with Carlotta practiced on the grounds, while a tethered balloon kept automatic instruments aloft gathering meteorological data.[49]

The balloon farm remained the couple's headquarters until 1910, when they retired to daughter Bessie's home in Atlanta, Georgia. During his twenty-one years "on the farm," Carl Myers became one of the best-known balloon builders in the nation. The steady flow of envelopes, baskets, and aeronautical accessories shipped from the balloon farm provided the family with a substantial income.

The Myers were also involved in a variety of special balloon projects during this period. Carl Myers, for example, earned national fame as a rainmaker. His technique involved sending aloft 8-to-10-foot aerostats filled with a mixture of hydrogen and oxygen, which were exploded at the desired altitude by means of a time fuse. During the season of 1891–1892, the aeronaut built at least seventy-four of these balloons, including a set of ten that were produced in five days to meet an emergency. The work, which was encouraged by the U.S. government, was conducted, for the most part, at Frankfort, New York, and Midland, Texas.[50]

Carlotta was also involved in a number of special projects after the turn of the century. While her husband served as manager of events at the Aeronautical Concourse of the Louisiana Purchase Exposition held in St. Louis in 1904, Carlotta operated the large captive balloon at the fair. Passengers, safely housed in a wicker basket with a door on one side, ascended to 900 feet for a bird's-eye view of the city.

The Myers also staged several balloon "races" at the exposition. On Monday, July 4, 1904, Carlotta, with Carl flying as a passenger, flew against T. A. (Tracy) Tisdell, a Brooklyn aeronaut who was often associated with the Myers. Carlotta and Tisdell split $250 for staging this race.

On August 28 Carl Myers and George Tomlinson launched two balloons in a well-publicized race whose announced goal was Washington, D.C. Myers landed that evening near St. Charles, Missouri, while Tomlinson flew on to a landing near Wyoming, Illinois, the next day. St. Louis newspapers attacked the races as staged publicity stunts for which no real prizes had been offered. Myers was eventually dismissed as superintendent of aeronautical events at the fair.

Much of Carl Myers's energy was devoted to experiments with small airships. His very first balloon ascent was made from Mohawk, New York, in 1878, aboard a small, unpowered, spindle-shaped gasbag. He described the experience to a conference on aerial navigation held at Chicago's World's Columbian Exposition in 1893.

> *No basket was attached, but I sat simply astride of a folded bank of cloth, hung like a saddle from the concentrating ring. Thus my feet extended below, while the balloon neck, tied down beside me, made us all one machine for cleaving the air. The air was calm and the ascent nearly perpendicular, reaching finally to two miles in height. Then, noting time and altitude, I pulled the valve wide open. The balloon responded immediately and the speed accelerated till the first downward mile was made in one minute.*
>
> *The neck of the balloon formed a sharp prow terminating with my body and extended toes. Retaining the slim form throughout, my craft cleft the air like a dart. I estimated my speed now to be more than a mile a minute. I could only judge by the sting of the air rushing past me. I was now within a thousand feet of the earth. . . . I dropped half my ballast and released the balloon neck. The speed was checked and as the balloon neck flew up, forming a concavity in the netting above me, there was a shock which seemed to lift me upwards. . . . Then the balloon slid downward into a cornfield without shock and ended my experiment.*[51]

Carl Myers was by no means the first aeronaut to experiment with very small airships. Charles Francis Ritchell, a native of Portland, Maine, had unveiled an airship with a barrel-shaped gasbag powered by a hand crank as early as 1878.

Ritchell was living in Corry, Pennsylvania, at the time he patented his airship. He was already well established as a mechanic and general inventor. His local fame was based on his invention of a boring machine used in the manufacture of brushes. An ordinary curry brush used in a stable required drilling 700 separate holes for the bristles. Ritchell revolutionized the process by drilling the correct number of holes at the proper angle in a single operation. While living in Corry, the inventor served as the manager of Frank S. Allen's brush block factory. A local newspaper, the *Corry Blade*, reported that "Mr. R. was the original inventor and patentee of all those boring machines; has had charge of starting the works, manufacturing the machines and finally brought the invention and the business to a success. It would seem that no other man could take hold of the concern and handle it as Mr. R. can."[52]

During his time in Corry, Ritchell was also responsible for developing a

Inventor CHAS. F. RITCHEL.

highly touted process for producing inlaid wood products. He moved to Bridgeport, Connecticut, in 1878, where he was employed by the firm of Ives, Blakeslee, and Company, manufacturers of mechanical toys. There his inventive genius was given free rein. "Whenever he wants money," reported one townsman, "he invents something, gets it patented and sells his rights."[53] There were, for example, his walking dolls, a child, a sheep, and a cat "that has only to be pulled to walk off as naturally as can be, and is without clockwork, and needs no winding."

The list of inventions that poured forth from Ritchell's fruitful mind seemed endless. By the mid-1880s he claimed to have obtained 150 patents on such pleasant and harmless devices as windup toy motors, corsets, and envelopes with the gum on the body rather than on the flap. Ritchell offered to construct a life-size automaton of Christopher Columbus that would "speak . . . as easily as Armour's pigs squeak" for the Columbian Exposition of 1893.[54] His "Brownie Miteascope" reduced live actors to the size of fairies when viewed by spectators looking through a system of lenses.[55]

Outside the entertainment field, Ritchell developed a number of machine and electrical techniques. He experimented with colors extracted from moth larvae and demonstrated a "mosquito exterminator" and a cockroach trap in which the hapless insect was captured in a wire sieve. In addition to his airship, Ritchell expressed his thoughts on the submarine and demonstrated a special belt for use by assassins which would inject poison into a victim. In November 1884 he unveiled a gas bomb designed for use in land or naval warfare.[56]

In spite of these many well-publicized inventions, Ritchell remained best known for his airship experiments. In later years the inventor claimed to have begun thinking about the problems of flight in 1869. Nine years later he was awarded U.S. patent no. 201,200 for an "improvement in flying machines." His craft was first flown in public at Philadelphia's Concert Hall in May 1878.

The barrel-shaped gasbag was made of rubberized fabric 25 feet long and 13 feet in diameter, with a weight of 66 pounds. The envelope was so small that it could be flown only with hydrogen produced by the standard method of adding sulphuric acid to iron filings. Seven broad, worsted bands extended over the top of the gasbag and attached to a 23-foot piece of nickle-plated, mandrel-drawn brass tubing 1½ inches in diameter. The airship framework was suspended from this keel.

The operator's frame was constructed of the same brass tubing as the keel. It was 11 feet long and 2½ feet wide and constructed in two sections. The pilot sat to the rear of the frame with his feet on two pedals and with a hand-operated wheel crank between his knees. The hand crank turned a four-bladed propeller set horizontally beneath the pilot. Twenty-four inches in diameter, it worked at a speed of 2,000 RPM. The four blades were cut from white holly, each with a surface area of 50 square inches.

A tractor propeller measuring 22 inches across was placed at the very front of the frame and was geared to the main horizontal propeller and hand crank drive system. It was controlled by the foot pedals. By pressing on the left treadle, the pilot threw the forward airscrew into gear, permitting it to operate at speeds of up to 2,800 RPM. When the pilot pressed his right foot forward the tractor propeller swung to the left; a push to the rear with his right heel pivoted it to the right.[57]

The finished machine weighed only 114 pounds and was normally flown 6 pounds heavy. No valves or ballast were employed. To rise, the pilot turned the hand crank forward to operate the lifting airscrew. To descend, he spun it in the other direction. The tractor propeller was employed for forward motion and for turning.

The inventor, who was too heavy to fly the craft himself, hired a lightweight young man for the early indoor exhibitions. Not long after the beginning of his successful run in Philadelphia, he was deluged with young women volunteers answering his advertisement for a female pilot. He finally selected Miss Mabel Harrington, who, suitably clad "in her new suit resembling a silk jacket and knee breeches," made a number of circles of the hall to the applause of an admiring crowd.[58]

The first outdoor trial of the Ritchell airship took place in Hartford, Connecticut, on June 12, 1878. The pilot, 96-pound Mark Quinlan, rose to 250 feet, sailed by the spire of the Colt Memorial Church and over the Connecticut River. Quinlan had ascended in a dead calm, but a strong wind sprang up while he was in the air, presaging the onset of a storm. The pilot was able to crank his way back to the ball park where the flight had begun, landing a few feet from his take-off spot before the rain began.[59]

The following day, June 13, Quinlan once again demonstrated that he had "the confidence and nerve enough to go up in a gale." A great deal of time was lost in properly balancing the machine that day. Loaded with nine pounds of stone ballast, the craft rested lightly on the ground, but could be lifted with a slight pressure of one finger. A local reporter described the take-off: "Then the word was given to 'Go.' Quinlan began turning the wheel, the horizontal fan revolved with a noise like a buzz-saw, and the machine darted vertically to a height of about two hundred feet. There a steady current of wind setting toward the southwest was encountered, and the machine was swept away by it."[60]

Operating the forward propeller at maximum speed, the aeronaut could not make headway against the wind. He dropped to 100 feet, then rose to 300,

Ritchell's airship

apparently searching for lighter winds, all the while being blown out of sight toward New Haven. After nearly an hour in the air, Quinlan gave up his attempts to return to Hartford and landed at Newington, 5 miles from the take-off point.

Ritchell began exhibiting his machine at Boston's Tremont Temple on June 24, 1878. The demonstration, arranged by William McMahon, who played a major role in introducing Edison's phonograph to the public, was a complete success. As the *Boston Herald* noted, "It is hardly an exaggeration to say that a more novel and interesting exhibition has never been given in this city."[61]

In addition to the indoor flights, Quinlan made an exciting ascension from Boston Common. Once in the air, the propeller gears jammed, allowing the balloon to rise dangerously high. Without a valve, the envelope swelled, breaking several of the bands from which the frame was suspended. Quinlan could not slit his envelope, for there was no netting in which the fabric could gather to form a parachute. He had little choice but to tie one hand and ankle to the frame, then drop beneath the craft to make repairs with a jackknife as his only tool. He finally descended at Farnumsville, 44 miles from the Common, after a flight of one hour and twenty minutes.[62]

Business was looking up for Ritchell. He was receiving orders for airships from other prospective exhibitors, including one Cuban syndicate. In cooperation with the D. P. Ells Company of Chicago, Ritchell marketed toy airships known as Patent Flying Wizards. Selling for fifty cents each, the Wizards were guaranteed to fly and "furnish splendid outdoor exercise and amusement to old and young, boys and girls."[63]

Ritchell, Quinlan, and the airship also found employment with W. C. Coup's Greatest Show on Earth in 1878. Coup, a former associate of P. T. Barnum, operated a touring show that included "the only traveling aquarium in the United States," a "giant devil fish," ten members of the "Royal Japanese Circus," performing greyhounds, a horse on stilts, the only banded proteus, or walking fish, and the Ritchell Flying Machine. Coup was careful to inform visitors that his extravaganza had been "endorsed by the entire Clergy of New York City."[64]

Ritchell continued to operate his original machine at Brighton Beach, near Coney Island, in 1879. A larger machine known as the *Peerless*, apparently constructed to carry two persons, broke loose and blew out to sea. The inventor was also said to be building a very small airship for W. C. Coup. Weighing only six pounds, it would be operated by a dwarf.[65]

By 1880, as the *New York Times* noted, "even flying machines lose their novelty."[66] Ritchell was now regaling editors with the prospect of an aerial voyage to the North Pole. He was also giving some thought to the possibility of a rotary wing, heavier-than-air machine and to the construction of an 83-foot-long version of his small airship. With such a craft, he predicted, flights of up to twenty-four hours would be possible. He was entirely pessimistic, however, about the ability to fly against the wind. "After all my experiments of building five complete flying machines and working them, I am convinced that if an engine could be made of 1,000 horse power weighing but one pound and no larger than your hat, it could not propel a balloon of 3,000-foot capacity against a wind current of ten miles an hour five miles an hour."[67]

Ritchell was never to alter his attitude. As late as 1905, long after the appearance of very large Zeppelin airships, he would still remark that the possibility of navigating an airship against the wind remained "just as ridiculous as a perpetual motion machine, and the latter will be invented just as soon as the former." Still an adherent of hand propulsion, he did believe that the crew of such a craft would be able to search out favorable winds blowing toward their destination. Ritchell's rather primitive notions of aeronautical propulsion guaranteed that he would make only minor contributions to lighter-than-air technology. He was, nonetheless, a superb showman who inspired other, more experienced, aeronauts to attempt their own airship building programs.

Carl Myers was the best-known experimenter who followed Ritchell's lead in constructing pedal-powered airships. In his hands, the "Skycycles," as he so aptly termed the little single-seat craft, were to delight thousands of aeronautical thrill seekers who attended his shows between 1879 and 1900. Loosely based on Ritchell's original machines, Myers added a few sensible touches. His craft were, for example, pedal-powered. The early Skycycles featured both a rudder at the rear and forward steering vanes. The rudder was abandoned in later versions. A shift of the pilot's weight was sufficient to climb or descend.[68]

Carlotta, who spent a great deal of time on the seat of a Skycycle, described the appearance and operation of one of the early "gas kites":

> *Looked at on its side it appears like a canoe bottom up. Viewed in front its section seems like that of a half sphere or dome. Its under side is flat, like a kite, and its interior space is filled with gas enough to nearly lift the entire machine and aeronaut. A balloon netting surrounds it, and its cords support a concentrating ring, exactly as with an ordinary balloon. Instead of the customary basket, there hangs from this ring a velocipede seat. In front of the operator, where the "steering bar" of a velocipede is, there are cranks for the hands instead. At the feet are ordinary velocipede cranks. All are geared so that moving one crank moves all, and together they revolve a screw shaft which projects to some distance in front like a bow sprit, supported by the netting stays. At the outer end of the shaft is*

The Myers Skycycle

> *a huge screw of cloth supported by two yards, like a ship's sail. This lies flat and motionless like the outspread wings of a soaring bird until revolved, when it instantly twists itself into a screw.*[69]

The exact number of Skycycles constructed is uncertain, but the couple flew exhibition dates with the machinery in thirteen states. The craft were so simple to operate that novice pilots were occasionally allowed to take a hop. Such was the case on August 3, 1895, when a *New York World* reporter pedaled his way from the Brooklyn Navy Yard across the East River and Manhattan to a landing in Yonkers.[70]

Carl Myers finally patented the Skycycle in 1897. Together he and Carlotta continued to operate the small airships into the early years of the new century. In 1900, for example, they made 120 Skycycle ascents during a single engagement at the St. Louis Coliseum. Myers's Electric Aerial Torpedo, also developed by 1900, was a smaller, unmanned version of the Skycycle intended to deliver a load of explosives against an enemy.[71]

While Ritchell and Myers were the best-known aeronauts operating human-powered airships, a few other balloonists followed suit. Arthur Barnard, of Nashville, Tennessee, for example, made a number of pedal-powered flights in 1897.

The Campbell Airship, one of the more controversial projects, was the product of a number of the major aeronauts of the period, including both the Myers and Allen families. The airship was conceived by Peter Carmount Campbell. A native of Rhinebeck, New York, born in 1832, Campbell had for many years operated a jewelry store in Brooklyn. A longtime aeronautical enthusiast, he had discussed his schemes with Horace Greeley, S. F. B. Morse, and others for many years before beginning construction of his airship in 1888. The craft was built around an oblong envelope 60 feet in length and 42 feet across. As with the Ritchell machine, a long metal rod beneath the gasbag served as a keel. The keel was directly tied to the bag at the center. A web of cords extended from the bar to the ends of the bag. A boatlike car was slung from the keel. Large enough to house a pilot and three passengers, the car sported two large, birdlike wings on each side. The wings were not designed to be flapped, but could be raised or lowered to control the direction of motion. A forward rudder was also employed.

A large, multibladed propeller was located on the underside of the car. Campbell originally hoped to power this fan, and the three smaller pusher propellers at the rear, with batteries. In order to save weight, however, the electrical power system was rejected in favor of foot pedals.[72]

In August 1888 Campbell contracted with Carl Myers for the construction of three small practical working models and a single large envelope and net for a full-scale airship. Campbell and the investors in his airship enterprise worked the rest of that summer and fall to prepare their craft for flight. By early December 1888 they had erected a gas generator in a vacant lot next to the Sea Beach Palace at Coney Island, New York, and were ready for a trial.[73]

James K. Allen was engaged to operate the craft. Allen, worried that the freezing temperatures might crack the fabric, waited until December 8, a warm day, to attempt an ascension. After several hours lost in generating hydrogen, Allen climbed aboard at 4:00 P.M., tested the pedals and rudder rope, and ordered his release. Ascending to 100 feet, the aeronaut heard Campbell calling him to return to earth for a photograph. Allen cranked himself back down without difficulty.

Reascending to 500 feet, Allen paused to gauge the strength and direction of the wind. He then flew toward Brooklyn, reversed course back to Coney Island, and spent the next half hour zigzagging across the skyline. "I never had a pleasanter sail in my life," Allen informed reporters. "This mechanism is nearly perfect, and the ship minds her rudder as quickly as does a sloop at sea. The labor of propulsion is not at all difficult. I wish I owned the ship. I could make a fortune out of her."[74]

The Campbell Airship was taken back into storage until the following spring. Campbell had engaged the services of E. D. Hogan to operate the craft in 1889. Hogan, born in 1852, was the brother of John Hogan, whose fatal hot air balloon accident was noted earlier. An experienced Canadian aeronaut with 200 ascensions to his credit, he had made his first flight from Jackson, Michigan, and had apparently met Campbell while making regular parachute drops from a hot air balloon over Rockaway Beach in the summer of 1888.

Campbell and Hogan made their first attempt to launch the airship on June 19, 1889. After five hours of effort, they were forced to admit defeat to a jeering crowd of 1,000 spectators. The failure to ascend was blamed on a leaky gas generator.

Charles Ritchell, who had followed Campbell's progress with interest, was incensed. Campbell, he charged, had directly copied his own airship plans. "I abandoned the machine," Ritchell complained to the press, "because it was only a toy, and couldn't be made of any practical use, but I want whatever glory there is in the invention and don't propose to let these people go on deceiving the public with a machine built after my ideas and which is just as useless and even more helpless than mine was."[75]

Campbell ignored Ritchell's comments and brought the airship out for another trial on July 16, 1889. On this occasion, Campbell and Hogan abandoned the old generator in favor of city gas. The ascension began well, with Hogan moving toward the southeast. When the ship had traveled roughly a mile from its take-off point at the Nassau Gas Company works, the large horizontal propeller beneath the car was seen to fall into the water.

Some observers thought they could see Hogan climbing the network as the airship was blown out of sight. Late that afternoon, the crew of a New York pilot boat reported seeing the balloon dragging in the water 74 miles off Long Island. The boat was unable to overtake the balloon. No trace of the aeronaut was ever found.[76]

The desire to navigate the air had infected a few of Hogan's countrymen north of the border. One of the best-publicized airship projects of the period was the work of two citizens of Montreal, Richard Cowan and Charles A. Pagé.

Pagé, "a very ingenious and skillful mechanic," began to consider the possibility of the navigable balloon during the Franco-Prussian War, when free balloons were employed by besieged Parisians to communicate with the outside world.[77] Pagé's invention lay fallow until 1878, when his work came to the attention of Richard Cowan, a wealthy retired Montreal merchant. It would seem that Charles Ritchell's limited success with his pedal-powered airship that summer was a factor of some importance in drawing Cowan's attention to the subject.

Fortunately for Pagé and Cowan, neither of whom had any aeronautical experience, Charles H. Grimley, a veteran of forty-one ascents, was on hand to take charge of practical details. Grimley had immigrated to the United States from Worcester, England, in 1870. After several years as a merchant's clerk, he turned to aeronautics.[78]

Grimley concentrated his activity in New York and Pennsylvania. He had experienced the usual number of near disasters, suffering from intense cold during an ascent to 10,000 feet at Monongahela City, Pennsylvania, on September 18, 1875; nearly drowning in the Hackensack River following a New York City launch; and dropping 40 feet to the ground after the collapse of his balloon during a Pittsburgh ascension in October 1875.

Grimley began a long Canadian tour in March 1878. Over the next eighteen months he made four ascents from Ottawa, the last on the occasion of the annual picnic of the St. George Society on August 20, 1878. Accompanied by William Gibbons of the *Ottawa Citizen*, Grimley flew for an hour and forty-five minutes, making a rough but safe landing in the upper branches of a 50-foot pine.

Grimley brought his 17,500-cubic-foot balloon *City of Ottawa* to Montreal in the fall of 1878. Built especially for the Canadian tour, the craft was constructed of Irish linen. It stood 50 feet tall and measured 35 feet in diameter. Montreal newsmen were enthralled by Grimley's tales of his years in the air and by his hopes to fly the Atlantic. The city turned out in force for his first Montreal ascent on September 28, 1878. Accompanied by Hiram A. Moulton of the *Montreal Daily Witness*, he narrowly escaped falling into the St. Marys River after an hour in the air. The officers of the Irish Protestant Benevolent Society, which had sponsored the event, were delighted.[79]

Cowan and Pagé displayed the partially completed car of their proposed airship at the Shamrock Lacrosse Gardens on the occasion of Grimley's flight. The relationship between the three men apparently began at this time. By the following spring the car was complete, and so was a brand-new 70,000-cubic-foot aerostat which Grimley had commissioned for Cowan and Pagé. The first flight of the new craft was scheduled for another Irish Protestant Benevolent benefit on June 21, 1879.

The Cowan-Pagé car was the subject of some attention in the Montreal newspapers. The passengers would travel in a cage of iron tubing 7 feet long, 7 feet high, and 4½ feet wide with a plank floor. Twin sidewheel propellers driven by hand cranks were attached to the side of the passenger car. The car was enclosed in a larger iron frame, 30 feet long, designed to be slung beneath the usual cigar-shaped gasbag. A single-piece, cruciform rudder-elevator was attached to the rear of this frame. Provision was also made for a valve on the balloon, although it was hoped that, through judicious use of the paddles and control unit, it would not be necessary to resort to valving gas or dumping ballast.[80]

For the preliminary experiments, Grimley's new balloon *Canada* would be substituted for the ideal cigar-shaped envelope. Four times the size of the old *City of Ottawa*, the new craft stood 80 feet tall and measured 50 feet in diameter.[81]

The premier ascent of the new balloon and car on June 21 was a disappointment to Cowan and Pagé. Delays in the inflation forced Grimley to cut the process short. He had originally intended to fly with Cowan, Pagé, and two reporters, a Mr. Creelman and Hiram Moulton. Now he asked Cowan and Pagé to give up their seats in favor of one of the journalists in order to foster publicity for the program. The craft was still too heavy however. Moulton was next to go, followed by the fore and aft sections of the car, the propelling wheels and mechanism. Grimley and Creelman would make a normal balloon voyage using valve and ballast.

It was a rough journey during which the *Canada* was blown over the Richelieu and St. Lawrence rivers toward a vast forested area beyond. On landing, they crashed through a series of treetops and were dragged across a large open area before their grapnel took hold. Grimley was knocked unconscious but neither man was seriously injured.[82]

Grimley moved on to Ottawa to entertain at the Dominion Exposition that July, while Cowan and Pagé were reassembling the car for another trial. During this waiting period, Charles Ritchell's airship became a subject of discussion in Canadian newspapers. One disappointed correspondent, who had attended Grimley's ascent on June 21, noted that while Ritchell's machine was smaller, it had proven far more trustworthy.

Anxious to redeem themselves, Cowan, Pagé, and Grimley scheduled another test flight for July 31, 1879, from the Shamrock Lacrosse grounds in Montreal. On this occasion all three men flew with three additional passengers, Mr. Moulton of the *Montreal Witness*, Mr. Browning of the *Montreal Herald*, and Mr. Harper of the *Montreal Star*.[83]

All navigating equipment was in operating order this time, and Moulton was ordered to take the helm soon after take-off. After enjoying the scenery from 1,500 feet, Grimley ordered his crew to man the sidewheel propellers. Grimley and Harper operated the windlass on one side while Pagé and Browning manned the other propeller.

Everything seemed to be going well until Pagé noted that the revolving propellers were cutting the suspending ropes connecting the car to the balloon. After some minor repairs they set the propellers in motion a second time, only to start up a bumping and jolting in the car. The *Canada* reverted to the status of a free balloon once again while Grimley climbed out of the car to effect still more repairs. In the third trial, the propellers functioned as planned. While the equipment proved inadequate to control the ascent or descent of the craft by itself, it did help to conserve gas and ballast.

The mechanism created one anxious moment when the valve and rip line became tangled around one of the cranks. If the ripping line were pulled, the entire top of the balloon would open, allowing all the gas to spill out. Afraid to attempt disentangling the lines, Grimley began searching for a landing place. After two and a half hours in the air, they came safely to earth near the village of St. Aimée.

Cowan and Pagé were more than satisfied with the outcome of the voyage. While their equipment had certainly created problems, it had proved effective in stabilizing altitude. Others were less certain. Correspondents writing to the *Montreal Star* continued to call attention to the fact that Ritchell's small machine had operated with greater success and fewer problems.

The flight of July 31, 1879, was apparently the last trial of the Cowan-Pagé car. Grimley returned to the United States where he continued an active aeronautical career. A trip over the Green Mountains from Montpelier, Vermont, to a Montreal suburb on July 4, 1884, drew national attention.[84] The airship experiment was not forgotten however. When the giant airship R-100 visited Canada half a century later, Montreal newspapers resurrected the story of this earlier airship trial.[85]

The small airships powered by pedals or hand cranks were a uniquely North American phenomenon. Aimed at solving the problem of aerial navigation, they were put to use as yet another means of entertaining the public. Developed before the availability of lightweight steam, internal combustion, or electrical engines, there was no chance for these small craft to evolve into genuine airships capable of moving against the wind and maneuvering at will.

Few Americans were thinking of large powered airships in the tradition of Rufus Porter or Solomon Andrews. Russell Thayer of Philadelphia was an exception. Thayer, an 1874 graduate of West Point, had resigned his commission after a period of service as an artillery instructor at the academy. By 1888 he had become a senior partner in the Philadelphia civil engineering firm of Thayer and Patterson.

On January 7, 1885, Thayer sent a letter to Secretary of War Robert Lincoln suggesting the military possibilities of a large airship powered by a jet of compressed air ejected from the rear of the craft. Impressed by the proposal, Lincoln invited Thayer to bring his proposal to the attention of the Ordnance Board. The board approved Thayer's plan and suggested that a congressional appropriation of $5,000 be sought to fund construction of a prototype. After another year of discussion, however, the War Department decided not to submit the request to Congress. Thus, while European governments began the process of investing in the earliest large dirigible airships, American military officials decided against entering the field.[86]

In spite of the efforts of Russell Thayer, Americans took little interest in the potential of the large airship. The small airships had flourished for exactly the same reason that the dangerous acrobatic performances beneath a hot air balloon had proven so popular. They were entertaining. That was the watchword of late-nineteenth-century aeronautics. The prosperity of a balloonist was much less dependent on his skill in the air than on his ability to develop new and thrilling stunts to please a crowd.

Chapter 16 Captain Tom and Sergeant Ivy

NO ONE PLEASED MORE CROWDS DURING THE CLOSING YEARS OF THE nineteenth century than Thomas S. Baldwin, a man whose career both epitomized and helped to shape the course of American aeronautics. Baldwin himself encouraged some mystery about his birth. At different points in his long career he gave a variety of birthplaces and dates, ranging from Marion County, Mississippi, to Decatur and Quincy, Illinois. He also gave birthdates between 1854 and 1861. Even Baldwin's middle name is uncertain, having been either Scott/Scot or Sackett/Sakett.

Baldwin's son Thomas seems to have provided the final answer to these problems in a letter written to aviation pioneer and historian Ernest L. Jones on February 3, 1950. His father's name, he noted, was Thomas Sackett Baldwin, and he had been born in Decatur, Illinois, on June 30, 1858, the son of Samuel Yates and Jane Baldwin.[1]

When Baldwin was very young, his father, a local physician, moved the family to Quincy, Illinois, where Tom and his brother Sam attended local schools. His parents died in a Civil War border raid. Refusing to accept the life of a bound orphan, Tom ran away and worked as a lamplighter, newsboy, railroad brakeman, and book salesman.[2]

As Baldwin recalled many years later, he had been amusing himself doing handsprings on a Hot Springs, Arkansas, sidewalk one day while on a bookselling trip. A circus performer named Bryant, observing the scene, suggested that Baldwin team up with him to form a two-man high-wire and trapeze act for a circus that W. W. Cole of Quincy was organizing.[3] Baldwin put in three months practicing on a high wire stretched between two 30-foot buildings. By 1870 he was touring the nation as a gymnast and acrobat.

Baldwin's years on the sawdust circuit certainly inured him to danger. Having broken his connection with the alcoholic Bryant, Baldwin entered into a partnership with Harry Victor, a trapeze performer. During a Columbus, Ohio, performance, both men suffered a 65-foot fall that took Victor's life. Tom continued on his own, earning local fame all across the western United States. One of his best-known stunts was a 700-foot tightrope walk from the San Francisco Cliff House to Seal Rocks in 1885.[4]

It was apparently in San Francisco that the young daredevil first met Park A. Van Tassel, one of the leading aerial impresarios of the period. Van Tassel, a native of Ohio born in 1852, had been captivated by a balloon ascension he had witnessed as a boy. Three decades later he became a specialist in introducing the balloon to small western towns that had not yet seen a man in flight.[5]

Van Tassel had begun by staging a flight for the citizens of Albuquerque, New Mexico, with a balloon that he had purchased from F. F. Martin of San Francisco for $850. Less than a year after sheriff Pat Garrett shot Billy the Kid in Fort Sumner, New Mexico, Park Van Tassel brought the territory into the air age.

The flight, which took place on July 4, 1882, was a great success. C. W. Talbott, superintendent of Albuquerque's new gasworks, supplied the lifting power. Van Tassel's balloon *City of Albuquerque* rose until it was "no larger than a derby hat" and flew to a safe landing near the Rio Grande. Returning to town, Van Tassel was greeted by a celebration at the Elite Saloon. The locals were not only honoring the aeronaut but celebrating the recent death of Johnny Ringo, a notorious gunfighter.[6]

Van Tassel also brought the first balloon to Utah with a flight of 6 hours 45 minutes from Salt Lake City in 1883.[7] On August 7, 1886, Van Tassel inflated his balloon from a city gas main at the corner of Eleventh and Broadway in Denver. He then loaded the craft onto a wagon for transportation to the lot from which Colorado's first balloon ascension was made.[8]

By 1887 when Van Tassel encountered Tom Baldwin in San Francisco, he had worked the West for what it was worth and was in the market for a new stunt to liven up his performances and draw new audiences. The parachute was one possibility. Parachuting, once all the rage in Europe and America, had died out after Robert Cocking's death in 1837.

The manager of the Brush Street Theater first informed Baldwin and Van Tassel that two local San Francisco gymnasts were experimenting with a parachute. The men had enjoyed little success in attempting to incorporate the device into their indoor act because of the low altitudes involved. Baldwin and Van Tassel bought the parachute from the discouraged experimenters and, on January 21, 1887, made their first trial in the Mechanics Pavillion. They drew the canopy into the rafters by means of a rope run over a pulley. After a few test drops with sand bags, Baldwin made three jumps and Van Tassel one.

The parachute itself was fairly simple. Sewn of Wamsutta muslin, the canopy measured from 18 to 21 feet in diameter and featured a hole at the top of 4 to 6 inches. The lower edge of the canopy was held open by a hoop, so that the canopy could not fully collapse. Baldwin would continue to jump with this hoop in place at least until the time of his fifth descent at Syracuse, New York, on September 22, 1887.

Van Tassel, weighing in at 200 pounds, was scarcely an ideal parachute jumper. Baldwin, however, was perfect for the job. He was the very picture of an athlete, as an admiring female journalist noted in England the next year:

> *A clean-limbed, well-built man, evidently possessed of enormous muscular strength, and in the best of health. He stands five feet ten inches in height, measures a close forty-eight inches around the chest, and weighs very nearly thirteen stone [182 lbs.]. . . .*
>
> *He has blue eyes, a fair complexion, light brown hair of a somewhat wiry texture, and a well-trimmed moustache of the same colour. He has a hearty, genial manner, with a dry humour, speaks with a strong, but by no means disagreeable American accent, has a grip of iron when he shakes hands with you, and is about as free from conceit and affectation as they make 'em.*[9]

Their preliminary tests a success, Baldwin and Van Tassel agreed on an acceptable split of the profits, then approached a San Francisco streetcar company which agreed to pay a dollar for every foot of altitude Baldwin reached prior to dropping. The great day came on January 30, 1887. Baldwin climbed to 1,000 feet over Golden Gate Park. The crown of the parachute had been lightly stitched to the netting and gathered around the lower hemisphere of the balloon with the trapeze bar in the basket. Baldwin grasped the bar and leaped into space, his weight pulling the parachute free. The 65-pound canopy did its work, lowering the newly hatched aeronaut safely to earth.

Having reintroduced the parachute, Baldwin and Van Tassel parted company. Their relationship had been stormy, and the parting was unhappy. In later years Van Tassel claimed that Baldwin had simply "stolen" the parachute and set up business on his own.[10]

Tom Baldwin returned to Quincy for a reunion with his brother Samuel Yates Baldwin, who would soon join him on the exhibition circuit under the stage name Professor Yates. Tom made his second parachute drop before a hometown crowd on July 4, 1887. On this occasion, as in San Francisco and on most subsequent ascents, he was flying a gas balloon. While safer to operate than a hot air aerostat, it would remain in the air much longer after the aeronaut had

A publicity photo commemorating T. S. Baldwin's first parachute jump over Quincy, Ill., July 4, 1887

departed, making recovery a problem. For this reason the Golden Gate jump had been from a tethered craft.

In Quincy, however, high winds made tethering too dangerous. Rather than disappoint the crowd, the aeronaut chose to make his first jump from a free balloon. Baldwin leaped from 4,000 feet, returning safely to earth in a wheat field only half a mile from the take-off point after a fall to earth of 3 minutes and 20 seconds. The balloon traveled 50 miles to a landing in Perry, Pike County, Illinois.

"The best men in Quincy" were so pleased with the performance that they presented Baldwin with a $450 bonus and a gold medal struck expressly for the occasion.[11] The young aeronaut was rapidly becoming a very famous man. Dozens of letters offering the promise of lucrative engagements were pouring into Quincy. With theatrical manager Richard Fitzgerald handling the business arrangements, Baldwin set off on an East Coast tour less than a month after the Quincy jump.

Forty thousand people watched as Baldwin jumped over Brooklyn on August 9, 1887. He was forced to cancel a Rockaway Beach ascent on August 29 because of inferior city gas, but was back in the air from the same spot the following day.[12] Then he was off to Syracuse for another jump on September 22, 1887, and a second from Potsdam, New York, on September 24. Baldwin returned to Quincy for his seventh jump on October 12, 1887.[13]

It had been a most successful year for the young aeronaut. News of his exploits had even reached London, where an Italian showman, G. A. Farini, became convinced that Baldwin would be a sensation.[14] Baldwin and Fitzgerald were also quick to recognize the possibility of rich pickings in England. Tom, his young wife, and Fitzgerald sailed for Southampton early in the summer of 1888. Fitzgerald had some initial difficulty in booking the act and exhibited a drinking problem as well, so the Baldwins dismissed him and entered into a partnership with Farini, who proved a much more able manager.

Baldwin's British tour more than fulfilled Farini's expectations. The aeronaut made more than a dozen jumps over London's Alexandra Palace between July 28 and September 14, 1888. His appearances were nothing short of a sensation. Accurate news accounts reported a regular attendance of from 30,000 to 50,000 at his exhibitions. The Prince of Wales, the Duchess of Teche, her daughter, the future Queen Mary, and other members of the royal family attended.[15]

In addition to his London flights, Baldwin ascended from Newcastle, Birmingham, Ashton, Manchester, and Glasgow. The Scottish ascent proved particularly exciting when Baldwin's parachute did not fully deploy. Fortunately, the aeronaut suffered only minor bruises.[16]

The young American completely eclipsed other aeronauts operating in Great

Britain that summer and fall. Leona Dare, the American performer, who was also touring the country, had previously drawn large crowds when she ascended hanging from the trapeze bar by her teeth. Now she found herself upstaged by the newcomer and his parachute.[17] Captain Templar, of the army balloon factory at Farnborough, was also impressed. London newspapers reported that he had placed an order for three Baldwin parachutes.

Not everyone shared the general enthusiasm. The parachute drops sparked a strong negative reaction as well. Letters to the *Times* suggested that Baldwin be forbidden to make any more jumps "before he pays the penalty for daring."[18] An editorial in the London *Chronicle* objected to such a "showy suicide" and upbraided its readers for failing to "shrink with horror from the picture of the dying gladiator butchered to make a Roman holiday."[19] The *St. James Gazette* remarked that the reason why people should jump from balloons was far from obvious, while the *Daily Chronicle* complained that Baldwin's show "belongs to the type of personal exhibition with which we can feel little sympathy."[20] Another journal struck a particularly elitist note by congratulating the aeronaut on providing a "treat" for the "intellectual masses of London."[21] At one point, the House of Lords considered the possibility of banning further jumps.[22]

British balloonists were not deluded by the criticism. They saw nothing more than the pounds, shillings, and pence pouring into the joint coffers of Messrs. Baldwin and Farini. Baldwin recognized the danger of competition, and, with Farini, obtained an English patent on the parachute. This action does not seem to have affected the rush to take up parachuting however.

The Spencer brothers, Percival, Stanley, and Arthur, were particularly impressed by the enormous drawing power of Baldwin's show and did not hesitate to borrow his idea. Arthur remained to carry on at home, while Percival and Stanley toured, respectively, the Far East and South America with balloons and parachutes.[23]

Baldwin returned home from Europe with a full purse and a gold medal from the Balloon Society of Great Britain. During another year of ascensions in the United States, Tom had initiated both his brother Sam and a new assistant, William Ivy, into the parachute-jumping business.

A native of Houston, Texas, born on July 31, 1866, Ivy had grown up on the rough-and-ready Texas frontier. Determined to become a circus acrobat, he had achieved local fame by walking a high wire stretched over a lake near San Pedro Springs, Texas. He joined the Thayer-Noyes Circus as an acrobat and made his first balloon flight aboard a homemade hot air craft from Evansville, Indiana, the following year.

Ivy pursued his career throughout the early 1880s, in spite of occasional

T. S. Baldwin performs at the Alexandra Palace, London.

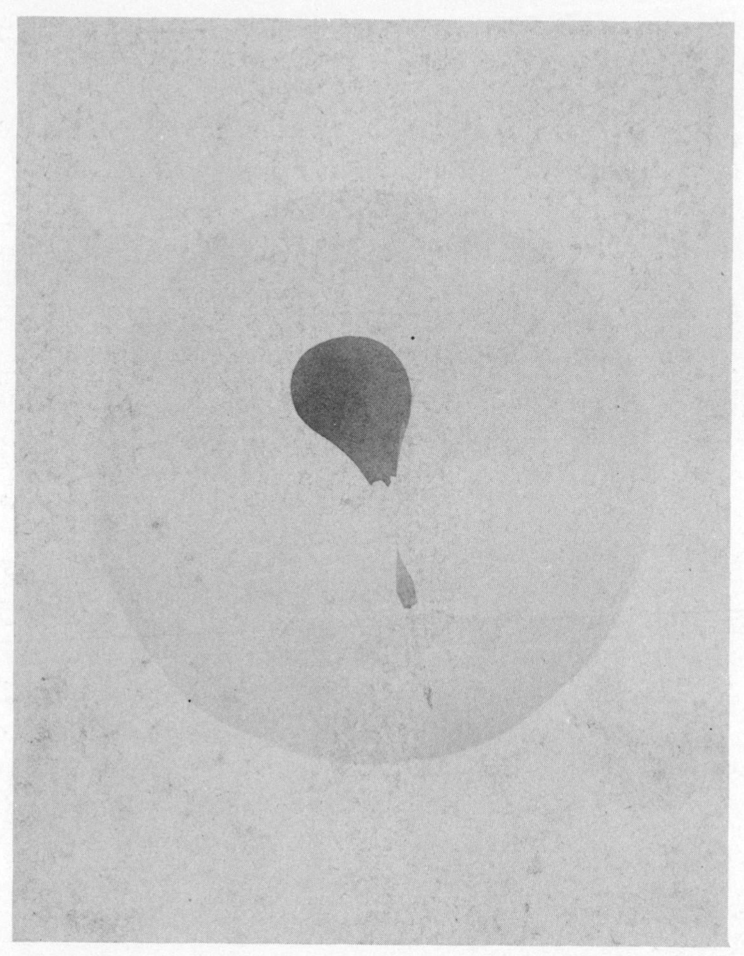

T. S. Baldwin dangles beneath his parachute in this rare photo.

injuries such as the broken ribs and ankle he suffered at Wichita Falls, Texas, in 1882, when a drunken cowboy rode his horse through the party of men that were holding the guy wires of his tightrope. Soon the young aerialist-acrobat met T. S. Baldwin. His act, which included not only ballooning and parachuting but also a 150-foot dive from the top of a tower into a net, was quickly incorporated into the Baldwin brothers' performance.[24] William Ivy soon took to calling himself Ivy Baldwin, and so he would remain for the rest of his long and active career.

By late summer 1890 the Baldwins were once again casting their eyes on what appeared to be greener pastures overseas. The Spencers, inspired by Baldwin's English exploits, seemed to be doing well. Other American balloonists had also begun to venture abroad by this time.

Rufus Gibbon Wells, for example, had built an enviable aeronautical reputation in Europe, America, and Asia. Wells was perhaps the most garrulous and well-traveled of all late-nineteenth-century balloonists. Charles Dollfus, the great French aeronautical historian, believed that Rufus Wells had been born in 1832 and had served a balloon apprenticeship under Eugène Godard in 1856. There is evidence to suggest that he knew A. J. DeMorat and may, in fact, have been the "Richard Wells" that flew a hot air balloon in a race with DeMorat in New Orleans on February 15, 1858. Wells lived in Europe between 1860 and 1872. During these years he engaged in balloon races in England, attempted to find employment with both the Prussians and the French during the Franco-Prussian War, and actually constructed a large balloon in which he hoped to fly General Bazaine to safety before the surrender of Metz.[25]

Visiting Rome in 1869, he ascended in a hot air balloon that did not carry a burner, only to find himself held aloft for four hours by solar heat. Pope Pius IX refused to allow Wells to wave the American flag during ascents from the Vatican. When the aeronaut blatantly ignored the order, he was banned from making further flights. Wells flew anyway, ascending from Rome and flying seven hours to Benevento, "where I was well received by the Italians, who praised me very much, because I had the courage to defy the Pope."[26]

Wells also flew from a Copenhagen garden to the Swedish coast and claimed to have reached an altitude of three miles during an ascent from Milan. Following his Italian tour, he visited Constantinople, where he received "some rich and splendid gifts."

He also claimed to have flown from Agra, Delhi, Lucknow, and Calcutta, India; Akyab, Burma; and Batavia in the Dutch East Indies. In a brief autobiographical account, Wells said that he had made a grand total of sixty ascensions in Mexico as well as flights in Rio de Janeiro and Buenos Aires. During a return visit to Europe in 1889, he certainly did fly William Hammer, Thomas Edison's business representative, and Abbot Lawrence Rotch, an American meteorologist, on a scientific ascent from Paris. In his later years Rufus Wells published accounts of his aerial adventures in English children's publications.[27]

While it is difficult, at best, to verify Wells's tales of his Asian and European ascents, it is certain that he preceded the Van Tassel troupe to Hawaii. Wells offered his first lecture on ballooning in Honolulu's Kawaiahao church in December 1879. His audience included Princess Liliuokalani and "a large respectable native audience."[28]

Wells claimed that he could not only fly the Pacific but could, in fact, circle the globe in twenty-five days in a balloon equipped with both a steam engine and a portable gas generator. He topped off the talk with "scientific" demonstrations of laughing gas and "electrobiology." He also offered to provide Hawaiians with their first view of a balloon in flight, "if given sufficient encouragement." That encouragement was apparently not forthcoming. Wells did not fly.[29]

Nor did Emil L. Melville, who at least attempted an ascent. Melville appeared in Honolulu in March 1889, fresh from a flight before a crowd of more than sixty thousand spectators at San Francisco's Ocean Beach. Melville had learned the balloon trade from his father, an aeronautical veteran of the Siege of Paris.

Young Melville promised the citizens of Honolulu that he would perform the usual acrobatic stunts while dangling beneath his 86-foot hot air balloon *America*. The much-used craft burst twice and was collapsed by high winds. Panic ensued when Melville's fire threatened to spread toward the spectators. The aeronaut was back for a second try on March 11. In spite of strong winds and two more rips in the envelope, Melville leaped onto the trapeze as the balloon crashed through a grove of algarroba trees. He drifted along for several hundred yards, just below the treetops, and finally leaped to the ground, allowing the empty balloon to rise to 200 feet.[30]

Park Van Tassel, his wife, and a new aeronaut, Joseph Lawrence, operating under the name Joseph Lawrence Van Tassel, succeeded in getting properly into the air from Hawaiian soil for the first time on November 2, 1889. Kapiolani Park, the slopes of Diamond Head, and surrounding areas were crowded with spectators as Van Tassel himself stepped onto the bar, climbed over a mile into the air and jumped. He landed safely 200 yards from the take-off point.

The Van Tassel troupe staged its next flight on King Kalakaua's fifty-third birthday, November 18, 1889. Strong winds were blowing across the Punch Bowl that day. Lawrence, who was to make this ascent, decided to wear a life preserver for the flight. He rose into the air, jumped over the corner of Richard and King streets, and found himself being carried higher into the air when wind gusts caught his parachute. The aeronaut finally landed at sea, two miles off the shore of Keehi Lagoon. When the tug *Eleu* reached the scene thirty minutes later, Lawrence was nowhere in sight. The balloon was discovered floating in the water, but no trace of the aeronaut or his parachute was ever found.[31] Van Tassel and his wife continued to tour the Far East, flying in India and Siam before returning to the United States. They faded from the scene after the turn of the century.

Van Tassel's most successful protégé, Tom Baldwin, began his own Far Eastern tour in 1890. Traveling with brother Sam and newcomer William Ivy, now known as Ivy Baldwin, he sailed from San Francisco aboard the SS *City of Peking* on November 1, 1890. Following a rough passage, complete with hurricane-force winds, they landed in Japan on November 21. The Baldwins were not the first aeronauts to reach the Land of the Rising Sun however. Percival Spencer was already on the scene.

Spencer had begun his tour through the Orient sometime before. His ascents were enthusiastically received. In Surabaya, on the island of Java, for example, where he flew in July 1890, local officials provided a launch site, complete with a fence and a band, free of charge. Spencer jumped to a safe landing from 3,000 feet. His balloon flew 30 miles to a landing in a jungle clearing. The natives, believing the craft to be a gift from God, cut sarongs from the fabric.[32]

After a series of ascents in China, Spencer reached Japan on October 4, 1890, and made his first flight from a Tokyo cricket ground later that month. Assisted by crew members from a visiting British man-of-war, he took off at 4:30 P.M., remained in the air for three minutes, and dropped safely to earth.[33] Spencer made a second Japanese ascent in October. The twin flights created a sensation. Proof of Japanese interest in Spencer's activity came on November 8, when newspapers announced that the emperor had ordered a command performance.[34]

The ascent for the imperial family had to be postponed while search parties combed the countryside for Spencer's balloon, which had been lost during a flight from Kobe. Finally, on November 15, the aeronaut ventured aloft from an open area near the Niju Bridge outside the palace gate. The flight was well attended in spite of a steady rain. Pupils at the Nobles School for Girls and Boys were released from class for the event. The emperor and his family watched from within the palace. Spencer, already drenched by the rain, was thoroughly soaked when he fell into the palace moat. One newspaperman quipped that Spencer could make good use of a waterproof parachute and a life preserver.[35]

Tom and Sam Baldwin arrived in Tokyo with William Ivy (Ivy Baldwin) on November 22. While Spencer's ascensions had taken the edge off Japanese interest in ballooning, the Baldwins could offer double ascents topped off by Ivy's leap into a safety net. Tom Baldwin was also careful to inform the Japanese that he had inspired Spencer's act.[36]

The Baldwins made their first Tokyo ascensions from Uyeno Park in December 1890. The proceedings began with Ivy's tower jump. "It was not a very graceful or edifying feat," noted the *Japan Weekly Mail*, "but the recklessness and peril of such a leap pleased the multitude."[37]

With the preliminaries out of the way, Sam supervised the inflation of the hot air balloon, a process that fascinated the Japanese.

> *The balloon differed materially from that employed by Spencer. The latter was of silk, its dimensions comparatively small, and the means to inflate it were tedious and elaborate. The Baldwins' balloon was a big and smutty affair, made of red and white strips of coarse, strong cotton, sewn alternately together. Observing it, as it lay upon the ground, the centre slightly raised by means of a rope run through pulleys, the spectators imagined that they were looking at the cover of the balloon only. But they were by and by undeceived. Conjecture was rife as to the means of inflating the balloon. At the Spencer performance an elaborate arrangement of barrels and pipes indicated the brewing of gas, but the Baldwins had provided themselves, apparently, with nothing more than a pile of fire-wood and a can of kerosene. In short, it is impossible to conceive anything rougher than their whole system. They had not so much as an instrument to open the oil can. The operation had to be performed eventually with a hatchet, and when it came to ladling out the kerosene, a considerable delay occurred while somebody ran to borrow a tumbler from the nearest restaurant.*[38]

With all of the equipment on hand at last, inflation began. Fifteen minutes later, Baldwin was whisked skyward at the end of a fifty-foot rope and climbed to the bar for his usual gymnastics, all culminating in a parachute drop.

From Japan the Baldwins moved on to China and Southeast Asia. They flew from Shanghai, Hong Kong, the Straits Settlements, Singapore, three towns in Java, Saigon, Rangoon, Calcutta, Madras, and other Indian towns. The group had also hoped to fly from Vladivostok but changed their minds when the Russians demanded sixty-five percent of the profits.[39] The Baldwin brothers then cut their Asian tour short in order to escort Tom's wife, who was about to give birth to Thomas Baldwin, Jr., back to Quincy.[40]

Tom Baldwin made a serious effort to settle down following his European and Asian travels. The Baldwins purchased a section of the local fairgrounds at the corner of Thirteenth and Maine in Quincy for $18,000 on September 21, 1891. Renamed Baldwin Park, the area included a racecourse, grandstands, baseball diamond, and bowling alley. Sam Baldwin took over managerial responsibility for the park, while Tom continued to fly and manufacture balloons. Because his weight had now climbed to over 200 pounds and he no longer sported the trim waistline of an athlete, Baldwin seldom made parachute jumps anymore. Ivy Baldwin ran the saloon at the park after his return to Quincy. In addition, he continued to jump and conducted aerial wedding parties aloft.[41]

Tom Baldwin was a settled family man now, but he continued to make

A Japanese print commemorating Baldwin's tour

occasional tours. After the turn of the century, he began operating a large tethered balloon from Los Angeles and San Francisco. Caught in a storm during a free flight from Denver with this balloon, the aeronaut and two passengers were forced to remain aloft at altitudes of over 14,000 feet for more than fourteen hours. The large aerostat was finally destroyed in 1906 during the San Francisco earthquake.[42]

Aeronauts that trained with Baldwin remained active into the twentieth century. Thomas Greenleaf, for example, had become associated with Baldwin at the time of the first Quincy jump, July 4, 1887. A native of Quincy, born on June 27, 1866, Greenleaf performed over a hundred jumps for the Baldwin brothers. In 1893 he arranged balloon and fireworks demonstrations at Chicago's Columbian Exposition. Greenleaf and Eugene Falk began operating as "The Baldwin Brothers," with Sam Baldwin as their booking agent. Baldwin Park remained the headquarters for the two stand-ins.

Eventually Greenleaf was joined in the air by his two daughters, Florence and Ruth, as well as by two newcomers, Walter Thomas and Ray Jones. Together the new troupe toured the nation as "The Flying Baldwins." Toby Thomas and William Beckman replaced Thomas and Jones in 1905. Three years later Greenleaf was permanently grounded by a fractured kneecap, but the troupe continued to operate until 1918.[43]

W. E. Blanchard, another early Baldwin associate, had also become well known before the turn of the century. Blanchard, who claimed descent from J. P. Blanchard, had been born in Boston in 1852, and claimed to have made his first ascension with Samuel Archer King at age sixteen. He had helped Baldwin and Van Tassel to prepare for their first San Francisco parachute jump. Blanchard began making jumps on his own soon thereafter, completing a total of thirty-five flights in the United States, Mexico, Chile, and Peru by the turn of the century.[44]

Baldwin's first protégé, William Ivy, also enjoyed a distinguished later career, the high point of which was his service as the first U.S. Army aeronaut since the Civil War.

Tom Baldwin had dispatched Ivy to Mexico in 1893 to operate a tethered passenger balloon. Returning to Quincy at the conclusion of the Mexico City engagement, Ivy found that a cyclone had put Baldwin Park temporarily out of business. With his new bride, Bertha Sherman Ivy, in tow, the young aeronaut moved on to Denver, where he starred at John Elitch's new beer garden in 1894.

Ivy made the acquaintance of Capt. William Glassford not long after his arrival in Colorado. Glassford, who had been involved in attempts to establish an active balloon unit within the Signal Corps since 1892, decided that Ivy

Baldwin was exactly the sort of man he needed in the army. Ivy, anxious to obtain regular employment in aeronautics, enlisted as a sergeant in the Signal Corps in 1894. He was immediately assigned to take charge of the single U.S. Army balloon, then stationed at Fort Logan, Colorado.

The small Signal Corps balloon unit, of which Ivy Baldwin had become a part, originated in 1890 when Congress transferred the weather service from the army to the Department of Agriculture. For some years Signal Corps officials had conducted balloon experiments as part of their meteorological responsibilities. Now Gen. Adolphus W. Greely, the chief signal officer, and Capt. R. E. Thompson, head of the corps's military signaling division, were forced to define a purely military role for the balloon if they were to continue in the field.

By 1891 Greely had decided to add a balloon to the army inventory for the first time since the Civil War. Following a careful review of current European practice in military aeronautics, the chief signal officer determined that a tethered balloon constructed of goldbeater's skin, carrying an insulated telephone line as a part of the tether, would best suit the army's purposes. T. S. C. Lowe's system of field generators for producing hydrogen was rejected in favor of steel cylinders in which the gas would be stored and transported under pressure.[45]

Recognizing the value of good public relations, Greely hoped to have his balloon system fully operational in time for demonstration flights at the 1893 World's Columbian Exposition in Chicago. William A. Glassford was dispatched to Paris to study military aeronautics in European armies and to purchase a balloon and equipment.

Glassford arrived in the city at a most inopportune moment. On the morning of June 25, 1892, a Frenchman named Grenier had been arrested at the home of Captain Borup, the U.S. military attaché in Paris. Grenier was charged with having sold state secrets, which Borup not only passed on to Washington but also made available to German and Italian attachés.

Borup was recalled, but the incident sparked a wave of anti-American feeling that made it difficult for Glassford to do business. Informing General Greely that "until the Borup matter has blown over, nothing new . . . is to be secured," Glassford shifted operations to England where he observed the use of balloons during army maneuvers in August 1892.[46] His subsequent report on English aeronautical practice served as the basis for Greely's balloon planning.

At the same time, Captain Thompson was preparing final specifications for the construction of a "balloon wagon," complete with cable reel, and three additional wagons designed to carry a total of 270 gas cylinders, each containing hydrogen at a pressure of 120 atm. The contract was awarded to the experienced firm of Siemens Brothers.

Glassford helped arrange the acquisition of the wagon and equipment but was unable to purchase a balloon in England. Back in Paris on October 24, 1892, he suggested to Greely that all attempts to obtain an English balloon be abandoned. A quick survey of French balloon builders led him to Henri Lachambre. Two months later, on December 16, Greely authorized Glassford to purchase a 13,000-foot goldbeater's skin aerostat from Lachambre for 8,000 francs ($1,970).

Glassford remained in Paris to superintend work on the balloon, which was to be named the *General Myer* in honor of the first chief signal officer. At the same time Greely dispatched Thompson to consult with the aging James Allen on the auxiliary equipment to be purchased for the balloon and to discuss the problems and lessons of the Civil War Balloon Corps.

The *General Myer*, with its accompanying wagons and equipment, reached the exposition grounds in Chicago in late May 1892. While Glassford remained on duty in Paris, Thompson traveled to Chicago, where, assisted by two sergeants, he prepared the balloon for its first flight.

A delay in the delivery of the gas cylinders allowed Greely and Thompson to consider the offers of S. A. King and Carl Myers to assist the novice army aeronauts. Both men had originally hoped to obtain construction contracts for the military balloon. The responsible officials had chosen Lachambre on the basis of his experience in producing military balloons of proven quality. Now King and Myers offered to manage the operation of the craft.

Greely and Thompson continued to vacillate into October. Now, with the fair due to close in a few days, and with the cylinders on hand at last, Thompson realized that he would, in fact, require professional assistance. While it was too late to bring King or Myers to Chicago, Tom Baldwin was already on hand. He had met Thompson during a visit to the fair earlier that summer and was now recalled for emergency duty.

Working against a deadline, Baldwin set up a generator which filled the *General Myer* three-quarters full of gas, then topped it off with additional hydrogen from the reserve cylinders. With the inflation complete, the *General Myer* made its first flight on October 21, 1893. A Sergeant Crichton went aloft on this ascent, while Captain Thompson remained on the ground to oversee the crew and communicate with the aeronaut via telephone. Subsequent flights were made on October 25, 28, and 30. No record of the number of flights was kept, but at least twenty ascents were made on the final day.

The Chicago experience was an enormous success. Thompson informed Greely that "the utility of the equipment is unquestionable." In spite of problems with the netting, which required Thompson to call on the services of a number of sailors to splice and repair the ropes, he seemed genuinely pleased with the

General Myer. "It is easy to operate," he noted, "and the arrangement of the suspension gear, and spreader, being such as to promote rotation, allows the basket to maintain its equilibrium."[47] As a result of these flights, Secretary of War Daniel S. Lamont could remark in his annual report that the balloon had "taken its place in the mechanism of war."[48]

The *General Myer* was transferred to Fort Riley, Kansas, soon after the close of the exhibition. While in service there, it was employed in combined maneuvers. Lt. J. E. Maxfield also used the craft to illustrate balloon procedures then current in the British Army.

Greely was disappointed that the balloon could not be returned to Chicago in time to contribute to the suppression of the Pullman strike in 1894. "The fact that the war balloon is simply indispensable for observing and checkmating the operations of marauding bands within the limits of great cities has not been recognized," he argued. "It need only be stated to be acknowledged, that in offensive city fighting, such as, for instance, that against the Communists [i.e., Communards] in Paris, there is absolutely no way of observing the movements by the enemy and of their lines of barricades except by means of balloons."[49]

In 1894 Greely ordered the balloon transferred to Fort Logan, Colorado. Glassford, who was now serving as signal officer for the Department of the Colorado, became responsible for the balloon he had purchased in Paris. Although Signal Corps funds were cut to the bone during this period, Glassford was able to enlist Ivy Baldwin to take charge of the tiny balloon unit. Baldwin spent his time drilling the band of noncoms and troops in balloon handling procedures. He practiced aerial photographic techniques and developed telephone communication procedures.[50]

Glassford recognized Ivy Baldwin's value to the army. "In all of this work," Glassford remarked, "Baldwin was invaluable for he knew the practical details of every part of the work from cutting the gores of a balloon-envelope, making the netting, inflation, and in all ascensions he was in the basket." Baldwin's years of experience gave confidence to the men. He was a natural teacher, always ready to answer questions and provide practical advice.[51]

The army officers involved were interested in the program and attempted to ease the path for Baldwin. Colonel Merriam, the commander of Fort Logan, detailed his best troops to the balloon company and provided wagons and other items required to equip a mobile balloon train.

Serious problems still remained to be overcome however. With no funds for the construction of a suitable balloon house, Baldwin improvised three high walls from bails of hay and straw. When the Secretary of War toured the site with Major General Davis, he was sufficiently impressed by Baldwin's skill at improvisation to approve construction of a large wooden building to shelter the balloon and additional money for an improved generator, compressor, and accessories.[52]

None of this was of much value without a balloon, however, and the hard Colorado winters had taken a fearful toll on the fragile goldbeater's skin of the existing envelope. The *General Myer* was finally torn to shreds during an inflation in a high wind.

The small appropriations made to the Signal Corps during these years scarcely allowed for the purchase of a new balloon. Unwilling to drop the program, Glassford scraped together a few hundred dollars for the purchase of silk with which Ivy and Bertha proceeded to build a new balloon for the army. Thus the first U.S. Army aircraft to go into combat in thirty-five years was a home-built affair constructed of domestic dress silk.[53]

Ivy's effort sparked a great deal of interest in ballooning in army circles during the years 1894–1898. Glassford himself outlined plans for a balloon unit designed to accompany the rapidly moving field telegraph teams that provided communications for an army in the field. The wagon transporting the balloon would be followed by an escort vehicle carrying the cable drum and four others filled with racks of compressed hydrogen. A second balloon, spare parts, and equipment would be packed aboard a sixth and final wagon. The balloon, the eyes of the army, would operate in cooperation with its nervous system, the telegraph.

Lt. Joseph Dickman introduced students at the U.S. Infantry and Cavalry School, Fort Leavenworth, to the potential of military ballooning in an 1896 lecture. As historian Russell Parkinson has noted, by the outbreak of the Spanish-American War, the appearance of the balloon at maneuvers or demonstrations was no longer a cause for comment.[54]

Greely was anxious to demonstrate the value of the balloon unit in combat. As the war with Spain approached in the spring of 1898, he ordered Ivy, the balloon, and the crew to Fort Wadsworth, New York, where they came under the command of Lt. Joseph E. Maxfield. With the approach of the crisis, funds for expansion became available at last.

On April 20, 1898, Maxfield released an invitation to bid on the construction of two 20,000-cubic-foot silk war balloons, complete with baskets and netting. In contracting for the *General Myer*, Greely and Glassford had chosen not to contact potential American balloon builders, assuming that they were less capable than more experienced European contemporaries like Lachambre. Now, with time a major factor, Glassford attempted to find an American supplier.

A number of native aeronauts were anxious to obtain government work.

Young Leo Stevens, for example, had offered his services to Pres. William McKinley as early as April 1898. His proposal apparently did not reach Maxfield however. Samuel Archer King and Carl Myers were in a much better position to meet the army's needs.

Before the release of the Maxfield specification, King had offered to sell the army a 35,000-cubic-foot, rubber-coated-cotton envelope for $800 or a 25,000-cubic-foot, oil-varnished aerostat for $600. Both King and Myers ignored the specific requirements of Maxfield's advertisement, bidding $2,000 and $3,080 respectively for cotton rather than silk balloons. James K. Allen offered to meet all the government requirements for $7,000.[55] Maxfield received a fourth bid from an unexpected source, J. Chenard, who represented the French balloon builder A. Varicle. Varicle was, as Charles Dollfus has noted, "quite a character."[56] Having amassed a small fortune as the inventor of the key-equipped sardine tin, he had become interested in aeronautics and learned to fly with Maurice Mallet, a leading French aeronaut.

In 1897 Varicle was preparing an aerial supply expedition to the Klondike. He planned to carry food and supplies over fabled Chilkoot Pass aboard two large balloons complete with sails, draglines, and manually operated propellers. His party left Paris on March 28, 1898, bound for Montreal, planning to travel overland to Juneau, where he would launch the balloon *Paris-Alaska* on the first flight into the goldfields.[57]

Chenard contacted Maxfield while Varicle was still crossing the Atlantic. The officer noted that the *Paris-Alaska* was too large for army service, but invited Varicle to bid on the existing contract. When the bids were opened, Maxfield accepted the French offer of two balloons, one of 500 and one of 600 cubic meters, for $4,250.

Varicle immediately returned to France to fill the order. The aeronaut did not intend to construct the balloons himself. Rather, he subcontracted the work to M. Mallet's newly established firm, L'Ecole Française des Aéronautes. Much less experienced than Gabriel Yon, the Godards, Lachambre, or other French balloon firms, Mallet was not the perfect choice for the job. Soon after the contract had been let, the American military attaché in Paris informed Greely, "From what I can gather I think it doubtful whether he would be able to furnish this material in the time stated."[58] Glassford, who was clearly the most experienced judge, also cabled that he would prefer an American-built balloon from Baldwin or some other producer who had not been included in the bidding. All of these communications arrived too late. Impressed by Varicle's enthusiasm—and price—Greely was now fully committed to the French contract. He did finally agree to order an additional cotton balloon from Carl Myers however.

When all the balloons and equipment were on hand, Greely planned to establish two combat balloon trains, each equipped with one silk Varicle balloon. The Myers balloon would accompany one train, Ivy Baldwin's the other. One train would be dispatched to Cuba, the other would accompany the planned expedition against the Spanish in Puerto Rico.

Maxfield was kept busy traveling back and forth between Fort Wadsworth in New York City, where Ivy Baldwin and his crew were operating the old silk aerostat, and Carl Myers's balloon farm at Frankfort, New York. He was also assembling the additional equipment to outfit both trains and was charged with responsibility for maintaining cable traffic between the United States and naval forces in Cuba.[59]

Ivy Baldwin and the members of Company B, Signal and Balloon Section, were ordered to the embarkation point at Tampa late in May. Maxfield remained in New York to supervise the varnishing of the two Varicle balloons, which had been received in record time but in an unfinished condition. Facing a lengthy drying process, Maxfield placed a subordinate, Lieutenant Wildman, in charge of the Varicles and followed his crew to Florida, where they were preparing to board the troopship USS *Rio Grande* with other members of the Fifth Corps.

Maxfield found Baldwin packing his balloon and equipment for transportation to Havana. As compressed hydrogen would obviously be unavailable in Cuba, the men had filled all of their available cylinders using a portable generator and compressor. As the safety of the relatively delicate balloon equipment would scarcely be a major concern of those planning the expedition, Maxfield and Baldwin packed their apparatus with extreme care.

Lieutenant Wildman arrived with the two Varicle balloons and the Myers aerostat while the rest of the Signal Corps contingent was completing preparations for departure. Fully aware of the difficulties he would face in caring for the one balloon already packed, Maxfield ordered Wildman to remain in Tampa to prepare the three remaining envelopes for service in Puerto Rico. Ultimately, the Cuban expedition succeeded so completely that the Puerto Rican campaign was unnecessary. The Varicle and the Myers balloons would never be flown in combat, though they would see service at American military posts after the war.

In spite of the care Maxfield and Baldwin had taken with the Colorado balloon, the aging envelope, now dubbed the *Santiago*, was in poor condition when it arrived in Cuba. The varnish had melted in the tropical heat, sticking sections of the rotted silk together. The crew was kept busy repairing small holes and large rips in the fabric. Maxfield was particularly discouraged.[60] "The *Santiago*," he noted, "was in such condition that, had the ascents to be made in time of peace, it would have been felt unsafe to use it."[61]

Far from the amenities of Fort Logan, the crew had nothing more than the balloon, a minimum number of gas cylinders, and other items of absolutely essential equipment. They were without camping equipment, shelter, or logistical support. In spite of these difficulties, Maxfield and Ivy were able to inflate the *Santiago* and make three flights on June 30, 1898, providing invaluable topographical information on the country in front of the army. General Shafter, commander of the U.S. forces in Cuba, was most impressed and, on the recommendation of his chief engineer, ordered the balloon and crew to the front the next day.[62]

High winds on the night of June 30 caused new rents to appear in the *Santiago*. Following early morning repairs, Maxfield rode to El Poso Hill, where the crew had been ordered to ascend. Fire was so heavy when he arrived that Maxfield's horse was immediately shot from under him. When the balloon and crew arrived a short time later, they went into action at the base of El Poso, directly in front of General Kent's division.

Maxfield and Lieutenant Colonel Derby, representing General Shafter, made the first ascents with Ivy Baldwin observing the movement of troops at El Caney and on the road leading toward San Juan Hill. Derby then ordered the *Santiago* and her crew to the front line, over Maxfield's objection that Spanish artillery zeroed in on the area would make it impossible for the balloon to survive.[63]

Reascending in a grove of trees along the Aguanadores River, the tether and telephone lines became so hopelessly tangled in the branches that the *Santiago* could rise no higher than 300 feet. It was an ideal target for everything from small arms fire to heavy artillery. The balloon observations were immediately useful. Strong Spanish entrenchments on the slopes of San Juan Hill were pinpointed for U.S. artillery. The aeronauts also spotted a hidden trail and ford across the Aguanadores that would permit General Kent to ease the congestion on the one road then in use and already under heavy fire.

Conditions aboard the *Santiago* and on the ground surrounding the tether point were growing hotter by the minute. Mauser bullets tore through the thin reeds of the basket and peppered the gasbag. Maxfield finally ordered the balloon down, but it was too late. Spanish artillery and small arms fire from the troops entrenched on the lower slopes of San Juan Hill, only 650 yards away, continued to pour onto the site. Maxfield feared that the envelope was so badly damaged that it would never be flown again. Nevertheless, he ordered the crew to remain on the scene until the balloon was packed safely into the basket and loaded onto a wagon.[64]

Fire had become so heavy that leaves fell from the trees in showers. Protected by an earthen bank, the balloon unit was relatively safe. Only one man,

Private Hayward, suffered a serious wound. Two others, Corporal Boone and Lt. W. S. Volkmor, were cited for particular gallantry and resourcefulness during the operation.[65]

The troops surrounding the balloon company would have very different memories of the morning. The *Santiago* had been the aiming point for every Spanish gun in the neighborhood. The men on the receiving end of this fire would remember the place as Hell's Corner and Bloody Ford. "The front had burst out with a roar like a brushfire," Stephen Crane wrote. "The balloon was dying, dying a gigantic and public death before the eyes of two armies. It quivered, sank, faded into the trees amid the flurry of battle that was suddenly and tremendously like a storm."[66]

First Lieutenant John J. "Black Jack" Pershing led his troops of the black Tenth U.S. Cavalry onto the scene as the balloon was descending. "When the Tenth Cavalry arrived at the crossing of the San Juan River," he recalled, "the balloon had become lodged in the treetops above and the enemy had just begun to make a target of it—no doubt correctly supposing that our troops were moving along this road and were near at hand. A converging fire from all the works within range opened upon us that was terrible in its effect; the Seventy-first New York, which lay in a sunken road near the ford, became demoralized and well-nigh stampeded."[67]

As the field hospital established near the ford spotted by the men aboard the *Santiago* began to fill with dead and wounded, it became clear that the U.S. troops would either have to capture the heights held by the Spanish or withdraw to safer positions. At about one o'clock that afternoon, word was passed that an assault would be mounted on Kettle Hill.

Theodore Roosevelt, riding back and forth under heavy fire in front of his Rough Riders, decided that their present position was so dangerous as to be untenable. Seeing no alternative, he ordered an immediate charge up the hill. At the same time, other units moved uphill against the Spanish San Juan blockhouses on the Rough Riders' left. By four-thirty that afternoon, the fight that had begun with the ascent of the *Santiago* had drawn to a close, with the victorious Americans still in possession of the all-important line of hills.[68]

There were few participants willing to extend congratulations to the aeronauts however. In the wake of the heavy fighting on July 1, officers and newsmen alike directed a steady rain of criticism on the balloonists. Many of the troops that had suffered through the fire brought down on their heads by the *Santiago* clearly wondered if the intelligence obtained had been worth the price. Leading war correspondents, including Richard Harding Davis and Stephen Bonsal, were particularly critical. "The history of our war balloon on the first day of the fight," wrote Bonsal in 1899, "would serve to discredit ballooning as an aid to an army, if it were not quite so apparent that the blame lay with those who were in charge of it, or those by whose direction it was sent out."[69]

Greely, Glassford, Maxfield, and all those responsible were, of course, concerned about the adverse public and professional reaction to their debut in combat. Greely, as chief signal officer and primary responsible official, did his best to explain and defend the role of the military balloon.[70]

In terms of equipment, the balloon unit was in far better shape at the conclusion of the war than at the outset. By early 1899 the Signal Corps counted four balloons and a gas generating plant in its inventory. The Signal School established at Fort Myer in Arlington, Virginia, at the conclusion of the war became the new home of two balloon companies. In 1899 Greely added a Siegsfeld *drachen* kite balloon, the latest in German aeronautical technology, to the army balloon roster.

In spite of all this, these were not good years for army aeronautics. The new quarters at Fort Myer were far from ideal, and appropriations for operations were so limited that it was seldom possible to inflate any of the balloons. The grand total of $144.75 made available in 1900 for "war balloons and expense in connection therewith" was, in all probability, just enough to allow a single inflation of the *drachen*, officially Signal Corps No. 5, at Benicia Barracks, California. The same balloon was still in use as late as 1908, when it was pressed into service as an unmanned pylon moored at Shooter's Hill, Alexandria, Virginia, for the distance trial of the first Wright military airplane.[71]

The attempts of Greely, Glassford, Thompson, Maxfield, and other Signal Corps officers to demonstrate the combat potential of the observation balloon laid the foundation for the development of military aeronautics in the United States. While the Signal Corps radically reduced its aeronautical effort in the years immediately following the Spanish-American War, the organization took the lead in the operation of aircraft and dirigibles after 1908.

Free and tethered balloons would continue to play a role in both the army and the navy for many years to come. During the First World War tethered observation balloons remained an essential means of gathering information on enemy troop movements and positions. Free balloons were employed in training airship pilots throughout the interwar years.

The bright future of military aeronautics was not apparent to Sergeant Ivy Baldwin however. Convinced that the army was no longer a place for a balloonist, he resigned at the conclusion of his second enlistment in 1900 and returned to his old position as chief aeronaut at Elitch Gardens in Denver. After an unsuccessful attempt to construct a battery-powered airship, he went back to high-wire work to supplement his income.

Ivy was back in the headlines when he walked across a 530-foot high wire over South Boulder Creek Canyon at Eldorado Springs. Shortly thereafter he moved to Alameda, California, where, like his old friend Tom Baldwin, he made the transition from balloon to airplane.[72]

"Uncle Tom" Baldwin came to the airplane by way of the airship. Having reinvigorated exhibition ballooning once through the reintroduction of the parachute, Baldwin recognized the need for yet another new departure in aeronautics to keep paying customers pouring through the gates.

The little pedal-powered airships pioneered by Ritchell and Myers had failed for lack of a power plant. The development of lightweight internal combustion engines during the first decade of the new century solved this problem. Led by Alberto Santos-Dumont, European aeronauts were the first to combine the new gasoline power plant with the old spindle-shaped gasbag originally suggested by Meusnier in 1784. Tom Baldwin was not far behind. While Baldwin's *California Arrow* was not the first U.S. airship, it was certainly the most famous.

Baldwin had built an unsuccessful, pedal-powered airship in 1892, followed by another dirigible propelled by a 24-horsepower automobile engine. While the second craft was able to carry an operator aloft, it could not be controlled in the air. The *California Arrow*, his first navigable airship, was constructed in San Jose, California, in 1904. The craft was 52 feet long and 17 feet in diameter, with a 7-horsepower Curtiss motorcycle engine that drove an eight-foot propeller at 150 RPM. Baldwin reported his first successful round-trip flight in the craft from Idora Park in Oakland, on August 4, 1904.

Baldwin, who weighed in at 200 pounds, was hardly the ideal pilot for an airship offering such minimal performance. He chose a young balloonist named A. Roy Knabenshue to fly the craft. Together Baldwin and Knabenshue breathed new life into aerial showbusiness with their small, one-man airships. They were soon joined by other daredevils who toured the county fair circuit demonstrating their small craft.[73]

Captain Tom Baldwin, whose *California Arrow* had launched the airship craze in the United States, continued his interest in aeronautics. Between 1904 and 1907 Baldwin airships made a total of 193 ascents, 174 of which were successful round trips. In addition to his exhibition work, Baldwin sold Signal Corps No. 1, the first powered U.S. military airship, to the U.S. Army on August 22, 1908. By the summer of 1910 Captain Tom had made the move to the airplane, flying the *Red Devil*, a specially designed exhibition machine, at Mineola, Long Island. Prior to the First World War, he served as head of the Curtiss Newport News Flying School, where American and Canadian pilots were trained for service in France. During the war Baldwin was commissioned as a major and served as balloon inspector for the army in Akron. Following his discharge, he served as chief of balloon production at the Goodyear plant until his death in May 1923.

The era of the small, powered, dirgible balloon in the United States was a short one. The period of real enthusiasm for the small dirigibles lasted scarcely a decade, but these frail craft were important, for they bridged the gap between the classic free balloon of the nineteenth century and the airplane of the twentieth. They kept the aeronautical exploits of their operators on the front pages of the nation's newspapers, preparing the way for the coming of the heavier-than-air flying machines. More important, they served as a training ground for the small band of daredevils that would spawn the first generation of American airplane pilots. The balloonists were no longer alone in the sky. There was a new way to fly.

Chapter 17 The Sport of Kings

THE ADVENT OF THE AIRPLANE BROUGHT THE ERA OF THE BALLOON showman to an end. The nation's leading professional aeronauts, men like Tom and Ivy Baldwin, Roy Knabenshue, Cromwell Dixon, and Lincoln Beachey, who had long earned their livelihood in the air, were quick to abandon aerostation for aviation. While a few balloonists continued to obtain bookings at small local fairs, no large celebration seemed complete without the sight of one of those frail new contraptions of wood, wire, and fabric sputtering across the sky.

Ironically, the death of the old tradition of aerial showmanship led to an unexpected renaissance of interest in the free balloon at the very moment when that craft seemed well on its way to extinction. The classic hydrogen balloon passed directly from the hands of the last generation of barnstorming "Professors" and "Captains" into the care of wealthy amateur sportsmen.

The sudden interest of American socialites in an activity that, until recently, had been the province of "county fair fakes" was a bit startling to some contemporary social commentators. As reporter Arnold Kruckman noted, "The most interesting aspect of the sport is the character of the people who engage in it." Scientists, he continued, might have been expected to take an interest in ballooning, "but the enthusiasm of prominent society people and notable businessmen is as unexpected as it is gratifying."[1]

The birth of sport ballooning becomes easier to understand if one examines the self-image entertained by turn-of-the-century captains of industry. American industrialists regarded themselves as a venturesome and forward-looking social élite, pioneers for a nation that had grown beyond its last geographic frontier. They prided themselves on being risk-takers in sport as well as in business. Such men and women echoed the sentiment of a contemporary journalist who remarked that "a sport which hazards nothing is no sport." Further:

When you skim swiftly over the landscape on the back of a high-spirited thoro-

bred isn't there always present the one unpreventable chance that an accident will happen to its delicate ankle or that it will rear back upon you in affright for some absurd reason? And even though a multitude of things may happen while you are devouring the distance in your gloriously dashing motor-car, isn't it worth it all just to experience the exhilaration of that mad rush? It is the same element of chance that makes yachting, sailing—any genuine sport appeal to all full-blooded people.[2]

With this attitude, ballooning emerged as a "royal sport":

It is a sport which by its very nature can only appeal to the kings and queens of man and womanhood. It makes its appeal to the healthy, wholesome, steady, cool, level-headed members of society. It takes more than the spirit of foolish dare-deviltry to become an aeronaut. The balloonist . . . needs brains. Plenty of brains. It takes the kind of brains that can administer the affairs of a huge business, or can govern a city or a state, or can command an army, or write a book, or compose music. It takes the kind of intellectual and physical stamina that has the good courage to do the thing that ordinary people would tremble even in thinking of doing. In short, the successful aeronaut is made of the kind of stuff out of which they make the most perfect specimens of American humanity.[3]

In the summer of 1905 a number of these "perfect specimens of American humanity" banded together to form the Aero Club of America, the organization that would spearhead the sport balloon movement in the United States. The decision to organize the new club was made by leading members of the Automobile Club of America who had attended a lecture on aeronautics by Charles Matthews Manly, Samuel Pierpont Langley's chief "aerodromic" assistant at the Smithsonian Institution. Inspired by Manly's vision of mankind's future in the air, the wealthy automobile enthusiasts founded the Aero Club of America with the vague objective of promoting the "development of aerial navigation."[4]

Uncertain how they might best achieve this goal, the charter members of the new club mounted the first large American exhibition illustrating the current state and future prospects of the flying machine. This was far from a simple task, since the group had little experience or knowledge of aeronautical matters. Money was no object for the wealthy sportsmen, however, and their show, staged in the 69th Regiment Armory in New York as part of the Annual Automobile Club exhibition in January 1906, was a great success.

The emphasis of this first exhibit was clearly on heavier-than-air flight. Leading aircraft experimenters, including Samuel Langley, Augustus Herring,

Octave Chanute, and the Wright brothers, provided engines, models, and full-scale gliders. John Brisbane Walker, an American sportsman and publisher of *Cosmopolitan Magazine*, lent his original Lilienthal hang glider. Hiram Maxim, Clément Ader, Gustave Whitehead, and other European and American experimenters provided photos of their machines on the ground and in the air.

A great many photos of balloons in flight adorned the walls. In addition, the club purchased a large French balloon to top off the exhibition. The show was thoroughly covered in the local press and was the subject of long articles in national periodicals like *Scientific American*. It not only helped to educate the public and popularize the subject of aeronautics, it established the reputation and credentials of the Aero Club of America. The shows were to become an annual event.

Thus encouraged, the members of the Aero Club of America began to set a proper course for their organization. There was little initial interest in sponsoring experiments with airplanes, an activity in which few of the members would be able to participate. As seasoned travelers, however, charter members like A. Lawrence Rotch, Courtlandt Field Bishop, J. C. McCoy, Augustus Post, and others were aware of the growing popularity of sport ballooning among wealthy Europeans. Aeronautical clubs had been organized in England and France to purchase and maintain sport balloons that could then be used to train the members as pilots. This was the pattern the leaders of the Aero Club of America decided to adopt.

The goals of the organization were coming into focus at last. As Courtlandt Field Bishop, an early leader of the club, informed a reporter for the *New York Herald*, "The first step in the propaganda undertaken by the Aero Club of America is the popularization of ballooning as a sport, especially among the leisure and more wealthy class."[5]

The first order of business was the acquisition of club balloons. Two first-rate envelopes, *Orient* and *Centaur*, were purchased from Count de la Vaux, a well-known European aeronaut who was touring the United States in 1906. *Centaur* was a particularly well-known balloon, having flown 1,193 miles from Paris to Kiev earlier in the year. Count de la Vaux also provided demonstration flights for leading members, including Homer W. Hedge, president of the Aero Club, and its treasurer, Charles Jerome Edwards. By the end of summer 1906 the group had completed a grand total of thirty-three ascensions from different points in the New York-New England area. As the city offered few suitable launch sites, the club maintained a flying field and gas supply at Pittsfield, Massachusetts.[6]

A. Leo Stevens provided professional guidance for the wealthy Aero Club

Inflating a balloon at the Pittsfield, Mass., launch site in 1906

members during these early years. Stevens, who had made his first ascent in his native Cleveland, had flown the first small, powered airship in the United States. When the interest of wealthy amateurs in ballooning first became apparent, Stevens was quick to take advantage of the situation. He emerged as a key Aero Club figure at the time of the first aero show, when his wide aeronautical contacts and sound advice proved indispensable.

Leo Stevens also initiated the members of the Aero Club into the mysteries of ballooning. The Pittsfield launch site became Stevens's preserve. His early pupils included some of the first Aero Club members to earn their pilot's license, including J. C. McCoy, Alan R. Hawley, Fitzhugh Whitehouse, and Homer W. Hedge. Between January and September 1907, Stevens, his pupils, and a number of visiting foreign members of the club completed thirty-three ascensions.[7]

A. Leo Stevens's balloon factory

A. N. Chandler was the first club member to purchase his own balloon, the *Initial*, which was frequently flown by Alan Hawley and others that year. Chandler was a central figure in the foundation of the nation's second balloon organization, the Aero Club of Philadelphia. Philadelphia was, of course, a city with a rich aeronautical tradition. As one reporter noted, "For many years ballooning has been a source of entertainment in Philadelphia and rarely has an ascension been made that has not been witnessed by hundreds of persons who within themselves often gave expression to the thought, 'I wish I were going up.'"[8]

At the same time, Philadelphians had always viewed ballooning as something akin to "a side show to a circus, [conducted] only by professionals, partly because of the fear possessing the public and in part because the professionals themselves wanted ballooning as a business from which to make a livelihood."[9]

Chandler, a Philadelphia businessman, yachtsman, and automobile enthusiast, was determined to change this attitude. Having been introduced to ballooning through the Aero Club of America, Chandler made his first hometown ascent with the *Initial* on May 12, 1906. The 35,000-cubic-foot French-built aerostat descended at South Amboy, New Jersey, 70 miles from its take-off point.[10]

The flight inspired other Philadelphia sportsmen to form their own club on May 24, 1906. Active flying with the *Initial* and with the Aero Club of America balloon *Orient* was soon under way at the Point Breeze Gas Works. Later that year a second club, the Philadelphia Aeronautical Recreation Society, was formed in the city with the avowed purpose of constructing the *Ben Franklin*, a large balloon which became a familiar sight aloft over the city.

As in New York, Leo Stevens was on the scene in Philadelphia, serving as a balloon instructor and professional advisor for the new group. Stevens was already well on his way to becoming the single most important figure in the spread of the Aero Club movement across the United States prior to 1910. As the magazine *Aeronautics* noted: "The development of ballooning as a sport among notable people in America was entirely due to the conscientious propaganda of Leo Stevens. He was the first man in this country to elevate it above the plane of the county fair fake and he has practically given his whole life to establishing it as a medium of giving pleasure in a safe and sane manner."[11]

Stevens and other members of the Aero Club of America were particularly important in establishing sport ballooning in New England. The Aero Club of New England, organized at Young's Hotel in Boston on November 21, 1907, was to become the most active balloon organization in the area. The movement to establish the club was sparked by Charles J. Glidden, an early member of the Aero Club of America.[12]

Glidden, one of the best-known leaders of the automobiling movement in the United States, exemplified the many ties that bound the sport balloon movement to early automobiling. The personal connection through leaders like Glidden and others was underscored by the fact that, like the Aero Club of America, many of the sport balloon organizations grew directly out of local automobile organizations.

Glidden was to emerge as one of the most influential amateurs in the Aero Club movement. His business took him from coast to coast, and everywhere in his travels he sparked enthusiasm for sport ballooning. Leo Stevens was the other important figure in New England. Named "aeronautical engineer" to the Aero Club of New England, he sold the group their two club balloons, the *Boston* and the *Massachusetts*.[13]

The first of the local northeastern groups, the Pittsfield Aero Club, was founded on November 15, 1907, but the members were unable to purchase their first balloon from Stevens until 1908. Named the *Heart of the Berkshires*, the lovely envelope had its name encased in a red heart on one side and the identification "Pittsfield" on the opposite.[14] The North Adams [Massachusetts] Aero Club followed on March 9, 1908. Like its predecessors, the group purchased a Stevens balloon, which became widely known as *North Adams No. 1*.[15]

Glidden and Stevens were also instrumental in founding active aero clubs in Hartford, Connecticut (1908); Springfield, Massachusetts (September 1908); Worcester, Massachusetts (November 18, 1908); and Rutland, Vermont (1909).[16]

By January 1909 the various New England aero clubs were thriving, with a total membership of nearly 400 and a grand total of seven balloons in operation. The Aero Club of New England was still flying the Stevens balloons *Massachusetts* (56,000 cu. ft.) and *Boston* (35,000 cu. ft.). The two aerostats that Stevens had sold to Pittsfield and North Adams, flown by Dr. Randall of the North Adams Club, were also 35,000-cubic-foot specimens. The *Springfield* and the *Hartford* were of the same size and operated by the clubs in those towns. A total of eighty-three ascents were completed by members of the various New England clubs in 1909.[17]

The Aero Club of St. Louis was the first of the western balloon organizations. Organized on January 7, 1907, to assist in preparations for the upcoming James Gordon Bennett Race, the group grew to become one of the most influential balloon clubs in the nation. The charter membership of 37 had grown to 300 in less than ten days.[18]

The founders of the St. Louis organization were an extraordinary group even by contemporary aero club standards. Fully one-quarter of the first 400 members claimed to be millionaires. There were railroad presidents, bank and trust company executives, merchant princes, leading physicians, surgeons, and lawyers, as well as one Catholic archbishop. Albert Lambert, secretary of the organization, and H. E. Honeywell, who, as director of the Franco-American Balloon Company, was the city's chief balloon builder, were the most influential members of the St. Louis group.

Honeywell and another St. Louis man, C. L. Bumbaugh, joined Leo Stevens as professional balloonists, becoming key figures in the midwestern balloon clubs. Honeywell's St. Louis-based Franco-American Balloon Company provided envelopes and supplies for amateur aeronauts. In addition, Honeywell provided professional guidance for a number of clubs, notably the San Antonio Aero Club of Texas. Organized in June 1908 by Dr. Frederick H. Fielding, a motorboat and automobile enthusiast who had once defeated Barney Oldfield in a local race, the San Antonio club remained small but active and continued to make use of Honeywell's services as copilot on special occasions.[19]

Bumbaugh did not sell balloons, but he did inspire the foundation of a number of local clubs. The Aero Club of Indiana was his most successful venture. The Indianapolis-based group was organized in 1908 by Carl G. Fisher, an official of the Prest-O-Lite Corporation, which manufactured a hand-held portable light fed by bottled gas, and the local dealer for Stoddard-Dayton automobiles. Fisher made his first flight with Bumbaugh on September 14, 1908. The citizens of Indianapolis had been treated to their first ascent in 1883 and since that time had witnessed the usual series of parachute jumps from hot air balloons. The Fisher-Bumbaugh flight in a large and impressive gas balloon was something very different, however, and attracted a great deal of attention. The following month the two men staged a stunt that made headlines throughout the state, flying in a Stoddard-Dayton automobile slung beneath the balloon *Chicago* in place of a basket. When the Aero Club was organized later that year, Fisher was elected president. The group quickly purchased the balloon *Indiana* and, with Bumbaugh's continued aid, began an active flying program.[20]

The Aéronautique Club of Chicago was another venture in which C. L. Bumbaugh played an important role. Charles Andrew Coey, the president of the Chicago group, was to become one of the most publicized society aeronauts of the period.[21] Bumbaugh introduced Coey to ballooning in 1908. Delighted with the new sport, Coey ordered an enormous new balloon from Bumbaugh. Named the *Chicago*, it stood ten stories tall and was perhaps the largest envelope flown in the United States during this period. Bumbaugh also had permission to fly the craft in fulfillment of other contracts, as in the case of the Indianapolis ascents.[22]

The Aero Club of Ohio, organized in Canton on December 10, 1907, was another early midwestern balloon organization. Frank S. Lahm, a wealthy native of the city who spent most of his life in Paris, was the prime moving force in this case. Born on April 25, 1846, Lahm was a successful businessman who introduced the Remington typewriter to France. He had taken up sport ballooning in France and decided to introduce the pastime to old friends in his hometown.[23]

At Lahm's request, Leo Stevens brought a 35,000-cubic-foot balloon to Canton in December 1907. The charter members of the Ohio club were a distinguished lot, including Walter Wellman, another Canton native, who had made his fortune in newspaper publishing and who would later become famous for his unsuccessful attempts to fly to the North Pole and across the Atlantic in dirigible airships.[24]

Ten thousand people turned out to witness the club's first activity, the Stevens flight. The event was an enormous success, and, on December 29, 1907,

C. L. Bumbaugh (in cap)

the group purchased its first balloon, the *Psyche*, from Stevens. By the end of July 1908 they had completed a total of six ascensions.[25]

Stevens went on to help another group of Ohioans form an aero club at Salem, Ohio, later in 1908. The citizens of Wilkinsburg, Pennsylvania, were inspired to organize a balloon club in 1909, when H. W. Thompson of Canton ended his ninety-third ascent in town. Dayton, Ohio, soon followed suit, organizing an aero club after a series of ascents by C. L. Bumbaugh. Columbus had a balloon group by 1910, while Max Fleishman of Cincinnati, a member of the Aero Club of America, represented the Queen City at balloon races and meets.[26]

Sport ballooning was spreading farther west by the spring of 1908. The Milwaukee Aero Club, organized on March 16, 1908, boasted a membership roster including "the best men of the State, all enthusiastic sportsmen and willing to give their time and energetic attention to all affairs of the club."[27] Henry B. Hersey, a local official of the U.S. Weather Bureau, was the club's only charter licensed pilot. When August Pabst purchased a 16,000-cubic-foot aerostat for the Milwaukee group in 1908, the number of local pilot trainees began to grow under Hersey's watchful eye.[28]

The Aero Club of the Northwest was founded in June 1908 under the initial leadership of Alec Sloane of the sports department of the St. Paul *Daily News*. This group grew rapidly to include forty "well known sportsmen" of the Minneapolis-St. Paul area who were the proud operators of two balloons purchased from Leo Stevens.[29]

The number of Aero Clubs continued to grow through the remaining months of 1908, as Buffalo, Denver, Baltimore, and San Francisco were added to the list. The growth continued into 1909. During that year Charles Glidden swung through the western states on his annual tour. Aero clubs popped up all along his path. Missouri (Kansas City), Michigan, Colorado, Utah, and Nebraska each sported a balloon organization by the end of the year. Clubs also sprouted in Cleveland; Jacksonville, Florida; Atlantic City; Washington, D.C.; Memphis, Tennessee; and Buffalo, New York, that year.[30]

Regional associations were formed as well. The California Balloon Club, for example, was organized to coordinate the activity of existing West Coast clubs "and to set before eastern balloonists the advantages of touring and winter ballooning in California."[31]

Specialized clubs also appeared. Collegiate groups became common with Columbia University, Amherst, Harvard, Cornell, the University of Pennsylvania, and Notre Dame all boasting student clubs by the end of 1909. The Columbia group was perhaps the most active, its membership ranging from socialite student Jay Gould to Grover Loening, a young man who would play a major role in shaping the infant U.S. aircraft industry. The possibility of a women's aero club was discussed as well. Mrs. A. Leo Stevens and Mrs. Albert C. Triaca, the wife of one of the most active members of the Aero Club of America, publicized their plans for a Women's Aero Club, which apparently were never realized.[32]

With or without their own club, society women were eager to join their husbands in the air. Some organizations, notably the Philadelphia Aeronautical Recreation Society, took special pains to attract women members. Stories of flights made by these women were frequently featured in the aeronautical journals of the period. Special competitions were also planned for women aeronauts. The Philadelphia Aeronautical Recreation Society, which boasted sixty-six female members, sponsored a Ladies Balloon Race on August 14, 1909, and offered the Eldridge-Zimmerman Cup to the nonprofessional woman pilot flying the greatest distance during a single ascent conducted during the years 1908–1910.[33]

The balloon movement had swept across the nation in less than three years. Of course, not all of the many clubs founded during this period survived the

A. H. Morgan (above) in a balloon he purchased from Leo Stevens (in basket, wearing dark hat), Cleveland, 1909.

initial boom of enthusiasm. The surprising thing is that so many organizations did grow and prosper before 1910. Competition was the lifeblood of sport ballooning. Races and tests of pilot skill brought members of various clubs together and gave substance to the movement.

By 1909 a number of standing prizes had been established to tempt daring balloonists. The North Adams [Massachusetts] *Herald*, for example, offered a silver cup and a $100 prize to the first aeronaut launching from North Adams and landing within five miles of Boston Common. Anderson and Price, the owners of twin hotels on Mount Washington and Mount Pleasant, established a $100 cup for a balloon voyage of over 150 miles ending within a mile of either hotel. Similar prizes were offered for flights to Poland Springs, New York; Fitchburg, New York; and the *La Patrie* office in Montreal.[34]

Then there were the club prizes. The "Sky Pilot Trophy" offered by the Aero Club of Ohio was typical. The silver loving cup was destined for the first aeronaut to fly more than 100 miles from a launch in Canton. Charles Daugherty of the Ohio group sponsored a separate trophy to be awarded to the first automobile driver who reached the next balloonist to ascend from Canton.[35]

There were opportunities aplenty for those pilots interested in racing their balloons. A few specialty contests were established. The North Adams Aero Club sponsored a point-to-point race, in which each pilot was required to land as close as possible to a preselected point. The distance races were the preferred contests of the period however. The object was simply to fly farther than anyone else in the event. Occasionally, as in the case of an Indiana Aero Club contest on June 5, 1908, the race was handicapped.[36]

The straight distance and endurance races were far more common. A number of these contests were sponsored by local clubs, including the Ohio, St. Louis, and Pacific Aero Clubs, and the Milwaukee, Columbus, and St. Paul groups. Some very creditable distances were flown during these competitions. F. H. Fielding and C. L. Bumbaugh flew 786.5 miles in 23 hours, 15 minutes to win an Aéronautique Club of Chicago event on July 4–5, 1908. The 74-mile trip (16 hours, 43 minutes) that won both the distance and endurance prizes for C. A. Coey at a St. Paul contest on July 18, 1908, was a more typical performance.[37]

The sport balloon phenomenon, of course, was not unique to the United States. The enthusiasm of wealthy Americans for aeronautics was inspired in large measure by the rise of aero clubs in England and on the Continent. In view of the new popularity of balloon racing in nations on both sides of the Atlantic, the establishment of international competitions was perhaps inevitable. The greatest of these contests, the James Gordon Bennett Races, comprise one of the most exciting chapters in the history of ballooning. Twenty-six Gordon Bennett races were staged in six different nations between 1906 and 1938. The United States won ten of these contests, Belgium seven, Poland four, Germany and Switzerland two each, and France one. A grand total of 351 pilots and copilots, the world's foremost aeronauts, participated in the competition. One hundred thirty-nine of these men competed in more than one of the contests. Ernest Demuyter held the grand record, however, flying in eighteen of the annual events.[38]

The Gordon Bennett races, spawned by the new enthusiasm for sport ballooning, kept the spirit and excitement of ballooning alive in the age of the airplane. Once a year the gas balloons recaptured the headlines as aeronauts wafted across two continents, overcoming enormous obstacles in their pursuit of pure sport in the air.

The races began in 1906 when James Gordon Bennett, an eccentric New York newspaper publisher who stood out even in an age dominated by colorful entrepreneurs, donated an ornate silver trophy to the Fédération Aéronautique Internationale (FAI).[39]

Bennett had established his first racing trophy, the Coupe Internationale d'Automobile, in 1910. When officials of the Automobile Club de France objected to Bennett's rules prior to the 1906 race, the American withdrew his prize. Fully aware of the new enthusiasm for ballooning, Bennett decided to establish an international balloon prize in order to maintain his image as a philanthropist of sport.[40]

The rules of the Coupe Internationale des Aéronautes were simple and straightforward. It was not a race but a distance contest. A provision for transforming the event into an endurance contest in the event of weather unsuitable for distance flying was never invoked.

The Gordon Bennett Cup competition was open to free balloons ranging from 22,000 to 80,000 cubic feet in capacity. Entries were to be chosen by the national member clubs of the FAI. The trophy would pass into the temporary possession of the winning club, which would also be responsible for hosting the next year's competition. A $2,500 cash prize awarded to the winning crew during the first three years was abandoned once the competition was firmly established.[41]

The trophy was to be permanently retired to any club winning the event for three consecutive years. As a result, there were, in fact, three James Gordon Bennett trophies. The original trophy was finally awarded permanently to Belgium in 1924, and the second to the United States in 1928. The third cup, first awarded to the United States in 1929, remained in use until the end of the competition.

The Gordon Bennett classic attracted international attention from the time of its first running on September 30, 1906. It featured sixteen balloons from seven nations. Two balloons represented the United States. The Brazilian-born expatriate, Alberto Santos-Dumont, the hero of three continents, would fly his airship *Deux Amériques* under the colors of the Aero Club of America. While there was nothing in the initial rules (later altered) to preclude the entry of an airship in the contest, Santos-Dumont's craft, which sported a six-horsepower engine, was the only powered machine ever to participate in the Gordon Bennett Race.[42]

The other American balloon was a standard, if aging, aerostat that Frank S. Lahm had purchased from the Count de la Vaux. Lahm, a friend of Leo Stevens and a founder of the Aero Club of Ohio, planned to fly the *United States* himself but was called home shortly before the contest to attend his daughter's wedding. Fortunately, Lahm's son was available to fly in his father's place.[43]

Frank Purdy Lahm was born in Mansfield, Ohio, on November 17, 1877, and grew up with relatives in the United States and with his father in France. Graduating from West Point in the class of 1901, he saw early service in China and the Philippines and as a language instructor at the military academy. He made his first balloon ascent during a 1904 visit with his father in Paris. During a return visit the following summer, Lahm made six more flights, including the night and solo ascents required for an FAI balloon license.

Lahm, now a lieutenant in the U.S. Sixth Cavalry, was detailed to the French Cavalry School at Saumur in October 1906. Arriving early for another visit with his father, the younger Lahm was persuaded to step in as the Aero Club of America representative in the upcoming James Gordon Bennett Cup competition.[44]

Lahm invited his friend the French aeronaut Levée to serve as copilot. At the last minute, patriotic members of the Aéro Club de France persuaded Levée to withdraw. Looking around for a competent American balloonist to accompany him, Lahm's attention was drawn to the presence of Henry Blanchard Hersey in Paris.

Hersey, a native of Williamstown, Vermont, born on July 28, 1861, was a graduate of Norwich University. Since 1885 he had been employed as a meteorologist with the U.S. Weather Bureau. Hersey had a distinct taste for adventure. He served with the Rough Riders, and arrived in Paris in the fall of 1906 fresh from service as a scientific advisor to the second Wellman Polar Expedition. An Ohio native and a friend of the senior Lahm, Walter Wellman had made two unsuccessful attempts to fly an airship to the North Pole. While Hersey was not yet a licensed balloon pilot, he was an aeronautical enthusiast and had a professional's eye for wind and weather.[45]

Lahm, Hersey, and the other contestants in the first James Gordon Bennett balloon classic began their journeys from the Tuileries Gardens in Paris on September 30, 1906. An enormous crowd of spectators had gathered in the garden and the neighboring Place de la Concorde hours before the scheduled 4:00 P.M. launch. Lahm recalled that Parisians crowded the nearby bridges over the Seine, hung out the windows, and darkened the roofs.

Several bands entertained the crowd with the music of Offenbach. The festivities were enlivened by the release of hundreds of small colored balloons and of inflated figures constructed of goldbeater's skin. Immediately prior to launch, a large flock of pigeons was released. As Lahm recalled, "All of this, combined with the natural gaiety of a French holiday gathering, made a fascinating and long to be remembered scene."[46]

One million cubic feet of coal gas had been generated to fill the sixteen balloons bobbing up and down on the grounds that afternoon. Each balloon was equipped with a barograph sealed by FAI officials which would provide a detailed record of the flight. The aeronauts also carried a three days' supply of food, a number of messages to be dropped every two hours so that those on the ground could follow their course, and a printed card bearing helpful phrases in English, French, German, and Russian: "Can we obtain a cart to carry this balloon to the railway station? Would you be so kind as to go and see if you can get us a cart? Will you take me to the house of the Mayor, as I wish to have this certificate signed by him in accordance with the Rules of the Race?"[47]

The aeronauts had drawn lots to determine the order of starting. The first balloon, the Italian *Elfe*, cast off at 4:00 P.M. James Gordon Bennett himself fired the starting pistol. The other balloons rose into the sky at five-minute intervals thereafter. As Lahm noted, "none drew more applause than Santos-Dumont, who, true to his mechanical ingenuity, went up with a six-horsepower motor buzzing away."[48]

Lahm and Hersey, twelfth in the draw, were led to the starting point by a group of French balloon troops and released at 4:55. While the two Americans were unaware of the fact, their compatriot Santos-Dumont was already out of the race. An hour into the flight his sleeve had become caught in the mechanism. He was forced to land with minor injuries.[49]

By 9:40 P.M. Lahm and Hersey had lost sight of all other balloons. They had followed their course as long as daylight lasted, even speaking to villagers on the ground. Passing over Lisieux at 11:00 P.M., their 300-foot dragline scraping over the rooftops, the aeronauts became aware of a change in wind direction. Now traveling northwest, they decided to chance a night crossing of the English Channel. Crossing the French coast at 11:17, they began what Lahm regarded as "the . . . most interesting part of our voyage":[50]

> *To describe the beauty of a Channel crossing would require the pen of a master. With a full moon shining overhead, an almost cloudless sky, the balmy air, and except for the gentle breaking of waves beneath us, not a sound to disturb the perfect calm, nothing could be more charming, nothing more delightful. With occasional reference to the compass and north star, we knew our direction was good, so had no uneasiness on that score. Sitting on the bottom of the car on the ballast bags, occasionally looking over to see if the guide-rope was clear of the water, if not, throwing out a scoopful of sand to send us up a few feet, we quietly ate our long-postponed dinner of sandwiches, chicken, eggs, fruit, coffee and other good things which we laid in before starting. Once a little sailing vessel slipped under us and disappeared in the night. This was the only sign of life we saw in the Channel.*[51]

At 2:30 A.M., the aeronauts caught their first glimpse of a lightship anchored off the English coast. Soon after, they passed the lights of Chichester on their left. They crossed Sussex, Hampshire, and Berkshire shrouded in an early morning fog. The *United States*, heated by the rising sun, climbed higher as the aeronauts caught a glimpse of Warwick Castle and Stratford-on-Avon. They were flying at 10,000 feet by 2:00 P.M. on October 22. Shortly thereafter they valved gas in an effort to find a current that might carry them north toward Scotland. Once again, however, they were running out of land. Their final descent was made on a farm not far from Hull. The Yorkshire squire on whose estate they landed later remarked that "an earthquake could not have caused more excitement."

Lahm and Hersey had flown 402.4 miles in just over 22 hours and earned the victory in the first James Gordon Bennett Coupe Internationale des Aéronautes. Nine balloonists had chosen to bring their flights to a close at the channel coast the night before. Three of the remaining crews had halted on the English side of the channel, while the remaining craft had struck the North Sea coast farther south than the Americans.[52]

The adulation that Lahm and Hersey received in France and England was but a foretaste of the enthusiasm that greeted them in the United States. They were universally acclaimed national heroes. Their dark-horse victory was an important factor in building interest in sport ballooning.

At the same time, the unexpected win presented very real problems for the members of the infant Aero Club of America, for with it came the responsibility of hosting the 1907 contest. Their organization was a new one, with only a handful of pilots, not one of whom had had any experience in planning an event of international significance.

The choice of a host city was their first problem. St. Louis officials were interested and, having staged the successful Louisiana Purchase Exposition only two years before, had a great deal of the required planning experience. Moreover, St. Louis seemed an ideal geographic starting point for a long balloon race. An aeronaut launching from the city would have to fly hundreds of miles in any direction before encountering a significant water hazard.

An Aero Club committee composed of president Courtlandt Field Bishop, secretary Augustus Post, directors J. C. McCoy and Alan R. Hawley, foreign representative Frank S. Lahm, and advisor Leo Stevens visited St. Louis in late December 1906. After talks with local officials and a trial ascent of the *Orient* by Hawley and McCoy, the committee approved the city as the site of the race. The St. Louis Aero Club was founded to assist in planning the race, which rapidly grew to become an aerial extravaganza with the Gordon Bennett race as its centerpiece.[53]

The new St. Louis Aero Club also established the first significant American aeronautical trophy in honor of the U.S. victory in the first James Gordon Bennett contest. Aptly named the Lahm Cup, the award consisted of a $1,500 trophy designed and produced by the jewelers Black, Starr and Frost, of New York. The cup was to go to the first member of the Aero Club of America to exceed the distance flown by Lahm and Hersey. Thereafter, the trophy would pass to the next member to exceed the first winner's distance. Contestants were required to post a $1.00 entry fee and notify the Aero Club of America at least twenty-four hours before a trial was to be made. The cup was to become the permanent possession of the first aeronaut to hold it for three successive years.[54]

A number of attempts were made to win the Lahm Cup between 1906 and 1912, only three of which were successful. Charles de Forest Chandler and James C. McCoy, vice-president of the Aero Club of America, were the first victors.

An aeronautical tourney, held to celebrate the hundredth anniversary of St. Louis in October 1909, was the occasion for several attempts on the Lahm Cup. Three aeronauts, Sylvester Louis von Phul (540 miles), H. H. McGill (523 miles), and H. E. Honeywell (488 miles), surpassed Chandler and McCoy's distance. But since none of these men had complied with the letter of the contest rules, all were disqualified.[55]

Several days later, on October 11, 1909, A. Holland Forbes, acting president of the Aero Club of America, and Max Fleishman, former mayor of Cincinnati and a leading midwestern balloonist, flew 697.7 miles from St. Louis to a landing 20 miles south of Richmond, Virginia. Unlike the earlier entrants, Forbes could hardly be accused of breaking the rules. As he informed a St. Louis reporter, "All I have to do to register for the Lahm Cup is to hand myself a dollar."[56]

Alan R. Hawley and Augustus Post, two leading lights of the Aero Club of America, were the final winners of the Lahm Cup. Like Chandler and McCoy, their victory came, appropriately enough, during the course of a James Gordon Bennett race. Their winning flight, covering 1,172.9 miles from St. Louis to the Canadian wilds, set a new world's distance record. Hawley, the official pilot, received permanent ownership of the lovely Lahm Trophy in 1913, no one having bettered his record during the preceding three years.[57]

The American victory in 1906, coupled with the organization of the Aero Club in St. Louis and other U.S. cities, as well as the establishment of the Lahm Cup and other balloon races and events, helped build an atmosphere of excitement that approached a peak with the second James Gordon Bennett competition. Held in St. Louis on October 21, 1907, the event was an enormous success. Local papers estimated that a crowd of three hundred thousand spectators gathered to watch the launch at Forest Park. Newspapers and aeronautical journals were filled with compliments for those who had planned the contest.[58] Leo Stevens, who was hired to superintend the inflation and launch of the balloons, drew special praise. As the official of the Aero Club of America noted, "Great credit . . . must be given to Mr. Stevens whose direction of the inflation was admirable. Like a general he deployed his soldiers and the filling went on like a well-planned attack. French clubs can no longer boast of superiority in the handling of a balloon race."[59]

The United States and Germany were represented by three balloons each. The Aéro Club de France fielded two craft and Great Britain one. It was a long race. All the balloons remained in the air for more than 24 hours, dropping messages all over the midwestern landscape. Ohio newspapers reported that "balloon neck" was suddenly a common complaint among citizens of the state who were scanning the sky for a glimpse of the balloons. The American Edgar W. Mix, who was flying with Albert LeBlanc aboard the *Ile de France*, a French entry, passed directly over the Ohio farm on which he had been born, and was able to drop two notes that were later found by his aunts.

Henry B. Hersey, now a licensed pilot who agreed to represent the Aero Club of America when Frank P. Lahm was unable to compete, excited less favorable comment. Passing over Macomb, Illinois, at an altitude of 300 feet, Hersey hailed the crew of a train standing on a siding, asking his location. "Macomb," came the reply, "but don't run your damned old balloon into this smoke stack!"[60]

The winning balloon, the *Pammern*, piloted by German aeronaut Oscar Erbslöh, flew 872.5 miles to a landing in Asbury Park, New Jersey, in precisely 40 hours. The American aeronauts were less successful than in 1906. J. C. McCoy and Charles de Forest Chandler finished fourth, landing the balloon *America* at Patuxent, Maryland, after a flight of 726.42 miles. Alan Hawley and Augustus Post finished fifth with an aerial voyage of 714.5 miles to Westminster, Maryland, aboard the *St. Louis*. Henry Hersey and his companion, Arthur T. Atherholt, finished eighth by piloting the *United States* to Caledonia, Ontario.[61]

American aeronauts were even less successful in the third Gordon Bennett competition, staged in Berlin on October 11, 1908. The problems began at the very outset with the destruction of the American balloon *Conqueror* flown by A. Holland Forbes and Augustus Post.

Both men were among the best-known of the new breed of aeronauts. Forbes, a New Yorker, and vice-president of the Aero Club of America, was often described as a "cowboy of the air," and had been identified by American journalists as the most "enthusiastic balloonist in the United States, or perhaps

A. H. Forbes's daughter Natalie christens her father's balloon in 1908.

in the entire world." His aide, Augustus Post, described Forbes as a "sportsman such as only the word intrepid will describe, yachtsman, automobilist, balloonist of experience."[62]

Forbes had invited thirty-five-year-old Augustus Post, a New York banker and a founder of the Aero Club of America, to fly as his assistant. Post, a veteran of the 1907 contest, was flattered for he knew that the choice of a companion in a competition of this sort was a very important matter.

> *The successful make-up of a team in a long distance balloon-race depends on many qualifications, mental almost more than physical. For many hours perhaps,*

> *two men, cut loose from the earth, sharing a perfect solitude, must have one mind and one motive, and must act instinctively with a precision that admits of no hesitation and no discussion. In any event, a long distance balloon race is bound to be a great memory, and your companion must be one with whom you are willing to share a great memory—and that is in itself something of a test of one's opinion of a man.*[63]

Forbes's balloon, the *Conqueror*, was the largest size permitted in the race. Built by Leo Stevens, it stood 80 feet tall, measured 50 feet in diameter, and had a gas capacity of 80,000 cubic feet. It featured a very long appendix that extended down into the basket, where it was closed by a rubber band. In this way the aeronaut could precisely control the loss of coal gas through expansion.[64]

Stopping off in Paris on their way to Berlin, Forbes and Post joined the elder Frank Lahm and J. C. McCoy, owner of the *America II*, which would also be competing in the Gordon Bennett race, in a visit to M. Mallet's balloon shop. There they saw the other U.S. entry, the *St. Louis II*, which Post described as "looking beautiful with her cloth covered with composition and varnish on the outside."[65]

Upon their arrival in Berlin, Forbes and Post were immediately involved in preparations for the flight. A voyage matching the current European distance record of 1,193.8 miles could carry an aeronaut launching from Berlin into Russia, Asia Minor, or North Africa. Forbes and Post therefore armed themselves with a variety of visas. Post recalled, "They made me take off my hat to a picture of the Tsar" when applying for travel permits at the Russian consul's office. They also carried phrase books that would enable them to offer such comments as, "Don't all talk at once," "you shall have a big tip," and "show me how many on your fingers" in a number of central European, Asian, and African dialects.

As Post noted, the basket of the *Conqueror* was filled with "clothing for every clime, eatables, and drinkables for three days and more, for it was not impossible that we might have to camp out on landing."[66] They carried drinks in patented bottles and soup packed in lime, "so that opening the can and pouring water on the packing heated the contents."[67]

Just after launch, the *Conqueror* was blown through a board fence, which tore away two full bags of ballast and ripped a third. Suddenly lightened, the balloon shot up to 2,900 feet so rapidly that coal gas was rushing through the long, narrow appendix "like the blowing off of a steam boiler."[68]

Suddenly, Post, who was making a log entry, felt a shudder "such as one sometimes feels when a wrinkle straightens out the balloon-envelope. . . . I looked up. The bottom of the balloon was beginning to shrivel like an old dry

Preparing to launch balloons for the Gordon Bennett race, Berlin, Germany

apple. It was plain what had happened. I said to Forbes, 'She's gone.'"[69]

The coal gas emptied through a huge rip in the side of the envelope. Post freed the appendix line, allowing the lower hemisphere of the balloon to gather in the top of the net, as the two aeronauts began a perilous descent toward the Berlin suburbs. They scrambled around the basket cutting ballast bags loose. Even so, they were careful to empty the bags first to avoid sending the missiles dashing down on the crowd. Forbes was startled to note that they were dropping faster than the empty bags. "The city was coming up to meet us," Post recalled. "It seemed as if some great giant was hurling buildings, streets, churches, up at us with all his might."[70] They finally smashed into a chimney and onto a mansard roof. Stunned but safe, Forbes saw to the safety of the instruments while Post snapped pictures of the envelope draped over the top of the house. German newspapers made much of this matter-of-fact behavior on the part of two men perched on a rooftop. The aeronauts were, it seems, "true American types."[71] Forbes and Post finally climbed into the house through the window of a lady's bedchamber. Forbes later received a note from the woman, who apologized for having been out when he called.

The aeronauts can hardly have been surprised to discover that they had finished twenty-third in a field of twenty-three, just behind the second American team of Arnold and Hewat. Their trip aboard the *St. Louis II* was every bit as exciting as their countrymen's. Arnold and Hewat chanced a crossing of the North Sea. The water anchor had to be cut loose as the rope had become tangled around Hewat's hand. Soon thereafter they found themselves immersed to the waist, bobbing about in the frigid water. Hewat retrieved a bottle of whiskey and an electric flashlight from the bottom of the car. Thus armed, they were able to signal a passing boat that effected a rescue.[72]

McCoy and Fogmann, flying the third U.S. entry, *America II*, had a less adventurous but scarcely more successful trip, finishing sixteenth with a flight of 127.1 miles in 30 hours and 43 minutes.[73]

The victory of the Swiss aeronaut M. Schaeck (808 miles in 73 hours) meant that the 1909 Gordon Bennett race would begin in Zurich. Edgar W. Mix, the only American competitor that year, was something of a controversial figure. Although a native of Ohio, Mix had lived in France for many years, had flown aboard the Aéro Club de France entry in the 1907 Gordon Bennett race, and would be flying the *America II*, a French (Mallet)-built balloon, in the 1909 competition. As the journal *Aeronautics* commented, "France really looks upon him as an adopted son. . . . It might have looked better to have sent a resident native with an American balloon."[74] Ultimately the choice was a wise one, leading to the second U.S. victory in the contest.

Seventeen balloons began the race at Zurich on October 3, 1909. Thirty-five hours and one minute after launch, Mix and his companion, M. Roussel, landed at Gustrovo, near Ostrolenka, northeast of Warsaw, Poland, after a flight of 696.5 miles.

Mix and Roussel had faced the usual problems, which in this case included serving as a target for a hunter near Warsaw and being stopped near Prague by a group of young men who insisted on holding onto the dragline. These hazards seemed tame indeed, however, when compared with the difficulties encountered by the winners of the 1910 race.[75]

Edgar Mix's victory brought the honor of hosting the fifth James Gordon Bennett competition back to the United States, where a friendly rivalry ensued between several cities anxious to serve as a launch site. Denver's bid was rejected on the basis of its altitude and the prevalence of high winds in the area. Kansas City was unable to demonstrate its ability to furnish quality balloon gas. Indianapolis, Baltimore, Philadelphia, and Washington were also contenders. Courtlandt Field Bishop, president of the Aero Club of America, finally announced that the balloons would rise from St. Louis.[76]

The fifth running of the Gordon Bennett race was the smallest to date, with only ten entries. The original U.S. competitors were selected through an elimination race from Indianapolis on September 17, 1910. The winner, A. R. Hawley, would fly the *America II* with Augustus Post as aide. The St. Louis aeronaut and balloon builder H. E. Honeywell would fly his own *Centennial*, assisted by Albert Lambert. J. H. Wade, Jr., the third-place winner, would pilot the *Buckeye* with his friend A. H. Morgan. Wade and Morgan were, according to *Aeronautics*, "two of Cleveland's wealthiest citizens." The two men made their first flight with Leo Stevens in 1907 and had been "keen" on ballooning since that time. When Wade and Morgan were forced to withdraw from the Gordon Bennett competition, their place was taken by the fourth-place Indianapolis balloon, *Million Population Club*, flown by S. Louis von Phul, assisted by J. M. O'Reilly.[77]

The ten balloons were launched from St. Louis on September 17 and moved northeast toward the Great Lakes, the traditional death trap for American aeronauts. Three teams, including Von Phul/O'Reilly and Honeywell/Joland (who had replaced Albert Lambert), landed when they struck the southern shore of the lakes. Seven other balloons crossed Lake Michigan and Lake Superior and continued their voyages into the Canadian wilderness, where several of them simply disappeared for a time.[78]

Lt. Hans Gericke and Samuel F. Perkins, the crew of the *Düsseldorf*, a German entry, remained in the air for 42 hours, 20 minutes, landing 17 miles north of Lake Kiskisink, Quebec, 1,127.5 miles from the take-off point. Although the two men had only the vaguest notion of their location, they had glimpsed a railroad just before descending and had assumed that they could walk back to it in five or six hours.[79]

Once on the ground, they began to realize the seriousness of the situation. The underbrush was so dense that, without a hatchet or cutting tools, they were lucky to hike a mile a day. They crossed three small rivers and two lakes on makeshift rafts. Each night they camped on the highest point they could find, building a huge bonfire. One man slept while the other tended the fire. "We heard wild beasts all around us," Perkins recalled, "and thought we could distinguish the howling of wolves, but nothing attacked our camp." Finally, at noon on Saturday, October 22, they heard a gunshot. Perkins scrambled up the nearest hill and began to wave a red flag and fire his .22 pistol into the air. Theodore Belwin, a local game warden, saw the signal and took the aeronauts to the nearest settlement in his canoe.[80]

Lt. L. Vogt and William F. Assman also had what one journalist aptly described as "an exciting time." Descending into Lake Nipissing, Ontario, they were rescued by Indians. Both were in such poor physical shape that they had to be hospitalized at Powassan, 755 miles from St. Louis.[81]

News of other balloons was slow in reaching the United States. One by one, however, the missing balloonists reappeared at small Canadian villages. Lieutenant Messner and M. Givaudau of the Swiss balloon *Azura* landed 32 miles northeast of Biscotasing, Ontario. The *Helvetia*, piloted by Col. M. Schaeck (winner of the 1908 contest) and Paul Armbruster, had flown 810 miles to Villa Maria, Ontario. The *Germania* brought Capt. Hugo von Abercron and August Blackert to earth at Coocoocash, 179 miles north of Quebec.[82]

By October 24 Hawley and Post, aboard the *America II*, were the only contestants still missing. In spite of this fact, newspapers were already heralding Gericke and Perkins as the unofficial winners.

Then, a week after the take-off from St. Louis, the missing aeronauts reappeared, having wandered for five days in the Canadian wilds before being rescued by trappers. They had won the competition with a flight of 1,172.9 miles in 44 hours, 25 minutes. The flight distance was measured precisely by the U.S. Coast and Geodetic Survey using the most accurate techniques available, for it was clear that Hawley and Post were very close to capturing the absolute distance record for free balloons. This careful survey indicated that they had fallen 20.9 miles short of the record.[83]

Much to the chagrin of the Europeans, the victory of Hawley and Post meant that the 1911 Gordon Bennett contest would also take place in the United States. Moreover, the Aero Club of America now had the first opportunity to capture permanent possession of the cup. This time Kansas City succeeded in its bid to become the host city. George M. Myers, president of the Kansas City Aero Club, working with municipal gas officials, was able to reduce the specific gravity of the local product to an acceptable level.

The elimination race to choose the U.S. team was run from Kansas City on July 5, 1911. Five of the seven balloons entered were the product of H. E. Honeywell's St. Louis-based French-American Balloon Company. The first-, second-, and third-place craft (*St. Louis IV*, Lt. F. P. Lahm, pilot, Lt. J. F. Hart, aide; *Million Population Club*, John Berry, pilot, P. J. McCullough, aide; *Miss Sapphire*, W. F. Assman, pilot, J. M. O'Reilly, aide) were all Honeywell balloons.[84]

The 1911 James Gordon Bennett competition, which took place on October 5, was the smallest ever staged. Only six competitors were involved. It was a short race as well. The winner, Lt. Hans Gericke, assisted by S. O. Duncker, flew a German entry, *Berlin II*, only 468 miles to a landing at Ladysmith, Wisconsin. Frank P. Lahm and his aide, J. H. Wade, piloted the *Buckeye* to second place with a flight of 365 miles. Third place went to the other German entry, Lt. Leopold Vogt and his aide, Lt. M. Schoeller, who flew the *Berlin I* 345 miles to Austin, Minnesota.[85]

The seventh race, staged in Stuttgart, Germany, on October 27, 1912, boasted a large field of entries, including balloons from Germany (2), France (2), the United States (2), Austria (3), Denmark (1), Italy (2), Belgium (2), and Switzerland (3). Maurice Bienaimé of France was the victor, with a 1,364-mile flight into Russia that set an all-time Gordon Bennett distance record. The second-place winner, Albert LeBlanc, flew his *Ile de France* 1,240 miles.[86]

It was a disastrous contest for the Americans involved. The team of John Watts (pilot) and Arthur Atherholt (aide) was disqualified at the start when their balloon, *Kansas City II*, split during inflation. The pair launched that evening in a borrowed German balloon, the *Düsseldorf*, but were no longer regarded as official competitors. Like so many 1912 contestants, Watts and Atherholt simply dropped out of sight after launch. Five days later, officials were searching Norway, Lapland, and Arctic Russia for some sign of the balloonists. In fact, Watts, a Kansas City man, and Atherholt, a Philadelphian, were being held incommunicado in a Russian prison.[87]

They launched from Stuttgart at 6:30 P.M., an hour after all the other balloons had gone. By ten o'clock the next night they were flying over Warsaw and noted the approach of a storm. They were finally forced to descend after 136 hours, their balloon now covered with snow and ice. Watts received a severe bump on the head in landing, while Atherholt "was a mass of bruises, with a stiff knee and a sharp pain in his side which was later found to be caused by a broken rib."

The injured aeronauts had no idea where they were, but were pleased to see "an odd-looking" fellow approach them not long after landing. Atherholt later recalled that their rescuer was "dressed in sheepskin down to his knees, long hair and a beard, wearing birchbark shoes and his legs wrapped in cloth, crossed with red straps, and a hatchet in his red belt."[88]

Presently they were joined by a larger party of locals driving sledges drawn by shaggy ponies. Tired, wet, sore, the Americans were herded none too ceremoniously to the village of Pskov, 40 miles from St. Petersburg, Russia.

After a meal of goat soup served in a wooden wash bowl, they went to sleep on a narrow board covered with skins. Rudely awakened at three-thirty the next morning, they were transported aboard droskies to another village consisting of fifty log cabins. Then followed a two-day hearing conducted in German, after which the aeronauts were held for several more days before being released to return, flat broke, to Stuttgart. The episode was apparently a result of the diplomatic crisis brewing in the Balkans. The sudden descent of a German balloon on Russian soil had, Atherholt surmised, "looked very suspicious to them."[89]

H. E. Honeywell and his aide, Herman F. Long of Kansas City, the crew of the *Uncle Sam*, fared little better. They flew to the northeast throughout the race, passing over Warsaw at 8:00 P.M. on their second night in the air. Caught in a cold rain, now flying over Russia, which Honeywell noted was "a desolate country compared to prosperous Germany," things had begun to look very dismal. The howling of wolves on the ground was a final touch. As Honeywell commented, "It was trouble from there on to the end."[90]

By 6:00 A.M. on the third day, the *Uncle Sam* "resembled an iceberg." They began dropping rapidly. Everything went overboard, but the aeronauts could not halt their descent. The flight ended in a grove of dead trees. "The bag was a total wreck." The Americans soon found themselves surrounded by "hundreds of peasants, some with guns and axes. . . . None could read their own language, much less talk ours. We drew a picture of a horse and wagon. They shook their heads. We then rubbed our stomachs, chewed our thumbs and pointed up and down the railroad track. One long haired fellow pointed west, holding up five fingers, by which we understood 'five verstes to town.'"[91]

After two more hours spent floundering through deep snow, they arrived at a railway station, where they discovered that they had flown 1,054 miles to capture third place in the Gordon Bennett race. Their troubles were far from over however. After returning to the landing site with a handcar to retrieve the remains of the *Uncle Sam*, they boarded a train for Germany at 7:30 that evening. The aeronauts were stopped short of the border, dragged off the train early the next morning, and sent back to a provincial center under guard. They were searched and their belongings were seized. No contact with the U.S. consul in St. Petersburg was permitted. They paid for their own meals at the station restaurant and slept on a desk in the room where they were locked up at night.

Three days after their arrest, a wire came from St. Petersburg ordering their release. Reporting the experience to American newsmen, Honeywell could only breathe a sigh of relief. "Thank God we got out and are home once more!"[92]

John Berry and his assistant, A. von Huffman, had a less adventurous and much shorter voyage aboard the *Million Population Club.* Berry was an especially colorful fellow. A native of Rochester, born in 1849, he made his first ascension at age fourteen. He moved to St. Louis where he became involved in an airship scheme. When Berry was shot by a coworker in an argument over this project, he sued and purchased his first balloon with the settlement.[93]

Berry flew his balloon to victory in the first National Balloon Race at Indianapolis on June 9, 1909. He competed in the 1911 Gordon Bennett race from Kansas City, and in the Stuttgart race landed on the shores of the Baltic Sea after a flight of 434 miles. Berry's son Albert was also well known in St. Louis aeronautical circles, having made the first parachute jump from an airplane. John Berry died in St. Louis on April 22, 1931.

The year 1913 was especially significant in the history of U.S. participation in the Gordon Bennett classic, for it was in this year that young Ralph Hazlett Upson made his debut in international competition. Born in New York City on June 21, 1888, Upson graduated from the Stevens Institute of Technology in 1910.

Upson dated his early interest in aeronautics to the visit of Frank S. Lahm, his mother's cousin, to his Glen Ridge, New Jersey, home during the winter of 1905–1906. Upson, a high school senior, was fascinated by Lahm's tales of the Wright brothers, whose work had not yet been made public. While a guest of Frank P. Lahm two years later, young Upson attended the first public flights of the Wright military airplane at Fort Myer, Virginia. He made his first balloon flight on September 13, 1906, while visiting Lt. Lahm in Paris while Lahm was training for the first Gordon Bennett competition. He made a second ascent with C. L. Bumbaugh in 1912.[94]

Upson had accepted a position with the small aeronautical accessories department of the Goodyear Tire and Rubber Company in Akron, Ohio, in 1910. At the time, Goodyear was offering a line of aircraft tires, shock absorbers, and rubberized fabric. The company's interest in lighter-than-air flight was a result of the production of rubberized fabric. Originally designed as a covering for aircraft wings, the material proved less satisfactory than standard doped aircraft fabric, but it was found to be ideal for balloon envelopes.[95]

Paul W. Litchfield, Goodyear's factory manager, was interested in expanding the aeronautical department but had decided that the company should not compete with its accessory customers in the manufacture of full-scale airplanes. The construction of lighter-than-air craft appeared to be a wide open field. This fact, coupled with the success of the Goodyear rubberized fabric as balloon cloth, led Upson and his small team into the field of lighter-than-air flights.

Their first project was a contract job, the construction of an envelope for a transatlantic airship designed by Melvin Vaniman, who later died in the crash of the machine. The experimental balloon *Goodyear*, the second company project, was far more significant.[96]

Finished in 1912, the *Goodyear* was the first of a series of famous aerostats to bear that name. It was a spherical balloon of 80,000-cubic-foot capacity constructed of two-ply rubberized fabric. A lovely yellow color, the balloon was the first designed by Upson and featured several novelties, including a maneuvering valve, as well as special dimensioning of the panels.[97]

Proud of his new craft, Upson turned the *Goodyear* over to the well-known St. Louis aeronaut C. L. Bumbaugh, who finished next to last with it in the 1912 U.S. Gordon Bennett elimination race. Upson, disappointed when another pilot finished last with the balloon in a race held later in the year, decided that "all pilots up to that time depended altogether too much on luck."[98]

Certain that he could do better, Upson immersed himself in the problems of ballooning and emerged with a "theory" of aerostation that he referred to as "scientific operation." "Most pilots up to that time," he noted, "had been content to drift along, economizing ballast as well as possible, exercising judgment as to altitude and landing place, but not having any systematically coordinated data on which to base judgment or satisfactory means of recording it."[99] Careful planning was the heart of Upson's "theory." Every item that went into the basket had a purpose. He also emphasized the need for a serious understanding of where the balloon was and what the weather conditions above, below, and around it were at all times.

As Upson admitted, "I was neither skillful or experienced." But if he lacked the intuition of an experienced aeronaut, neither was he hampered by useless "rules of thumb" that had long been accepted. "For instance," he noted, "there is a tradition which has grown almost universal among balloonists to the effect that if you would save ballast, never touch the valve until landing."[100] To Upson's way of thinking, "not knowing how to use a valve would be the only reason for adopting such a rule."[101] Upson's competitors would later be surprised to discover that he could valve gas constantly, maintain a precise altitude, and still finish a race with more ballast than anyone else.

On May 13, 1913, Upson took the *Goodyear* aloft from Akron with associates R. A. D. Preston and William Morgan as passengers. His only experience prior to this solo had been the two flights with Lahm and Bumbaugh. Neither passenger had ever flown before. This trip, which covered 19 miles in one hour and fifteen minutes, was the first of several that enabled Upson to obtain his balloon license later that year.

Upson and Preston entered and won the National Championship Balloon Race at Kansas City on July 4, 1913, thus earning the right to represent the Aero Club of America in the James Gordon Bennett race, to be held in Paris on October 12, 1913. This would be the ultimate test of "scientific operation."

The key to success for Upson and Preston was a clear understanding of meteorological conditions. They were in constant touch with weather offices in

Ralph Upson in his flying togs, 1913(?)

France and England prior to launch. Once in the air, they laid careful plans that would take full advantage of existing conditions. The team began reading the winds from the moment of take-off. "The general situation was a large 'high' spreading over nearly the whole of Europe, and an equally large 'low' trying hard to press in from the west. The only sure inference was that as soon as the low arrived within feeling distance we would have pronounced surface winds from the south and later from the west, with the usual variations corresponding to the altitude."[102]

Moving toward the southwest just after launch, the two Americans made a crucial decision. On the basis of the existing data, they assumed that all the other competitors would attempt to move due west with the changing winds, traveling as close to Brest, on the Atlantic coast, as possible. They, on the other hand, would fly low, catching surface winds that would sweep them directly north, away from the pack, over the Channel, and as far to the north of England as they could go before reaching the North Sea.

The race proceeded as though Upson and Preston had scripted the events. Nineteen other balloons moved west toward Brest, H. E. Honeywell flying the farthest, 299.46 miles in 29 hours and 35 minutes, to win second place.

The crew of the *Goodyear*, meanwhile, hugged the ground, turned north over Caen, and crossed the Channel coast at Cherbourg. Stopwatch and maps in hand, attempting to predict the changing wind conditions as they flew, they reached England at Southampton and continued northeast to a landing at Bridlington on the North Sea. They were the clear winners, having covered 383.16 miles in 43 hours and 10 minutes.[103]

The victory of Upson and Preston created more excitement in the United States than any balloon event since Lahm's original win in the 1906 contest. Frank A. Sieberling, president of the Goodyear Tire and Rubber Company, was especially proud of his employees and his balloon: "The achievement was only accomplished through the scientific handling of the balloon by these young aeronauts. They were competing with men of experience, and under foreign conditions that from the beginning were considered a big handicap. These conditions, however, did not prevent our boys from exerting themselves to the utmost, and putting into practice all the knowledge they had ever known about the flying game."[104]

Congratulatory telegrams poured in from all across Europe and America. Sieberling wired, "Warmest congratulations upon your brilliant achievement. Your countrymen are proud of you and the Goodyear clan supremely happy." Courtlandt Field Bishop telegraphed "heartiest congratulations" from the "Club," while James Gordon Bennett added his "hearty congratulations" in a telegram from Biarritz.[105]

The Aero Club of America struck a medal in honor of the aeronauts and directed special praise to the Goodyear firm "for its superior balloon as well as for the sportsmanship and patriotism shown by sending it to bring back to America the Gordon Bennett Cup."[106]

Fêted in London and New York, Upson and Preston received full honors in Akron as well, where thousands turned out to greet the returning heroes. They were presented with loving cups during the special full day of celebration. Aero Club of America officials held such high hopes for Upson and Preston in the 1914 race that the two men were appointed as U.S. contestants even before the annual elimination race staged from St. Louis on July 11.

But the guns of August 1914 precluded the Gordon Bennett competition that October. A much deadlier international contest was already well under way.

Chapter 18 Racing with the Wind

AS THE UNITED STATES WAS DRAWN CLOSER TO WAR, AMERICAN BAL-loonists were eager to contribute their skills to the national defense effort. They were fully aware that, as the official historian of the U.S. Air Service noted, "The United States Army had almost no balloon service previous to our declaration of war."[1]

At the close of hostilities, the Balloon Section consisted of 446 officers and 6,365 noncommissioned officers and enlisted men. All but 14 of the officers were on flying status. From February 16, 1918, when the 2nd Balloon Company moved into position near Royamieux in the Toul Section, until the end of the war, the officers and men of the Balloon Section made a total of 5,866 ascents in France, spending a total of 6,832 hours in the air. They assisted in 932 artillery adjustments, were attacked by enemy aircraft on 89 occasions, and suffered 35 balloons lost in action to the enemy aircraft and 12 destroyed by German artillery.

American balloon observers were forced to make 116 parachute jumps under enemy fire. They spotted 12,018 enemy shell bursts, reported the presence of 11,856 enemy airplanes and 2,649 German balloons, and observed 400 enemy batteries in action. They reported 22 German infantry movements, 1,113 cases of enemy traffic on roads or railroads, 2,941 smokes, fires, or flares behind enemy lines, and 597 explosions.[2]

In large measure this record of unqualified success was due to the unflagging efforts of prewar American balloonists. Col. Charles de Forest Chandler, a professional officer who had been close to the sport balloon movement since the foundation of the Aero Club of America, commanded the U.S. Air Service balloon effort. Chandler, the holder of Aero Club balloon pilot certificate no. 8, had made numerous flights from the Pittsfield, Massachusetts, site, and had served as an aide to J. C. McCoy during both his Lahm Cup Trophy flight and the 1907 Gordon Bennett contest.

Albert Lambert, the St. Louis pharmaceutical executive and a founder of the St. Louis Aero Club, was another major prewar figure who made a substantial contribution to the creation of the Balloon Section. Shortly after the declaration of war, Lambert organized and financed a training school for military balloon observers. The St. Louis school supplemented the efforts of the army's only existing school at Fort Omaha, Nebraska. The St. Louis program was officially recognized by the U.S. government in the spring of 1917 and transferred to new quarters at the aptly named Camp John Wise, near San Antonio, that May. Lambert was commissioned a major and continued to command the training school until the end of the war.[3]

Henry B. Hersey, Lahm's companion in the first Gordon Bennett race and one of the mainstays of the Aero Clubs in the Minneapolis-St. Paul area, was also quick to offer his services. Commissioned a lieutenant colonel, he commanded the training school at Fort Omaha and was later placed in charge of all balloon training activity in France. Hersey earned the sincere respect of his men when, at age fifty-five, this veteran of the Rough Riders became one of the first officers to make a parachute jump from a balloon.

Other well-known sport balloonists, including H. E. Honeywell, Joseph O'Reilly, William Assman, Paul McCullough, E. S. Cole, Harlow B. Spencer, James Bemis, and John Berry, served as balloon instructors. Ralph Upson, R. A. D. Preston, and their colleagues at Goodyear spent the war designing and building balloons and airships for the army and navy. B. F. Goodrich, the other Akron-based rubber giant, was involved in the same work, having obtained the services of balloon professionals like Stanley Vaughn to assist in the effort.

Tom Baldwin, commissioned a captain on April 25, 1917, and later promoted to major, was assigned as chief U.S. Army balloon inspector with headquarters in Akron. Discharged in October 1919, he continued as manager of balloon production and inspection with Goodyear. Baldwin died on May 17, 1923.

Sport ballooning enjoyed a revival with the return of peace. Such well-known prewar aeronauts as Ralph Upson, H. E. Honeywell, William Assman, and Paul McCullough were now joined by young veterans of the U.S. Army Air Service. Between 1918 and 1920 the Aero Club of America issued 850 balloon pilot certificates, most of them to men who had learned to fly in military programs. Ernest S. Cole, John S. McKibben, and Warren Rasor were among the former Air Service balloon instructors that built solid reputations in sport ballooning after 1919.

Albert Lambert, fresh from his military duties, organized the first postwar balloon competition on October 11, 1919. Charles Glidden, Alan R. Hawley, and George Myers were among the leading U.S. sport balloonists who assisted in planning the event, the National Balloon Race, held in St. Louis.

The fourteen initial entries included three military teams that would compete for the experience but would not accept prizes. Upson finished first in the St. Louis race, flying 1,050 miles with Ward T. Van Orman as aide. Honeywell flew 920 miles with Kansas City balloonist Harry Worthington assisting. Newcomer E. S. Cole and his aide, Lt. Leo C. Ferrenback, finished third with an 860-mile voyage. Because of the desperate situation in postwar Europe, FAI officials chose not to schedule a Gordon Bennett balloon competition in 1919. Upson's victory in the 1919 elimination race was still an important event, however, for it marked the competitive debut of his aide, Ward Tunte Van Orman.[4]

Van Orman, born in Lorain, Ohio, on September 2, 1894, was destined to succeed Upson as both Goodyear's "company" balloonist and the most successful American competitive aeronaut of his generation. A graduate of Cleveland's Case Institute of Technology, Ward Van Orman went to work for Goodyear on June 6, 1917, where he was placed in charge of installing aircraft engines aboard the coastal patrol blimps being constructed for the U.S. Navy. In addition, he instructed student blimp pilots in engine technology.[5]

Van Orman was introduced to free ballooning by the master himself. Ralph Upson, now Goodyear's chief aeronautical engineer, was in search of a new aide to accompany him on the upcoming National Balloon Race from St. Louis in 1919. As part of the selection process, he invited both Van Orman and Commander Hoyt, commanding officer of the Akron Air Station, to accompany him on a balloon voyage from Akron to New York where they would witness the triumphal parade planned in honor of General Pershing.

The trip was a spectacular success. The trio landed in Andover, New York, so close to their destination that even Upson must have been surprised. He was certainly impressed with Van Orman, who was invited to accompany him in the St.Louis race that October and in the next year's National Balloon Race from Birmingham, Alabama, on September 26, 1920.[6]

The balloons were launched from the Sloss-Sheffield Steel Plant a half mile north of Birmingham, which had also been selected as the site of the 1920 James Gordon Bennett contest. Honeywell was declared the winner of the 1920 Nationals and the number one U.S. entry in the international race, after a 700-mile flight to Thamesville, Ontario, aboard the *Kansas City II*. Captain R. F. Thompson of the Army placed second (690 miles) and Upson third (620 miles).[7]

The Gordon Bennett competition followed on October 23, 1920. Belgian pilot Ernest Demuyter won the event with a 1,100-mile flight. Honeywell and his aide, Jerome Kingsbury, finished second, flying roughly one thousand miles to Tongue Mountain, Lake George, New York. Upson finished fifth with a 651-

mile trip to Amherstburg, Ontario, aboard the *Goodyear II*. Thompson finished in sixth place with a 589-mile trip.[8]

The 1921 elimination returned to Birmingham on May 21. Upson was selected as the chief U.S. competitor. His teammate, C. Andrus, was the chief forecaster of the U.S. Weather Bureau. Ward T. Van Orman, the only other American contestant in the 1921 race, flew the *City of Akron* with Willard F. Sieberling, son of Goodyear president Frank A. Sieberling, as his aide.[9]

The 1921 Gordon Bennett race began with a rocky start from Brussels, Belgium, on October 18. Ascending in a rain squall, one of the two Italian entries flew straight through a grove of trees. Demuyter's grappling hook caught in the clothing of one of his ground handlers at take-off. He went aloft with the helpless man dangling beneath the basket, and was finally able to pull the man up to the safety of the car. The extra weight of the soldier was, however, a major factor in Demuyter's poor finish as twelfth in a field of fourteen.

Both Upson and Van Orman were worried about the third American contestant, Bernard von Hoffman, who was flying the *City of St. Louis*, a brand-new aerostat, which was leaking gas at an alarming rate. Hoffman and his aide, J. S. McKibben, were eventually disqualified in spite of a very long flight that carried them over the English Channel and across Great Britain to a watery landing 12 miles off Cork in the Irish Sea. Quickly rescued by a ship, the aeronauts fell victim to an FAI rule which stated that all Gordon Bennett landings had to be made on land, where the distance could be precisely measured.[10]

Van Orman and Sieberling finished eighth with a flight of 15 hours, five minutes to the estate of Squire Byrum, 6 miles north of Exeter and 371.38 miles from Brussels. Upson placed third, behind the Swiss pilot Armbruster and the Englishman Spencer. It was to be the last Gordon Bennett flight of Upson's career.[11]

The 1922 National Balloon Race, intended, as usual, to select the American contestants for that year's international contest in Switzerland, marked the real entry of both the U.S. Army and U.S. Navy into competitive ballooning. Both of the service air arms were eagerly seeking the favorable publicity that would attract public support and increased congressional appropriations for their programs. Participation in balloon and airplane races, long-distance flights, the appearance of military aircraft at state fairs, and the formation of crack military aerobatic teams were all a part of this new emphasis on public relations.

The 1922 Nationals, held at Milwaukee on May 31, featured three army balloons, piloted by Maj. Oscar Westover and Lieutenants Neeley and Weeks. Lt. William F. Reed and Lt. Cdr. Joseph Pugh Norfleet represented the navy.

H. E. Honeywell, Ward Van Orman, Warren Rasor, Ralph Upson, J. S. McKibben, John Berry, Bernard von Hoffman, and Roy F. Donaldson were the civilian entries.[12]

Reed and his aide, Mit Mullenaux, flew the race as the navy's official contenders for the 1922 Gordon Bennett competition. Norfleet, a pioneer naval aviator who had acquired his balloon training in Europe during the war, was flying as an experimental entry in order to compare the performance of a helium balloon with that of the aerostat filled with standard coal gas. As all official contestant balloons were required to be filled with gas drawn from a single source, Norfleet was disqualified from participating for the honor of representing the United States in Switzerland in 1922. Norfleet discovered that the only discernible advantage was that he and his aide could smoke and cook in the basket.[13]

As Reed noted in a report to his naval superiors, "On the afternoon of May 31st there was perhaps the widest range of winds to select from of any balloon race ever held in this country."[14] Most of the pilots chose to fly to a landing in Indiana, Illinois, or Missouri. The winner, Maj. Oscar Westover, flew 866 miles in quite a different direction, landing near Lake Johns, Quebec.[15] Westover, who had spent most of the war on assignment to the Bureau of Aircraft Production, graduated as a balloon observer from the Balloon School at Ross Field, California. While serving as chief of the Balloon and Airship Division, Office of the Chief of the Air Service, he flew as Frank Lahm's aide in the 1921 elimination race.[16]

Westover and his aide, Carlton Bond, finished sixth in the 1922 Gordon Bennett race, flown from Geneva on August 6. Their trip was cut short when Hungarian peasants caught their dragline and hauled them to earth. Lieutenant Reed placed thirteenth in the field of nineteen. H. E. Honeywell was more successful, placing second with a flight of 657.82 miles, right behind the Belgian winner, Ernest Demuyter.[17]

Encouraged by their success in 1922, both the army and the navy sent large teams to the fourteenth National Balloon Races staged at Indianapolis on July 4, 1923. The now familiar civilian entries were present as well.[18] (See page 566.)

Tragedy struck in 1923 for the first time in the history of the Nationals. The envelope of Navy balloon A6698, flown by Lt. L. J. Roth, aided by Lt. T. B. Null, was picked up by a Lake Erie tug 25 miles south-southeast of Port Stanley, Ontario. After a full-scale air search by two DH-4s, a Loening Air Yacht from Selfridge Field, and the flying boats *Buckeye* and *Nina* of Aeromarine Airways, the basket was found floating on the lake with Roth's body strapped in place. Lieutenant Null was never found.[19]

In his report to Air Service superiors, Lt. R. S. Olmsted stressed the impor-

Entries in the 1923 National Balloon Race
(in order of finishing)

Pilot	Aide	Balloon	Landing
1. Lt. R. S. Olmsted	Lt. J. W. Shoptaw	Army S6	Lake Marilla, New York 449.5 miles
2. Lt. J. B. Lawrence	Lt. F. W. Reichelsderfer	Navy A6700	Glen Campbell, Pennsylvania 398.1 miles
3. H. E. Honeywell	P. J. McCullough	*St. Louis*	Brocton, New York 397.2 miles
4. C. E. McCullough	C. R. Bond	*American Legionnaire*	Frankfort Springs
5. Capt. L. T. Miller	Lt. C. M. Brown	Army S5	Ford City, Pennsylvania 400 miles (approx.)
6. Lt. F. B. Culbert	Lt. T. D. Quinn	Navy A6699	Alliance, Ohio 310 miles
7. Lt. Comdr. J. P. Norfleet	Lt. J. B. Anderson	Navy A6704	Mount Eaton, Ohio 300 miles (approx.)
8. J. A. Boettner	J. M. Yolton	*Goodyear II*	Freemont, Ohio 250 miles (approx.)
9. Roy F. Donaldson	P. A. Erlach	*City of Springfield*	8 miles NE of Bryan, Ohio 169 miles
10. Ralph Upson	C. Andrus	*Detroit*	Wapakoneta, Ohio 150 miles
11. Lt. J. B. Jordan	M. F. Moyer	Army S7	Macedonia, Ohio 150 miles (approx.)
12. W. T. Van Orman	H. V. Thaden	*City of Akron*	5 miles N of Hartford City, Indiana 75 miles

The 1922 Gordon Bennett classic, Geneva, Switzerland

tance of a lightweight radio on which he had received both weather reports and news of other competitors. "In the future," he reported, "all racing balloons should be radio equipped."[20]

The tragedy of the 1923 Nationals was compounded at that year's Gordon Bennett race. Sixteen balloons launched from Brussels in the middle of a violent thunderstorm on September 23. Belgian officials were anxious to postpone the event, but found that FAI rules did not permit a cancellation under the existing conditions.[21]

The problems besetting the American team began at the start of the race. H. E. Honeywell's *St. Louis*, unable to stand the strain of high winds, burst just prior to launch. The Navy aerostat A6699, flown by J. B. Lawrence and F. W. Reichelsderfer, almost struck a group of balloons at take-off. The navy aeronauts, who flew with an impressive array of both flight and scientific instruments, were forced down at Pitten, Holland, finishing sixth with a total distance of only 110.36 miles.[22]

The army balloon S6, with Olmsted and Shoptaw on board, was launched at 4:00 P.M. and immediately collided with the *Ville de Bruxelles*, ripping the netting of the Belgian craft and forcing it to withdraw from the event.[23] The Americans, unaware that they had been disqualified as a result of the accident, flew on. Three hours later, flying over Nistelrode, Holland, the S6 was struck by lightning. Both Olmsted and Shoptaw were killed.[24]

Ernest Demuyter's third victory was marred by the appalling death toll of the 1923 race. Like the S6, the Spanish balloon *Polar* was struck by lightning, killing the pilot and severely injuring his aide. The *Geneva*, a Swiss entry, suffered a similar fate with the loss of both crew members. The Spanish balloon *Espheria* hit a power line on landing, injuring the pilot and aide. Major Baldwin and Captain Dunville, of the British entry *Banshee II*, were injured when their craft landed on a roof. Allen and Berry of the second British balloon, *Margaret*, descended in the sea near Skogen, Denmark, but were rescued.[25] The opinion of the American trade journal *Aviation* was echoed around the globe. "We hope that there will never be another race permitted where the odds are so heavy against these fine sportsmen of the air."[26]

By 1924 Ward Van Orman was particularly anxious for a victory. A perennial competitor who was universally recognized as one of the world's most experienced aeronauts, Van Orman had been plagued by difficulties which kept him out of the running in recent balloon contests.

Van Orman and his aide, Carl K. Woolam, had designed an entirely new balloon for the 1924 Nationals, which were scheduled to begin at Kelly Field, San Antonio, Texas, on April 23. The resulting *Goodyear III*, a single-ply envelope of rubberized fabric, weighed only 385 pounds, less than half the weight of a standard balloon of the same dimensions. Van Orman was particularly proud of his lightweight radio receiver. With its 200-foot antenna wrapped around the equator of the balloon, the set was capable of receiving signals from distances of up to 2,500 miles.[27]

During the course of the race, Van Orman and Woolam, who had been a balloon instructor at Goodyear's Wingfoot Lake facility from 1917 to 1920, had arranged for regular weather reports to be broadcast by radio stations across the nation. In addition, Goodyear officials had scheduled a special broadcast to the hometown aeronauts over Cleveland's WTAM.

Floating 1,500 feet over the Texas prairie, Van Orman and Woolam listened to messages from company officials, a special concert by Smith's Metropolitan Orchestra, and the live voices of their wives. "I could not believe my ears," Van Orman remarked, "when from out of that quietude came my wife's voice broadcasting a message of 'God's speed' from the Willard Storage Battery station in Cleveland. I can only state that the miracle of radio in bringing Edith's voice

The windy conditions plaguing the 1923 Gordon Bennett, held in Brussels, are evident in this photo.

to me over such a distance caused shivers to run up and down my spine."[28]

As Van Orman and Woolam traveled north over Oklahoma, the radio became their most essential item of equipment. Broadcast weather reports informed the aeronauts that they were heading directly into a storm. Rising to altitudes of from 15,000 to 17,000 feet for over three hours, then descending to make a run around the edge of the storm, the aeronauts flew the *Goodyear III* 1,100 miles to a landing near Rochester, Minnesota, and victory.[29]

As Ladislas D'Orcy of *Aviation* noted, it was a good year for America's veteran balloonists: "Van Orman's victory is popular, as he has been a prominent figure in all of the balloon races since the war. . . . Honeywell, the 'grand old man' of ballooning who has been in every race that anyone in America can remember, always coming in among the leaders, has again added to his reputation by securing second place."[30] It was Honeywell's 550th flight.

Honeywell, Van Orman, and U.S. Army Major Norman W. Peck, the third-place winner in the Nationals, finished far back in the field of the 1924 Gordon Bennett race (6th, 14th, and 16th out of 17, respectively). Ernest Demuyter won the event, which was flown from Brussels on June 15.[31]

For the first time in the history of the race, a pilot had won the trophy for three consecutive years. Demuyter took permanent possession for the Aero Club of Belgium. There was some question whether or not the event would continue beyond the retirement of the original cup. Early in 1925, however, the Belgians scheduled the annual race as usual and offered a new prize trophy.[32]

By the time of the 1925 Nationals, Ward Van Orman had perfected the system that would lead to an unprecedented string of victories in the future. In effect, he applied new technology to improve upon the lessons he had learned from Ralph Upson.

Launching from St. Joseph, Missouri, on May 1, 1925, with C. K. Woolam, Van Orman carried an efficient six-tube heterodyne receiver developed with the assistance of David Sarnoff and the Radio Corporation of America. This was combined with an experimental radio compass. Like Upson, Van Orman had become a devoted student of meteorology. He depended on this knowledge and on his ability to obtain up-to-date weather information to plot a winning course. Upson was not alone in his reliance on the radio and meteorology however. All five army and civilian teams competing in the 1925 Nationals carried radios.[33]

In addition to competing for a place on the U.S. Gordon Bennett team, the aeronauts were also vying for the P. W. Litchfield Trophy for the first time. Established by Paul W. Litchfield, vice-president and factory manager of the Goodyear Tire and Rubber Company, to celebrate his twenty-fifth anniversary with the firm, the handsome trophy carried with it a $1,000 cash prize.[34]

Van Orman and Woolam won both the trophy and the first-place seat on the 1925 U.S. Gordon Bennett team with a flight of 585 miles to Reform, Alabama. Lt. William Flood placed second, and H. E. Honeywell third. Honeywell had actually flown farther than Flood, but was reduced to third place because he had been late in inflation and launching.[35]

Van Orman always felt that he had lost the 1925 Gordon Bennett on a technicality. Launching from Brussels on June 7, he and Woolam crossed France and set out over the English Channel, hoping to make a landfall and landing in Brest. For once, fortune seemed to be working against Van Orman. Once over the channel, they were carried north of the coast and straight out into the Atlantic. Passing over Brest on their left, they were so close they could hear dogs barking and carts rattling through the streets, but were unable to locate a current blowing inland.

At seven o'clock on the evening of June 8, they passed the tiny island of

Quessaut, six miles to the south. Both men were painfully aware that their next possible landfall was the North American coast, 3,300 watery miles away. The *Goodyear III* had sufficient ballast for roughly 1,500 miles, but, knowing the vagaries of the Atlantic winds, Van Orman planned to come down at sea somewhere to the east of a line drawn from Cape Torinana, Spain, to Land's End, England, inside the heavily traveled shipping lanes.

Now the radio only added to the gloomy atmosphere. "At 10:00 P.M. sharp we tuned in the London weather report," Van Orman recalled. "The most discouraging bit of information we had ever received came in over our radio that night. The forecast indicated the winds were going to blow towards the west for three more days. Then to add a touch of irony to our dire straits, the London station then presented the Savoy Hotel dance orchestra. But we were in no mood for dance music nor did we feel like dancing at that particular moment."[36] There was silence in the basket now. As Van Orman considered the narrow range of options open to them, Woolam worked his way through the cognac ration. Without saying a word, the aide pulled out the flare pistol, took aim at the gasbag, and pulled the trigger. Van Orman watched in horror as the flare arched within two feet of the hydrogen-filled envelope.

A bit more sober now, Woolam offered to jump from the basket to extend Van Orman's flight. The pilot, deeply moved, replied, "If we have to go we will go together."[37] Soon thereafter, Van Orman spotted the lights of a small steamer struggling through twenty-four-foot seas to reach the Channel. With a Morse code book in one hand and a five-cell flashlight in the other, he was able to attract the attention of the bridge crew. Incredibly, his message was, "We are going to land on board."[38]

Minutes before, Van Orman was facing a watery grave. Now, with rescue at hand, he refused to give up his chance for a victory in the race. Certain that he had flown farther than any of the other competitors, Van Orman knew that he would be disqualified under strict rules which forbade a landing on water. If he could set down on the pitching deck, however, he might circumvent the rule book.

The landing was nothing short of miraculous. Van Orman dropped from 2,000 feet to within 30 feet of the ocean. Tossing a sea anchor out of the basket, he prepared for the toughest landing of his career. Almost half a century after the event, Van Orman remarked, "It seems to me to this day as if some unseen hand held our basket at just the right height so that as we struck the outside rail of the ship's forward deck, six sailors found it possible to grab onto the basket and hold it to the rail."[39] Van Orman immediately pleaded with the captain to bring the ship into the wind so that the envelope would be blown across the deck and away from the hot smokestack. The two men then leaped onto the deck, yanking the ripping line as they went.

The aeronauts were safe, but the last-minute stratagem for winning the race failed to impress the FAI officials. The race was awarded to A. Veenstra and his aide, F. Quersin, of Belgium, who had flown 835.7 miles. Van Orman would never forgive the judges for what he regarded as an unfair decision. As late as 1978 he could still remark: "I have always believed in good sportsmanship and the good loser. Countless things are more important than victory. Yet unfairness, either on the part of participants or officials, cannot be tolerated. To the extent that indifference to bias or wrongdoing is tolerated—in any organization or segment of society—the entire institution deteriorates in character and importance to that degree."[40]

The official records of the FAI indicate that Ward Tunte Van Orman won three James Gordon Bennett Cups (1926, 1929, 1930). Van Orman knew better. That providential landing on a pitching deck—at night, in high seas—should have earned him a victory. He had flown the farthest without getting his feet wet, and that should have been enough. But it was not, and Ward Van Orman faced the flying season of 1926 more determined than ever to achieve his first victory in the international classic.

As usual, the first test of the season came with the 1926 Nationals held at Little Rock, Arkansas, on April 29. As Ralph Upson noted in *Aviation*, this promised to be a good year for America's amateur aeronauts. In addition to the elimination race and the Litchfield Trophy, the Thomas Baldwin Trophy and the Detroit News Trophy had been established for balloons of 35,000 cubic feet and under.[41]

Little Rock staged an entire air meet to honor the coming of the Nationals to town that year. A wide range of military and civil aircraft were demonstrated. Crop dusters, parachute jumpers, aerobatic teams, and a National Guard air race enlivened the proceedings.

A field of nine balloons left Little Rock Airport between 5:00 and 5:40 P.M. on April 29. (See page 573.)

Van Orman and his aide, Walter H. Morton, took first place with an 815-mile trip to a point south-southeast of Petersburg, Virginia. John (Jack) Boettner, a Goodyear-Zeppelin Co. airship pilot and instructor and a veteran of the 1923 Nationals, finished second. Boettner also won the Detroit News Trophy for 1926. Capt. Hawthorne C. Gray, a balloon officer from Scott Field, Illinois, placed third.[42]

Sixteen balloonists gathered in Antwerp for the sixteenth running of the Gordon Bennett classic on May 30, 1926. Once again the launch was made

Entries in the 1926 National Balloon Race
(in order of starting)

Pilot	Aide	Balloon	Comments
1. Lt. James F. Powell	James Earley	Army S5	
2. Capt. Lawrence E. Stone	Capt. Guy R. Oatman	Army S19	
3. Walter H. Hamm	Robert P. Lehr	*Goodyear Southern California*	
4. Herbert V. Thaden	Charles D. Williams	*The Detroit*	
5. Ward T. Van Orman	Walter H. Morton	*Goodyear IV*	First place, 815 miles. Landed SSE of Petersburg, Virginia
6. Capt. Hawthorne C. Gray	Lt. Douglas Johnson	Army S23	Third place
7. Sverd W. Rasmussen	Edward J. Hill	*Detroit Adcrafter*	
8. Lt. William A. Gray	Lt. Ronald Kieburtz	Army S20	
9. John A. Boettner	Herbert W. Mayson	*Akron NAA*	Second place

during heavy weather. The field was reduced to fourteen when two balloons suffered minor accidents at take-off.

The remaining entries were scattered across Denmark, Holland, and Sweden by a southwest wind. Van Orman found that his radio was more useful than ever both in obtaining weather data and in pinpointing his position. He and his aide, Charles Morton, finally came to a safe landing near Salnesburg, Sweden.

Van Orman was at first disappointed in having made only 864 kilometers (a little over 540 miles). Morton was especially displeased at what he regarded as "a terrible run in the International Race." Their spirits rose abruptly the day after landing when they received a telegram from the Goodyear branch manager in Stockholm. "You have won the International Balloon Race," it read, "even doubling the distance of your nearest foreign competitor, Lieutenant Demuyter."[43]

It was a great moment for Van Orman, and it had been a long time coming.

Hawthorne C. Gray and Ralph Johnson, two American entries in the 1926 Gordon Bennett

Returning to Antwerp, he met Albert, king of the Belgians, and was treated to a string of banquets and receptions. The returning hero, bearing the Gordon Bennett Cup back to the United States for the first time in thirteen years, was greeted with a reception that matched the one offered Ralph Upson in 1913.

Van Orman's victory in 1926 was made especially sweet by the fact that Capt. Hawthorne Gray had finished second with a distance of 371.38 miles. Their twin victories marked the beginning of a six-year period of American Gordon Bennett triumphs.

Van Orman, who had an automatic place on the 1927 U.S. Gordon Bennett team, nevertheless participated in and won the Nationals that year. Flying once again with his aide, Charles Morton, he launched from Akron on May 30 and flew to Bar Harbor, Maine. Two newcomers, E. J. Hill and A. G. Schlosser of Detroit, earned the second-place spot on the team. Capt. W. E. Kepner and Lt. W. O. Earickson, piloting an army balloon, finished third.[44]

Preparations for the Gordon Bennett race from Ford Airport, September 1927

A great deal of publicity was devoted to preparations for the international competition to be staged in Detroit in September 1927. Henry Ford had made Ford Airport available for the event. Throughout the late summer of 1927, crews were busy installing the pipes that would carry gas to the various launch sites and erecting the grandstands and other facilities for the expected crowds.[45]

The aeronauts were preparing as well. Van Orman had flown a new balloon, *Goodyear IV*, in the Nationals. An 80,000-cubic-foot "racing bag," it incorporated all the improvements Goodyear had developed in almost fifteen years of experience in the construction of balloons. A single-ply envelope, it was perhaps the lightest racing balloon ever built.[46]

Van Orman equipped his basket with the usual array of navigation instruments, including a marine sextant, a radio compass, and his 17-pound RCA receiver. In addition, he had developed two new instruments, a vertimeter and an alarm altimeter. The vertimeter was a "vertical speedometer" so sensitive that it would measure the speed of a rise or fall as small as four feet. Van Orman tested the device by operating it aboard the elevator of a New York skyscraper. Less of a problem to develop, the alarm altimeter sounded when a desired altitude had been reached, thus obviating the need for the pilot or aide to monitor the instrument constantly.[47]

The day of the race was a gala day in Detroit. The crowd of 50,000 spectators included many of the leaders of American aeronautics. Gen. Mason F. Patrick, chief of the Army Air Corps, was present, as were William P. McCracken, assistant secretary of commerce for aeronautics; Edward P. Warner, assistant secretary of the navy for aeronautics; Porter Adams, president of the National Aeronautic Association (NAA); and Carl F. Schory, secretary of the NAA. Gov. Fred Green of Michigan, Mayor John Smith of Detroit, E. M. Statler of the Statler Hotels, and Henry and Edsel Ford were also on hand.[48]

As each of the twelve balloons took off, the Ford band struck up the appropriate national anthem. As the last balloon, a U.S. Army entry, took off, the pilots of Selfridge Field passed in review, led by Maj. Thomas G. Lamphier.[49]

Van Orman and Morton were fifth into the air. Because of existing weather patterns, the aeronauts had decided that the longest flight would be obtained by moving straight south, aiming for a landing in Florida. They were stopped short when they encountered a violent thunderstorm fifty-one hours after launch. "The decision," Van Orman recalled, "was quite simple. We had victory in our grasp if we wished to pay the price, but the price possibly would be our lives. It was out of reason."[50]

Goodyear IV returned safely to earth a half mile north of Adrian, Georgia. Unfortunately, the envelope of the balloon was allowed to escape because of carelessness on the part of the spectators. The disappointment of the loss of the balloon was compounded by the news that they had finished only third behind Hill and Schlosser of the United States and Kaulen and Dahl of Denmark.[51]

Still, Van Orman and Morton had made a wise decision in landing. Since the invention of the balloon, thunderstorms had been the bane of the aeronaut. If nothing else, the tragedy of the 1928 National Balloon Race would demonstrate this basic truth.

The race began from Bettis Field, near McKeesport, Pennsylvania, on May 30, 1928. Fourteen balloons were launched between 5:01 and 6:05 that morning. Traveling over the rough country east of Pittsburgh at 6:00 P.M., in the *Goodyear V*, Van Orman noted the sky darkening to the west. He instructed Morton to don a parachute and prepare for a storm.

About five minutes later it seemed to us that some unseen hand grasped the top of the balloon and started to draw it rapidly upward. Our speed accelerated and our vertimeter, designed to register vertical currents of up to thirty miles per hour, had reached its limit. The hydrogen gas was pouring out of the bottom of the appendix at a furious rate. The balloon started to spin, caused by the diamond shaped mesh of the net. Due to this spinning action, the basket started to wind in crazy circles produced by centrifugal force.[52]

Neither aeronaut, with their total of thirty-seven years of experience in the air, had ever experienced anything like this. In spite of a torrential rain, they were shooting up to 12,000 feet, where a hailstorm pelted the balloon so severely that they feared the danger of a puncture. Then they were on their way down, with Morton frantically scooping ballast over the side. Passing through 8,000 feet, Van Orman began to slash the ballast bags with his knife.

This was to no avail. They struck the ground with a force that shattered their instruments, then rebounded back to 10,000 feet, only to begin a second drop. Another smash into the ground, and they were back up to 2,000 feet. Morton was sprawled on the bottom of the basket as Van Orman peered helplessly over the edge just in time to see the team of Bill Kepner and W. O. Earickson drop past at terrific speed. They were so close that Van Orman could hear Earickson screaming "Alley-oop!"

At that moment, over the roar of the storm, Van Orman saw a lightning flash only a few feet away. He turned to look at the gasbag, fearing an explosion, then passed out. "Five hours later I regained consciousness to the sensation of flies buzzing about my eyes and nose. I swished my hand but to no avail. Then it struck me that there were no flies at all, but drops of rain bouncing off my face. As I ever so slowly recovered my senses, I realized that I was lying with my head on the ground and my feet and body in the capsized basket which now rested on its side on the ground."[53] Van Orman called for Morton, and, receiving no response, presumed that his aide had walked away in search of help. After freeing himself from the equipment that held him in place, Van Orman pushed himself upright and saw Morton sprawled dead in the bottom of the basket.

Van Orman, his leg broken, was unable to walk. He gave up on an attempt to crawl in search of a farmhouse when exhaustion claimed him after less than fifty feet. Only five minutes later he heard a voice and directed his flashlight toward the sound. Rescue arrived in the form of a farmer venturing out before dawn to investigate the strange events of the night before.

Recuperating in a Greensburg, Pennsylvania, hospital, Van Orman was able to reconstruct the disaster. The lightning had struck *Goodyear V* at the equator, and passed down the fabric to strike Morton. The balloon had exploded just as Van Orman looked up. The lower three quarters of the envelope had been completely destroyed.[54]

A thousand feet away at the time of the explosion, Lt. Paul Evert and his aide, Lt. V. G. Ente, watched from the basket of Army No. 3. Ente recalled that he had screamed, "My God! Look there, Evert!" Evert turned in time to see *Goodyear V* falling out of the sky "at rocket speed." The bag, which had just exploded, was "strung out in the form of a colossal necktie." "Oh," Evert remarked, "I'll bet those fellows are done for!" These were his last words. A moment later he was killed by a second lightning bolt.[55]

Two men were dead, two balloons were down. At the moment, Kepner and Earickson were positive that they were next. The night of May 31, 1928, was one neither man would ever forget. In all, eleven balloons had been trapped in the storm. In addition to the two dead aeronauts and the injured Van Orman, James P. Cooper, Carl Woolam's aide aboard the *City of Cleveland*, was burned and bruised. All in all, May 31, 1928, was the worst day in the history of American ballooning.

Some good did come from the events of the day. Working with Arthur Austin at the Ohio Insulator Laboratories at Barberton, Ohio, Van Orman launched a study of the problems of lightning strikes on aircraft of all types. After a year of experimentation, the team developed a set of sixteen wires which would surround the basket, providing protection from lightning. The invention, an adaptation of the Faraday cage, was installed around the basket of a new aerostat, the *Goodyear VIII*.[56]

Van Orman was convinced that the device was the solution to the problem. Others were not so sure. Capt. William Flood and Lt. V. G. Ente chose to fly without a radio in 1929, fearing that the apparatus had been the cause of the accident. Van Orman could only "question the logic and good sense of scrapping their primary means of communication. For me, I would sooner dispense with the balloon basket, itself—and rations, firearms and clothing—than give up the life-saving radio during a race!"[57]

In addition to providing a possible solution to the problem of lightning strikes, Van Orman and Austin paid a great deal of attention to the presence of the hazardous vertical currents that created havoc in 1928.

Bill Kepner and W. O. Earickson survived the storm and went on to win the 1928 Gordon Bennett. They flew 460 miles from Detroit to a spot 3 miles southeast of Kenbridge, Virginia, to finish ahead of the Danish father and son team of Kaulen.[58]

Fully recovered from his accident, Ward Van Orman was ready for the 1929 Nationals held in Pittsburgh.

Entries in the 1929 National Balloon Race
(in order of starting)

Pilot	Aide	Balloon	Comments
1. Capt. W. Flood	Lt. V. G. Ente	Army No. 1	
2. Capt. E. W. Hill	Lt. Robert Heald	Army No. 2	Third Place Landed near Newcomb Lake, New York
3. Lt. L. A. Lawson	Lt. Edgar Fogelsonger	Army No. 3	
4. Lt. Thomas G. W. Settle	Ens. Wilfred Bushnell	Navy No. 1	First Place 900 miles (approx.) to Prince Edward Island
5. Lt. Jack C. Richardson	Lt. Maurice Bradley	Navy No. 2	
6. Dr. George Milegalles	Walter Chambers	*Pittsburger*	
7. H. E. Honeywell	Roland J. Gaupel	*St. Louis Chamber of Commerce*	
8. E. J. Hill	A. G. Schlosser	*Detroit Balloon Club*	
9. Sverd W. Rasmussen	Tracy Southworth	*Detroit Balloon Club*	
10. W. A. Klikoff	Thorvald Larson		
11. Ward T. Van Orman	Alan L. MacCracken	*Goodyear VII*	Second place Landed at Harkness, New York
12. A. C. Palmer	Lt. W. O. Earickson	*American Business Club of Akron*	

As the 1928 Gordon Bennett winner, Captain Kepner was automatically named to the 1929 American team. Lt. Thomas (Tex) G. W. Settle and Van Orman, who won the first and second place in the Pittsburgh contest, would round out the U.S. entries. This was perhaps the most distinguished James Gordon Bennett team ever fielded by any nation. Van Orman was the nation's most distinguished civilian aeronaut, a worthy successor to H. E. Honeywell and Ralph Upson. In addition, he was a leading engineering contributor to Goodyear's airship program.[59] Settle and Kepner, on the other hand, were among the leaders of the lighter-than-air groups in their respective services.

Nine balloons left St. Louis on September 28, 1929, for the seventeenth running of the James Gordon Bennett classic. The Americans swept the field. Van Orman and Alan MacCracken finished first with a 341-mile flight to a landing 3 miles southeast of Troy, Ohio. Kepner and Capt. J. F. Powell came in a close second, traveling 338 miles to North Neptune, Ohio. Tex Settle and Ensign Bushnell were third, landing 304 miles from St. Louis, near Eaton, Ohio. Having retired the second Gordon Bennett Trophy in 1928, the Americans were the first to win the third cup.[60]

In 1930 the Nationals traveled to Houston, where fifteen balloons were launched on July 4. (See page 581.)

Van Orman, as the 1929 winner, flew to his third and final Gordon Bennett victory from Cleveland on September 1, 1930. Traveling with Alan MacCracken in the *Goodyear VIII*, he flew 542 miles to a point near Canton Junction, Massachusetts. Demuyter finished second, Hill and Schlosser third, and Blair and Trotter fourth.[61]

The 1930 race had been run with only six balloons, three of them American. It was obvious that four straight years of U.S. victories was discouraging to all but the most determined European competitors. Moreover, having retired one trophy, the Americans needed only one more victory to retire a second. These facts, combined with the expense of shipping aeronauts and balloons to the United States in order to participate in yet another American sweep, proved so unattractive that there were no European entries for the 1931 race.[62]

The U.S. Nationals were held in 1931 as usual. Tex Settle won the contest with a short 215-mile trip from Akron to Marella, New York, on July 19. With only three balloons, all American, scheduled to participate in the Gordon Bennett, the race was cancelled. U.S. officials, fearing that the very survival of the twenty-five-year-old contest was in danger, broke tradition and agreed to send the race to Basel, Switzerland in 1932.[63]

The 1932 Nationals, held at Omaha on May 30, provided the sort of excitement everyone had come to expect of the annual elimination race. Lt. Wilfred J. Paul and M. Sgt. John Bishop, the winning team, battled 30 hours of thunderstorms, snow, rain, and fatigue in a flight that carried them 901.4 miles to a point 13 miles north of Hutton, Saskatchewan. Two Akron men, Ronald J. Blair and Frank Trotter, finished second with a trip of 19 hours, 25 minutes to Tyvan, Saskatchewan. Capt. William Flood, flying Army No. 1, finished in third place, turning in what he described as "one of the most exciting balloon rides of his career."[64]

Entries in the 1930 National Balloon Race
(in the order of starting)

Pilot	Aide	Balloon	Comments
1. Lt. T. G. W. Settle	Lt. R. G. Meyer		Third place 700 miles to Dover, Tennessee
2. Geo. Hineman	Milford Vanik		
3. Lt. W. D. Budie	Lt. J. P. Kidwell		
4. S. T. Moore	Lt. W. O. Earickson		
5. E. J. Hill	A. G. Schlosser	*Detroit-Times*	Second place 775 miles to Russellville, Kentucky
6. Capt. K. Axtater	Lt. Holmes		
7. Lt. W. R. Turnbull	C. M. Brown		
8. R. J. Blair	F. A. Trotter	*Goodyear-Zeppelin*	First place 850 miles to Greenville, Kentucky
9. Lt. W. Bushnell (USN)	Lt. J. A. Greenwald (USN)		
10. Lt. R. R. Dennett (USN)	Lt. C. F. Miller (USN)		
11. Sverd W. Rasmussen	Tracy Southworth		
12. W. A. Klikoff	R. S. Cunningham		
13. D. R. C. Legalle	R. W. Ebert		
14. C. H. Roth	W. P. Carey		
15. Lt. Otto Wiencke	W. N. Fox		

This year, however, the outcome of the Nationals had little to do with the selection of the U.S. James Gordon Bennett entries. Van Orman, the 1930 winner, now flying the *Goodyear VII*, had already entered with Ronald Blair as his aide. Tex Settle, the second U.S. competitor, would fly with Lt. Wilfred Bushnell.

As in 1929, the Americans swept the field of the twentieth Gordon Bennett race, which began at Basel, Switzerland, on September 25,1932. Settle and Bushnell flew the balloon *U.S. Navy* across Czechoslovakia and into Poland, climbing as high as 10,000 feet by 11:00 A.M. on their second day in the air.

Ward T. Van Orman addresses the crowd at the 1930 Nationals.

Settle recalled that the most humorous incident of the voyage occurred at this point: "About this time we discovered that we had had a rather serious accident. There was a full thermos bottle of hot chocolate in my duffle bag and this had exploded, presumably from the lowered atmospheric pressure. Hot chocolate was distributed over my spare white shirt, underwear and socks, all of which were soaked. Later on we used these garments for ballast. The natives finding them must have thought, 'These balloonists are truly peculiar people.'"[65]

They continued their voyage at low altitudes, making use of their dragline to conserve ballast. Passing over a Polish village late on the afternoon of September 26, the dragline ran over a thatched roof and made a cracking sound as it whipped around a post. The sight of a group of young women scurrying out of sight led Bushnell to quip, "They see the Navy coming."[66]

After crossing Warsaw at 8,000 feet, they dropped back down to continue "draglining" over the countryside. With only enough ballast left for landing, they were finally forced to descend the next morning. They had won the 1932

Inflation procedure, 1930 Nationals

contest with a flight of 963 miles in 41 hours, 20 minutes.[67]

Van Orman and Blair had adopted a very different tactic. Rather than hugging the ground, they chose to venture into the upper atmosphere. Sucking oxygen from pressurized bottles, they remained at 24,000 feet for the last nine hours of the flight. With the thermometer hovering at 15 degrees above zero and snow drifting into the basket, Blair asked, "Do you know, Van, your face is blue?" "I don't like to worry you," replied Van Orman, "but yours is indigo."[68] The pair finished in second place, having covered a distance only 15 miles short of Settle's winning trip.

Both Settle and Van Orman returned across the South Atlantic as the guests of Hugo Eckner aboard the *Graf Zeppelin*. As appealing as an airship voyage from Danzig to Brazil was, it presented at least one problem for the chain-smoking Van Orman. Just before take-off, Captain Ernst Lehman took him aside and warned him that absolutely no smoking was permitted aboard the hydrogen-filled airship. Lehman then handed him a small package with instructions to open it when he got the urge to smoke. "The urge," Van Orman recalled, "came about thirty seconds after he turned and retreated to the control car."[69] It was a beautiful harmonica which Van Orman would treasure for the rest of his life. The urge to smoke didn't leave the aeronaut during the trip, but he had mastered *Juanita* by the time they were halfway across the South Atlantic.

Settle's victory in 1932 marked the end of American hegemony in the Gordon Bennett competition. The 1933 race took place in Chicago on September 2, in conjunction with the Century of Progress Exposition. The balloon *Opel*, which was to have been flown by the German rocket experimenter Fritz von Opel, escaped prior to take-off and was later found torn to shreds.[70] Six other balloons were successfully launched. Two of these, the Danish and Belgian entries, stopped short on the western shore of Lake Michigan, while the French balloon came down in northern Ohio after a flight of only 154.38 miles.

Settle and his aide, Lt. C. H. Kendall, brought their balloon down near Branford, Connecticut, after a relatively uneventful flight of some 750 miles, which would ultimately prove good enough for second place. The other two competing crews, Francis Zekhyn Hynek and Zebigniew Burzynsky of Poland and Ward Van Orman and Frank Trotter of the United States, had a more difficult time of it.[71]

Van Orman had not been anxious to compete in 1933. His wife had died the year before, leaving him with three small children to raise on his own. Paul W. Litchfield, however, was determined to get a Goodyear entry into the competition. When Litchfield offered to insure the aeronaut through Lloyd's of London, Van Orman reluctantly agreed to fly. Studying the unsettled weather conditions, the aeronaut planned to overfly the Great Lakes, aiming for a landing somewhere in Newfoundland and the opportunity to capture the U.S. distance prize.

In spite of a dunking in Lake Michigan, they crossed safely into Canada, only to be caught in a thunderstorm. Van Orman recalled the scene:

> *The most notable thing at first was the darkness. We were flying at 8,000 feet, yet everything was black as ink. It got worse and worse. Suddenly the squall came with a roaring rush. When it struck, we had never seen a balloon act as ours acted—not even in the Pittsburgh storm. It hit us while we were high and we careened all over, bounced all over. We rushed up and fell and we were only on the outskirts of it! Had we been in the center of that disturbance, I'm certain we would have been dead at the start.*[72]

They fought the storm for six hours. Although furnished with parachutes, the

Goodyear VII

men were unable to jump because of the incredible winds. The problem was compounded by Van Orman's continual valving of gas in an effort to control the balloon.

The balloon finally plunged into the top of a forest at a speed of 55 MPH. As the aeronauts were thrown back and forth across the basket, the balloon smashed through five tree trunks and into a sixth which ripped the envelope and ended the flight. Hanging twenty-five feet in the air, with the remnants of the balloon draped over the basket to protect them from the rain, the two men shook hands, split a can of peaches, and fell asleep in their dangling perch.

The next morning Van Orman went to work with a tiny sextant in an attempt to figure their position. The sophisticated navigational equipment carried in the basket was either broken or lost. Using a water-filled peach can to determine the sun's angle, and after eight hours of waiting for the breeze to die down or the sun to reemerge from behind the leaves, Van Orman concluded that they had fallen into the Timagami forest of Quebec, 26 miles from the nearest railroad.

"Twenty-six miles may not sound far," Van Orman remarked, "but in tall, razor-sharp underbrush, including tens of thousands of fallen tree trunks, we learned that our best distance could not be better than two miles on foot a day."[73] The two aeronauts spent fourteen days hiking out of the wilderness. Each day it rained with what Van Orman could only describe as "abandon."

> *At the end of the first week, we were both overcome with ptomaine poisoning. Each day from then on, our progress steadily diminished as over-powering fatigue and nausea nearly conquered us. Frank and I ultimately reached the point where we'd flop on the ground for an hour, then proceed on for a city block. Our meager plan was to measure our diminishing strength against the distance to that damn railroad.*[74]

Van Orman compared the experience to "a bad dream, a strange unreal nightmare."[75] He brooded over his children's welfare. The matches were soaked. Rainwater caught in the balloon fabric tasted of rubber and made the nausea worse.

On Sunday, September 10, eight days after take-off, they heard an aircraft overhead, but the flares that they had carefully saved brought no response. They were resting more frequently now, very ill with ptomaine poisoning from canned pork and beans, and suffering from blisters as well as the bumps and bruises sustained in the crash.

They had given up on the railroad and were now following a high-tension line. Providentially, they found an old line shack, complete with blankets and dried beans. They had found a telephone line as well, which they were able to cut, hoping that the downed line would attract the notice of the outside world.[76] Their ruse worked. As linemen began working their way toward the stranded aeronauts in search of the reason for the loss of phone service, Van Orman and Trotter were able to stay alive by shooting grouse with a shotgun found in the cabin.

Meanwhile, the wilderness of the northern United States and Canada was being scoured by searchers intent on rescuing the missing U.S. aeronauts and the Poles, Hynek and Burzynsky, who had also flown into oblivion. Finally, two full weeks after the take-off, a six-foot Canadian lineman burst through the door of the shack where Van Orman and Trotter were regaining their strength. His first words were, "What the hell are you fellows doing here?"[77]

Having suffered the tortures of "high water and a touch of hell," as Van Orman put it, the American did not even have the satisfaction of winning the race. Hynek and Burzynsky had trudged even farther, 90 miles, out of the Canadian wilds. So ended the 1933 James Gordon Bennett race: Hynek and Burzynsky in first place, Settle and Kendall in second, Van Orman and Trotter in third.[78]

The old guard of American sport ballooning had passed from the scene. Van Orman, like Ralph Upson before him, would spend the rest of his life as a working aeronautical engineer. He was a pioneer in air navigation, aeronautical radio technology, puncture-proof gasoline tanks, rescue equipment, and pressure suits. Honeywell, Berry, and others had retired from the field because of age. Kepner and Settle had turned their attention from balloon racing to the development of balloons designed to penetrate the substratosphere.

As a result, less experienced aeronauts rather suddenly inherited leadership in American sport ballooning. The 1934 Nationals, for example, featured a very small field of only five balloons. The competition was won by two naval officers, Lt. Charles H. Kendall and Lt. Howard T. Orville, who flew a U.S. Navy balloon 206.4 miles from Birmingham, Alabama, to Commerce, Georgia. The *Goodyear VII*, *Buffalo-Courier-Express*, Army No. 1, and Army No. 2 were all forced down within less than 180 miles. That year's Gordon Bennett competition, held in Warsaw, Poland, on September 23, 1934, was won by the Polish team of Hynek and Pamaski. George Hineman and Milford Vanik, flying the *Buffalo-Courier-Express*, finished only eleventh, while Kendall and Orville came in twelfth in a field of sixteen.[79]

Nor did 1935 bring any improvement for the United States. Once again, the Gordon Bennett was staged in Warsaw, on October 16, 1935. The only American competitors, Lt. Raymond Tyler and Lt. H. T. Orville (USN), finished dead last. Zebigniew Burzynsky, Hynek's original partner, flew to victory.[80] That year marked the end of U.S. participation in the James Gordon Bennett Cup competition. The race continued for three more years, 1936–1938. Two of these victories went to Belgium and one (1938) to Poland. The 1939 race, scheduled for Warsaw, was cancelled as a result of the invasion of Poland.

Sport ballooning was dying in the United States after 1935. The National Balloon Races continued for one more year, but the winners no longer represented their nation in the international championship.[81] The foundation of the Cleveland Balloon Club did draw a few new figures into the sport. The older clubs had long since disappeared. Most of these groups had failed to reestablish themselves after the vicissitudes of the First World War. Since that time, most sport balloonists had been either wealthy private individuals, officers in the military services, or aeronauts flying under the sponsorship of a major corporation like Goodyear, a newspaper, or a city government like Akron. In essence, the Cleveland Balloon Club was the last gasp of sport gas ballooning in the United States.

The Cleveland group was the result of the efforts of Milford F. Vanik and George Hineman, each of whom owned a 35,000-cubic-foot balloon in which they had been flying for several years. These two men interested a number of friends in the sport, including George Vanik, Tony and Edmund Fairbanks, John E. Copeland, J. D. Hartishorne, and H. K. Damels. Together these eight men organized the Cleveland Balloon Club on January 19, 1932.

The club immediately purchased a vintage 80,000-cubic-foot navy gasbag originally manufactured in 1920. The group had paid $200 to save the craft from its two previous owners, who had already begun to cut it in half to make two balloons.

On the occasion of the club's first flight, its new acquisition began to split at the equator seams. As the crew brought the balloon back to earth, the dragline whipped around some telephone wires and snapped four telephone poles. The result was a $150 bill from the phone company for repairs.

During the club's first "midnight flight," the repaired balloon was nearly rammed by a Beach Staggerwing. The six-man crew was then dunked into Lake Erie before continuing to a landing on the grounds of the Detroit Ladies Polo Club. By 1935 the Cleveland group had developed a small coterie of new pilots, the most active of whom was Anthony Fairbanks. Milford Vanik and Fairbanks flew the club balloon *Great Lakes Exposition* to second place in the 1936 Nationals from Denver on July 3.[82] (See page 589.)

The 1936 Nationals marked the end of the sport gas balloon era in the United States. Random flights continued to be made. In August 1936, for exam-

Entries in the 1936 National Balloon Race
(in order of starting)

Pilot	Aide	Sponsor	Comments
1. Capt. Haynie McCormick	Capt. John A. Tarro	U.S. Army	Fifth Place
2. Lt. Comdr. Francis H. Gilmer	Reginald H. Wald	U.S. Navy	Third Place 64 miles
3. Frank Trotter	V. L. Smith	Goodyear Tire & Rubber Co.	First place 115 miles, landing at Sterling, Colorado
4. Lt. R. F. Taylor	Lt. F. F. Flaherty	U.S. Navy	Fourth place 55 miles
5. Milford Vanik	Anthony Fairbanks	Cleveland Balloon Club	Second place 100 miles, landing at Akron, Colorado
6. R. S. Cunningham		Detroit Balloon Club	No flight; balloon ripped by high winds prior to launch

ple, a "grudge match" pitted the first- and second-place winners in that year's Nationals against one another in a flight from Cleveland Municipal Stadium. Goodyear's Frank Trotter was victorious for the second time.

Of course, ballooning continued in the army and navy, both as a means of training blimp pilots and as part of the now all-important effort to investigate conditions at very high altitudes where combat air crews would soon be operating. But there would never be another National Elimination Race nor another American entry in the Gordon Bennett. When Tony Fairbanks expressed an interest in entering the 1939 race, Charles Logsdon of the NAA informed him of a War Department decision "that in view of the unsettled conditions in Europe . . . a balloon team from the United States should not venture to Poland this year."[83]

It was a wise decision. On September 1, 1939, the Wehrmacht crossed the Polish border. This action was not only the first step in a conflagration that would soon engulf the entire world, but also brought to an end one of the most colorful eras in the history of the balloon.

Chapter 19 Destination Stratosphere

LATE ON THE AFTERNOON OF NOVEMBER 4, 1927, JACK FISHER, A BLACK field hand working a patch of ground near Orlindo, Tennessee, was startled to see a small package drifting to earth on a parachute. Running over to investigate, he found a set of radio batteries bearing a tag that requested the finder to contact Capt. Hawthorne C. Gray at Scott Field, Belleville, Illinois.[1]

Shortly thereafter, a Tennessee farmer saw a large balloon pass low over his head. The man called out, but there was no response as the balloon drifted off toward the White County line. Will Gordon was the next man to spot the balloon some three miles from the little town of Doyle. It was now so low that Gordon remarked he could have caught the line dangling from the basket.[2] Then it was gone, up and over a neighboring ridge. It was dark now, and none of these Tennessee farm folk chose to go scouring the countryside in search of the visitor from the sky.

Early the next morning, however, a curious youngster found the balloon in a wood near Sparta, Tennessee. The basket was held fast in the upper branches of one tree, while the envelope was draped over the top of another. Like Gordon, the boy called up to the basket, then scrambled up the tree to peer over the edge of the wicker car. A body was curled up in the bottom of the basket. Clad in a fleece-lined flying suit, his face hidden behind an oxygen mask, a parachute strapped on his back, Capt. Hawthorne Charles Gray, United States Army Air Corps, had returned to earth. He had flown higher than any other human being in history, but he had not lived to tell the tale.

Hawthorne Gray had ventured into the stratosphere in a pioneering attempt to solve a number of pressing technical and scientific problems. Since the turn of the century, engineers, scientists, and the general public had become ever more interested in conditions in the upper atmosphere. There were suddenly a great many reasons to attain extreme altitudes.

There were, for example, urgent engineering problems to be solved. The development of the turbosupercharger after 1920 made it possible to operate aircraft engines in the thin, frigid air of the substratosphere for the first time. The small band of U.S. Air Service test pilots at McCook Field, Dayton, Ohio, battled for the title "Icicle King" as they coaxed their supercharger-equipped Packard Le Père biplanes to altitudes approaching 40,000 feet. Breathing oxygen through a pipestem, operating their open cockpit aircraft without benefit of pressure garments or electrically heated flying clothing, these pilots returned to earth semi-conscious, bearing tales of the incredibly difficult physical conditions encountered nearly eight miles above the surface of the earth. A great many questions would have to be answered, a great many problems solved, before military or commercial flying operations could be conducted in the inhospitable region of the substratosphere.

As pilots and engineers were beginning to ponder the difficulties of high-altitude flight, the attention of the scientific community was also focusing on the upper layers of the atmosphere. The electrical properties of the upper atmosphere had been a subject of discussion during the late nineteenth century. With the advent of radio, the Kennelly-Heaviside and Appleton layers (ionosphere) were identified and their importance in wireless communication noted.

The discovery of cosmic rays in 1912 was even more intriguing to physicists. Victor F. Hess had conducted the first serious electroscope studies from the basket of a free balloon on August 7, 1912. The electroscope, an instrument used to detect and measure radiation, indicated, in Hess's words, "that a radiation of very great penetrating power enters our atmosphere from above."[3]

As physicist Bruno Rossi has noted, this discovery "was the beginning of one of the most extraordinary adventures in the history of science":

> *Subsequent investigations of the "radiation from above" opened up the new and bewildering world of high-energy physics. Here scientists found particles of subatomic dimensions, with energies thousands, millions, billions, trillions of times greater than the energy of particles emitted by radioactive materials found on earth. Here for the first time they witnessed processes in which particles of matter are created out of energy, and then promptly disappear in giving birth to other particles. Beyond this, Hess's discovery revealed new vistas in astrophysics and cosmology. The mysterious radiation was found to carry important messages concerning the physical conditions of the distant regions of space through which it had traveled on its way to the earth. And finally, in an effort to explain the origin of the radiation, physicists developed a number of novel ideas about the nature of the events that take place in stars and in the masses of dilute gas that fill interstellar space.*[4]

Thus, by 1920, the upper reaches of the atmosphere seemed to hold the answers to some very intriguing questions. But how were experiments to be conducted at the roof of the ocean of air? The test pilots of the day who ventured much over 30,000 feet had to devote their complete attention to maintaining some control over an airplane that was wallowing through thin air in which it had never been intended to operate.

But the balloon was another matter. A large aerostat, properly designed, would rise far beyond the altitude of the most high-flying airplane of the day. Moreover, the aeronaut, while obviously concerned with conserving sufficient ballast and gas for a safe descent, would be relatively free to tend his scientific instruments and make any required observations.

Balloonists had blazed a trail into the upper air during the nineteenth century. As early as 1804, the French scientist Joseph Louis Gay-Lussac reached an altitude of 23,000 feet during the course of a scientific flight that had begun from Paris. He was followed by Pascal Andreoli and Carlo Brioschi, Italian meteorologists, who rose to 20,000 feet aboard a hot air balloon in 1808.

This record stood for over half a century, until September 5, 1862, when Henry Coxwell and James Glaisher almost died at an altitude of just under 30,000 feet. While Glaisher lay in a helpless stupor against the side of the basket, Coxwell climbed onto the load ring to free a tangled valve line. The two men barely survived their adventure and were widely hailed as exemplars of courage in the pursuit of scientific knowledge, but their experience was an object lesson to those who sought to probe the upper atmosphere.

Théodore Sivel and Joseph Crocé-Spinelli ignored the lesson, and paid with their lives. Equipped with an experimental breathing device consisting of oxygen-filled bladders and simple mouthpieces, these two men, accompanied by the great French aeronaut Gaston Tissandier, ascended aboard the balloon *Zenith* to approximately 28,000 feet in April 1875. All three men blacked out. When Tissandier regained consciousness, he found his companions dead.

The *Zenith* tragedy marked the end of the "heroic age" of nineteenth-century high-altitude ballooning. Even so, the German meteorologist Arthur Berson ascended to 30,000 feet in 1894. Seven years later, Berson flew to an all-time record of 35,424 feet with companion Reinhard Süring in 1901. By this time, however, unmanned, instrumented sounding balloons were returning meteorological data that could earlier be obtained only by means of dangerous manned ascents.

The first meteorological balloon sondes, or "registering balloons," were flown in France in 1892. Relatively large craft of several thousand cubic feet, these balloons carried registered instruments that provided a constant record of altitude and temperature. The balloons were furnished with standard open appendixes, so that they descended very slowly through the normal loss of gas. As a result, the aerostats were commonly retrieved at distances of up to 700 miles from the launch point.

The German meteorologist Assmann solved this difficulty through the introduction of closed rubber balloons designed to burst at altitude, dropping the instruments to earth by parachute much closer to the take-off site. These balloons offered the additional advantage of fairly constant speeds of ascent and descent, which helped to insure more accurate temperature readings.

A. Lawrence Rotch, director of the Blue Hill Meteorological Observatory in Massachusetts, introduced the Assmann balloons to America during the 1904 Louisiana Purchase Exposition in St. Louis. Between 1904 and 1908 the Blue Hill staff supervised the launch and retrieval of some eighty instrumented balloons. Rotch reported that a typical flight lasted two to three hours, during which the balloon would reach an altitude of eight to ten miles. The recovery rate was good. Fifty-three of the fifty-six balloons launched in 1908 were recovered with good instrument records.[5]

The use of sounding balloons was not restricted to meteorology. Charles Greeley Abbot, director of the Smithsonian Astrophysical Observatory, pressed the small aerostats into service carrying specially designed pyrheliometers into the upper atmosphere. Abbot was attempting to study the nature of solar energy and its impact on the earth. Because of the extent to which portions of the spectrum were absorbed during passage through the earth's atmosphere, it was important to obtain readings from as high an altitude as possible.

The first of the instruments was launched aboard a standard balloon from a Weather Bureau station at Omaha, Nebraska, in July 1914. Abbot claimed that the scientific package, which was retrieved in Iowa, attained an altitude of 15 miles. Later Smithsonian balloons reached 20-mile altitudes.

In addition to the data on solar energy, Abbot's instruments provided an accurate record of temperature variation with altitude. The scientist noted a strange anomaly in this temperature data: "The reader will notice how curiously the temperatures alter with altitude. It would have seemed more natural to find them continually falling with increasing altitude, till at last they fell to nearly absolute zero. On the contrary, the temperature of the atmosphere ceases to fall at about 7 to 10 miles of elevation, and sometimes is found even to rise a few degrees as the recording apparatus goes higher."[6]

Charles Abbot referred to this mysterious region of static temperature as the isothermal layer. In fact, the instrumented balloons had entered the stratosphere, an atmospheric region first identified by the French meteorologist Teisserenc de Bort on the basis of his own experiments with sounding balloons in 1899.

For the newspaper-reading public of the decade 1926–1936, the stratosphere was far more than a layer of sky that began roughly ten miles above the surface of the earth at the equator. The word stratosphere conjured up images similar to those that "darkest Africa" had evoked in the nineteenth century or the names Arctic and Antarctic in the early twentieth. Like an unexplored continent, it was a region to be visited by only the most intrepid explorers, members of well-financed and well-equipped expeditions. In fact, no one took a greater interest in the exploration of the stratosphere than the National Geographic Society, the same organization that had supported Robert Peary's first trip to the North Pole. As in the case of geographic exploration, international rivalries were played out in the stratosphere, with the United States and the Soviet Union trading the world's absolute altitude record back and forth while the world press urged them on.

More than any other figure, Capt. Hawthorne Gray ushered in the age of the great stratosphere expeditions. A native of Pasco, Washington, born on February 16, 1889, Gray was the son of Capt. William P. Gray, a well-known steamboat skipper, and the grandson of a member of Marcus Whitman's mission to the Indians of the Northwest. After graduating from the University of Idaho, he enlisted as an infantry private for service in the Mexican Punitive Expedition of 1916, and was commissioned a second lieutenant on June 2, 1917. Gray was transferred to the Army Air Service with the rank of captain in 1920. By 1924 he had graduated from the balloon school at Ross Field, the airship school at Scott Field, and the primary flying school at Brooks Field. Assigned as engineering officer to Scott Field in Belleville, Illinois, Gray quickly became a key figure in army lighter-than-air circles. His second-place finish in the 1926 Gordon Bennett race identified him in the public mind as one of the best-known of the nation's military balloonists.

Hawthorne Gray's career as a stratosphere balloonist began with a flight to 28,510 feet from Scott Field on March 9, 1927. As on his subsequent high-altitude ascents, science played a small role in the planning for the first flight. It was an outright attempt to capture the world's absolute altitude record. While Gray's three ventures into the upper atmosphere did offer an opportunity to test high-altitude clothing, oxygen systems, instruments, and other items of equipment required for aircraft operations in the substratosphere, Air Corps publicity releases emphasized the fact that this was an attempt to set a record.

Gray's instruments (a statoscope, a 50,000-foot altimeter, a thermograph, and a rate of climb indicator) were flight oriented and not designed to provide scientific information. There was little doubt that the most important instruments carried aboard the balloon were twin barographs sealed by officials of the NAA. The record of altitude that appeared on the strips of paper inside these instruments would serve as the basis for official recognition of Gray's position as the world's highest man.

The balloon employed, Army No. S-30-241, was a single-ply rubberized silk envelope coated with aluminum paint. The 70,000-cubic-foot bag, loaded with supplies and instruments, weighed 726 pounds. The ballast consisted of 4,520 pounds of sand packed in bags suspended from a special rack on the load ring. In an effort to make one final push to an absolute ceiling, Gray proposed to discard the useless weight of the rack once all the sand had been dropped. Ross Asbill, foreman of the fabric shop at Scott Field, had sewn special small parachutes so that Gray could drop heavy objects, including the ballast rack and empty oxygen bottles, without endangering those below.

Gray's oxygen system consisted of three 200-inch tanks mounted high on the side of the basket so that they could be shoved overboard with less effort. A 100-inch tank was held in reserve for use as a jump bottle in the event the aeronaut was forced to abandon his balloon at altitude. The three primary oxygen cylinders fed through two pressure valves. Gray received oxygen through a rather primitive oxygen mask covering his mouth.

When Gray emerged from a Scott Field hangar on the afternoon of March 9, 1927, the 183-pound aeronaut, clad in 57 pounds of clothing, appeared, according to one observer, like "a large brown bear preparing for winter hibernation."[7] Gray himself remarked: "Over the heaviest of woolen underwear I wore two woolen shirts, a sweater and a winter uniform. On top of that was a flying suit, leather on the outside, reindeer fawn skin on the inside, and two thicknesses of heavy woolen blanket cloth between."[8] His boots were fleece-lined, high-topped leather moccasins laced at the front and back. A leather helmet, upper face cover, mouth mask, goggles, and mittens completed his outfit.

Gray's first record attempt began promptly 1:18 P.M., when, after checking the instruments, the distribution of ballast, and his oxygen system, the aeronaut tuned in a jazz melody on his seven-tube Atwater-Kent radio and gave the command, "O.K., let her go." One chase plane followed Gray aloft, while two Signal Corps theodolites established on a three-mile north-south baseline tracked Gray from the time of take-off until he disappeared from view moments before landing.

The aeronaut's problems began soon after lift-off. He had planned a leisurely ascent that would permit time to check radio reception and conduct instrument observations. In fact, he had time to do little more than heft the individual ballast bags into the car, slit each one with a knife, and dump the contents over the side. It was an exhausting task that required Gray to dispose of 2¼ tons of ballast by hand. As he passed through 25,000 feet, the oxygen system was also proving less than satisfactory, pouring out a stream of frigid gas painful to breathe.

Gray passed out at 27,000 feet. The balloon continued to climb to 28,510 feet, then began to drop rapidly. The aeronaut regained consciousness at 17,000 feet, only to discover that he was falling at the frightening rate of 1,200 feet per minute. He immediately resumed dumping ballast, abandoning each of his three knives as they became dull, ripping the seams out of the bags by hand. Gray slowed his descent to 600 feet per minute, finally tearing through a set of telephone lines to make a hard landing near Ashley, Illinois, some 40 miles southwest of Scott Field.

The chase plane landed and returned Gray to the post hospital, where he made a rapid recovery. While the aeronaut had set a new U.S. altitude mark, he fell far short of the 35,424-foot record set by Süring and Berson. He had, nevertheless, created much favorable publicity for the Army Air Corps. More important, he could now provide some sound advice on the subject of flying clothing, oxygen systems, and other items of equipment for high-altitude flights.[9]

His helmet, for example, had been almost completely unsatisfactory. In the future, he planned to adapt a gas mask to his needs. Such a mask would provide a tighter seal on his face to conserve oxygen and make breathing easier. Oxygen would pass from the cylinder through an asbestos-covered cannister where the gas was heated by electrical coils operated by a two-volt battery. The oxygen exited the cannister, which was strapped on the aeronaut's chest, and entered the mask through flat pipes located near the inside of the goggles to prevent fogging at altitude.

Thus equipped, Gray was ready for a second altitude attempt from Scott Field on May 4, 1927. The aeronaut had a very specific goal for this, his 107th ascent. He intended not only to beat Süring and Berson's balloon altitude record but to climb above the absolute world mark of 40,820 feet set by French aviator Jean Callizo the previous year.

Gray had no illusions about the dangers involved. He knew that somewhere between 40,000 and 50,000 feet the air pressure would be so reduced that his lungs would not function and gases would begin to bubble out of his blood. Gray would be venturing very close to the absolute ceiling, where a human being could no longer survive without artificial pressurization.

The flight began well. Gray rose rapidly toward his maximum altitude, anxious to clinch the record before oxygen shortage became a problem. Each item of equipment functioned properly. He was able to relax on the way up,

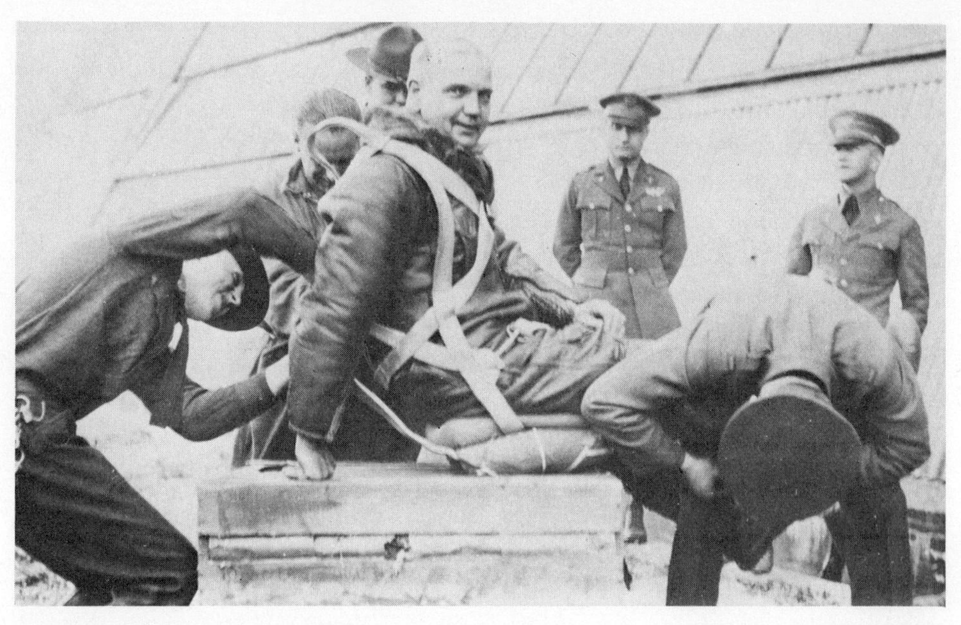

Capt. Hawthorne C. Gray prepares for his flight of November 4, 1927, Scott Field, Illinois.

listening to jazz tunes and watching his instruments. Within forty-five minutes he was approaching his ceiling.

> *At 42,000 feet, having been kept alive by compressed oxygen for the last four miles, I was listening to a jazz orchestra playing in St. Louis, the music coming in clear and loud on my radio, without a single trace of static. That was the only connecting link with the world I had left. Far below, cruising along the top of the cloud banks at 13,000 feet, two escort planes, one with a movie photographer aboard and the other with the post surgeon as passenger, hovered and watched me, though I could not pick them out of the mist. Below them the clouds covered the land, except for an occasional rift. Once through such a crevasse, I caught a magnificent view of the Mississippi and Missouri, tracing their winding course for miles and miles to the north and south.*[10]

The sky, Gray noted, "was magnificent in the depth of coloring, which was a deep, almost cobalt, blue."[11]

His sand ballast was gone at 40,000 feet. In order to climb higher, he parachuted the first of his twenty-five pound oxygen cylinders overboard. This

H. C. Gray, in helmet, just prior to launch, November 4, 1927

sent him up another few thousand feet, high enough to insure his hold on the world record.

> *At that height, though still distended, I knew the gas bag above contained less than one-eighth of the gas I had started with. As the balloon had climbed into lighter air and the pressure against it was removed, the gas had rushed out through the big appendix in the bottom, keeping the silvered fabric from bursting. So long as I stayed up, the balloon would be full, but once I started down the gas would begin to contract under the increasing air pressure, so that if I could keep all the gas I had, there still would be less than 10,000 cubic feet when I reached the ground.*[12]

After bouncing up and down for a few minutes at 42,240 feet, Gray valved gas to start back down. Low on gas, low on ballast, he faced the classic dilemma of the free balloonist. As his descent picked up speed, he began dumping everything disposable overboard. Keeping one eye on the statoscope, which indicated his rate of fall, Gray knew he was in trouble, and increased his attempts to lighten the load by parachuting equipment.

> *The parachutes were designed to fall at sixteen feet a second, the same rate as the large chutes used by flyers, but the bag was falling so much more rapidly that when I dropped things over the sides they appeared to fly straight up in the air. It was queer to see twenty-five pound steel bottles apparently flying upwards. Two more oxygen tanks, the storage battery used to run the electric heater in my oxygen mask, my radio batteries and loud speaker, and finally the wooden framework which supported the ballast bags, with all the empty bags still attached, were released to lighten the balloon.*[13]

Passing through 8,000 feet, Gray could see trees and marshlands rushing up to meet him. Unable to slow his drop below 1,800 feet per minute, twice the safe landing speed, he climbed onto the side of the basket, balanced himself with one hand on the load ring, and jumped. Drifting to earth beneath his parachute canopy, Gray could see S-30-241, free of his weight, rise at last, as the Air Service camera plane circled. He landed safely 110 miles from Belleville one hour and twenty minutes after take-off.

The precious barograph instruments were found undamaged with the balloon, ten miles from Gray's landing spot. They indicated that the flight had reached a crest at 42,240 feet. Ultimately, the effort to capture the altitude record failed however. In spite of all the arguments advanced by officials of the NAA, the FAI ruled that because Gray did not land with his balloon, the record would be disallowed.[14]

As before, the flight of May 4, 1927, provided important information on oxygen equipment, clothing, and instruments. Gray, however, was more determined than ever to capture the altitude title and immediately began planning a third flight.

Army Air Corps officials, anxious to avoid the charge of publicity seeking, issued statements to the effect that "Captain Gray's flight was not made with the intention of breaking any records but for the purpose of studying atmospheric conditions at high altitude."[15] Newspapers were ready to accept this position. The *New York Times* commented that meteorologists would be more interested in the results of Gray's effort than military officials. "It is," remarked the *Times*, "another case of the Army Air Service laying science under obligation."[16] Successful Air Corps propaganda notwithstanding, it was apparent that the third flight of Hawthorne Gray was nothing more than an all-out effort to set an official world altitude record.

The aeronaut lifted off on his third altitude flight from Scott Field on November 4, 1927. After bidding good-bye to his wife and son, Gray took off at 2:24 P.M., accompanied by four chase planes. Fifty-mile-per-hour winds were blowing at 1,600 feet that day, and much higher at Gray's altitude. The aircraft pilots, who had had little difficulty maintaining contact with the balloon on previous occasions, now lost sight of Gray when he flew into a cloud bank over McLeansboro, Illinois. The aircraft continued south, hoping to spot the balloon again, but were finally forced to scatter and land at Henderson and Madisonville, Kentucky, late that afternoon. Gray, they knew, had to be on the ground by this time, probably someplace farther south. At the time, however, they could only hope for the best and wait for Gray to send word of his location.

The young boy who found the mortal remains of Hawthorne Charles Gray in the trees near Sparta, Tennessee, early on the morning of November 5, was quick to inform Sheriff Hawk Templeton. Templeton arrived on the scene soon thereafter with a number of phone linemen to retrieve the balloon from the trees and an undertaker to take charge of the body. There was little doubt about who this was. Nashville papers had headlined the news of Gray's disappearance in the previous evening edition. Positive identification came from papers on the body. Noting the instruments and papers in the basket, Sheriff Templeton posted a guard at the crash site, then started back for Sparta with the body.

Postmaster S. C. Dodson wired the news to Washington with a request for instructions on where the body should be shipped. The news found its way onto the Associated Press wire as well. That afternoon two of the army chase planes were headed directly for Sparta, while the other two flew to Blackwood Field in Nashville to wait for orders. Only one aircraft, flown by Sgt. S. J. Sampson with Capt. C. J. Bryan as his observer, actually arrived on the scene. The other airplane bound for Sparta was forced down short of its destination, with no injury to the crew.[17]

Col. John Paegelow, a First World War balloon veteran commanding Scott Field, gradually took charge of the situation. Gray's body was sent on to Washington for burial at Arlington National Cemetery. A board of three, including a flight surgeon and two balloon officers, was appointed from the Scott Field staff to investigate the causes of the disaster.

On the basis of the instruments found in the basket and a hastily scribbled log that Gray had kept during the flight, the details of the tragedy came into focus. Station KNOX had been playing "Kashmiri" at take-off. The ascent was much more leisurely than on either previous flight. In March Gray had reached 28,510 feet in forty minutes. This time he required an hour to reach that point. In May he attained 42,240 feet in one hour and five minutes. He required almost two hours to reach the same ceiling on November 4.

Gray's slower pace is not difficult to explain. On both previous flights he had dumped ballast with wild abandon, attempting to rise as quickly as possible before the oxygen ran low. The result had been far too rapid a descent with no ballast left to slow his fall. It seems that Gray was now sufficiently comfortable with his oxygen apparatus that he was willing to take a chance on a very slow climb, which would permit him to conserve ballast for landing.

He was enjoying the radio during the leisurely ascent. KSD was coming in loud and clear at 12,000 feet. As Gray noted, he was beginning to feel "symptoms of rickets" as a result of the falling pressure. At 15,000 feet he turned the batteries on his electrical oxygen heater from two to four volts. The electrical heater for the lenses of a pair of experimental goggles was also switched on as the lenses began to fog. The thermometer read zero as the balloon passed through 19,000 feet at 3:05 P.M. Gray had switched channels now and was enjoying a saxophone solo of "Traumeri." At 3:10 P.M. he was at 23,000 feet, listening to "Thinking of You" on KNOX. Five minutes and a thousand feet later, "Just Another Day Wasted Away" was spilling out into that cobalt blue sky from WLW in Cincinnati. It was even colder, 25 degrees below, at 3:17. The Pied Piper was entertaining over Chicago's WLS.

Gray's fate may well have been sealed with an entry penciled into the log at 3:17: "Clock frozen." Now he was literally reduced to guessing his time in the air and the amount of oxygen remaining in the tanks.

Gray's first tank was exhausted before he reached 34,000 feet. He opened the cock on the second tank, closed the valve on the junction box of the empty cylinder, cut the hoses and lashings, and hefted it overboard to float to earth. As the cylinder tumbled free, it struck the long trailing radio antenna, ripping it from the basket. Gray was absolutely alone now. "No more music."

The journal entries continued in a shakier hand now. Gray was in the stratosphere, and the temperature was rising a bit and stabilizing. At 39,000 feet, the bag was fully distended. A thousand feet higher, the sand ballast completely gone, Gray noted that the sky was a very deep blue and the sun was bright.[18]

This was the last entry in the log, but the barograph continued to run, providing an exact record of the balloon's movements. The craft was oscillating between 42,200 and 42,100 feet. At about 4:05 there was a sudden jump up to the peak altitude, 42,470 feet. We can assume that this jump represents the point at which Gray dropped the radio batteries that Jack Fisher, the field hand, saw descending to earth a few minutes later.

Gray was obviously still alive. He valved gas on the next down oscillation, and was on his way back to earth at last. The descent was slow and steady until he reached 39,000 feet at 4:28 P.M. Now the balloon started down much more rapidly. Gray, probably realizing for the first time that he was running low on oxygen, had valved gas again. The aeronaut's oxygen should, according to his own calculations, have run out at 4:38 P.M., at which time he was still descending through air far too thin to breathe. Gray must have fallen unconscious by this time, for the barograph track shows a slow, steady descent toward a landing in that grove of trees near Sparta, Tennessee, at 5:20 P.M.

The Scott Field board of inquiry had few doubts about the ultimate cause of Gray's death. "There is one thought—if the clock had only not stopped."[19] Gray's determination to achieve the record explained the slow ascent. The aeronaut's confidence in his own experience with the oxygen system led him to continue his climb after the failure of the clock. When he realized that his oxygen was running out, he was either unable to jump or unwilling to do so, knowing that it would cost him the record. Gray's determination to gain official recognition as the world's highest man had ultimately cost him his life.

The trappings of a hero were heaped on Gray. After burial in Arlington, he was posthumously awarded the Distinguished Flying Cross "for heroism while participating in aerial flights." "His courage," noted the citation, "was greater than his supply of oxygen."[20] In 1938 the Army Air Corps named a facility at Fort Lewis, Washington, in his honor. None of this would be of much help to Gray's widow and three sons, who received his Air Corps insurance and $2,700, the equivalent of six months' pay.[21]

Hawthorne Gray's death underscored the central problem facing high-altitude balloonists. To fly above 40,000 feet without a pressure garment or the protection of a pressurized vessel was to invite death. It was a problem that attracted the attention of would-be stratosphere balloonists in the United States and Europe.

Tex Settle was one of the first to suggest the construction of a fully enclosed, pressurized gondola as a means of venturing to very high altitudes. Thomas Greenhow Williams Settle was a native of Washington, D.C., born on January 4, 1895. Graduating second in his class from the Naval Academy in 1918, Tex Settle served on cruisers and destroyers before completing postgraduate work at Annapolis in aviation communications engineering in 1924. He then served in a variety of posts aboard the dirigibles *Shenandoah* and *Los Angeles*, earning a reputation as one of the rising leaders of the navy lighter-than-air contingent.

As noted earlier, Settle was also the navy's ace balloon racer. He was a member of the navy team in the Nationals from 1927 until 1930. He won the Nationals in 1929, placed third in the Gordon Bennett race that year, and won that race in 1932. From 1925 to 1929 he was an instructor in lighter-than-air flight at Lakehurst, New Jersey, then reported to Akron as inspector of naval aircraft with the Goodyear-Zeppelin Corporation, where he remained during the construction of the *Akron*.

While serving at Lakehurst in the mid-1920s, Settle had studied Hawthorne Gray's experience and concluded that similar high-altitude flights would require a sealed gondola. Working with C. P. Burgess of the Navy Bureau of Aeronautics, he prepared what may well have been the first design for such an altitude pressure vessel. Nicknamed the "flying coffin," the gondola would have stood seven feet tall with rounded ends. The outside width would have been three feet, leaving just enough room for one man, his life-support system, instruments, and flight controls.

Settle and Burgess interested Rear Adm. William A. Moffett, chief of the Bureau of Aeronautics, in their project, and received initial approval to have the flying coffin constructed at the Naval Aircraft Factory in Philadelphia. Settle's gondola finally fell victim to the Schneider Trophy however. With congressional budget analysts uncomfortable over the cost of developing the special high-speed racing aircraft required for the great float-plane competition, navy officials bowed to pressure and canceled a number of "unconventional projects," including the flying coffin.[22]

Tex Settle was not the only would-be stratosphere balloonist thinking in terms of a sealed gondola. Auguste Piccard, a professor of physics at the University of Brussels, was the man most responsible for the introduction of the pressurized balloon gondola.

Piccard was one of Europe's leading cosmic ray investigators. Knowing that the mysterious radiation originated beyond the earth's atmosphere, Piccard believed it was clear, as did other physicists, that the phenomena could best be studied above the atmosphere, which absorbed a portion of the radiation. Piccard considered the possibility of using rockets or an airplane to lift his instruments into the air, but the choice of the balloon was obvious from the beginning. Able to fly higher and stay longer at altitude, the balloon would also provide a stable instrument platform free from vibration or electrical interference of engines.

Piccard also studied the possibility of developing a pressure suit for the scientist-aeronaut who would fly the balloon. Like Settle, however, he decided that a sealed gondola would be safer and provide a better environment for men and instruments.

This first high-altitude gondola represented the solution of a number of engineering problems. A hand-wrought, hand-welded aluminum sphere measuring 82 inches in diameter with an empty weight of 300 pounds, the first Piccard gondola was designed to keep two human beings alive at altitudes in excess of 40,000 feet for up to ten hours. Breathable air was provided by a Draeger apparatus placed under a table in the small cabin. Developed for use aboard German submarines during the First World War, the apparatus released 2 quarts of pure oxygen into the cabin every minute while at the same time scrubbing and recirculating 20 gallons of cabin air by filtering it through alkalai.

The gondola, which was built at Liège, was fitted with two large portholes for entry and exit and eight small observation ports. In order to deal with the problems of solar heating, Piccard painted half of the sphere black, left the other side bare metal, and installed an electrically driven propeller to turn either the absorbing or reflecting surface toward the sun to control the internal temperature.

With the design in hand, Piccard approached the National Fund for Scientific Research (FNRS), which granted £3,000 for the project. Piccard then contracted with the Riedinger firm of Augsburg, Germany, for construction of the balloon. The 500,000-cubic-foot gasbag was constructed of rubberized cotton, heavier in the upper hemisphere than in the lower. Fully inflated at take-off, the balloon would have been capable of lifting a small locomotive. Piccard, however, planned to launch with only 100,000 cubic feet of gas, which would expand to inflate the envelope fully at altitude with no loss of hydrogen.

After several false starts, Piccard and an assistant, Paul Kipfer, made the first ascent aboard *FNRS* from Augsburg on May 27, 1931. Soon after lift-off, Kipfer discovered their precious oxygen escaping through a hole in the gondola. They were able to close the hole with a mixture of petroleum jelly and oakum, only to discover the temperature rising rapidly within the gondola. As the heat built up within it, the two scientists stripped off most of their clothes and hoped for the best.

They reached their maximum altitude of 51,775 feet (15,781 meters) barely a half hour into the flight, but were unable to descend because of a fouled valve line. There was little choice but to wait for the gas to cool and the balloon to descend as the sun set. They finally touched down on a glacier in the Bavarian Alps at 9:30 P.M. that evening and walked to safety the next day. The original pressurized gondola remained on the ice until recovered for exhibition in a museum nearly a year later. Piccard ascended with Max Cosyns to 53,152 feet (16,000 meters) the following year, and later, beginning in 1937, applied the lessons learned in the construction of balloon gondolas to the design of the first deep-sea bathyscaphe.[23]

Piccard's flight was celebrated in newspapers around the globe. The man himself became one of the great public figures of twentieth-century science. Physically, he personified the popular image of the scientist. Tall, thin, with a wild fringe of fine hair surrounding a balding pate, Piccard was lionized when he came to the United States on a lecture tour in 1933. From the moment of his first talk at the inaugural meeting of the Institute of Aeronautical Sciences in New York, reams of newsprint were devoted to his tour.

The success of Piccard's American visit inspired the sponsors of the Century of Progress, a world's fair being planned to celebrate Chicago's 100th birthday, to stage an American stratosphere flight from the fairgrounds. Auguste Piccard was interested in the proposal, and the Chicago *Daily News* and the National Broadcasting Company were willing to underwrite the venture. Two American Nobel laureates, Arthur H. Compton of the University of Chicago and Robert Millikan of the California Institute of Technology, offered enthusiastic support and were anxious to provide the cosmic ray experiments that were, at least officially, the raison d'être for the flight. The Union Carbide and Carbon Company would donate hydrogen, while the Dow Chemical Company agreed to construct the gondola at no cost. Goodyear-Zeppelin would design and build the envelope at cost. In a few short weeks, the first American stratosphere exhibition was well on its way.

The press of business in Europe forced Auguste Piccard to withdraw from the flight almost at the outset. As a substitute leader, he suggested his twin brother, Jean Félix Piccard, a chemist, who was then employed as head of organic research for the Hercules Powder Company in Wilmington, Delaware.

Jean Piccard had graduated from the Swiss Institute of Technology in 1907 with a degree in chemical engineering. Further studies in Zurich led to a doctorate in organic chemistry and postgraduate work in Munich. He served in the lighter-than-air unit of the Swiss Army before accepting a position at the University of Chicago in 1916. While teaching in Chicago he met and married Jeannette Ridlon in 1919.

A graduate of Bryn Mawr with a University of Chicago masters in organic chemistry, Jeannette Piccard became her husband's scientific partner and collaborator. The pair spent the years 1919–1926 on the faculty of the University of Lausanne, then returned to the United States when Piccard accepted a post as director of research at MIT. At the time of twin brother Auguste's two flights, Jean was serving as head of research in organic chemistry for the Hercules Powder Company.[24]

Jean eagerly accepted Auguste's suggestion that he take charge of the Chicago venture. He immediately began work on the design of the gondola that

Dow had agreed to construct and paid his first visit to Akron to discuss the envelope with Goodyear-Zeppelin engineers.

The gondola, of course, was the central problem. Auguste Piccard had originally considered steel, duralumin, and aluminum for the 1931–1933 gondolas. Steel walls could be constructed thinner than duralumin to obtain similar tensile strength, but steel was ultimately rejected because the capsule walls would have to be so thin to conserve weight that the gondola might be dented by an accidental blow on the ground or by the weight of a man standing on it. Welding was required to guarantee an airtight sphere, and duralumin was difficult to weld. Great advances had been made in the technique of welding aluminum, however, as a result of the brewing industry's requirement for large aluminum beer vats. Piccard, therefore, had selected a soft, relatively pure aluminum.

In designing the first American stratosphere gondola, Jean and Auguste Piccard were interested in taking advantage of recent metallurgical advances in the United States. Dowmetal seemed to be especially promising. The various Dowmetal alloys, produced by the Dow Chemical Company, contained 95 percent magnesium, but had the same advantages over pure magnesium that duralumin had over aluminum. Unlike duralumin, however, Dowmetal could be welded. The decision was sealed when Dow agreed to provide the gondola free of charge for its publicity value. The finished shell, 7 feet in diameter with walls 3½ millimeters (.138 inches) thick, weighed only 192 pounds.[25] Jean Piccard was more than satisfied. "Enclosed in such a shell," he remarked, "one does not need much courage to trust his life to the perils of the stratosphere."[26]

Meanwhile work was moving forward on an envelope for the *Century of Progress* at Goodyear-Zeppelin in Akron. Auguste and Jean Piccard paid their first visit to Goodyear on March 8, 1933. They were secretive, admitting to newsmen only that they were considering a flight and were discussing the venture with rubber company officials. Goodyear-Zeppelin agreed to build a 600,000-cubic-foot balloon at cost.

With Auguste Piccard withdrawing from the venture, the Chicago sponsors were left with the problem of selecting a pilot to fly with Jean Piccard, who was not a licensed balloonist. As Jeannette Piccard noted many years later, "this arrangement was fine with Jean, for as long as he was commander of the flight, who was pilot was immaterial to him."[27] Ward Van Orman, whom Piccard had met during the recent visit to Akron, was apparently first choice. Van Orman refused. At best, he regarded the Piccard brothers with amused tolerance. They exhibited far too many eccentricities for the serious-minded engineer. "My own personal dislike for . . . Piccard started when it came to my attention that he wore inverted chicken baskets as helmets," Van Orman commented. "The day he visited the Goodyear plant he bounced around all over the lot and insisted on drinking from the laboratory spigots instead of the drinking fountains in the hall."[28]

These were, Van Orman admitted, minor eccentricities, but, facing up to ten hours in the cramped quarters of the gondola, "nothing is trifling."[29] In addition to his personal dissatisfaction with Piccard as a balloon companion, Van Orman was not at all certain that the cosmic ray investigations were worth the danger involved. "At the time," he commented, "I was asked quite often if I considered the adventure of questionable scientific significance. I didn't like the word questionable. It always was a matter of balancing benefits to be received against the risk involved. I never questioned the benefit gained from stratosphere flights. It was entirely a question of value in terms of knowledge gained versus human life risked. . . . I considered the price too high."[30]

Van Orman suggested Tex Settle as an alternate. Settle, who had already given a great deal of thought to stratosphere ballooning, leaped at the chance of a seat aboard the *Century of Progress*. "I'm damned enthusiastic about it," he informed newsmen.[31] Van Orman, one of Settle's best friends, watched this performance in wonder. "It was the only incredibly stupid remark I ever can recall him making to anybody," Van Orman noted, "for Settle was an uncommonly brilliant fellow."[32]

After observing the Piccards for a time and listening to Van Orman's advice, Settle finally decided to insist on making a solo flight. Van Orman, concerned about Settle's safety, voiced his objections to a solo ascent to the Chicago sponsors of the flight. Settle, misreading his friend's motives, offered to withdraw if Van Orman wished to fly. Van Orman immediately withdrew all of his objections and wished his friend the best of luck. Tex Settle would fly alone.

Jean and Jeannette Piccard were a bit perplexed, but continued their efforts to make the flight a success. Even Van Orman was forced to admit that Jean "made himself tremendously useful in the days leading up to ascent, helping the rest of us in every capacity." He was, the engineer commented, "a real sportsman."[33]

In fact, the decision to allow Settle to fly solo was based largely on rapidly developing events in the Soviet Union and in Belgium, where two rival stratosphere expeditions were being fitted out. When Settle insisted on flying alone, the Chicago backers asked Jean Piccard if he could guarantee to set a new world altitude mark higher than that likely to be attained by their new competitors. Piccard replied that this was unlikely. "With all the scientific equipment and two men, I cannot guarantee it."[34] For those financing the flight, the decision between a record and sound science was not a difficult one to make. Settle would fly alone and push hard for maximum altitude.

The Chicago investors were quite right in assessing the international situation. The stratosphere was about to become the arena where nations would vie for the prestige and honor of holding the world's absolute altitude record. It was a competition made to order to capture the public fancy: brave men risking life and limb ten miles above the earth's surface to uphold national honor. This was the stuff of headlines.

W. G. Quissenberry, a United Press correspondent, caught the spirit of the moment. "Six world powers have begun a race for supremacy in the stratosphere—that freezing region ten miles above the earth where airplanes can attain the speed of bullets."[35] News reports spoke of rockets and remotely controlled aircraft traveling through the stratosphere to deliver death to an enemy a continent away. No longer solely the domain of science, the stratosphere had become the "high ground" of which military strategists so often spoke. The race was on to explore it.

For the moment, the Soviet Union and Belgium seemed to have the best chance to beat the United States to a record altitude. In Belgium Max Cosyns announced plans to return to the stratosphere aboard the *FNRS*. Ernest Demuyter would ascend with his famed *Belgica* tied to the larger balloon, helping to control the ascent and to conserve ballast. He would cut loose at 30,000 feet to allow the *FNRS* to enter the stratosphere alone.[36] These plans did not reach immediate fruition. The Russians proved to be more persistent and successful contenders.

Throughout the early 1930s the leaders of the Soviet Union attempted to use aviation as a means of establishing their position in the world pecking order. Stalin's airmen staged spectacular mass flybys, built and flew the largest airplane in the world, and made well-publicized intercontinental flights, all aimed at drawing world attention to spectacular advances made by the Russian people under Soviet leadership.

Soviet participation in the race for the stratosphere was an important part of this program of aeronautical propaganda. Soviet engineers were inspired to construct a stratosphere balloon as a result of Piccard's success. Planning began in December 1932, under the direct sponsorship of the Aviation Branch of the Ministry of War. The Scientific Investigation Institute for the Rubber Industry (NIIRP) worked out the theoretical problems and supervised construction of the envelope *USSR*. The balloon was enormous, far and away the largest built to date. With a capacity of 859,688 cubic feet, the bag would measure 118 feet in diameter when fully distended.

The gondola was constructed under the supervision of an engineer named Chisheski. The construction was of sheet aluminum, tied together with 5,000 aluminum rivets. Two large entrance and exit ports were provided, along with nine small portholes. The gondola carried a full array of cosmic ray instruments.

George Prokofiev, age thirty-one, would serve as chief pilot of the *USSR*. He would be accompanied by M. Birnbaum of the Central Military Aviation Department, and M. Godsunoff, a mechanic and balloon constructor.

The first attempt to inflate the giant "stratostat" on the night of September 23–24, 1933, failed when early morning dew collecting on the bag added too much additional weight for a successful ascent. The inflation was finally completed by 6:00 A.M. on September 30. Two normal balloons ascended the sides of the monster, inspecting the bag and valve while waiting for a lift-off at dawn.

Seven hours later, the three newest heroes of the Soviet Union were safely back on the ground near Kolomna, some 70 miles from Moscow. They had reached a record altitude of 60,695 feet.[37]

Once the Piccard balloon and gondola had been mated and tested in Akron, it was on to Chicago, where the flight would be staged from Soldiers Field. The Compton and Millikan instruments were on hand now and installed under the watchful eye of University of Chicago scientists.

There was a great deal of public curiosity about these cosmic rays. Van Orman, Settle, and the other engineers were grateful to Jean Piccard for keeping the press off their backs. As Van Orman remarked: "Jean was far more interested in cosmic rays than I was. What's more, his chicken hat made him a reporter's dream. Newsmen figured all scientists were nuts anyway. Jean Piccard looked the part. He seemed to know what he was talking about, and he kept the reporters busy for hours."[38]

Inflation began on the night of August 3–4, 1933. White canvas ground cloths were spread over Soldiers Field as 700 steel hydrogen cylinders were connected to the inflation sleeves running to the balloon. The rubberized-cotton envelope would receive 125,000 cubic feet of hydrogen, which would completely fill the balloon at altitude.

The gondola hung from ropes attached to a reinforced belt of material (catenary band) located in the middle of the lower half of the envelope. To facilitate ground handling, small iron rings had been attached to a second belt in the middle of the upper half. Ropes led from the ground through a ring on the bag and back to the ground. One end of the rope was attached to sandbags. The other end was held by one of the marines from the Chicago marine recruiting unit serving as ground crew.

By 2:00 A.M. on August 4, the *Century of Progress* had risen from Soldiers Field and was held in place, caught by searchlights. As thousands of Chicagoans watched, the gondola was wheeled into the launch area and suspended from the

Sailors guarding the Century of Progress *gondola prior to the Soldiers Field flight*

load ring.

Settle encountered his first problem when he tested the gas valve at 2:15. As in a normal balloon, the valve line passed all the way through the envelope. Goodyear engineers, realizing that at take-off the hydrogen would be gathered in a bubble at the top of the bag with the lower fabric hanging in folds and creases, had coated the valve line with graphite so that if it were caught in such a fold, it would still slip free. When Settle valved gas, however, it became clear that the line was not slipping. The gas could be heard whistling through the giant valve on top of the balloon long after the pilot had released the line.

The project leaders were now caught on the horns of a dilemma. To launch without proper control of the valve would be to doom the flight to failure, perhaps a disastrous failure. At the same time, Settle was unwilling to pull the rip line and dump 125,000 cubic feet of hydrogen so close to thousands of spectators, some of whom might be smoking. Settle decided to go.[39] As the pilot was climbing into the gondola, Jean Piccard offered a final bit of advice. "Don't crack the valve."[40]

Settle lifted off at 3:00 A.M. Scarcely ten minutes later, flying over the deserted Chicago railroad yards, he did crack the valve. This time it stuck open, and Settle began dumping sand and lead, realizing that he was about to land.

Watching from Soldiers Field, Maj. Chester Fordney, who had been in charge of the ground crew, commandeered a car and four marines and sped off toward the yards. When he arrived the *Century of Progress* was spread over the Chicago, Burlington, and Quincy tracks at Fourteenth and Canal streets. The only danger facing Settle at the moment came from spectators who were approaching the balloon to cut samples of fabric from the bag.

Fordney and his marines quickly restored order. As a Chicago *Daily News* reporter commented, "In the ensuing three minutes the mob was treated to a gala performance of language and action that have won reputations for potency from the Halls of Montezuma to the Shores of Tripoli."[41] Fordney and his men had saved the day. The giant balloon was rolled up and placed under guard by the leathernecks while the gondola was sent to a warehouse for safekeeping.

The envelope, folded and placed on the railroad car while still wet, was in danger of rotting. The Piccards, who had been promised the balloon and gondola after a successful altitude flight by Settle, had no intention of seeing their property ruined by mildew. They placed a call to Chicago's Rosenwald Museum, promising to present the gondola to that institution in exchange for a working party and an area large enough to spread and dry the bag. As a result of this agreement, the *Century of Progress* gondola, a veteran of one aborted flight and two successful stratosphere ascents, today rests permanently in the Chicago Museum of Science and Industry. With full approval of the Chicago backers and his Navy Department superiors, Tex Settle was quick to reclaim the envelope and gondola for a second attempt at the record. The Piccards had little choice but to agree.

This time Settle was determined to stage the flight under the best possible conditions. He accepted a Goodyear-Zeppelin offer to begin the ascent from its Akron launch facility, far from the crowds that had thronged Soldiers Field. The company would also repair and refurbish the envelope while Dow was putting the gondola through another series of final checks. Union Carbide was also persuaded to provide an additional 700 hydrogen cylinders.

Settle had also recognized the difficulties involved in a solo attempt. Still convinced that Jean Piccard was not a suitable companion, Settle offered Maj. Chester Fordney the opportunity of a seat in the gondola. Fordney was a wise choice. In addition to his personal courage and ability to perform under pressure, the marine, a University of Michigan engineer with a moderate scientific back-

Right, Chester Fordney; left, T. G. W. Settle.

ground, had been in charge of the mathematics exhibit at the Chicago fair. Now he would take charge of the scientific instruments to be carried aboard the balloon.

The scientific program was considerable. Two instruments would be employed to measure the conductivity of gas for cosmic rays. A cosmic ray telescope would determine the direction from which the rays were coming. Fordney would also operate a polariscope to study the polarization of light at high altitudes. He would have to collect air samples for the University of Illinois. Single-celled organisms and fruit flies carried aboard the craft would enable Department of Agriculture and University of Chicago scientists to study genetic mutations. Fordney would also operate an infrared camera and a Gaertner-quartz spectrograph to study the ozone layer.

The second flight of the *Century of Progress* began on November 20, 1933.

It was a hazy Indian summer day as the giant balloon was walked out of the airship hangar at the Akron airport. Settle and Fordney, both clad in street clothes, were concerned about a light ground wind rising toward 8 MPH. Their scientific experiments could best be conducted with a high sun, so they hoped to postpone launch until late morning.

By 9:30 A.M., however, the decision was made to ascend before the winds rose to an unsafe level. The flight began with all hatches open and Settle on the "roof garden" between the gondola and load ring dumping ballast. They rode the low altitude winds until 12:45 when, passing over East Liverpool, Ohio, they buttoned up the gondola and began a steady rise of two to four meters a second. They had also set the oxygen pressure system in operation. It consisted of a submarine regenerator and liquid oxygen evaporator to raise the pressure and provide fresh oxygen. Two momson lungs were also packed aboard. If it became necessary to abandon the gondola, Settle and Fordney would don the lungs before taking to their parachutes.

The gondola that kept the two aeronauts alive had been carefully designed by the Piccards. The men stood on a three-foot circular deck, surrounded by shelves holding the instruments, experiments, and equipment. The deck and shelving were tied to stanchions suspended directly from the load ring. Thus the thin Dowmetal walls of the sphere bore little of the structural weight.

While Fordney performed the required experiments, Settle was busy with the radio. Call letters for the balloon were W9XZ. A three-watt transmitter was carried aboard, its antenna strung up in the rigging above the load ring. A 60-foot receiving antenna trailed beneath the gondola. Throughout the early portions of the flight, W9XZ was on the air, allowing the two aeronauts to converse with their backers, Frank Knox of the Chicago *Daily Mail*, Niles Trammell, an NBC vice-president, and Adm. William J. King, head of the Navy Bureau of Aeronautics.

The *Century of Progress* reached its ceiling at 2:10 in the afternoon and remained there for some two hours. A natural descent began at about 4:15 as the gas cooled. Settle was forced to drop ballast to slow the fall. The ballast system chosen by the Piccards made use of a special hopper with a stopcock on the bottom and a glass cover on the top. To drop ballast the aeronaut opened the glass cover and poured a small amount of lead shot into the hopper. With the cover closed, the stopcock was turned, and the shot dropped free. It was a slow but precise means of ballasting. The size of the lead shot had been carefully selected so that, even falling at terminal velocity, the tiny pellets would not injure the eyes of those looking up from the ground.

Descending through 26,500 feet, the aeronauts opened the hatches and

The Century of Progress *launches from Akron.*

began to drop disposable equipment by parachute. By 5:40 P.M., they were drifting steadily over the New Jersey countryside at an altitude of 800 feet. The *Century of Progress* came to rest in a tidal marsh seven miles southwest of Bridgeton, New Jersey. After a careful study of the barograph records, Settle and Fordney were awarded the official world altitude mark of 61,237 feet. They had flown 542 feet higher than the Soviet crew.[42]

The Soviets were quick to attempt to recapture the altitude title. On January 30, 1934, only seventy days after Settle and Fordney took off from Akron, *Osoaviakhim-1* carried three Soviet airmen to a new record altitude, then dropped them to their deaths. *Osoaviakhim* was a new balloon, with a capacity of 24,920 cubic meters, constructed of rubberized cambric muslin. It was the first gondola to be constructed of stainless, nonmagnetic steel welded to a tubular frame. In consequence, it was very thin-walled. Designed by E. E. Chertovski, the new vessel was much better equipped and instrumented than the *USSR* and would fly with a three-man crew.

One hundred eyelets for the ground handling ropes were located 14 meters down the bag from the valve. The gondola was suspended from a second ring of 64 eyelets located below the equator, 27 meters from the pole. Thirty-two "goosefeet" diamonds tied to these eyelets led to 16 diamonds in the second layer and 8 final risers which descended to the gondola. The hoop was a ring woven of flax and hemp secured with steel clamps, which blew open several long sleeves full of ballast housed outside the gondola.

In addition to recapturing the world altitude record, the Soviet aeronauts were expected to perform a lengthy series of scientific experiments. These ranged from the usual cosmic ray studies to magnetic observations, aerial photography, photometric observations, radio tests, actinometric research on solar radiation, the study of bacilli in the air, and the determination of the electrical properties of the atmosphere, including its electrical gradient, conductivity, and degree of ionization. The scientific instruments were automated to the degree possible. One Soviet scientist went so far as to speculate that the day of the fully instrumented, unmanned stratostat was close at hand.

The crew, Paul F. Fedossyenko, I. D. Usyskin, and A. B. Vasenko, were perhaps the best-trained stratosphere aeronauts of the period. The Soviets had instituted a vigorous selection process, including extended periods in an altitude chamber. This was followed by familiarization in a simulator and parachute training.

The *Osoaviakhim* rose from the Moscow airport at 9:04 A.M. on January 30, 1934. Radio reception was good throughout the morning as the three men munched apples while watching their instruments and reporting their progress to ground listening stations. They achieved their peak altitude of 72,178 feet at 12:23. The trio had now only to effect a safe landing to clinch the record. In fact, that was already impossible. The inexorable balance between ballast and gas was about to deal with the Soviet airmen as it had with Hawthorne Gray.

The crew had dropped over 793 pounds of ballast to reach 72,178 feet. As a result of expansion heating, they had lost a large quantity of hydrogen. At least 1,500 pounds of ballast should have been available to check the descent. They had at their disposal only 925 pounds. There was no hope of a safe landing now and little chance of the sort of stable descent that would permit an escape by parachute.

Continuous solar heating held them aloft, forcing occasional use of the valve, which only compounded the situation. They were also encountering radio problems and difficulties with a sticky altimeter. The final communications from the balloon were garbled:

> 4:05 "*We are dropping fast; considerable discharge of the. . . .*"
> 4:07 "*The bright sunlight. . . . The gondola. . . . Beautiful sky. . . . The ground. . . . This. . . . The sky. . . . The balloon. . . . It. . . .*"[43]

The Soviet airmen faced their final problems as the sun sank and the hydrogen cooled. Dropping far too rapidly into the thicker air of the troposphere, the envelope began to tear. Some contemporary experts believed that the rip may have begun at one of the inflation sleeves.

Violent buffeting set in, making it impossible to jettison the remaining ballast, let alone open the twin entry ports to dump heavy equipment or prepare to use the parachutes. The cables supporting the gondola finally snapped, dropping the sphere and its occupants free. There was no chance of survival. The gondola smashed to earth at 4:23 P.M. near the village of Potish-Ostrog, 16 kilometers east of the Kadoshkino station on the Moscow-Kazan Railroad. The three dead aeronauts were given a state funeral. Stalin, Molotov, and Voroshilov, chief of Soviet military affairs, presided as the ashes were interred in the Kremlin wall.[44]

Capt. Albert W. Stevens had been watching all of this stratosphere activity from his billet at Wright Field, Dayton, Ohio. Stevens was an extraordinary officer. As his longtime friend and commander Orvil Anderson recalled, he was a man "who had never developed the art of responding to cant's."[45]

Stevens was capable of total commitment to the job at hand and would labor for hours without food and for days without sleep. Another colleague,

Benjamin Kelsey, recalled Stevens's habit of catching catnaps stretched out on top of a row of filing cabinets while working for extended periods without leaving the base.[46]

Captain Stevens was an enthusiast, and in the summer of 1932 his enthusiasm, drive, and energy were directed toward carrying the U.S. Army Air Corps into the stratosphere. Surveying his fellow officers at Wright Field, Stevens selected Maj. William Ellsworth Kepner as his most likely ally in selling the notion to Air Corps officials. Kepner, an Indiana native, had served in the Marine Corps from 1909 to 1913 before transferring to the infantry. He entered the Air Service in 1920 and quickly became involved in army lighter-than-air activity. Like Gray, he was a frequent contestant in major balloon races.

Major Kepner's reaction to Stevens's proposal was mixed. "I agreed to help get it started, but I did not agree to go, until I gave it more consideration."[47] Kepner voiced his fear that "after Gray's death it would be difficult to get the service to spend $60,000 on the project."[48] Recognizing the validity of Kepner's argument, Stevens put up $1,000 of his own money as a starter, crediting half of the amount to Kepner in order to make it appear that the project had at least two enthusiastic sponsors.

Stevens then approached Dr. Karl Arnstein and the staff of Goodyear-Zeppelin in the spring of 1933. He asked that they conduct a study of the altitude that would be attained by very large balloons of up to 3,000,000 cubic feet in capacity. The engineers determined that such a large craft would reach 79,000 feet with a one-ton payload and sufficient ballast to land safely once the superheat to which the gas would be subjected in the stratosphere had dissipated.[49]

With this study and some cost estimates in hand, Stevens approached the National Geographic Society, which was immediately interested in the project. Following a series of discussions, Gilbert M. Grosvenor, president of the society, agreed to fund the venture. The Army Air Corps would provide the personnel, equipment, and expertise.

Grosvenor established a thirteen-man advisory committee to plan a scientific program for the flight. Composed of representatives of government scientific agencies, corporate industrial laboratories, and university faculties, this group called for studies in four basic areas. Cosmic radiation experiments topped the list. The crew would also determine the position and distribution of the ozone layer, collect air samples, and compare altitude measurements as determined by photographs with the altitude computed from the pressure and temperature of the air at the moment the photo was taken. Albert Stevens, who would serve as chief scientific observer, was responsible for coordinating this effort and for working with individual experimenters in the design of their packages.

Meanwhile Goodyear-Zeppelin was busy with the design of the balloon *Explorer I*. The 3,000,000-cubic-foot envelope was five times the volume and almost three times the surface area of the *Century of Progress*. When fully inflated, it would measure 179 feet in diameter. The balloon, filled to 7.5 percent of its capacity at take-off, would stand 307 feet high. The now usual twin catenary support bands would provide attachment points for the ground handling lines and the gondola. The enormous appendix measured 8 feet in diameter and 15 feet in length. Twin valves were set at the pole. One of these could be operated by a crewman standing on top of the gondola at low altitudes. The other valve was pneumatically operated from within the capsule. The rubberized cotton gasbag also featured a standard ripping panel for use on landing or for an emergency deflation.[50]

As in the case of the *Century of Progress*, *Explorer I* would feature a gondola constructed of Dowmetal. The shell measured 100 inches in diameter and incorporated all the improvements devised since the original Piccard flight.[51]

With work on the balloon, gondola, and experiments under way, Kepner, who had been assigned as pilot and commanding officer of the flight by Brig. Gen. Oscar Westover, assistant chief of the Army Air Corps, went in search of a launch site. Accompanied by Capt. Orvil A. Anderson, who as operations officer would be responsible for rigging, inflating, and launching the balloon, Kepner undertook a two-week flight through the western United States inspecting one area after another.

When asked by the secretary of the Denver Chamber of Commerce to describe the area they were looking for, Kepner replied, "A hole 400 feet deep with vertical walls; a 500 foot square grassy meadow in the bottom, with a 20,000 volt electrical power line; a railroad and a first class truck highway running through it; and, if possible, I would also like a good trout stream running through it."[52] Denver; North Platte, Nebraska; the Great Meteor Crater near Winslow, Arizona; Cheyenne, Wyoming; and Salt Lake City, Utah, were each rejected in turn. Lander, Wyoming, appeared promising at first. The two airmen spent the night there, burning tires filled with oil to study the local winds, which proved unsatisfactory.

Rapid City, South Dakota, was the next stop on the tour. Here Kepner and Anderson visited an area known locally as Moonlight Valley. "We walked out of a wooded area," Kepner recalled, "and there it was. . . . I could hardly believe it. The walls were vertical 400 feet high on two sides; the third side almost a 50 degree slope; and the fourth side well protected. The bottom was 600 feet in diameter—a log cabin on the bottom, and a huge pile of sawdust that was most useful. A fair road that we traveled by auto down to the bottom, and, of all things, a beautiful trout stream tucked into one corner."[53] The local government offered to construct the required roads and to install power lines. Kepner and Anderson were satisfied. Moonlight Valley was already well on its way to becoming the Stratobowl.

George H. Dern, the secretary of war, had taken an interest in the project and gave assurances that the *Explorer I* team would receive any assistance it required. By June 1, 1933, when Anderson moved into the Stratobowl, three airplanes and a detachment of troops from the Seventh Army Corps area were already assigned to the project. As Rapid City and South Dakota state officials made good on their promise to build roads and power lines to the site, the Stratobowl was quickly transformed into a small city complete with a fire department, telephone switchboard, and a radio station manned by NBC technicians. The South Dakota National Guard began the process of stacking 1,500 hydrogen cylinders in the launch area. Meteorologists were on hand now, as were the scientists who would assist Stevens in loading the instruments into the gondola.

Kepner flew back to Washington during the first week in July in order to brief Grosvenor and Westover. For the first time, Westover raised the possibility of carrying a third man, Anderson, on the flight. Both officials pointed out that there would be a great deal of work for only two men to perform. "I agreed that I had thought some of the same things," Kepner recalled, "but Steve and I had always talked of it as a two-man flight. I felt sure if he could do the work on the technical side, I could fly the balloon. After some further discussion, they left it to me to decide but impressed on me that they both considered it would be a wise decision to take Anderson. I thought it over and on July 27, when it was certain we were going, I told Anderson that he was going."[54]

Back at the Stratobowl, preparations for the launch were well under way. On July 6, 1934, Kepner and Anderson conducted a training session with a standard 35,000-cubic-foot army balloon modified with ground handling lines tacked to the spot where the upper catenary band had been placed on the *Explorer*. They inflated, moved, and launched the small craft using the procedures developed for the giant balloon.

July 9 was a day for ceremony, as the wife of Gov. E. Y. Berry of South Dakota christened the gondola, which was now ready to fly. It was only a matter of waiting for perfect weather over the 700-to-800-mile range to the east. Stevens had no intention of going with anything less than ideal weather. "Photography," he noted, "was to play an important part in our work during the proposed twelve hours aloft; and for satisfactory photography we must have cloudless skies and good visibility. Only a broad area of high atmospheric pressure could assure such conditions, and we were determined to wait for such a 'high' if it took all summer."[55]

While Kepner poured over the daily weather records with the members of the camp meteorological staff, headed by V. E. Jakl of the U.S. Weather Bureau, serious scientists, including Lyman J. Briggs, director of the National Bureau of Standards, and W. F. G. Swann, head of the Bartol Research Foundation of the Franklin Institute, devised a dehumidifier for use in the gondola.

At noon on July 27, 1934, Kepner announced that the long-awaited high-pressure area was in the offing. The *Explorer* would be inflated that night and launched at dawn. Reporters and other nonessential personnel were barred from the work area as the envelope was pulled out of its packing crate for the first time and spread on the carefully prepared sawdust and canvas bed. The special "gondola house" also buzzed with activity as scientists and technicians installed the required batteries and ran final checks.

Inflation was complete by 2:00 A.M. Three hours later the gondola had been wheeled into place and attached to the balloon. Stevens and Anderson were helped into the gondola, while Kepner took his place on the roof where he could better supervise the ground crew. Immediately after lift-off, Anderson joined Kepner on top of the gondola, while Stevens remained inside operating the pneumatic valve. When it became apparent that the external valve line would not have to be employed, Anderson returned inside, and Stevens joined Kepner "on the slippery top of the gondola." The two men lowered a 125-pound spectrograph 500 feet below the balloon on a ¼-inch rope. "It took us more than half an hour to finish the task," Stevens recalled, "for we had to be extremely careful that the heavy and valuable instrument did not get away from us—or take us with it."[56]

Kepner and Stevens then joined Anderson inside and completed the task of sealing the craft. Passing through 15,000 feet, the trio opened radio contact with General Westover and Dr. La Gorce, vice-president of the National Geographic Society. Anderson then began the slow process of dumping 400 pounds of ballast through the hopper. An hour later they paused at 40,000 feet, where they started the geiger counters, which would detect the direction of cosmic rays. By 1:00 P.M. they had ceased valving gas and risen to 60,000 feet. Stevens was giving instrument readings over the radio when all three men heard a distinct ripping sound. Gazing through the three-inch upper port they saw a large tear in the lower surface of the envelope. The balloon was not yet spherical and their calculations indicated that if they continued to ascend they could reach 75,000 feet.

This would clearly have been foolish. The hydrogen, which was rapidly expanding due to solar superheating, would soon reach the rip, placing increased pressure on the area.

Heavy valving was required to compensate for the heating and begin the premature descent. Crowding around the tiny porthole, they could see the rip enlarging. Stevens continued to operate his experiments. The air sample bottles, which were to have been opened at 75,000 feet, were cracked at 60,000 instead.

Kepner was in touch with the ground, relaying news of their situation. They cracked the hatches and climbed onto the roof at 18,000 feet. As Kepner recalled, "It was a grand feeling to be outside where we could, if necessary, use our individual parachutes. We were no longer in a trap! We were free!"[57]

Anderson did have a problem. His parachute had opened in the gondola, and he emerged carrying yards of silk and shroud line in his arms. Anderson, with some trepidation, asked Kepner if he thought the chute would open. At that moment the discussion became academic. The entire bottom half of the envelope tore away, allowing the air to create a volatile mix with the hydrogen. The balloon exploded. Anderson tossed his canopy into space and jumped. It worked.

Stevens, meanwhile, was unable to exit the gondola because of the pressure of the air flowing around the falling sphere. With the photographer wedged halfway out, Kepner put his foot on Stevens's chest and pushed him free, then jumped himself.

Explorer I smashed into a plowed field owned by Mr. Reuben Johnson, four miles north of Loomis, Nebraska, at 3:40 P.M. Kepner, whose parachute had opened within a few hundred feet of the ground, was the first to reach the shattered sphere. His companions were safely on the ground within minutes.

Kepner trudged half a mile to a farmhouse and called the Air Corps camp at the Stratobowl. The NBC engineers routed his call onto the air so that, as Kepner recalled, "I was broadcasting the story of our flight and that all three crewmen had jumped safely."[58]

A great deal of scientific information was salvaged. The spectrograph, which had been cut free to come down on its own parachute, was safely recovered. Much of the filmed record of instrument readings was also intact and salvaged through careful processing in Stevens's photographic laboratory at Wright Field.

The joint board of review appointed to study the crash accepted the aeronaut's explanation for the failure of the envelope. In packing, the section of lightweight two-ounce fabric below the lower catenary had adhered to the heavier three-ounce fabric above the band. Excessive shear resistance between the two areas had produced multiple radial tears, one of which enlarged to form the original rip.

Thanks to the fact that both balloon and gondola were insured through

Explorer I, *inflated and ready for launch*

Above, The Explorer I *envelope, just prior to its collapse; opposite, The* Explorer I *gondola falling free, and the impact.*

Lloyd's of London, the financial loss to the National Geographic Society and the Army Air Corps was contained. This, coupled with the fact that so much promising scientific data had been salvaged, led to immediate discussions of a second try to be made in 1935.

Certainly the press and public viewed the flight of *Explorer I* in a most favorable light. The story was reported as high adventure. "Never," noted an editorial in the *New York Times*, "was there drama like this." Air Corps officials received high marks for their soundness of judgment in selecting the crew. "As Kepner, Anderson, and Stevens have proved, the satisfaction of scientific curiosity calls for the physical courage and the fiber that we associate with true argonauts."[59]

Two weeks after the much-heralded flight of *Explorer I*, Dr. Max Cosyns

was back in the stratosphere, on August 18, 1934. Flying with a Belgian student, Neree van der Elst, Cosyns flew a new balloon named the *Piccard-Cosyns* to an altitude of 16,000 meters. This was purely a scientific venture. Cosyns had emphasized from the outset that he was not making an attempt to capture the altitude record. The flight, which began in the Ardennes, ended with a safe landing 1,000 miles away in Zenaveljie, Yugoslavia.[60]

The American Piccards, Jean and Jeannette, were also anxious to make a stratosphere flight in the reconditioned *Century of Progress* balloon. True to their word, the backers of the fair had turned the balloon and gondola over to the Piccards following the successful Settle-Fordney flight. Convinced that the emphasis of the first ascent had been on altitude, not science, the Piccards were now determined to control all aspects of what would surely be the last ascent of their veteran stratosphere aerostat.

This would, they decided, be a family venture. Jean Piccard would fly as scientific observer, while Jeannette would pilot the balloon. Many years later, when asked why she had not simply hired a balloonist, Jeannette Piccard countered with the query, "How much loyalty can you count on from someone you hire?"[61]

The first step in their program, therefore, was to obtain balloon training for Jeannette. During their search for a suitable balloon instructor, a friend in Washington advised the Piccards to find Ed Hill, the winner of the 1927 Gordon Bennett race. After a considerable search, Hill was located working in a Detroit truck factory. The Piccards were impressed. A short, stocky man, no taller than Jeannette Piccard, Hill exhibited a confident smile and eyes that Jeannette Piccard noted "could suddenly turn gay with a sparkling twinkle."[62]

Flight training began with Hill's 35,000-cubic-foot balloon. Aptly named *Patches*, the craft had once escaped unmanned from ground handlers and had to be brought down with a shotgun. The instruction was furnished at cost. Jeannette Piccard was required to complete six flights for a license. Two of these would be made under Hill's control, and two under the student's direction with Hill in the basket. Two solo ascents would also be necessary, one of these a night flight.

Jeannette Piccard's instruction in ballooning began at the Ford Airport in Dearborn on the evening of May 15, 1935.

> *We spread Ed's big balloon on the field . . . near the big gas lines that had been put in for the Gordon Bennett race and which were still there. Ed showed us how to walk with stocking feet on the deflated balloon lying on a big canvas sheet, how to insert the valve and spread the net over the balloon, how to run the valve*

rope and rip-panel rope through the balloon and coil them under the appendix.

Then we fastened sand bags to the net all around the balloon. We moved the bags down the net square by square as the gas flowed and the balloon rose in the air. That night the air was still, the sky star-studded. Ed and two of his friends handled most of the sand bags, Jean and I only a few. One of my bags got as high as my shoulders before I got around to lifting it down. I almost couldn't do it. Ed jumped to help me.

"Mustn't let them get out of reach," he said firmly. "If you do, it will be there to stay for the time of the flight. It will cut down on your lift. It's lost ballast. Happened to me once."[63]

From the moment of take-off into the soft night air of Detroit, it was a flight Jeannette Piccard would always recall with special fondness.[64] During a second practice inflation two weeks later a sudden squall ripped the balloon free from the netting. Hill made a wild jump for the rip line and was carried into the air "as a child lifts a small toy."[65] Finally yanking the line before dropping safely to earth, Hill watched as the dying balloon fell into a pocket of trees behind the Dearborn Inn. Repairing the shredded envelope became an unexpected part of Jeannette Piccard's course of instruction.

When news of the proposed Piccard stratosphere expedition reached the press, interest centered on the fact that the balloon would be piloted by a woman, the mother of three sons. Was she not afraid? Jeannette Piccard had a ready answer to this question.

Even if one were afraid to die there is so much of interest in a stratosphere trip that one does not have time to be afraid. It is too absorbing, too interesting. . . . When one is a mother, though, one does not risk life for a mere whim. One must back up emotions by cold reason. One must have a cause worthy of the danger. In times of war one sacrifices self and children for country. In times of peace the sacrifice is made for humanity. If one can forward by so much as even a little bit the sum total of man's knowledge one will not have lived in vain. If we do not add something to the knowledge of cosmic rays by our trip to the stratosphere this summer, we had better not go. We had better stay on the ground, be hewers of wood and drawers of water.[66]

The fact that a woman would pilot the balloon created some hostility in official quarters as well. The attitude of one potential funding agency, the National Geographic Society, was typical, as Jeannette Piccard recalled: "The National Geographic Society would have nothing to do with sending a woman—*a mother*—

in a balloon into danger. And it turned out that the National Geographic and the United States Army were going to make a flight. Perhaps they didn't want any competition."[67]

Nor did Goodyear-Zeppelin offer much support. The Piccards returned the *Century of Progress* to the company for repairs, but had great difficulty in diverting any effort from workers who were concentrating on building the new *Explorer* bag. In Jeannette Piccard's words, "We had quite a bit of opposition in Akron all along the way."[68] Even Dow Chemical asked that the company logo be removed from the gondola and requested that the Piccards refrain from using the trade name Dowmetal in the publicity relating to the upcoming flight.[69]

The Piccards proceeded with their plans in spite of all these handicaps. They planned a major modification for the envelope, asking Goodyear-Zeppelin to replace the small appendix of the *Century of Progress* balloon with a much larger one to be held open by magnesium rings at the top and bottom. They hoped that this would allow air to enter the bag during the ascent, creating a pear shape that would open the folds and creases at the outset, thus avoiding the difficulties that had plagued the *Century of Progress* and *Explorer I* flights.

The Piccard scientific program was also under way during the summer of 1934. The Piccards had paid their first visit to Dr. W. F. G. Swann of the Franklin Institute's Bartol Research Foundation at Swarthmore, Pennsylvania, in late November 1933. Swann, one of the nation's leading cosmic ray researchers, had provided instruments for the Settle flight and was currently working on an apparatus to be carried aboard *Explorer I*. The scientist agreed to collaborate with the Piccards as well by preparing four direction geiger counters and a burst instrument. The directional apparatus would only register rays running parallel to the axis of the counter. The burst apparatus was employed to study the simultaneous bursting of lead atoms bombarded by cosmic radiation.

During a subsequent meeting in Washington with Robert Millikan, then the president of Cal Tech, Millikan agreed to prepare three cosmic ray instruments. One of these would be unshielded, one would be shielded by 700 pounds of lead, and the third by 540 pounds of lead dust.[70]

The Piccards returned to Swarthmore following Jeannette's first solo flight on June 16, 1934, only to be informed that Swann's instruments would first be flown on *Explorer I*. When the apparatus was destroyed in the crash, Swann immediately went to work on a new set of instruments for the Piccard expedition. Millikan was having difficulty as well, and was finally able to turn over a single ionization chamber shielded with 700 pounds of lead dust, which would be dropped as ballast when the experiment was complete.

Funding was also a problem for the Piccards. Cut off from the National

Geographic Society and government agencies, they finally obtained support from private firms, including the People's Outfitting Company and Grunow Radio. Private citizens, notably C. P. Burgess, Gustavus Cook, and Sonin Krebe, also made substantial contributions. Instruments and technical assistance was provided by the Bartol Foundation at no charge. Henry Ford also made the facilities of Ford Airport available as a launch site.

The Piccards were back in Dearborn on August 16, 1934. The gondola, repainted and refurbished, arrived from Midland, Michigan, soon thereafter. The balloon was trucked onto the airport grounds from Akron on September 13. Ed Hill, now serving as volunteer flight director, organized the corps of workers swarming over the gondola. Two radio experts from Ford built and installed a 56-megacycle two-way radio. Pressure tests were conducted on September 29–30 and October 10.

Special care had to be taken in soldering the connection to sixteen screws mounted on a panel in the cabin. Once in flight, a battery applied to any of these screws would detonate a blasting cap that would open an external ballast bag. Lead dust would also be dropped through a now standard airlock hopper in the gondola. The valve system was tested as well. In the original *FNRS* gondola, the valve line had passed through a pressure seal consisting of a U-joint filled with mercury. The *Explorer I* had featured a pneumatic valve system. The *Century of Progress* capsule had a simpler arrangement in which the valve line fed through a hole in the wall that was sealed with a packing gland. The line was treated to improve the seal.

Canceled once because of deteriorating weather, launch preparations were set in motion for the second time on the evening of October 22, 1934. Hill and the Piccards originally intended to inflate the balloon with pure hydrogen, not allowing atmospheric air to enter the bag until the process was complete. Once inflation was under way, however, the bottom half of the envelope was sucked into the upper hemisphere, raising the possibility that the fabric might be damaged as that of the *Explorer I* had been. Capt. Albert Stevens, who was on the scene to give advice and assistance, had a worried look on his face and remarked to the Piccards, "That is just the way our balloon looked in Rapid City."[71] The large appendix was opened, allowing air to stream in beneath the upper bubble of hydrogen and fill out the gasbag.

Jeannette Piccard was standing on top of the gondola, with Jean and their pet turtle inside. Following the presentation of a bouquet by the Piccard boys to their mother, the raising of a small flag over the gondola, and the playing of the *Star Spangled Banner*, the balloon weighed off for the third time at 6:51 A.M.

The take-off from the field was slow and uncertain. Rising away from the

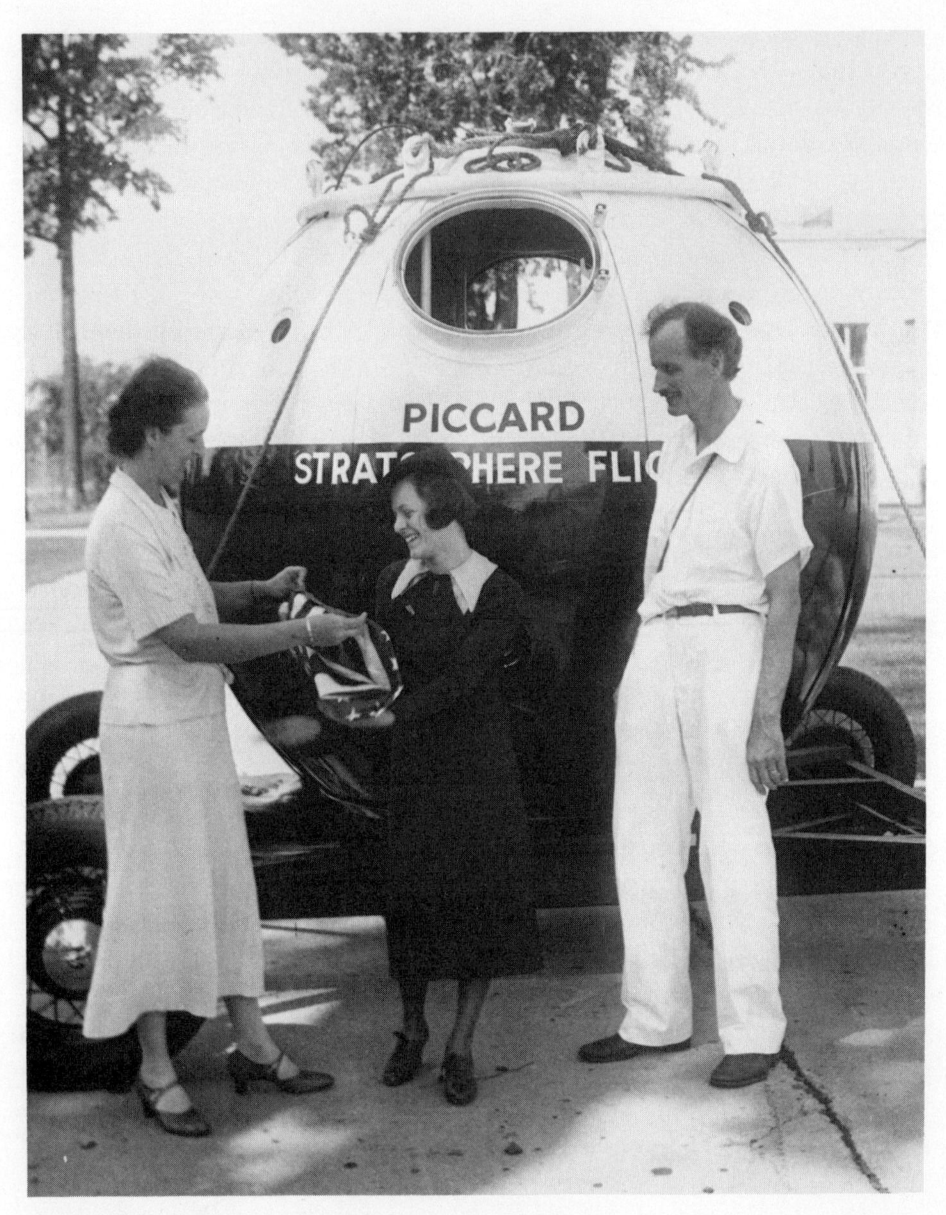

Jean and Jeannette Piccard with their veteran gondola.

field, a plume of smoke poured from the appendix as the expanding hydrogen forced the cool atmospheric air out of the envelope. One by one, they blew open the external ballast bags, gradually rising to an altitude of 57,579 feet as they moved toward the east. Precise navigation became a problem when the contents of one bag, clumped together in the thin air, struck the drift ring hanging beneath the gondola and knocked it off. The ring had been of little assistance at any rate, as most of the flight was conducted over clouds.

Descending over the Ohio countryside eight hours after launch, the Piccards dumped the 700-pound lead dust shield through the hopper as planned. They barely had time to haul a battery up from the floor and toss it out the hatch. A safe landing was made near Cadiz, Ohio, at 2:45 P.M.[72]

Although the Piccards did not approach a record altitude, they had now joined the select fraternity of those who had flown into the stratosphere. As such, they were now public figures, the subject of newspaper feature stories and radio interviews. Jeannette Piccard was a particular favorite. The first woman to enter the stratosphere, she was cited as a courageous example for American girls. Asked repeatedly if she would venture aloft again, her answer was invariably, "Oh! Just give me a chance."[73]

The Soviet Union opened the stratosphere season of 1935 with their first ascent since the tragedy of *Osoaviakhim*. The balloon *USSR-1-bis* was launched from the Moscow airport on June 26, 1935. With M. Christopzille as commander, M. Prilutski as pilot, and Alexander Varigo of the Leningrad Observatory as scientific observer, the craft reached an altitude of almost ten miles before returning safely to earth two and a half hours later. The voyage marked the beginning of the systematic exploration of the upper atmosphere by Soviet aeronauts more concerned with physics than records.[74]

The National Geographic Society-Army Air Corps Explorer team was also anxious to return to the field in 1935. Karl Arnstein of Goodyear-Zeppelin had designed a new balloon, *Explorer II*, for this latest assault on the upper atmosphere. Care was taken to avoid the problems that had led to the loss of the first balloon. The new envelope would feature a lower section constructed of heavier rubberized cloth. The inside surface would also be dusted with powder to reduce the possibility of adhesion.[75]

The *Explorer II* was to be the first helium stratostat, in order to negate any possibility of an explosion such as the one that had destroyed its predecessor. Since helium is roughly twice the weight of hydrogen, *Explorer II* was expanded to 3,700,000 cubic feet. The 1934 flight had also provided lessons for the engineers at Dow Chemical who were responsible for the *Explorer II* gondola. They expanded the size of the portholes in which Albert Stevens had so nearly become fatally wedged to 20 by 22 inches, 2 inches larger in each dimension. The instruments were permanently bolted to the walls rather than placed on shelves.

A parachute with an 80-foot canopy, designed by Lt. Col. E. L. Hoffman, was provided to slow the descent of the gondola in the event of another catastrophic failure of the balloon. Once the free gondola had reached a lower altitude, the crewmen could jump with their personal parachutes. A hinged, 14-foot arm constructed of Dowmetal tubing mounted between the two portholes carried an 18-inch fan on the far end. Driven by a battery within the gondola, the fan was used to turn the gondola through 360 degrees for instrument observations.[76]

The scientific program was much more ambitious than in 1934. Cosmic ray studies, spectrographic observations of sunlight, the collection of upper air samples, temperature and barometric measurements, observations of sun and earth brightness, wind direction and velocity studies, the determination of the electrical conductivity of the atmosphere at various altitudes, radio tests, fruit fly experiments, experiments in altitude determination, and the solution of balloon navigation problems were all on the program. Extensive still- and motion-picture footage would document activity aboard the gondola and record the scene outside the portholes.[77]

Only two men would fly the new balloon. Kepner, given the choice of remaining with the Explorer team or attending the Tactical School at Maxwell Field, had chosen the latter. Capt. Randolph P. Williams replaced Kepner as operations officer, alternate pilot, and weather officer.

The Stratobowl camp was back in full operation when the balloon and gondola arrived in May 1935. Oscar Westover estimated that 3,700 men and women were involved in the project, and many of them were present in Rapid City for the flight. Gilbert Grosvenor noted the remarkable spirit that permeated the Stratobowl: "The Stratobowl camp was . . . a city of scientists and technicians operated with the aid of military men intent on peacetime accomplishments. And it reflected the geography and the cosmopolitanism of science. Men from 21 states and the District of Columbia, representing 21 colleges and universities and nearly a score of professions worked together in this little city built overnight in the hills, each laboring ceaselessly for many weeks that two men might fly for eight hours."[78]

The crew had hoped to launch *Explorer II* in June, but poor weather conditions prevented inflation until July 12, 1935. As three riggers were attaching the gondola at 3:00 A.M. that morning, the bag burst, dropping 2⅔ acres of rubberized cotton over the sphere. A steel guard rail around the gondola protected the workers until balloon troops from Scott Field were able to extricate them.

The Explorer II *gondola*

There was no choice but to ship the giant envelope back to Akron for study and repair. The camp was cleared, with all military personnel being sent back to their duty stations while the investigation was under way. Goodyear-Zeppelin engineers, Bureau of Standards officials, and Army Air Corps officers discovered that the collapse had resulted from the failure of the rip panel. Goodyear-Zeppelin fabricated a new top and new upper catenary band. The lower section of the balloon was repaired, and the standard rip panel was replaced by a new section of the fabric attached to a thin steel cable.[79]

The Stratobowl camp had been reestablished by early September. Preparations for the flight were complete by October 1. Once again, the wait was on for good weather. It was late in the season now. Three snowstorms swept through the area by November 10. The temperature dipped below zero on a number of occasions. National Guard and army troops shivered in summer-weight tents and clothing. All the troops worked to keep the snow scraped off the inflation area.

A break in the weather finally occurred on November 10, 1935. The crew labored through that evening and into the early hours of November 11. With 20,000 cubic feet of helium in the envelope, a loud ripping sound echoed across the base of the Stratobowl. Inspection revealed a 17-foot tear immediately beneath the equator. Stevens and Anderson, recognizing that this was their last opportunity to fly before spring, authorized immediate repairs. With the emergency stitching complete two hours later, the preparations for launch continued.

Anderson and Stevens finally lifted off at 7:01 A.M. They reached their maximum altitude of 72,395 feet at 10:50 that morning. It was a clear day, and the view was spectacular for the first men in the history of the world to rise high enough to see the curvature of the earth. Looking toward their take-off point, the history of the American West lay spread at their feet. The Cheyenne and Belle Fourche rivers, Wounded Knee, Devils Tower, the Little Big Horn, the Black Hills, the Bad Lands were all visible. A single photograph taken at peak altitude became one of the most widely published images of the decade.

Anderson and Stevens landed safely at 3:14 P.M. on November 11. The flight had been a complete success. New data had been gathered for the cosmic ray physicists. The National Geographic Society and the U.S. Army Air Corps garnered a wealth of the most favorable publicity and a world altitude record that would stand for twenty years.

The flight of *Explorer II* marked the end of the great era of stratosphere ballooning. The investment in money and man-hours required to reach 72,000 feet had been enormous. Moreover, the large and heavy envelope was clearly very close to the ultimate in rubberized fabric balloon technology. The possibility

Explorer II *in the air*

of mounting another such costly expedition in the depths of a world-wide depression was slim, particularly in view of the fact that the record now stood so high.

As governments and those few private agencies capable of funding a large-scale stratosphere expedition withdrew from the field, those scientists interested in the exploration of the atmosphere returned to experiments with sounding balloons. Soviet experimenters were early leaders in this area. The Institute of Aerology in Slutsk, near Leningrad, and the Radio Institute cooperated to send automated radio-equipped balloons to altitudes of up to 8.7 miles by the mid-1930s. In February 1936 the Borispol Aerological Observatory, near Kiev, claimed to have sent an unmanned balloon sonde to an incredible 132,840 feet.[80]

American meteorologists, finding the instrumented Soviet meteorographs (radio-equipped instrument packages) too expensive to purchase, built their own. The staff of the Blue Hill Observatory near Boston drew special attention to the balloon-borne "radio bullets" which they sent into the stratosphere.[81]

American physicists were also active in the area of sounding balloon research. Jean Piccard and John Ackerman of the University of Minnesota concentrated their attention on improving the equipment for automated atmospheric research beginning in the spring of 1936. Most of their time was spent on the development of meteorological instruments, transmitters, and receivers equipped with direction-finding antennae. Whereas other experimenters had emphasized the construction of ultra-lightweight packages with low-power transmitters and special receivers, the University of Minnesota team aimed at a transmission range of 200 to 300 miles and a signal that would be received by amateur radio operators.

Piccard and Ackerman also gave a great deal of thought to new balloon technologies. Dissatisfied with the latex rubber sounding balloons then in common use, they undertook the first experiments with plastic film balloons. At the time, cellophane was the only material available. While far from ideal because of its tendency to crack during cold weather inflations, the two scientists enjoyed a real measure of success with cellophane balloons. A balloon launched from the University of Minnesota stadium on June 24, 1936, remained in the air for ten hours, covering 613 miles. The tests continued with the improved instruments and transmitters on November 14 and 18. Thomas Johnson of the Bartol Foundation, which had been so helpful in sponsoring the scientific portions of the Piccards' earlier stratosphere flights, provided funding and assistance for this early plastic balloon work as well.[82]

In spite of the success of the automated aerostats, Piccard continued to argue for the importance of manned altitude flights. In search of a means of circumventing the high cost of balloon construction, he began to focus his attention on the possibility of taking to the air aboard a cluster of small sounding balloons.

Dubbed *Pleiades*, the project culminated in a single low-altitude test flight from Soldiers Field, Rochester, Minnesota, on July 18, 1937. Piccard obtained funding for the venture from the Kiwanis Club of Rochester, Minnesota. The aeronautical engineering department of the University of Minnesota also provided advice, while the mechanical engineering department tested the materials to be flown. Hydrogen cylinders were lent by the Commercial Gas Company of Minneapolis. The *Pleiades* system consisted of 92 latex balloons and a gondola constructed of alclad sheets. Since a valve line and rip cord were obviously out of the question, Piccard was forced to devise a new method of controlling lift. The final answer was to group the balloons in two equal clusters, one tied just above the gondola so that the aeronaut could puncture or cut free any desired number of these balloons. The upper group was tied 50 feet above the car, linked to the lower balloons through a small charge of TNT. Piccard could cut the lower balloons loose to descend and blow all of the upper group away at once at the moment of landing to avoid being dragged across the landscape.

Piccard chose 350-gram Dewy and Almy balloons for the upper cluster. The lower group would be 350-gram aerostats from the same manufacturer. When inflated and linked, the composite furnished 528 pounds of lift over and above the 111-pound weight of the balloons. Inflating and linking 92 separate balloons into a single system would obviously require a closely coordinated ground crew. Jeannette Piccard was given the task of drilling 160 volunteers from the university and the Rochester area into a single team.

As the U.S. Weather Bureau in St. Paul predicted, the early morning hours of July 18, 1937, were virtually windless. Jean Piccard rose into the air at 12:06 A.M., dropping zinc punchings as ballast to maintain the speed of his ascent as he climbed out of the spotlights. The aeronaut established radio contact with the ground immediately after take-off. He cut the first balloon away two hours later, having reached his agreed ceiling of two miles.

More balloons were cut loose at 5:00 A.M. as the sun began to rise. Flying toward a thick fog bank that marked the point at which the warm air over the Mississippi River met the cold air of the prairies, Piccard decided to land. Swooping low over a farmhouse about a hundred miles from Rochester, the aeronaut began to shoot balloons from the lower cluster in order to speed his descent. "I was careful," he noted, "not to shoot in the direction of the farm house." He was also careful to "shoot the balloons by a glancing blow, otherwise one produces only a very small hole which is quite insufficient for rapid deflation. If properly hit, however, the balloons burst easily."[83]

Piccard made a soft landing on grass and immediately detonated the TNT to release the upper cluster. The explosive package was surrounded by protective layers of sand and excelsior. Piccard leaped from the car as a flaming mass of this material dropped onto the floor, causing extensive damage and bringing the last of the great prewar scientific balloon voyages to an end.

Chapter 20 Ballooning in the Space Age

JEAN PICCARD'S CELLOPHANE BALLOONS AND PLEIADES PROJECT WERE experiments that bore important fruit during the turbulent years following War War II. That conflict had seen the operational altitude of combat aircraft climb into the substratosphere. The future for which Hawthorne Gray and others had been preparing was at hand. But if the contrails tracing a path through German and Japanese skies marked the prophecy fulfilled, the war also contributed a few prophetic moments of its own. On September 8, 1944, the first V-2 plunged into the heart of London. Launched from Holland, the rocket had arced to a peak altitude of 60 miles before descending on England. While few were aware of the fact at the time, the space age had begun.

The push to occupy the "high ground" that had impelled McCook Field test pilots to probe the stratosphere two decades earlier was on once again. The time was ripe for Jean Piccard to continue his pursuit of upper atmosphere balloon research. Piccard had spent the war years working on navy aeronautical programs at the University of Minnesota and had served as an aeromedical consultant at the Mayo Clinic. In 1945 he had been sent to Germany with the simulated rank of colonel to assist in the evaluation of enemy wartime research. In 1946 Piccard was eager to reenter the stratosphere aboard a vehicle dubbed *Helios*. An outgrowth of the 1937 *Pleiades*, *Helios* would consist of 80 to 100 plastic balloons that would carry Jean and Jeannette Piccard, safely housed in a sealed gondola, to an altitude of 100,000 feet.[1]

Piccard was searching for a means of funding the project when he was introduced to Otto C. Winzen, a twenty-eight-year-old engineer who was visiting the University of Minnesota aeronautical laboratory on other business. Winzen, born in 1918 in Cologne, Germany, had immigrated to the United States in 1937, studied aeronautical engineering at the University of Detroit, and spent part of World War II in an internment camp. Prior to the meeting with Piccard, Winzen had been employed as chief engineer of the Minnesota Tool and Manufacturing Corporation. Intrigued by Piccard's discussion of *Helios*, he signed on to assist in the design of the life-support system for the gondola.[2]

Together Piccard and Winzen approached George Hoover, then a research project officer with the Office of Naval Research (ONR). Hoover had a reputation for funding interesting projects that more conservative naval officers would reject as outrageous. Word of Hoover's penchant for unusual proposals had apparently reached Winzen, for his first words upon meeting the officer were, "I hear you'll buy anything!"[3]

Hoover did buy *Helios*, writing a modest contract to support the project. The navy did have a few stipulations however. Jeannette Piccard would not be allowed to fly. Jean Piccard would have to ascend with a navy pilot.

A second and more important problem, in view of the outcome, resulted from Piccard's decision not to accept ONR funding for the University of Minnesota since Jeannette Piccard, because of university nepotism regulations, could not play a full role in planning the project. At this point General Mills entered the picture. During World War II the firm had developed a fully equipped engineering facility to produce bombsights and other optical equipment. This work had given General Mills, in the words of Malcolm Ross, a later navy balloonist, a capacity to "do research and development on a number of things other than Betty Crocker and Wheaties."[4] Piccard's idea, Winzen's enthusiasm and engineering talent, and George Hoover's navy contract convinced General Mills officials that their shop capacity might be put to profitable use on Project *Helios*.

With Piccard and Winzen both ensconced at General Mills, work on the gondola and accessories moved forward rapidly. Finding a suitable plastic for the balloons was a more serious problem. Piccard's original cellophane was clearly impractical, as were several other films investigated. Polyethylene was finally selected, and experiments were begun to determine the best means of fabricating a balloon from sheets of this plastic only 1/1000 of an inch (one mil) thick.

Ten years later Otto Winzen could still recall the difficulties encountered in flying the first experimental *Helios* balloons: "On the first plastic balloons the adhesive tape served as much to seal the many leaks as they did to support the load. And yet, the first big *Skyhook* balloon, which the author launched on 25 September 1947, reached 100,000 feet, despite millions of tiny pinholes in the film, leaky seams covered with adhesive tapes, and makeshift launch method."[5]

This first large postwar plastic balloon, progenitor of hundreds of later *Skyhook* scientific research balloons funded by ONR, had a volume of less than 100,000 cubic feet and carried a payload of only 70 pounds. The next two launches were outright failures, followed by the first of many "balloon conferences" called to iron out the difficulties.

A real breakthrough occurred on one of the next experimental *Helios* balloon flights. George Hoover had approved the request of two Brookhaven National Laboratory physicists, J. Hornbostel and E. O. Salant, to fly a pair of cosmic ray plates on one of the first plastic balloon ascents. Technically, the flight was a near disaster. As Winzen recalled, the balloon "refused to come down for three days, and it became apparent that we had a long way to go to develop reliable control and radio equipment and insulated containers which would function in this new environment for extended periods."[6]

In spite of these difficulties, Hornbostel and Salant were delighted with their returned plates. "It will take years to analyze the wealth of cosmic ray events in these plates," they remarked to Winzen. "From that day on," Winzen noted, "cosmic ray research and plastic balloons became inseparable. Project *Skyhook* was on the way."[7]

Skyhook was indeed on its way. *Helios* was not. The manned project had never been a popular program with ONR officials. The gondola shell and other items of equipment were complete, but serious technical problems still remained to be solved before the primitive plastic balloons could be man-rated. Moreover, as a result of the success of the unmanned Brookhaven experiment, the potential of single plastic balloons for research in physics was apparent. Unable to justify the complications of the manned *Helios*, ONR funders chose to drop the project and contract with the General Mills Mechanical Division aeronautical team for the first *Skyhook* scientific balloon flights.

General Mills and the University of Minnesota were rapidly becoming the focal point for all U.S. balloon activity, but there was one other major group operating in the field. A New York University team of meteorologists and physicists working under the leadership of Athelston Spilhaus had also been drawn into the balloon business. Spilhaus, Charles B. Moore, Jr., J. R. Smith, and other members of this group had been charged with developing acoustical equipment and balloons, presumably for use in detecting Soviet atomic tests, although in published articles Spilhaus emphasized the meteorological application of such aerostats.[8]

The NYU team borrowed heavily from Japanese balloon technology in an effort to produce a constant-altitude craft capable of remaining at a specific height for a very long period of time. The Japanese balloons had been constructed of paper and silk and carried incendiary bombs completely across the Pacific to start forest fires in the Pacific Northwest. The problem, of course, was that a standard free balloon would simply rise until it burst. Japanese engineers overcame this difficulty through the use of automatic, barometrically operated ballasting systems. The NYU group paid careful attention to this wartime experience, developing their own automatic ballast-dropping technique.

The Spilhaus team had also noted Jean Piccard's pioneering work with plastic balloons and, like the General Mills engineers, had decided on polyethylene as the most promising film. By 1947 they were enjoying a good measure of success. One of their balloons, launched that year from the White Sands proving ground in New Mexico, was recovered in Norway.

The NYU group had employed a specialist to fashion their envelopes from what one participant fondly characterized as "carrot bag grade polyethylene."[9] Having discovered the extent of the General Mills operation, however, it made sense to abandon their own balloon-building program and purchase their aerostats in Minneapolis.

Both of the original key figures, Piccard and Winzen, had left General Mills by 1949. Piccard had been drawn away from the balloon business by the press of other scientific interests. Winzen left the firm under a cloud in 1949 to found Winzen Research, Inc. With the help of his wife, Vera, who played a key role as vice-president and chief of production at Winzen, the firm quickly grew to become a leader in the field.

Charles B. Moore, Jr., a member of the original NYU balloon group, replaced Winzen as head of the General Mills aeronautical laboratories. Other NYU alumnae, including J. R. Smith and H. A. Smith, the group's original polyethylene balloon builder, quickly followed Moore onto the General Mills payroll. J. R. Smith would be elevated to leadership of this group following Moore's departure for Arthur D. Little in 1953.

The third Minneapolis-based balloon firm, G. T. Schjeldahl, was founded in Northfield, Minnesota. A fourth manufacturer, Raven Industries, was founded by Paul Edward Yost, J. R. Smith, Joseph Kaliszewski, and Dwayne Thon, all of whom had left General Mills in a personnel dispute.[10]

From the efforts of these and other pioneers, the plastic balloons that Jean Piccard conceived in 1935 began to play a major role in both science and national defense. To the general public, science seemed to be the prime beneficiary of the new technology. It is certainly true that the advent of a lightweight, reasonably low-cost means of lofting instrument payloads to altitudes in excess of 100,000 feet led to a renaissance of scientific ballooning.

Cosmic ray physicists were the first to make use of plastic balloons. During the decade 1947–1957, literally hundreds of cosmic ray instruments and plates would be carried aloft beneath billowing yards of polyethylene. Dr. James Van Allen of the University of Iowa's department of physics achieved yet another technical breakthrough with the *Rockoon*, developed under a grant from the

ONR in 1952. By launching a sounding rocket from a balloon at an altitude of 70,000 feet, Van Allen and his colleagues were able to attain altitudes of up to 300,000 feet. These early experiments suggested the possible existence of trapped radiation in near-earth space. The existence of the Van Allen belt was confirmed by the instruments flown aboard the first U.S. earth satellites.[11]

Astronomers also profited from the introduction of plastic research balloons. Solar and planetary specialists have found balloon research especially useful. *Stratoscope I*, launched on August 19, 1957, carried a 12-inch telescope to take photographs of the sun from an altitude of 80,000 feet. The ONR and the National Center for Atmospheric Research (NCAR) sent *Coronascope I*, another solar instrument package, to 80,000 feet on March 10, 1960.

Stratoscope II, a 36-inch telescope designed to make infrared measurements of Mars, was flown on January 13, 1963. A second *Stratoscope II* balloon package studied the planet Venus and six red giant stars. The NCAR *Coronascope II* was sent aloft on May 3, 1964, aboard a 32 million-cubic-foot balloon. An X-ray telescope prepared by MIT staff members flown aboard a 34 million-cubic-foot aerostat on October 16, 1970, remained above 148,000 feet for over ten hours. Neutron spectroscopy, micrometeorite counts, and X-ray, gamma ray, infrared, and ultraviolet experimental packages have also been flown aboard balloons. Geophysicists and earth scientists were also quick to take advantage of the new balloons for earth resources and aerial photographic work and to conduct auroral and zodiacal light studies. Biologists and aerospace medical specialists have sent plants and animals into the upper atmosphere aboard balloons.

By the early 1950s the plastic aerostats had been put to a variety of more practical scientific uses as well. Predictably, the meteorologists were among the early users of these craft. Under the sponsorship of the National Science Foundation, the National Center for Atmospheric Research was organized in Boulder, Colorado, to coordinate the efforts of over thirty-five universities engaged in atmospheric and cosmic research. Since the early 1960s the NCAR has taken the lead in encouraging technical innovation in balloon construction, instrumentation, telemetry, tracking, and other aspects of balloon technology. NCAR launch facilities from Palestine, Texas, to Fort Churchill, Canada, have supported the ascent of some of the largest plastic balloons ever constructed.

This scientific balloon work kept the major balloon manufacturers busy, but it did not make them rich. During their early years, both Winzen and Raven were forced to turn their facilities to the production of every sort of polyethylene product from boat covers to food packaging. For these struggling young companies, secret contracts with the intelligence community often meant the difference between financial success or failure. Carefully hidden from view, each of the major balloon companies maintained a proverbial "back room" operation where work was pursued on lucrative intelligence contracts. The balloon builders had enlisted for service in the cold war.

The links between the intelligence community and the plastic balloon industry were forged during the years immediately following World War II. The NYU group was an early recipient of intelligence funding in support of its work on a constant-altitude aerostat. Officials of Radio Free Europe were also quick to recognize the potential of the balloon as a delivery system for propaganda pamphlets. Intelligence planners soon found a variety of other imaginative applications for the silent, high-altitude plastic balloons. They were employed to gather information of great value on atmospheric radiation levels, helping to monitor atomic weapons testing. During the ensuing decades, U.S. defense organizations, such as the Advanced Research Projects Agency, have employed both free and tethered balloons as sonar platforms and communications relays as well as for air defense surveillance, atmospheric testing, and other applications.

The plastic balloon played a particularly important—and controversial—role as a reconnaissance vehicle. Drifting silently above the altitude of most operational aircraft of the 1950s, the balloon served as an ideal means of sending cameras and other sensors across a hostile border.

Moby Dick, a USAF Strategic Air Command program begun in 1956, was the best-known of these balloon reconnaissance programs. Thanks to the availability of recently declassified documents, the story of *Moby Dick* can be told in some detail.

The USAF publicly unveiled *Moby Dick* with considerable fanfare on January 9, 1956, when a new type of plastic meteorological balloon was launched from stations in Okinawa, Alaska, and Hawaii. As Otto Winzen noted in an official statement, it was to be "a project which aims at exploring the meteorological problems including high-altitude circulation and the jet stream. Object of this work are flights of many days duration, therefore necessitating heavier payloads."[12]

In fact, this portion of the total program, which was identified as *White Cloud*, was nothing more than a cover for the reality of *Moby Dick*. The following day, January 10, 1956, eight balloons were launched from widely dispersed European bases with absolutely no publicity. Weather permitting, ten more balloons were launched each day until January 17, when the daily quota was doubled. The number climbed to thirty a day on January 25 and was upped to forty on January 28. Although each balloon carried a placard indicating in Russian that a monetary reward was offered for the safe return of this meteorological package, the real mission had little to do with the weather. In the words of the project

A Moby Dick *balloon launch*

final report, *Moby Dick* was designed "to obtain photographic and electronic reconnaissance of the Union of Soviet Socialist Republics and its satellites."[13]

The project had begun with the activation of Strategic Air Command's 1st Air Division at Offutt Air Force Base on April 15, 1955. Training began at bases in the United States that May. General Mills was the prime contractor for the large balloons, which were designed to cross Europe and Asia in eight to ten days before being snatched out of the sky over the Bering Sea or Alaska by C-119 pickup planes. The balloons would be tracked based on known weather patterns and predicted wind speeds. A locator beacon was switched on once the balloon was thought to have entered the recovery area.

The training operation, *Moby Dick Hi*, stretched beyond the scheduled completion date because of equipment failures requiring further testing. The major decision during this period involved terminating the contract for the large balloons originally chosen and substituting a smaller aerostat. The original balloon, which flew at an altitude of 75,000 to 80,000 feet, was less vulnerable than the small balloon, which would have a ceiling of only 45,000 to 60,000 feet. The high failure rate with the original balloons during training forced the change. The decision altered the operations plan, however, for the smaller balloons would move much more quickly, arriving in the recovery zone twice as fast as originally scheduled.

Vulnerability tests were conducted as well. The Air Defense Command was informed of launch times and presumed trajectories and ordered to destroy as many of the test balloons as possible. While the balloons were found to be visible from the ground and by radar, and were intercepted and destroyed by fighters, the promise of the information to be gained was judged worth the risk.

Once operations began, both balloon types on hand were employed. The basic configuration consisted of the balloon, load straps to support the cargo, a rotator and bar assembly to position cameras and instruments, and the equipment package. A 24-foot parachute was rigged between the balloon and the gondola to prevent a free-fall drop of the equipment in case of damage to the balloon. A radio-sonde transmitted the level of altitude in the event of a successful launch. Four packages of chaff were attached to the bar assembly to provide a positive radar locator should it become necessary to drop the package at sea. A saltwater-activated, battery-operated transmitter would then signal the location of the equipment for twenty-four hours. Like the Japanese transpacific balloons, the *Moby Dick* aerostats were automatically ballasted. Care was taken to insure that the cameras would not be switched on over friendly territory.

Operations moved forward from January 10, 1956, until the termination of the launches on March 1, 1956. During this period, 516 balloons were launched.

Of these, 399 went into operation, while 117 were known failures. Also, 12 balloons were recovered without having crossed the Soviet border. Of these 387 operational craft, 243 simply disappeared. In all, 123 balloons were tracked, but only 67 entered a recovery area. Forty-four returned balloons were on hand on March 5, 1956.

Suspicions about the vulnerability of the balloons rose as the operation proceeded. Steps were taken to increase the altitude at which the balloons flew and to reduce radar visibility as the success rate declined late in the program. Past project analysis revealed that overballasting had been a major problem, cutting the life of the balloons short by as much as two days, dropping them directly onto Soviet soil.

The electronic packages had not been prepared in time, so all missions were strictly photographic. Pictures were returned by 40 balloons. The number of usable exposures totaled 13,813. The photos charted 1,388,745 gross statute miles of Sino-Soviet territory. This was roughly equivalent to 8 percent of the total land area of these nations.

Air Force analysts offered a mixed assessment of the program. "The type of photography obtained affords an excellent source for pioneer reconnaissance," remarked the authors of the final project report, "but due to limiting factors, detailed analysis of new intelligence is difficult."[14]

The real problem was the immediate and vociferous reaction on the part of the Soviets. It is uncertain when the Russians first became aware of the balloons floating over their nation. The Air Force believed that the initial flights caught the Soviets by surprise and attributed the early success rate to the absence of countermeasures. The protests began early in February 1956. Soviet officials quoted in the *New York Times* accused the United States "of conducting a policy of going to the 'brink' of war by sending large numbers of espionage balloons over this country."[15] The comment was a reference to Secretary of State John Foster Dulles's recent remark on "brinksmanship" in a *Life* magazine article.

In this case, however, Dulles was far more conciliatory than Air Force officials. The State Department had played a major role in the preparations for *Moby Dick*, assuring West German, Turkish, and other friendly governments that no reconnaissance would be conducted over their territory. State Department officials had also worked closely with the military to prepare the cover-up scheme. Now with a potential crisis brewing over the Soviet protests, Dulles remarked that he was "disposed to respect the protests of countries which—like Russia—have objected to flights by American weather balloons over their territory."[16] Pentagon officials, on the other hand, stated that they had no intention of stopping the flights "just because the Russians object."[17]

The Russians attempted to force the issue with a lavish press conference at the Spirdonovka Palace. The area outside the palace where the remains of as many as fifty *Moby Dick* balloons were exhibited was floodlit from trucks carrying portable generators. Leonid F. Illychiv, the Soviet Foreign Ministry press chief, presided over the affair, while army experts explained the recovered equipment and displayed photo enlargements of the Russian-Turkish border area from film found on board one of the recovered gondolas.

In spite of convincing Soviet proof to the contrary, Secretary of the Air Force Donald A. Quarles continued to argue that these were meteorological balloons, not spy vehicles. Once again, the State Department position was very different. Ambassador Charles E. Bohlen, responding to an official Soviet protest, noted that the United States would "seek to avoid" sending any more camera-carrying balloons over Soviet territories. By February 11, 1956, American newspapers were featuring headlines such as "Balloons Create Washington Rift."[18] Not one to brook opposition, Dulles finally prohibited military officials from making any further comment on the balloon situation.

The controversy cooled after February 3, when the attrition rate of the lower-altitude balloons became apparent. A complete stand-down was ordered on February 6 while Air Force authorities studied the situation. Two choices were discussed. One possibility was to begin mass launches of the remaining balloons of both types at maximum altitude in an effort to saturate Soviet air defenses and end the project as soon as possible. Alternatively, they could continue the launches using only the remaining 400 high-altitude balloons. Headquarters USAF favored the latter approach, but before the decision could be implemented, the Soviet protest campaign led to abandonment of the original *Moby Dick* program.

It was not the end of the U.S. intelligence community's attempts to reconnoiter the Soviet heartland using plastic balloons. The Soviets issued a new protest against balloon overflights in September 1958. The new balloons were, according to a State Department source, part of a purely meteorological program operated by the Air Force Cambridge Research Laboratory (AFCRL). The craft had been launched from the West Coast with cameras "for photographing cloud formations and other weather phenomena."[19]

The Soviets were as unimpressed with this explanation as they had been two years earlier during the *Moby Dick* confrontation. Once again, the remains of balloons and equipment were unveiled at a press conference on October 11, 1958. Colonel Taransov, a Russian expert, remarked that the equipment was much improved over that recovered in 1956. He claimed that the cameras would operate continuously for 18 to 20 hours, with each frame covering an area of

800 to 1,200 square kilometers.[20]

The Soviets charged the United States with violating the 1956 promise to cease such operations and called attention to the danger the balloons posed to high-flying aircraft.

By 1958 the high-flying plastic balloons had become obsolete as a means of conducting a photo reconnaissance of the Asian heartland. Lockheed U-2 aircraft, operating from Turkish bases, could provide precise information on specific locations, something that was impossible with an unmanned balloon. The U-2 leaped into the headlines with the capture of Francis Gary Powers in 1960. The furor over *Moby Dick* was scarcely more than a prelude to the U-2 debacle. Ultimately, the spy satellite would prove to be the perfect means of conducting the sort of photographic reconnaissance pioneered with balloons and airplanes.

The effort to meet the needs of the multitude of users that have appeared since 1947 has led to the production of ever larger balloons, built of superior materials and capable of carrying heavier cargoes to greater altitudes. By 1965 the original 10,000-cubic-foot aerostat of 1947 had given way to enormous craft with gas capacities of up to 3,000,000 cubic feet. The size of the largest balloons had reached 1.5 million cubic meters by 1972. Such a balloon measured some 750 feet tall and was constructed with 24.8 miles of heat-welded seams. It was capable of carrying a seven-ton payload to relatively low altitudes or lighter instrument packages up to 50 kilometers.

But this was only the beginning. Plastic balloons of up to 35,000,000 cubic feet were making regular scientific ascents by 1972. In June 1976 a 52.6 million-cubic-foot envelope was launched from the NCAR facility at Palestine, Texas. A 70 million cubic footer was sent up in October of that year. This balloon expanded to an incredible 580 feet in diameter at altitude. As historian Roger Pineau has pointed out, such a balloon would theoretically contain 70 Washington monuments.

These giant plastic balloons were so tough they carried their instruments through jet stream winds of up to 155 miles per hour and into temperatures as low as −86C. Operating above 99.9 percent of the earth's atmosphere, they were exposed to the full force of solar and cosmic radiation. The reliability rate of 96 percent achieved with these craft was indeed remarkable.

The advent of superpressure balloons during the 1960s brought another revolution to aerostatic technology. The giant plastic balloons, like almost all of their predecessors, were zero pressure craft. That is, they were constructed with an open appendix through which gas vented automatically with increasing altitude. The pressure inside and outside the envelope therefore remained equal.

This made it very difficult to maintain a constant altitude or to keep a balloon in the air for extended periods.

Automatic ballasting offered one possible solution to this problem. The Japanese used this technique during World War II to send sealed balloons constructed of paper and silk, equipped with pressure valves, across the Pacific, carrying incendiary bombs intended to start forest fires in the western United States. Complex barometric ballasting systems were employed to keep the balloons from rising so high that they burst.

The American sounding balloon firm of Dewy and Almy had attempted to develop constant-altitude balloons in 1943, encasing a normal rubber balloon in a nonextensible shroud. The experiment clearly left much to be desired, and those meteorologists and scientific researchers that sought to use constant-altitude aerostats were forced to continue to use an automatic ballasting system.

The superpressure revolution, which provided an answer to all of these difficulties, began at the AFCRL in 1958. The research lab had been the center of much defense balloon research in the 1950s. Two members of this research team, Maj. Thomas Haig and Vincent Lundy, first pointed out that the recent development of mylar plastic films, combined with advances in electronic miniaturization, had made constant-altitude craft possible at last. These balloons could be launched to remain aloft at specified altitudes for weeks or months at a time. Moreover, artificial earth satellites could be used to track and interrogate many balloons in the atmosphere in order to obtain a simultaneous picture of atmospheric conditions all over the globe. The result was the GHOST (Global Horizontal Sounding Technique) balloon system. The GHOST ballooning became reality after Lundy's move to NCAR in 1961.

The new mylar superpressure balloons were completely sealed, with no gas vents. This material was able to sustain relatively high internal pressures as the craft rose into the atmosphere. By carefully calculating the weight of the balloon and payload, the altitude at which the balloon would achieve equilibrium and float could be calculated. As long as the excess pressure was maintained inside the balloon, it would remain at that altitude in spite of temperature changes, which would only increase or decrease the degree of superpressure.[21]

With the success of the GHOST balloon, meteorologists at last achieved their dream of semipermanent platforms floating high in the atmosphere. In 1966 one of these craft launched from Christchurch, New Zealand, circled the globe in ten days at an altitude of 42,000 feet. By 1973 NASA had orbited scientific instrument packages aboard sealed balloons at altitudes of up to 78,000 feet. Other GHOST balloons have remained aloft for up to a year at a time.

The two-centuries-old balloon reached its ultimate altitude when *Echo I* was placed in earth orbit on August 12, 1960. *Echo*, constructed by the G. T. Schjeldahl Company, was an aluminum-coated mylar balloon measuring 100 feet in diameter. Inflated in space by residual air, benzoic acid, and anthraquinone, the craft served as a reflector for bouncing radio signals from coast to coast.

If the plastic balloons performed yeomanly service in science and national defense, the revival of manned stratosphere flights after 1957 monopolized public attention. Just as the new balloon technology provided an inexpensive means of lofting experimental packages to extraordinary altitudes, so it made possible manned flights to the very edge of space. This activity was no longer undertaken to capture altitude records. Nor was pure science the only motivation. Except in the case of certain astronomical studies, most scientific research could now be performed by unmanned instrumented balloons.

The high-altitude balloon flights of the 1950s and 1960s were inspired by the necessity to achieve a better understanding of the problems to be faced by future military pilots and, eventually, the first human beings to venture beyond the atmosphere. The emphasis was on research in aerospace medicine and on the development of life-support systems, pressure garments, high-altitude parachutes, instruments, and other items of equipment that would be required for operations under the harsh conditions of the space flight environment.

As with the plastic balloon itself, this new campaign to send human beings to the edge of space began with a U.S. Navy contract to General Mills. C. B. Moore had brought his interest in constant-altitude balloons and a taste for adventure when he migrated from New York to Minneapolis. Moore made his first flight aboard a General Mills balloon rigged to carry a human being on November 3, 1949. Moore recalls that his boss at the General Mills Mechanical Division, Frank Jewett, failing to see any immediate benefit for the company, treated the young meteorologist as a hobbyist, insisting that he fly after hours on his own time. The experiments remained strictly unofficial to avoid insurance or liability problems for General Mills. After a year of effort, Mechanical Division officials brought Moore's manned balloon work to the attention of a federal agency, which offered funding to develop these balloons as training devices for blimp pilots and for use in various intelligence operations.[22]

M. Lee Lewis, a naval officer assigned to Moore's General Mills balloon team before becoming the ONR balloon project officer in 1951, and Malcolm Ross, who followed him in both posts, were the men most responsible for focusing attention on the high-altitude potential of Moore's early experimental manned plastic balloons. Lewis and Ross cooperated to argue for sending human beings to the edge of space aboard the plastic giants.

These two men, with a handful of others, including George Hoover and

Comdr. Marion Buaas, developed a plan for manned balloon research flight that was rejected by the chief of naval research, Rear Adm. Calvin Bolster, as being "too experimental."[23] The situation had changed by 1954 when Ross and Capt. W. C. Fortune, head of the ONR Air Branch, presented the program to Adm. Frederick Furth, the new head of the ONR, who quickly approved. As Ross later recalled, "with this nod, the program that became *Strato-Lab* was created."[24]

Ross's first step was to order the old *Helios* shell, which was in storage at Lakehurst, New Jersey, to the Winzen laboratories in Minneapolis for refurbishing. The crew also began to test large plastic balloons to destruction in a hangar at South Weymouth, Massachusetts. At the same time, pressure suits, instruments, and other items of equipment were being evaluated.

Ross later recalled the coordination required to set Project *Strato-Lab* in motion:

> *We called in the best talents of the Army, Navy, and the Air Force to consult in these areas. The interest of experts in the various fields was quite high. The entire effort was coordinated with the General Science Committee which no longer exists, but then functioned within the Department of Defense to look over scientific efforts. An advisory panel was established to make decisions on systems and various balloon shapes and types. [It was] comprised of Professors Ney and Winckler of the University of Minnesota, Charles B. Moore (who had been with General Mills, but had since moved to Arthur D. Little Co.), Lee Lewis and myself. Tremendous activity was instituted in record time to prepare for the first flight, but with all the pushing, it took six months to get things in motion.*[25]

By the summer of 1956 a program of manned training flights was also under way. The most important of these was an open basket ascent by Ross and Lewis to 40,000 feet on August 10, 1956. The flight marked the first return of a U.S. manned balloon to the stratosphere since 1935. During a second open basket flight on September 24, 1956, H. Froelich and K. Long reached an altitude of 42,000 feet.

The *Strato-Lab* gondola was, as *U.S. News and World Report* noted, "a manned space laboratory."[26] The old *Helios* shell had been refurbished and completely tested at China Lake, California, and in Minneapolis. Once again, the Stratobowl in the Black Hills had been chosen as the launch site. The *Strato-Lab I* flight, which began on November 8, 1956, saw Ross and Lewis reach an altitude of 76,000 feet aboard a 2,000,000 cubic-foot-balloon before a valve failure caused a rapid descent that proved very difficult to slow. The two naval officers emerged safely from the gondola with a record altitude in hand and much new scientific and technical information.

The manned balloon flights that began with *Strato-Lab I* in November 1956 had a special meaning. The orbiting of *Sputnik I* in October 1957 underscored the importance of space as a cold war arena in which the United States and the Soviet Union would compete for military position, international prestige, and scientific honors. Manned space flight clearly lay in the immediate future, and the high-altitude balloon flights offered an opportunity to test men and equipment under nearly equivalent conditions. As early as 1958, Otto C. Winzen could report that "it is now generally recognized in scientific circles that the manned balloon capsule is the prototype for the manned space cabin. It serves not only for the study of the human factors of space flight, but for the selection and training of space pilots and as a test bed for the multitude of accessories and components which will eventually go into the construction of the sealed cabin which will carry the first man into space."[27]

This was the emphasis of the high-altitude flights of the 1950s and 1960s. The manned balloon capsule was, in the words of Malcolm Ross, "an important stepping stone, or bridge, to manned earth satellites or space ships."[28] Winzen Research, Inc., was the balloon firm most heavily involved in the early manned flights. Between June 1957 and October 1958, the firm constructed five manned capsules and balloons employed in an epic series of manned flights.

U.S. Manned High-Altitude Balloon Ascents, 1957–1958

Flight	Date	Pilot	Altitude	Duration
USAF *Manhigh I*	June 2, 1957	Capt. J. W. Kittinger	96,000 ft.	7 hours
USAF *Manhigh II*	August 19–20, 1957	Maj. D. G. Simons	101,500 ft.	32 hours
USN *Strato-Lab II*	October 18, 1957	Lt. Comdrs. M. D. Ross & M. L. Lewis	86,000 ft.	10 hours
USN *Strato-Lab III*	July 26–27, 1958	Comdr. M. D. Ross & Lt. Comdr. M. L. Lewis	82,000 ft.	34½ hours
USAF *Manhigh III*	October 8, 1958	Lt. Clifton McClure	100,000 ft.	12 hours

The *Strato-Lab* ascents were, of course, a continuation of the original ONR program. The USAF *Manhigh* flights were operated under the supervision of the staff of the Aero Medical Field Laboratory at Holloman Air Force Base, New Mexico. Under the leadership of Col. John Paul Stapp, this laboratory had conducted a series of highly publicized experiments to measure the human reaction to extreme conditions faced during high-altitude, high-speed flight operations. Stapp himself had become one of the most famous aerospace figures of the 1950s as the result of his rides on rocket-powered sleds in order to study the effect of windblast and G-forces encountered during ejection.[29]

The Holloman team had conducted a series of balloon flights with animals as test subjects. Stapp insisted on the necessity for carrying the program into the stage of manned flights. In his words, "The animals did nothing up there but breathe, eat, and defecate. They didn't talk on the radio or shift around in a 180-pound mass or fidget in a pressure suit or try to grab scientific observations out of those saucer-shaped portholes, or do any of the things you will have to do when you go up."[30]

Maj. David G. Simons, Stapp's project officer for *Manhigh*, had hoped to make the first flight himself. Stapp rejected his request. Simons, a physician and flight surgeon, would remain on the ground to run the flight, which would be made by Capt. Joseph W. Kittinger, a test pilot who had become something of a parachute expert. Simons made the second *Manhigh* ascent.[31]

Other altitude flights sponsored by the U.S. Navy and Air Force followed over the next several years.

Selected U.S. Manned High-Altitude Balloon Ascents, 1959–1961

Flight	Date	Pilot	Altitude	Comments
USAF *Excelsior I*	November 16, 1959	Capt. J. W. Kittinger	76,400 ft.	Parachute jump from maximum altitude
USN *Strato-Lab IV*	November 28, 1959	Lt. Comdr. M. D. Ross & C. B. Moore	81,000 ft.	Observations of Venus with 16-inch telescope and spectrograph
USAF *Excelsior II*	December 11, 1959	Capt. J. W. Kittinger	74,700 ft.	Parachute jump at maximum altitude
USAF *Excelsior III*	August 16, 1960	Capt. J. W. Kittinger	102,000 ft.	Unofficial world altitude record. Parachute jump from 102,800 ft.
USN *Strato-Lab V*	May 4, 1961	Lt. Comdrs. M. D. Ross & V. C. Prather	113,740 ft.	World record altitude

David G. Simons prior to the second Manhigh flight

With astronauts and cosmonauts orbiting the earth during the early 1960s, the second era of manned high-altitude ascents drew to a close. *Manhigh*, *Strato-Lab*, and *Excelsior* had accomplished the goal of preparing the way for human beings to enter space. But the price had been high. M. Lee Lewis, a *Strato-Lab* pioneer, died in a 1959 ground accident involving a gondola suspension system. V. C. Prather drowned when his pressure suit filled with water after a landing at sea following the flight of *Strato-Lab V*. Prather and Malcolm Ross had launched from the carrier *Antietam* aboard a 10-million-cubic-foot aerostat, the largest balloon ever to carry men aloft.

Lt. Clifton McClure, a twenty-eight-year-old volunteer, was, in the words of a *Washington Star* news headline, "Nearly Baked in AF Balloon Error," during the flight of *Manhigh III* on October 8, 1958. McClure, who was neither a pilot nor a balloonist, had endured temperatures of from 150 to 160 degrees above 80,000 feet for several hours. He was so severely dehydrated that he had to be hospitalized. Air Force spokesman Dr. Knox Millsaps commented, "Something went wrong with our calculating about insulation and temperatures.[32]

For the most part, balloon science in the 1950s and 1960s was conducted with unmanned instrumented balloons. The astronomers were virtually the only scientists that continued to find important uses for manned aerostats. On May 31, 1954, Charles Dollfus, the great French aeronaut, ascended to 7,000 meters to assist his astronomer son Audoin in searching for the presence of water vapor in the Martian atmosphere. Five years later, on April 22, 1959, Audoin reached 42,000 feet beneath a large string of weather bureau balloons in order to study Venus. Malcolm Ross and R. Cooper conducted solar coronal studies from the open basket of a balloon at 38,000 feet on August 7, 1959. Important telescopic and spectrographic observations were also conducted during the *Strato-Lab V* flight.[33]

The new balloon technology that had been put to work in the service of science and national defense also enabled aeronauts to take up the two-centuries-old challenge of the Atlantic once again. In the age of supersonic transports, manned journeys to the moon, and sophisticated planetary probes, the lure of an Atlantic balloon voyage may at first appear more than a little puzzling. For the twenty-three men and women who hazarded their lives in the venture between 1955 and 1978, the sheer challenge was motivation enough. Millions of dollars were spent. Five aeronauts, four men and a woman, lost their lives before the elusive goal of a transatlantic balloon voyage was realized.[34]

For every balloonist who lifted off on an attempted Atlantic crossing, there were hundreds who dreamed of doing so, and a few made preliminary flights in order to prepare for the great voyage itself. Kurt Stehling and two naval officers,

Launching Manhigh II

Transatlantic Balloon Summary 1955–1978

Take-Off	Aeronauts	Balloon	Launch Site
December 12, 1958	Arnold Eiloart Tim Eiloart Colin Mundie Rosemary Mundie	*Small World:* Hydrogen; 53,000 cu. ft.	Tenerife, Canary Islands
August 10, 1968	Mark Winters Jerry Kostar	*Maple Leaf:* Helium; 35,000 cu. ft.	Halifax, Nova Scotia
September 20, 1970	Malcolm Brighton Rodney Anderson Pamela Anderson	*Free Life:* Helium and hot air; 73,000 cu. ft.	East Hampton, L.I., New York
August 7, 1973	Bob Sparks	*Yankee Zepher*: Helium and hot air; 73,000 cu. ft.	Bar Harbor, Maine
February 18, 1974	Tom Gatch	*Light Heart:* Helium; composite	Harrisburg, Pennsylvania
August 6, 1974	Bob Berger	*Spirit of Man*: Helium	Lakehurst, New Jersey
January 6, 1975	Malcolm Forbes Tom Heinsheimer	*Windborne:* Helium	Santa Ana, California
August 21, 1975	Bob Sparks Haddon Wood (stowaway)	*Odyssey:* Helium, hot air; 100,000 cu. ft.	Mashpee, Massachusetts
June 25, 1976	Karl Thomas	*Spirit of 76:* Helium; 77,000 cu. ft.	Lakehurst, New Jersey
October 5, 1976	Ed Yost	*Silver Fox:* Helium; 60,000 cu. ft.	Milbridge, Maine
September 9, 1977	Maxie Anderson Ben Abruzzo	*Double Eagle:* 101,000 cu. ft.	Marshfield, Massachusetts
October 10, 1977	Dewey Reinhard Charles Stephenson	*Eagle:* Helium, 86,000 cu. ft.	Bar Harbor, Maine
August 10, 1978	Ben Abruzzo Maxie Anderson Larry Newman	*Double Eagle II*	Presque Isle, Pennsylvania

Distance	Duration	Landing	Comments
1,200 miles	94.5 hours	1,500 miles E of Caribbean Islands	Sailed to safety in sail-equipped gondola-boat
70 miles	20 hours	35 miles SE of Halifax	Rescued
1,400 miles(?)	30 hours(?)	East of Newfoundland	Lost at sea
850 miles	23.5 hours	45 miles NE of St. Johns	Rescued
1,400 miles(?)	18 hours(?)	(?)	Lost
12 miles	1 hour	Barnegat Bay, New Jersey	Rescued
Aborted launch	—	—	—
125 miles	18 hours	125 miles S of Cape Cod	Rescued
550 miles	33 hours	375 miles NE of Bermuda	Rescued
2,475 miles	107.5 hours	200 miles E of Azores	Rescued
2,440 miles	66 hours	3 miles from NW Iceland	Rescued
200 miles	46 hours	50 miles SE of Halifax	Rescued
3,120 miles	137 hours	Miserey, France	First successful transatlantic balloon voyage

Capt. Gordon Benson and Comdr. Ben Levitt, for example, sailed away from the small airport at Brawley, California, at 8:30 on the morning of April 30, 1962. Their balloon was the *Kathleen*, a 170,000-cubic-foot plastic Winzen aerostat. Their goal was the first-ever nonstop balloon crossing of the United States in preparation for an Atlantic flight. In spite of meticulous planning, the support of the Decker Foundation of Philadelphia, and the cooperation and assistance of Winzen Research, Inc., the flight ended with a safe landing in the Mexican desert. Stehling would go on to make a number of noteworthy ascensions, including a scientific flight on October 30, 1970, covering 380 miles, and another on October 26, 1971, during which he traveled 640 miles, but his dreams of an Atlantic trip were dead.[35]

With the successful flight of *Double Eagle II*, American balloonists seemed compelled to go in search of new challenges. The United States had still not been crossed nonstop by a balloon. Publisher Malcolm Forbes called attention to this goal in 1973, when he achieved a balloon crossing of the United States, stopping each night to rejoin the thirty-man crew trailing him across the continent with an airplane, a bus, several cars, and a mobile home.[36]

Vera Simons's *Da Vinci Trans-America* flight also drew a great deal of attention. As Vera Winzen, Mrs. Simons had been one of the important early pioneers in the area of plastic balloons. Not only had she played a major role in the production of Winzen balloons, she had also been a central figure in planning the series of air force and navy manned research flights of the 1950s and 1960s. She received her gas balloon pilot's license in 1957, was herself an early plastic balloon aeronaut, represented the United States at the 30th Annual International Gas Balloon Races in Holland, and was the recipient of a gold medal for her contributions to balloon research.

The series of four *Da Vinci* flights was planned by Vera Simons and Rudolf J. Englemann, a National Oceanic and Atmospheric Administration scientist. The project had several goals. As manned high-altitude flights were being phased out, Simons and Englemann hoped to spark interest in the balloon as a vehicle for scientific research in the lower atmosphere. Atmospheric structure, turbulence, pollutant levels, and the suspension of fine particles in clean air were all objects of concern. In addition, Simons, an artist, sought to use the unique potential of the balloon to gather landscape and cloud images to be used in producing works of art. She was also anxious to explore the potential of a balloon playing colored lights on the clouds and the ground to create artistic effects.

The first *Da Vinci* flight was made from Los Cruces to Wagon Mound, New Mexico, in November 1974. This 12-hour ascent proved the utility of a thoroughly instrumented, low-altitude manned sampling platform. *Da Vinci II* was a

Double Eagle II

successful 24-hour trip from St. Louis to Griffin, Indiana, on June 8–9, 1976. The aeronauts remained aloft for 24 hours, traveling the length of the St. Louis plume, an air pollution band. *Da Vinci III*, a flight from St. Louis to Morehead, Kentucky, also included extensive air pollution studies.

Da Vinci IV, or *Da Vinci Trans-America*, by far the most ambitious flight of the program, was an attempt to conduct a similar mix of scientific and artistic activities during a nonstop, coast-to-coast flight. Simons and Englemann, accompanied by flight surgeon Lawrence L. Hyde and NBC-TV cameraman Randy Birch, lifted off from Tillamook, Oregon, on September 26, 1979. The flight ended with a crash landing in a thunderstorm near Van Wert, Ohio, on October 2. Vera Simons suffered a compound fracture of her left leg. The other aeronauts escaped without injury.[37]

Maxie L. Anderson and his son Kris finally completed the first successful nonstop crossing of North America with a landing near the small town of Matane, Quebec, on the Gaspé Peninsula, on May 12, 1980. They had launched 99 hours and 54 minutes earlier from Fort Baker, California. Their balloon, the *Kitty Hawk*, was a plastic, helium-filled aerostat standing 75 feet tall. The pair traveled 2,417 miles (*Da Vinci Trans America* covered 2,003), but still fell 160 miles short of the East Coast. This, and the fact that the Andersons had landed in Canada, encouraged others to continue efforts to achieve an absolute coast-to-coast flight within the boundaries of the United States.[38]

This goal was finally achieved by John Shoecraft and Frederick Gorrell who piloted *Super Chicken III* from Costa Mesa, California, to Blackbeard's Island, off the Georgia coast, only 200 yards from the surf. It was a remarkably quick trip during which the huge polyethylene balloon traveled 2,515 miles in 55 hours and 25 minutes from October 9 to October 12, 1981. Shoecraft had made two previous attempts to fly coast to coast. On September 22, 1980, he and companion Ron Ripps had been forced to parachute from *Super Chicken I* during a thunderstorm over Columbus, Ohio. The *Super Chicken II* flight was aborted on December 4, 1980, when a helium leak developed near Liberal, Kansas.[39]

With the crossing of the Atlantic and the North American continent accomplished facts, a Pacific crossing seemed the next logical goal. The Pacific had, of course, been crossed many times by the Japanese wartime balloons, but a manned flight would be a very different and much more difficult matter.

Yet success was achieved here as well. At 7:36 A.M. on November 13, 1981, *Double Eagle V* swept in to a rough landing in a wilderness area not far from Covelo, California. The gondola disconnected from the balloon and fell sharply on one side, pitching the four occupants into a heap. For Ben Abruzzo, Larry Newman, Ron Clark (all of Albuquerque, New Mexico), and Rocky Aoki,

The Kitty Hawk *transcontinental attempt*

Japanese-born founder of the Benihana restaurant chain, it was the end of a flight that had begun 5,070 miles to the east in Nagashima, Japan.

It had been a difficult trip, dangling beneath the 26-story balloon for 84 hours and 31 minutes. Thunderstorms, icing, helium leakage, and a rough landing were among the obstacles to be overcome. In describing the trip to newsmen, Larry Newman remarked that he had felt "the closest anyone flying a balloon will come to meeting his maker."[40]

But Ben Abruzzo, who had now flown both the Atlantic and the Pacific, was still in search of new goals. At the conclusion of the *Double Eagle V* voyage, he announced his intention of attempting a nonstop, round-the-world flight. When asked what drove him to attempt such things, Abruzzo answered, "For the adventure, the achievement, the challenge. You do it for the horizons that are before you and to keep moving forward."[41]

The great transoceanic and transcontinental flights of the 1960s and 1970s were only the highly visible peak of a sport balloon revival. Sport ballooning had been virtually nonexistent in the United States during the first two postwar decades. Hydrogen remained as dangerous as ever, while helium costs were so high as to virtually prohibit gas ballooning. The Balloon Club of America, founded by Don Piccard, Robert McNair, and Peter Wood prior to 1950, was a focal point for the tiny band of postwar U.S. aeronauts. In addition to Piccard and the other founders, Tony Fairbanks, Francis Shields, Jerry Burns, Eleanor Vodala, and Connie Wolf were among the premier American balloonists associated with this Swarthmore, Pennsylvania, group.[42]

For most Americans, however, the balloon seemed to be a thing of the past. Just as it appeared that the oldest of all flying machines would vanish from the American scene forever, the astonishing revival of the hot air balloon began. Throughout the twentieth century, various inventors had toyed with the notion of developing a fuel-fed burner system sufficiently powerful to keep a man-carrying balloon in the air. One of these men, veteran balloonist James A. Contos of Akron, Ohio, actually built and flew a hot air balloon with a burner fueled by a kerosene-propane mixture as early as November 18, 1948.[43]

But the real origins of the modern hot air sport balloon, like so much else in the recent history of American aerostation, can be traced to ONR contracts with General Mills and Raven Industries. While the exact nature of these early contracts remains classified, it seems probable that federal agencies were supporting the development of a silent entry craft to be used in making short hops across an unfriendly border. It would be impossible to inflate and launch a gas balloon in reasonable secrecy, while a hot air craft could be inflated and launched in minutes.

Raven Industries was the organization most responsible for the reemergence of hot air ballooning. Founded by Ed Yost and four disgruntled General Mills ex-employees in 1956, the firm struggled through several difficult early years, pursuing research contracts and manufacturing plastic products. Yost, a former Alaskan bush pilot-mechanic, had served as a chase pilot on General Mills balloon projects and brought his enthusiasm for the possibility of one-man sport balloons with him to Raven. He flew his first home-built hot air balloon, an 8,000-cubic-foot polyethylene envelope with a plumber's blowtorch attached, in 1953. Within weeks, he had constructed a 27,000-cubic-foot envelope fired by multiple torches which could lift a human being.[44] Yost and Raven continued this research with a $47,000 grant from the ONR.

The early envelopes had a lifespan of only five flights, and the search was soon on for improved envelope materials and a replacement for the torches. The result, the prototype of the modern sport balloon, made its first short flight at Bruning, Nebraska, on October 10, 1960. Refinements in the pressure-fed propane burner were still required, but all the essentials were now in place.

Raven executives still expected that most of the money to be made from hot air balloons would come from intelligence and defense sources. As one founder remarked, "No one was more astonished than we were when people began to buy the things."[45]

In spite of their doubts about the salability of hot air sport balloons, Raven executives hired Don Piccard, aeronaut son of Jean and Jeannette Piccard, to serve as test and demonstration pilot. By the spring of 1962 Piccard was hard at work peddling a one-man "thermal balloon" dubbed the *Vulcoon* from coast to coast. While the Federal Aviation Agency had yet to lay down guidelines for airworthiness certification, the 30,000-cubic-foot craft was being sold to sportsmen and commercial firms looking for advertising gimmicks.

The *Vulcoons* were constructed of a specially laminated silicon rubber and ripstop nylon. The gores were sewn in alternating strips of white and orange. Each balloon stood 50 feet high and measured 40 feet in diameter. A cylindrical fabric "flame shield" was provided at the base.

The early baskets were suspended from aircraft cables. Propane tanks fed a stainless steel burner through a needle valve, which was used for flame control, and a blast control for full power. *Vulcoons* featured a quick opening crown for immediate deflation in a wind.

The balloon had a useful lift of 240 pounds (roughly a pilot and his parachute). With 80 pounds of propane aboard, a skilled pilot could fly for as long as four hours.

The complete *Vulcoon* came with an inflation unit, instructions, and flight training for $3,500 F.O.B. Raven's Flight Center at Sioux Falls, South Dakota. New envelopes could be purchased for $1,600.

Rather than inflating the craft on the ground with cold air from a fan, as is common practice today, Raven advised its customers to use a special "ground burner." This consisted of a propane-fired, four-cycle engine with a double-headed, stainless-steel burner in the mouth of a "squirrel cage" blower. This unit was connected to the balloon by a flexible duct.[46]

By 1962 a number of these *Vulcoons* had already been sold, and the first hot air balloon race for the Jean Piccard Trophy had been staged at the St. Paul Winter Carnival. The Piccard Race required all participating pilots to land as close as possible to a given target. Don Piccard was full of ideas for other balloon sporting events. He was also full of enthusiasm for the "fun" potential of the new sport, as he explained in a letter to Louis Casey of the National Air and Space Museum in May 1962:

> *When a small club has one one-man aircraft and it is a glider, the ground crew (tow pilot or driver and wing holder) must just sit while the craft is airborne. If it is a Vulcoon that is flying, they chase it in a car or light plane, with extra fuel tanks for additional flights. The chase by car can be a real sport in itself and is not to be undervalued as an exciting time when the balloon is moving fast and may often be "lost."*
>
> *So, whether it is a quiet Sunday afternoon "picnic" flight or a full blown national race, here we have an economical and ideally suited sporting machine.*[47]

As the veteran of several balloon chases down back-country Virginia roads in the dwindling twilight, the author can attest to the fact that this phase of the hot air balloon experience has not changed during the past two decades.

The sport balloon craze spread from the United States to Europe when Yost and Piccard visited England with a 60,000-cubic-foot balloon, flying the English Channel in 3 hours and 17 minutes during the last week in March 1963 with the great French aeronaut-historian Charles Dollfus. Back in the United States, the first Hot Air Balloon National Race was held in 1963, by which time there were three manufacturers in business: Raven, Piccard (Don Piccard was now operating his own firm), and Barnes, founded by Tracy Barnes, who, as a physics student at the University of Minnesota, had built and flown his first balloon two years before.

The Balloon Federation of America, founded in 1969 by Ed Yost, Don Kersten, and Peter Pellegrino "to promote, develop, and aid the art of ballooning," provided a focus for the rapidly growing sport in America. Within four years, the organization had grown to 100 balloonist and 200 nonpilot members.

Today, thirty years after the introduction of the hot air balloon, the sport is firmly entrenched and seems destined for continued growth. Each year thousands of enthusiasts trek to Albuquerque, New Mexico, Indianola, Iowa, or to smaller local balloon festivals for an opportunity to watch a mass fly-off of beautifully decorated aerostats rising slowly into the air. The balloon, now two centuries old, remains very much on the scene. The American balloon lovers of the past, from Benjamin Franklin and Sally Jay to Ralph Upson and Ward Van Orman, would be pleased, but probably not surprised.

Notes

Chapter 1

1. David Grayson Allen et al., *The Diary of John Quincy Adams* (Cambridge, Mass., 1981), vol. 1, pp. 187–188.

2. Unidentified article, Upcott Scrapbook, Ramsey Rare Book Room, National Air and Space Museum Library, Smithsonian Institution, Washington, D.C. (hereafter, NASM Library).

3. B. Franklin to Joseph Banks, Aug. 30, 1783, in Abbot Lawrence Rotch, *Benjamin Franklin and the First Balloons* (Worcester, Mass., 1907), p. 5.

4. Unidentified article, Upcott Scrapbook.

5. B. Franklin to J. Banks, Aug. 30, 1783, in Rotch, *Benjamin Franklin*, p. 5.

6. Allen et al., *Diary of John Quincy Adams*, vol. 1, pp. 187–188.

7. B. Franklin to J. Banks, Aug. 30, 1783, in Rotch, *Benjamin Franklin*, p. 5.

8. *Ibid.*

9. Carl Van Doren, *Benjamin Franklin* (New York, 1938), p. 700; I. Bernard Cohen, "Benjamin Franklin and Aeronautics," *Journal of the Franklin Institute* (August 1941), pp. 103–104.

10. Baron Grimm, *Correspondance littéraire, philosophique, et critique* (Paris, 1877–1882), vol. 13, p. 349; I. Minis Hays, *Calendar of the Papers of Benjamin Franklin in the Library of the American Philosophical Society* (Philadelphia, 1908), p. 468.

11. John Adams to Elbridge Gerry, Dec. 14, 1782, Adams Family Papers, Massachusetts Historical Society.

12. "Extrait d'une lettre d'Annonay du 21 Juin 1783," *Affiches, Annonces, et Divers* (Paris), July 10, 1783, p. 110.

13. B. Franklin to J. Banks, July 27, 1783, in A. H. Smyth, ed., *The Writings of Benjamin Franklin* (New York, 1907–1910), vol.9, p.73.

14. James Woodman, *Nazca: Journey to the Sun* (New York, 1977).

15. Joseph Needham, *Science and Civilization in China* (Cambridge, Eng., 1965), vol. 4.2, pp. 163; 555–599.

16. *Ibid.*

17. Clive Hart, *The Dream of Flight: Aeronautics from Classical Times to the Renaissance* (London, 1972), pp. 50–51; Needham, *Science and Civilization*, vol. 4.2, p. 573.

18. Hart, *Dream*, pp. 110–111.

19. Javier Merino Arroyo, "Ballooning in Mexico," *Ballooning* (Spring 1975), pp. 14–17.

20. Edgar Cardosa, "Bartolomeu de Gusmao," Air BP 53 (n.p., n.d.), pp. 11–14; Viscond de Faria, *Le Précurseur des Navigateurs Aériens Bartolomeu Laurenco de Gusmao L'Homme Volant Portugais né au Brésil* (Paris, 1910).

21. Joanne Sonnichsen, "Kria Kutnoi's 1731 Balloon Flight," *Ballooning* (Jan.–Feb. 1980), pp. 44–45.

22. Gaston Tissandier, *Histoire des Ballons et*

des Aéronautes Célèbres, 1783–1800 (Paris, 1887), p. 33.

23. Allen et al., *Diary of John Quincy Adams*, vol. 1, p. 187.

24. David Schoenbrun, *Triumph in Paris: The Exploits of Benjamin Franklin* (New York, 1976), p. 392.

25. *Ibid.*

26. Charles Coleman Sellers, *Benjamin Franklin in Portraiture* (New Haven, 1962), pp. 136–137.

27. *Ibid.*

28. Tissandier, *Histoire*, p. 33.

29. B. Franklin to J. Banks, Aug. 30, 1783, in Rotch, *Benjamin Franklin*, p. 5.

30. "Excerpt from a Report Submitted to the Académie des Sciences," *The Romance of Ballooning: The Story of the Early Aeronauts* (New York, 1971), p. 18.

31. *Ibid.*

32. B. Franklin to J. Banks, Oct. 8, 1783, in Rotch, *Benjamin Franklin*, pp. 7–9.

33. For a catalogue of period balloon artifacts, see Roger Pineau, *Ballooning, 1782–1972* (Washington, D.C., 1972), p. 72.

34. B. Franklin to R. Price, Sept. 16, 1783, in Smyth, *Writings*, vol. 9, p. 85.

35. Grimm, *Correspondance*, vol. 13, p. 351.

36. Lyman H. Butterfield, ed., *The Diary and Autobiography of John Adams* (Cambridge, Mass., 1961), vol. 3, p. 169.

37. A. B. Le Roy to B. Franklin, Library of the American Philosophical Society, Franklin Papers, vol. 44, p. 146.

38. B. Franklin to J. Banks, Oct. 8, 1783, in Rotch, *Benjamin Franklin*, p. 8.

39. Allen et al., *Diary of John Quincy Adams*, vol. 1, p. 192.

40. B. Franklin to R. Price, Sept. 16, 1783, in Smyth, *Writings*, vol. 9, p. 85.

41. M. Deschamps to W. T. Franklin, abstract in Hays, *Calendar*, CVIII, p. 79.

42. B. Franklin to J. Banks, Nov. 21, 1783, in Rotch, *Benjamin Franklin*, p. 12.

43. Allen et al., *Diary of John Quincy Adams*, vol. 1, p. 192.

44. "Extract of a Letter from Paris, October 16," *General Evening Post* (London), Nov. 11, 1783; see also Cohen, "Franklin and Aeronautics," p. 124.

45. John Jay to Gouverneur Morris, July 20, 1783, John Jay Papers, Columbia University.

46. John Jay to William Livingston, Sept. 12, 1783, in Richard B. Morris, *John Jay: The Winning of the Peace* (New York, 1980), pp. 585–586.

47. Peter Jay Munro to John Jay, Oct. 16, 1783, MS. 42.315.132, Museum of the City of New York. The author thanks Jacqueline Hayott of the Department of Prints, Photographs, and Paintings for her assistance in obtaining copies of Peter Jay Munro's balloon correspondence.

48. John Jay to Peter Jay Munro, Oct. 20, 1783, Jay Papers.

49. B. Franklin to J. Banks, Nov. 21, 1783, in Rotch, *Benjamin Franklin*, p. 9.

50. *Ibid.*

51. *Ibid.*

52. *Ibid.*

53. *Ibid.*

54. *Ibid.*

55. *Ibid.*

56. *Ibid.*

57. *Ibid.*

58. B. Franklin to R. Price, Sept. 16, 1783, in Smyth, *Writings*, vol. 9, p. 85.

59. B. Franklin to J. Banks, Dec. 1, 1783, in Rotch, *Benjamin Franklin*, p. 14.

60. *Norwich [Conn.] Packet*, Feb. 26, 1784. All quotes are from this article. Biographical information on Daniel Lathrop Coit is from a manuscript volume, "Memorabilia," Connecticut Historical Society.

61. See also D. L. Coit to Dr. Joshua Coit, and reply, in "Memorabilia."

62. Sarah Livingston Jay to John Jay, Dec. 2, 1783, Jay Papers.

63. *Ibid.*

64. *Ibid.*

65. *Ibid.*

66. Peter Jay Munro to John Jay, Dec. 7, 1783, MS. 42.315.123, Museum of the City of New York.

67. B. Franklin to J. Banks, Nov. 21, 1783, in Rotch, *Benjamin Franklin*, p. 12.

68. B. Franklin to J. Ingenhousz, Jan. 16, 1784, in Smyth, *Writings*, vol. 9, p. 156.

69. B. Franklin to J. Banks, Nov. 21, 1783, in Rotch, *Benjamin Franklin*, p. 12.

70. B. Franklin to H. Laurens, Dec. 6, 1783, in Smyth, *Writings*, vol. 9, p. 13; B. Franklin to J. Ingenhousz, Jan. 16, 1784, *ibid.*, p. 156.

71. *Ibid.*

72. B. Franklin to [Sir Edward Newenham], June 20, 1783, *ibid.*, p. 12.

73. B. Franklin to J. Ingenhousz, Jan. 16, 1784, *ibid.*, p. 156.

74. B. Franklin to J. Banks, Nov. 21, 1783, in Rotch, *Benjamin Franklin*, p. 12.

75. Allen et al., *Diary of John Quincy Adams*, vol. 1, pp. 187–188.

76. Sarah Livingston Jay to John Jay, Dec. 4, 1783, Jay Papers.

77. John Adams to Abigail Adams, Sept. 7, 1783, Adams Family Papers, Massachusetts Historical Society.

78. John Quincy Adams to Peter Jay Munro, Nov. 10, 1784, Adams Family Papers.

79. *Ibid.*

80. E. C. Genêt, *Memorial on the Upward Forces of Fluids and Their Applicability to Several Arts, Sciences, and Public Improvements* (Albany, 1825), p. 25.

81. Caroline Amelia De Wendt, *Journal and Correspondence of Miss Adams, Daughter of John Adams, Written in France and England* (New York, 1841), vol. 1, pp. 18–19.

82. Hester Lynch Piozzi, *Observations and Reflections Made in the Course of a Journey Through France, Italy, and Germany* (Dublin, 1789), vol. 1, pp. 22–24; see also Howard C. Rice, *Thomas Jefferson's Paris* (Princeton, N.J., 1976), p. 26.

83. De Wendt, *Journal*, pp. 18–19.

Chapter 2

1. John Jay to Gouverneur Morris, July 20, 1783, John Jay Papers, Columbia University. The author owes an enormous debt to Ms Ene Simit, Associate Editor of the Jay Papers, for providing photocopies of this and other unpublished letters in the Jay Collection.

2. John Jay to Robert Livingston, Sept. 12, 1783, in H. P. Johnson, ed, *The Correspondence and Published Papers of John Jay* (New York, 1891), vol. 3, pp. 78–80.

3. Robert Livingston to John Jay, Nov. 29, 1783, Jay Papers.

4. Robert Morris to John Jay, Nov. 27, 1783, Jay Papers.

5. William Vans Murray to Henry Maynaidier, Feb. 8, 1789, Maynaidier Papers, Manuscript Division, Maryland Historical Society.

6. John Coakley Lettsom to Benjamin Rush, Feb. 28, 1784, Rush Manuscript XXVIII, p. 4, Historical Society of Pennsylvania.

7. Thomas Jefferson to Francis Hopkinson, Feb. 28, 1784, in Julian Boyd, ed., *The Papers of Thomas Jefferson* (Princeton, N.J., 1952) (hereafter, Boyd, *Jefferson Papers*) vol. 6, p. 542.

8. James McClurg to Thomas Jefferson, Apr. 12, 1784, in Boyd, *Jefferson Papers*, vol. 15, p. 610.

9. Thomas Jefferson to Philip Turpin, Apr. 28, 1784, Cincinnati Historical Society. The letter was reprinted in Marie Dickworth, ed., *Two Unpublished Letters of Thomas Jefferson Found in Ohio* (Oxford, Ohio, 1941), pp. 9–16.

10. George Washington to Louis Le Bèque du Portail, Apr. 4, 1784, in John C. Fitzpatrick, ed., *The Writings of George Washington* (Washington, D.C., 1938), vol. 27, p. 3.

11. *Salem [Mass.] Gazette*, Nov. 6, 1783.

12. *Ibid.*, Nov. 27, 1783.

13. *Ibid.*

14. "Air Balloons," *Essex Journal and Massachusetts and New Hampshire General Advertiser* (Boston), Jan. 12, 1784.

15. "Extract of a Private Letter from Paris," *Independent Ledger and American Advertiser* (Boston), Jan. 12, 1784.

16. *Ibid.*, Jan. 19, 1784; *Massachusetts Spy or Worcester Gazette*, Jan. 22, 1784.

17. *Massachusetts Spy or Worcester Gazette*, May 6, 1784.

18. "A Description of a Machine Proper to be Navigated through the Air," *Massachusetts Centinel*, May 12, 1784.

19. *Essex [Mass.] Journal*, June 9, 1784; *Massachusetts Spy or Worcester Gazette*, May 13, 1784; *Providence Gazette and Country Journal*, Mar. 26, 1785; Apr. 12, 1784; *Gazette of the State of Georgia*, Aug. 25, 1785; Oct. 14; Dec. 30, 1784.

20. *Connecticut Courant*, Mar. 16, 1784.

21. *Massachusetts Spy or Worcester Gazette*, Jan. 29, 1784.

22. *Gazette of the State of Georgia*, Oct. 14, 1784.

23. "The Air Balloon," *Independent* (Philadelphia), Apr. 17, 1784.

24. *Massachusetts Centinel*, Jan. 15, 1785; see also *The Balloon Almanac for the Year of Our Lord 1786* (Philadelphia, 1786).

25. *Gazette of the State of Georgia*, Oct. 14, 1784.

26. *Massachusetts Spy or Worcester Gazette*, Jan. 15, 1784.

27. *Connecticut Courant*, June 8, 1784.

28. *Gazette of the State of Georgia*, Sept. 30, 1784.

29. *Massachusetts Centinel*, July 14, 1784.

30. *Ibid.*, June 16, 1784.

31. *Boston Magazine* (Feb.–Sept. 1784).

32. *Massachusetts Spy or Worcester Gazette*, July 22, 1784.

33. *Pennsylvania Gazette*, Dec. 8, 1786.

34. *Freeman's Journal* (Philadelphia), July 28, 1784.

35. *Ibid.*

36. *Maryland Gazette*, Sept. 16, 1784 (from New York, Aug. 27).

37. *Providence Gazette and Country Journal*, Apr. 2, 1785.

38. *Freeman's Journal*, July 21, 1784.

39. *Journal de Paris*, May 13, 1784. One English translation appeared in the *Gentleman's Magazine* (London) (June 1784), pp. 433–434; another translation is given in Joseph Jackson, "The First Balloon Hoax," *Pennsylvania Magazine of History and Biography*, vol. 35 (1911).

40. Francis Hopkinson to Thomas Jefferson, Mar. 12, 1784, in Boyd, *Jefferson Papers*, vol. 7, p. 20.

41. Foulke biographical file, NASM Library.

42. Thomas Bond to B. Franklin, Apr. 27, 1780, Franklin Papers, the American Philosophical Society Library (hereafter, APSL).

43. J. Foulke to B. Franklin, Oct. 12, 1781, APSL.

44. *Ibid.*

45. *Ibid.*; see also Foulke to William Temple Franklin, Nov. 19, 1781, APSL.

46. G. Fox to William Temple Franklin, Oct. 12, 1780, APSL.

47. B. Franklin to the American Philosophical Society, Dec. 26, 1783, APSL.

48. American Philosophical Society, *Early Proceedings of the American Philosophical Society* (Philadelphia, 1884), p. 125.

49. Francis Hopkinson to Thomas Jefferson, Mar. 12, 1784, in Boyd, *Jefferson Papers*, vol. 7, p. 20.

50. Unpublished notes by Benjamin Franklin, Franklin Papers, American Philosophical Society, L(i), p. 51.

51. *Freeman's Journal*, June 2, 1784.

52. Francis Hopkinson to Thomas Jefferson, Mar. 31, 1784, in Boyd, *Jefferson Papers*, vol. 7, p. 57.

53. Original note and ticket in the Ramsey

Rare Book Room, NASM Library. See also Horace Mather Lippincott, "Dr. John Foulke, 1780, A Pioneer in Aeronautics," *The General Magazine and Historical Chronicle*, vol. 34, no. 1 (October 1934), pp. 525–531.

54. *Ibid.*

55. George Washington to Sir Edward Newenham, Nov. 25, 1785, in Fitzpatrick, *Writings of Washington*, vol. 28, p. 323.

56. *Ibid.*

57. Francis Hopkinson to Thomas Jefferson, Mar. 31, 1784, in Boyd, *Jefferson Papers*, vol. 7, p. 57.

58. Francis Hopkinson to Thomas Jefferson, May 12, 1784, *ibid.*, vol. 7, pp. 245–246.

59. *Ibid.*

60. Francis Hopkinson to B. Franklin, May 24, 1784, in Jared Sparks, ed., *The Works of Benjamin Franklin* (Boston, 1840), vol. 6, p. 452.

61. *Ibid.*

62. *Pennsylvania Gazette*, May 12, 1784.

63. *Ibid.*

64. Thomas Jefferson to James Monroe, May 21, 1784, in Boyd, *Jefferson Papers*, vol. 7, p. 280.

65. *Independent Gazetteer*, May 29, 1784.

66. *Freeman's Journal*, July 21, 1784.

67. *Pennsylvania Gazette*, June 30, 1784; see also undated Philadelphia articles in the Franklin Scrapbook, American Institute of Aeronautics and Astronautics (AIAA) History Collection, Manuscript Division, Library of Congress. See also *Independent Gazetteer*, June 26, 1784.

68. Minutes of the American Philosophical Society, June 11, 1784, APSL.

69. *Ibid.*, June 19, 1784; see also Whitfield Bell, Jr., "John Morgan," *Bulletin of the History of Medicine*, vol. 22, no. 5 (Sept.–Oct. 1948), pp. 543–562.

70. *Ibid.*

71. *Pennsylvania Gazette*, June 30, 1784.

72. Information on the early life of Peter Carnes has been drawn from a variety of sources in: Prince George's County Courthouse, Upper Marlboro, Md.; the Maryland Historical Society, Baltimore, Md.; and the Maryland Hall of Records, Annapolis, Md. For a good summary of these record groups, see *A Brief History of the George Washington House, Bladensburg, Md.* (prepared by John Wolton for the Prince George's County Jaycees, Jan. 8, 1974). A copy of this report is contained in the NASM Library and in the Maryland Room, McKeldon Library, University of Maryland. R. Lee Van Horn, *Out of the Past: Prince Georgians and Their Land* (Riverdale, Md., 1976), is also a helpful compilation of local legal records. The author also owes a debt to William A. Ayleshire of the Prince George's County Historical Society for data on Peter Carnes.

73. John Belton O'Neal, *Biographical Sketches of the Bench and Bar of South Carolina*, 2 vols. (Charleston, S.C., 1859), n.p.

74. *Idem*, *The Annals of Newberry* (1858), pp. 12–19.

75. *Ibid.*

76. *Ibid.*, p. 20.

77. *Maryland Journal and Baltimore Advertiser*, June 22, 1784.

78. *Ibid.*, Mar. 5, 12, 19, 21; Apr. 16; May 14, 21, 1784.

79. *Ibid.*, June 15, 1784.

80. *Ibid.*; see also *Virginia Journal* (Alexandria), June 24, 1784.

81. For information on Bladensburg and its leading citizens, see John Pendleton Kennedy, *Memoirs of the Life of William Wirt* (Philadelphia, 1856). Scattered notes on the area are also to be found in the William Wirt Papers, Maryland Historical Society, and in the Maryland Room, McKeldon Library, University of Maryland.

82. *Virginia Journal*, June 10, 1784.

83. *Maryland Journal and Baltimore Advertiser*, June 22, 1784.

84. *Virginia Journal*, June 10, 1784.

85. *Ibid.*, July 1, 1784.

86. Letterbook, Johannot Johnson and Co.,

MS. 497–9, Manuscript Division, Maryland Historical Society.

87. *Virginia Journal*, July 1, 1784.

88. *Maryland Journal and Baltimore Advertiser*, June 15, 22, 25, 1784.

89. A thorough search of state and local records, including a survey of the very extensive genealogical materials in the library of the Maryland Historical Society, has failed to unearth any information on Edward Warren.

90. *Pennsylvania Packet*, June 17, 1784.

91. "American Aerostatic Balloon," *Independent Ledger and American Advertiser*, Aug. 2, 1784 (article datelines Philadelphia, June 21).

92. "Air Balloon," *Pennsylvania Gazette*, July 29, 1784; see also *Connecticut Journal*, July 28, 1784.

93. *Maryland Gazette*, July 29, 1784; *Virginia Gazette*, July 29, 1784; *Maryland Journal*, June 25, 1784; *Connecticut Journal*, July 28, 1784; *Gazette of the State of Georgia*, Aug. 12, 1784; *Freeman's Journal*, July 21, 1784; *Pennsylvania Gazette*, July 21, 1784.

94. John Page, "The Balloon," ms copy, Virginia State Library.

95. For information on Carnes's legal problems and later career, see Wolton, *History of the George Washington House* (see above, n. 72); and Raymond D. Hall, "The Search for Peter Carnes," *Richmond County History*, vol. 10, no. 2 (Summer 1978), pp. 5–11.

96. *Augusta Chronicle and Gazette of the State*, Aug. 2, 1794.

97. *Massachusetts Centinel*, June 30, 1784.

98. *Pennsylvania Gazette*, July 28, 1784.

99. *Ibid.*; *Maryland Gazette*, July 1, 1784.

100. Bell, "John Morgan," p. 562.

Chapter 3

1. John Jeffries Diary, Nov. 25, 1783, original in Houghton Library, Harvard University, MS. Am 1220 (hereafter, Jeffries Diary).

2. Jeffries Manuscript, MSS. Am 1220.7, 11, 12, Houghton Library, Harvard University.

3. Jeffries Diary, 1781–1784; see Sept. 23, 1781, for example.

4. Jeffries Diary, July 4, 1781. Other entries indicate a close relationship between Jeffries and Copley.

5. Jeffries Diary; see Mar. 30, 1779, for example.

6. *Ibid.*, July 27, 1781.

7. Jeffries's diaries for the years 1781–1784 are filled with similar references.

8. Jeffries Diary, Mar. 25, 1783.

9. Jeffries Diary, Sept. 1, 1782.

10. *Ibid.*

11. Jeffries Diary, Jan. 3, 1783.

12. L. T. C. Rolt, *The Aeronauts: A Dramatic History of the Great Age of Ballooning* (New York, 1966), p. 61. See also Blanchard files and correspondence in the Tissandier Collection, Manuscript Division, Library of Congress.

13. J. Banks to B. Franklin, in Sparks, *Works of Benjamin Franklin*, vol. 4, p. 32.

14. Rolt, *Aeronauts*, p. 61.

15. On early British aeronautics, see J. E. Hodgson, *The History of Aeronautics in Great Britain* (Oxford, 1924).

16. *London Morning Herald*, Sept. 17, 1784. The author used a copy in the Upcott Scrapbook, NASM Library, in which the exact date was not given.

17. *London Morning Post*, Feb. 6, 1784.

18. *Ibid.*

19. *Ibid.*

20. *Ibid.*, July 21, 1786.

21. For information on Tytler, see Sir James Fergusson of Kilkerran, *Balloon Tytler* (London, 1972).

22. Rolt, *Aeronauts*, p. 69.

23. Jeffries Diary, Sept. 15, 1784.

24. *Ibid.*

25. *Gazette of the State of Georgia*, Dec. 23, 1784.

26. Jeffries Diary, Oct. 16, 1784.

27. John Jeffries, A *Narrative of Two Aerial Voyages of Doctor Jeffries With Mons. Blanchard* (London, 1786), p. 11.

28. *Ibid.*, p. 10.

29. "Reminiscences of Dr. Jeffries," undated news article, Jeffries biographical file, NASM Library.

30. Kurt R. Stehling, *Bags Up!* (Chicago, 1975), p. 107.

31. Rolt, *Aeronauts*, p. 84.

32. Léon Coutil, *Jean-Pierre Blanchard: Biographie et Iconographie* (Paris, 1911).

33. Jean Pierre Blanchard, *Journal and Certificates of the Fourth Aerial Voyage of Mr. Blanchard* (London, 1784). This volume contains a detailed description of Blanchard's inflation procedure.

34. Jeffries Diary, Nov. 1–29, 1784; Jeffries, *Voyages*; Jeffries, manuscript account, MS. Am 1210.6, "Remarks and Observations on Balloon Ascent," Houghton Library; *London Chronicle*, Dec. 8, 1784.

35. *Ibid.*, quote from MS. Am 1210.6.

36. MS. Am 1210.6.

37. Photostat of note from Amherst College Library Collection.

38. Jeffries Diary, Nov. 30, 1784; MS. Am 1210.6.

39. *Ibid.*

40. Jeffries Diary, Dec. 2, 1784.

41. MSS. Am 1210.6; 1210.4. See also *London Morning Post*, Dec. 4, 1785.

42. Jeffries Diary, Nov.–Jan. 1784.

43. Repeated references to "honourable," "amiable," and "Little Captain" in Jeffries Diary for the period.

44. Jeffries Diary, Dec. 3, 1784.

45. Jeffries, *Voyages*, pp. 40–41.

46. *Ibid.*; see also Jeffries Diary, Jan. 5, 1785.

47. MS. Am 1210.4; see also Jeffries, *Voyages* and Diary.

48. Claudia Kidwell, "Apparel for Ballooning with Speculations on More Commonplace Garb," *Costume* (London, 1977), pp. 73–85, provides much insight into the matter of Jeffries's attire. Some of the items have survived in the collection of Harvard University. For more detail, see Jeffries Diary and MSS. Am 1210.5 and 1220.14.

49. Jeffries Diary, Jan. 7, 1785; see also unidentified articles on the Channel flight in the Upcott Scrapbook, NASM Library. Good, though often undated, clippings are also to be found in box 177, AIAA History Collection, Manuscript Division, Library of Congress.

50. Jeffries Diary, Jan. 7, 1785; see also Jeffries's open letter from Calais, original in the British Museum, copy in the Jeffries file, NASM Library.

51. *Kentish Gazette*, Jan. 15, 1785.

52. *Ibid.*; *Gazette of the State of Georgia*, Apr. 28, 1785.

53. *Ibid.*

54. *Ibid.*

55. *Ibid.*

56. Jeffries, *Voyages*, p. 87.

57. *Ibid.*; Jeffries Diary, Jan. 7, 1785.

58. *Ibid.*

59. Jeffries, *Voyages*, p. 48.

60. *Ibid.*

61. Jeffries Diary, Jan. 8–12, 1785.

62. *Ibid.*, Jan. 16, 1785; *Gazette of the State of Georgia*, Apr. 25, 1785.

63. Jeffries Diary, Jan. 13, 1785.

64. *Ibid.*

65. *Ibid.*, Jan. 23, 1785.

66. *Ibid.*

67. *Ibid.*, Jan. 26, 1785.

68. *Ibid.*, Jan. 14, 1785.

69. *Ibid.*, Feb. 18, 1785.

70. *Ibid.*, Jan. 22, 1785.

71. De Wendt, *Journal*, vol. 2, p. 47.

72. Jeffries Diary, Feb. 3, 1785.

73. *Ibid.*, Mar. 5, 1785.

74. *Ibid.*, June 2, 1785.

75. *Ibid.*, May 23, 1785.

76. All quotations from *Lloyd's Evening Post*, June 28–30, 1786; see also correspondence in MSS. Am 1220.15–18.

77. Oliver Wendell Holmes, *Medical Essays, 1842–1882* (Boston, 1892), p. 349.

Chapter 4

1. *New Jersey Gazette*, Feb. 21, 1785.

2. *Gazette of the State of Georgia*, Mar. 10, 1785.

3. "Extract of a Letter from New York," July 15, 1785, Upcott Scrapbook, NASM Library.

4. *Gazette of the State of Georgia*, Feb. 24, 1785.

5. *Providence Gazette and Country Journal*, Apr. 2, 1785.

6. *Ibid.*

7. "American Notes," Upcott Scrapbook; *Norwich [Conn.] Packet*, Feb. 24, 1785.

8. Franklin Bowditch Dexter, ed., *The Literary Diary of Ezra Stiles* (New York, 1901), vol. 3, p. 157. The author has also consulted a Yale University microfilm of the original diary. My thanks to the manuscript division, Sterling Library, Yale University.

9. *Ibid.*, pp. 112, 129.

10. *Ibid.*, pp. 157–161.

11. *Ibid.*

12. Rev. James Madison to Thomas Jefferson, Apr. 28, 1784, in Boyd, *Jefferson Papers*, vol. 7, p. 133.

13. Rev. James Madison to Thomas Jefferson, Mar. 27, 1786, *ibid.*, vol. 9, p. 355.

14. *Ibid.*; see also "Letters of Reverend James Madison, President of the College of William and Mary, to Thomas Jefferson," *William and Mary Quarterly* (April 1925), pp. 77–95.

15. "Letters from William and Mary College, 1798–1801," *The Virginia Magazine of History and Biography* (April 1921), pp. 129–179.

16. *Ibid.*

17. *Ibid.*

18. *Ibid.*

19. "The Royal Balloon" and other undated articles in the Deeker file, Tissandier Collection, Manuscript Division, Library of Congress.

20. *New York Advertiser*, June 7, 1789; *New York Gazetteer or Daily Evening Post*, July 7, 1789; *New York Packet*, Aug. 6, 1789.

21. *New York Advertiser*, Aug. 8, 1789; *New York Packet*, Aug. 11, 1789; *New York Journal*, Aug. 13, 1789; *New York Daily Advertiser*, Aug. 11, 1789.

22. *New York Journal and Weekly Register*, Aug. 20, 1789.

23. *Ibid.*

24. Dexter, *Literary Diary*, vol. 3, p. 367.

25. *New York Packet*, Sept. 23, 1789; *New York Journal*, Sept. 24, 1789.

26. *Ibid.*

27. *New York Journal*, Nov. 15, 1789.

28. *New York Daily Advertiser*, June 3, 1793; June 20, 1793; see also references to James West in George C. D. Odell, *Annals of the New York Stage* (New York, 1927).

29. *Federal Gazette*, Dec. 17, 1792. Lewis Leary, "Phaeton in Philadelphia: Jean Pierre Blanchard and the First Balloon Ascension in America," *The Pennsylvania Magazine of History and Biography* (1943), pp. 49–60; and Carroll Frey, *The First Aerial Voyage in America* (Philadelphia, 1943), are the two best accounts of Blanchard's activity in Philadelphia. Frey's volume contains a complete reprinting of Blanchard's *Journal of My Forty-Fifth Ascension and the First in America* (Philadelphia, 1793).

30. *American Apollo* (Boston), Dec. 28, 1792; *Dunlop's American Advertiser*, Dec. 24, 25, 1792; "To Mr. Blanchard," Dec. 25 and 31, 1792;

Jan. 3, 14, 22, 26, 1793; *National Gazette*, Jan. 2, 1793.

31. *Federal Gazette*, Jan. 5, 1793.

32. *National Gazette*, Jan. 5, 1793.

33. "On Balloons," *National Gazette*, Jan. 2, 1793.

34. *Ibid.*

35. Unidentified article, Blanchard biographical file, NASM Library.

36. "Reflections on Balloons," *National Gazette*, Jan. 30, 1793.

37. *Ibid.*

38. *National Gazette*, Jan. 19, 1793.

39. "The Echo," *American Mercury* (Hartford), Feb. 25, 1793.

40. Charles Francis Adams, Jr., ed., *Letters of John Adams Addressed to His Wife* (Boston, 1841), vol. 2, p. 119.

41. *Dunlop's American Daily Advertiser*, Dec. 28, 1793.

42. Jean Pierre Blanchard, *Voyage*, p. 15.

43. Quoted in Frey, *First Aerial Voyage*, p. 18.

44. See introductory essay in Frey, *First Aerial Voyage*.

45. Lyman H. Butterfield, ed., *The Autobiography of Benjamin Rush* (Princeton, N.J., 1948), p. 303.

46. *Ibid.*, p. 304; Blanchard, *Voyage*.

47. Blanchard, *Voyage*, p. 14.

48. Butterfield, *Rush Autobiography*, p. 304.

49. Archibald Henderson, "Washington and Aeronautics," *The Archive* (May 1932), p. 8.

50. Blanchard, *Voyage*, p. 20.

51. *Ibid.*, p. 21.

52. Quoted in Frey, *First Aerial Voyage*, p. 18.

53. Lyman H. Butterfield, ed., *The Letters of Benjamin Rush* (Princeton, N.J., 1951), vol. 2, p. 627.

54. *Ibid.*

55. *Ibid.*

56. *Columbian Centinel* (Boston), Jan. 12, 19, 30; Apr. 2, 1793.

57. *National Gazette*, Jan. 30, 1793; *General Advertiser*, May 14, 1793.

58. *General Advertiser*, May 14, 1793; *National Gazette*, May 25, 1793.

59. *General Advertiser*, June 6, 1793; *Federal Gazette*, June 25, 1793.

60. *National Gazette*, May 25, 1793.

61. *Federal Gazette*, June 29, 1793.

62. *Ibid.*, June 29; Aug. 28, 29, 1793; *National Gazette*, July 3; Sept. 21, 1793.

63. J. H. Powell, *Bring Out Your Dead: The Great Plague of Yellow Fever in Philadelphia in 1793* (Philadelphia, 1949), is the most complete study of the catastrophe.

64. *General Advertiser* (Aurora, Pa.), Nov. 26, 1794.

65. *Charleston City Gazette*, Nov. 14, 1794; see also George C. Rogers, Jr., ed., "The Letters of William Smith to Edmond Rutledge," *The South Carolina History Magazine* (January 1969), pp. 38–59.

66. *General Advertiser* (Aurora, Pa.), Dec. 30, 1794.

67. *Ibid.*

68. *Federal Quarterly* (Boston), July 30, 1795; see also assorted Blanchard correspondence in the Tissandier Collection.

69. Jeffries MS. Am 1220.15–17; all in the Houghton Library, Harvard University.

70. Jeffries Diary, Jan. 30; Feb. 23–25; Apr. 13, 1793; passim to Nov. 27, 1795.

71. *General Advertiser* (Philadelphia), Apr. 28, 1794.

72. *Columbian Gazetteer* (New York), Oct. 10, 1793. For a detailed treatment of the museum, see Robert M. and Gale S. McClurg, "Tammany's Remarkable Gardiner Baker: New York's First Museum Proprietor, Menagerie Keeper, and Promoter Extraordinary," *New York Historical Society Quarterly*, vol. 42, no. 2 (April 1958), pp. 143–169.

73. William A. Duer, *Reminiscences of an Old New Yorker* (New York, 1867), p. 9.

74. Isaac Newton Phelps Stokes, *Iconography of Manhattan Island, 1498–1909* (New York, 1915–1928), entry for June 13, 1796.

75. *Minerva* (New York), July 9, 1796; *Columbian Museum*, July 8, 1796; *Argus* (New York), June 27, 1796; see also Gardiner Baker, *The Principles, History, and Use of Air-Balloons* (New York, 1796).

76. Baker, *Principles*, p. 43.

77. *Ibid.*

78. *Diary* (New York), Sept. 17, 19, 1796; *New York Journal*, Sept. 20, 1796.

79. *Minerva and Mercantile Evening Advisor*, Oct. 31; Nov. 14; Dec. 7, 1796.

80. *Ibid.*

81. *Ibid.*

82. *Ibid.*

83. *Minerva*, Nov. 2, 1796.

84. *Ibid.*

85. Handbill, "Balloon's Misfortune and Hope for Relief," New York Historical Society; *Diary* (New York), May 17, 1797.

86. *Argus*, Sept. 29, 1796.

87. *New York Gazette and General Advertiser*, Mar. 4, 1797.

88. *The Diary and Mercantile Advisor*, May 11, 1797.

Chapter 5

1. *Columbian Centinel* (Boston), Oct. 9, 1796.

2. Odell, *Annals*, vol. 2, p. 116.

3. *New York Daily Advertiser*, July 11, 1800.

4. Odell, *Annals*, p. 116.

5. *Ibid.*, p. 117.

6. *New York Daily Advertiser*, Oct. 14, 1800.

7. *Columbian Centinel* (Boston), Dec. 20, 1800.

8. *Ibid.*, July 24, 1805.

9. *Connecticut Courant*, July 16, 1834.

10. *Ibid.*, Nov. 9, 1819.

11. *Nile's Weekly Register*, Aug. 31, 1816, p. 44.

12. *Ibid.*, June 10, 1820, p. 312.

13. Anon., *Parlour Magic* (Philadelphia, 1834), p. 95.

14. *Ibid.*, p. 126.

15. *Ibid.*, p. 95.

16. *Ibid.*, p. 127.

17. "Making Small Fire Balloons," *Scientific American* (September 19, 1863).

18. "Balloon," *Atkinson's Saturday Evening Post* (Philadelphia), Sept. 15, 1838.

19. *New York Evening Post*, Sept. 1, 1825.

20. *Atkinson's Saturday Evening Post*, Sept. 15, 1838.

21. *Ibid.*

22. Odell, *Annals*, vol. 2, p. 272.

23. See entry for Oct. 28, 1799, in *A List of Patents Granted by the United States from April 10, 1790, to December 31, 1836* (Washington, D. C., 1872).

24. *Connecticut Courant*, June 2 and 30, 1800.

25. *Ibid.*, June 2, 1800.

26. *Ibid.*, June 30, 1800.

27. *New York Daily Advertiser*, July 2, 1800.

28. *Connecticut Courant*, Sept. 1, 1800; May 10 and 11, 1801.

29. *The Times* (Hartford), Aug. 17, 1819; *New York Mercantile Advertiser*, Aug. 4, 1819.

30. *New York Evening Post*, Aug. 3, 1819.

31. David B. Lee to Thomas Jefferson, Apr. 5, 1822, in Boyd, *Jefferson Papers*, vol. 22, pp. 602–603.

32. Lee was later to become involved in a bitter dispute with James Bennett, who had petitioned Congress for financial support. Bennett hoped to build a flying machine, which Lee believed was a direct copy of his own design. The controversy was a tempest in a teapot which everyone ignored. It was, however, the first time that Congress considered a bill relating to aerial navigation.

33. D. B. Lee to Thomas Jefferson, Apr. 5, 1822.

34. *Ibid.*

35. Undated clipping, Scrapbook 216, AIAA History Collection, Manuscript Division, Library of Congress; *United States Gazette and True American* (Philadelphia), Sept. 3, 1819.

36. *Union* (Philadelphia), Sept. 8 and 21, 1819; J. Thomas Scharff and Thomas Westcott, *A History of Philadelphia* (Philadelphia, 1884), p. 598; quotation, Lee to Jefferson, Apr. 5, 1822.

37. Scharff and Westcott, *History*, p. 598.

38. *Ibid.*

39. *United States Gazette*, quoted *ibid.*

40. *Connecticut Mirror*, Nov. 1, 1819; *Connecticut Courant*, Nov. 2, 1819; John Pintard, *The Letters of Jon Pintard to His Daughter Elizabeth Noel Pintard Davidson* (New York, 1940), vol. 1, pp. 210–211, 237.

41. *Connecticut Mirror*, Nov. 1, 1819.

42. *Ibid.*

43. *Ibid.*; *New York Evening Post*, Oct. 28, 1819.

44. *New York Evening Post*, Oct. 28; Nov. 18, 20, 23, 1819; see also undated article, AIAA Scrapbook; Cincinnati *Advertiser*, Dec. 14, 1819.

45. Anna Boyd McHenry to John McHenry, no date, MS. 647, Manuscript Division, Maryland Historical Society.

46. *Ibid.*

47. *Federal Gazette* (Baltimore), Apr. 21, 24, 27, 31, 1820.

48. *Ibid.*, Apr. 20; May 15 and 19, 1820.

49. *New York Spectator*, Oct. 17, 1820.

50. *Columbian Centinel*, Oct. 17 and 20, 1820; *Federal and Baltimore Republican*, Oct. 20, 1820; *Connecticut Mirror*, Nov. 1, 1820.

51. Odell, *Annals*, p. 604.

52. *New York Post*, May 21, 1821; Odell, *Annals*, p. 604.

53. *Ibid.*

54. William Johnson, *Reports of Cases Argued and Determined in the Supreme Court of Judicature and in the Court for the Trial of Impeachments and the Correction of Errors* (New York, 1882), pp. 381–383.

55. Robert T. Schendal, "Flight Safety," *Ballooning* (Sept.–Oct. 1980), p. 8.

56. *Boston Intelligencer and Evening Gazetteer*, Sept. 1, 1822.

57. Wilfrid de Fonvielle, *Histoire de la Navigation Aérienne* (Paris, 1907), p. 195.

58. E. Roch, *Essais sur les Voyages Aériens d'Eugène Robertson en Europe, aux Etats-Unis et aux Antilles* (Paris, 1831), chaps. 1–4, passim. This volume is virtually the only source of information on Robertson's American tour. All quotations are from the present writer's translation.

59. *Ibid.*, p. 19.

60. Information on the New York pleasure gardens is drawn from what must be regarded as the greatest of all U.S. city resource books, I. N. Phelps Stokes, *The Iconography of Manhattan Island, 1498–1909* (New York, 1915–1928). Information on the Philadelphia pleasure gardens is from Scharff and Westcott, *History*.

61. *New York Sun*, July 9, 1825.

62. *Albany Argus*, July 15, 1825; see also *Nile's Weekly Register*, July 16, 1825, p. 320.

63. *Ibid.*; Roch, *Essais*, pp. 18–23.

64. *New York Evening Post*, July 11, 1825.

65. *Ibid.*

66. *Ibid.*

67. Roch, *Essais*, pp. 24–27.

68. Advertisements and handbills in Eugène Robertson file, Tissandier Collection.

69. Roch, *Essais*, pp. 33–44; *American Journal of Science* (June 1827), pp. 166–168.

70. Roch, *Essais*, pp. 33–44.

71. *Ibid.*

72. *Ibid.*

73. Roch, *Essais*, p. 50.

74. *Ibid.*, pp. 51–62.

75. *Ibid.*, pp. 63–69.

76. *New York Statesman*, Sept. 24, 1828; Roch, *Essais*, pp. 70–74; see also *New York Statesman*, Dec. 29, 1826.

77. *Daily Chronicle* (New York), Oct. 28, 1828; Roch, *Essais*, pp. 75–83.

78. *Ibid.*

79. *New York Sun*, July 7, 1834; Eugène Robertson, *Relation du Premier Voyage Exécuté dans la République Mexicaine* (Paris, 1835), pp. 1–8; *Saturday Chronicle and Mirror of the Times* (Philadelphia), Sept. 16, 1837.

80. Jose Villas Gomez, *Breve Historia de la Aviacion en Mexico* (Mexico City, 1971), pp. 26–27.

Chapter 6

1. *New York American*, Oct. 24, 1825.

2. *National Gazette and Literary Register* (Philadelphia), Oct. 25, 1825.

3. *New York Statesman*, Oct. 25, 1825.

4. *National Gazette and Literary Register*, July 24, 1828.

5. *Ibid.*, Oct. 25 and 29, 1828; *Nile's Weekly Register*, Oct. 25, 1828.

6. *National Gazette and Literary Register*, Oct. 29, 1828.

7. Most of the biographical data on Durant has been drawn from obituary articles, such as *New York Evening Post*, Mar. 3, 1873, as well as undated clippings annotated by Emma Durant in the NASM Library biographical file.

8. *Independent* (New York), Jan. 2, 1851; collection of additional reviews in the Durant biographical file, NASM Library.

9. Undated obituary articles in the NASM Library biographical file.

10. Peter Parley, "History of Balloons," *Parley's Magazine for 1837* (New York, 1837), pp. 230–240.

11. Emma Durant to Paul E. Garber, Aug. 16, 1932, NASM Library biographical file.

12. Undated advertising clippings in the Charles Ferson Durant Scrapbook, box 233, AIAA History Collection, Manuscript Division, Library of Congress (hereafter, Durant Scrapbook).

13. *New York American*, Sept. 9, 1830.

14. *Bell's Life in New York*, Sept. 14, 1830.

15. *Daily Centinel*, Sept. 10, 1830.

16. *Ibid.*; *New York Evening Post*, Sept. 10, 1830.

17. "The Aeronaut's Address"; copy in Durant Scrapbook.

18. *New York Evening Post*, Sept. 11, 1830; *New York Statesman*, Sept. 10, 1830; *Bergen County Gazette*, Sept. 11, 1830; *New York Gazette*, Sept. 10, 1830; *National Gazette and Literary Register*, Sept. 14, 1830.

19. *New York Journal of Commerce*, Sept. 23, 1831.

20. *New York Commercial Advertiser*, Sept. 22, 1831.

21. *National Gazette and Literary Register*, Sept. 25, 1831.

22. *Long Island Star*, Sept. 22–23, 1831.

23. *New York Standard*, Aug. 26, 1831; see also *New York Daily Advertiser*, Aug. 13, 1831; *New York Commercial Advertiser*, Aug. 23, 1831; *New York Journal of Commerce*, Aug. 24–25, 1831; *New York Gazette*, Aug. 26, 1831.

24. *National Gazette and Literary Register*, Sept. 8, 1831; *Mercantile Advertiser*, Sept. 8, 1831.

25. *New York Gazette*, Sept. 12, 1831.

26. *New York American*, Sept. 8, 1831.

27. *New York Evening Post*, Sept. 8, 1831.

28. *Ibid.*

29. *New York Standard*, Sept. 9, 1831.

30. *New York Evening Journal*, Sept. 10, 1831.

31. *New York Gazette*, May 17; June 5, 1832; *New York Commercial Advertiser*, June 5, 1832.

32. *New York Advertiser*, May 30, 1833; *New York Gazette*, May 30, 1833.

33. *New York Traveler, Times and Journal*, June 1, 1833.

34. *Ibid.*, *New York American*, May 25, 1833; *Morning Post*, May 28, 1833; *New York Gazette*, May 29, 1833; *New York Journal of Commerce*, May 30, 1833; *New York Commercial Advocate*, May 30, 1833; *New York Daily Advocate*, May 31, 1833.

35. *Atkinson's Saturday Evening Post*, June 8, 1833.

36. *New York Gazette*, June 14, 1833.

37. Undated article, Durant Scrapbook.

38. *Portland Gazette*, June 17, 1833.

39. *Ibid.*

40. Undated article, Durant Scrapbook.

41. *New York Standard*, June 13 and 15, 1833; *Courier* (Charleston, S.C.), June 16–20, 1833; *New York Gazette*, June 12, 1833; *New York Daily Advocate*, June 12, 1833; *New York American*, June 11, 1833.

42. *American* (Baltimore), June 24, 1833.

43. *Gazette* (Philadelphia), Sept. 4, 1833; *Boston Gazette*, June 19, 1833; *American* (Baltimore), June 24, 1833; New York *Traveler*, Sept. 5, 1833; *Atkinson's Saturday Evening Post*, June 24, 1833.

44. *Ibid.*

45. *Ibid.*

46. Albany *Microscope*, n.d., in Durant Scrapbook.

47. Albany *Advocate*, quoted in *Mercantile Advertiser and New York Advocate*, July 18, 1833; *Daily Albany Argus*, July 25, 1833; Albany *Evening Journal*, July 31 and Aug. 3, 1833.

48. Albany *Evening Journal*, Aug. 17, 1833.

49. *New York Gazette*, July 3, 1833; *Boston Evening Transcript*, July 19 and Aug. 28, 1833; *Palladium* (Morristown, N.J.), July 24, 1833.

50. *Daily Albany Argus*, Aug. 9–10, 1833; Albany *Microscope*, Aug. 3, 1833.

51. Albany *Evening Journal*, Aug. 16, 1833.

52. *New York Gazette*, Sept. 7, 1833; Baltimore *Journal of Commerce*, Sept. 9, 1833; Albany *Daily Advocate*, Sept. 10, 1833.

53. Baltimore *American*, Sept. 27, 1833; *Baltimore Chronicle*, Sept. 16 and 23, 1833; Easton (Md.) *Gazette*, Sept. 26, 1833.

54. Baltimore *Gazette*, Sept. 26, 1833.

55. Hagerstown (Md.) *Free Press*, Oct. 3, 1833.

56. *Ibid.*

57. Baltimore *Gazette*, Oct. 2, 1833; *Baltimore Chronicle*, Oct. 3, 1833; *Sentinel* (Baltimore), Oct. 4, 1833; scattered additional articles in Durant Scrapbook.

58. "To Mr. C. F. Durant," *Sentinel* (Baltimore), Oct. 4, 1833.

59. *Baltimore Visitor*, Oct. 11, 1833.

60. *Baltimore Chronicle*, Oct. 11, 1833.

61. *Republican and Commercial Advertiser*, Oct. 15, 1833; *Baltimore Chronicle*, Oct. 16, 1833; see other articles in Durant Scrapbook.

62. Boston *Courier*, July 8, 1834; *Traveller* (Boston), July 11; Aug. 1, 1834; *Commercial Gazette*, Aug. 1, 1834; *National Palladium* (Boston), Aug. 1, 1834; *Daily Journal* (Lowell, Mass.), July 21, 1834; *Columbian Centinel* (Boston), July 29, 1834; *Commercial Gazette*, Aug. 1, 1834; Boston *Atlas*, Aug. 1, 1834; *Boston Evening Transcript*, Aug. 1, 1834.

63. *Traveller* (Boston), Aug. 19, 1834; *Boston Gazette*, Aug. 20–26, 1834; *Courier* (Charleston, S.C.), Aug. 18, 1834; Boston *Morning Post*, Aug. 26, 1834; *Columbian Centinel*, Aug. 26, 1834. See also undated articles in Durant Scrapbook.

64. *Daily Advocate* (Boston), Sept. 7 and 15, 1834; *Boston Gazette*, Sept. 11, 1834; *Observer*, Aug. 30, 1834; *Boston Centinel*, Sept. 15, 1834; *Boston Evening Transcript*, Sept. 15, 1834; *Daily Advertiser and Patriot* (Boston), Sept. 15, 1834.

65. *Baltimore Visitor*, Sept. 28, 1834.

66. *Ibid.*

67. *Ibid.*

68. *Ibid.*

69. *Gazette* (Baltimore), July 2, 1833.

70. *Evening Star* (Baltimore), July 2, 1834.

71. Baltimore *Patriot*, Aug. 12, 1834.

72. *Nile's Weekly Register*, Apr. 5, 1834, p. 85.

73. *New York American*, May 9, 1834; *National Intelligencer*, May 3, 1834; *Nile's Weekly Register*, May 3, 1834, p. 148.

74. *Nile's Weekly Register*, May 10, 1834; *American Sentinel* (Baltimore), June 24, 1834; Baltimore *Patriot*, Mar. 8, 1834; *Evening Star* (Baltimore), Mar. 10; June 28, 1834; *Gazette*, May 26, 1834; *Baltimore Chronicle*, June 9, 1834; *Atkinson's Saturday Evening Post* (Philadelphia), June 28, 1834; Baltimore *American*, Apr. 8, 1834; *New York Gazette*, June 30, 1834.

75. *Ibid.*

76. *Baltimore Chronicle*, May 29 and 31; June 2, 1834. The statement that Ash was an artist is based on his identification as a "Painter" in the 1834 Baltimore Directory as well as on scattered references to him as an artist in contemporary advertisements.

77. *Baltimore Chronicle*, May 31; June 3–4, 1834; *Atkinson's Saturday Evening Post*, June 7, 1834.

78. *Baltimore Chronicle*, July 8 and 12, 1834.

79. *National Intelligencer*, July 22, 1834; Baltimore *Patriot*, July 22, 1834.

80. *National Intelligencer*, quoted in Baltimore *Patriot*, Aug.1,1834; *Boston Morning Post*, Aug. 14; Sept. 18, 1834; *Gazette* (Baltimore), Aug. 1; Sept. 29, 1834. Other unidentified articles can found in the AIAA Scrapbooks, AIAA History Collection, Manuscript Division, Library of Congress. These scrapbooks also contain three advertisements announcing the ascension of one Joseph Ames from Analoston Garden on Mason's Island in the District of Columbia. The ads are not dated, but the style suggests a date before 1840. The author has been unable to locate any other information on the Ames ascents.

81. *Baltimore Chronicle*, June 17, 1834; *Gazette* (Baltimore), June 18, 1834; undated articles under Woodall, box 220, AIAA Scrapbooks.

82. *Ibid.*

83. *Gazette* (Baltimore), May 26; June 26, 1834.

84. *Ibid.*

85. *Baltimore Chronicle*, July 3, 1834; *Evening Star*, July 7 and 11, 1834.

86. Baltimore *Patriot*, Aug. 6–7, 1834; *Gazette* (Baltimore), July 4–5, 7, 1834; *Baltimore Chronicle*, July 7; August 7, 1834; *New York Evening Star*, July 11, 1834; *Nile's Weekly Register*, June 21, 1834; *National Gazette* (Philadelphia), Aug. 8, 1834; *New England Galaxy* (Boston), Aug. 9, 1834.

87. Baltimore *Patriot*, July 24 and 26, 1834; other undated articles, author's collection.

88. *Journal of Commerce* (New York), July 31, 1834; *New York Gazette*, July 31, 1834; *New York Commercial Advertiser*, Aug. 1, 1834.

89. *National Gazette* (Philadelphia), Aug. 13, 1834; Baltimore *Gazette*, Aug. 12, 1834; Baltimore *Patriot*, Aug. 12, 1834; *New York American*, Aug. 15, 1834; *Evening Star*, Aug. 13 and 19, 1834.

90. *Baltimore Chronicle*, Aug. 12, 1834.

91. *Providence Journal*, Aug. 22, 1834.

92. Undated article, box 220, AIAA Scrapbooks.

93. Anonymous article (Aug. 13, 1834), under Women in Aeronautics, box 216, AIAA Scrapbooks.

94. *United States Gazette*, July 2; Aug. 6 and 9; Sept. 17, 1834; *Atkinson's Saturday Evening Post*, July 5; Sept. 20; Oct. 4, 1834; *New York Evening Star*, Oct. 8, 1834.

95. *New York Gazette*, Nov. 5, 1834; Philadelphia *Liberalist*, Oct. 11, 1834; *Baltimore Chronicle*, Oct. 30, 1834; *United States Gazette*, Oct. 8, 1834.

96. *United States Gazette* (Philadelphia), May 9, 1835; *Pennsylvanian*, May 12, 1835; *National Gazette and Literary Register*, May 7, 1835; *New York American*, May 12, 1835.

97. *Jeffersonian*, Sept. 7, 1835.

98. Undated clipping, box 220, AIAA Scrapbooks; John Wise, *Through the Air: A Narrative of Forty Years Experience as an Aeronaut* (Philadelphia, 1873), p. 260.

99. *Journal of Commerce* (Baltimore), Aug. 27, 1834; *Gazette* (Baltimore), Aug. 27, 1834; *National Gazette* (Baltimore), Aug. 26, 1834

100. *Gazette* (Baltimore), Aug. 27, 1834; *Baltimore Chronicle*, Aug. 27, 1834; other undated articles under Elliott, box 217, AIAA Scrapbooks.

101. *Baltimore Chronicle*, Oct. 31, 1834.

102. *Ibid.*

103. *Baltimore Chronicle*, Nov. 11, 1834; *Evening Star* (Baltimore), Nov. 12, 1834.

104. *Baltimore Chronicle*, Jan. 17, 1835; *New York Gazette*, Jan. 17, 1835; *National Gazette and Literary Register* (Philadelphia), Jan. 20, 1835; *Atkinson's Saturday Evening Post*, Jan. 10, 1835.

105. Undated article, under Elliott, AIAA Scrapbooks.

106. Norfolk *Beacon*, Sept. 13, 1834; Boston *Atlas*, Sept. 17, 1835; Baltimore *Patriot*, Sept. 13, 1834.

107. Boston *Atlas*, Sept. 17, 1835; *Gazette* (Baltimore), Oct. 15, 1835; *Baltimore Chronicle*, Oct. 16, 1835.

108. *New York Evening Star*, Feb. 8, 1835; undated articles under Mills, box 219, AIAA Scrapbooks.

109. *Gazette* (Baltimore), Nov. 11, 1834.

110. *Ibid.*

111. *Ibid.*

112. *National Gazette and Literary Register* (Philadelphia), Feb. 7, 1835.

113. *Ibid.*

114. *Baltimore Chronicle*, Aug. 16, 1834; *Gazette* (Baltimore), Aug. 19; Sept. 13, 1834; Baltimore *Patriot*, Aug. 19, 1834.

115. *Baltimore Chronicle*, Mar. 23, 1835.

116. *New York American*, June 5, 1835, from the Richmond *Whig*.

117. *Boston Weekly Messenger*, Sept. 18, 1834; assorted undated clippings under Warren, box 220, AIAA Scrapbooks.

118. *Daily Albany Argus*, Sept. 21; Oct. 11, 1837.

119. Wise, *Through the Air*, p. 260.

Chapter 7

1. The best short treatment of Cincinnati history remains the WPA guidebook, *A Guide to the Queen City and Its Neighbors* (Cincinnati, 1943).

2. Rebekah Gest to Erasmus Gest, Dec. 4, 1834, Erasmus Gest Papers, Ohio Historical Society, Columbus, Ohio.

3. *Commercial Register*, Nov. 29, 1834; Cincinnati *Gazette*, Dec. 4, 1834.

4. *Ibid.*

5. *Literary Gazette*, Dec. 20, 1834.

6. *Commercial Register*, Dec. 16, 1834.

7. *Cincinnati Gazette*, Dec. 18, 1834; Jan. 1, 1835.

8. *Commercial Register*, Dec. 23–24, 29, 1834.

9. *Literary Gazette*, Mar. 7, 1835.

10. See also Tom D. Crouch, "Thomas Kirkby: Pioneer Aeronaut in Ohio," *Ohio History* (Winter 1970), pp. 56–62.

11. *Cincinnati Gazette*, Apr. 7, 1835.

12. Erasmus Gest Papers, box 1, Ohio Historical Society; see also Maurer Maurer, "Richard Clayton, Aeronaut," *Historical and Philosophical Society of Ohio Bulletin* (April 1955), pp. 143–150; Tom D. Crouch, "Up, Up, and—Sometimes—Away," *The Cincinnati Historical Society Bulletin* (Summer 1970), pp. 109–132.

13. "Mr. Clayton," *The Western Monthly Magazine* (October 1835), pp. 235–236.

14. "Aerial Navigation," *ibid.* (June 1838), p. 359.

15. "Mr. Clayton's Account of his Aerial Voyage," *National Gazette and Literary Register* (Philadelphia), July 18, 1835; *Daily National Intelligencer*, Apr. 28; May 22; July 18, 1835; for additional information, see *New York American*, Apr. 28; July 24, 1835; *Jersey City Gazette*, May 13, 1835; assorted undated articles in AIAA Scrapbooks under Clayton; and Robert M. Dresser to Comfort C. Dresser, May 10, 1835, Richard C. Gimball Collection, Library, USAF Academy, Colorado Springs, Colorado (hereafter, Gimball Collection).

16. *National Gazette and Literary Register*, July 18, 1835.

17. *Ibid.*

18. *Ibid.*

19. Rolt, *Aeronauts*, p. 118.

20. *New York Evening Star*, Jan. 13, 1836; undated articles under Clayton, box 217, AIAA Scrapbooks.

21. *Ibid.*

22. Undated clippings under Clayton, box 217, AIAA Scrapbooks.

23. *Ibid.*

24. Louisville *Journal*, Aug. 9, 1837; *Daily Cincinnati Gazette*, Aug. 4, 1837; *New York American*, Aug. 18, 1837; *Daily Albany Argus*, Aug. 18, 1837.

25. *Daily Pittsburgh Gazette*, Aug. 30; Sept. 1, 5, 1837.

26. *Ibid.*, Sept. 3, 1837.

27. *Ibid.*; *Boston Weekly Messenger*, Sept. 21, 1837.

28. *Ibid.*; *Daily Pittsburgh Gazette*, Sept. 5, 12, 14, 16, 1837.

29. *Ibid.*

30. *Ibid.*

31. *Ibid.*

32. Baltimore *Patriot*, Apr. 16, 1838; *New York Daily Express*, Apr. 18, 1838; *New York Evening Star*, June 22, 1838.

33. Tom D. Crouch, *The Giant Leap: A Chronology of Ohio Aerospace Events and Personalities, 1815–1969* (Columbus, Ohio, 1971).

34. Maurer, "Richard Clayton," p. 150.

35. *Christian Watchman* (Boston), Sept. 11, 1835; *Traveller*, July 17, 1835.

36. *Ibid.*

37. *Ibid.*

38. *Ibid.*

39. *Hartford Daily Times*, Oct. 30, 1841.

40. *New York Daily Express*, Aug. 18, 1837.

41. Most of the available biographical data on Paullin is contained in "William Paullin," *Appleton's Cyclopedia of American Biography* (New York, 1900), vol. 4, p. 682.

42. Unidentified article under Paullin, box 219, AIAA Scrapbooks.

43. *Ibid.*

44. *Daily National Intelligencer*, July 3, 1837.

45. *The United States Gazette*, July 28, 1837.

46. *Ibid.*

47. *Ibid.*

48. Unidentified articles, box 217, AIAA Scrapbooks.

49. *Ibid.*

50. See "John Wise," *Appleton's Cyclopedia*; see also Pearl I. Young, "John Wise and His Balloon Ascensions in the Middle West," *Wingfoot Lighter-Than-Air Society Bulletin* (October 1967), p. 2.

51. Wise, *Through the Air*, p. 249.

52. *Ibid.*, p. 253.

53. *Ibid.*

54. *Ibid.*, p. 255; *National Gazette and Literary Register*, May 5, 1835.

55. *Ibid.*

56. Wise, *Through the Air*, pp. 260–261; *Philadelphia Liberalist*, May 23, 1835.

57. Wise, *Through the Air*, pp. 262–269.

58. *Ibid.*, pp. 270–284.

59. *Ibid.*

60. *Ibid.*

61. *Ibid.*

62. *Ibid.*

63. *Ibid.*

64. *Ibid.*

65. Young, "John Wise."

66. Wise, *Through the Air*, pp. 285–291.

67. *Ibid.*, pp. 295–302.

68. *Globe*, Aug. 20, 1838.

69. Wise, *Through the Air*, pp. 304–310.

70. *Ibid.*

71. *Ibid.*

72. *Ibid.*

73. *Ibid.*, pp. 311–314.

74. *Ibid.*, pp. 315–318.

75. *Ibid.*, pp. 320–323.

76. *Ibid.*

77. "William Paullin," *Appleton's Cyclopedia.*

78. Wise, *Through the Air*, p. 305.

79. *Ibid.*

80. *Ibid.*

81. *Ibid.*

82. My thanks to Rev. Nathaniel Lauriat of Hartford, Connecticut, for biographical information on his distinguished ancestor. Reverend Lauriat's paper, "Louis Anselm Lauriat—Pioneer New England Balloonist," presented at the First Northeast Aero-Historians Conference in Hartford, October 15–17, 1965, was particularly useful. My thanks also, as always, to the generous nature of Mr. Harvey Lippincott, archivist with the United Technologies Corp., for introducing me to Reverend Lauriat.

83. Unidentified article under Lauriat, box 218, AIAA Scrapbooks.

84. Lauriat, "Louis Anselm Lauriat."

85. *National Gazette and Literary Register*, July 22, 1835.

86. *Jersey City Gazette*, Sept. 12, 1835.

87. *Ibid.*, Sept. 26, 1835.

88. Unidentified articles, box 218, AIAA Scrapbooks.

89. Rochester *Democrat*, n.d., box 218, AIAA Scrapbooks.

90. *Boston Daily Times*, June 20, 1836.

91. *Lowell Courier Extra*, July 5, 1836.

92. *New York Express*, Oct. 14, 1836; *Daily Albany Argus*, Aug. 3, 1837; *Providence Journal*, Aug. 22, 1837; *Boston Post*, July 14, 1838.

93. *Boston Notions*, June 17, 1839.

94. Poem in the collection of Rev. Nathaniel Lauriat.

95. *Universal Trumpet and Magazine*, Oct. 10. 1840; *Boston Times*, Aug. 23, 1842.

96. George A. Fuller's series of articles, "American Aeronauts in Canada in the Nineteenth Century," which began running in the *Wingfoot Lighter-Than-Air Society Bulletin* in April 1970, remains the finest treatment of early Canadian ballooning. My own treatment of the subject relies on these articles.

97. *Ibid.*, passim.

98. Wright and Lapham to L. A. Lauriat, Mar. 10, 1840, in the collection of Rev. Nathaniel Lauriat; copy in the author's collection.

99. Lauriat, "Louis Anselm Lauriat."

100. *Ibid.*

101. L. A. Lauriat to S. Lauriat, July 28, 1842, in the collection of Rev. Nathaniel Lauriat; copy in the author's collection.

102. Lauriat, "Louis Anselm Lauriat."

103. *Sacramento Union*, Sept. 1, 1858.

104. *Boston Evening Transcript*, June 5, 1838.

105. Robert H. Goodell, "Mathias Zahm's Diary," *Papers of the Lancaster County Historical Society*, vol. 47, no. 4 (1943), pp. 65–89. The Zahm diary has been my guide to all of John Wise's Lancaster ascents for the period. See also *Supplement to the Courant* (Hartford), Oct. 29, 1842; Wise, *Through the Air*, pp. 323–358.

106. Wise, *Through the Air*, pp. 329–330.

107. *Ibid.*, pp. 330–335.

108. *Ibid.*, p. 330.

109. *Ibid.*, p. 329.

110. *Ibid.*

111. *Ibid.*, p. 358.

112. *Ibid.*, pp. 365–366.

113. *Ibid.*

114. *Ibid.*, pp. 367–368.

115. *Ibid.*, pp. 373–375.

116. Tom D. Crouch, "The History of American Aviation, 1822–1905," *Aviation Quarterly* (Spring 1976), pp. 10–11.

117. *Ibid.*

118. Wise, *Through the Air*, p. 375.

119. *Ibid.*, p. 426.

120. *Ibid.*

121. *Ibid,*

122. *Ibid.*, p. 430.

123. *Ibid.*

Chapter 8

1. Clarence S. Brigham, *Poe's Balloon Hoax* (Metuchen, N.J., 1932).

2. Young, "John Wise."

3. Wise, *Through the Air*, pp. 382–408.

4. *Ibid.*, pp. 419–452; unidentified articles under John Wise, AIAA Scrapbooks; see also Crouch, *Giant Leap*, pp. 6–9.

5. Goodell, "Mathias Zahm's Diary," pp. 66–89.

6. Wise, *Through the Air*, pp. 463–469; *Newburyport Daily Herald*, Feb. 12, 1855.

7. *Ibid.*

8. *Ibid.*

9. Wise, *Through the Air*, p. 464.

10. *Ibid.*, p. 467.

11. *Ibid.*, p. 463.

12. Unidentified articles under Mme. Delon, box 217, AIAA Scrapbooks.

13. Charles Dollfus, "On Ballooning Firsts in America," *Buoyant Flight* (December 1963), p. 8; Crouch, *Giant Leap*, p. 9.

14. Nashville, *Gazette*, July 16, 1858; see also Pontiac (Michigan) *Examiner*, Nov. 30, 1859.

15. Joseph Jenkins Cornish III, *The Air Arm of the Confederacy: A History of Origins and Usages of War Balloons by the Southern Armies during the American Civil War* (Richmond, 1963), is an excellent guide to balloonists operating in the South prior to the Civil War.

16. Wise, *Through the Air*, p. 460.

17. *Ibid.*, p. 342.

18. *Ibid.*, p. 341.

19. *Ibid.*, p. 416.

20. *Ibid.*

21. *Ibid.*, p. 414.

22. Crouch, *Giant Leap*, p. 8.

23. Hartford *Courant*, July 3, 1897. This biographical article on Brooks is the basic source of information on the aeronaut's early career. The author thanks Ms. Phyllis Kihn of the Connecticut Historical Society for her help in unraveling Brooks's career; see Phyllis Kihn, "Silas M. Brooks, Aeronaut, 1824–1906," *Connecticut Historical Society Journal* (April 1972), pp. 41–55.

24. Kihn, "Brooks"; see also a poster, "Grand Exhibition of Voightlander Views," in the collection of the Connecticut Historical Society.

25. *Ibid.*

26. Kihn, "Brooks," p. 44.

27. *Hartford Courant*, June 5, 1854; Hartford *Times*, July 15, 1854.

28. Howard L. Scamehorn, *Balloons to Jets: A Century of Aeronautics in Illinois* (Chicago, 1957), pp. 5–9.

29. *Ibid.*; Kihn, "Brooks."

30. Ann Holtgren Pellegreno, *Iowa Takes to the Air: Volume One, 1845–1918* (Iowa City, Iowa, 1980), pp. 5–7.

31. Muscatine (Iowa) *Daily Evening Enquirer*, Oct. 6 and 11, 1856.

32. *Daily Gate City* (Keokuk, Iowa), Oct. 28, 1856.

33. *Hartford Daily Courant*, Aug. 11; Sept. 26, 1857.

34. *Daily National Intelligencer*, Oct. 11, 1853; *Hartford Daily Courant*, Jan. 3, 1866.

35. *Ibid.*

36. For a strange sidelight on the early history of aeronautics in California, see Mary Lou Henry, "C. A. A. Dellschau and the Sonora Aero Club," *Ballooning* (Mar.–Apr. 1980), pp. 23–26; Jerome Clarke and Loren Coleman, "Mystery Airships of the 1800's," *Fate* (May–

July 1973), pp. 84–94, 96–104; correspondence of the author with Elizabeth Glassman, May 7, 1980; and assorted materials on the Dellschau Collection from the Menil Foundation, author's collection.

37. Gomez, pp. 34–35.

38. "Balloon Travelling," *Scientific American* (July 21, 1855), p. 355.

39. "Balloon Ascension—Snow Storm," *Hartford Daily Courant*, Sept. 21, 1850.

40. "The Ascension," *Hartford Daily Courant*, May 21, 1855; Crouch, *Giant Leap*, p. 8.

41. Assorted unidentified articles under Pusey, box 219, AIAA Scrapbooks.

42. *Ibid.*

43. "Samuel Archer King," *Appleton's Cyclopedia*, p. 546; Samuel Archer King, *The Balloon: Noteworthy Aerial Voyages from the Discovery of the Balloon to the Present Time* (New York, 1879), p. 1; assorted material in the S. A. King biographical file, NASM Library.

44. *Ibid.*; S. A. King news articles, box 218, AIAA Scrapbooks.

45. *The Aeronaut*, assorted undated issues in the S. A. King biographical file, NASM Library. I have been unable to locate an extended run of this magazine; Clarence P. Wynn, "Professor Samuel Archer King: A Memorial," *Fly* (January 1915), p. 358; T. Chalmers Fulton, "Professor Samuel Archer King: A Tribute of Regard from His Pupil," *ibid.* (February 1909), p. 12. The unique manuscript chronology of early U.S. aeronautics prepared by Ernest Jones, himself a pioneer aviator and aviation publicist, also contains much useful information on King. The original of this chronology is in the Albert F. Simpson Research Center, Air University, Maxwell Air Force Base, Alabama. A microfilm copy is held by the NASM Library.

46. James Grant Wilson and John Fiske, eds., *The Cyclopedia of American Biography* (New York, 1888), vol. 9, p. 210.; F. Stansbury Haydon, *Aeronautics in the Union and Confederate Armies with a Survey of Military Aeronautics prior to 1861* (Baltimore, 1941), pp. 39–41.

47. Account in King biographical file, NASM Library.

48. *Ibid.*

49. The two volumes of James Allen Scrapbooks and the S. A. King article in the AIAA History Collection, Manuscript Division, Library of Congress, provide the fullest account of the early careers of Allen and King and are the basis for my treatment. In particular, see Worcester *Daily Bay State*, July 28, 1857; Worcester *Daily Transcript*, July 28, 1857; Manchester (New Hampshire) *Mirror*, July 27, 1857; Manchester *Daily American*, July 27, 1857; Paterson (New Jersey) *Daily Guardian*, Aug. 5, 1858; Providence (Rhode Island) *Daily Post*, June 14, 1858; Boston *Ledger*, June 24, 1859; *Boston Journal*, July 5, 1858; *Boston Herald*, July 5, 1858; Boston *Transcript*, July 5 and 7, 1859; Boston *Saturday Evening Gazette*, July 16, 1858; New Bedford (Mass.) *Mercury*, Aug. 26–27, 1859; Norwich (Conn.) *Morning Bulletin*, Sept. 30, 1859. See also other articles cited in Haydon, *Aeronautics*, p. 41.

50. "Strange Coincidences," *Aeronaut*, n.d., King biographical file, NASM Library.

51. *Ibid.*

52. "Ernest Pétin," *L'Aéronaute* (August 1878), pp. 255–260.

53. "Aerial Navigation," *Scientific American* (April 17, 1852), p. 255.

54. "Going to Europe in a Balloon," *ibid.*, (July 13, 1852), n.p.

55. *Hartford Daily Courant*, July 21, 1852.

56. *Ibid.*; see also "The Giant Balloon," *Scientific American* (July 17, 1852), p. 349.

57. Samuel Orcutt, *A History of the Old Town of Stratford and the City of Bridgeport, Connecticut* (Boston, 1886), pt. 2, p. 819.

58. "Mr. Pétin's Balloon Ascension," Springfield *Republican*, Oct. 1, 1852.

59. *Ibid.*

60. Cornish, *Air Arm*, p. 13; *Tägliche Deutsche Zeitung* (New Orleans), Dec. 19, 22, 28, 1852; Feb. 6, 1853; Apr. 24, 1853.

61. *Tägliche Deutsche Zeitung*, Apr. 24, 1853; see Gomez, *Aviacion*, for some information on Pétin in Mexico.

62. Charles Dollfus and H. Bouché, *Histoire de l'Aéronautique* (Paris, 1932), p. 94.

63. *Ibid.*; Gaston Tissandier, *Histoire des Ballons et des Aéronautes Célèbres, 1783–1800* (Paris, 1887), p. 53.

64. "To Their Honors . . . ," an advertising brochure in the Tissandier Collection, Manuscript Division, Library of Congress.

65. "Balloon Excursion by Moonlight," *Scientific American* (November 29,1851), p. 80.

66. "Performing Somersets from a Balloon," *ibid.* (December 18, 1852), p. 80.

67. "To Their Honors. . . ."

68. *Ibid.*

69. *New York Daily Tribune*, Oct. 23, 1854.

70. "To Their Honors. . . ."

71. "Rapid Traveling," *Scientific American* (January 27, 1855), n.p.

72. *Cincinnati Daily Gazette*, Sept. 26, 1855.

73. *Ibid.*

74. *Ibid.*, Oct. 3, 1855.

75. *Ibid.*

76. *Ibid.*

77. *Ibid.*; Hartford *Daily Courant*, Oct. 9, 1855; *Gazette and Chronicle* (Dorchester), Oct. 27, 1855; *Daily Union* (Pittsburgh), Nov. 18, 1855.

78. "Ballooning," *Scientific American* (August 2, 1856), p. 368; *Daily Mirror* (Manchester), July 5, 1856.

79. *Ibid.*

80. *Boston Post*, July 22, 1856; *Boston Evening Transcript*, Aug. 4, 1856; *La Patrie* (Montreal), Sept. 10, 1856; *Le Pays* (Montreal), Sept. 18, 1856; *Pilot* (Montreal), Sept. 23, 1856; cf. George E. Fuller, "American Aeronauts in Canada," *Wingfoot Lighter-Than-Air Society Bulletin* (Sept.–Oct. 1970), p. 8.

81. *Daily Union* (Pittsburgh), Nov. 18, 1856.

82. Crouch, *Giant Leap*, pp. 8–9.

83. *Ibid.*

Chapter 9

1. *Philadelphia Ledger*, Oct. 5, 1858.

2. Crouch, *Giant Leap*, p. 9.

3. *Frank Leslie's Illustrated Newspaper*, July 4, 1857.

4. "Perilous Balloon Ascension," Detroit *Tribune*, undated article, box 219, AIAA Scrapbooks.

5. *Frank Leslie's Illustrated Newspaper*, July 4, 1857.

6. Undated article, box 219, AIAA Scrapbooks.

7. *Cincinnati Daily Gazette*, Oct. 13, 1858.

8. Munson Baldwin, *With Brass and Gas: An Illustrated and Embellished Chronicle of Ballooning in Mid-Nineteenth Century America* (Boston, 1967), p. 53.

9. Cincinnati *Daily Gazette*, Oct. 16, 1858.

10. *Ibid.*, Oct. 16 and 19, 1858.

11. Baldwin, *With Brass and Gas*, p. 77.

12. Cincinnati *Daily Gazette*, Oct. 19, 1858.

13. *Ibid.*

14. *Ibid.*

15. Cleveland *Leader*, Oct. 20, 1858.

16. *Illinois State Journal*, July 3, 1858.

17. Cleveland *Leader*, Oct. 20, 1857.

18. *Frank Leslie's Illustrated Newspaper*, Oct. 23, 1858.

19. *Ibid.*; see also related unidentified articles in the Gimball Collection.

20. *Niagara Mail*, Aug. 14, 1850; *St. Catharines Journal*, Sept. 5, 1850; *Halifax Nova Scotian*, Oct. 14, 1850.

21. Crouch, *Giant Leap*, p. 6.

22. Baldwin, *With Brass and Gas*, pp. 1–27; Adrian (Michigan) *Expositor*, June 15, 1855.

23. Baldwin, *With Brass and Gas*, pp. 1–27.

24. "A Man Lost in the Clouds," *Harper's Weekly* (Oct. 2,1858); *Frank Leslie's Weekly Illustrated Newspaper*, Oct. 9, 1858.

25. Wise, *Through the Air*, p. 483.

26. *Frank Leslie's Illustrated Newspaper*, Oct. 9, 1858.

27. Unidentified articles under Thurston, box 219, AIAA Scrapbooks.

28. *Ibid.*

29. *Hartford Daily Courant*, Oct. 15, 1855.

30. A. J. B. DeMorat to the Bureau of Topographical Engineers, Aug. 3, 1862, MS. D536, Letters Received, Bureau of Topographical Engineers, National Archives.

31. Cornish, *Air Arm*, offers the best treatment of mid-nineteenth-century ballooning in the South. This pamphlet has served as an invaluable guide to contemporary news accounts. For DeMorat, see *Tägliche Deutsche Zeitung* (New Orleans), Dec. 19, 1857.

32. *Hartford Daily Courant*, June 19, 1857; unidentified articles, AIAA Scrapbooks.

33. Cornish, *Air Arm*, p. 15.

34. *Ibid.*

35. *Ibid.*

36. New Orleans *Picayune*, Feb. 16, 1858; James Allen Scrapbooks, AIAA History Collection.

37. *Ibid.*

38. *Ibid.*

39. *Ibid.*

40. Cornish, *Air Arm*, p. 17.

41. *Ibid.*

42. Wise, *Through the Air*, p. 376.

43. Savannah *Daily Morning News*, Feb. 28; Mar. 9–10, 12, 19, 1860.

44. *Ibid.*, Apr. 26; June 8 and 21, 1860.

45. Crouch, *Giant Leap*, pp. 9–10.

46. Erie (Pa.) *Dispatch*, May 17, 1859; *Frank Leslie's Illustrated Newspaper*, July 14, 1859.

47. Wise, *Through the Air*, p. 507.

48. John A. Haddock, *Mr. Haddock's Account of His Hazardous and Exciting Voyage in the Balloon "Atlantic" with Prof. J. La Mountain* (Philadelphia, 1872), p. 3.

49. "Great Balloon Trip Across Country," *Philadelphia Evening Journal*, July 6, 1859 (hereafter, "Great Balloon Trip").

50. "Mr. La Mountain's Balloon Journal," *Frank Leslie's Illustrated Newspaper*, July 16, 1859.

51. Undated articles under Wise, La Mountain, and Haddock, boxes 218–220, AIAA Scrapbooks.

52. Wise, *Through the Air*, pp. 488–490.

53. Baldwin, *With Brass and Gas*, pp. 86–87.

54. "Great Balloon Trip."

55. *Ibid.*

56. *Ibid.*; see also Wise, *Through the Air*, p. 490, and "The Aerial Trip from St. Louis to the Atlantic Seaboard," *St. Louis Republican*, July 20, 1859.

57. "Mr. La Mountain's Account of the Balloon Voyage," unidentified article under La Mountain, box 218, AIAA Scrapbooks (hereafter, "La Mountain's Account").

58. This account of the flight has been reconstructed from various accounts cited above.

59. "La Mountain's Account."

60. "Great Balloon Trip."

61. *Ibid.*; see also Wise, *Through the Air*, pp. 509–510.

62. "La Mountain's Account."

63. *Ibid.*

64. "Great Balloon Trip."

65. John Wise to the editor of the *New York Tribune*, in Baldwin, *With Brass and Gas*, pp. 138–139.

66. *Ibid.*, pp. 140–141; *Daily Herald* (Newburyport, Mass.), July 9, 1859.

67. *Ibid.*; see also "Aerial Trip from St. Louis" (above, n. 56).

68. "La Mountain's Account."

69. Baldwin, *With Brass and Gas*, pp. 142–144.

70. O. A. Gager to John Wise, *ibid.*, p. 147.

71. *Ibid.*, pp. 138–139; unidentified articles in boxes 218–220, AIAA Scrapbooks. These

articles also allude to the controversy and have been incorporated into this account.

72. Baldwin, *With Brass and Gas*, p. 148; St. Louis *Democrat*, July 2, 1859; Buffalo *Courier*, July 4, 1859; Albany *Atlas*, July 4, 1859; John Wise et al., *Full Particulars of the Greatest Aerial Voyage on Record, from St. Louis, Mo., to Adams, New York, in Nineteen Hours* (New York, 1859).

73. Haddock, *Balloon Voyage*, p. iv.

74. *Ibid.*; see also unidentified articles in AIAA Scrapbooks.

75. *Ibid.*

76. My account of the voyage relies on Haddock's *Balloon Voyage*; "The Great Balloon Voyage: Narrative of Mr. La Mountain," *New York Daily Tribune*, Oct. 7, 1859; and "Voyage of the Atlantic Balloon," *Frank Leslie's Illustrated Newspaper*, Oct. 22, 1859.

77. "Great Balloon Voyage"; "Voyage of the Atlantic Balloon."

78. *Ibid.*

79. Haddock, *Balloon Voyage*, pp. 9–10.

80. *Ibid.*, passim.

81. *Ibid.*, p. 12.

82. *Ibid.*; see also La Mountain accounts cited above.

83. Haddock, *Balloon Voyage*, pp. 12–13.

84. *Ibid.*

85. *Ibid.*

86. *Ibid.*; see also La Mountain accounts cited above.

87. "The Great Balloon Voyage," *Mercury* (New Bedford, Mass.), July 7, 1859; "The Perils of Ballooning," *Boston Daily Evening Transcript*, Oct. 7, 1859; unidentified articles on the Wise reception, box 220, AIAA Scrapbooks.

88. Wise, *Through the Air*, pp. 529–537; see also clippings on the Buffalo flight in the Wise biographical file, NASM Library.

89. "The Death of Mr. Connor, the Aeronaut," *The New York Illustrated Evening News*, May 19, 1860; Wise, *Through the Air*, pp. 537–543; see also two unidentified articles in box 220, AIAA Scrapbooks.

90. Wise, *Through the Air*, p. 543.

91. Biographical material on Lowe has been drawn from T. S. C. Lowe, "My Balloons in Peace and War: Memoirs of Thaddeus S. C. Lowe," unpublished manuscript in the collection of the NASM Library. See also Lowe biographical files in the NASM Library and in the AIAA History Collection.

92. Lowe, "My Balloons," p. 9.

93. *Bytown Tribune* (Ottawa), May 29; June 19, 1858.

94. John Wise to T. S. C. Lowe, May 17, 1859, in T. S. C. Lowe file, boxes 80–84, AIAA History Collection, Library of Congress (hereafter, Lowe File).

95. Lowe, "My Balloons," pp. 5–6, 7–18; *Harper's Weekly*, Sept. 24, 1859; assorted articles under Lowe, box 218, AIAA Scrapbooks.

96. *Rome [N.Y.] Sentinel*, n.d., AIAA Scrapbboks; see also Crouch, *Giant Leap*, p. 12.

97. Lowe, "My Balloons," p. 12.

98. Letters to T. S. C. Lowe from Ferdinand Gross, Oct. 21, 1859; Vernon Henry Vaughn, Oct. 26, 1859; Samuel Jackson, Nov. 19, 1859; Volney S. Anderson, May 7, 1860; and W. Hendricks, n.d., all in Lowe File.

99. *New York Tribune*, Nov. 18, 1859; *New York Herald*, Nov. 19, 1859; Natchez *Courier*, Nov. 16, 1859; *Frank Leslie's Illustrated Newspaper*, May 19, 1860; unidentified articles and advertisements, box 218, AIAA Scrapbooks.

100. Unidentified article in box 218, AIAA Scrapbooks.

101. Lowe, "My Balloons," pp. 15–17.

102. *Ibid.*

103. *Ibid.*, pp. 19–25; *New York Illustrated News*, May 19, 1860; unidentified articles and advertisements, box 218, AIAA Scrapbooks.

104. *Philadelphia Inquirer*, July 4, 1860.

105. Lowe, "My Balloons," p. 125.

106. John Wise to T. S. C. Lowe, Sept. 9, 1860, in Lowe File.

107. *Ibid.*

108. Lowe, "My Balloons," pp. 32–34.

109. *New York Tribune*, May 10, 1861; Cincinnati *Daily Commercial*, Apr. 2, 22, 30; May 9–10, 1861.

110. Cincinnati *Daily Commercial*, Apr. 30, 1861.

111. *Ibid.*

112. *Ibid.*

113. *Ibid.*

Chapter 10

1. My discussion of Zeppelin's visit to America is based on Rhoda Gilman, "Zeppelin in Minnesota: A Study in Fact and Fable," *Minnesota History* (Fall 1965), pp. 278–285; and *idem*, "Zeppelin in Minnesota: The Count's Own Story," *Minnesota History* (Summer 1967), pp. 265–279.

2. Hugo Eckner, *Count Zeppelin: The Man and His Work* (London, 1938), p. 53.

3. F. von Zeppelin to Count von Zeppelin, July 24, 1863, quoted in Gilman, "The Count's Own Story," pp. 271–272.

4. *Ibid.*

5. *Ibid.*, p. 273.

6. *Ibid.*, pp. 273–274.

7. *Ibid.*

8. *Ibid.*

9. St. Paul *Pioneer Press*, Aug. 20; Oct. 15, 1863; Gilman, "A Study in Fact and Fable," pp. 284–285.

10. F. von Zeppelin to Count von Zeppelin, Aug. 19, 1863, in Gilman, "The Count's Own Story," p. 276.

11. *Ibid.*

12. *Ibid.*

13. Gilman, "A Study in Fact and Fable," p. 285.

14. For one particularly well known heavier-than-air proposal of the period, see Richard O. Davidson, *A Description of the Aerostat: A Practical View of Aerial Navigation* (New York, 1841).

15. For additional details on U.S. attitudes toward heavier-than-air flight, see Tom D. Crouch, "The History of American Aviation, 1822–1905," *Aviation Quarterly* (Spring 1976), pp. 8–13.

16. Thomas Jefferson to David Lee, Apr. 27, 1822, quoted in Jeremiah Milbank, *The First Century of Flight in America* (Princeton, N.J., 1943), p. 72. Jefferson's attitude toward flight as expressed in this letter typifies that of many Americans of the era.

17. For an overview of early European thinking on the airship, see Douglas Robinson, *Giants in the Sky* (Seattle, Wash., 1973); on Meusnier, see C. C. Gillispie, *Science and Polity in France at the End of the Old Regime* (Princeton, N.J., 1980), pp. 538–544.

18. E. C. Genêt, *Memorial . . .* (Albany, 1825), p. 18.

19. *Ibid.*, p. 75.

20. *Ibid.*, p. 80.

21. Thomas Jones, "Notice of a Work . . . ," *Franklin Journal and American Mechanic's Magazine* (July 1826), pp. 41–45.

22. Benjamin Silliman, "Remarks on Genêt and Pascalis," *ibid.* (January 1827), pp. 33–37. For more information on Genêt, see *Vindication of Mr. E. C. Genêt's Memorial on the Upward Forces of Fluids in Two Letters to Professor Silliman to Which Are Added Remarks on Aerostation* (New Haven, 1827).

23. *Ibid.*

24. *United States Gazette*, June 17, 1828, quoted in John Wise, *A System of Aeronautics* (Philadelphia, 1850), pp. 86–87.

25. John H. Pennington, *Aerostation, or Steam Aerial Navigation* (Baltimore, 1838).

26. Wise, *System*, p. 85.

27. John H. Pennington, *A System of Aerostation* (Washington, D.C., 1842).

28. "Pennington's Aerial," *Scientific American* (November 14, 1846), p. 59.

29. "Flying Cotton: John H. Pennington," *ibid.* (January 9, 1847), p. 125.

30. "Pennington and Co.," a poster in the collection of the Rare Book Division, Library of Congress; John H. Pennington to Joseph Gates, Oct. 31, 1853, pasted in the copy of *Steam Aerial Navigation* in the Rare Book Room of the Library of Congress.

31. *Ibid.*

32. Pennington to Mehan, Oct. 31, 1853.

33. *Ibid.*

34. *Ibid.*

35. *Ibid.*

36. *Ibid.*

37. Rufus Porter, "Epitome of Experience and Practice," *Aerial Reporter* (April 27, 1854), p. 2. For many years writers on Porter had assumed that only one issue of the *Aerial Reporter* had survived. Thanks in large measure to the bibliographic sleuthing of Mimi Scharf of the NASM Library, the author uncovered a complete run of the newsletter in the library of the Maryland Historical Society.

38. Jean Lipman, *Rufus Porter Rediscovered* (New York, 1980), offers a good biographical treatment with an emphasis on Porter the artist.

39. Porter, "Epitome."

40. My discussion of Porter's inventions is based on Lipman, *Rufus Porter*, as well as on Porter's various autobiographical offerings.

41. Porter, "Epitome."

42. "Inventor's Institute," *Scientific American* (March 6, 1847).

43. Porter, "Epitome."

44. Roger Burlingame, *March of the Iron Men* (New York, 1938).

45. Rufus Porter, "Early Aspirations," *Aerial Reporter* (July 17, 1852), p. 2.

46. *Ibid.*

47. Rufus Porter, "Motives," *ibid.* (December 18, 1852), p. 2.

48. Rufus Porter, "Utility of Aerial Navigation," *ibid.* (August 14, 1852), p. 2.

49. Rufus Porter, "Transportation of Ponderous Merchandise," *ibid.* (July 17, 1852), p. 2.

50. Porter, "Early Aspirations."

51. Rufus Porter, "Plan for an Observatory Balloon," *Mechanic's Magazine* (American edition) (September 27, 1834), pp. 215–216.

52. Rufus Porter, "Travelling Balloon, or Flying Machine," *ibid.* (November 8, 1834), pp. 273–275.

53. Rufus Porter, "Travelling Balloon," *New York Mechanic* (March 13, 1841).

54. Rufus Porter, "The Travelling Balloon," *Scientific American* (September 18, 1845), p. 1.

55. B.G.N., "Aeronautic Steam Car," *Mechanic's Magazine* (American Edition) (August 1834), pp. 142–144.

56. "Aerial Navigation," *Scientific American* (October 30, 1847), p. 46.

57. *Opinions of Thirty-Two Scientific Gentlemen on the Invention of Muzio Muzzi* (New York, 1845).

58. *Ibid.*

59. Rufus Porter, "Aerial Navigation: Signor Muzio Muzzi's Travelling Balloon," *Scientific American* (October 28, 1845), n.p. Porter reprinted this article on several occasions, holding Muzzi up as an exemplar of the foolishness of his predecessors in the field of aerial navigation.

60. Rufus Porter, "The Travelling Balloon," *Scientific Mechanic* (December 18, 1847), p. 1.

61. Rufus Porter, "The Travelling Balloon," *ibid.* (December 25, 1847), p. 1.

62. "Former Experiments," *Aerial Reporter* (June 19, 1852), p. 2; "Aerial Navigation," *Scientific American* (November 20, 1869), p. 325; "The Travelling Balloon," *ibid.* (December 18, 1845), p. 2.

63. "Travelling Balloon," *Scientific American* (March 3, 1849), p. 125.

64. Boston *Bee*, quoted in "Former Experiments."

65. *New York True Sun*, quoted in "Former Experiments."

66. *New York Evening Post* (Feb. 22, 1849), quoted in Thomas B. Settle, *Rufus Porter and His Aeroport*, a 1980 catalogue essay for a Hudson River Museum exhibition.

67. Porter, "Epitome."

68. Rufus Porter, "Opposition of the Press," *Aerial Reporter* (October 9, 1852), p. 1.

69. "A Flying Machine," Philadelphia *Bulletin*, quoted *ibid.*

Chapter 11

1. James Marshall, quoted in David Lavender, *California: Land of New Beginnings* (New York, 1972), p. 150. J. S. Holiday, *The World Rushed In: The California Gold Rush Experience* (New York, 1981), has been the author's guide to Gold Rush sources.

2. Holiday, *World*, p. 39.

3. *Ibid.*

4. *New York Herald*, Jan. 11, 1849.

5. Rufus Porter, *Aerial Navigation: The Practicality of Travelling Pleasantly and Safely from New York to California in Three Days Fully Demonstrated* (New York, 1849). This pamphlet was reprinted with Porter's *Aerial Steamer, or Flying Ship* (Washington, D.C., 1850) in Rhoda Gilman, ed., *A Yankee Inventor's Flying Ship* (St. Paul, Minn., 1969).

6. Porter, *Aerial Navigation*, in Gilman, *Yankee Inventor*, pp. 30–31.

7. *Ibid.*, pp. 29–30.

8. *Ibid.*

9. *Ibid.*, pp. 36–37.

10. *Ibid.*, p. 36 and rear cover.

11. Copies of both prints are housed in the Prints and Photographs Division of the Library of Congress.

12. "Navigating the Air," *Scientific American* (September 29, 1849), p. 181.

13. "New Flying Ship," *Gleason's Pictorial Drawing Room Companion* (September 22, 1851), p. 208.

14. *Ibid.*

15. *Ibid.*; "Navigating the Air," *Scientific American* (September 29, 1849), p. 48.

16. Rufus Porter, *An Aerial Steamer, or Flying Ship* (Washington, D.C., 1850), in Gilman, *Yankee Inventor*.

17. Rufus Porter, "A Fruitless Application," *Aerial Reporter* (July 31, 1852), p. 2.

18. *Ibid.*

19. Rufus Porter, "Unsuccessful Proposition," *ibid.* (August 28, 1852), p. 1.

20. *Ibid.*

21. *Ibid.*

22. Rufus Porter, "The Flying Ship," *ibid.* (November 24, 1852), p. 2; "Report of Progress," *ibid.* (May 28, 1853), p. 1.

23. *Aerial Reporter*, nos. 1–20, Maryland Historical Society.

24. William Markoe to William Paullin, Dec. 16, 1856, Markoe Letterbook, Minnesota Historical Society, quoted in Rhoda Gilman, "Pioneer Aeronaut: William Markoe and His Balloon," *Minnesota History* (December 1962), p. 168.

25. Gilman, "Markoe," p. 168.

26. *Ibid.*; see also Gilman, *Yankee Inventor*, p. 4.

27. "Report of Progress," *Aerial Reporter* (December 18, 1852), p.1.

28. *Ibid.* (July 17, 1852), p. 1.

29. *Ibid.* (July 31, 1852), p. 1.

30. *Ibid.*

31. *Ibid.*

32. *Ibid.* (September 25, 1852), p. 1.

33. *Ibid.* (October 23, 1852), p. 1.

34. *Ibid.* (November 6, 1852), p. 1.

35. *Ibid.* (December 18, 1852), p. 1.

36. *Ibid.*; "Aerial Navigation," *Washington Republic*, Mar. 26, 1853.

37. "Aerial Navigation," *National Intelligencer*, Apr. 5, 1853.

38. "The Flying Ship," *Washington Evening Star*, Apr. 13, 1853.

39. *Ibid.*

40. "Report of Progress," *Aerial Reporter* (April 23, 1853), p. 1.

41. "Report of Progress," *ibid.* (July 9; August 13; October 12, 1853; April 4, 1854).

42. Rufus Porter to William Markoe, Aug. 23, 1853, in Gilman, *Yankee Inventor*, p. 4.

43. "Report of Progress," *Aerial Reporter* (April 27, 1854), p. 1.

44. Gilman, *Yankee Inventor*, p. 5.

45. Gilman, "Markoe," p. 168.

46. John Seymour to the editor of the *New York Sun*, undated clipping in the Gimball Collection.

47. John Seymour to Rufus Porter, n.d., MS. XF2–1 2429, Gimball Collection.

48. "Aerial Navigation," *Scientific American* (November 20, 1869), p. 325; *ibid.* (November 25, 1854), p. 85.

49. *Ibid.*; "Another Steam Balloon," *ibid.* (May 10, 1856).

50. "Memorial of the Citizens of the State of Tennessee . . . November 5, 1854," machine copy in the author's collection; original, National Archives and Records Service (NARS). My thanks to Prof. Bryan Leary, Georgia Institute of Technology, for providing me a copy of this document.

51. "News about Aerial Navigation," *Scientific American* (November 13, 1849), n.p.

52. *Ibid.*

53. "New Plan of Aerial and Terra Firma Locomotion," *Scientific American* (April 7, 1849), p. 228.

54. "Aerial Navigation—A New Old Plan," *ibid.* (October 9, 1852), p. 29; "Captain Taggert's Flying Machine," *ibid.* (July 13, 1850), p. 340; "Another Flying Machine," *ibid.* (January 13, 1849), p. 132; "Aerial Navigation," *ibid.* (November 20, 1869), p. 325.

55. Biographical sources on the early years of Solomon Andrews include William C. McGinnis, *History of Perth Amboy* (Perth Amboy, N.J., 1959); and W. Northey Jones, *The History of St. Peter's Church in Perth Amboy, New Jersey* (Perth Amboy, N.J., 1923). The Andrews autobiographical fragments in *The Art of Flying* (New York, 1865) and *The Aereon, or Flying Ship* (New York, 1866) are also useful. Family papers are held by the New Jersey State Library and the Rutgers University Library.

56. Andrews, *Aereon*, pp. 2–3.

57. *Ibid.*

58. *Ibid.*

59. *Ibid.*

60. Solomon Andrews to the secretary of the Mansfield Lodge, I.O.O.F., Aug. 10, 1848; *Inventor's Institute—To All Persons Interested in the Progress of Invention* (Perth Amboy, N.J., 1848); both reprinted in McGinnis, *History*, chap. 3.

61. *Ibid.*

62. Advertisements, reprinted *ibid.*

63. "Aerial Speculation," *Scientific American* (July 29, 1848), p. 359.

64. Andrews, *Aereon*, p. 5.

65. *Ibid.*; John Toland, *Great Dirigibles: Their Disasters* (New York, 1972), contains a popular, but solid and generally trustworthy, account of Andrews and his project. John McPhee, *The Deltoid Pumpkin Seed* (New York, 1973), provides a much sketchier and less satisfactory portrait of the origins of the Aereon.

66. McGinnis, *History*, n.p.; Andrews, *Aereon*, pp. 5–7; "A New Flying Machine," *Scientific American* (September 19, 1863); "The New Flying Machine," *ibid.* (October 24, 1863); "Aerostation," *New York Herald*, undated article in box 217, AIAA Scrapbooks.

67. Discussion of the Aereon is based on material cited above. Toland, *Great Dirigibles*, is also helpful.

68. *Ibid.*

69. Undated articles, AIAA Scrapbooks.

70. "Aerial Navigation," *New York Herald*, Sept. 8, 1866; "Aerial Navigation," *New York Herald Tribune*, Sept. 8, 1866; other undated articles, AIAA Scrapbooks.

71. *Ibid.*; see also Andrews, *Aereon*, p. 7.

72. Andrews, *Aereon*, pp. 6–9.

73. *Ibid.*; see also A. D. Bache, J. C. Woodruff, and J. Henry to E. M. Stanton, July 22, 1864, NARS, RG 107, Records of the Office of the Secretary of War, Letters Received (entry 33), vol. 131, file no. B1841; copy in the Joseph Henry Papers, Smithsonian Institution Archives.

74. *Ibid.*

75. Robert Schenck to Solomon Andrews, Mar. 22, 1865, in Andrews, *Aereon*, p. 9.

76. "The Minute Books and Certificate of Incorporation of the Aerial Navigation Company, Dr. Solomon Andrews, President, New York, 1865," box 174, AIAA Collection. This notebook contains the complete history of the Aerial Navigation Company.

77. *Ibid.*

78. *Ibid.*

79. Undated articles, box 217, AIAA Scrapbooks; see also above, n. 70.

80. *Ibid.*

81. *Ibid.*

82. *Ibid.*

83. *Ibid.*

84. *Ibid.*

85. "Life of Frederick Marriott," *San Francisco Newsletter*, December 20, 1884; see also C. S. Kirkpatrick, Notes on Frederick Marriott, Manuscript Division, Library of Congress; and Richard A. Hernandez, "Frederick Marriott: A Forty-Niner Banker and Editor Who Took a Flyer in Pioneering American Aviation," *Journal of the American West* (October 1963), pp. 401–424.

86. *Ibid.*

87. Idwal Jones, "The Flying Editor," *Westways*, vol. 6, no. 2, pp. 14–15.

88. Hernandez, "Marriott," p. 415.

89. *Ibid.*

90. Kirkpatrick Notes; Hernandez, "Marriott," pp. 415–417.

91. *Daily Alta California*, Sept. 29, 1869.

92. Marvin Martin, "Marriott and His Flying Avitor," *Popular Aviation* (November 1935), p. 290.

93. *Ibid.*

94. *Annual Report of the Aeronautical Society of Great Britain for 1868* (London 1869), pp. 85–89.

95. Martin, "Marriott," p. 290.

96. Hernandez, "Marriott," p. 419.

Chapter 12

1. Bell Irvin Wiley, *The Life of Billy Yank: The Common Soldier of the Union* (Indianapolis, Ind., 1951), p. 20.

2. *Providence Post*, Apr. 19, 1861; undated clipping, James Allen Scrapbooks, vol. 1, p. 45, box 228, AIAA History Collection, Manuscript Division, Library of Congress (hereafter, Allen Scrapbooks).

3. See esp. Frederick Stansbury Haydon, *Aeronautics in the Union and Confederate Armies, with a Survey of Military Aeronautics prior to 1861* (Baltimore, 1941). This volume remains, after more than four decades, the finest work on the subject. My manifold debts to Haydon are obvious.

4. Haydon, *Aeronautics*, pp. 1–39, remains the best introduction to the subject of early military aeronautics.

5. Frederick E. Beaseley to Joel Poinsett, Sept. 8, 1840, Quartermaster Consolidated File, box 87, Letters Received, "Balloons," National Archives (hereafter, QCF).

6. Copies of all correspondence relating to Sherbourne's project are included in the "Balloon History—USA" file, NASM Library.

7. Wise, *Through the Air*, pp. 386–391.

8. *Ibid.*

9. *Ibid.*; see also undated articles under John Wise, box 20, AIAA Scrapbooks.

10. Haydon, *Aeronautics*, p. 59.

11. *Daily Guardian* (Paterson, N.J.), Aug. 5, 1858.

12. For the early history of aerial photogra-

phy, see *Boston Journal*, July 5; Oct. 15–16, 1860; *Boston Transcript*, Oct. 13 and 15, 1860; *Providence Journal*, July 12; Aug. 17 and 21; Oct. 6, 1860; *American Journal of Photography* (September 1860), pp. 105–106; quotation from unidentified clipping, Allen Scrapbooks, vol. 1, p. 43.

13. *Scientific American* (May 3; September 21, 1861); *American Journal of Photography* (February 1861), pp. 105–106; *Photographic News* (London) (November 1860), p. 347.

14. *New York Tribune*, June 22, 1861; *Philadelphia Ledger*, Sept. 10, 1861.

15. Haydon, *Aeronautics*, p. 331.

16. *Providence Daily Post*, n.d., Allen Scrapbooks, vol. 1, p. 45.

17. *Ibid.*

18. "Letter from James Allen," *Providence Daily Post*, Apr. 26; May 30; June 10, 1861; *Cincinnati Commercial*, June 10, 1861; *Providence Journal*, June 10, 1861; *Providence Press*, June 10, 1861; *New York Tribune*, June 15, 1861; *Washington National Republican*, June 10, 1861; *Scientific American* (June 22, 1861).

19. "War Signals," *Providence Daily Post*, Apr. 19, 1861.

20. *Ibid.*, Haydon, *Aeronautics*, p. 46.

21. Allen Scrapbooks, vol. 1, p. 45.

22. Haydon, *Aeronautics*, p. 60.

23. *Ibid.*

24. *Ibid.*; Wise provides a simplified account in *Through the Air*, pp. 554–557.

25. "La Mountain Petition, May 1, 1861," QCF.

26. *Ibid.*

27. *Ibid.*

28. "La Mountain Memorandum, May 29, 1861," QCF.

29. *New York Commercial Advertiser*, June 11, 1861; *Philadelphia Ledger*, June 13, 1861.

30. O. A. Gager to B. F. Butler, June 22, 1861, B. F. Butler Papers, Manuscript Division, Library of Congress.

31. *Ibid.*; see also La Mountain notes, QCF.

32. *Ibid.*; see also La Mountain notes, QCF.

33. T. S. C. Lowe, "My Balloons in Peace and War," pp. 41–42, unpublished memoirs, NASM Library.

34. *Ibid.*, pp. 55–57.

35. *Ibid.*

36. *Ibid.*; see also the bulk of the early Lowe correspondence with Halstead, Henry, and others in the Lowe biographical file, boxes 80–84, AIAA History Collection, Manuscript Division, Library of Congress (hereafter, Lowe File). These manuscripts, together with those in RG 98 and 168, NARS, constitute the Lowe Papers.

37. Lowe File; Lowe, "My Balloons," pp. 58–61.

38. *The War of Rebellion: A Compilation of the Official Records of the Union and Confederate Armies* (Washington, D.C., 1899), vol. 3, p. 254 (hereafter, *OR*).

39. Lowe, "My Balloons," pp. 60–61.

40. *Boston Transcript*, June 20, 1861.

41. *New York Herald*, June 20, 1861.

42. *Washington Evening Star*, June 20, 1861; see also undated articles under Lowe, box 81, AIAA Scrapbooks.

43. Lowe, "My Balloons," pp. 58–60.

44. *Ibid.*

45. *Ibid.*, pp. 62–65; *OR*, vol. 3, pp. 256–257; Haydon, *Aeronautics*, p. 186.

46. *Ibid.*

47. Lowe, "My Balloons," p. 63.

48. *Ibid.*

49. *Ibid.*

50. *Ibid.*

51. *Ibid.*

52. Henry L. Abbot, "Early Experiences with Balloons in War," *Professional Memoirs of the Corps of Engineers* (Washington, D.C., 1912), vol. 4, p. 680; see also Lowe, "My Balloons," p. 64; and *OR*, vol. 3, pp. 255–256.

53. Haydon, *Aeronautics*, pp. 50–51.

54. Abbot, "Early Experiences," p. 681.

55. *Lancaster Daily Evening Express*, July 17, 22, 24, 1861; Wise, *Through the Air*, p. 554; *London Times*, July 22, 1861; *New York Herald*, Aug. 20, 1861.

56. *Ibid.*

57. *Ibid.*

58. Lowe, "My Balloons," pp. 67–70.

59. *Lancaster Daily Express*, July 29, 1861; *New York Tribune*, July 26, 1861; *Detroit Free Press*, July 28, 1861; *Philadelphia Press*, July 26, 1861; *Washington Star*, July 30, 1861.

60. Haydon, *Aeronautics*, p. 75.

61. *Ibid.*

62. Lowe, "My Balloons," p. 65.

63. *Ibid.*, pp. 71–72.

64. Haydon, *Aeronautics*, p. 197.

65. Lowe, "My Balloons," pp. 72–73.

66. Joseph Henry to A. W. Whipple, Aug. 2, 1861, in Lowe, "My Balloons," pp. 72–73.

67. Haydon, *Aeronautics*, p. 198.

68. H. Bache to A. W. Whipple, Aug. 2, 1861, in Lowe, "My Balloons," pp. 73–74.

69. J. C. Woodruff to T. S. C. Lowe, Aug. 28, 1861, in Lowe, "My Balloons," pp. 74–75.

70. T. S. C. Lowe to A. W. Whipple, Aug. 29, 1861, in Lowe, "My Balloons," pp. 75–76.

71. See, for example, F.-J. Porter to T. S. C. Lowe, Sept. 11, 1861, in Lowe, "My Balloons," pp. 79–80.

72. W. F. Smith to F.-J. Porter, Sept. 23, 1861; J. F. Questen to T. S. C. Lowe, Sept. 24, 1861, in Lowe, "My Balloons," pp. 81–82; F. Beaumont, "On Balloon Reconnaissance as Practiced by the American Army," *Professional Papers of the Corps of Royal Engineers* (London, 1863); see also G. E. Grover, "On the Uses of Balloons in Military Operations"; and *idem*, "On Reconnoitering Balloons," *ibid.*; all three reprinted in *Military Ballooning, 1862*, undated pamphlet issued by the Corps of Royal Engineers.

73. Haydon, *Aeronautics*, pp. 211–216.

74. W. A. Glassford, "Prolegomenon, or the Balloon During the Civil War," *Journal of the Military Service Institution of the United States*, 18 (1896), p. 261.

75. E. P. Alexander, quoted in Cornish, *Air Arm*, p. 18.

76. J. E. Johnston to P. G. T. Beauregard, Aug. 22, 1861, quoted in a catalogue of manuscripts for sale by Stanley V. Henkels, October 1915, Manuscript Division, Library of Congress.

77. *Providence Daily Post*, June 28, 1861.

78. Cornish, *Air Arm*, pp. 16–18.

79. E. L. Jones compiled the list of balloons and equipment; see his unpublished "Chronology," microfilm copy, NASM Library. See also Haydon, *Aeronautics*, passim; Lowe File; and RG 97, NARS.

80. *Ibid.*

81. Haydon, *Aeronautics*, p. 356.

82. *Ibid.*

83. *Ibid.*, pp. 275–276; 377–386. Most of the original documents relating to Starkweather and all other aeronauts are now in the Lowe File and RG 97.

84. E. L. Mason to T. S. C. Lowe, Sept. 5, 1861; Jan. 3, 1862, Lowe File.

85. E. L. Mason to T. S. C. Lowe, June 2, 1862, Lowe File.

86. E. Seaver to F.-J. Porter, Nov. 17, 1861; T. S. C. Lowe to J. N. Macomb, January 16, 1862; F.-J. Porter to T. S. C. Lowe, Nov. 17, 1861; all in Lowe File.

87. Unidentified clipping, Allen Scrapbooks, vol. 1, pp. 47–49.

88. T. S. C. Lowe to M. Meigs, Mar. 7, 1863, Lowe File.

89. Haydon, *Aeronautics*, p. 267.

90. Lowe, "My Balloons," passim.

91. Beaumont, "On Balloon Reconnaissance," p. 96.

92. Lowe File, passim, 1861–1863.

93. *Detroit Free Press*, Aug. 27, 1861; *New York Times*, Aug. 6–7, 1861; *New York Trib-*

une, Aug. 6, 1861; *New York Commercial Advertiser*, Aug. 5, 1861; *New York World*, Aug. 6, 1861; *New York Evening Post*, Aug. 6, 1861; Philadelphia *Press*, Aug. 6, 1861; Baltimore *American*, Aug. 6, 1861; *Scientific American* (August 17, 1861). See also J. La Mountain to B. F. Butler, July 31, 1861, Butler Papers, Library of Congress.

94. J. La Mountain to B. F. Butler, Aug. 13, 1861, Butler Papers.

95. This account of La Mountain's activity during and after the period of Butler's departure from Fort Monroe is based on assorted documents in the Butler Papers and the QCF.

96. "Col. R. B. Marcy, Chief of Staff, Headquarters, Army of the Potomac, Report of the Meeting with La Mountain and T. S. C. Lowe, Sept. 21, 1861," QCF.

97. *Washington Evening Star*, Oct. 5, 1861; *Washington Sunday Chronicle*, Oct. 6, 1861; Baltimore *American*, Oct. 7, 1861; *New York Herald*, Oct. 6, 1861; *Philadelphia Press*, Oct. 7, 1861; *Boston Transcript*, Oct. 10, 1861; *Providence Post*, Oct.11, 1861.

98. *Washington Evening Star*, Oct. 19, 1861.

99. J. La Mountain to W. Franklin, Oct. 21, 1861.

100. *Boston Journal*, Nov. 16, 1861.

101. J. N. Macomb, "Report, November 16, 1861"; J. La Mountain to W. Franklin, Dec. 9, 11, 21, 1861; Franklin Note, Dec. 9, 1861; all in QCF.

102. *Ibid.*

103. Haydon, *Aeronautics*, p. 135.

104. T. S. C. Lowe to A. V. Colburn, Jan. 18, 1862; T. S. C. Lowe to G. McClellan, Dec. 18, 1862; T. S. C. Lowe to G. McClellan, undated; copy of a *New York Herald* article, "The New Aeronautic Department Under Prof. La Mountain"; all in QCF.

105. *Ibid.*

106. *Ibid.*

107. *Ibid.*

108. Undated notes, QCF.

109. *Ibid.*

110. T. S. C. Lowe to S. Williams, Feb. 22, 1862, QCF.

111. Assorted undated documents, QCF.

112. W. H. Helme to J. Macomb, Sept. 1861, QCF.

113. J. Macomb to W. H. Helme, Oct. 23, 1861; W. H. Helme to J. Macomb, Dec. 18, 1861; J. H. to G. McClellan, Oct. 18, 1861; all in QCF.

Chapter 13

1. T. S. C. Lowe to J. Dahlgren, Nov. 16, 1861, Lowe File; *New York Herald*, Nov. 13, 1861; Haydon, *Aeronautics*, pp. 256–257.

2. *Ibid.*

3. Miscellaneous reports, Lowe File.

4. J. Hooker to T. S. C. Lowe, Nov. 30, 1861, Lowe File.

5. W. Small to J. Hooker, Dec. 9, 1861, QCF.

6. J. Starkweather to T. S. C. Lowe, Jan. 7, 13, 18; Mar. 10; Apr. 3, 10, 12, 15, 19, 1862; all in Lowe File.

7. The story of Steiner's service in the West is found in: T. S. C. Lowe to J. Steiner, Feb. 12, 1862; J. Steiner to T. S. C. Lowe, Mar. 1, 8, 13; Apr. 15; May 7; June 9, 16, 28, 1862; quotation from J. Steiner to T. S. C. Lowe, Mar. 1, 1862; all in Lowe File.

8. Steiner to T. S. C. Lowe, Mar. 13, 1862, Lowe File.

9. Steiner to T. S. C. Lowe, Apr. 15, 1862, Lowe File.

10. *Ibid.*

11. Steiner to Lowe, June 9, 1862, Lowe File.

12. Lowe, "My Balloons," p. 95.

13. *Ibid.* pp. 86–102; see also material for this period in both the Lowe File and the QCF.

14. Lowe, "My Balloons," pp. 101–102.

15. *Ibid.*, p. 102.

16. Haydon, *Aeronautics*, p. 340.

17. W. J. Handy to T. S. C. Lowe, Dec. 1, 1909, in Lowe, "My Balloons," pp. 105–108.

18. Joel Cook, *Siege of Richmond* (Philadelphia, 1862), p. 32.

19. Handy to Lowe, Dec. 1, 1909.

20. Lowe, "My Balloons," pp. 112–130.

21. S. Millett Thompson, *13th Regiment, New Hampshire Volunteer Infantry in the War of the Rebellion* (Boston, 1888), p. 163.

22. *Ibid.*

23. "The Yankee Balloon," copied from the *Detroit Free Press*, in Lowe, "My Balloons," pp. 103–105.

24. *Ibid.*

25. John R. Bryan, "Balloon Used for Scout Duty," *Southern Historical Society Papers*, vol. 33 (1905), p. 32.

26. G. A. Townsend, *Rustics in Rebellion* (Chapel Hill, N.C., 1950), p. 150.

27. Bryan, "Balloon," p. 33.

28. *Ibid.*

29. *Ibid.*; see also G.A. Townsend, *The Sword Over the Mantle* (New York, 1960), p. 44.

30. *Ibid.*

31. Bruce Catton, *Mr. Lincoln's Army* (New York, 1951), p. 115.

32. *Ibid.*

33. George Armstrong Custer, "War Memoirs," *Galaxy Miscellany and Advertiser* (November 1876), p. 686.

34. *Ibid.*

35. *Ibid.*

36. *Ibid.*

37. *Ibid.*, p. 687.

38. *Ibid.*, pp. 687–688.

39. It should be noted that while both Lowe and Allen often gave the names of officers flying with them, this was not always the case. Having flown with a great many generals, both aeronauts may simply not have seen fit to mention an as yet unknown young lieutenant. It is far more difficult to understand why Lowe, writing his memoirs in 1911, at which time Custer's name was known to all Americans, failed to mention his own early connection with the famed "Yellow Hair."

40. Lowe, "My Balloons," pp. 109–116.

41. *Ibid.*, pp. 114–115.

42. Anonymous, "Three Months with the Balloons in America," *St. James Magazine* (London, 1863), pp. 96–105.

43. *Ibid.*

44. Unidentified Richmond newspaper clipping, May 26, 1862, quoted in Lowe, "My Balloons," p. 115.

45. *Ibid.*

46. Beaumont, "Balloon Reconnaissance."

47. *Ibid.*

48. *OR*, vol. 3, pp. 280–281.

49. *Ibid.*

50. *Ibid.*; Lowe, "My Balloons," pp. 119–120.

51. S. P. Heintzelman to T. S. C. Lowe, July, 1862, Lowe File.

52. A. W. Greely, "Balloons in War," *Century Magazine* (June 1900), p. 42.

53. Beaumont, "Balloon Reconnaissance."

54. Glassford, "Prolegomenon," p. 261.

55. James Longstreet, "Our March against Pope," in *Battles and Leaders of the Civil War* (New York, 1950), p. 512.

56. J. H. Ensterly, "Captain Langdon Cheeves, Jr., and the Confederate Silk Dress Balloon," *South Carolina Historical and Genealogical Magazine* (January 1944), p. 8.

57. *Ibid.*

58. Cornish, *Air Arm*, p. 39. Cornish is the major authority on the Cevor balloon as on most other aspects of Confederate aeronautics.

59. *Ibid.*, p. 43.

60. *Ibid.*, p. 48.

61. *Ibid.*

62. *Ibid.*

63. Lowe, "My Balloons," p. 134.

64. A. A. Humphreys to C. Lowe, Aug. 13, 1862, in Lowe, "My Balloons," p. 134.

65. A. M. Stewart, *Camp, March, and Battle Field* (Philadelphia, 1865), p. 248.

66. Lowe, "My Balloons," p. 136.

67. Haydon, *Aeronautics*, pp. 280–281.

68. *Ibid.*; Haydon provides by far the best administrative history of the Balloon Corps.

69. *OR*, vol. 3, pp. 292–293; Lowe, "My Balloons," pp. 136–138.

70. Lowe, "My Balloons," p. 137.

71. *Ibid.*

72. *Ibid.*

73. *Ibid.*

74. *Ibid.*

75. Bruce Catton, *Glory Road: The Bloody Route from Fredericksburg to Gettysburg* (New York, 1952), provides an excellent treatment of the military campaigns involved here.

76. Lowe, "My Balloons," pp. 140–141.

77. *Ibid.*, p. 142.

78. Joseph P. Cullen, *Where a Hundred Thousand Fell* (Washington, D.C., 1966), p. 21.

79. Lowe, "My Balloons," p. 142.

80. *Ibid.*

81. Glassford, "Prolegomenon," p. 263.

82. G. Harding to E. M. Stanton, Aug. 29, 1863, in Lowe File.

83. Undated clipping, Allen Scrapbooks, vol. 1, p. 53.

84. Glassford, "Prolegomenon," p. 265.

85. Lowe, "My Balloons," p. 142.

86. *Ibid.*, pp. 142–147.

87. Haydon, *Aeronautics*, p. 326.

88. The story of Lowe's "caloric" signal balloons is contained in: T. S. C. Lowe to A. Burnside, Dec. 22, 1862; T. S. C. Lowe to D. Butterfield, Feb. 16, 1863; Butterfield to Lowe, n.d.; T. S. C. Lowe to A. Myer, n.d. (all in Lowe File); T. S. C. Lowe to S. Williams, Mar. 21, 1863, *OR*, vol. 3, p. 298; Lowe to Williams, Mar. 20, 1863, *ibid.*, p. 297; Lowe to Williams, Mar. 21, 1863, *ibid.*, p. 298.

89. *Ibid.*

90. *Ibid.*

91. Oliver to T. S. C. Lowe, Mar. 17, 1863, Lowe File.

92. B. England to H. Halleck, Mar. 18, 1863, Lowe File; W. Paullin to General Crossman, Mar. 7, 1863, QCF.

93. T. S. C. Lowe to S. Williams, Mar. 30, 1863, in Lowe, "My Balloons," pp. 148–150.

94. J. Allen to T. S. C. Lowe, Apr. 1, 1863, with endorsement by E. Allen, Lowe File; a copy of Allen's letter is in Lowe, "My Balloons," pp. 150–152.

95. *Ibid.*

96. Haydon, *Aeronautics*, p. 266.

97. Charges by Jacob C. Freno, Dec. 22, 1863, copies in both QCF and Lowe File; Haydon, *Aeronautics*, p. 267. Haydon believed that Lowe had answered all of the charges to the satisfaction of the War Department. This was clearly not the case.

98. T. S. C. Lowe to O. Howard, Jan. 31, 1863; T. S. C. Lowe to E. Canby, Jan. 22, 1863; copies in Lowe File. See also W. S. Kitchner, asst. adj. gen., comments on Lowe's answer to Freno's charges, dated Apr. 27, 1863, Lowe File.

99. T. S. C. Lowe to M. C. Meigs, with enclosures from J. Allen, E. Allen, J. Starkweather, J. O'Donnell, et al., Mar. 7, 1863, Lowe File.

100. *Ibid.*

101. Mrs. L. Lowe to T. S. C. Lowe, n.d., in RG 94, NARS.

102. Mrs. L. Lowe to T. S. C. Lowe, Feb. 10, 1863, *ibid.*

103. E. Canby, "Notes on Lowe's Request for Reimbursement," Dec. 3, 1863, Lowe File.

104. S. Williams, "Special Order No. 95," in Lowe, "My Balloons," p. 153.

105. C. B. Comstock to P. H. Waterman, Apr. 25, 1863, QCF.

106. *Ibid.*

107. *Ibid.*

108. C. B. Comstock to T. S. C. Lowe, Apr. 12, 1863, Lowe File.

109. T. S. C. Lowe to D. Butterfield, Apr. 12, 1863, Lowe File.

110. T. S. C. Lowe to C. B. Comstock, Apr. 13, 1863, Lowe File; see also Lowe, "My Balloons," p. 159.

111. *Ibid.*, with notes and instructions from Comstock and S. Williams ("By Command of General Hooker"), Apr. 15, 1863, Lowe File.

112. Miscellaneous undated documents, Lowe File.

113. Lowe, "My Balloons," pp. 159–169.

114. *Ibid.*, pp. 170–188; original communication in Lowe File.

115. G. F. R. Henderson, *Stonewall Jackson and the American Civil War* (London, 1919), vol. 2, pp. 307, 410–419.

116. J. Sedgwick to T. S. C. Lowe, Sept. 3, 1863, Lowe File.

117. *Ibid.*

118. Lowe, "My Balloons," p. 189.

119. *Ibid.*

120. J. Allen to T. S. C. Lowe, June 6, 1863, Lowe File.

121. *Ibid.*

122. *Ibid.*; see also J. Allen to G. K. Warren, June 11, 1863, Lowe File.

123. *Ibid.*

124. *Ibid.*

125. *Ibid.*

126. *Ibid.*

127. *Ibid.*

128. Undated article in Allen Scrapbooks, vol. 1, p. 49.

129. *Ibid.*

130. *Ibid.*

131. W. J. Rhees, "Reminiscences of Ballooning in the Civil War," *Chatauquan*, vol. 27 (1890), p. 261.

132. Greely, "Ballooning," p. 4.

133. Haydon, *Aeronautics*, p. 85.

134. T. S. C. Lowe to E. M. Stanton, May 26, 1863, in *OR*, vol. 3, pp. 252–319.

135. T. S. C. Lowe to A. J. Myer, July 13, 1863, Lowe File.

136. G. Harding to E. M. Stanton, Aug. 29, 1863, Lowe File.

137. E. Canby to T. S. C. Lowe, Dec. 3, 1863, Lowe File.

138. *Ibid.*

139. *Ibid.*

140. W. S. Ketchem, "Notes on Lowe's Claims," Apr. 27, 1864, Lowe File.

141. T. S. C. Lowe to J. Henry, July 15, 1864, Lowe File.

Chapter 14

1. "William Paullin," *Appleton's Cyclopedia*, p. 682.

2. Undated article, John La Mountain, box 218, AIAA Scrapbooks.

3. "La Mountain's Death," undated article, box 218, AIAA Scrapbooks.

4. *Ibid.*

5. "Army Balloon Ascensions," poster, under Lowe, box 82, AIAA Scrapbooks.

6. "Up in a Balloon," *Frank Leslie's Illustrated Newspaper*, Oct. 21, 1865, p. 127; "By Air and Water," *New York Herald*, July 10, 1866; *Harper's Weekly* (November 25, 1866), p. 745.

7. "Thaddeus S. C. Lowe," *Dictionary of American Biography*.

8. *Evening Standard* (New Bedford, Mass.), Sept. 13, 1864; *Delaware Inquirer* (Wilmington), July 2 and 9, 1864; "The Bicentennial Celebration," *Mercury* (New Bedford, Mass.), Sept. 15, 1864; "Balloon Ascension," *Daily Inquirer*, June 25, 1864.

9. *Ibid.*

10. "The Balloon Ascensions and Concert,"

Daily Courier (Lowell, Mass.), July 24, 1865; *Kennebec Journal*, Sept. 1, 1865; undated article in Allen Scrapbooks, vol. 1, p. 55.

11. *Providence Journal*, June 15; July 6, 1866; *Providence Press*, June 18, 1866; *Providence Daily Post*, June 18, 1866.

12. *Ibid.*

13. *Ibid.*

14. Emperor Dom Pedro to T. S. C. Lowe, n.d., Lowe File.

15. Lowe, "My Balloons," p. 197.

16. *Morning Herald* (Providence, R.I.), Apr. 23, 1868; *Providence Journal*, Oct. 24, 1867; undated clippings in English and Spanish, Allen Scrapbooks, vol. 1, pp. 62–65.

17. E. S. Allen to T. S. C. Lowe, July 14, 1867, in Lowe File.

18. *Ibid.*

19. *Ibid.*

20. *Ibid.*

21. Emperor Dom Pedro to T. S. C. Lowe, n.d., Lowe File.

22. H. G. Warren to E. L. Jones, Jan. 23, 1951, microfilm copy in the Jones Chronology, NASM Library; H. G. Warren, *Paraguay: An Informal History* (Norman, Okla., 1949), p. 236; T. S. C. Lowe, *The Latest Developments in Aerial Navigation* (Los Angeles, 1910).

23. Robert A. Chodesciorcy to J. Allen, n.d., Allen Scrapbooks, vol. 2, n.p.

24. "Noted Aeronaut Dead," undated article, Allen Scrapbooks, vol. 2, n.p.

25. *Boston Evening Transcript*, June 21, 1872.

26., Undated articles, Allen Scrapbooks, vol. 2, n.p.

27. "Ballooning," *San Francisco Morning Call*, undated clipping, Allen Scrapbooks, vol. 2, p. 86; *Evening Bulletin*, Oct. 25, 1878; undated articles, Allen Scrapbooks, vol. 2, pp. 86–88.

28. "Aloft at Anchor," *Daily Evening Call* (San Francisco), Feb. 22, 1874; "Castle in the Air," *Daily Evening Bulletin*, Feb. 21; Apr. 20, 1874; *San Francisco Chronicle*, Mar. 16, 1874; *Daily Call* (San Francisco), Apr. 21, 1874; undated articles, Allen Scrapbooks, vol. 2, pp. 86–88.

29. "Ballooning at Night," *San Francisco Chronicle*, July 8, 1874.

30. *Dayton Journal*, June 21, 1877.

31. Assorted articles, Allen Scrapbooks, vols. 2–3, passim.

32. See, for example, *Frank Leslie's Illustrated Newspaper*, Oct. 6, 1888; *Illustrated Police News*, Sept. 14, 1889; "Up in the Air," *Sunday Telegram* (Providence, R.I.), July 5, 1891; undated articles, Allen Scrapbooks, vols. 2–3.

33. "Post Sky Bulletins," *Boston Post*, Nov. 23, 1891; undated article, Allen Scrapbooks, vol. 3, n.p.

34. "The Sky Couriers," *Boston Post*, June 1, 1892.

35. *Ibid.*

36. "Up in the Sky," *Sunday Telegram* (Providence, R.I.), July 5, 1891.

37. *Haverhill Gazette*, July 12, 1883.

38. "Up in a Balloon, Boys," unidentified clipping, July 1889, Allen Scrapbooks, p. 135.

39. See also "Up in the Sky"; "In Upper Air," *Boston Globe*, July 5, 1889.

40. "Up in the Air"; "Out of the Balloon," *New Bedford Evening Journal*, July 6, 1891; "He Wasn't in It," unidentified article, Allen Scrapbooks, vol. 3, p. 168.

41. "Fishermen Save Aeronaut Allen," *Boston Post*, July 7, 1906.

42. Unidentified articles, Allen Scrapbooks, vol. 3, p. 172.

43. "Allen Would Brave Perils of the Ocean," *Evening Bulletin* (Providence, R.I.), n.d., Allen Scrapbooks, vol. 3, p. 176.

44. John Wise, "Aerial Navigation," *Scientific American* (January 8, 1870), p. 33; "The Balloon as an Aid to Meteorological Research," *ibid.* (November 26, 1870), p. 341; and "Balloon Varnishes," *ibid.* (June 11, 1870), p. 381.

45. Wise, *Through the Air*, pp. 565–570.

46. *Ibid.*, p. 588; M. L. Amick, *History of Donaldson's Balloon Ascensions* (Cincinnati, 1875), n.p.

47. Robert H. Stepanek, "The Transatlantic Flight of the Daily Graphic," *American Avia-*

tion Historical Society Journal (Summer 1981), pp. 143–149.

48. On the origins of the *Daily Graphic*, see the following *Scientific American* articles: "To Europe in a Balloon" (July 19, 1873), p. 33; "The Proposed Transatlantic Balloon Voyage" (July 19, 1873), p. 37; "To Europe by Balloon" (July 26, 1873), p. 49; "Balloonamania" (September 13, 1873), p. 160; "The Voyage of the Graphic Balloon" (September 20, 1873), p. 176; "Failure No. 2 of the Balloon to Europe" (October 25, 1873), p. 256.

49. *Ibid.*; see also *Balloon Graphic*, Aug. 9 and 16, 1873, in John Wise, box 220, AIAA Scrapbooks; *Evening Herald* (New York), July 14, 1873.

50. *Balloon Graphic*, Aug. 9, 1873.

51. "The Transatlantic Flight," Allen Scrapbooks, vol. 2, p. 148.

52. "The Rotten Balloon," *New York Times*, Sept. 13, 1873.

53. *Ibid.*

54. *Ibid.*

55. *Ibid.*

56. Unidentified article, John Wise, box 220, AIAA Scrapbooks.

57. *Ibid.*

58. King, *Balloon*, pp. 11–13.

59. All quotations and information on W. H. Donaldson from Amick, *Donaldson*, pp. 86–199; unidentified article, Washington H. Donaldson, box 20, AIAA Scrapbooks.

60. Unidentified article, John Wise, box 220, AIAA Scrapbooks.

61. *Ibid.*

62. *Ibid.*

63. The author wishes to thank Ms. Phyllis Kihn of the Connecticut Historical Society for her assistance in unscrambling the careers of George and Silas Brooks.

64. *Hartford Daily Courant*, July 9 and 11, 1862.

65. "The Celebration," *ibid.*, June 7, 1862.

66. *Hartford Daily Courant*, July 11, 1862.

67. "Edwin C. Bassett: An Old Resident of Hartford with an Interesting History," Connecticut Historical Society Obituary Scrapbooks, vol. 16, p. 66, CHS Library, Hartford, Conn.

68. *Hartford Daily Courant*, June 15, 1863; *Hartford Evening Press*, June 22; July 6; Aug. 16, 20, 28; Sept. 1, 26; Oct. 1, 6, 12, 26, 1863; June 4, 1864; Aug. 11; Sept. 13, 1865; July 7, 1869; "How the Fourth was Celebrated in Hartford," July 6, 1863; "A Cat's Tale," July 6, 1863; "Balloon Ascension at Middletown," Aug. 27, 1863; "Narrative of Professor Brooks," Sept. 5, 1863; "The Balloon Ascension," July 5, 1866; "That Balloon," July 18, 1866; "Balloon Ascensions," Aug. 7, 1866; "The Balloon Ascensions," *Springfield Daily Republican*, Sept. 29, 1863.

69. *Ibid.*

70. "Fourth of July—The Balloon Ascension," *Hartford Evening Post*, July 5, 1882; see also "The Balloon Ascension Today," *Hartford Daily Courant*, July 4, 1882; and *ibid.*, July 6, 1882.

71. "Independence Day," *Hartford Evening Post*, July 3, 1883; "Animals in a Balloon," *ibid.*; "The Fourth of July in Hartford," July 5, 1883; "The Fourth," July 3, 1885; "The Balloon Ascended," July 5, 1885; "The Glorious Fourth," July 6, 1885; "The Balloon Ascended," July 5, 1887; "A Fourth of July Balloon," *Hartford Times*, July 6, 1886; "Old Fashioned Fourth," *Hartford Evening Post*, July 5, 1894; "Gas Was Too Heavy," *Hartford Daily Courant*, July 5, 1894.

72. "Perilous Balloon Ascension," *Boston Journal*, July 10, 1862; "Loss of the Balloon," *Boston Herald*, July 10, 1862.

73. "Aeronautics," *Boston Journal*, July 11; Sept. 9, 1867; King, *Balloon*, pp. 1–3.

74. "The Buffalo Balloon Ascension," *Boston Journal*, July 12, 1869.

75. *Ibid.*

76. "Facing Death in the Clouds," *Yankee Blade*, Apr. 15, 1871.

77. "An Aerial Journey," *Daily Patriot* (Washington D.C.), Oct. 8, 1872.

78. "Grand Balloon Ascension," *Aeronaut*

(Buffalo, N.Y.), July 4, 1874; "The Buffalo," *Inquirer* (Buffalo, N.Y.), July 10, 1874; "The Buffalo Balloon Ascension," unidentified article, July 10, 1874, S. A. King, box 218, AIAA Scrapbooks.

79. U.S. Congress, House of Representatives, "Report of the Assistant in Charge of the Study Room," App. 4, *Report of the Chief Signal Officer, Executive Document No. 1, Part 2, Vol. IV*, 48th Cong. 2nd Sess., 1884, serial 2283, p. 6; see also Russell J. Parkinson, "United States Signal Corps Balloons, 1871–1902," *Military Affairs* (Winter 1960–1961), pp. 190–202; and Hunter Dupree, *Science in the Federal Government: A History of Policies and Activities to 1940* (Cambridge, Mass., 1957), for a history of the weather service. Russell Parkinson has been my guide through the Signal Corps records.

80. U.S. Congress, House of Representatives, "Paper M," *Report of the Chief Signal Officer, Executive Document No. 1, Part 2,* Sess. 1872, serial 1558, p. 754.

81. *Ibid.*, *Vol. IV*, 45th Cong., 2nd Sess., 1877, serial 1798, pp. 541–544. This flight has assumed special significance in philatelic history since King carried mail bearing the first balloon stamps to be issued in the U.S.

82. Henry C. De Ahna, letter to the editor of the *Philadelphia Public Ledger*, July 7, 1873, De Ahna, AIAA Scrapbooks; *Philadelphia Evening Herald*, Aug. 27, 1873; *New York Times*, Aug. 27, 1873.

83. Unidentified article, John Light, box 218, AIAA Scrapbooks.

84. *New York Times*, July 22, 1878; *New York Herald*, Aug. 3, 1880.

85. King, *Balloon*, pp. iii-ix.

86. *Ibid.*, pp. x-xvi; "Aeronautic Potentialities," *New York Graphic*, Apr. 7, 1879; *ibid.*, July 7, 1879; "Transatlantic Ballooning," *New York Herald*, Apr. 7, 1879; "The Captive Balloon," *ibid.*, July 8, 1879; "To Europe by Balloon," *ibid.*, May 30, 1879; "The Balloon at the Beach," *New York Graphic*, June 30, 1879; "Aerial Navigation," *New York Graphic*, Apr. 25, 1879; "In Mid-Air," *New York Herald*, July 7, 1879.

87. *Ibid.*

88. Rhoda Gilman, "Balloon to Boston," *Minnesota History* (Spring 1970), pp. 17–22, is the best secondary account of the flight; see also "Aerial Navigation," *New York Herald*, Sept. 8, 1881; "Balloon Experiments," *ibid.*, Aug. 16, 1881; *Boston Evening Journal*, Sept. 12, 13, 24, 1881; *Minneapolis Tribune*, Sept. 6–15, 1881; St. Paul and Minneapolis *Pioneer Press*, Sept. 6–15, 1881.

89. St. Paul and Minneapolis *Pioneer Press*, Sept. 13, 1881; see also U.S. Congress, House of Representatives, "Report upon the Balloon Ascension Made from Minneapolis, Minn., September 12, 1881, by Winslow Upton," App. 66., *Report of the Chief Signal Officer, Executive Document No. 1, Part 2, Vol. IV*, 47th Cong., 2nd Sess., 1882, Serial 2098, pt. 1.

90. Parkinson, "U.S. Signal Corps Balloons," p. 192.

91. W. H. Hammon, "Report on the Observations Made in Four Balloon Ascents," *Proceedings of the American Association for the Advancement of Science, 1890* (Washington, D.C., 1891), pp. 94–96.

92. Parkinson, "U.S. Signal Corps Balloons," passim.

93. *Ibid.*, p. 192.

94. *Ibid.*

95. Unidentified article, S. A. King, box 218, AIAA Scrapbooks.

96. Wynn, "Professor Samuel Archer King," p. 358.

97. Unidentified obituary notices, S. A. King, box 218, AIAA Scrapbooks.

Chapter 15

1. "Hot Air Ballooning, Weehawken, N.J.," *Scientific American* (September 5, 1891), p. 147.

2. Author's interview with Captain Eddie Allen, Apr. 30, 1982.

3. "The Dangers of Ballooning," *Scientific American* (July 16, 1876), p. 97.

4. "Blown to Pieces High in the Air," J. E. Baldwin, box 217, AIAA Scrapbooks.

5. "Balloon Accidents," *Scientific American* (August 3, 1862), p. 69.

6. "The Balloon on Fire," *Frank Leslie's Illustrated Newspaper*, Aug. 1, 1868.

7. "A Balloon Ascension Suicide," *Day's Doings*, Oct. 21, 1871, p. 16.

8. "Terrible Fall from a Balloon," unidentified clipping, John Short, box 219, AIAA Scrapbooks.

9. "Falling from the Clouds," unidentified clipping, box 218, AIAA Scrapbooks.

10. "Down from the Clouds to Death," *Day's Doings*, undated, Dennison, AIAA Scrapbooks.

11. "Falling from a Balloon," *Sun*, Sept. 21, 1877.

12. "Terrible Fall of an Aeronaut," unidentified clipping, E. L. Stewart, box 219, AIAA Scrapbooks.

13. "Balloon Disaster," *Winchester Herald*, July 23, 1879; other unidentified clippings, S. W. Colegrove, box 217, AIAA Scrapbooks.

14. "The Balloon Caught Fire"; "An Aeronaut Killed"; "An Aeronaut's Fate"; unidentified clippings, Sam Black, box 217, AIAA Scrapbooks.

15. "Fell from the Clouds," unidentified clipping, Bartholomew, box 217, AIAA Scrapbooks.

16. *Ibid.*

17. *Ibid.*

18. "Aeronaut Hart's Fall," unidentified clipping, box 218, AIAA Scrapbooks.

19. "An Aeronaut Hurt," E. E. Craig, box 217, AIAA Scrapbooks.

20. "He Was Drowned," F. H. Vandergrift, box 219, AIAA Scrapbooks.

21. "Killed by a Fall from a Balloon," Annie Harkness, box 218, AIAA Scrapbooks.

22. "She Broke Her Leg," Clara Pattee, box 219, AIAA Scrapbooks.

23. "Aeronaut's Fearful Fall," Nellie Wheeler, box 220, AIAA Scrapbooks.

24. This list of additional aeronauts has been extracted from articles in the AIAA Scrapbooks and assorted undated clippings in the NASM collection. The author also thanks Robert Rechs, who furnished his list of American balloonists active during the nineteenth century.

25. *Ibid.*

26. "Gymnastic Balloonists," *Scientific American* (August 17, 1872), n.p.

27. Assorted advertising materials, McClellan, box 219, AIAA Scrapbooks.

28. Undated articles, Eddie Allen Scrapbooks, Eddie Allen, Batavia, N.Y. The author's thanks to Captain Allen's daughter, Mrs. Arlene Stambach, of Mechanicsburg, Pa., who not only introduced me to her father but made family records available as well.

29. *Ibid.*; see also "The 100th Anniversary of the Flying Allens," *Ballooning* (Autumn 1974), pp. 20–24.

30. Unidentified clipping, Allen Family Scrapbook.

31. *Ibid.*

32. "*Flying Allens*," p. 22.

33. Author's conversation with Captain Eddie Allen, August 1982.

34. "Flying Allens," pp. 23–24.

35. Preston R. Bassett, "Carlotta, the Lady Aeronaut of the Mohawk Valley," *New York History* (April 1963), pp. 145–172, is the best source of biographical information on both Carl and Mary Myers.

36. *Ibid.*

37. *Watertown [N.Y.] Daily Times*, July 4, 1882.

38. Mary Myers, *Skylarking in Cloudland* (Mohawk, N.Y., 1883), pp. 5–13.

39. *Ibid.*, pp. 18–22.

40. *Ibid.*, pp. 23–27.

41. *Ibid.*, pp. 24–25.

42. *Ibid.*, p. 44.

43. *Ibid.*, pp. 112–113

44. *Ibid.*, pp. 47–49.

45. *Ibid.*, pp. 92–94.

46. Carl E. Myers, "Sphere of the Gas Balloon," *Fly* (April 1909), p. 1.

47. Bassett, "Carlotta," p. 164.

48. Mary Myers, *Skylarking*, pp. 58–62.

49. Bassett, "Carlotta," pp. 165–168.

50. H. Paul Draheim, "Rainmaking is Old Stuff in the Utica-Frankfort Area," *Wingfoot Lighter-Than-Air Society Bulletin* (Summer 1960), p. 4; Pearl I. Young, "Balloons at the St. Louis Fair, 1904," *ibid.* (March 1964), p. 2.

51. *Ibid.*; Carl E. Myers, "Dirigible Balloons with the Screw in Front," *American Magazine of Aeronautics* (January 1908), pp. 29–31.

52. *Corry [Pa.] Blade*, undated, 1876, in a collection of articles on C. F. Ritchell presented to the NASM Library by the Connecticut Aviation Historical Society (hereafter, CAHA Scrapbook).

53. "Inventor's Dreams," *Sunday Register* (New Haven, Conn.), July 24, 1881.

54. "Chicago's Columbus," *Philadelphia Record*, June 3, 1892.

55. "Saw the Miteascope," *Standard*, Jan. 16, 1901; "Live People Appear as Fairies and Brownies in This Marvel of Pantomime Production," undated article, CAHA Scrapbook.

56. "Under Ocean Waves," *Evening Register*, Oct. 3, 1893; "Death in 15 Minutes," *Sunday Register*, Oct. 7, 1883; "Ritchell's Deadly Bombs," *Daily Register*, Nov. 23, 1884; "A Pretty Good Suggestion," *New York World*, Jan. 5, 1891; U.S. Patent No. 346,416, "Corset"; No. 378,536, "Wave-Guide Matching Machine"; No. 352,068, "Toy Motor"; assorted undated articles, CAHA Scrapbook.

57. The best descriptions of the Ritchell machine and its operation are found in U.S. Patent No. 201,200; "Voyages in Cloudland," *Fireside Almanac* (July 15, 1878), p. 278; "Ritchell's Flying Machine," *ibid.*, undated, CAHA Scrapbook; *New York Daily Graphic*, June 26, 1878; "Professor Ritchell's Flying Car," *Hartford Daily Courant*, June 14, 1878; assorted unidentified articles, CAHA Scrapbook.

58. "The Flying Machine," *Boston Globe*, June 26, 1878; "Sailing on the Flying Machine," *Philadelphia Times*, June 17, 1878; "Elevation of Women," unidentified article, CAHA Scrapbook.

59. Unidentified articles, CAHA Scrapbook.

60. "The Flying Machine," *Boston Post*, June 25, 1876; see also "Ritchell's Flying Machine," *Fireside Almanac*.

61. "The Flying Machine," *Boston Herald*, June 26, 1876.

62. "Voyages in Cloudland," *Fireside Almanac* (July 15, 1878), p. 278; unidentified articles, CAHA Scrapbook.

63. Advertisements, CAHA Scrapbook.

64. *Ibid.*

65. *Ibid.*

66. "A New Flying Machine," *New York Times*, Mar. 28, 1880.

67. "Flying to the North Pole," *New York Daily Tribune*, Mar. 28, 1880; "A New Aerostat," *Harper's Weekly* (Oct. 27, 1883); unidentified articles, CAHA Scrapbook.

68. Carl E. Myers, "Half a Lifetime with the Hydrogen Balloon," *Fly* (June 1913), pp. 16–17; *idem*, "Dirigible Balloons," pp. 29–31.

69. Mary Myers, *Skylarking*, pp. 52–53.

70. *Ibid.*

71. *Ibid.*

72. "That Airship Wouldn't Ascend," *Press* (N.Y.), June 30, 1889; "The Balloon Wouldn't Rise," *New York Sun*, June 26, 1889; "The Loss of the Campbell Airship," *Scientific American* (July 27, 1889), p. 54.

73. Carl E. Myers, "Gas Balloon," p. 1; "Campbell's Airship Goes," *Sun*, Dec. 15, 1889.

74. "Campbell's Airship Goes."

75. "That Airship Wouldn't Ascend."

76. "Loss of the Campbell Airship."

77. Biographical information on Cowan and Pagé from "Montreal Airship 51 Years Ago," *Witness and Canadian* (Homestead), May 21, 1930; *Montreal Witness*, May 14, 1879.

78. *Ibid.*; "Cloudland," *Daily Citizen* (Oc-

tawa), Aug. 28, 1878; "The Balloon Ascension," *Montreal Witness*, Sept. 30, 1878.

79. "The Late Balloon Ascension," *Montreal Witness*, Oct. 24, 1878; "Testimonial to an Aeronaut," *Montreal Herald*, Oct. 24, 1878; *Montreal Witness*, July 26, 1879; Richard Thomas to C. H. Grimley, Oct. 5, 1878, R. Cowan, box 217, AIAA Scrapbooks.

80. "A Canadian Invention," *Ottawa Citizen*, Aug. 26, 1878; *Montreal Gazette*, Sept. 30, 1878; "Will Artificial Wings Fly?" *Montreal Star*, June 21, 1879.

81. "An Immense Balloon," *Montreal Witness*, March 28; Apr. 9, 16, 26; May 27; June 3, 1879; *Montreal Herald*, May 19; June 19, 1879; *L'Aurore* (Montreal), May 8, 1879; *Montreal Star*, June 14, 1879; *Montreal Witness Balloon Bulletin*, June 21, 1879.

82. *Montreal Star*, June 23–24, 1879; *Montreal Herald*, June 23–24, 1879; unidentified articles, R. Cowan and C. Pagé, boxes 217–219, AIAA Scrapbooks.

83. "Up in a Balloon," *New York Herald*, June 23, 1879; "The Late Balloon Ascension," *Montreal Witness*, July 3, 1879; C. H. Grimley to R. Cowan, June 1, 1879, in R. Cowan, AIAA Scrapbooks; "The Balloon Ascension," *Montreal Witness*, June 23, 1879; "Sailing through the Air," *Ottawa Daily Press*, Aug. 2, 1879; "Voyage of the Balloon *Canada*," *Montreal Herald*, Aug. 2, 1879; "Journeying in the Clouds," *New York Herald*, Aug. 12, 1879; unidentified articles, Grimley and Cowan, boxes 217–218, AIAA Scrapbooks.

84. "Strange Experience in a Balloon," *Harper's Weekly* (July 10, 1884), p. 469.

85. *Ibid.*

86. Russell J. Parkinson, "U.S. Signal Corps Balloons," pp. 189–202; "A Proposed War Balloon," *Scientific American Supplement* (1885), p. 7943; *London Graphic*, June 13, 1885, p. 594; notes, Russell Thayer biographical file, NASM Library.

Chapter 16

1. Thomas Baldwin to E. L. Jones, Feb. 3, 1950, microfilm copy of the Jones Chronology, NASM Library.

2. Harold E. Morehouse, "Thomas S. Baldwin," notes in the author's collection; Austin Gregory, "America's Foremost and Most Famous Aeronaut, Captain Thomas Scott Baldwin," *Aeronautics* (June 1908), pp. 36–41; "Captain Thomas Baldwin," *Aircraft* (Mar. 10, 1910), vol. 1, p. 123; Edith Day Robinson, "The Personal Side of Talked of Aviators," *ibid.* (November 1910), vol. 1, p. 9; Augustus Post, "Captain Thomas Scott Baldwin," *Curtiss Flyleaf* (Jan.–Feb. 1918), pp. 12–14; Ladislas D'Orcy, "The Passing of a Great Aeronautical Pioneer," *Aviation* (May 28, 1923), pp. 584–585.

3. Morehouse, "Thomas S. Baldwin"; T. S. Baldwin biographical file, NASM Library; Jones Chronology.

4. Undated clippings, T. S. Baldwin, box 5, AIAA Scrapbooks (hereafter, Baldwin Scrapbook).

5. "Used Parachutes Fifty Years Ago," *New York Times*, Sept. 14, 1930; Frank Cliff, "The Balloon Jumpers," *Long Beach Press Magazine* (August 12, 1928), p. 4.

6. Dick Brown, "New Mexico's Ballooning Heyday," *Ballooning* (Summer 1976), pp. 62–63.

7. Cliff, "Balloon Jumpers," p. 4.

8. Sandy Branham, "Over Ninety Years of Ballooning in Colorado," *Ballooning* (Summer 1976), pp. 71–72.

9. "A Genuine Long Drop," *Evening News* (London), July 31, 1888.

10. Duke Schroer, *A True History of the Daring Aeronaut, Thomas Baldwin and His Thrilling Parachue Leaps from a Balloon* (Quincy, Ill., 1887), pp. 1–4; Charles J. Murphy, *Parachute* (New York, 1930); Don D. Eklund, "Baldwin and the Parachute," *Wingfoot Lighter-Than-Air Society Bulletin* (June 1968), pp. 2–5.

11. Schroer, *True History*, pp. 4–10.

12. *Ibid.*, pp. 11–19.

13. *Ibid.*, pp. 20–29.

14. Unidentified clippings, Baldwin Scrapbook, "Chats with Celebrities," *Sunday Times* (London), Sept. 16, 1888; Griffith Brewer to Lester Gardiner, Sept. 30, 1943, box 5, AIAA History Collection.

15. Assorted articles, Baldwin Scrapbook.

16. *Ibid.*; see also *Echo* (London), July 3, 1888; *Evening News* (London), Aug. 26, 1888; *Liverpool Mercury*, Sept. 5, 1888; *Newcastle Chronology*, Sept. 5, 1888; *Hull Express*, Sept. 6, 1888; *York Evening Press*, Sept. 14, 1888.

17. *Morning Advertiser* (London), July 30, 1888.

18. *Times* (London), Aug. 5, 1888.

19. Assorted clippings, Baldwin Scrapbook.

20. *Ibid.*

21. *Ibid.*

22. *Ibid.*

23. Brewer to Gardiner, Sept. 30, 1943.

24. Irwin Harrison, "Around the World by Parachute," *Rocky Mountain News* (Denver), Sept. 11, 1927; "Ivy Baldwin Clings to Memories of Quincy's Air Pioneer," *Quincy [Ill.] Herald-Whig*, Aug. 14, 1949; Harold Morehouse, "Ivy Baldwin," NASM Library, Baldwin File; "Ivy Baldwin," *Chirp* (October 1948), p. 7.

25. Rufus G. Wells, "An American Balloon Trip over Rome," *Aeronautics* (October 1910), pp. 111–112; Charles Dollfus, "Firsts," p. 8; Rufus G. Wells, scattered articles in *The Boy's Own Paper*, 1888–1889.

26. Wells, "Balloon Trip."

27. *Ibid.*; Dollfus, "Firsts"; Wells articles, *Boy's Own Paper* (above, n. 25).

28. William J. Horvat, *Above the Pacific* (Fallbrook, Cal., 1966), p. 14.

29. *Ibid.*

30. *Ibid.*, pp. 14–16.

31. *Ibid.*, pp. 15–17; *Commercial Advertiser* (Honolulu), Oct. 31, 1889.

32. "Spencer the Balloonist," *Japan Weekly Mail*, July 12; Oct. 11, 1890.

33. "Mr. Spencer's Parachute Descent," *ibid.*, Oct. 25, 1890.

34. "Ballooning before the Emperor," *ibid.*, Nov. 8, 1890.

35. *Japan Weekly Mail*, Nov. 15, 1890; "Ballooning in Tokyo," *ibid.*, Nov. 21, 1890.

36. *Japan Weekly Mail*, Nov. 22, 1890.

37. "The Baldwin Brothers at Uyeno," *ibid.*, Dec. 1, 1890.

38. *Ibid.*

39. Harrison, "Around the World by Parachute"; "Ivy Baldwin Clings to Memories" (above, n. 24).

40. *Ibid.*

41. *Ibid.*; Jones Chronology; Baldwin File, NASM Library.

42. "Ivy Baldwin," *Chirp*; Russell J. Parkinson, "Signal Corps Balloons, 1892–1907," *Wingfoot Lighter-Than-Air Society Bulletin* (July–Aug. 1962), pp. 2–4; *idem*, "Varicle Balloons and the Spanish American War," *ibid.* (February 1969), pp. 2–6; Charles Dollfus, "Hispano-American War and Varicle," *ibid.* (December 1968), pp. 6–8.

43. *Quincy* [Ill.] *Whig-Herald*, Sept. 4, 1949; Jones Chronology.

44. Cliff, "Balloon Jumpers."

45. Parkinson, "U.S. Signal Corps Balloons," pp. 189–202.

46. Russell J. Parkinson, "A Signal Corps Balloon for the Columbian Exposition, 1893," *Wingfoot Lighter-Than-Air Society Bulletin* (January 1966), pp. 2–6.

47. *Ibid.*; *Army and Navy Journal* (November 24, 1894), p. 201; *Aeronautics* (1893–1894), pp. 98–111.

48. A. W. Greely, "Balloons in War," *Century Magazine* (June 1900), p. 45.

49. *Ibid.*

50. Ivy Baldwin to Lester Gardiner, July 10, 1946, Ivy Baldwin File, NASM Library; Harrison, "Around the World by Parachute"; "Ivy Baldwin Clings to Memories"; "Mail Reunites Army Buddies"; all in Ivy Baldwin File, NASM Library.

51. William A. Glassford, "Memorandum on the Santiago Captive Balloon," *Aeronautics* (February 1908), p. 14.

52. *Ibid.*; Greely, "Balloons"; William A. Glassford, "Our Army and Aerial Warfare," *Aeronautics* (January 1908), p. 18.

53. *Ibid.*

54. Parkinson, "U.S. Signal Corps Balloons," pp. 185–186.

55. Leo Stevens to Pres. William McKinley, Apr. 15, 1898, RG 111, Office, Chief Signal Officer, Entry 9, Doc. File No. 6961, NARS. My thanks to Russell Parkinson for this and the following references.

56. J. E. Maxfield to A. Greely, Apr. 21, 1898, RG 111, OCSO, Entry 9, Doc. File No. 6985; Apr. 25, 1898, *ibid.*

57. Dollfus, "Hispano-American War."

58. Alex Rogers, Capt., 4th Cavalry, Military Attaché, Paris, to A. Greely, May 5, 1898, RG 111, OCSO, Entry 9, Doc. File No. 6985; W. Glassford to A. Greely, May 6, 1898, *ibid.*; Adolphus Greely, *Reminiscences of Adventures and Service: A Record of Sixty-Five Years* (New York, 1927), p. 183.

59. Parkinson, "Varicle Balloons"; *idem*, "U.S. Signal Corps Balloons"; *idem*, "Columbian Exposition." My debts to Dr. Parkinson are obvious.

60. Greely, "Balloons"; Glassford, "Our Army"; "Mail Reunites Army Buddies."

61. J. E. Maxfield, quoted in Jones Chronology.

62. Glassford, "Memorandum"; *idem*, "Our Army"; Ivy Baldwin, "Under Fire in a War Balloon at Santiago," *Aeronautics* (February 1908), pp. 13–14.

63. *Ibid.*

64. *Ibid.*

65. *Ibid.*

66. Stephen Crane, quoted in Frank Freidel, *The Splendid Little War* (Boston, 1958), p. 150.

67. John J. Pershing, quoted *ibid.*, pp. 151–152.

68. For a general description of the events of the day, see *ibid.*

69. Stephen Bonsal, *The Fight for Santiago* (New York, 1899), p. 300; Richard Harding Davis, *The Cuban and Puerto Rican Campaigns* (New York, 1898), pp. 211–213.

70. Glassford, "Memorandum"; *idem*, "Our Army."

71. Parkinson, "U.S. Signal Corps Balloons," pp. 201–202.

72. Morehouse, "Ivy Baldwin"; Ivy Baldwin biographical file, NASM Library.

73. The story of the one-man, gasoline-powered airship in America is best told in the various biographies of pioneer aviators prepared by Harold Morehouse. The original manuscript copies of these are in the author's collection; typed copies are to be found in the NASM Library files. Tom D. Crouch, "The Gas Bag Era," *Aviation Quarterly* (Winter 1977), pp. 291–301, offers a short history of the airship era. Douglas Robinson, "Dr. August Graeth and the First Airship Flight in the U.S.," *American Aviation Historical Society Journal* (Summer 1976), pp. 84–92; Pearl I. Young, "Airships and Balloons at the St. Louis Fair," *Wingfoot Lighter-Than-Air Society Bulletin* (Feb., Mar., Apr., 1964), pp. 2–4, 2–3, 2–5; James Horgan, "Airship Races," *ibid.* (February 1968), pp. 2–5; and A. Roy Knabenshue, "Chauffeur of the Skies," unpublished manuscript autobiography in the Knabenshue biographical file, NASM Library, also provide information on various aspects of the pioneer airship era in the U.S.

Chapter 17

1. "Aeronaut Leo Stevens," *Fly* (January 1907), p. 12.

2. Arnold Kruckman, "The Sport of Kings—Ballooning," *Aeronautics* (July 1907), pp. 10–11.

3. *Ibid.*

4. "Something about the Aero Clubs," *Fly* (November 1908), pp. 10–11; "The Promotion of Aerial Navigation," *New York Herald*, Jan. 27, 1907; undated articles, Aero Club of America file, NASM Library.

5. "Promotion of Aerial Navigation."

6. "Aeronaut Leo Stevens"; see also Leo Stevens Scrapbook, box 238, AIAA History Collection.

7. "Our Aero Club," *Aeronautics* (August 1907), p. 3.

8. Thomas Edwin Eldridge, "Ballooning in Philadelphia," *Fly* (November 1908), p. 11; Joseph Jackson, "The Aero Club of Philadelphia," *Aeronautics* (July 1907), pp. 16–17.

9. Eldridge, "Ballooning in Philadelphia."

10. Jackson, "Aero Club."

11. Kruckman, "Sport of Kings."

12. "New Aero Clubs in America" *Aeronautics* (December 1907), p. 6; Charles J. Glidden, "A Talk on Ballooning," *Fly* (May 1909), p. 3.

13. "Club Notes," *Aeronautics* (August 1908), p. 29; "Aero Club of New England," *ibid.* (July 1908), p. 29; *ibid.*, (September 1908), p. 152.

14. "Pittsfield Aero Club," *Aeronautics* (December 1907), p. 6; *ibid.* (May 1908), p. 40.

15. "North Adams Aero Club," *Aeronautics* (December 1907), p. 38; *ibid.* (May 1908), pp. 38–39.

16. "Many Ascents for New England," *Aeronautics* (January 1910), p. 27; "The Aero Club of New England," *ibid.* (November 1909), p. 201.

17. *Ibid.*; Alfred R. Shrigley, "Aero Club of New England," *Fly* (May 1909), p. 14.

18. J. W. Kearney, "The Aero Club of St. Louis," *Aeronautics* (July 1907), pp. 21–22; "Dirigible Balloon and Heavier-Than-Air Contests at St. Louis," *ibid.* (September 1907), pp. 17–19; "Aero Club of New England," *ibid.* (August 1908), p. 34.

19. "The San Antonio Aero Club," *Aeronautics* (July 1908), p. 28; "July Balloon Racing," *ibid.* (August 1908), p. 19.

20. Goethe Link, "Aeronautics in Indiana," *Fly* (January 1909), p. 14; "Aero Club of Indiana," *ibid.* (March 1909), p. 13; "Ascensions," *Aeronautics* (October 1908), pp. 41–43; "Aero Club of Indiana," *ibid.* (January 1909), p. 89.

21. Robert Spangler, "C. A. Coey of Chicago," *Fly* (December 1908), pp. 8–12.

22. "Ascensions."

23. The story of the Aero Club of Ohio has been drawn from undated items in the Frank S. Lahm Papers, MS. 53–151, 53–91, Manuscript Division, Library of Congress (hereafter, Lahm Papers). See also Johnson Sherwick, "The Aero Club of Ohio," *Fly* (January 1909), p. 15; "New Aero Clubs in America," *Aeronautics* (January 1908), p. 32; "Aero Club of Ohio," *ibid.* (April 1908), p. 38; (August 1908), p. 34; "Aero Club of America," *Aeronautics* (November 1908), p. 27; Frank S. Lahm biographical file, NASM Library.

24. *Ibid.*

25. *Ibid.*

26. *Ibid.*; "Note," *Aeronautics* (May 1908), p. 32.

27. "Milwaukee Aero Club," *Aeronautics* (April 1908), p. 39; "Club Notes," *ibid.* (August 1908), p. 29; "Milwaukee Aero Club," ibid. (November 1908), p. 27.

28. *Ibid.*

29. "New Aero Clubs," *Aeronautics* (July 1908), p. 28.

30. Information on these new aero clubs has been drawn from a variety of notes in *Aeronautics* and *Fly*, 1908–1909.

31. E. O. Paulson, "Aero Club of California," *Fly* (March 1910), p. 16; "Aero Club of California," *Aeronautics* (March 1909), p. 90.

32. Harold Augustus Content, "Columbia University Aero Club Letter," *Fly* (January 1909), p. 16. On ladies aero club activity, see Thomas Edwin Eldridge, "Why Ladies Are and Should Be Interested in Ballooning," *Fly* (December 1908), p. 17; Triaca-Stevens note, undated clipping, Stevens Scrapbook; and "New Organizations," *Aeronautics* (September 1908), p. 32.

33. Cora Thompson, "My Initial Trip to the Clouds," *Aeronautics* (July 1908), p. 24.

34. "North Adams Aero Club," *Aeronautics* (August 1908), p. 33; "New Aero Prizes," *ibid.* (January 1908), p. 35; "New Aero Club Prizes," *ibid.* (February 1908), p. 71.

35. "Aero Club of Ohio," *Aeronautics* (August 1908), p. 45.

36. "July Balloon Racing," *Aeronautics* (August 1908), p. 19.

37. *Ibid.*; "After the Chicago Balloon Race," *Aeronautics* (September 1908), p. 45.

38. Jan Boesman, *Gordon Bennett Balloon Race* (The Hague, 1976), is the only published account offering satisfactory coverage of all the races.

39. Julia Lamb, "The Commodore Enjoyed Life—But New York Society Winced," *Smithsonian* (November 1978), pp. 132–144, is a good popular treatment of Bennett's career and the basis for my own account.

40. *Ibid.*

41. "A History of the James Gordon Bennett International Balloon Race," file A2000420, NASM Library.

42. "Winning the Blue Ribbon of the Air," undated article, Lahm biographical file, NASM Library.

43. Frank P. Lahm, "The First Annual Aeronautic Cup Race," in *Navigating the Air* (New York, 1907), pp. 34–47; Lahm biographical files, NASM Library.

44. Various items from the Lahm Papers; Lahm, "Cup Race."

45. "Henry Blanchard Hersey," in *Who's Who in American Aeronautics* (New York, 1924), p. 56; "Gordon Bennett Race," *Aeronautics* (October 1907), p. 8.

46. Lahm, "Cup Race," p. 35.

47. *Ibid.*; see also "The Memories of a Pioneer," *Flying* (August 1957), pp. 41–51; Boesman, *Gordon Bennett*, p. 12.

48. Lahm, "Cup Race," p. 35.

49. *Ibid.*, p. 38.

50. *Ibid.*, p. 39.

51. *Ibid.*

52. *Ibid.*, pp. 39–45.

53. "Gordon Bennett International Aeronautic Cup Race," *Aeronautics* (July 1907), pp. 19–20; *ibid.* (August 1907), pp. 8–9; "Gordon Bennett International Race," *ibid.* (September 1907), pp. 19–20; "Gordon Bennett Race," *ibid.* (October 1907), pp. 4–8; "Gordon Bennett, 1907," *ibid.* (December 1907), pp. 6–20.

54. James J. Horgan, "Competition for the Lahm Cup," *Wingfoot Lighter-Than-Air Society Bulletin* (July–Aug. 1965), pp. 4–5; *Aero Club of America, 1911* (New York, 1911), p. 52; *Official Program of the Second Competition for the Gordon Bennett Aeronautical Cup* (St. Louis, 1907); *St. Louis Globe Democrat*, Oct. 18–20, 27, 1910; *St. Louis Post-Dispatch*, Oct. 13, 20, 24, 1907; *St. Louis Republic*, Oct. 9, 1909; Oct. 27, 1910; Charles de Forest Chandler, "The Winning of the Lahm Cup," *Aeronautics* (December 1907), pp. 19–20.

55. *Ibid.*

56. *St. Louis Post-Dispatch*, Oct. 13, 1909.

57. Horgan, "Lahm Cup," p. 5.

58. *New York Herald*, undated clipping, 1907, Gordon Bennett File, NASM Library.

59. "Gordon Bennett Race," *Aeronautics* (December 1907), p. 6.

60. Henry B. Hersey, "The Trip of the *United States*," *Aeronautics* (December 1907), pp. 16–17.

61. "Gordon Bennett, 1907," *Aeronautics* (December 1907), pp. 6–11.

62. Augustus Post, "A Fall from the Sky," *Century Magazine* (October 1910), p. 935; "A. Holland Forbes: Yachtsman and Aeronaut," *Aircraft* (April 1910), p. 74.

63. Post, "Fall," p. 935.

64. *Ibid.*; "October Balloon Racing," *Aeronautics* (November 1908), p. 19; "The Gordon Bennett Race for 1908," *Aeronautics* (London), (October 1908), p. 76; *ibid.* (November 1908), pp. 81–82; "The Gordon Bennett in October," *American Aeronaut* (Winter 1908), p. 127.

65. Post, "Fall," p. 937.

66. *Ibid.*, p. 939.

67. *Ibid.*

68. *Ibid.*, p. 941.

69. *Ibid.*

70. *Ibid.*

71. *Ibid.*

72. "October Balloon Racing," *Aeronautics* (November 1908), pp. 19–20.

73. *Ibid.*; Boesman, *Gordon Bennett*, pp. 19–24.

74. "Gordon Bennett Balloon Race," *Aeronautics* (November 1909), p. 177.

75. "Gordon Bennett Balloon Race," *Flight* (October 9, 1909), p. 630; "Gordon Bennett Balloon Race," *ibid.* (October 30, 1909), p. 692; "The Gordon Bennett Balloon Race at Zurich," *Aeronautics* (London), (November 1909), pp. 137–138.

76. "With the Lighter-Than-Air," *Aeronautics* (January 1910), p. 26; "Gordon Bennett Balloon Race," *ibid.* (September 1910), p. 91; "The International Balloon Race," *ibid.* (November 1910), pp. 168–169; "Balloon Race Nearly Breaks World Record," *ibid.* (December 1910), pp. 216–218; "Gordon Bennett Balloon Race," *Flight* (January 8, 1910), p. 29; *ibid.* (October 29, 1910), p. 890; *ibid.* (November 5, 1910), p. 916; "The Gordon Bennett Aviation Cup," *Aeronautics* (November 1910), p. 168; Assman and Hawley biographical files, NASM Library.

77. *Ibid.*

78. *Ibid.*

79. *Ibid.*

80. Samuel F. Perkins, "Trip of the Düsseldorf," *Aeronautics* (December 1910), pp. 217–218.

81. Undated clippings, Assman and Hawley biographical files, NASM Library.

82. *Ibid.*; Boesman, *Gordon Bennett*, pp. 30–31.

83. *Ibid.*

84. "Gordon Bennett Balloon Race," *Aircraft* (November 1911), p. 310; "National Balloon Race," *Aeronautics* (July 1911), pp. 32–33; "St. Louis Wins Balloon Elimination," *Aeronautics* (August 1911), p. 67.

85. "The Kansas City International Balloon Contest," *Aeronautics* (November 1911), pp. 159–160; "Germany's Victory Preserves Balloon Classic," *Aero* (October 21, 1911), p. 56; "Balloon Race Distances Verified," *Aero* (November 3, 1911), p. 125; "Germany Wins Gordon Bennett Balloon Race," *Fly* (November 1911), p. 17.

86. "G. B. Balloon Elimination," *Aeronautics* (September 1912), p. 95; Boesman, *Gordon Bennett*, p. 33.

87. "Seeking Atherholt's Balloon in Norway," *Philadelphia Evening Bulletin*, Nov. 16, 1912; "Balloonists Twice Arrested in Russia," *Kansas City Journal*, Nov. 17, 1912; undated clippings, Gordon Bennett 1912 file, NASM Library.

88. "Balloonists Twice Arrested in Russia."

89. *Ibid.*

90. H. Eugene Honeywell, "My Voyage in the International Balloon Race," *Aeronautics* (December 1912), p. 164.

91. *Ibid.*

92. *Ibid.*

93. Undated clippings, John Berry biographical file, NASM Library.

94. Ralph Upson, "Answer to Questionnaire, Feb. 26, 1956," MS. in Upson biographical file, NASM Library.

95. *Ibid.*

96. *Ibid.*

97. Ralph Upson, *Free and Captive Balloons* (New York, 1926), pp. 42–49.

98. Upson, "Questionnaire."

99. *Ibid.*

100. *Ibid.*

101. *Ibid.*

102. Upson, p. 44.

103. *Ibid.*; "The Gordon Bennett Balloon Race, 1913," *Flying* (November 1913); R.A.D. Preston, "How We Won the Gordon Bennett," *Aeronautics* (November 1913), pp. 167–168; "Gordon Bennett Balloon Race," *ibid.*, p. 166; R. H. Upson, "Spherical Ballooning," *Theta Nu Epsilon Quarterly* (n.d.), Upson biographical file, NASM Library; "Balloon Racing, a Game of Practical Meteorology," *Monthly Weather Review* (January 1912), pp. 6–7.

104. Undated article, Ralph Upson Scrapbook, Upson biographical file, NASM Library.

105. Original telegrams, *ibid.*

106. *Ibid.*

Chapter 18

1. Maurer Maurer, ed., *The US Air Service in World War I* (Washington, D.C., 1978), vol. 1, p. 137.

2. *Ibid.*, pp. 137–143; for details of tactical operations, see pp. 379–387.

3. James J. Horgan, "Wartime Ballooning in St. Louis," *Wingfoot Lighter-Than-Air Society Bulletin* (September 1965), pp. 2–3; see also Albert Bond Lambert Aeronautical Papers, Missouri Historical Society; and Albert Bond Lambert and William Robertson, "Early History of Aeronautics in St. Louis," *Missouri Historical Society Proceedings* (June 1928), p. 244.

4. "Thrilling Sport of Ballooning Revived," *Aerial Age* (September 15, 1919), p. 9.

5. Ward T. Van Orman, *The Wizard of the Winds* (St. Cloud, Minn., 1978), is the source of all my biographical information on Van Orman. It is, perhaps, the finest autobiography of an American balloonist.

6. *Ibid.*, pp. 3–20.

7. "Honeywell Wins Air Race," *Aerial Age Weekly* (October 11, 1920), p. 137.

8. "Four Nations Enter Seven Balloons in Race," *ibid.* (October 10, 1920), p. 167; "Air Race Won by Belgian Balloon," *ibid.* (November 8, 1920), p. 246; Van Orman, *Wizard*, pp. 31–46.

9. "National Balloon Race," *Aerial Age Weekly* (March 28, 1921), p. 51; "Nine Balloons in Elimination Race," *ibid.* (May 30, 1921), p. 268; Van Orman, *Wizard*, pp. 57–60.

10. "Gordon Bennett Entries," *Aerial Age Weekly* (June 27, 1921), p. 365; "US Balloon Team Selected," *ibid.* (June 30, 1921), p. 316; "Dinner to Gordon Bennett Pilots," *ibid.* (September 5, 1921), p. 608; "Gordon Bennett Balloon Races," *ibid.* (September 5, 1921), p. 622; "Meeting of the FAI," *ibid.* (October 17, 1921), p. 136; Van Orman, *Wizard*, pp. 61–70.

11. Upson, *Balloons*, pp. 48–59; Ralph Upson, "Inside Story of the Gordon Bennett Race," *Leslie's Weekly* (November 5, 1921), pp. 628–631; "The Lessons of the Gordon Bennett Race," *Aviation* (October 31, 1921), pp. 374–376; "Ralph Upson Wins National Balloon Race," *ibid.* (June 26, 1921), pp. 633–634; "The Great Balloon Flight," *St. Nicholas Magazine* (June 1922), pp. 313–319; Van Orman, *Wizard*, pp. 71–76.

12. "Gordon Bennett Balloon Race," *Aerial Age Weekly* (January 9, 1922), p. 424; "Gordon Bennett Balloon Race, 1922," *ibid.* (January 22, 1922), p. 244.

13. Lt. William F. Reed, "Report on Flight of Free Balloon # 5860 from Milwaukee on 31 May 1922," RG 86, Secretary of the Navy, General Correspondence file 26983–1282:6, NARS.

14. *Ibid.*

15. *Ibid.*

16. Oscar Westover biographical file, NASM Library.

17. "Gordon Bennett Balloon Race," *Flight* (August 24, 1922), p. 481; "G.B. Team Sails," *Aviation* (July 31, 1922), p. 129; "The Gordon Bennett Cup Race," *ibid.* (August 21, 1922), p. 225; "DeMuyter Wins Gordon Bennett Cup," *ibid.* (September 18, 1922), p. 421; Van Orman, *Wizard*, pp. 61–76.

18. "Army Team, National Balloon Race," *Aviation* (June 25, 1923), pp. 121–123; "Results of the National Balloon Race," *ibid.* (July 16, 1923), p. 71; "Army Team Wins National Balloon Race," *ibid.* (July 30, 1923), pp. 121–123; Van Orman, *Wizard*, pp. 77–80.

19. *Ibid.*

20. Olmsted's Report, quoted in "Army Team Wins," p. 121.

21. "Tragic Sport," *Aviation* (October 8, 1923); "Belgium Again Wins Gordon Bennett," *ibid.*, pp. 438–439; "Final Gordon Bennett Results," *ibid.*, p. 549; Van Orman, *Wizard*, pp. 77–80.

22. *Ibid.*

23. *Ibid.*

24. *Ibid.*

25. *Ibid.*

26. "Tragic Sport."

27. "National Balloon Race," *Aviation* (May 5, 1924), p. 484; *ibid.* (May 12, 1924), p. 501; H. V. Thaden, "Impressions of the 1924 National Balloon Race," *ibid.*, pp. 510–511; Van Orman, *Wizard*, pp. 81–86.

28. Van Orman, *Wizard*, p. 81.

29. *Ibid.*

30. "National Balloon Race," *Aviation* (May 12, 1924), p. 501.

31. Van Orman, *Wizard*, pp. 87–91; Boesman, *Gordon Bennett*, pp. 54–56.

32. "The 1925 Gordon Bennett Cup," *Aviation* (January 2, 1925), p. 129.

33. "National Balloon Race," *Aviation* (February 16, 1925), p. 188; "National Balloon Race, 1925," *ibid.* (March 2, 1925), p. 245; "The National Elimination Balloon Race," *ibid.* (May 18, 1925), p. 548; Van Orman, *Wizard*, pp. 93–96.

34. "National Elimination Balloon Race."

35. *Ibid.*

36. Van Orman, *Wizard*, pp. 99–100.

37. *Ibid.*, p. 100.

38. *Ibid.*, p. 101.

39. *Ibid.*

40. *Ibid.*, pp. xv-xvi; see also "Gordon Bennett Results," *Aviation* (July 13, 1925), p. 43.

41. Ralph H. Upson, "Balloon Racing," *Aviation* (April 12, 1926), p. 547.

42. "The Little Rock Air Meet," *Aviation* (April 19, 1926), p. 598; "The National Balloon Race," *ibid.* (May 10, 1926), p. 708; Van Orman, *Wizard*, pp. 107–110.

43. Van Orman, *Wizard*, p. 117; "The Gordon Bennett Cup," *Aviation* (June 28, 1926), p. 982.

44. "The Gordon Bennett Balloon Races," *Aviation* (August 29, 1927), p. 479; Van Orman, *Wizard*, pp. 119–123.

45. "Gordon Bennett Balloon Race," *Aviation* (September 15, 1927), p. 282; "America Wins Gordon Bennett Balloon Race," *ibid.* (September 26, 1927), p. 726; Van Orman, *Wizard*, pp. 125–132.

46. "Gordon Bennett Balloon Races," *Aviation* (August 29, 1927), p. 479.

47. Van Orman, *Wizard*, pp. 125–126.

48. "America Wins Gordon Bennett."

49. *Ibid.*

50. Van Orman, *Wizard*, pp. 125–126.

51. *Ibid.*

52. *Ibid.*

53. *Ibid.*, pp. 138–140.

54. *Ibid.*, pp. 141–142.

55. *Ibid.*, p. 142.

56. *Ibid.*, pp. 143–144; see also "Gordon Bennett Victory Goes to American Team," *Air Travel News* (August 1928), pp. 7–8.

57. Van Orman, *Wizard*, pp. 145–146.

58. "Gordon Bennett Victory."

59. Van Orman, *Wizard*, passim.

60. "Elimination Race Teams Announced," *Aviation* (April 27, 1929), p. 1440; "Navy Balloon No. 1 Flies to Prince Edward," *ibid.* (May 11, 1929), p. 1616.

61. Van Orman, *Wizard*, pp. 153–162.

62. "Goodyear Zeppelin," *Air Transportation* (July 12, 1930), p. 2; "50,000 See National Air Races," *ibid.* (September 6, 1930), p. 1; "Gordon Bennett Balloon Race Results," *Airway Age* (October 30, 1930), p. 1327; "Balloon Prize Goes to Blair-Trotter," *New York Times*, Aug. 5, 1930; "Last Balloon Lands," *ibid.*, July 7, 1930.

63. "Navy Balloon in National Race," *Aero Digest* (August 1931), pp. 84–85.

64. *Washington Evening Star*, June 26, 1932.

65. T. G. W. Settle, "Winning the Gordon Bennett Cup," *National Aeronautic Magazine* (December 1932), p. 15.

66. *Ibid.*, p. 18.

67. *Ibid.*

68. Van Orman, *Wizard*, p. 188.

69. *Ibid.*, pp. 190–191.

70. "Air Races," *Aviation* (September 1933), p. 298; "Balloons," *ibid.* (October 1933), p. 318; "Racing for Records," *ibid.*, p. 326.

71. Van Orman, *Wizard*, pp. 193–203.

72. *Ibid.*, pp. 195–196.

73. *Ibid.*, p. 197.

74. *Ibid.*

75. *Ibid.*

76. *Ibid.*, pp. 197–202.

77. *Ibid.*, pp. 203–204.

78. *Ibid.*, p. 203.

79. "National Balloon Races," *Aviation* (September 1934), pp. 206–207; "Balloons," *ibid.*, p. 37; "The Gordon Bennett Race," *Flight* (October 4, 1934), p. 1037.

80. "Lighter-Than-Air," *Aviation* (September 1935), p. 54; "Balloons," *ibid.* (October 1935), p. 501.

81. *Ibid.*

82. Michael Fairbanks, *Down One Diamond* (Swarthmore, Pa., 1979), pp. 3–30.

83. *Ibid.*, p. 29.

Chapter 19

1. "Gray's Radio Batteries Found," *New York Times*, Nov. 8, 1927.

2. "Asphyxiated beyond the Clouds," *Nashville Banner*, Nov. 5, 1927.

3. Victor F. Hess, quoted in Bruno Rossi, *Cosmic Rays* (New York, 1964), p. 2.

4. *Ibid.*

5. "Blue Hill Observatory Creates a Stratosphere Radio Bullet," *Boston Evening Transcript*, Oct. 30, 1935.

6. Charles Greeley Abbot, *The Sun and the Welfare of Man* (New York, 1938), pp. 43–45.

7. "More about Captain Gray's Altitude Flight," *Air Corps Newsletter* (Apr. 16, 1927), p. 107; "Captain H. C. Gray Breaks American Altitude Record for Balloons," *ibid.* (Mar. 31, 1927), p. 80; "Request for Homologation of World Records, Class A, Altitude, Captain Hawthorne C. Gray, March 9, 1927," NAA Files, NASM Library.

8. H. C. Gray, "Eight Miles—Straight Up!" *Popular Mechanics Magazine* (August 1927), p. 180.

9. "Captain Gray Breaks American Altitude Record"; "More about Captain Gray's Altitude Flight"; Gray, "Eight Miles Up."

10. Gray, "Eight Miles Up," pp. 177–178.

11. *Ibid.*

12. *Ibid.*

13. *Ibid.*

14. "Request for Homologation of World Records, Class A, Spherical Balloons, Seventh Category, Altitude, Captain H. C. Gray, Scott Field, Belleville, Ill.," NAA Files, NASM Library.

15. "Captain Gray's Last Flight," *Air Corps Newsletter* (Jan. 19, 1928).

16. "Meteorology at Great Altitudes," *New York Times*, May 19, 1927.

17. "Gray Not Heard From," *Commercial Appeal* (Memphis), Nov. 5, 1927; *Knoxville Journal*, Nov. 7, 1927; "Asphyxiated beyond the Clouds," *Nashville Banner*, Nov. 5, 1927; "Neglect of Clock Killed Balloonist," *New York Times*, Sept. 2, 1927.

18. This account of Gray's last flight is based on the aeronaut's logbook dated November 4, 1927, NASM biographical files; my interpretation has been guided by the Board of Inquiry Report in "Captain Gray's Last Flight."

19. "Captain Gray's Last Flight," p. 6.

20. "Army Flying Cross Awarded to Five," *New York Times*, Feb. 21, 1928.

21. *Knoxville Journal*, Nov. 7, 1927.

22. Material on the "flying coffin" appeared in J. Gordon Vaeth, "When the Race for Space Began," *U.S. Naval Institute Proceedings* (August 1961), pp. 70–71.

23. Biographical information on the early career of Auguste Piccard has been drawn from: *Au-Dessus des Nuages* (Paris, 1933); C. G. Philp, *The Conquest of the Stratosphere* (Lon-

don, 1937); Auguste Piccard, "Ballooning in the Stratosphere," *National Geographic Magazine* (March 1933), pp. 353–384; "Prisoners of the Air," *Popular Mechanics Magazine* (August 1931), pp. 177–185; "Ten Miles Up in a Balloon," *Flight* (June 5, 1931), p. 511; "Professor Piccard's Balloon Ascent," *ibid.* (June 12, 1931), p. 571; "Professor Piccard's New Venture," *ibid.* (June 3, 1932), p. 498; scattered material in the Piccard Family Papers, Manuscript Division, Library of Congress (hereafter, Piccard Papers).

24. Piccard Papers.

25. Jean Piccard, "Construction of Welded Gondolas for Stratosphere Balloons," *Welding Engineer* (June 1933), p. 489.

26. *Ibid.*; see also "The Construction of the Stratosphere Balloon 6/7/33," box 71, Piccard Papers.

27. Paul Maravelas, "Jeannette Piccard Interviewed," *Ballooning* (July–Aug. 1980), pp. 15–19.

28. Van Orman, *Wizard*, p. 161.

29. *Ibid.*

30. *Ibid.*

31. *Ibid.*

32. *Ibid.*

33. *Ibid.*

34. Maravelas, "Jeannette Piccard," p. 16.

35. Philp, *Conquest*, p. 52.

36. *Ibid.*

37. G. Prokofiev, "Ascents into the Stratosphere," *Aircraft Engineering* (September 1933), p. 232; "The Soviet Stratostat," *Flight* (December 21, 1933), p. 1287; Philp, *Conquest*, pp. 69–72.

38. Van Orman, *Wizard*, p. 172.

39. Vaeth, "Race for Space," pp. 73–75.

40. Maravelas, "Jeannette Piccard," p. 16.

41. Vaeth, "Race for Space," p. 76.

42. *Ibid.*; unidentified clippings, Stratosphere Balloon file, NASM Library; Van Orman, *Wizard*, pp. 174–175; Philp, *Conquest*, pp. 73–80.

43. Reysa Bernson, "Ascension and Catastrophe of the Russian Stratosphere Balloon 'Osoaviakhim I', *National Advisory Committee for Aeronautics Misc. Papers No. 39*; E. E. Hildreth, Memo for the Chief, Materiel Division, Wright Field, July 11, 1934, Stratosphere Balloon file, NASM Library.

44. *Ibid.*

45. Orvil Anderson, "Ballooning in the Stratosphere," *Air Power Historian* (January 1957), pp. 3–14.

46. Benjamin Kelsey, conversation with the author.

47. William E. Kepner and James Scriven, Jr., "The Saga of *Explorer I*: Man's Pioneer Attempts to Reach Space," *Aerospace Historian* (Fall 1971), p. 124.

48. *Ibid.*

49. Karl Arnstein, "The Design of the Stratosphere Balloon *Explorer*," in *The National Geographic Society-U.S. Army Air Corps Stratosphere Flight of 1934 in the Balloon Explorer* (Washington, D.C., 1935), p. 95.

50. *Ibid.*; technical details of the *Explorer I* have also been drawn from other articles in this volume.

51. A. W. Winston, "The Design and Construction of the Gondola for *Explorer*," *Stratosphere Flight of 1934*.

52. Kepner, "*Explorer I*," p. 125.

53. *Ibid.*

54. *Ibid.*, p. 126.

55. Albert W. Stevens, "Exploring the Stratosphere," *Stratosphere Flight of 1934*, pp. 401–402.

56. *Ibid.*, p. 405.

57. Kepner, "*Explorer I*," p. 127.

58. *Ibid.*; see also unidentified clippings in *Explorer I* file, NASM Library.

59. *New York Times*, July 30, 1934.

60. Philp, *Conquest*, pp. 99–111.

61. Maravelas, "Jeannette Piccard," p. 17.

62. Jeannette Piccard, "He Taught Me How

to Fly," unpublished manuscript, box 77, Piccard Papers.

63. *Ibid.*

64. *Ibid.*

65. *Ibid.*

66. Jeannette Piccard, untitled manuscript, "Speech and Writing File," box 77, Piccard Papers.

67. Maravelas, "Jeannette Piccard," p. 19.

68. *Ibid.*

69. Jeannette Piccard to Claudia Oakes, April 21, 1979, author's collection. My thanks to Ms. Oakes.

70. Assorted manuscript records, box 77, Piccard Papers.

71. "The Piccard Stratosphere Flight," a manuscript report by Ray Cunningham, author's collection. My thanks, once again, to Ms. Oakes.

72. *Ibid.*

73. Transcript of a radio interview with Jeannette Piccard, box 77, Piccard Papers.

74. Philp, *Conquest*, pp. 130–134.

75. Karl Arnstein and F. D. Swann, "The Design of the Stratosphere Balloon *Explorer II*," in *The National Geographic Society-U.S. Army Air Corps Stratosphere Flight of 1935 in the Balloon Explorer II* (Washington, D.C., 1936), pp. 240–245; see also various articles in the *Black Hills Engineer* (September 1936).

76. A. W. Winston, "The Design and Construction of the Gondola for *Explorer II*," in *Stratosphere Flight of 1935*.

77. Albert W. Stevens, "Man's Farthest Aloft," in *Stratosphere Flight of 1935*, pp. 173–217.

78. Gilbert Grosvenor, "The Stratosphere Expeditions," *Black Hills Engineer* (September 1936), pp. 49–50.

79. Arnstein and Swann, "*Explorer II*," pp. 240–245.

80. Philp, *Conquest*, pp. 139–149.

81. "Blue Hill Observatory Creates a Stratosphere Radio Bullett" (above, n. 5).

82. John D. Ackerman and Jean F. Piccard, "Upper Air Study by Means of Balloons and the Radio Meteorograph," *Journal of the Aeronautical Sciences* (June 1937), pp. 332–337.

83. The material on *Pleiades* has been drawn from "The Voyage of the *Pleiades*," a manuscript dated 1937, box 62, Piccard Papers.

Chapter 20

1. The story of Project *Helios* is found in various documents contained in box 77, Piccard Papers.

2. Biographical information on Otto C. Winzen is drawn from the author's conversations with Winzen in 1974. The correspondence between F. C. Durant III and Winzen, found in the Winzen biographical file, NASM Library, was also useful. My thanks to Vera Simons, who consulted the Winzen diaries for the answers to specific questions.

3. Author's conversation with George Hoover, Nov. 15, 1982.

4. Shirley Thomas, "Malcolm D. Ross," *Men in Space* (Philadelphia, 1972), vol. 4, p. 147.

5. Otto C. Winzen, "Ten Years of Plastic Balloons," *Proceedings of the VIIIth International Astronautical Congress, Barcelona, Spain, 1957* (Wien, 1958), p. 438.

6. *Ibid.*

7. *Ibid.*

8. Athelston Spilhaus, C. S. Schneider, and C. B. Moore, "Controlled Altitude Free Balloons," *Journal of Meteorology* (August 1948), pp. 130–137.

9. Author's conversations with J. R. Smith, Nov. 16, 1982, and C. B. Moore, Nov. 16, 1982.

10. Author's conversations with C. B. Moore, J. R. Smith, Ed Yost, Vera Simons, Nov. 15–20, 1982. See also *Minnesota: Balloon Capital of the World*, a pamphlet prepared by the public relations department of General Mills, Inc., for the Eleventh Annual Breakfast for Editors, Oct. 1, 1960, author's collection.

11. J. A. Van Allen and M. B. Gottlieb, "The Inexpensive Attainment of High Altitudes with

Balloon-Launched Rockets," Research Note, State University of Iowa, n.d., author's collection; see also Winzen, "Ten Years of Plastic Balloons"; Otto C. Winzen, "Twenty-Five Years of Plastic Research Balloons," *Proceedings of the XXIIIrd International Astronautical Congress, 1972*; *idem*, "Plastic Balloons in the Rocket Age," *Missiles and Rockets* (March 1957), pp. 50–52.

12. "First Air Division (Meteorological Survey), Final Report, Project 119L," n.d., copy in the author's collection. My thanks to Glen Sweeting, who provided a copy of this report, which was recently declassified by R. W. Koch under the Freedom of Information Act.

13. *Ibid.*, n.p.

14. *Ibid.*, n.p.

15. "Russians Display Balloons of US," *New York Times*, Feb.10, 1956; other unidentified articles in "High Altitude Balloons, Science" file, NASM Library.

16. *Ibid.*

17. *Ibid.*

18. *New York Times*, Feb.11, 1956.

19. "US Reassures Russia, Asks about Balloons," *Washington Post Herald*, Oct. 11, 1958.

20. "Russia Again Protests at US Balloon Flights," *ibid.*, Sept. 7, 1958.

21. *GHOST: A Technical Summary* (Boulder, Colo., 1969); "Horizontal Sounding Balloons," *NCAR Quarterly* (Summer 1967), n.p.

22. Author's conversation with C. B. Moore, Nov. 16, 1982; see also Roger Pineau, *Ballooning, 1782–1972* (Washington, D.C., 1972), p. 77. Pineau's chronology of ballooning is still the finest available.

23. Thomas, "Malcolm Ross," p. 149.

24. *Ibid.*

25. *Ibid.*, pp. 149–150.

26. *U.S. News and World Report* (November 2, 1956), p. 20.

27. Otto C. Winzen, "From Ballooon Capsules to Space Cabins," *Proceedings, IXth International Astronautical Congress, Amsterdam, 1958* (Wien, 1959), p. 562.

28. Thomas, "Malcolm Ross," p. 148.

29. Malcolm D. Ross, "The Role of Manned Balloons in the Exploration of Space," *Aerospace Engineering* (August 1958), p. 52.

30. Capt. Joseph W. Kittinger, Jr., with Martin Caiden, *The Long, Lonely Leap* (New York, 1961), p. 73.

31. *Ibid.*; David G. Simons, *Man High* (Garden City, N.Y., 1960).

32. "Officer Nearly Baked in AF Balloon Error," *Washington Star*, Oct. 10, 1958.

33. On manned balloon astronomy, see Kurt R. Stehling, "Balloon Astronomy, A Case for More," *Smithsonian* (June 1971), pp. 28–33; Martin and Barbara Schwarzschild, "Balloon Astronomy," *Scientific American* (May 1959), pp. 52–57.

34. "Summary of Attempted Atlantic Balloon Crossings," *Ballooning* (Nov–Dec. 1977), pp. 29–31, has served as the basis for this table.

35. Kurt R. Stehling, *Bags Up! Great Balloon Adventures* (Chicago, 1975), pp. 41–58, narrates the voyage of the *Kathleen*.

36. Donald Dale Jackson, *The Aeronauts* (Alexandria, Va., 1980), p. 155.

37. Da Vinci Trans-America Press Kit, author's collection; "Balloonists Forced Down by Storm in Northwest Ohio," *Washington Post*, Oct. 2, 1979; "Balloon Soars above Rockies; Oxygen, Cold Affect Crew," *ibid.*, Sept. 28, 1979; "Why Balloonists Risk the Wild Wind," *ibid.*, Oct. 28, 1979; "Balloonists Down in Ohio, Crew Reported Safe," *Washington Star*, Oct. 2, 1979.

38. "Flight of the Kitty Hawk," Press Release #5–80–043, U.S. Army Western Regional Recruiting Command; "Father and Son Achieve the Goal: First Trip across Continent," *New York Times*, May 13, 1980; "Balloon Lands in Canada after Crossing Continent," *Washington Post*, May 13, 1980.

39. "Balloon Pilots Recall Sleeplessness, Cold," *Newark [N.J.] Star-Ledger*, Oct. 13, 1981; "Balloon Soars East in Cool Jet Stream," *Los Angeles Times*, Oct. 10, 1981; assorted unidentified clippings, file A2003915, NASM Library.

40. "Falling Apart, Balloon Set Mark," *New*

York Times, Nov.14, 1981; "Double Eagle V Drops In," *Round Valley News*, Nov. 26, 1981; assorted clippings, file A2003975, NASM Library.

41. "Things Were Very, Very Bad," *Time* (November 23, 1981).

42. Fairbanks, *Down One Diamond*, pp. 31–108.

43. Pineau, *Ballooning*, p. 77.

44. Dick Wirth, *Ballooning: The Complete Guide to Riding the Winds* (New York, 1980), pp. 48–49.

45. Author's conversations with J. R. Smith, Nov. 15, 1982, and Don Piccard, Nov. 14, 1982.

46. "Vulcoon Data Sheet," Hot Air Balloon file, NASM Library.

47. Don Piccard to Louis Casey, May 8, 1962, copy in author's collection.

A Select Bibliography of Materials on the History of the Balloon in the United States

Books

Abbot, Charles Greeley. *The Sun and the Welfare of Man*. New York, 1938.

Adams, Charles Francis, Jr., ed. *Letters of John Adams Addressed to His Wife*. Boston, 1841.

Aero Club of America. *Navigating the Air*. New York, 1907.

Aeronautical Society of Great Britain. *Annual Report of the Aeronautical Society of Great Britain for 1868*. Greenwich, Eng., 1869.

Allen, David Grayson, et al., eds. *The Diary of John Quincy Adams*. Cambridge, Mass., 1981.

American Philosophical Society. *Early Proceedings of the American Philosophical Society*. Philadelphia, 1884.

Amick, M. L. *History of Donaldson's Balloon Ascensions*. Cincinnati, 1875.

Andrews, Solomon. *The Aereon, or Flying Ship*. New York, 1866.

———. *The Art of Flying*. New York, 1865.

Appleton's Cyclopedia of American Biography. New York, 1900.

Baker, Gardiner. *The Principles, History and Use of Air-Balloons*. New York, 1796.

Baldwin, Munson. *With Brass and Gas: An Illustrated and Embellished Chronicle of Ballooning in Mid-Nineteenth Century America*. Boston, 1967.

Blanchard, Jean Pierre. *Journal and Certificates of the Fourth Aerial Voyage of Mr. Blanchard*. London, 1784.

———. *Journal of My Forty-Fifth Ascension and the First in America*. Philadelphia, 1943.

Block, Eugene B. *Above the Civil War: The Story of Thaddeus Lowe, Balloonist, Inventor, and Railway Builder*. Berkeley, Cal., 1966.

Boesman, Jan. *Gordon Bennett Balloon Race*. The Hague, 1976.

Bonsal, Stephen. *The Fight for Santiago*. New York, 1899.

Boyd, Julian, ed. *The Papers of Thomas Jefferson*. Princeton, N.J., 1952.

Brigham, Clarence S. *Poe's Balloon Hoax*. Metuchen, N.J., 1932.

Burke, James. *Connections*. Boston, 1978.

Butterfield, Lyman H., ed. *The Autobiography of Benjamin Rush*. Princeton, N.J., 1948.

———. *The Diary and Autobiography of John Adams*. Cambridge, Mass., 1961.

———. *The Letters of Benjamin Rush*. Princeton, N.J., 1951.

Catton, Bruce. *Glory Road: The Bloody Route from Fredericksburg to Gettysburg*. New York, 1952.

———. *Mr. Lincoln's Army*. New York, 1951.

Cavallo, Tiberius. *The History and Practice of Aerostation*. London, 1785.

Chandler, Charles de Forest, and Frank Lahm. *How Our Army Grew Wings: Airmen and Aircraft before 1914*. New York, 1943.

Cornish, Joseph Jenkins, III. *The Air Arm of the Confederacy: A History of Origins and Usages of War Balloons by the Southern Armies during the American Civil War*. Richmond, Va., 1963.

Cottrell, Leonard, *Up in a Balloon*. New York, 1970.

Coutil, Léon. *Jean-Pierre Blanchard: Biographie et Iconographie*. Paris, 1911.

Crouch, Tom D. *The Giant Leap: A Chronology of Ohio Aerospace Events and Personalities, 1815–1969*. Columbus, Ohio, 1971.

Cullen, Joseph P. *Where a Hundred Thousand Fell*. Washington, D.C., 1966.

Davidson, Richard O. *A Description of the Aerostat: A Practical View of Aerial Navigation*. New York, 1841.

Davis, Richard Harding. *The Cuban and Puerto Rican Campaigns*. New York, 1898.

De Faria, Viscond. *Le Précurseur des Navigateurs Aériens Bartolomeu Laurenco de Gusmao L'Homme Volant Portugais né au Brésil*. Paris, 1910.

De Fonvielle, Wilfrid. *Adventures in the Air*. London, 1877.

———. *Histoire de la Navigation Aérienne*. Paris, 1907.

De Wendt, Caroline Amelia. *Journal and Correspondence of Miss Adams, Daughter of John Adams, Written in France and England*. New York, 1841.

Dexter, Franklin Bowditch, ed. *The Literary Diary of Ezra Stiles*. New York, 1901.

Dickworth, Marie, ed. *Two Unpublished Letters of Thomas Jefferson Found in Ohio*. Oxford, Ohio, 1941.

Dollfus, Charles. *The Orion Book of Balloons*. New York, 1961.

Dollfus, Charles, and Henri Bouché. *Histoire de l'Aéronautique*. Paris, 1942.

Drattell, Alan and Jeanne O'Neil, *Journey to Deptford*. Deptford, N.J., 1976.

Duer, William A. *Reminiscences of an Old New Yorker*. New York, 1867.

Dupree, Hunter. *Science in the Federal Government: A History of Policies and Activities to 1940*. Cambridge, Mass., 1957.

Dwiggins, Don. *Riders of the Winds: The Story of Ballooning*. New York, 1973.

Eckner, Hugo. *Count Zeppelin: The Man and His Work*. London, 1938.

Emme, Eugene M. *Aeronautics and Astronautics: An American Chronology of Science and Technology in the Exploration of Space, 1915–1960*. Washington, D.C., 1961.

Fairbanks, Michael. *Down One Diamond*. Swarthmore, Pa., 1979.

Fergusson of Kilkerran, Sir James. *Balloon Tytler*. London, 1972.

Fitzpatrick, John C., ed. *The Writings of George Washington*. Washington, D.C., 1938.

Freidel, Frank. *The Splendid Little War*. Boston, 1958.

Frey, Carroll. *The First Aerial Voyage in America*. Philadelphia, 1943.

Garrison, Paul. *The Encyclopedia of Hot Air Balloons*. New York, 1978.

Genêt, E. C. *Memorial on the Upward Forces of Fluids and Their Applicability to Several Arts, Sciences and Public Improvements*. Albany, 1825.

GHOST: A Technical Summary. Boulder, Colo., 1969.

Gibbs-Smith, Charles H. *Ballooning*. London, 1948.

Gillispie, C. C. *Science and Polity in France at the End of the Old Regime*. Princeton, N.J., 1980.

Gilman, Rhoda, ed. *A Yankee Inventor's Flying Ship*. St. Paul, Minnesota, 1969.

Gomez, Jose Villas. *Breve Historia de la Aviacion en Mexico*. Mexico City, 1971.

Greely, Adolphus. *Reminiscences of Adventure and Service: A Record of Sixty-Five Years*. New York, 1927.

Haddock, John A. *Mr. Haddock's Account of His Hazardous and Exciting Voyage in the Balloon "Atlantic" with Prof. John La Mountain*. Philadelphia, 1872.

Haining, Peter. *The Dream Makers*. London, 1972.

Hart, Clive. *The Dream of Flight: Aeronautics*

from Classical Times to the Renaissance. London, 1972.

Haydon, F. Stansbury. *Aeronautics in the Union and Confederate Armies, with a Survey of Military Aeronautics Prior to 1861*. Baltimore, 1941.

Hays, I. Minis. *Calendar of the Papers of Benjamin Franklin in the Library of the American Philosophical Society*. Philadelphia, 1908.

Henderson, G. F. R. *Stonewall Jackson and the American Civil War*. London, 1919.

Hodgson, J. E. *The History of Aeronautics in Great Britain*. Oxford, 1924.

Hoehling, Mary. *Thaddeus Lowe: America's One-Man Air Corps*. New York, 1958.

Honour, Alan. *Ten Miles High, Two Miles Deep*. New York, 1957.

Horvat, William J. *Above the Pacific*. Fallbrook, Cal., 1966.

Jackson, Donald Dale. *The Aeronauts*. Alexandria, Va., 1980.

Jeffries, John. *A Narrative of Two Aerial Voyages of Doctor Jeffries With Mons. Blanchard*. London, 1786.

Johnson, H. P., ed. *The Correspondence and Published Papers of John Jay*. New York, 1891.

Johnson, William. *Reports of Cases Argued and Determined in the Supreme Court of Judicature and in the Court for the Trial of Impeachments and the Correction of Errors*. New York, 1882.

Kennedy, John Pendleton. *Memoirs of the Life of William Wirt*. Philadelphia, 1856.

King, Samuel Archer. *The Balloon: Noteworthy Aerial Voyages from the Discovery of the Balloon to the Present Time*. New York, 1879.

Kittinger, Joseph W., Jr., with Martin Caiden. *The Long, Lonely Leap*. New York, 1961.

Lavender, David. *California: Land of New Beginnings*. New York, 1972.

Lipman, Jean. *Rufus Porter Rediscovered*. New York, 1980.

Longstreet, James. "Our March against Pope," in *Battles and Leaders of the Civil War*. New York, 1950.

Lowe, T. S. C. *The Latest Developments in Aerial Navigation*. Los Angeles, 1910.

McCarry, Charles. *Double Eagle*. Boston, 1979.

McGinnis, William C. *History of Perth Amboy*. Perth Amboy, N.J., 1959.

McPhee, John. *The Deltoid Pumpkin Seed*. New York, 1973.

Maurer, Maurer, ed. *The US Air Service in World War I*. Washington, D.C., 1978.

Mikesh, Robert C. *Japan's World War II Balloon Bomb Attacks on North America*. Washington, D.C., 1973.

Milbank, Jeremiah. *The First Century of Flight in America*. Princeton, N.J., 1943.

Morris, Richard B. *John Jay: The Winning of the Peace*. New York, 1980.

Murphy, Charles J. *Parachute*. New York, 1930.

Myers, Mary. *Skylarking in Cloudland*. Mohawk, N.Y., 1883.

Needham, Joseph. *Science and Civilization in China*. Cambridge, Eng., 1965.

Nørgaard, Erik. *The Book of Balloons*. New York, 1971.

Northey, L. *The History of St. Peter's Church in Perth Amboy, New Jersey*. Perth Amboy, N.J., 1923.

Odell, George C. D. *Annals of the New York Stage*. New York, 1927.

O'Neal, John Belton. *The Annals of Newberry*, 1858.

———. *Biographical Sketches of the Bench and Bar of South Carolina*. Charleston, S.C., 1859.

Opinions of Thirty-Two Scientific Gentlemen on the Invention of Muzio Muzzi. New York, 1845.

Orcutt, Samuel. *A History of the Old Town of Stratford and the City of Bridgeport, Connecticut*. Boston, 1886.

Parkinson, Russell J. *Politics, Patents and Planes: Military Aeronautics in the United States, 1863–1907*. Ph.D. dissertation, Duke University, 1963.

Parlour Magic. Philadelphia, 1838.

Pellegreno, Ann Holtgren. *Iowa Takes to the Air: Volume One, 1845–1918*. Iowa City, Iowa, 1980.

Pennington, John H. *Aerostation, or Steam Aerial Navigation*. Baltimore, 1838.

———. *A System of Aerostation, or Steam Aerial Navigation*. Washington, D.C., 1842.

Philp, C. G. *The Conquest of the Stratosphere*. London, 1937.

Piccard, Auguste. *Between Earth and Sky*. London, 1950.

———. *In Balloon and Bathyscaphe*. London, 1956.

Pineau, Roger. *Ballooning, 1782–1972*. Washington, D.C., 1972.

Pintard, John. *The Letters of Jon Pintard to his Daughter Elizabeth Noel Pintard Davidson*. New York, 1940.

Piozzi, Hester Lynch. *Observations and Reflections Made in the Course of a Journey through France, Italy, and Germany*. Dublin, 1789.

Poole, Lynn. *Ballooning in the Space Age*. New York, 1958.

Porter, Rufus. *Aerial Navigation: The Practicality of Travelling Pleasantly and Safely from New York to California in Three Days Fully Demonstrated*. New York, 1849.

Powell, J. H. *Bring Out Your Dead: The Great Plague of Yellow Fever in Philadelphia in 1793*. Philadelphia, 1949.

Rice, Howard C. *Thomas Jefferson's Paris*. Princeton, N.J., 1976.

Robertson, Eugène. *Relation du Premier Voyage Exécuté dans la République Mexicaine*. Paris, 1835.

Robinson, Douglas. *Giants in the Sky*. Seattle, Wash., 1973.

Roch, E. *Essais sur les Voyages Aériens d'Eugène Robertson en Europe, aux Etats-Unis et aux Antilles*. Paris, 1831.

Rolt, L. T. C. *The Aeronauts: A Dramatic History of the Great Age of Ballooning*. New York, 1966.

The Romance of Ballooning: The Story of the Early Aeronauts. New York, 1971.

Rossi, Bruno. *Cosmic Rays*. New York, 1964.

Rotch, Abbot Lawrence. *Benjamin Franklin and the First Balloons*. Worcester, Mass., 1907.

———. *Sounding the Ocean of Air*. London, 1900.

Scamehorn, Howard L. *Balloons to Jets: A Century of Aeronautics in Illinois*. Chicago, 1957.

Scharff, J. Thomas, and Thomas Westcott. *A History of Philadelphia*. Philadelphia, 1884.

Schoenbrun, David. *Triumph in Paris: The Exploits of Benjamin Franklin*. New York, 1976.

Schroer, Dick. *A True History of the Daring Aeronaut, Thomas Baldwin and His Thrilling Parachute Leaps from a Balloon*. Quincy, Ill., 1887.

Scrivner, John H. *The Military Uses of Balloons and Dirigibles in the United States, 1793–1963*. Norman, Okla., 1963.

Sellers, Charles Coleman. *Benjamin Franklin in Portraiture*. New Haven, 1962.

Simons, David G. *Man High*. Garden City, N.Y., 1960.

Smyth, A. H., ed. *The Writings of Benjamin Franklin*. New York, 1907–1910.

Sparks, Jared, ed. *The Works of Benjamin Franklin*. Boston, 1836–1840.

Stehling, Kurt. *Bags Up! Great Balloon Adventures*. Chicago, 1975.

Stehling, Kurt, and William Beller. *Skyhooks*. Garden City, N.Y., 1962.

Stewart, A. M. *Camp, March, and Battle Field*. Philadelphia, 1865.

Stokes, Isaac Newton Phelps. *Iconography of Manhattan Island, 1498–1909*. New York, 1915–1928.

Thomas, Shirley. *Men in Space*. Philadelphia, 1972.

Thompson, S. Millett. *13th Regiment, New Hampshire Volunteer Infantry in the War of the Rebellion*. Boston, 1888.

Tissandier, Gaston. *Histoire des Ballons et des Aéronautes Célèbres 1783–1800*. Paris, 1887.

Toland, John. *Great Dirigibles: Their Disasters*. New York, 1972.

Townsend, G. A. *Rustics in Rebellion*. Chapel Hill, N.C., 1950.

———. *The Sword Over the Mantle*. New York, 1960.

Turner, Hatton. *Astra Castra: Experiments and Adventures in the Atmosphere*. London, 1865.

Upson, Ralph. *Free and Captive Balloons*. New York, 1926.

Van Doren, Carl. *Benjamin Franklin*. New York, 1938.

Van Horn, R. Lee. *Out of the Past: Prince Georgians and Their Land*. Riverdale, Md., 1976.

Van Orman, Ward T. *The Wizard of the Winds*. St. Cloud, Minn., 1978.

The War of Rebellion: A Compilation of the Official Records of the Union and Confederate Armies. Washington, D.C., 1899.

Warren, H. G. *Paraguay: An Informal History*. Norman, Okla., 1949.

Wiley, Bell Irvin. *The Life of Billy Yank: The Common Soldier of the Union*. Indianapolis, Ind., 1952.

Wilson, James Grant, and John Fiske, eds. *The Cyclopedia of American Biography*. New York, 1888.

Wirth, Dick. *Ballooning: The Complete Guide to Riding the Winds*. New York, 1980.

Wise, John. *A System of Aeronautics*. Philadelphia, 1850.

———. *Through the Air: A Narrative of Forty Years Experience as an Aeronaut*. Philadelphia, 1873.

Woodman, James. *Nazca: Journey to the Sun*. New York, 1977.

Articles

Abbot, Henry L. "Early Experiences with Balloons in War." *Professional Memoirs of the Corps of Engineers* (Washington, D.C., 1912), p. 680.

Ackerman, John D., and Jean F. Piccard. "Upper Air Study by Means of Balloons and the Radio Meteorograph." *Journal of the Aeronautical Sciences* (June 1937), pp. 332–337.

Anderson, Orvil. "Ballooning in the Stratosphere." *Air Power Historian* (January 1957), pp. 3–14.

Anonymous. "Aerial Navigation." *Scientific American* (October 30, 1847), p. 46.

———. "Aerial Navigation." *Scientific American* (April 17, 1852), p. 255.

———. "Aerial Navigation." *Scientific American* (November 20, 1869), p. 325.

———. "Aerial Navigation." *Western Monthly Magazine* (June 1838), p. 359.

———. "Aerial Speculation." *Scientific American* (July 29, 1848), p. 359.

———. "Aero Club of America." *Aeronautics* (November 1908), p. 27.

———. "Aero Club of Indiana." *Fly* (March 1909), p. 13.

———. "Aero Club of New England." *Aeronautics* (July 1908), p. 29.

———. "Aero Club of New England." *Aeronautics* (November 1909), p. 201.

———. "Aeronaut Leo Stevens." *Fly* (January 1907), p. 12.

———. "Air Races." *Aviation* (September 1933), p. 298.

———. *American Journal of Photography* (September 1860), pp. 105–106.

———. *American Journal of Photography* (February 1861), pp. 105–106.

———. "Army Team, National Balloon Race." *Aviation* (June 25, 1923), pp. 121–123.

———. "Army Team Wins National Balloon Race." *Aviation* (July 30, 1933), pp. 121–123.

———. "Balloon Accidents." *Scientific American* (August 3, 1862), p. 69.

———. "Balloonamania." *Scientific American* (September 13, 1873), p. 160.

———. "Balloon Excursion by Moonlight." *Scientific American* (November 29, 1851), p. 80.

———. "Balloon Racing, a Game of Practical Meteorology." *Monthly Weather Review* (January 1912), pp. 6–7.

———. "Balloons." *Aviation* (October 1933), p. 318.

———. "Balloon Traveling." *Scientific American* (July 21, 1855), p. 355.

———. "Captain Gray's Last Flight." *Air Corps Newsletter* (January 19, 1928), p. 6.

———. "Captain H. C. Gray Breaks American Altitude Record for Balloons." *Air Corps News* (March 31, 1927), p. 80.

———. "Captain Thomas Baldwin." *Aircraft* (March 10, 1910), p. 123.

———. "Club Notes." *Aeronautics* (August 1908), p. 29.

———. "The Dangers of Ballooning." *Scientific American* (July 16, 1876), p. 97.

———. "Dirigible Balloon and Heavier-Than-Air Contests at St. Louis." *Aeronautics* (September 1907), pp. 17–19.

———. "Ernest Pétin." *L'Aéronaute* (August 1878), pp. 255–260.

———. "Extrait d'une Lettre d'Annonay du Juin 1783." *Affiches, Annonces, et Divers* (July 10, 1783), p. 110.

———. "Failure No. 2 of the Balloon to Europe." *Scientific American* (October 25, 1873), p. 256.

———. "First Atlantic Crossing." *Time* (August 28, 1978).

———. "Flight of the Eagle." *Newsweek* (August 23, 1978).

———. "Flying Cotton, John H. Pennington." *Scientific American* (January 9, 1847), p. 125.

———. "Former Experiments." *Aerial Reporter* (June 19, 1852), p. 2.

———. "G.B. Team Sails." *Aviation* (July 31, 1922), p. 129.

———. "Germany's Victory Preserves Balloon Classic." *Aero* (October 21, 1911), p. 56.

———. "Germany Wins Gordon Bennett Balloon Race." *Fly* (November 1911), p. 17.

———. "The Giant Balloon." *Scientific American* (July 17, 1852), p. 349.

———. "Going to Europe in a Balloon." *Scientific American* (July 13, 1852), n.p.

———. "Goodyear Zeppelin." *Air Transportation* (July 12, 1930), p. 2.

———. "The Gordon Bennett Aviation Cup." *Aeronautics* (November 1910), p. 168.

———. "Gordon Bennett Balloon Race." *Flight* (August 24, 1922), p. 481.

———. "The Gordon Bennett Balloon Race at Zurich." *Aeronautics* (November 1909), pp. 137–138.

———. "Gordon Bennett Entries." *Aerial Age Weekly* (June 27, 1921), p. 365.

———. "Gordon Bennett International Aeronautic Cup Race." *Aeronautics* (July 1907), pp. 19–20.

———. "Gordon Bennett Race." *Aeronautics* (December 1907), p. 6.

———. "Gordon Bennett Victory Goes to American Team." *Air Travel News* (August 1928), pp. 7–8.

———. "The Great Balloon Flight." *St. Nicholas Magazine* (June 1922), pp. 313–319.

———. "Gymnastic Balloonists." *Scientific American* (August 17, 1872), n.p.

———. "History of Balloons." *Parley's Magazine* (New York, 1837), pp. 230–240.

———. "Honeywell Wins Air Race." *Aerial Age Weekly* (October 11, 1920), p. 137.

———. "Hot Air Ballooning, Weehawken, N.J." *Scientific American* (September 5, 1891), p. 147.

———. "Inventor's Institute." *Scientific American* (March 6, 1847), n.p.

———. "July Balloon Racing." *Aeronautics* (August 1908), p. 19.

———. "The Kansas City International Balloon Contest." *Aeronautics* (November 1911), pp. 159–160.

———. "Letters from William and Mary College, 1798–1801." *The Virginia Magazine of History and Biography* (April 1921), pp. 129–179.

———. "Letters of the Reverend James Madison, President of the College of William and Mary, to Thomas Jefferson." *William and Mary Quarterly* (April 1925), pp. 77–95.

———. "Letters of William Smith to Edmund

Rutledge." *The South Carolina History Magazine* (January 1969).

———. "The Little Rock Air Meet." *Aviation* (April 19, 1926), p. 598.

———. "The Loss of the Campbell Airship." *Scientific American* (July 27, 1889), p. 54.

———. "Making Small Fire Balloons." *Scientific American* (September 19, 1863), n.p.

———. "A Man Lost in the Clouds." *Harper's Weekly* (October 2, 1858), n.p.

———. "Many Ascents for New England." *Aeronautics* (January 1910), p. 27.

———. "Meeting of the FAI." *Aerial Age Weekly* (October 17, 1921), p. 136.

———. "The Memories of a Pioneer." *Flying* (August 1957), pp. 41–51.

———. "Milwaukee Aero Club." *Aeronautics* (April 1908), p. 39.

———. "More about Captain Gray's Altitude Flight." *Air Corps News* (April 16, 1927), p. 107.

———. "Mr. Clayton." *The Western Monthly Magazine* (October 1935), pp. 235–236.

———. "National Balloon Race." *Aerial Age Weekly* (March 28, 1921), p. 51.

———. "Navigating the Air." *Scientific American* (September 29, 1849), p. 181.

———. "Navy Balloon in National Race." *Aero Digest* (August 1931), pp. 84–85.

———. "Navy Balloon No. 1 Flies to Prince Edward." *Aviation* (May 11, 1929), p. 1616.

———. "New Aero Clubs in America." *Aeronautics* (December 1907), p. 6.

———. "New Aero Prizes." *Aeronautics* (January 1908), p. 35.

———. "New Flying Ship." *Gleason's Pictorial Drawing Room Companion* (September 22, 1851), p. 208.

———. "News about Aerial Navigation." *Scientific American* (November 13, 1849), n.p.

———. "The 1925 Gordon Bennett Cup." *Aviation* (January 2, 1925), p. 129.

———. "North Adams Aero Club." *Aeronautics* (January 1910), p. 38.

———. "The 100th Anniversary of the Flying Allens." *Ballooning* (Autumn 1974), pp. 20–24.

———. "Our Aero Club." *Aeronautics* (August 1907), p. 3.

———. "Pennington's Aerial." *Scientific American* (November 14, 1846), p. 59.

———. "Performing Somersets from a Balloon." *Scientific American* (December 18, 1852), p. 80.

———. "Pittsfield Aero Club." *Aeronautics* (December 1907), p. 6.

———. "Prisoners of the Air." *Popular Mechanics Magazine* (August 1931), pp. 177–185.

———. "Professor Piccard's New Venture." *Flight* (June 3, 1932), p. 498.

———. "The Proposed Transatlantic Balloon Voyage." *Scientific American* (July 19, 1873), p. 37.

———. "Racing for Records." *Aviation* (October 1933), p. 326.

———. "Rapid Traveling." *Scientific American* (January 27, 1855), n.p.

———. "The San Antonio Aero Club." *Aeronautics* (July 1908), p. 28.

———. "Something about the Aero Clubs." *Fly* (November 1908), pp. 10–11.

———. "The Soviet Stratostat." *Flight* (December 21, 1933), p. 1287.

———. "St. Louis Wins Balloon Elimination." *Aeronautics* (August 1911), p. 67.

———. "Strange Experiences in a Balloon." *Harper's Weekly* (July 10, 1884), p. 469.

———. "Summary of Attempted Atlantic Crossings." *Ballooning* (November–December 1977), pp. 29–31.

———. "Things Were Very, Very Bad." *Time* (November 23, 1981).

———. "Three Months with the Balloons in America." *St. James Magazine* (London, 1863), pp. 96–105.

———. "Thrilling Sport of Ballooning Revived." *Aerial Age* (September 15, 1919), p. 9.

———. "To Europe by Balloon." *Scientific American* (July 26, 1873), p. 49.

———. "To Europe in a Balloon." *Scientific American* (July 19, 1873), p. 33.

———. "Tragic Sport," *Aviation* (October 8, 1923).

———. "Travelling Balloon." *Scientific American* (December 18, 1845), p. 2.

———. "Travelling Balloon." *Scientific American* (March 3, 1849), p. 125.

———. "The Voyage of the Graphic Balloon." *Scientific American* (September 20, 1873), p. 176.

Arnstein, Karl. "The Design of the Stratosphere Balloon *Explorer*." *The National Geographic Society-U.S. Army Air Corps Stratosphere Flight of 1934 in the Balloon Explorer* (Washington, D.C., 1935), p. 95.

Arroyo, Javier Merino. "Ballooning in Mexico." *Ballooning* (Spring 1975), pp. 14–17.

Baldwin, Ivy. "Under Fire in a War Balloon at Santiago." *Aeronautics* (February 1908), pp. 13–14.

Bassett, Preston R. "Carlotta, the Lady Aeronaut of the Mohawk Valley." *New York History* (April 1963), pp. 145–172.

Beaumont, Frederick F. E. "On Balloon Reconnaissance as Practiced by the American Army." *Professional Papers of the Corps of Royal Engineers* (London, 1863).

Bell, Whitfield, Jr. "John Morgan." *Bulletin of the History of Medicine* (September–October 1949), pp. 543–562.

B.G.N. "Aeronautic Steam Car." *Mechanic's Magazine* [American edition] (August 1834), pp. 142–144.

Branham, Sandy. "Over Ninety Years of Ballooning in Colorado." *Ballooning* (Summer 1976), pp. 71–72.

Brown, Dick. "New Mexico's Ballooning Heyday." *Ballooning* (Summer 1976), pp. 62–63.

Bryan, John R. "Balloon Used for Scout Duty." *Southern Historical Society Papers*, 33 (1905), p. 32.

Clarke, Jerome, and Loren Coleman. "Mystery Airships of the 1800's." *Fate* (May–July 1973), pp. 84–94.

Cliff, Frank. "The Balloon Jumpers." *The Long Beach Press Magazine* (August 12, 1928), pp. 4–5.

Cohen, I. Bernard. "Benjamin Franklin and Aeronautics." *Journal of the Franklin Institute* (August 1941), pp. 103–120.

Crouch, Tom D. "The Gas Bag Era." *Aviation Quarterly* (Winter 1977), pp. 291–301.

———. "The History of American Aviation, 1822–1905." *Aviation Quarterly* (Spring 1976), pp. 8–13.

———. "Thomas Kirkby: Pioneer Aeronaut in Ohio." *Ohio History* (Winter 1970), pp. 56–62.

———. "Up, Up, and—Sometimes—Away." *The Cincinnati Historical Society Bulletin* (Summer 1970), pp. 109–132.

Custer, George Armstrong. "War Memoirs." *Galaxy Miscellany and Advertiser* (November 1876), p. 686.

Dollfus, Charles. "Hispano-American War and Varicle." *Wingfoot Lighter-Than-Air Society Bulletin* (December 1968), pp. 6–8.

———. "On Ballooning Firsts in America." *Buoyant Flight* (December 1963), p. 8.

Donty, Esther M. "The Greatest Balloon Voyage Ever Made." *American Heritage* (June 1955).

D'Orcy, Ladislas. "The Passing of a Great Aeronautical Pioneer." *Aviation* (May 28, 1923), pp. 584–585.

Draheim, H. Paul. "Rainmaking is Old Stuff in the Utica-Frankfort Area." *Wingfoot Lighter-Than-Air Society Bulletin* (Summer 1960), p. 4.

Eiloart, Arnold. "Braving the Atlantic by Balloon." *National Geographic Magazine* (July 1959).

Eklund, Don D. "Baldwin and the Parachute." *Wingfoot Lighter-Than-Air Society Bulletin* (June 1968), pp. 2–5.

Eldridge, Thomas Edwin. "Ballooning in Philadelphia." *Fly* (November 1908), p. 11.

———. "Why Ladies Are and Should Be Interested in Ballooning." *Fly* (December 1908), p. 17.

Ensterly, J. H. "Captain Langdon Cheeves, Jr., and the Confederate Silk Dress Balloon." *South Carolina Historical and Genealogical Magazine* (January 1944), p. 8.

Fuller, George E. "American Aeronauts in Canada." *Wingfoot Lighter-Than-Air Society Bulletin* (April–October 1970), p. 8.

Fulton, T. Chalmers. "Professor Samuel Archer King: A Tribute of Regard from His Pupil." *Fly* (February 1909), p. 12.

Gammon, Clive. "Across the Sea to Glory." *Sports Illustrated* (August 28, 1978).

Gilman, Rhoda. "Balloon to Boston." *Minnesota History* (Spring 1970), pp. 17–22.

———. "Pioneer Aeronaut: William Markoe and His Balloon." *Minnesota History* (December 1962).

———. "Zeppelin in Minnesota: A Study in Fact and Fable." *Minnesota History* (Fall 1965), pp. 278–285.

———. "Zeppelin in Minnesota: The Count's Own Story." *Minnesota History* (Summer 1967), pp. 265–279.

Glassford, William A. "Memorandum on the Santiago Captive Balloon." *Aeronautics* (February 1908), p. 14.

———. "Our Army and Aerial Warfare." *Aeronautics* (January 1908), p. 18.

———. "Prolegomenon, or the Balloon During the Civil War." *Journal of the Military Service Institution of the United States*, 18 (1896), p. 261.

Goodell, Robert H. "Mathias Zahm's Diary." *Papers of the Lancaster County Historical Society*, 47, no. 4 (1943), pp. 65–89.

Gray, H. C. "Eight Miles—Straight Up!" *Popular Mechanics Magazine* (August 1927), pp. 177–181.

Greely, A. W. "Balloons in War." *Century Magazine* (June 1900).

Gregory, Austin. "America's Foremost and Most Famous Aeronaut, Captain Thomas Scott Baldwin." *Aeronautics* (June 1908), pp. 36–41.

Grosvenor, Gilbert. "The Stratosphere Expeditions." *The Black Hills Engineer* (September 1936), pp. 49–50.

Hall, Raymond D. "The Search for Peter Carnes." *Richmond County History* (Summer 1978), pp. 5–11.

Hammon, W. H. "Report on the Observations Made in Four Balloon Ascents." *Proceedings of the American Association for the Advancement of Science* (Washington, D.C., 1891), pp. 94–96.

Henderson, Archibald. "Washington and Aeronautics." *The Archive* (May 1932), p. 8.

Henry, Mary Lou. "C. A. A. Dellschau and the Sonora Aero Club." *Ballooning* (March–April 1980), pp. 23–26.

Hersey, Henry B. "The Trip of the United States." *Aeronautics* (December 1907), pp. 16–17.

Honeywell, H. Eugene. "My Voyage in the International Balloon Race." *Aeronautics* (December 1912), p. 164.

Horgan, James J. "Competition for the Lahm Cup." *Wingfoot Lighter-Than-Air Society Bulletin* (July–August 1965), pp. 4–5.

———. "Wartime Ballooning in St. Louis." *Wingfoot Lighter-Than-Air Society Bulletin* (September 1965), pp. 2–3.

Jackson, Joseph. "The First Balloon Hoax." *Pennsylvania Magazine of History and Biography* (1911), p. 51.

Jones, Idwal. "The Flying Editor." *Westways*, vol. 6, no. 2, pp. 14–15.

Jones, Thomas. "Notice of a Work . . ." *The Franklin Journal and American Mechanic's Magazine* (July 1826), pp. 41–45.

Kearney, J. W. "The Aero Club of St. Louis." *Aeronautics* (July 1907), pp. 21–22.

Kepner, William E., and James Scriven, Jr. "The Saga of Explorer I: Man's Pioneer Attempts to Explore Space." *Aerospace Historian* (Fall 1971), pp. 123–128.

Kidwell, Claudia. "Apparel for Ballooning, with Speculations on More Commonplace Garb." *Costume* (London, 1977), pp. 73–85.

Kihn, Phyllis. "Silas M. Brooks, Aeronaut,

1824–1906." *Connecticut Historical Society Journal* (April 1972), pp. 41–55.

Kittinger, Joseph W., and Volkmar Wentzel. "The Long, Lonely Leap." *National Geographic Magazine* (December 1960), pp. 854–873.

Kruckman, Arnold. "The Sport of Kings—Ballooning." *Aeronautics* (July 1907), pp. 10–11.

Lahm, F. P. "The Memories of a Pioneer." *Flying* (August 1957), pp. 41–51.

Lamb, Julia. "The Commodore Enjoyed Life—But New York Society Winced." *Smithsonian* (November 1978), pp. 132–144.

Lambert, Albert, and William Robertson. "Early History of Aeronautics in St. Louis." *Missouri Historical Society Proceedings* (June 1928), p. 244.

Leary, Lewis. "Phaeton in Philadelphia: Jean Pierre Blanchard and the First Balloon Ascension in America." *The Pennsylvania Magazine of History and Biography* (1943), pp. 49–60.

Link, Goethe. "Aeronautics in Indiana." *Fly* (January 1909), p. 14.

Lippincott, Horace Mather. "Dr. John Foulke, 1780, A Pioneer in Aeronautics." *The General Magazine and Historical Chronicle* (October 1934), pp. 525–531.

McClurg, Robert M. and Gale S. "Tammany's Remarkable Gardiner Baker: New York's First Museum Proprietor, Menagerie Keeper, and Promoter Extraordinary." *New York Historical Society Quarterly* (April 1958), pp. 143–169.

Maravelas, Paul. "Jeannette Piccard Interviewed." *Ballooning* (July–August 1980), pp. 15–19.

Martin, Marvin. "Marriott and His Flying Avitor." *Popular Aviation* (November 1935), p. 290.

Maurer, Maurer. "Richard Clayton, Aeronaut." *Historical and Philosophical Society of Ohio Bulletin* (April 1955), pp. 143–150.

Myers, Carl E. "Dirigible Balloons with the Screw in Front." *American Magazine of Aeronautics* (January 1908), pp. 29–31.

———. "Half a Lifetime with the Hydrogen Balloon." *Fly* (June 1913), pp. 16–17.

———. "Sphere of the Gas Balloon." *Fly* (April 1909), p. 1.

Palmer, Harry R., Jr. "Lighter-Than-Air Flight in America, 1784–1910." *Journal of the American Aviation Historical Society* (Fall 1982).

Parkinson, Russell J. "A Signal Corps Balloon for the Columbian Exposition, 1893." *Wingfoot Lighter-Than-Air Society Bulletin* (January 1966), pp. 2–6.

———. "Signal Corps Balloons, 1892–1907." *Wingfoot Lighter-Than-Air Society Bulletin* (July–August 1962), pp. 2–4.

———. "United States Signal Corps Balloons, 1871–1902." *Military Affairs* (Winter 1960–1961), pp. 189–202.

———. "Varicle Balloons and the Spanish-American War." *Wingfoot Lighter-Than-Air Society Bulletin* (February 1969), pp. 2–6.

Perkins, Samuel F. "Trip of the Düsseldorf." *Aeronautics* (December 1910), pp. 217–218.

Piccard, Auguste. "Ballooning in the Stratosphere." *National Geographic Magazine* (March 1933), pp. 353–384.

Piccard, Jean. "Construction of Welded Gondolas for Stratosphere Balloons." *The Welding Engineer* (June 1933), pp. 488–489.

Porter, Rufus. "Aerial Navigation: Signor Muzio Muzzi's Travelling Balloon." *Scientific American* (October 28, 1845).

———. "Plan for an Observatory Balloon." *Mechanic's Magazine* [American edition] (September 27, 1834), pp. 215–216.

———. "Travelling Balloon, or Flying Machine." *Mechanic's Magazine* [American edition] (November 8, 1834), pp. 273–275.

———. "Travelling Balloon." *New York Mechanic* (March 13, 1841).

———. "The Travelling Balloon." *Scientific American* (September 18, 1845), p. 1.

———. "The Travelling Balloon." *Scientific Mechanic* (December 18, 1847), p. 1.

———. "The Travelling Balloon." *Scientific Mechanic* (December 25, 1847), p. 1.

Post, Augustus. "Captain Thomas Scott Baldwin." *The Curtiss Flyleaf* (January–February 1918), pp. 12–14.

———. "A Fall from the Sky." *The Century Magazine* (October 1910), pp. 935–947.

Preston, R. A. D. "How We Won the Gordon Bennett." *Aeronautics* (November 1913), pp. 167–168.

Prokofiev, G. "Ascents into the Stratosphere." *Aircraft Engineering* (September 1933), p. 232.

Rhees, W. J. "Reminiscences of Ballooning in the Civil War." *Chatauquan*, 27 (1890), p. 261.

Robinson, Douglas. "Dr. August Graeth and the First Airship Flight in the U.S." *American Aviation Historical Society Journal* (Summer 1976), pp. 84–92.

Robinson, Edith Day. "The Personal Side of Talked of Aviators." *Aircraft* (November 1910), p. 9.

Ross, Malcolm D. "The Role of Manned Balloons in the Exploration of Space." *Aerospace Engineering* (August 1958), p. 52.

Schendal, Robert T. "Flight Safety." *Ballooning* (September–October 1980), pp. 8–9.

Schwarzschild, Martin and Barbara. "Balloon Astronomy." *Scientific American* (May 1959), pp. 52–57.

Settle, T. G. W. "Winning the Gordon Bennett Cup." *National Aeronautic Magazine* (December 1932), p. 15.

Sherwick, Johnson. "The Aero Club of Ohio." *Fly* (January 1909), p. 15.

Silliman, Benjamin. "Remarks on Genet and Pascalis." *The Franklin Journal and American Mechanic's Magazine* (January 1827), pp. 33–37.

Sonnichsen, Joanne. "Kria Kutnoi's 1731 Balloon Flight." *Ballooning* (January–February 1980), pp. 44–45.

Spangler, Robert. "C. A. Coey of Chicago." *Fly* (December 1908), pp. 8–12.

Spilhaus, Athelston, C. S. Schneider, and C. B. Moore. "Controlled Altitude Free Balloons." *Journal of Meteorology* (August 1948) pp. 130–137.

Stehling, Kurt R. "Balloon Astronomy: A Case for More." *Smithsonian* (June 1971), pp. 28–33.

Stepanek, Robert H. "The Transatlantic Flight of the *Daily Graphic*." *American Aviation Historical Society Journal* (Summer 1981), pp. 143–149.

Thompson, Cora. "My Initial Trip to the Clouds." *Aeronautics* (July 1908), p. 24.

Upson, Ralph H. "Balloon Racing." *Aviation* (April 12, 1926), p. 547.

Vaeth, J. Gordon. "When the Race for Space Began." *U.S. Naval Institute Proceedings* (August 1961), pp. 63–71.

Wells, Rufus G. "An American Balloon Trip over Rome." *Aeronautics* (October 1910), pp. 111–112.

Winzen, Otto C. "From Balloon Capsules to Space Cabins." *Proceedings, IXth International Astronautical Congress, Amsterdam, 1958* (Wien, 1959).

———. "Plastic Balloons in the Rocket Age." *Missiles and Rockets* (March 1957), pp. 50–52.

———. "Ten Years of Plastic Balloons." *Proceedings of the VIIIth International Astronautical Congress, Barcelona, Spain, 1957* (Wien, 1958), p. 438.

Wise, John. "Aerial Navigation." *Scientific American* (January 8, 1870), p. 33.

———. "The Balloon as an Aid to Meteorological Research." *Scientific American* (November 26, 1870), p. 341.

———. "Balloon Varnishes." *Scientific American* (June 11, 1870), p. 381.

Wynn, Clarence P. "Professor Samuel Archer King: A Memorial." *Flying* (January 1915), p. 358.

Young, Pearl I. "Airships and Balloons at the St. Louis Fair." *Wingfoot Lighter-Than-Air Society Bulletin* (February–March–April 1964).

———. "Balloons at the St. Louis Fair, 1904." *Wingfoot Lighter-Than-Air Society Bulletin* (March 1964), p. 2.

———. "John Wise and His Balloon Ascensions in the Middle West." *Wingfoot Lighter-Than-Air Society Bulletin* (October 1967), p. 2.

Index

Boldface numbers refer to illustrations.

A

C

D

E

G

H

I

J

K

L

M

Q

R

S

T

W

Y

Z

Y

Z